Lecture Notes in Artificial Intelligence   8371

Subseries of Lecture Notes in Computer Science

LNAI Series Editors

Randy Goebel
University of Alberta, Edmonton, Canada
Yuzuru Tanaka
Hokkaido University, Sapporo, Japan
Wolfgang Wahlster
DFKI and Saarland University, Saarbrücken, Germany

LNAI Founding Series Editor

Joerg Siekmann
DFKI and Saarland University, Saarbrücken, Germany

Sven Behnke   Manuela Veloso
Arnoud Visser   Rong Xiong (Eds.)

RoboCup 2013:
Robot World Cup XVII

 Springer

Volume Editors

Sven Behnke
University of Bonn
Autonomous Intelligent Systems
Bonn, Germany
E-mail: behnke@cs.uni-bonn.de

Manuela Veloso
Carnegie Mellon University
Computer Science Department
Pittsburgh, PA, USA
E-mail: mmv@cs.cmu.edu

Arnoud Visser
University of Amsterdam
Intelligent Robotics Lab
Amsterdam, The Netherlands
E-mail: a.visser@uva.nl

Rong Xiong
Zhejiang University
Institute of Cyber-Systems and Control
Hangzhou, China
E-mail: rxiong@iipc.zju.edu.cn

ISSN 0302-9743   e-ISSN 1611-3349
ISBN 978-3-662-44467-2   e-ISBN 978-3-662-44468-9
DOI 10.1007/978-3-662-44468-9
Springer Heidelberg New York Dordrecht London

Library of Congress Control Number: 2014945369

LNCS Sublibrary: SL 7 – Artificial Intelligence

Typesetting: Camera-ready by author, data conversion by Scientific Publishing Services, Chennai, India

Printed on acid-free paper

Springer is part of Springer Science+Business Media (www.springer.com)

Preface

RoboCup fosters research and development of intelligent robots by providing a standardized test bed through competitions and by organizing scientific meetings. This book documents some of the highlights of RoboCup 2013, which was held June 24-30, 2013, at Genneper Parks in Eindhoven, The Netherlands, and contains the proceedings of the 17th RoboCup International Symposium, which took place on July 1 at Eindhoven University of Technology.

The RoboCup 2013 competitions had 2,661 participants from 45 countries. About half of these competed in RoboCupJunior Soccer, Rescue, and Dance leagues, where the focus lies on mathematical and technical education for middle and high school students. The university-oriented Major leagues were held in the areas of: RoboCup Soccer, with five leagues spanning from simulated robots to full-scale humanoid robots; RoboCup Rescue, where the support of intelligent robots to first responders is investigated; RoboCup@Home, where service robots are tested in everyday environments; and the RoboCup@Work and Festo Sponsored Logistics leagues, which are inspired by industrial applications.

RoboCup 2013 was attended by 112 journalists from 22 countries, resulting in, press coverage in 75 countries. There were at least 60,000 individual viewers of RoboCup TV, 80,000 visitors to the RoboCup website and over 75 million tweets about RoboCup 2013. One of the highlights of RoboCup 2013 was certainly the final game of the Middle-Size soccer competition, which attracted 5,000 spectators from the general public. All together over 40,000 people visited RoboCup 2013, notably among them Her Majesty Queen Máxima.

Due to the complex nature of the RoboCup research and competitions, the RoboCup International Symposium offers a unique venue for exploring various and intimate connections of theory and practice across a wide spectrum of research fields. The experimental, interactive, and benchmark character of the RoboCup initiative presents the opportunity to disseminate novel ideas and promising technologies, which are rapidly adopted and field-tested by a large, and growing, community. To document the state of the art and as a basis for further development, this book contains 11 papers of teams winning individual RoboCup league competitions.

The RoboCup 2013 Symposium introduced a new special track on open-source hard- and software components to encourage the sharing of concrete implementations among RoboCup researchers. We had 19 contributions, from which some were invited based on suggestions of the leagues and the rest were submitted after an open call. Review of the special-track contributions was based on the technical innovation as well as the component use and benefit for the RoboCup community.

The main part of the book consists of the contributions selected for presentation at the 17th RoboCup International Symposium. From 78 regular paper submissions that were each reviewed by three members of the Program Committee, 14 papers (18 %) were selected for oral presentation. In addition, 20 regular papers (26 %) were presented as posters. We hope that all the careful reviews were of benefit for authors of both accepted and rejected papers.

The RoboCup 2013 Symposium had two invited keynote speakers, namely, Raffaello d'Andrea (ETH Zürich, Switzerland) "Why RoboCup Matters," and Chad Jenkins (Brown University, Providence, USA) "Towards a World Wide Web for Robotics." Both excellent presentations gave the more than 360 symposium participants a broader perspective on the use of robots for warehouse logistics and service tasks.

The single-track, single-day symposium had a novel and successful format. The program included a marathon of one-minute introductions of the special-track contributions, which were presented throughout the day as posters and demonstrations. The accepted poster papers, in addition to their scheduled poster presentations, had an opportunity for brief oral overviews. The regular papers were presented in topic-based sessions mixed well to appeal to the varied research emphases in the multiple RoboCup leagues.

The Award Committee selected three papers, printed first in the book:

- Best Paper Award for its *Application* contribution: Jos Elfring, Simon Jansen, René van de Molengraft, and Maarten Steinbuch "Active Object Search Exploiting Probabilistic Object–Object Relations"
- Best Paper Award for its *Engineering* contribution: Felix Wenk and Thomas Röfer "Online Generated Kick Motions for the NAO Balanced Using Inverse Dynamics"
- Best Paper Award for its *Theoretical* contribution: Oliver M. Cliff, Joseph T. Lizier, X. Rosalind Wang, Peter Wang, Oliver Obst, and Mikhail Prokopenko "Towards Quantifying Interaction Networks in a Football Match"

We would like to thank the 60 Program Committee members and the additional reviewers for their valuable expertise that ensured the quality of the technical program and the members of the Award Committee for their work during the symposium. We also thank all authors and participants for their high-quality contributions and vivid discussions. Last, but not least, we are very grateful to the chairs of RoboCup 2013, René van de Molengraft and Roel Merry, who organized one of the best RoboCup events ever, and all members of the Organizing Committee who helped to make RoboCup 2013 a big success.

Finally, we would like to share that we, the co-chairs of the RoboCup 2013 Symposium, had great pleasure working together and jointly making all the decisions with the unified goal of producing an informative and exciting event for all our symposium authors and participants. We thank the entire RoboCup community and friends!

March 2014

Sven Behnke
Manuela Veloso
Arnoud Visser
Rong Xiong

Organization

Symposium Co-chairs

Sven Behnke	University of Bonn, Germany
Manuela Veloso	Carnegie Mellon University, USA
Arnoud Visser	University of Amsterdam, The Netherlands
Rong Xiong	Zhejiang University, China

Program Committee

H. Levent Akın	Boğaziçi University, Turkey
Hidehisa Akiyama	Fukuoka University, Japan
Luís Almeida	University of Porto, Portugal
Jacky Baltes	University of Manitoba, Canada
Ansgar Bredenfeld	Dr. Bredenfeld UG, Germany
Stefano Carpin	University of California Merced, USA
Stephan Chalup	University of Newcastle, Australia
Weidong Chen	Shanghai Jiao Tong University, China
Xiaoping Chen	University of Science and Technology of China, China
Eric Chown	Bowdoin College, USA
Bernardo Cunha	University of Aveiro, Portugal
Klaus Dorer	Offenburg University of Applied Sciences, Germany
Amy Eguchi	Bloomfield College, USA
Alessandro Farinelli	Verona University, Italy
Bernhard Hengst	University of New South Wales, Australia
Todd Hester	University of Texas at Austin, USA
Dirk Holz	University of Bonn, Germany
Luca Iocchi	Sapienza University of Rome, Italy
Nobuhiro Ito	Aichi Institute of Technology, Japan
Jianmin Ji	University of Science and Technology of China, China
Gal Kaminka	Bar-Ilan University, Israel
Alexander Kleiner	Linköping University, Sweden

Gerhard Kraetzschmar   Bonn-Rhein-Sieg University of Applied
                            Sciences, Germany
Michail Lagoudakis   Technical University of Crete, Greece
Gerhard Lakemeyer   RWTH Aachen University, Germany
Tim Laue   DFKI Bremen, Germany
Pedro U. Lima   Instituto Superior Técnico, Portugal
Luis F. Lupián   La Salle University, Mexico
Hitoshi Matsubara   Future University Hakodate, Japan
Norbert Michael Mayer   National Chung Cheng University, Taiwan
Çetin Meriçli   Carnegie Mellon University, USA
Tekin Meriçli   .   Boğaziçi University, Turkey
Zhao Mingguo   Tsinghua University, China
Eduardo Morales   National Institute of Astrophysics, Optics and
                            Electronics, Mexico
Tadashi Naruse   Aichi Prefectural University, Japan
Itsuki Noda   National Institute of Advanced Industrial
                            Science and Technology, Japan
Paul G. Plöger   Bonn-Rhein-Sieg University of Applied
                            Sciences, Germany
Daniel Polani   University of Hertfordshire, UK
A. Fernando Ribeiro   University of Minho, Portugal
Thomas Röfer   DFKI Bremen, Germany
Raúl Rojas   Freie Universität Berlin, Germany
Javier Ruiz-Del-Solar   University of Chile
Uluc Saranli   Middle East Technical University, Turkey
Sanem Sariel-Talay   Istanbul Technical University, Turkey
Jesus Savage   National Autonomous University of Mexico,
                            Mexico
Elizabeth Sklar   Brooklyn College, USA
Domenico G. Sorrenti   University of Milano-Bicocca, Italy
Matthijs Spaan   Delft University of Technology,
                            The Netherlands
Mohan Sridharan   Texas Tech University, USA
Gerald Steinbauer   Graz University of Technology, Austria
Peter Stone   University of Texas at Austin, USA
Luis Enrique Sucar   National Institute of Astrophysics, Optics
                            and Electronics, Mexico
Komei Sugiura   National Institute of Information and
                            Communications Technology, Japan
Tomoichi Takahashi   Meijo University, Japan

Yasutake Takahashi   University of Fukui, Japan
Tijn Van Der Zant   University of Groningen, The Netherlands
Ubbo Visser   University of Miami, USA
Alfredo Weitzenfeld   University of South Florida, USA
Feng Wu   University of Southampton, UK
Changjiu Zhou   Singapore Polytechnic, Singapore

Fig. 1. Symposium invited speakers Raffaello d'Andrea (left) and Chad Jenkins (right)

Fig. 2. Impressions from RoboCup 2013. Images by Bart van Overbeeke.

Fig. 3. Impressions from RoboCup 2013. Images by Albert van Breemen.

Table of Contents

Best Paper Award for Its *Theoretical* Contribution

Towards Quantifying Interaction Networks in a Football Match   1
 Oliver M. Cliff, Joseph T. Lizier, Rosalind X. Wang, Peter Wang,
 Oliver Obst, and Mikhail Prokopenko

Best Paper Award for Its *Application* Contribution

Active Object Search Exploiting Probabilistic Object – Object
Relations   13
 Jos Elfring, Simon Jansen, René van de Molengraft, and
 Maarten Steinbuch

Best Paper Award for Its *Engineering* Contribution

Online Generated Kick Motions for the NAO Balanced Using Inverse
Dynamics   25
 Felix Wenk and Thomas Röfer

Champion Teams

Team Water: The Champion of the RoboCup Middle Size League
Competition 2013   37
 Song Chen, Zhe Zhu, Ye Tian, Charles Lynch, Di Zhu, Ye Lv,
 Xinxin Xu, Wanjie Zhang, Liang Zhao, Xiaoming Liu, Miao Wang,
 Peng Zhao, Chenyu Wang, Wei Liu, Feichan Yang, Binbin Li,
 Jieming Zhou, Zongyi Zhang, and Xueyan Wang

RoboCup 2013 Humanoid Kidsize League Winner   49
 Daniel D. Lee, Seung-Joon Yi, Stephen G. McGill, Yida Zhang,
 Larry Vadakedathu, Samarth Brahmbhatt, Richa Agrawal, and
 Vibhavari Dasagi

Learning to Improve Capture Steps for Disturbance Rejection in
Humanoid Soccer   56
 Marcell Missura, Cedrick Münstermann, Philipp Allgeuer,
 Max Schwarz, Julio Pastrana, Sebastian Schueller,
 Michael Schreiber, and Sven Behnke

RoboCup 2013: Best Humanoid Award Winner JoiTech 68
Yuji Oshima, Dai Hirose, Syohei Toyoyama,
Keisuke Kawano, Shibo Qin, Tomoya Suzuki, Kazumasa Shibata,
Takashi Takuma, and Minoru Asada

B-Human 2013: Ensuring Stable Game Performance 80
Thomas Röfer, Tim Laue, Arne Böckmann, Judith Müller, and
Alexis Tsogias

ZJUNlict: RoboCup 2013 Small Size League Champion 92
Yue Zhao, Rong Xiong, Hangjun Tong, Chuan Li, and Li Fang

The Walking Skill of Apollo3D – The Champion Team in the
RoboCup2013 3D Soccer Simulation Competition 104
Juan Liu, Zhiwei Liang, Ping Shen, Yue Hao, and Hecheng Zhao

The Decision-Making Framework of WrightEagle, the RoboCup 2013
Soccer Simulation 2D League Champion Team 114
Haochong Zhang and Xiaoping Chen

Clustering and Planning for Rescue Agent Simulation 125
Ahmed Abouraya, Dina Helal, Fadwa Sakr, Noha Khater,
Salma Osama, and Slim Abdennadher

Increasing Flexibility of Mobile Manipulation and Intuitive
Human-Robot Interaction in RoboCup@Home 135
Jörg Stückler, David Droeschel, Kathrin Gräve, Dirk Holz,
Michael Schreiber, Angeliki Topalidou-Kyniazopoulou,
Max Schwarz, and Sven Behnke

How to Win RoboCup@Work? The Swarmlab@Work Approach
Revealed 147
Sjriek Alers, Daniel Claes, Joscha Fossel, Daniel Hennes,
Karl Tuyls, and Gerhard Weiss

Oral Presentations

Analyzing and Learning an Opponent's Strategies in the RoboCup
Small Size League 159
Kotaro Yasui, Kunikazu Kobayashi, Kazuhito Murakami, and
Tadashi Naruse

An Entertainment Robot for Playing Interactive Ball Games 171
Tim Laue, Oliver Birbach, Tobias Hammer, and Udo Frese

Self-calibration of Colormetric Parameters in Vision Systems for
Autonomous Soccer Robots 183
António J.R. Neves, Alina Trifan, and Bernardo Cunha

BRISK-Based Visual Feature Extraction for Resource Constrained
Robots   195
    Daniel Jaymin Mankowitz and Subramanian Ramamoorthy

Compliant Robot Behavior Using Servo Actuator Models Identified by
Iterative Learning Control   207
    Max Schwarz and Sven Behnke

Unexpected Situations in Service Robot Environment: Classification
and Reasoning Using Naive Physics   219
    *Anastassia Küstenmacher, Naveed Akhtar, Paul G. Plöger, and
    Gerhard Lakemeyer*

Person Following by Mobile Robots: Analysis of Visual and Range
Tracking Methods and Technologies   231
    *Wilma Pairo, Javier Ruiz-del-Solar, Rodrigo Verschae,
    Mauricio Correa, and Patricio Loncomilla*

Fast Monocular Visual Compass for a Computationally Limited
Robot   244
    Peter Anderson and Bernhard Hengst

Reusing Risk-Aware Stochastic Abstract Policies in Robotic Navigation
Learning   256
    *Valdinei Freire da Silva, Marcelo Li Koga,
    Fábio Gagliardi Cozman, and Anna Helena Reali Costa*

Motivated Reinforcement Learning for Improved Head Actuation of
Humanoid Robots.   268
    *Jake Fountain, Josiah Walker, David Budden,
    Alexandre Mendes, and Stephan K. Chalup*

Efficient Distributed Communications for Multi-robot Systems   280
    João C.G. Reis, Pedro U. Lima, and João Garcia

Poster Presentations

Distributed Formation Control of Heterogeneous Robots with Limited
Information   292
    Michael de Denus, John Anderson, and Jacky Baltes

RoSHA: A Multi-robot Self-healing Architecture   304
    Dominik Kirchner, Stefan Niemczyk, and Kurt Geihs

Routing with Dijkstra in Mobile Ad-Hoc Networks   316
    *Khudaydad Mahmoodi, Muhammet Balcılar, M. Fatih Amasyalı,
    Sırma Yavuz, Yücel Uzun, and Feruz Davletov*

Shape Based Round Object Detection Using Edge Orientation
Histogram 326
  Hamid Mobalegh, Lovísa Irpa Helgadóttir, and Raúl Rojas

RoboCup Logistics League Sponsored by Festo: A Competitive Factory
Automation Testbed 336
  Tim Niemueller, Daniel Ewert, Sebastian Reuter,
  Alexander Ferrein, Sabina Jeschke, and Gerhard Lakemeyer

Parallel Computation Using GPGPU to Simulate Crowd
Evacuation Behaviors - Planning Effective Evacuation Guidance at
Emergencies - 348
  Toshinori Niwa, Masaru Okaya, and Tomoichi Takahashi

Vision Based Referee Sign Language Recognition System for the
RoboCup MSL League... 360
  Paulo Trigueiros, Fernando Ribeiro, and Luís Paulo Reis

Unsupervised Recognition of Salient Colour for Real-Time Image
Processing 373
  David Budden and Alexandre Mendes

Improved Particle Filtering for Pseudo-Uniform Belief Distributions in
Robot Localisation 385
  David Budden and Mikhail Prokopenko

Robust and Efficient Object Recognition for a Humanoid Soccer
Robot 396
  Alexander Härtl, Ubbo Visser, and Thomas Röfer

Automatic Generation of Humanoid's Geometric Model Parameters 408
  Vincent Hugel and Nicolas Jouandeau

Modification of Foot Placement for Balancing Using a Preview
Controller Based Humanoid Walking Algorithm...................... 420
  Oliver Urbann and Matthias Hofmann

Iterative Snapping of Odometry Trajectories for Path Identification 432
  Richard Wang, Manuela Veloso, and Srinivasan Seshan

Evaluation of Recent Approaches to Visual Odometry from RGB-D
Images 444
  Sergey Alexandrov and Rainer Herpers

Spontaneous Reorientation for Self-localization 456
  Markus Bader and Markus Vincze

Multi-sensor Mobile Robot Localization for Diverse Environments 468
  Joydeep Biswas and Manuela Veloso

High-Level Commands in Human-Robot Interaction for Search and
Rescue 480
  Alain Caltieri and Francesco Amigoni

Optimizing Energy Usage through Variable Joint Stiffness Control
during Humanoid Robot Walking 492
  Ercan Elibol, Juan Calderon, and Alfredo Weitzenfeld

A System for Building Semantic Maps of Indoor Environments
Exploiting the Concept of Building Typology 504
  Matteo Luperto, Alberto Quattrini Li, and Francesco Amigoni

Semantic Object Search Using Semantic Categories and Spatial
Relations between Objects 516
  Patricio Loncomilla, Marcelo Saavedra, and Javier Ruiz-del-Solar

Special Track on Open-Source Hard- and Software

HELIOS Base: An Open Source Package for the RoboCup Soccer 2D
Simulation 528
  Hidehisa Akiyama and Tomoharu Nakashima

The Open-Source TEXPLORE Code Release for Reinforcement
Learning on Robots 536
  Todd Hester and Peter Stone

NUbugger: A Visual Real-Time Robot Debugging System 544
  Brendan Annable, David Budden, and Alexandre Mendes

The 2012 UT Austin Villa Code Release 552
  *Samuel Barrett, Katie Genter, Yuchen He, Todd Hester,
  Piyush Khandelwal, Jacob Menashe, and Peter Stone*

Ground Truth Acquisition of Humanoid Soccer Robot Behaviour 560
  Andrea Pennisi, Domenico D. Bloisi, Luca Iocchi, and Daniele Nardi

Humanoid TeenSize Open Platform NimbRo-OP 568
  *Max Schwarz, Julio Pastrana, Philipp Allgeuer, Michael Schreiber,
  Sebastian Schueller, Marcell Missura, and Sven Behnke*

Modular Development of Mobile Robots with Open Source Hardware
and Software Components... 576
  Martino Migliavacca, Andrea Bonarini, and Matteo Matteucci

Sharing Open Hardware through ROP, the Robotic Open Platform 584
  *Janno Lunenburg, Robin Soetens, Ferry Schoenmakers,
  Paul Metsemakers, René van de Molengraft, and Maarten Steinbuch*

Enabling Codesharing in Rescue Simulation with USARSim/ROS 592
  Zeid Kootbally, Stephen Balakirsky, and Arnoud Visser

FUmanoids Code Release 2012 600
  Daniel Seifert and Raúl Rojas

Extensions of a RoboCup Soccer Software Framework 608
  Stephen G. McGill, Seung-Joon Yi, Yida Zhang, and Daniel D. Lee

Inexpensive Image Processing Solution for the RoboCupJunior Soccer
Scenario 616
  Marek Šuppa and Ján Ďurkáč

Hector Open Source Modules for Autonomous Mapping and Navigation
with Rescue Robots 624
  Stefan Kohlbrecher, Johannes Meyer, Thorsten Graber,
  Karen Petersen, Uwe Klingauf, and Oskar von Stryk

SimSpark: An Open Source Robot Simulator Developed by the
RoboCup Community 632
  Yuan Xu and Hedayat Vatankhah

Visualizing and Debugging Complex Multi-Agent Soccer Scenes in Real
Time 640
  Justin Stoecker and Ubbo Visser

On B-Human's Code Releases in the Standard Platform League –
Software Architecture and Impact 648
  Thomas Röfer and Tim Laue

Five Years of SSL-Vision – Impact and Development 656
  Stefan Zickler, Tim Laue, José Angelo Gurzoni Jr., Oliver Birbach,
  Joydeep Biswas, and Manuela Veloso

Integration of the ROS Framework in Soccer Robotics: The NAO
Case 664
  Leonardo Leottau Forero, José Miguel Yáñez, and
  Javier Ruiz-del-Solar

Development of RoboCup@Home Simulation towards Long-term Large
Scale HRI.. 672
  Tetsunari Inamura, Jeffrey Too Chuan Tan, Komei Sugiura,
  Takayuki Nagai, and Hiroyuki Okada

Author Index 681

Towards Quantifying Interaction Networks
in a Football Match

Oliver M. Cliff, Joseph T. Lizier, Rosalind X. Wang,
Peter Wang, Oliver Obst, and Mikhail Prokopenko

CSIRO Information and Communication Technologies Centre, Adaptive Systems,
P.O. Box 76, Epping, NSW 1710, Australia

Abstract. We present several novel methods quantifying dynamic interactions in simulated football games. These interactions are captured in directed networks that represent significant coupled dynamics, detected information-theoretically. The model-free approach measures information dynamics of both pair-wise players' interactions as well as local tactical contests produced during RoboCup 2D Simulation League games. This analysis involves computation of information transfer and storage, relating the information transfer to responsiveness of the players and the team, and the information storage within the team to the team's rigidity and lack of tactical flexibility. The resultant directed networks (interaction diagrams) and the measures of responsiveness and rigidity reveal implicit interactions, across teams, that may be delayed and/or long-ranged. The analysis was verified with a number of experiments, identifying the zones of the most intense competition and the extent of interactions.

1 Introduction

Many team games, real and virtual, are characterised by rich interactions occurring dynamically and shaping the course of the contest both locally and globally. The interactions across the teams are created by opposing objectives of competing players and tactical schemes. The interactions within a team are usually constrained by cooperation and shared plans. Generally, the interactions are directed (e.g., a defender is marking an opponent's forward), varying in strength over time and/or space, and typically do not result from direct messaging or communications — rather they manifest some tacit correlations that often are delayed in time and/or are long-ranged over the play-field.

While a significant number of patterns emerging during a game may be evident even without an in-depth analysis, most of the interactions may appear intractable to an external observer who does not have an access to the logic and neural processing of the players. One then may formulate a general problem: how can an external observer identify most generic interaction networks that link together autonomous players, without re-constructing the players' behaviour and using only the positional data, such as planar coordinates and their changes? The problem is difficult as some of the dependencies between players are not discernible simply by correlating their dynamic locations over time — one needs to take into the account a possibly directed nature of such correlations, where dynamics of one of the players affects the positioning of another.

In general, as mentioned by Vilar et al. [1], "quantitative analysis is increasingly being used in team sports to better understand performance in these stylized, delineated,

S. Behnke et al. (Eds.): RoboCup 2013, LNAI 8371, pp. 1–12, 2014.

complex social systems". One of the older examples is "sabermetrics" — the specialised analysis of baseball through objective evidence, e.g. baseball statistics measuring in-game activity [2]. Another recent example is described by Fewell et al. [3] who analysed basketball games as networks, where players are represented as nodes and passes as edges: the resulting network captures ball movement, at different stages of the game. Their work studies network properties (degree centrality, clustering, entropy and flow centrality) across teams and positions, and attempts to determine whether differences in team offensive strategy can be assessed by their network properties. Strategic networks analysed by Fewell et al. consider only explicit interactions (such as passes) within a team, and not implicit (delayed and/or long-ranged) interactions, across teams.

Another very recent investigation by Vilar et al. [1] proposed a novel method of analysis that captures how teams occupy sub-areas of the field as the ball changes location. This study was important in focussing on the local dynamics of team collective behavior rather than individual player capabilities: when applied to football (soccer) matches, the method suggested that players' numerical dominance in some local sub-areas is a key to "defensive stability" and "offensive opportunity". While the method rigorously used an information-theoretic approach (e.g. the uncertainty of the team numerical advantage across sub-areas was determined using Shannon's entropy), it was not aimed at and did not produce interaction networks, either explicit or implicit.

Construction of interaction networks for (possibly competing) teams is not unique to sport, but arguably its utility can be leveraged quite strongly in team games, such as football, basketball and so on. RoboCup 2D Soccer Simulation League is a well-known benchmark domain for Artificial Intelligence that specifically targets soccer with its realistic and challenging multi-agent dynamics, characterised by autonomous decision-making under constraints, set by tactical plans and teamwork (collaboration) as well as opponent (competition) [4,5,6,7,8,9,10,11], and so we use this domain in our study.

Information dynamics is a recent methodology for analysis of complex systems in general and swarm behavior in particular. In this paper we describe a novel application of information dynamics to the RoboCup 2D Simulation. In particular, we develop an approach to build several interaction diagrams, given data from a number of games, followed by a tactical analysis. The interaction diagrams reveal a few interesting dependencies between pairs of players that are useful for game analysis, while the tactical analysis extends these findings to formation-level interactions (e.g., between defensive line-up of team Y with the attacking line-up of team X, etc.).

2   Motivation and Approach

2.1   Information Dynamics

A recently developed framework of *information dynamics* studies the phenomenon of computation in a systematic way: it uncovers and analyses information-theoretic roots of the most basic computational primitives: *storage, transmission*, and *modification of information* [12,13,14,15].

The *active information storage* quantifies the information storage component that is directly in use in the computation of the next state of a process [15]. More precisely, it is the average mutual information between the semi-infinite past of the process

$x_n^{(k)} = \{x_{n-k+1}, \ldots, x_{n-1}, x_n\}$ (as $k \to \infty$) and its next state. The *local information storage* (or pointwise mutual information) is then a measure of the amount of information storage in use by the process at a particular time-step $n + 1$:

$$a_X(n + 1) = \lim_{k \to \infty} \log_2 \frac{p(x_n^{(k)}, x_{n+1})}{p(x_n^{(k)})p(x_{n+1})}. \tag{1}$$

In practice, one deals with finite-k estimates $a_X(n + 1, k)$, as well as the finite-k estimates $A_X(k)$ of the average active information storage $A_X = \langle a_X(n + 1) \rangle_n$.

Transfer entropy [16] is designed to detect asymmetry in the interaction of subsystems by distinguishing between "driving" and "responding" elements. The *local information transfer*, based on *transfer entropy*, captures information transmission [12] from source Y to destination X, at a particular time-step $n+1$. It is defined as the information provided by the source y_n about the destination's next state x_{n+1} that was not contained in the past of the destination $x_n^{(k)}$:

$$t_{Y \to X}(n + 1) = \lim_{k \to \infty} \log_2 \frac{p(x_{n+1} \mid x_n^{(k)}, y_n)}{p(x_{n+1} \mid x_n^{(k)})}. \tag{2}$$

It is important to realise that information transfer between two variables does not require an explicit communication channel, it rather indicates a high degree of directional synchrony or nonlinear correlation between the source and the destination. It characterises a degree of *predictive* information transfer, i.e., "if the state of the source is known, how much does that help to predict the state of the destination?" [12].

Sometimes it is useful to condition the local information transfer on another contributing process W, considering the *local conditional transfer entropy* [13]:

$$t_{Y \to X \mid W}(n + 1) = \lim_{k \to \infty} \log_2 \frac{p(x_{n+1} \mid x_n^{(k)}, y_n, w_n)}{p(x_{n+1} \mid x_n^{(k)}, w_n)}. \tag{3}$$

In this study we used the average information transfer $t_{Y \to X \mid W} = \langle t_{Y \to X \mid W}(n+1) \rangle_n$. One may, however, utilise local values as well in order to trace the information dynamics over time, e.g. identifying its peaks during specific moments.

2.2 Pair-Wise Information Dynamics and Interaction Diagrams

In order to estimate strength of directed coupling between two agents we compute the average transfer entropy between them during any given game. For a game g with N time steps, between two teams $\mathbb{X}$ and $\mathbb{Y}$, the local transfer entropy at each time step $n \leq N$ is calculated between each source variable Y_i (a change in the 2D positional vector of agent i from team $\mathbb{Y}$) and destination variable X_j (a change in the 2D positional vector of agent j from team $\mathbb{X}$), given the change in current 2D ball position b:

$$t^g_{Y_i \to X_j \mid b}(n) .$$

Dynamics of the ball is conditioned upon in order to compute the transfer entropy in context of the game, which is greatly affected by the ball trajectories. Then, the average transfer entropy for each source-destination pair over the entire match is calculated as

$$T^g_{Y_i \to X_j | b} = \frac{1}{N} \sum_{n=0}^{N-1} t^g_{Y_i \to X_j | b}(n) \, . \tag{4}$$

Information-Sink Diagrams. Once the game's average transfer entropy, $T^g_{Y_i \to X_j | b}$, is determined for each pair Y_i, X_j, we identify the *source* agent $\hat{Y}_i(X_j, g)$ from the opposing team that transfers maximal information to a given agent X_j:

$$\hat{Y}_i(X_j, g) = \arg\max_{Y_k \in Y} T^g_{Y_k \to X_j | b} \, . \tag{5}$$

Over a number of games G, we select the source agent $\hat{Y}_i(X_j)$ that transfers maximal information to X_j most frequently, as the mode of the series $\{\hat{Y}_{i_1}(X_j, 1), \ldots, \hat{Y}_{i_G}(X_j, G)\}$. Then, we consider the average information transfer between these two agents $\hat{Y}_i = \hat{Y}_i(X_j)$ and X_j across all games:

$$T_{\hat{Y}_i \to X_j | b} = \frac{1}{G} \sum_{g=1}^{G} T^g_{\hat{Y}_i \to X_j | b} \, . \tag{6}$$

Intuitively, the movement of the source agent $\hat{Y}_i = \hat{Y}_i(X_j)$ affected the agent X_j more than movement of any other agent in team Y. That is, the agent X_j was responsive most to movement of the source agent $\hat{Y}_i$. Crucially, when we use the notion of *responsiveness* to another (source) agent, we do not load it with such semantics as being dominated by, or driven by that other agent. Higher responsiveness may in fact reflect either useful reaction to the opponent's movements (e.g., good marking of the source), or a helpless behaviour (e.g., constant chase after the source). Vice versa, generating a high responsiveness from another agent may result in either a useful dynamic (e.g., positional or even tactical dominance over the responding agent), or a wasteful motion (e.g., being successfully marked by the responding agent). In short, responsiveness captured in the maximal transfer $T_{\hat{Y}_i \to X_j | b}$ detects a directed coupling from the source agent $\hat{Y}_i$ to the responding agent X_j and should not be interpreted in general as a simple index for comparative performance. It is, however, a useful identifier of the opponents' source player that was affecting a given agent X_j most.

Given a series of games, we identify the pairs "source-responder" by finding the source agent for each of the agents on both teams (always choosing the source among the opponents). The identified pairs can be visualised in an "information-sink" interaction diagram $\hat{D}(Y, X)$ that depicts a directed graph with 20 nodes representing players (typically excluding goalkeepers), with the edges representing all source-responder pairs, where a single edge is incoming to every agent from the corresponding source.

Figure 1a shows the information-sink interaction diagram $\hat{D}(\text{Oxsy, Gliders})$ built for several hundred games between Oxsy and Gliders (cf. Results section).

Information-Base Diagrams. Similarly, having obtained the average transfer entropy during a game, $T^g_{Y_i \to X_j | b}$ for all pairs, we identify the *responder* agent $\check{X}_j(Y_i, g)$ that "received" maximal information from a given agent Y_i. Formally, for any game g:

$$\check{X}_j(Y_i, g) = \arg\max_{X_k \in X} T^g_{Y_i \to X_k | b} \, . \tag{7}$$

Over a number of games G, we select the responder agent $\check{X}_j(Y_i, g)$ to whom maximal information was transferred by Y_i most frequently, as the mode of the series $\{\check{X}_{j_1}(Y_i, 1), \ldots, \check{X}_{j_G}(Y_i, G)\}$. Finally, we consider the average information transfer between these two agents Y_i and $\check{X}_j = \check{X}_j(Y_i, g)$ across all games:

$$T_{Y_i \to \check{X}_j | b} = \frac{1}{G} \sum_{g=1}^{G} T^g_{Y_i \to \check{X}_j | b} . \tag{8}$$

The pairs $(Y_i, \check{X}_j)$ identified for each agent treated as a source are combined in an "information-base diagram" $\check{D}(\mathbb{Y}, \mathbb{X})$.

The intuition in this case is the same as in the previous subsection — the difference is that now we identify the highest responder agent, having selected a source. In general, of course, the pair $(\hat{Y}_i, X_j)$ defined for the information-sink diagrams and the pair $(Y_i, \check{X}_j)$ defined for the information-base diagrams may differ. That is, the agent Y_i may be the most informative source $\hat{Y}_i$ for the agent X_j, among all possible sources in Y, but the agent X_j may be not the best responder $\check{X}$ to the agent Y_i among all possible responders in $\mathbb{X}$, and vice versa.

While an information-sink diagram reflects more where the information tends to be transferred to, an information-base diagram tends to depict where the information is transferred from. Neither of the diagrams presents a complete "story", highlighting only a small part of the overall information dynamics. There are more comprehensive diagrams, where the edges would represent in the descending order the highest information transfers for all the pairs, retaining a given number of such links, or keeping the edges for the information amounts above a certain threshold, etc. — in these instances, some agents may have no incoming or outgoing links at all. Nevertheless, we believe that the interaction diagrams presented here are valuable, being particularly simple and easy to interpret. Specifically, for an information-sink diagram every agent has an incoming edge, and for an information-base diagram every agent has an outgoing edge.

Figure 2a shows the information-base interaction diagram $\check{D}(\text{Oxsy}, \text{Gliders})$ built for several hundred games between Oxsy and Gliders (cf. Results section).

2.3 Tactical Analysis

Building up on the information dynamics measures, it is possible to investigate group behavior in complex systems, such as swarms. For instance, recent studies by Wang et al. [17] quantitatively verified the hypothesis that the collective memory within a swarm can be captured by *active information storage*. Higher values of storage are associated with higher levels of dynamic coordination, while information cascades that correspond to long range communications are captured by *conditional transfer entropy* [12,13]. In other words, information transfer was shown to characterise the communication aspect of collective computation distributed within the swarm.

In applying information dynamics to the RoboCup 2D Simulation League we make the following conjecture:

> *a higher information transfer* $t_{Y \to X|W}$ *from the source Y (e.g. dynamics of player Y) to the destination X (e.g., dynamics of another player X), in the context of some other dynamics W (e.g., the movement of the ball W), is indicative of a higher responsiveness of the process/player X to the process/player Y.*

That is, the "destination" player Y responds, for example, by repositioning, to the movement of the "source" player Y. This may apply to many situations on the field, for instance, when one team's forwards are trying to better avoid opponent's defenders, we consider the information transfer $t_{\mathbb{Y}_{def} \to \mathbb{X}_{att}}$ from defenders $Y_i \in \mathbb{Y}_{def}$ to forwards $X_j \in \mathbb{X}_{att}$, where the involved probability distributions are obtained for different relative positions on the soccer field. Vice versa, the dynamics of the opponent's defenders, who are trying to better mark our team's forwards, are represented in the information transfer $t_{\mathbb{X}_{att} \to \mathbb{Y}_{def}}$ from forwards $X_j \in \mathbb{X}_{att}$ to defenders $Y_i \in \mathbb{Y}_{def}$. These two examples specifically consider a coupling between the attack line $\mathbb{X}_{att}$ of our team and the defense line $\mathbb{Y}_{def}$ of opponent's team (henceforth we keep denoting opponent's lines (attack, midfield or defense) by $\mathbb{Y}_{line}$ and our team's lines by $\mathbb{X}_{line}$).

We further contrast these two transfers in the coupled lines:

$$\Delta(\mathbb{X}_{att}, \mathbb{Y}_{def}) = t_{\mathbb{Y}_{def} \to \mathbb{X}_{att}} - t_{\mathbb{X}_{att} \to \mathbb{Y}_{def}} \,. \tag{9}$$

When our forwards are more responsive on average to the opponent's defenders than the opponents defenders are to our forwards, $t_{\mathbb{Y}_{def} \to \mathbb{X}_{att}} > t_{\mathbb{X}_{att} \to \mathbb{Y}_{def}}$, and the relative responsiveness $\Delta(\mathbb{X}_{att}, \mathbb{Y}_{def}) > 0$. It is also possible to combine relative responsiveness scores for each of the coupled lines in the overall tactical relative responsiveness (including, for example, relative scores for midfielders $\mathbb{X}_{mid}$ and $\mathbb{Y}_{mid}$):

$$\Delta(\mathbb{X}, \mathbb{Y}) = \Delta(\mathbb{X}_{att}, \mathbb{Y}_{def}) + \Delta(\mathbb{X}_{def}, \mathbb{Y}_{att}) + \Delta(\mathbb{X}_{mid}, \mathbb{Y}_{mid}) \,. \tag{10}$$

Here all the transfers to team $\mathbb{X}$ are added up, and the transfers from team $\mathbb{X}$ are subtracted away. When each of the transfers is conditioned on some other contributor W (e.g., all the dynamics are computed in the context of the ball movement), the overall tactical relative responsiveness $\Delta(\mathbb{X}, \mathbb{Y}|W)$ is also placed in this specific context, W.

In principle, competitive situations result in quite vigorous dynamics within the involved lines and overall formations, and the team that manages to achieve a higher degree of tactical relative responsiveness does often perform better. While this is not a hard rule, we may correlate the scores of relative responsiveness (e.g., line-by-line) with the game scores, and identify the lines which impacted on the games more.

Our tactical analysis also involves computation of the active information storage within the teams. We characterise team's rigidity $A_{\mathbb{X}}$ as the average of information storage values for all players of the team. We also determine the relative rigidity $A(\mathbb{X}, \mathbb{Y}) = A_{\mathbb{X}} - A_{\mathbb{Y}}$ for the teams (or their coupled lines). The hypothesis here is that

a higher rigidity $A_{\mathbb{X}}$ within the team is indicative of a higher dependence of players on each other, or a higher redundancy within the team's motion.

The average information storage, or rigidity $A_{\mathbb{X}}$, is high whenever one can predict the motion of some players based on the movements of their other teammates. In these cases, the players are not as independent of each other as a truly complex or swarm behavior would warrant, making the tactics less versatile. Obviously, this may be counter-productive, since an opponent team can counteract by only partially observing the 'rigid' team's dynamics, and deducing the rest. Consequently, the relative rigidity $A(\mathbb{X}, \mathbb{Y})$ should be anti-correlated with the team $\mathbb{X}$ performance against team $\mathbb{Y}$.

3 Results

To compute the measures described in previous sections, produce interaction diagrams and correlate tactical responsiveness with team performance, we carried out multiple iterative experiments matching Gliders2013 up against some well-known teams, such as Oxsy [18] and Marlik [19]. The correlation scores (Pearson product-moment correlation coefficients) reported below were tested for statistical significance, and corrected for multiple comparisons.

3.1 Interaction Diagrams

Figure 1 presents the information-sink interaction diagram $\hat{D}(\text{Oxsy, Gliders})$ and the information-base interaction diagram $\check{D}(\text{Oxsy, Gliders})$, built over almost 500 hundred games between Oxsy and Gliders. Analogously, Fig. 2 shows the information-sink interaction diagram $\hat{D}(\text{Marlik, Gliders})$ and the information-base interaction diagram $\check{D}(\text{Marlik, Gliders})$, built over nearly 450 hundred games between Marlik and Gliders.

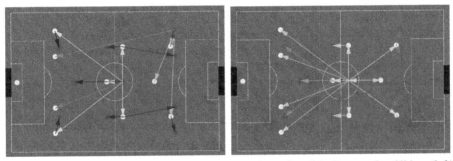

(a) Information-sink diagram for Gliders (left) and Oxsy (right)   (b) Information-sink diagram for Gliders (left) and Marlik (right)

Fig. 1. Interaction-sink diagrams. Arrows represent highest information transfer between players. MATLAB copper colormap is used to indicate the strength of transfer, varying smoothly from black (weakest) to bright copper (strongest). Example interactions: two arrows in the left diagram from Oxsy's central mid-fielder, positioned in the centre circle, to Gliders' left and right defenders indicate that these defenders respond mostly to the central mid-fielder's motion.

Several interesting observations can be made. In general, the diagrams are highly symmetric with respect to left and right wings. The diagrams represent interactions averaged over many games, and so the symmetry demonstrates that the employed methods are robust to noise present in individual games. Also, the information-sink diagrams do differ from information-base diagrams, as expected. We begin a more detailed analysis with the information-sink interaction diagrams 1a and 1b:

- Gliders' defenders mostly respond to opponent's central mid-fielder;
- Gliders' mid-fielders mostly respond to opponent's central mid-fielder;
- Gliders' forwards mostly respond to Oxsy's defenders or Marlik's central mid-fielder;

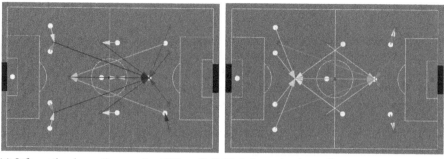

(a) Information-base diagram for Gliders (left) (b) Information-base diagram for Gliders (left) and Oxsy (right)   and Marlik (right)

Fig. 2. Interaction-base diagrams. Arrows represent highest information transfer between players. MATLAB copper colormap is used to indicate the strength of transfer, varying smoothly from black (weakest) to bright copper (strongest). Example interactions: four arrows in the right diagram from Marlik's central mid-fielder, positioned in the centre circle, to all Gliders' defenders indicate that the defenders respond mostly to the central mid-fielder's motion.

- Oxsy's wing forwards mostly respond to Gliders' side defenders, while Oxsy's centre-forward does not mostly respond to Gliders' centre-backs;
- Oxsy's side defenders mostly respond to Gliders' wing forwards, while Oxsy's centre-backs do not mostly respond to Gliders' centre-forward;
- Marlik's forwards mostly respond to Gliders' central mid-fielder;
- Marlik's defenders mostly respond to Gliders' side-wingers.

Now we turn our attention to the information-base interaction diagrams 2a and 2b:

- Gliders' defenders mostly transfer information to Oxsy's wing forwards, but not to their centre-forward;
- practically every Oxsy's player transfers infromation to Gliders' centre-forward;
- Gliders' defenders mostly transfer information to Marlik's centre-forward, but not to their wing-forwards;
- Gliders' centre-forward is transferred information from many Marlik's players, but not from their side defenders;
- Gliders's wing forwards are tightly coupled with Marlik's side defenders.

Even such a brief analysis helps to point out that in the contest with Oxsy, Gliders have a problem with their centre-backs not actively checking the opponent's centre-forward, but a similar problem also exists in Oxsy's own defense. Not surprisingly, most goals are scored in these games through the centre and not via the wing attacks and crosses. In addition, it appears that a lot of Gliders' motion is tuned to opponents' central mid-fielder which highlights a high degree of redundancy that may need to be exploited. In the games against Marlik it is evident that the opponents central mid-fielder plays a key role in both defense and attack, which again presents an opportunity to exploit such an overload. At the same time, it appears that a lot of interactions occur on the flanks of Marlik's defense (defenders mark forwards who try to evade), while Marlik's wing forwards are not marked by Gliders's side defenders.

3.2 Tactical Analysis

In this subsection, we correlate scores of relative responsiveness (either line-by-line or overall), as well as rigidity, with the game scores, and identify the lines which impacted on the games more. That is, we compute a correlation coefficient between a series of game scores and a series of information values per game.

The analysis of the games between Gliders and Oxsy shows that a sufficiently high correlation ($\rho_1 = 0.425$) exists between the game score and only one relative responsiveness $\Delta(\text{Gliders}_{def}, \text{Oxsy}_{att})$. That is, the games between these two teams are decided mostly in the opposition between Gliders' defenders and Oxsy's forwards. Specifically one may conjecture that whenever the Oxsy's forwards are more responsive in evading the defense, Oxsy tend to win, and whenever Gliders' defenders are more agile in closing on to the forwards, Gliders tend to win.

However, the main information transfer component of $\Delta(\text{Gliders}_{def}, \text{Oxsy}_{att})$, correlated with the performance, is $t_{\text{Oxsy}_{att} \rightarrow \text{Gliders}_{def}}$, at 0.553 ("our responsiveness helps our scoreline"), while the correlation with $t_{\text{Gliders}_{def} \rightarrow \text{Oxsy}_{att}}$ is just 0.089 ("opponents responsiveness does not hurt our scoreline"). This means that on average the relative agility of Gliders' defenders is correlated with the scoreline more than the response of Oxsy's forwards. This is not a causal inference, but simply a correlation observation.

The dynamic contests between Gliders' forwards and Oxsy's defenders, or between the midfield players, do not seem to be greatly correlated with the scoreline on average (the scoreline is correlated with $\Delta(\text{Gliders}_{att}, \text{Oxsy}_{def})$ at just 0.099, and with $\Delta(\text{Gliders}_{mid}, \text{Oxsy}_{mid})$ at just 0.216). The transfer components of these characteristics do not show any higher correlations either. The overall tactical relative responsiveness $\Delta(\text{Gliders}, \text{Oxsy})$ is correlated with the scoreline at a credible level of 0.310.

As expected, the relative rigidity $A(\text{Gliders}, \text{Oxsy}) = A_{\text{Gliders}} - A_{\text{Oxsy}}$ is observed to be highly anti-correlated with the scoreline: $\rho = -0.641$. The main contributing part is found to be the rigidity of the mid-fielders: the correlation of rigidity $A(\text{Gliders}_{mid}, \text{Oxsy}_{mid})$ with the performance is also quite high at -0.503, and the major component of this comes due to the rigidity of Oxsy's mid-fielders: correlation of $A(\text{Oxsy}_{mid})$ is 0.377 (it is positive as the scoreline is presented as Gliders vs Oxsy, so that higher game scores for Gliders are correlated with higher Oxsy's rigidity).

The tactical analysis of the games between Gliders and Marlik produces mostly concurring observations. In this pair, the outcome is mostly decided in the contest between Gliders' attack and Marlik's defense: the correlation between relative responsiveness $\Delta(\text{Gliders}_{att}, \text{Marlik}_{def})$ is low but statistically significant: 0.157. Interestingly, however, both individual components are anti-correlated with the scoreline: the transfer $t_{\text{Marlik}_{def} \rightarrow \text{Gliders}_{att}}$ is anti-correlated at -0.210, and the transfer $t_{\text{Gliders}_{att} \rightarrow \text{Marlik}_{def}}$ is anti-correlated at -0.366.

This poses an interesting question: why two individual components of the relative responsiveness are both anti-correlated with the scoreline, but their combination is positively correlated, albeit at a low level? One possible explanation is as follows. Both involved groups (Gliders forwards and Marlik defenders) are in almost constant interdependent motion that often confounds the players. When Marlik defenders respond to Gliders forwards' attempts to find free spots, they effectively mark and/or block the forwards, resulting in lower scores for Gliders team — hence, the negative correlation between $t_{\text{Gliders}_{att} \to \text{Marlik}_{def}}$ and the scoreline, which is seen from the Gliders' perspective ("opponents responsiveness hurts our scoreline"). However, when Gliders forwards respond to Marlik defenders' attempts to mark them, they may abandon good scoring positions, also resulting in lower scores for Gliders team — hence, the negative correlation between $t_{\text{Marlik}_{def} \to \text{Gliders}_{att}}$ and the scoreline ("our responsiveness also hurts our scoreline"). Nevertheless, when the scoreline is correlated with the relative responsiveness, rather than the individual components of the latter, the result is positive but low. This means that the remaining difference is still slightly important because of the interdependence of motion: when Gliders forwards reposition, they attract Marlik defenders again, and the 'circle' repeats, until one side gains a brief advantage ("when our responsiveness is higher than opponents responsiveness, it helps our scoreline"). In short, it is not the level of our responsiveness that is positively correlated with the scoreline, but the level of relative responsiveness.

There are no surprises with the analysis of rigidity: the relative rigidity $A(\text{Gliders}, \text{Marlik}) = A_{\text{Gliders}} - A_{\text{Marlik}}$ is highly anti-correlated with the scoreline: $\rho = -0.505$. It is interesting that (relative) rigidities of separate groups (defense, mid-field, attack) were not found to be correlated significantly with the outcomes: the dependence is detected only at the overall team level, being arguably an emergent property in this contest.

In summary, the findings demonstrate applicability of the information dynamics measures to analysis of football matches, revealing the areas of most intense competition and the extent of interactions. The latter aspect is evident when one compares the interpretations of relative responsiveness in the games against Oxsy and against Marlik. The higher responsiveness of Gliders' defenders to Oxsy's forwards was found to be positively correlated with the scoreline, while the higher responsiveness of Gliders' forwards to Marlik's defenders was anti-correlated (as was the responsiveness of Marlik's defenders to Gliders' forwards). The difference shows that in the first case responses were productive and the interaction was clearly directional, while in the second instance, the responses were strongly interdependent and the interaction was quite circular. These observations are also supported by the interaction diagrams: both information-sink and information-base diagrams 1a and 2a for Gliders vs Oxsy show that Gliders' defenders respond strongly to Oxsy's forwards, while the information-sink and information-base diagrams 1b and 2b for Gliders vs Marlik highlight the extent of cross-coupling between Gliders' forwards and Marlik's defenders. The anti-correlation of rigidity in both experimental set-ups is also encouraging: this measure can be suggested as a simple robust measure of tactical flexibility, at the emergent team level.

4   Conclusion

The paper proposed an approach for constructing interaction networks that reveal significant coupled dynamics produced during team games, or other activities that are characterised by concurrent cooperation and competition. The approach uses a novel application of information dynamics analysing pair-wise interactions and group-level tactics of RoboCup 2D Simulation League games. The input data needed for the analysis contain only positional data, such as planar coordinates and their changes, followed by computation of corresponding probability distributions and local information transfer measures. The model-free approach does not include any re-construction of the players' behaviour, being purely data-driven. Also, the method is not aimed at explicit interactions (such as passes) within a team (cf. [3]), but rather at implicit interactions, across teams, that may be delayed and/or long-ranged.

The interaction networks were exemplified with two sub-types highlighting different "slices" of the directed interactions: information-sink and information-base diagrams. In an information-sink diagram every node (every player) has an incoming edge, while in an information-base diagram every node has an outgoing edge. These diagrams were computed for two experimental set-ups that matched our team (Gliders) against two well-known teams, Oxsy [18] and Marlik [19], showing interesting player-to-player interactions, and pointing out weak spots and areas to be exploited.

The follow-up tactical analysis involved computation of information transfer and storage, and two hypotheses. The first one related positive information transfer from players Y to players X as an indication of responsiveness of the latter, suggesting to compute relative responsiveness between the opposing lines of two teams. The second hypothesis connected the information storage within the team with the team's rigidity, harming the fluidity and tactical richness of the team. This relation yielded the score for relative rigidity between the opposing teams. Both measures, relative responsiveness and rigidity, were correlated with the game results, and the obtained observations supported the hypotheses. In addition, the results pointed to important couplings that were particularly intense, and the main areas where the game outcomes were mostly decided.

This approach has been further successfully applied to opponent modelling and selecting the best available tactics in an opponent-specific way — this topic is a subject of future research. We hope that the proposed methods would be useful not only in the RoboCup leagues, but also in various analyses of team games, whether virtual or real.

Acknowledgments. This analysis was carried out using Gliders2013 [20], a follow up on Gliders2012 [11] — a simulated soccer team for the RoboCup soccer 2D simulator [21]. Gliders2012 and Gliders2013 reached semi-finals in RoboCup tournaments of 2012 and 2013. The team code is written in C++ using agent2d: the base code developed by Akiyama et al. [22], fragments of released source code of Marlik [19], as well as Gliders' in-browser basic soccermonitor (GIBBS): a log-player for viewing 2D Simulation League logs over web browser [23].

We thank Ivan Duong, Edward Moore and Jason Held for their contribution to Gliders2012 and GIBBS, as well as David Budden for his help with developing new self-localisation method. Team logo was created by Matthew Chadwick. Some of the Authors have been involved with RoboCup Simulation 2D in the past, however the code of their previous teams (Cyberoos and RoboLog, see, e.g., [10,24]) is not used in Gliders.

References

1. Vilar, L., Araujo, D., Davids, K., Bar-Yam, Y.: Science of winning soccer: Emergent pattern-forming dynamics in association football. Journal of Systems Science and Complexity 26, 73–84 (2013)
2. Grabiner, D.J.: The Sabermetric Manifesto, The Baseball Archive
3. Fewell, J., Armbruster, D., Ingraham, J., Petersen, A., Waters, J.: Basketball teams as strategic networks. PLoS One 7(11), e47445 (2012)

4. Noda, I., Stone, P.: The RoboCup Soccer Server and CMUnited Clients: Implemented Infrastructure for MAS Research. Autonomous Agents and Multi-Agent Systems 7(1-2), 101–120 (2003)
5. Riley, P., Stone, P., Veloso, M.: Layered disclosure: Revealing agents' internals. In: Castelfranchi, C., Lespérance, Y. (eds.) Intelligent Agents VII. LNCS (LNAI), vol. 1986, pp. 61–72. Springer, Heidelberg (2001)
6. Stone, P., Riley, P., Veloso, M.: Defining and using ideal teammate and opponent models. In: Proc. of the 12th Annual Conf. on Innovative Applications of Artificial Intelligence (2000)
7. Butler, M., Prokopenko, M., Howard, T.: Flexible synchronisation within RoboCup environment: A comparative analysis. In: Stone, P., Balch, T., Kraetzschmar, G.K. (eds.) RoboCup 2000. LNCS (LNAI), vol. 2019, pp. 119–128. Springer, Heidelberg (2001)
8. Reis, L.P., Lau, N., Oliveira, E.C.: Situation based strategic positioning for coordinating a team of homogeneous agents. In: Hannebauer, M., Wendler, J., Pagello, E. (eds.) Reactivity and Deliberation in MAS. LNCS (LNAI), vol. 2103, pp. 175–197. Springer, Heidelberg (2001)
9. Prokopenko, M., Wang, P.: Relating the entropy of joint beliefs to multi-agent coordination. In: Kaminka, G.A., Lima, P.U., Rojas, R. (eds.) RoboCup 2002. LNCS (LNAI), vol. 2752, pp. 367–374. Springer, Heidelberg (2003)
10. Prokopenko, M., Wang, P.: Evaluating team performance at the edge of chaos. In: Polani, D., Browning, B., Bonarini, A., Yoshida, K. (eds.) RoboCup 2003. LNCS (LNAI), vol. 3020, pp. 89–101. Springer, Heidelberg (2004)
11. Prokopenko, M., Obst, O., Wang, P., Held, J.: Gliders2012: Tactics with action-dependent evaluation functions. In: RoboCup 2012 Symposium and Competitions: Team Description Papers, Mexico City, Mexico (June 2012)
12. Lizier, J.T., Prokopenko, M., Zomaya, A.Y.: Local information transfer as a spatiotemporal filter for complex systems. Physical Review E 77(2), 026110 (2008)
13. Lizier, J.T., Prokopenko, M., Zomaya, A.Y.: Information modification and particle collisions in distributed computation. Chaos 20(3), 037109 (2010)
14. Lizier, J.T., Prokopenko, M., Zomaya, A.Y.: Coherent information structure in complex computation. Theory in Biosciences 131, 193–203 (2012)
15. Lizier, J.T., Prokopenko, M., Zomaya, A.Y.: Local measures of information storage in complex distributed computation. Information Sciences 208, 39–54 (2012)
16. Schreiber, T.: Measuring information transfer. Phys. Rev. Lett. 85(2), 461–464 (2000)
17. Wang, X.R., Miller, J.M., Lizier, J.T., Prokopenko, M., Rossi, L.F.: Quantifying and tracing information cascades in swarms. PLoS One 7(7), e40084 (2012)
18. Marian, S., Luca, D., Sarac, B., Cotarlea, O.: Oxsy 2011 team description paper. In: RoboCup 2011 Symposium and Competitions: Team Description Papers, Istanbul (2011)
19. Tavafi, A., Nozari, N., Vatani, R., Yousefi, M.R., Rahmatinia, S., Pirdir, P.: MarliK 2012 Soccer 2D Simulation Team Description Paper. In: RoboCup 2012 Symposium and Competitions: Team Description Papers, Mexico City, Mexico (June 2012)
20. Prokopenko, M., Obst, O., Wang, P., Budden, D., Cliff, O.: Gliders2013: Tactical Analysis with Information Dynamics. In: RoboCup 2013 Symposium and Competitions: Team Description Papers, Eindhoven (June 2013)
21. Chen, M., Dorer, K., Foroughi, E., Heintz, F., Huang, Z., Kapetanakis, S., Kostiadis, K., Kummeneje, J., Murray, J., Noda, I., Obst, O., Riley, P., Steffens, T., Wang, Y., Yin, X.: Users Manual: RoboCup Soccer Server — for Soccer Server version 7.07 and Later (February 2003)
22. Akiyama, H.: Agent2D Base Code (2010), http://www.rctools.sourceforge.jp
23. Moore, E., Obst, O., Prokopenko, M., Wang, P., Held, J.: Gliders2012: Development and competition results. CoRR abs/1211.3882 (2012)
24. Obst, O., Boedecker, J.: Flexible coordination of multiagent team behavior using HTN planning. In: Bredenfeld, A., Jacoff, A., Noda, I., Takahashi, Y. (eds.) RoboCup 2005. LNCS (LNAI), vol. 4020, pp. 521–528. Springer, Heidelberg (2006)

Active Object Search Exploiting Probabilistic Object–Object Relations

Jos Elfring*, Simon Jansen, René van de Molengraft, and Maarten Steinbuch

Faculty of Mechanical Engineering, Eindhoven University of Technology,
5600 MB Eindhoven, The Netherlands
j.elfring@tue.nl

Abstract. This paper proposes a probabilistic object-object relation based approach for an active object search. An important role of mobile robots will be to perform object-related tasks and active object search strategies deal with the non-trivial task of finding an object in unstructured and dynamically changing environments. This work builds further upon an existing approach exploiting probabilistic object-room relations for selecting the room in which an object is expected to be. Learnt object-object relations allow to search for objects inside a room via a chain of intermediate objects. Simulations have been performed to investigate the effect of the camera quality on path length and failure rate. Furthermore, a comparison is made with a benchmark algorithm based the same prior knowledge but without using a chain of intermediate objects. An experiment shows the potential of the proposed approach on the AMIGO robot.

1 Introduction

Domestic robots are expected to operate in human populated environments. In order to successfully complete tasks, *e.g.*, delivering objects or safe navigation, accurate descriptions of such environments are required. However, due to the limited sensing range of robots and the fact that many of the objects involved get moved regularly, a complete description will never be available.

In this work the focus will be on active object search. More specifically, a robot will be given a task such as 'Find the book "Little Red Riding Hood"'. The robot then needs a strategy to find this object. A very naive approach would be to start an exhaustive search for the book through all the rooms. Even though the robot might very well succeed, this approach is not desired since many human users will have lost their patience long before the robot finishes its task. Two different situations can be distinguished. Firstly, the position of the target object can be known in advance. The search task then simplifies to a navigation task. Secondly, in the more challenging and realistic scenario, the location of the target object is not known in advance. In this second case humans have a high success rate and the search is efficient. Most probably this is due to

---

* Corresponding author.

S. Behnke et al. (Eds.): RoboCup 2013, LNAI 8371, pp. 13–24, 2014.

the fact that humans use experience, *e.g.*, books are often seen on a table or in the book case. In [17] it was shown both empirically and mathematically that exploiting indirect object relations increases performance. Representing human experience in a robot understandable way and exploiting it during search are some of the the the main challenges in active object search.

If it comes to acquiring the common sense (or background) knowledge two possibilities are (i) learning or (ii) pre-programming it. The latter seems time consuming and will inevitably be incomplete. The first one however seems non-trivial since this knowledge is typically encoded in a human rather than robot understandable format. A popular alternative is to combine both. Various works [11,12] use human generated ontologies containing common sense knowledge, *e.g.*, 'coke is a drink' or 'a fridge is able to cool products'. In addition, this knowledge is combined with human generated knowledge available on the web. Among the popular alternatives are LabelMe [13], the Open Mind Common Sense (OMICS) [9] project and Flickr.com. In [15] LabelMe is used to get typical object-room relations, *e.g.*, 'kitchens typically contain a sink' or 'offices typically contain a bookshelf'. Afterwards, these statements are translated into probabilities. An approach leading to similar knowledge but using OMICS instead is presented in [11] whereas [10] uses Flickr for this purpose. Once this knowledge is represented by probabilistic models, the next step is exploiting it.

Already in the seventies, indirect object search was proposed as a way to limit the search space of the robot. In [3], the idea is to use an intermediate object that is easier to observe, *e.g.*, because it is larger, to find a target object. In case of the book, the robot could for example first try to find the, much larger, bookcase. This idea is the foundation of the strategy proposed in this work.

In [11], a decision-theoretic approach is presented for searching an object in large-scale indoor environments. Besides common sense knowledge, possibly complemented by robot observations, the robot also has a floor plan of the building available. Each room in the floor plan gets a probability of containing the target object based on the common sense knowledge. Both this probability and a utility function that incorporates the distance from the robot to the room are used to select the search location. An approach like this seems inevitable in large-scale environments, however, inside the rooms the robot still needs to perform an exhaustive search. Rather then selecting the most probable room, this work focuses on the search *inside* the most probable room.

In [4], a floor plan is not assumed to be available. After entering a room, it is classified autonomously and after classification a probability of containing the target object is calculated. Floor plans seldom change and for that reason the availability of a floor plan is considered reasonable. For that reason, this work aims at extending [11] rather than [4].

In [16,15] the focus is on place classification. Object-room relations are used for place classification using object detections. Similar knowledge is required and for that reason their approach for summarizing knowledge from LabelMe in probabilistic models is considered relevant in the context of this work. However, the problem investigated in this work is different and, for that reason, requires a different approach.

In [1] the focus is on organizing an efficient search given information about spatial relations between objects in an environment, *e.g.*, 'the book is on the table in room 1'. A decision theoretic strategy selection method is adopted for finding the object using the relations. The need of the spatial relations is considered to be too restrictive in the context of this work.

The application in [7,8] is finding target objects in a supermarket. Object-object attribute relations are used, rather then object-room relations, *e.g.*, 'an avocado will probably be located somewhere where other fruit can be found' or 'pizza can be found in a freezer'. A maximum entropy model steers the robot through the supermarket using these relations. The focus in this work is on the co-occurrence of objects rather than object-object attribute relations, since these are considered more useful in home or office like environments where objects are, contrary to supermarkets, not necessarily grouped by type.

Finally, [10] solves a similar problem with a different starting point, *i.e.*, a map containing objects is available. Based on a floor plan and the object locations, the best path is planned to a target object not included in the map based on the likelihoods of finding the object at each location in the map. This work differs from [10] by using a chain of intermediate objects and not assuming prior knowledge regarding object locations.

In order to find an object in large-scale indoor environments object-room relations are crucial for an efficient search. For that reason, this work builds upon the approach presented in [11]. However, [11] does not solve the problem of finding an object *inside* rooms. Indirect search using a single intermediate object is applied frequently but this work shows that it is even better to use a chain of intermediate objects, *e.g.*, books often lie at a nightstand, nightstands in turn are placed next to beds. Our first contribution is a strategy that allows using a chain of intermediate objects for active object search. A second contribution is a sensitivity analysis with respect to perception capability related parameters, *e.g.*, better camera or perception modules. Finally, we have performed an experiment to demonstrate the approach on a real robot.

The remainder of this paper is organized as follows. Section 2 explains how the object relations are learned and exploited during search. Section 3 explains the active object search strategy and Section 4 presents the findings obtained during an extensive performance analysis and comparison in simulation. Section 5 presents experimental results and Section 6 summarizes the conclusions and provides an outlook to future work.

2 Probabilistic Object Relations

2.1 Learning Object Relations

We are interested in two types of object relations. First of all, object-room relations and secondly, object-object relations. For the object-room relations we adopt the approach of [11]. In [11], human generated facts about typical combinations of rooms and objects stored in the OMICS database are matched to well-defined ontological concepts. Then the conditional probability of some object given a room is calculated by counting database entries and applying Lidstone's law with a λ of 0.5, according to Jeffrey-Perk's law, to compensate for unseen combinations.

To model object-object relations, the approach proposed in this work is to define the condition probabilities according to Lidstone's law [11,10]:

$$p\left(o_i \mid o_j\right) = \frac{N\left(o_i, o_j\right) + \lambda}{N\left(o_j\right) + \lambda n_{\text{obj}}}, \tag{1}$$

where both o_i and o_j are objects, $N\left(o_j\right)$ is the number of times object o_j is observed, $N\left(o_i, o_j\right)$ is the number of times objects o_i and o_j are observed in the same camera frame and n_{obj} is the number of objects. The dimensionless parameter λ can be used for smoothing. Furthermore, fading can be added to limit the sample size to grow unbounded [6], which is relevant in case of, e.g., time-dependence. By counting the number of times objects are observed, visibility of objects is explicitly taken into account. Objects that are easily recognized get more relations and will for that reason be used as intermediate object more often.

In this work, more than 50.000 labeled images provided by LabelMe [13] are used to calculate the probabilities in (1). A robot can refine these *general* relations by updating the probabilities based on observed *instances* in its own environment and using (1). This way the general set of object–object relations changes according to the specifics of the robot's environment.

2.2   Exploiting Object Relations

In order to exploit the conditional probabilities defined in (1) for active object search, they must be reversed. The probability of finding the target object o_T if a set of objects $\mathcal{O}$ is observed, is of interest, *i.e.*, if the target object is observed, what is the probability of observing the object currently in sight. Using Bayes' law:

$$p\left(o_j \mid o_T\right) = \frac{p\left(o_T \mid o_j\right) p\left(o_j\right)}{\sum_j p\left(o_T \mid o_j\right) p\left(o_j\right)}. \tag{2}$$

$\mathcal{O}$ is a set containing all objects observed by the robot and the summation over j represents the summation over all members of $\mathcal{O}$.

For selecting the room to which to navigate, the decision-theoretic approach introduced in [11] is adopted. For determining to which object to navigate inside that room, an expected utility, inspired by [11,7], that combines the probability of successfully finding the target object with the travel cost is maximized:

$$o^* = \underset{o_i \in \mathcal{O}}{\operatorname{argmax}} \left[p\left(o_i \mid o_T\right) + w \cdot V\left(o_i\right)\right], \tag{3}$$

where again o_T is the object that has to be found, o_i is an object in the set $\mathcal{O}$, w acts both as a relative weight and a scaling term and the travel cost is defined as the reciprocal of the arc tangent of the Euclidian distance $d_{r,o}$ from robot to object:

$$V(o) = \frac{1}{\arctan(d_{r,o})}. \tag{4}$$

The weighted sum in (3) balances between distance and probability and avoids oscillating behavior. The expected utility for some object o is represented by $\hat{U}(o)$ and given by the expression between squared brackets in (3).

3 Active Object Search Strategy

Section 2.2 introduced (3) for calculating which object from the set of observed objects $\mathcal{O}$ maximizes the expected utility. The next step is incorporating this in an active object search strategy. The first step of this strategy is to determine the room in which the object is expected to be according to the approach of [11]. As before, $\mathcal{O}$ is the set containing the objects observed by the robot. $\mathcal{F}$ is the set containing objects marked as dead end, initially $\mathcal{F} = \emptyset$. $\mathcal{T}$ is the set containing the (intermediate) target object(s), initially $\mathcal{T} = \{o_T\}$. If $|\mathcal{T}| > 1$, all members are considered in (3). The proposed search strategy is as follows:

1. $\mathcal{O} \leftarrow \mathcal{O} \backslash \mathcal{F}$, then determine o^* using (3).
2a. If $\hat{U}(o^*) \geq threshold$, start navigating towards object o^*. If during navigation new objects appear, add them to $\mathcal{O}$ and go to step 1 to see if the navigation goal has to be updated.
2b. If $\hat{U}(o^*) < threshold$, the observed objects have no sufficient relation with the target object and an intermediate object needs to be added. The object o_s with strongest relation to any of the objects in $\mathcal{T}$ according to Bayes' law is added to $\mathcal{T}$: $\mathcal{T} \leftarrow \mathcal{T} \cup \{o_s\}$. With $\mathcal{T}$ updated, go to step 1.
3a. If the robot reaches an intermediate object it is added to $\mathcal{F}$ since it did not lead to the target object. Now go back to step 1 and try to find a route via another object. If such route is not available, start a random search and update $\mathcal{O}$ if new objects are found.
3b. If the robot during navigation finds the target object, the task is succeeded
4. If the search fails, try the next best room.

For an illustrative example consider Figure 1. Both $p(o_T \mid A)$ and $p(o_T \mid D)$ are defined 0.45, $p(o_T \mid B) = 0.1$ and $p(o_T \mid C) = 0$. In Figure 1(a) the robot observes objects A, B and C. The robot moves via B to A because of the travel cost in (3). In Figure 1(b) object B is set as a dead end. After falsifying object A, the robot only observes C, however, objects C and o_T do not have a direct link. As a result, the most probable neighbor of the target object, i.e., object D since A is a dead end already, is added to the target set. This object has a relation with object C, hence the robot moves towards C, see Figure 1(c). While doing so, object D gets observed and the robot starts moving towards D. Before arriving at D, the target object is observed and the search is succeeded. This example shows how a target object can be found using a chain of intermediate objects.

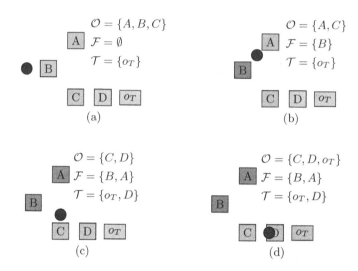

Fig. 1. Illustrative example of the search strategy. The robot is represented by the blue dot, green objects are in sight, red objects are dead ends. The target object o_T is assumed to have a 0.45 change being either next to D or A. Via a chain of intermediate objects, the target object is found.

4   Simulation Results

A large number of simulations is performed to (i) investigate the effect of the robot's sensors on the success of the search and (ii) compare the proposed strategy with a benchmark strategy. Simulations enable a high number of trials under the exact same conditions, thereby allowing for a fair comparison. Section 4.1 explains how we have updated the general relations using simulated data. After that, Section 4.2 briefly explains a benchmark search strategy and Section 4.3 presents the simulation results.

4.1   Simulation Set-Up

For all simulations, the general relations from LabelMe are refined by adding a large number of observations in different but similar simulated rooms. One simulated room, referred to as recipe room, was created. Based on a typical living room ten main objects were placed inside this room, *e.g.*, a dining table, two couches, etc. Furthermore twenty smaller objects were defined with locations relative to one or more main objects, *e.g.*, chairs stand next to the dining table, keys can be on any of four tables. All positions are disturbed by Gaussian noise. The recipe room was used to create 30 random rooms in which the object–object relations were learned. The simulated robot has a 360° viewing angle and the target object was always present in the room. Table 1 shows some of the typical condition probabilities of observing two objects in the same camera frame learned from the recipe room. A typical path towards the keys resulting from the proposed strategy is, *e.g.*, from the dining table towards the three seater via the chair, then from three seater to the coffee table on which the keys are found.

Table 1. Example probabilities learned from simulation

	Coffee table	Side table	Two seater
Book	0.20	0.33	0.11
Three seater	0.23	0.11	0.12
Keys	0.21	0.31	0.09
Remote control	0.01	0.19	0.13

For validation purposes, a second data set was obtained by redesigning the recipe room, *i.e.*, the dining table moved to the other side of the room and couches in a different formation. As a result, the smaller objects changed position too.

In order to be able to analyze the effect of the robot's sensors the view scale is varied. The view scale *VS*, in [14] referred to as effective range of detection, can be interpreted as the simulated resolution of the camera and is calculated as:

$$VS = \frac{d_{max}}{S_{avg}},$$ (5)

where d_{max} is the maximum distance from which an object with average size S_{avg} is detectable. *VS* is assumed independent of the object class and will be varied during the simulations. In [14] *VS* lies around 12, a Microsoft Kinect with object recognition using SIFT feature points has *VS* around 5 for typical household items on our robot.

4.2   Benchmark Search Strategy

For comparison a search strategy only exploiting direct neighbors was implemented, *i.e.*, a strategy that does not allow for a chain of intermediate objects. The direct neighbors in both strategies are the same. The strategy is as follows:

1. Check the probabilities of the objects neighboring the target object
2. Check the camera image for objects
3a. If none of the neighboring objects is in field of view, move to the center of the room while looking for objects
3b. If at least one neighboring object is found, navigate towards the most probable neighboring object
4a. If the robot during navigation finds the target object, the task is succeeded
4b. If the neighboring object is reached without observing the target, try another neighboring object
4c. If no new neighboring objects are found, start a random search

4.3   Results

Both the benchmark strategy and the strategy introduced in this work are tested on failure rate. The failure rate is the percentage of search tasks that would eventually end up with a random object search. The camera *VS* during the learning of the relations was arbitrarily set to 4. The effect of *VS* during learning will be investigated later. After that, the robot was asked to find the remote control, which has various possible locations as was already shown in Table 1. To investigate the effect of the *VS* the failure rate was calculated for 27 different values of *VS* in 25 different rooms per *VS*. Each room was entered from each corner for both methods, hence the results are an average of 5400 simulations. The results are shown in Figure 2.

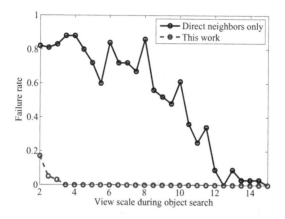

Fig. 2. Failure rate as a function of *VS* during the active object search

This figure shows a failure rate that rapidly goes to zero for the proposed approach. Only for very low values of *VS*, *i.e.*, a very limited number of object detections due to the very limited sensing range, the search failed. For more realistic values of *VS* the robot was always able to find a chain of intermediate objects leading to the target object, which means that the robot was performing a directed search. For the benchmark strategy only exploiting direct relations the failure rate is much worse. This is due to the fact that direct neighbors of the remote control, being the coffee table, side table, two seater and tv, are only observable in a relatively small portion of the room. The noisy line is caused by the sensitivity to the configuration of the limited number of direct neighbors. As *VS* increases, objects can be detected from larger distances and the failure rate drops. Increasing the number of 25 different rooms per configuration will lower the noise, increasing the room size will enlarge the differences.

Figure 3 shows the path lengths for the same set of simulations ignoring those that failed. For comparison, the path length using a very naive zigzag search strategy was added.

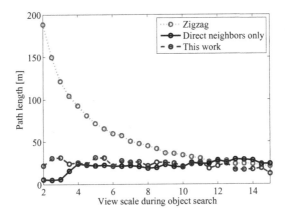

Fig. 3. Path length as a function of VS during the active object search

For low values of VS the benchmark approach seems to outperform the others, however, this method fails most of the time. The trails that succeeded started from a corner which appeared to be very close to the target object and as a result the average path length during the successful trials is very low. In case of the numerous failures, a random search has to be started and that will require many additional meters as can be seen from the red line. Such random search is however not yet combined with the benchmark strategy to allow for a fair comparison. The benchmark strategy does not incorporate the distance from robot to an object and for that reason, it performs worse than the naive zigzag strategy for very high values of VS. The proposed approach appears to be insensitive for the performance of the camera and is likely to outperform the benchmark approach over the full range of possible VS values. For very high values of VS, objects can be seen from many places in the room and for that reason, the differences are getting smaller.

To further investigate the effect of VS on the performance, the experiment above was repeated with relations that were learned by a robot with $VS = [5, 10, 15]$. The object search is again performed for 27 values of VS and starting from each corner of the room. Again each setting was simulated 25 times hence the average results of 8100 simulations are shown in Figure 4.

Two conclusions can be drawn from this figure. First of all the value of VS during learning does not have an effect on the path length during the active object search. As a result, it should be possible to share the relations among different robots with different perceptual capabilities without negatively influencing the performance. Secondly, the total path length decreases for VS above 9 or 10. Due to the different scale on the vertical axis, this conclusion can hardly be made from Figure 3 that partly shows the same information.

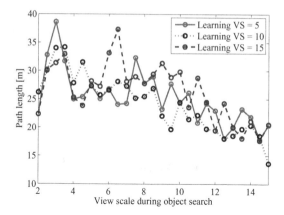

Fig. 4. Effect of *VS* during learning on the path length as a function of *VS* during the active object search

5   Experiment

In order to test the active search strategy on a real robot, our AMIGO robot was asked to find a lamp. The relations used in this experiment are learned from the LabelMe data set. Existing LabelMe tools [13] are used to unify the annotations, *e.g.*, nightstand and bedstand are synonyms. Relations are only considered if $N(o_i, o_j) \geq 5$ to avoid atypical object–object relations that might appear in the LabelMe database. The experiment is carried out in an atypical lab environment and for that reason, the room selection strategy is omitted in this experiment and the number of objects is somewhat lower than in a real world environment.

All perception on the robot uses RGB-D data obtained via a Kinect. By detecting horizontal planes at heights in a certain range, tables are detected. Objects are recognized using color information and/or using both 2D and 3D templates [5], anchoring, tracking and data association is performed using [2].

Figure 5 shows both a photograph of and a occupancy grid map of the lab. The numbers in the list below refer to the positions in the lab indicated in Figure 5(a):

1. Inside the lab, AMIGO detects a table and the bed. The table has a direct relation with the target object and for that reason, AMIGO navigates towards the table. On top of the table, no lamp was observed. Bed has no relations with lamp and for that reason, the most probable neighbors of lamp, being table and ceiling, are added as intermediate object. A ceiling is not observed, whereas the bed is, hence AMIGO navigates towards the bed.
2. While navigating towards the bed, the small table next to the bed is observed.
3. While navigating towards this table, the lamp is found on top of it.

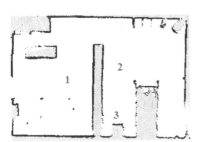 

(a) Occupancy grid map of the lab in which the search takes place.   (b) Photograph of the lab with AMIGO from the entrance.

Fig. 5. Lab environment in which the experiment is carried out

6   Conclusion and Future Work

This work introduced an active object search strategy exploiting probabilistic object-object relations learned based on their co-occurrence in camera images. The strategy is particularly useful inside rooms and is combined with the strategy introduced in [11], which calculates in which room to expect an object. The strategy is extensively analyzed in a large number of simulations. Simulations showed (1) the advantage of using a chain of intermediate objects over using direct neighbors only, (2) better perceptual capabilities lead to shorter paths during the search and (3) the perceptual capabilities of a robot during the learning of the relations do not influence the performance during the search. An experiment involving a real robot has shown the potential of the approach in the real world.

Future work can be to further validate the performance on real robots. Furthermore, the dependency of object–object relations on the room type should be investigated.

Acknowledgment. The research leading to these results has received funding from the European Union Seventh Framework Program FP7/2007-2013 under grant agreement no. 248942 RoboEarth.

References

1. Aydemir, A., Sjöö, K., Folkesson, J., Pronobis, A., Jensfelt, P.: Search in the real world: Active visual object search based on spatial relations. In: 2011 IEEE International Conference on Robotics and Automation, pp. 2818–2824 (2011)
2. Elfring, J., van den Dries, S., van de Molengraft, M., Steinbuch, M.: Semantic world modeling using probabilistic multiple hypothesis anchoring. Robotics and Autonomous Systems 61(2), 95–105 (2013)
3. Garvey, T.D.: Perceptual strategies for purposive vision. Ph.D. thesis, Stanford University, Stanford, CA, USA, aAI7613006 (1976)

4. Hanheide, M., Gretton, C., Dearden, R.W., Hawes, N.A., Wyatt, J.L., Pronobis, A., Aydemir, A., Göbelbecker, M., Zender, H.: Exploiting probabilistic knowledge under uncertain sensing for efficient robot behaviour. In: Proceedings of the 22nd International Joint Conference on Artificial Intelligence (IJCAI 2011), Barcelona, Spain (July 2011)

5. Hinterstoisser, S., Holzer, S., Cagniart, C., Ilic, S., Konolige, K., Navab, N., Lepetit, V.: Multimodal templates for real-time detection of texture-less objects in heavily cluttered scenes. In: 2011 IEEE International Conference on Computer Vision (ICCV), pp. 858–865 (November 2011)

6. Jensen, F.V., Nielsen, T.D.: Bayesian Networks and Decision Graphs. Springer (2007)

7. Joho, D., Burgard, W.: Searching for objects: Combining multiple cues to object locations using a maximum entropy model. In: 2010 IEEE International Conference on Robotics and Automation (ICRA), pp. 723–728 (May 2010)

8. Joho, D., Senk, M., Burgard, W.: Learning search heuristics for finding objects in structured environments. Robotics and Autonomous Systems 59(5), 319–328 (2011); special Issue ECMR 2009

9. Kochenderfer, M.J., Gupta, R.: Common sense data acquisition for indoor mobile robots. In: Nineteenth National Conference on Artificial Intelligence, AAAI 2004, pp. 605–610. AAAI Press / The MIT Press (2003)

10. Kollar, T., Roy, N.: Utilizing object-object and object-scene context when planning to find things. In: IEEE International Conference on Robotics and Automation, ICRA 2009, pp. 2168–2173 (May 2009)

11. Kunze, L., Beetz, M., Saito, M., Azuma, H., Okada, K., Inaba, M.: Searching objects in large-scale indoor environments: A decision-theoretic approach. In: 2012 IEEE International Conference on Robotics and Automation (ICRA), pp. 4385–4390 (May 2012)

12. Pronobis, A., Jensfelt, P.: Large-scale semantic mapping and reasoning with heterogeneous modalities. In: Proceedings of the 2012 IEEE International Conference on Robotics and Automation (ICRA 2012), Saint Paul, MN, USA, pp. 3515–3522 (May 2012)

13. Russell, B., Torralba, A., Murphy, K., Freeman, W.: Labelme: A database and web-based tool for image annotation. International Journal of Computer Vision 77, 157–173 (2008)

14. Tsotsos, J.K., Shubina, K.: Attention and Visual Search: Active Robotic Vision Systems that Search. In: 5th International Conference on Computer Vision Systems Conference (March 2007),
http://biecoll.ub.uni-bielefeld.de/volltexte/2007/1

15. Viswanathan, P., Southey, T., Little, J., Mackworth, A.: Place classification using visual object categorization and global information. In: 2011 Canadian Conference on Computer and Robot Vision (CRV), pp. 1–7 (May 2011)

16. Viswanathan, P., Southey, T., Little, J., Mackworth, A.: Automated place classification using object detection. In: 2010 Canadian Conference on Computer and Robot Vision (CRV), May 31-June 2, vol. 2, pp. 324–330 (2010)

17. Wixson, L.E., Ballard, D.H.: Using intermediate objects to improve the efficiency of visual search. International Journal of Computer Vision 12, 209–230 (1994)

Online Generated Kick Motions for the NAO Balanced Using Inverse Dynamics

Felix Wenk[1] and Thomas Röfer[2]

[1] Universität Bremen, Fachbereich 3 – Mathematik und Informatik,
Postfach 330 440, 28334 Bremen, Germany
`fwenk@informatik.uni-bremen.de`
[2] Deutsches Forschungszentrum für Künstliche Intelligenz,
Cyber-Physical Systems, Enrique-Schmidt-Str. 5, 28359 Bremen, Germany
`thomas.roefer@dfki.de`

Abstract. One of the major tasks of playing soccer is kicking the ball. Executing such complex motions is often solved by interpolating key-frames of the entire motion or by using predefined trajectories of the limbs of the soccer robot. In this paper we present a method to generate the trajectory of the kick foot online and to move the rest of the robot's body such that it is dynamically balanced. To estimate the balance of the robot, its Zero-Moment Point (ZMP) is calculated from its movement using the solution of the Inverse Dynamics. To move the ZMP, we use either a Linear Quadratic Regulator on the local linearization of the ZMP or the Cart-Table Preview Controller and compare their performances.

1 Introduction

To play humanoid soccer, two essential motion tasks have to be carried out: walking over a soccer field and kicking the ball. Both motions have to be both flexible and robust, i.e. the robot has to be able to walk in different directions at different speeds and kick the ball in different directions with different strengths, all while maintaining its balance to prevent falling over.

To design and execute motions to kick the ball, different methods have been developed. A seemingly obvious approach to motion design is to manually set up some configurations, i.e. sets of joint angles called key-frames, which the robot shall assume during the motion, and then interpolate between these key-frames while the motion is executed. This quite popular method has been used to design kick motions by a number of RoboCup Standard Platform League (SPL) teams including B-Human [12] and Nao Team HTWK [13].

Because it completely determines the robot's motion, the interpolation between fixed sets of joint angles precludes any reaction to changing demands or to external disturbances. Therefore, Czarnetzki et al. [3] specify key-frames of the motion of the robot's limbs in Cartesian space instead of joint space. This leaves the movement of the robot under-determined, so the Cartesian key-frame approach can be combined with a controller to maintain balance.

S. Behnke et al. (Eds.): RoboCup 2013, LNAI 8371, pp. 25–36, 2014.
© Springer-Verlag Berlin Heidelberg 2014

To design more flexible kick motions, Müller *et al.* [11] model the trajectories of the robot's hands and feet in Cartesian space using piecewise Bezier curves. Depending on the position of the ball relative to the kicking robot and the desired kick direction, the Bezier curves are modified such that the kick foot actually hits the ball in the desired direction. The modification of the Bezier curves is constrained such that the resulting curve is always continuously differentiable, which results in a smooth trajectory. In addition, a balancing controller is included to maintain static balance by tilting the robot such that its center of mass (COM) stays within the support polygon, i.e. the contour of the support foot. To make sure the robot gets properly tilted, the angular velocity measured by gyroscopes in the robot's torso is used as feedback.

In this work, the approaches mentioned in this introduction are combined to a motion engine that generates and executes dynamically balanced kick motions, but neither requires prior modeling of trajectories of limbs nor needs prerecorded key-frames to interpolate. Instead of searching the path of the kick foot to the ball like Xu *et al.* [15], the trajectory of the kick foot is an interpolation between a number of reference poses inferred from the ball position, the kick direction and the kick strength. The points are interpolated using a spline that is continuously differentiable twice so that the trajectory of the kick foot has no sudden jumps in acceleration [6]. The limbs of the robot, which are not part of the kick leg and whose motion is therefore not determined by the kick foot trajectory, are moved to maintain dynamic balance, i.e. to keep the Zero-Moment Point (ZMP) [14] within the support polygon. To achieve this, the ZMP is calculated via the solution of the Inverse Dynamics based on an estimate of the motion of the joints. Two different methods to move the ZMP are implemented and compared: first a Linear Quadratic Regulator (LQR) [9,4], which modifies the joint angles directly using a linearization of the ZMP depending on the motion of the joints, and second a Preview Controller [10] which generates a trajectory for the COM depending the current and the preview of the future ZMP [8]. The COM trajectory is then translated to joint angles using inverse kinematics.

The rest of the paper is organized as follows. The generation of the trajectory of the kick foot is treated in Sect. 2. Section 3 covers the motion of the robot to maintain dynamic balance. The latter includes the calculation of the balance criterion, the ZMP, and therefore the solution of the Inverse Dynamics. Experiments and their results including a comparison of the two balancing methods make up Sect. 4. Section 5 finally concludes this work.

2   Generating the Kick Foot Motion

To generate the kick foot motion, a number of reference points have to be calculated. At first it has to be determined which part of the contour of the foot should hit the ball. Because the contour of the foot is round and the kick foot will not be rotated during the kick, the kick is approximately a collision between two spheres. So the tangent at the contact point on the foot contour has to be orthogonal to the kick direction. To calculate the tangent, we approximate the contour of the front of the foot with a cubic Bezier curve.

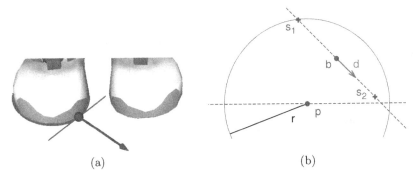

(a)   (b)

Fig. 1. a) The kick foot is to collide with the ball at the point where the foot contour's tangent is orthogonal to the kick direction. b) Geometric construction of the strikeout s_1 and swing s_2 reference points.

A cubic Bezier curve of dimension d is described by four control points $p_0, p_1, p_2, p_3 \in \mathbb{R}^d$, which jointly determine the resulting curve

$$C(t) = \sum_{i=0}^{3} B_{i,[\alpha,\beta]}^3(t) \cdot p_i \quad \text{with } t \in [\alpha, \beta], \alpha, \beta \in \mathbb{R} \ . \tag{1}$$

$B_{i,[\alpha,\beta]}^n(t) = \binom{n}{i} \cdot \left(\frac{t-\alpha}{\beta-\alpha}\right)^i \cdot \left(\frac{\beta-t}{\beta-\alpha}\right)^{n-1}$ is the i-th Bernstein polynomial of degree n defined on the interval $[\alpha, \beta]$ [6]. The derivative of such a curve is

$$\dot{C}(t) = \frac{3}{\beta - \alpha} \cdot \sum_{i=0}^{2} (p_{i+1} - p_i) B_{i,[\alpha,\beta]}^2(t) \ . \tag{2}$$

As shown in Fig. 1a, the ball contact point on the contour C_{foot} is then

$$C_{\text{foot}}(t_{\text{bc}}) \quad \text{with } \dot{C}_{\text{foot}}(t_{\text{bc}}) \cdot d = 0 \ , \tag{3}$$

where $d \in \mathbb{R}^2$ is the normalized kick direction vector in the plane of the soccer field. Since the support foot (hopefully!) does not move during a kick, its coordinate system serves as the reference for all coordinates related to the kick foot trajectory. The resulting curve describes the trajectory of the origin of the kick foot. Since $C_{\text{foot}}(t_{\text{bc}})$ is relative to the origin of the kick foot, the ball contact reference point b satisfies $b + C_{\text{foot}}(t_{\text{bc}}) = p_{\text{ball}}$, where $p_{\text{ball}} \in \mathbb{R}^2$ is the ball position on the pitch.

The points $s_1 \in \mathbb{R}^2$ and $s_2 \in \mathbb{R}^2$ to strike out and swing out the kick foot are the intersections of the line in kick direction through the ball reference point with a circle of radius r around the pelvis joint at $p \in \mathbb{R}^2$ of the support leg. The construction is pictured in Fig. 1b. p, s_1 and s_2 are all in a plane parallel to the field. The height of the plane is a parameter to be tuned by the user. r is determined manually so that the kick foot can reach all points on the circle.

If s_2 gets too close to the support leg side of the pelvis or even enters it, it is pulled back on the kick direction line as pictured in Fig. 1b, so that a small safety margin to the support leg remains and collisions between the kick foot and the support leg are avoided.

Aside from the ball position and the kick direction, the user of the kick engine also specifies the duration T of the kick foot motion and the speed v the kick foot should have when the ball is hit.

This information is used to determine the durations of the individual pieces of the curve. The curve must pass through the six points r_j with $0 \leq j < 6$ in order. The i-th curve segment starts at r_i, ends at r_{i+1} and takes the duration Δ_i. The segment around the ball from $r_2 = b - \lambda_1 d$ to $r_3 = b + \lambda_2 d$ shall be passed in $\Delta_2 = \frac{\lambda_1 + \lambda_2}{v}$. It turned out that a good choice for λ_1 and λ_2 is if they sum up to a little more than one diameter of the ball. The speed at which to move the kick foot back to the end position of the trajectory is set to $\frac{v}{4}$, so the duration for the last phase is $\Delta_4 = \frac{4 \cdot \|r_5 - r_4\|}{v}$. The other durations are calculated such that they are proportional to the square root of the length between their end points and sum up to $T - \Delta_2 - \Delta_4$. This is also called centripetal parameterization [6].

The durations Δ_i with $0 \leq i < 5$ are equivalent to a knot sequence of a cubic B-Spline curve with the clamped end condition [6]. The clamped end condition requires that the derivative of the curve is 0 at the ends, i.e. in this case $\dot{B}(0) = \dot{B}(T) = 0$, meaning physically that the kick foot does not move. B-Spline curves are a generalization of Bezier curves. As a cubic Bezier curve, a cubic B-Spline curve is a linear combination of control points d_j, which are also called de Boor points [6,2].

$$B(u) = \sum_{j=0}^{L} N_j^3(u) d_j \qquad (4)$$

$N_j^n(u)$ are called the B-Spline basis functions or just B-Splines and are recursively defined as

$$N_j^n(u) = \frac{u - u_{j-1}}{u_{j+n-1} - u_{j-1}} N_j^{n-1}(u) + \frac{u_{j+n} - u}{u_{j+n} - u_j} N_{j+1}^{n-1}(u) \qquad (5)$$

$$\text{with } N_j^0(u) = \begin{cases} 1 & \text{if } u_{j-1} \leq u \leq u_j \\ 0 & \text{otherwise} \end{cases}.$$

L in (4) is the index of the last control point. For a B-Spline curve of degree n, this is $L = K + 1 - n$, where $K + 1$ is the number of knots. A B-Spline curve is defined over the interval $[u_{n-1}, u_L]$, i.e. in the cubic case over $[u_2, u_L]$. Each reference point is associated with a knot. Since we want the curve to start at r_0 and end at r_5, this means that $B(u_{i+2}) = r_i$. This implies that $L = 7$, so there are 8 control points and $K + 1 = L + n = 10$ knots in the knot sequence $u_0, \ldots, u_9$. The clamped end condition defines the knots that are not associated to reference points to be $u_0 = u_1 = u_2$ and $u_9 = u_8 = u_7$. Given the durations between the reference points and that we want the curve to be defined on the interval $[0, T]$, the knots are

$$u_2 = 0 \quad \text{and} \quad u_{i+2} = u_{i+2-1} + \Delta_{i-1} \quad \text{for } 0 < i < 5 . \tag{6}$$

Due to the clamped end condition the first two control points collapse with the first reference point and the last two control points collapse with the last reference point, so $d_0 = d_1 = r_0$ and $d_6 = d_7 = r_5$. By inserting the association between knots and reference points $B(u_{i+2}) = r_i$ into (4) and by replacing the known control points with the corresponding reference points, we get a system of four equations for the remaining four control points. Figure 2 shows two different curves calculated by using this scheme. Since the B-Spline curves are infinitely often continuously differentiable between the knots and $n - r$ times continuously differentiable at a knot occurring r times in the knot sequence [6], the resulting curve is always at least $3 - 1$ times continuously differentiable.

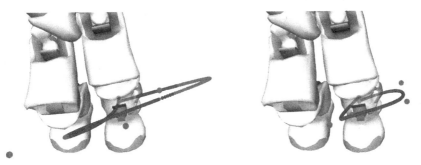

Fig. 2. Trajectories (blue) of the kick foot with the corresponding de Boor points (red dots). On the left, the speed of the kick foot is too large, so that some parts of the trajectory (red) are unreachably far away from the hip. On the right this is fixed by lowering the speed of the foot, resulting in a completely reachable trajectory.

If the speed v of the kick foot is too high relative to the kick duration T, the segment leading up to the ball contact segment may get too long, so that some parts of the curve are not reachable by the kick foot. To check whether this is the case, we calculate the point $B(u_{max})$ on that segment with the maximum (squared) distance from the pelvis joint of the kick leg and test whether this is larger than the maximum distance allowed d_{max}, i.e. whether $B(u_{max}) \cdot B(u_{max}) > d^2_{max}$. If this is true we define the error function

$$e(v) = \| B(u_{max}) \cdot B(u_{max}) - d^2_{max} \| \tag{7}$$

by making the speed v variable and keeping all other parameters of the curve construction fixed. Knowing that the current v is too large, we calculate the largest acceptable kick foot speed as $v' = \text{argmin}_v e(v)$ using (7). The effect of this procedure is pictured in Fig. 2, where an infeasible curve with a high kick foot speed is turned into a feasible one.

3 Dynamically Balanced Kick Foot Motion

Now that the kick foot trajectory is generated, it needs to be executed while the robot is dynamically balanced.

3.1 Balance Estimation

To determine how balanced the robot is, we estimate the difference between the current ZMP p_0 and a demanded ZMP $p_{0,d}$. The ZMP is defined as the point in the support polygon, i.e. the sole of the support foot, at which the (vertical) reaction force the ground exerts on the foot must act, such that all (horizontal) torques are canceled out [14]. If p_0 does not exist and the then-fictional ZMP lies outside of the foot sole, the robot will begin to rotate about the edge of the sole and probably fall over eventually. With $f_{r,z}$ being the vertical component of the reaction force, τ_{xy} the total torque of the robot relative to the projection of the support foot's origin on the ground, n the normal vector on the ground and $\times$ denoting the cross product of two vectors, the ZMP p_0 satisfies

$$p_0 \times f_{r,z} + \tau_{xy} = 0 \quad \text{and therefore} \quad p_0 = \frac{n \times (-\tau_{xy})}{n \cdot f_{r,z}} = \frac{1}{f_{r,z}} \begin{bmatrix} \tau_y \\ -\tau_x \\ 0 \end{bmatrix} . \quad (8)$$

If the ZMP exists as in Fig. 3, the robot is said to be dynamically balanced.

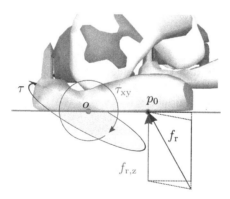

Fig. 3. The vertical component $f_{r,z}$ of the reaction force f_r acts on the Zero-Moment point p_0 to cancel out the horizontal component τ_{xy} of the total torque τ relative to the origin of the sole of the support foot

To calculate the ZMP, we need to know $f_{r,z}$ and τ_{xy}. To infer these from the robot motion, we estimate the angle θ_j, angular velocity $\dot{\theta}_j$ and angular acceleration $\ddot{\theta}_j$ of each joint j and solve the Inverse Dynamics problem using a variation [5] of the Recursive Newton-Euler Algorithm (RNEA) [7]. As the joint angle measurements are too noisy for simple numerical differentiation, we obtain the estimates for $\dot{\theta}_j$ and $\ddot{\theta}_j$ by numerically differentiating the target joint angles instead. Using Kalman filters [9] for each non-leg joint and one Kalman filter for each support leg, we filter the differences $\Delta\theta_j$, $\Delta\dot{\theta}_j$, and $\Delta\ddot{\theta}_j$ between the actual and commanded joint motions similar to the work of Belanger [1]. These are added to the commanded joint motions obtained by numerical differentiation. This allows us to use very small process variances in the filters, so that the estimates do not jump wildly while still having the estimates respond properly to changes of the joint motions due to our own software.

The motions of the joints of a single leg are filtered in a combined state instead of using individual filters, because we use the rotation of the gyroscope in the NAO's torso as a measurement model to correct the angular velocities of the joints of the support leg.

Since the support foot has no velocity and acceleration during the kick, it serves as the root of the kinematic tree, which is used by the RNEA to calculate the torques and forces across the joints. We also assume an imaginary 'joint' with zero degrees of freedom between the ground and the support foot. Because the support foot does not move, the torque and force exerted on the support foot via the imaginary joint must be the reactions to the torque and force of the support foot and thus $-\boldsymbol{\tau}$ and $\boldsymbol{f}_r$ in (8). Since we do not need to know the forces transmitted by the real joints of the robot, we use a simplified version of the RNEA by Fang $et\ al.$ [5], which only propagates the aggregated momenta and forces from the leaves of the tree to the root. From Fang $et\ al.$ [5], we also implemented taking the derivates

$$-\frac{\partial}{\partial\theta_j}\boldsymbol{\tau},\ -\frac{\partial}{\partial\dot{\theta}_j}\boldsymbol{\tau},\ -\frac{\partial}{\partial\ddot{\theta}_j}\boldsymbol{\tau},\ \frac{\partial}{\partial\theta_j}\boldsymbol{f}_r,\ \frac{\partial}{\partial\dot{\theta}_j}\boldsymbol{f}_r,\ \frac{\partial}{\partial\ddot{\theta}_j}\boldsymbol{f}_r \quad \text{for each joint } j. \quad (9)$$

Using these partial derivatives of (9), the partial derivatives of the ZMP $\boldsymbol{p}_0$ from (8) with respect to the joint motion are calculated by

$$\frac{\partial}{\partial\phi_j}\boldsymbol{p}_0 = \frac{\left[\boldsymbol{n}\times\left(-\frac{\partial}{\partial\phi_j}\boldsymbol{\tau}\right)_{xy}\right][\boldsymbol{n}\cdot\boldsymbol{f}_{r,z}] - \left[\boldsymbol{n}\times(-\boldsymbol{\tau})_{xy}\right]\left[\boldsymbol{n}\cdot\left(\frac{\partial}{\partial\phi_j}\boldsymbol{f}_r\right)_z\right]}{(\boldsymbol{n}\cdot\boldsymbol{f}_{r,z})^2} \quad (10)$$

with $\phi_j \in \left\{\theta_j,\dot{\theta}_j,\ddot{\theta}_j\right\}$ and j again being a joint of the robot.

$\boldsymbol{p}_{0,d}$ is moved from its initial position between both feet on the ground to a position within the sole of the support foot before the kick foot is moved, stays there while the kick foot is moved, and is moved back to its initial position when the kick foot reached its final position.

3.2   Balancing with Linear Quadratic Regulation

We implemented a balancer by directly modifying the joint angles of the support leg. To control the bearing of the robot's torso, both the pitch joints of hip and ankle as well as the roll joints of hip and ankle are coupled by a factor β, so that if the pitch pair is modified by $\Delta\theta'_{\text{pitch}}$, the hip joint of the pair will be modified by $\Delta\theta_{\text{hippitch}} = \Delta\theta'_{\text{pitch}}$ and the ankle joint by $\Delta\theta_{\text{anklepitch}} = \beta\Delta\theta'_{\text{pitch}}$. The roll pair is analogous. With these joint pairs, the change of the ZMP is determined by applying the chain rule using (10):

$$\frac{\partial}{\partial\phi'_{\text{pitch}}}p_0 = \frac{\partial}{\partial\phi_{\text{hippitch}}}p_0 + \beta \cdot \frac{\partial}{\partial\phi_{\text{anklepitch}}}p_0 \quad \text{with } \phi \in \{\theta, \dot{\theta}, \ddot{\theta}\} \qquad (11)$$

$\frac{\partial}{\partial\phi'_{\text{roll}}}p_0$ is analogous.

We use (11) to construct a linear model to be used in a discrete-time, finite-horizon LQR [4]. If the robot was perfectly balanced, the ZMP p_0 would coincide with the demanded ZMP $p_{0,d}$ and the error $e = p_0 - p_{0,d}$ would be zero. Assuming the demanded ZMP is constant, the linearization of the error over the time interval between time T_t and T_{t+1} with respect to the motions of the joint pairs is $e_{t+1} = e_t + \Delta e$ with

$$\Delta e = \dot{\theta}'_{\text{roll}} \overbrace{\frac{\partial p_0}{\partial\theta'_{\text{roll}}}\Delta T}^{a_1} + \ddot{\theta}'_{\text{roll}} \overbrace{\left(\frac{\partial p_0}{\partial\theta'_{\text{roll}}}\frac{1}{2}(\Delta T)^2 + \frac{\partial p_0}{\partial\dot{\theta}'_{\text{roll}}}\Delta T\right)}^{a_2}$$

$$+ \dddot{\theta}'_{\text{roll}} \overbrace{\left(\frac{\partial p_0}{\partial\theta'_{\text{roll}}}\frac{1}{6}(\Delta T)^3 + \frac{\partial p_0}{\partial\dot{\theta}'_{\text{roll}}}\frac{1}{2}(\Delta T)^2 + \frac{\partial p_0}{\partial\ddot{\theta}'_{\text{roll}}}\Delta T\right)}^{b_1}$$

$$+ \dot{\theta}'_{\text{pitch}} \overbrace{\frac{\partial p_0}{\partial\theta'_{\text{pitch}}}\Delta T}^{a_3} + \ddot{\theta}'_{\text{pitch}} \overbrace{\left(\frac{\partial p_0}{\partial\theta'_{\text{pitch}}}\frac{1}{2}(\Delta T)^2 + \frac{\partial p_0}{\partial\dot{\theta}'_{\text{pitch}}}\Delta T\right)}^{a_4}$$

$$+ \dddot{\theta}'_{\text{pitch}} \overbrace{\left(\frac{\partial p_0}{\partial\theta'_{\text{pitch}}}\frac{1}{6}(\Delta T)^3 + \frac{\partial p_0}{\partial\dot{\theta}'_{\text{pitch}}}\frac{1}{2}(\Delta T)^2 \frac{\partial p_0}{\partial\ddot{\theta}'_{\text{pitch}}}\Delta T\right)}^{b_2}, \qquad (12)$$

and $\Delta T = T_{t+1} - T_t$. This leads to the linear model $x_{t+1} = Ax_t + Bu_t$ with

$$x = \begin{bmatrix} e & \dot{\theta}'_{\text{roll}} & \ddot{\theta}'_{\text{roll}} & \dot{\theta}'_{\text{pitch}} & \ddot{\theta}'_{\text{pitch}} \end{bmatrix}^\top \quad u = \begin{bmatrix} \dddot{\theta}'_{\text{roll}} & \dddot{\theta}'_{\text{pitch}} \end{bmatrix}^\top \qquad (13)$$

and

$$A = \begin{bmatrix} 1 & 0 & a_{1,x} & a_{2,x} & a_{3,x} & a_{4,x} \\ 0 & 1 & a_{1,y} & a_{2,y} & a_{3,y} & a_{4,y} \\ 0 & 0 & 1 & \Delta T & 0 & 0 \\ 0 & 0 & 0 & 1 & 0 & 0 \\ 0 & 0 & 0 & 0 & 1 & \Delta T \\ 0 & 0 & 0 & 0 & 0 & 1 \end{bmatrix} \quad B = \begin{bmatrix} b_{1,x} & b_{2,x} \\ b_{1,y} & b_{1,y} \\ \frac{1}{2}(\Delta T)^2 & 0 \\ \Delta T & 0 \\ 0 & \frac{1}{2}(\Delta T)^2 \\ 0 & \Delta T \end{bmatrix}. \qquad (14)$$

With this model, the controls to be applied at time T_t are

$$u_t = -\left(R + B^\top P_k B\right)^{-1} B^\top P_k A x_t \qquad (15)$$

with

$$P_k = Q + A^\top P_{k-1} A - A^\top P_{k-1} B \left(R + B^\top P_{k-1} B\right)^{-1} B^\top P_{k-1} A, \qquad (16)$$

where $P_0 = Q$, k is the horizon parameter, and Q and P are the diagonal, positive-definite matrices of the quadratic cost function $C(\boldsymbol{x}, \boldsymbol{u}) = \boldsymbol{x}^\top Q \boldsymbol{x} + \boldsymbol{u}^\top R \boldsymbol{u}$.

We calculate the error $\boldsymbol{e}_t$ and the linearization of the ZMP in each execution cycle. We also keep track of θ'_t, $\dot{\theta}'_t$ and $\ddot{\theta}'_t$. From that, we build the model of (13) and (14) and compute $\boldsymbol{u}_t$, which is the rate of change of the acceleration and which is constant until the next motion cycle. The next angles to be set for the hip and ankle joints are computed from $\theta'_{t+1} = \theta'_t + \Delta T \dot{\theta}'_t + \frac{1}{2}(\Delta T)^2 \ddot{\theta}'_t + \frac{1}{6}(\Delta T)^3 \dddot{\theta}'_t$ according to factor β of the joint pairing. The remaining joints of the support leg are not changed.

To determine the angles of the kick leg, we calculate the pose of the torso using forward kinematics, evaluate the kick foot trajectory to the new kick foot pose, calculate the kick foot pose relative to the torso and use inverse kinematics to solve for the angles of the kick leg joints.

3.3   Balancing with a Cart-Table Controller

In addition to the LQR we implemented the Cart-Table controller by Kajita *et al.* [8]. Instead of a linearization of the ZMP, we use a truly linear model and use an extension of a linear-quadratic controller [10], which responds more quickly to changes of the demanded ZMP, because it also considers the N future demands of the ZMP instead of only the currently demanded ZMP.

According to the Cart-Table model, the ZMP depends on the motion of the COM $\boldsymbol{c}$. With the 2-dimensional COM restricted to a plane of height c_z, the 2-dimensional ZMP $\boldsymbol{p}_0$ on the ground is

$$\boldsymbol{p}_0 = \boldsymbol{c} - \frac{c_z}{g}\ddot{\boldsymbol{c}} \ . \tag{17}$$

With the ZMP, which is calculated using the inverse dynamics method explained above, and a preview series of ZMPs that will be demanded, we calculate the control output $\dddot{\boldsymbol{c}}$, the jerk of the COM. With an initial position the COM this effectively defines a trajectory for the COM. Using the kick foot pose from the evaluation of the kick foot trajectory, we implemented a simple numerical method to move the robot's torso relative to the support foot, such that the COM is moved to the position demanded by the controller and the kick foot to the pose relative to the support foot as demanded by the kick foot trajectory.

The downside of the Cart-Table model is that it only works at a constant COM height, which has to be approached before the actual balancing can start. Also, the effects of the angular motion of the robot around the COM are not modeled. But the advantages outweigh the downsides. As we will see in Sec. 4, the ZMP as modeled by the Cart-Table model is not too different from the ZMP calculated using the inverse dynamics. The computation is also much faster, because the gain matrices of the controller, including a Riccati equation such as (16), can be calculated once in advance and then repeatedly used while balancing.

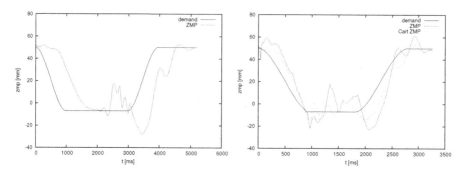

Fig. 4. ZMP tracking in y-direction of the LQR (left) and the Cart-Table controller (right). The bathtub-like plots are the ZMP references to be tracked by the robot, which leans to the side, kicks, and straightens up again. For the Cart-Table controller, the ZMP calculated using the Cart-Table model is also plotted as a dashed line.

4   Results

To evaluate the controllers, we performed different kicks and plotted the resulting estimated ZMPs. A typical example is pictured in Fig. 4. As expected, the Cart-Table controller responds more quickly to changes to the ZMP. As a consequence, the overall kick time is shorter, since the motion engine has to wait less for the robot to become balanced before and after the kick foot is moved. No matter which controller is used, the ZMP gets considerably perturbed while the kick foot is moved. To assess how well the Cart-Table model fits, we also plotted the ZMP as it would result from that model. Although the Cart-Table does not consider rotational effects, the Cart-Table-ZMP matches the ZMP calculated using inverse dynamics quite well. Overall, if the fixed height of the COM is not an issue, the Cart-Table controller appears to be the better choice.

To evaluate the kick foot trajectory, we let the NAO perform kicks in different directions using the Cart-Table controller: kicking straight ahead, and at $25°$, $50°$, and $80°$ angles. The results for the $50°$ kicks and the $80°$ kicks are plotted in Fig. 5. The trajectory duration (900ms) and the demanded kick foot speed $(1\frac{m}{s})$ were the same for all trials.

On average, the demanded angle is almost reached by the $80°$ kicks. The kicks performed with the left foot scatter with standard deviation $\sigma_{80,\text{left}} = 9°$ around the mean $\mu_{80,\text{left}} = 83°$, the right kicks with $\sigma_{80,\text{right}} = 10°$ around $\mu_{80,\text{right}} = 75°$. The results of the $50°$ kicks are a lot worse. On average, the kicks reach $\mu_{50} = 32°$. Only some left kicks come close to the demanded angle, but also show a high standard deviation of $\sigma_{50,\text{left}} = 13°$ compared to $\sigma_{50,\text{right}} = 7°$.

We observed similar results for the $25°$ kicks. As the $50°$ kicks, on average they only reached roughly half of the demanded angle. The straight kick was the most accurate with $\mu_{0,\text{left}} = -3.6°$ and $\mu_{0,\text{right}} = -3°$, scattering only with $\sigma_{0,\text{left}} = 3.7°$ and $\sigma_{0,\text{right}} = 3.1°$.

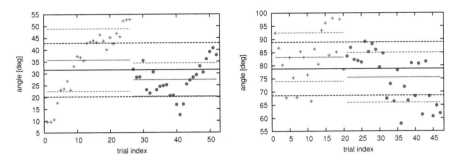

Fig. 5. The measured angles of the 50° kicks (left) and the 80° kicks (right). The means are plotted as solid lines, the dashed lines are one standard deviation away from the corresponding mean. Angles of kicks performed with left are plotted with crosses on the left side of each diagram, kicks performed with right on the right sides with circles.

We also measured the distances reached by the kicks. The straight kick reached 268cm on average, with 51cm standard deviation. By reducing the trajectory duration to 750ms and increasing the kick foot speed to $1.3\frac{m}{s}$, the average distance increased to 395cm, with 82cm standard deviation. The maximum distance reached was 560cm.

5 Conclusion

In the previous sections, we presented a method to calculate a smooth trajectory for the kick foot given the ball position, the kick direction, the desired duration for the kick and the speed the kick foot should have when hitting the ball, and two methods to execute the trajectory by a dynamically balanced robot.

The results are mixed. On the one hand, the angles reached for 80° and the straight kick are acceptable, while the kicks for 25° and 50° undershoot considerably. This is probably due to our approximation of the foot contour being not accurate enough.

We also observed large variations both in the angles and distances reached. As to why these variations happen, we currently suspect that minor variations in the ball positions have major effects on the resulting trajectories. In addition, small deviations from the trajectory probably cause large deviations of the kick direction if the ball is hit with a highly curved region of the foot.

We conclude that one can use the methods described to perform kicks of a soccer robot such as the NAO without prior modeling of trajectories of the robot's limbs or relying on key-framing methods, in particular if we can fix the variation issue.

Acknowledgement. We would like to thank the members of the team B-Human for providing the software framework for this work. This work has been partially funded by DFG through SFB/TR 8 "Spatial Cognition".

References

1. Belanger, P.: Estimation of angular velocity and acceleration from shaft encoder measurements. In: Proceedings of the 1992 IEEE International Conference on Robotics and Automation, vol. 1, pp. 585–592 (May 1992)
2. de Boor, C.: On calculating with B-splines. Journal of Approximation Theory 6(1), 50–62 (1972),
   http://www.sciencedirect.com/science/article/pii/0021904572900809
3. Czarnetzki, S., Kerner, S., Klagges, D.: Combining key frame based motion design with controlled movement execution. In: Baltes, J., Lagoudakis, M.G., Naruse, T., Ghidary, S.S. (eds.) RoboCup 2009. LNCS (LNAI), vol. 5949, pp. 58–68. Springer, Heidelberg (2010)
4. Doya, K.: Bayesian Brain: Probabilistic Aproaches to Neural Coding. Computational Neuroscience Series. MIT Press (2007)
5. Fang, A.C., Pollard, N.S.: Efficient synthesis of physically valid human motion. ACM Trans. Graph. 22(3), 417–426 (2003),
   http://doi.acm.org/10.1145/882262.882286
6. Farin, G.: Curves and Surfaces for CAGD: A Practical Guide. The Morgan Kaufmann Series in Computer Graphics and Geometric Modeling (2002)
7. Featherstone, R.: Rigid Body Dynamics Algorithms. Springer (2008)
8. Kajita, S., Kanehiro, F., Kaneko, K., Fujiwara, K., Harada, K., Yokoi, K., Hirukawa, H.: Biped walking pattern generation by using preview control of zero-moment point. In: Proceedings of the IEEE International Conference on Robotics and Automation, ICRA 2003, vol. 2, pp. 1620–1626 (September 2003)
9. Kalman, R.E.: A new approach to linear filtering and prediction problems. Transactions of the ASME–Journal of Basic Engineering 82, 35–45 (1960),
   http://www.cs.unc.edu/~welch/kalman/media/pdf/Kalman1960.pdf
10. Katayama, T., Ohki, T., Inoue, T., Kato, T.: Design of an optimal controller for a discrete-time system subject to previewable demand. International Journal of Control 41(3), 677–699 (1985)
11. Müller, J., Laue, T., Röfer, T.: Kicking a ball – modeling complex dynamic motions for humanoid robots. In: Ruiz-del-Solar, J., Chown, E., Plöger, P.G. (eds.) RoboCup 2010. LNCS (LNAI), vol. 6556, pp. 109–120. Springer, Heidelberg (2010)
12. Röfer, T., Laue, T., Müller, J., Bösche, O., Burchardt, A., Damrose, E., Gillmann, K., Graf, C., de Haas, T.J., Härtl, A., Rieskamp, A., Schreck, A., Sieverdingbeck, I., Worch, J.H.: B-Human Team Report and Code Release 2009 (2009), only available online:
   http://www.b-human.de/file_download/26/bhuman09_coderelease.pdf
13. Tilgner, R., Reinhardt, T., Borkmann, D., Kalbitz, T., Seering, S., Fritzsche, R., Vitz, C., Unger, S., Eckermann, S., Müller, H., Bellersen, M., Engel, M., Wünsch, M.: Team research report 2011 Nao-Team HTWK Leipzig (2011),
   http://robocup.imn.htwk-leipzig.de/documents/report2011.pdf
14. Vukobratović, M., Borovac, B.: Zero-moment point — thirty five years of its life. International Journal of Humanoid Robotics 01(01), 157–173 (2004),
   http://www.worldscientific.com/doi/abs/10.1142/S0219843604000083
15. Xu, Y., Mellmann, H.: Adaptive motion control: Dynamic kick for a humanoid robot. In: Dillmann, R., Beyerer, J., Hanebeck, U.D., Schultz, T. (eds.) KI 2010. LNCS (LNAI), vol. 6359, pp. 392–399. Springer, Heidelberg (2010)

Team Water: The Champion of the RoboCup Middle Size League Competition 2013

Song Chen, Zhe Zhu, Ye Tian, Charles Lynch, Di Zhu, Ye Lv, Xinxin Xu,
Wanjie Zhang, Liang Zhao, Xiaoming Liu, Miao Wang, Peng Zhao, Chenyu Wang,
Wei Liu, Feichan Yang, Binbin Li, Jieming Zhou, Zongyi Zhang, and Xueyan Wang

Beijing Information Science & Technology University,
No. 12, East Qinghexiaoying Rd., 100192, China
water.bistu@gmail.com
http://jdgcxy.bistu.edu.cn/robocup/blog

Abstract. This paper addresses the problems encountered and the advancements made to the robots of Team Water, the champion of the RoboCup Middle Size League Competition in 2013. This paper describes the robot from both the hardware point of view and the software. The hardware description includes details of the design of an omnidirectional wheel and an improved ball handling device. The software part focuses on the improvements on the masking function and the localization algorithm.

1 Introduction

The middle size league is an important part of RoboCup (Robot World Cup) which mainly focuses on advancing MAS (Multi-Agent System)[1] and DAI (Distributed Artificial Intelligence). During the competition, each team dispatches five autonomous robots on an 18 meters long and 12 meters wide soccer field. The winner is the team who scores more goals. There are many technical challenges. First, the robot collects information of the field in real time and follows that up with decisions, while the robot keeps moving simultaneously. Second, due to the special design of the omnidirectional camera, the image got from the camera is distorted. The program must correct the image efficiently and provide the robot with precise location information. Third, because of the fierce antagonism during the match among robots, the robot demands a high tensile structure. At last, on our course towards the goal of beating a human soccer team in 2050, the intelligent cooperation among robots is a big challenge to the researchers.

Team Water of BISTU devoted itself to the RoboCup middle size league since 2003, when the team designed its first generation robot with two differential wheels and a coach program. In 2005, Team Water turned its focus to improve performance of the robot's hardware and software. Consulting the paper on the control of omni-wheels[2], the second generation robot was designed with a three wheeled omnidirectional chassis and a 360 degree mirror. The image processing and motion strategy was also redesigned and implemented on the new robot, and the coach

S. Behnke et al. (Eds.): RoboCup 2013, LNAI 8371, pp. 37–48, 2014.

program improved. In 2010 and 2013, Team Water won the championship of the RoboCup middle size league. In the past year, the team improved the performance of the omnidirectional wheel, ball handling mechanism, mask function and localization module.

Remainder of this paper is organized as follows. Section 2 details on the improvements to the hardware, including the omnidirectional wheel and ball handling mechanism. Section 3 addresses the improvements to software, including the mask function and the localization functions. Section 4 describes the results of our improvement at the competition. Section 5 concludes the paper.

2   Hardware System

Robots of Team Water are shown in the Fig. 1. The camera captures the images and sends them to the onboard computer. The onboard computer makes decisions after receiving information that come from the sensors and the electrical circuits of the control system. Then the onboard computer sends the decisions to the servo.

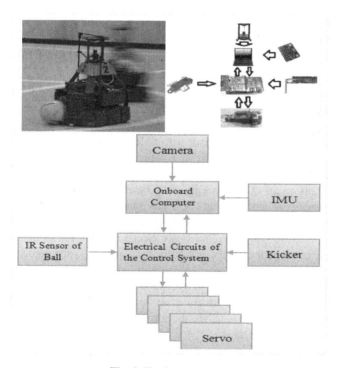

Fig. 1. Hardware system

Hardware defects have been fixed after the RoboCup in 2012 where we missed the championship. Most of the improvement has been made in the Omni-directional wheels and ball handling mechanism.

2.1   Omnidirectional Wheel

Omni-directional Wheels have small rollers to allow the wheels to move freely in any direction. They can rotate along the primary diameter, just as any other wheels. The smaller rollers along the outer of the diameter allow rotation along the orthogonal direction. With these two rotations combined, the robot can move along any direction no matter in which direction it's heading.

The Omni-directional Wheels used in our robots before 2013 have many defects. To start with, the wheel hubs are made of plastic, so they can break easily when crashed by other robots, as shown in Fig. 2(Left). Moreover, the thin rubber wrapped around the rollers is easy to wear and fall off after one or two competitive games, as shown in Fig. 2(Right). Additionally, it makes disassembling of the wheel hub almost impossible because all the parts of the wheel hub are glued together, which makes the replacement of small rollers impossible, resulting in a great waste.

Fig. 2. Left: A break caused by hits. Right: Worn rollers

To solve those problems, we design a new kind of Omni-directional Wheel, as shown in Fig. 3(Left). We use aluminum alloy, which is stronger than plastic, to make the wheel hubs, the shafts and the bearings of the small rollers so that the wheels can withstand more powerful hits. The diameter of the roller bearing has been increased in order to support more weight. Moreover, new rollers that have thicker and tougher rubber wrapped around them are used, and with this additional rugged tread they are able to provide larger friction force to actuate the robots. A new structure with rollers fixed on the wheel independently of each other has been proposed, as shown in Fig. 3(Middle) and Fig. 3(Right), to improve the time and economy efficiency when the rollers need to be replaced.

Fig. 3. Left: the new Omni-directional Wheel. Middle: Roller. Right: Wheel hub

2.2   Ball Handling Mechanism

The ball handling mechanism is comprised of a servo motor, two spring shock absorbers and components which are used to fix and link. The two ball handling mechanisms are fixed on each side of the kicker. The robot can dribble the ball better with the servo motors rotating. While colliding with other robots, the ball handling mechanism can control the ball and protect the possession of the ball from other robots [3]. Moreover, with the help of the ball handling mechanism, the robot can perform some movements like making a sudden stop or turning with speed.

In 2011, to pursue a better effect of dribbling, we managed to use active ball handling mechanism through fixing servo motors on each side of the kicker. Then, the robot could dribble the ball with more motions. It also decreased the failure rate of dribbling and increased the success rate of attack. Based on the result of tests in the 2011 and 2012 RoboCup competitions, the solution performed as well as we expected (Table 1). Some defects, however, were exposed as well. Due to the fierce fighting among robots during a match, the huge impact damaged the ball handling mechanism. Because a precise relative position between the ball handling mechanism and the ball were required, any deformations of the damaged ball handling mechanism would influence dribbling performance. Even more, sometimes the robot might not draw the ball into the ball handling mechanism or could not dribble the ball stably, finally result in losing the possession of the ball and failing to execute the robot's motion strategy.

Table 1. The average times of losing ball per game

Before 2011	In 2011	In 2012
30	20	14.5

In order to resolve these defects, we considered two schemes. One was to use stronger materials and increase the components' thickness. Another was to add new components to absorb the force of the impact. Considering the available mounting space and the weight of the robot, we decided to use scheme two.

We fixed spring shock absorbers which can absorb the force of the impact from fighting to protect the ball handling mechanism. The suspending cushion structure and the axial cushion structure convert the fixed ball handling mechanism into a floating structure. Therefore, with the help of two absorbers, the ball handling mechanism can effectively absorb the impact coming from both the axial direction and radial direction. Finally, this design protects the ball handling mechanism from damage.(Fig.4)

Fig. 4. Spring shock absorber

3   Software System

The software system of Water's robot is comprised of the robot module and the coach module (Fig.5). The software system starts with the robot image processing module which receives images of the robot's 360 degree of view from the camera, and then processes the images to calculate the coordinates of the ball and the robot's position and send to the coach module. The coach module is going to make decisions based on the coordinates received from the image processing module. Then, the coach module sends these decisions, such as the destination coordinates of the robots' motion, the

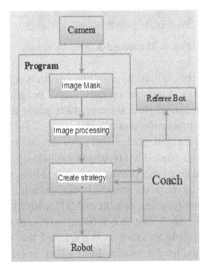

Fig. 5. The flow chart of robot's match system

robots' role in the game and so on, to each robot. Based on the commands from the coach module and the output of the image processing module, the robot itself will plan motion its strategy through its motion decision module.

This year, in order to improve the match system's efficiency and precision in localization, some improvements were made to the image process module and localization function. Following are the details.

3.1   Mask

Last year, the image processing module using OpenCV[4] segment the input image received from the camera by color directly (Fig.6).

According to the new rules, audiences have been allowed to stand around the field at their will. The colorful clothes audience members wear means much more interference for the robots. Therefore, detection approach that is based on simple color division and shape recognition is not reliable. More interference means that there is more information that the image processing module has to process, which is a heavy burden for the robot.

Fig. 6. The image divided by color

In order to improve the reliability and recognition efficiency of the image processing module, a mask function is introduced before the image processing module. Thus, the image received from the camera is going to be preprocessed by the mask function to reduce computation complexity.

Based on the location information received from the last image, the mask function generates a set of mask data named MD1 which excludes pixels that are outside the match field (as determined from the previous image). Meanwhile, the mask function also generates another set of mask data named MD2 which excludes pixels occluded by the robot. Then, the mask function combines MD1 with MD2 and gets the final set of mask data named MDF. Next, the mask function compares the picture from the camera (Fig.7) with MDF (Fig.8). This step excludes the image's pixels of robot occlusion and objects outside the match field. This result is sent to the image processing module as input.

 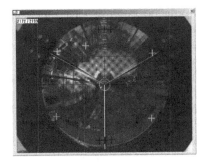

Fig. 7. The 360 degree picture camera takes   **Fig. 8.** The picture the mask function processes

3.2   Localization

Before 2012, based on the paper of Brainstormers Tribots[5], the localization function for determining the robot's location in the field first binarizes the image, in which the pixels including in the color threshold of green are defined as the field, the pixels included in the color threshold of white are defined as lines, and the other pixels are bypassed. The image devolved by the first step is named BD1. Then, the localization function scans the BD1 by a set of radials which set its origin at the center of BD1 with a 2.5 degrees difference in their direction. At the same time, the localization function also scans the BD1 by lines of 20 pixels' displacement from left to right and from top to bottom (Fig.9). When a scan line detects white pixels located between green pixels, the function is going to work out the weighted center of the white pixels. After scanning, all centers are saved in an array named WA1. Then, the function is going to restore the WA1 to a new array which can compare with the match field by distortion correction. This new array, without distortions, is named LTA. Next, we compare the LTA with the recent period's localization data, to exclude pixels outside the match field, and finally get an array named FLA. The FLA is continuously compared with the field template (Fig.10) by the point to point ICP method[6]. Finally, it solves for the robot's location, named FRL, and orientation, named FRA, relative to the field. Due to the fierce antagonism between robots during the match, the robot's orientation would suddenly change or the camera's image becomes blurry and the robot is going to locate itself in the wrong location. It becomes clear that the localization function which is only based on images is not reliable.

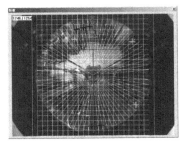 

Fig. 9. Grid method to scan the white line of image   **Fig. 10.** Field template

In order to improve localization precision, efficiency and reliability, the team manages to ask help from AI area. With high integration and low cost IMUs, however, being popularized by their use in cellphones and other electronic products in recent years, such as MPU 6050[7], we decide to use the IMUs to assist the robot localize itself, which is cheaper and easier to realize. More sensors are added to help the robot to determine its location. The localization function fuses the angle got from the IMUs (Fig. 11) with the FRA, then determines the robot's orientation.

When fusing, the localization function first gets the initial angle named A1' from the IMUs. Then, the localization function gets an angle named A1, from IMUs, by subtracting AVGDA from A1'. Subsequently, the A1 and the FLA together are compared with the field template by the ICP method and the Steepest Descent method[8]. This process works out a final angle name A2. DA results from subtracting A1 from A2. AVGDA is the average of the sum of the DAs from all the previous iterations.

To further improve the precision of robot's location, the localization function also collects data from the odometer of the servo motors. The data from the odometer is called A3. Using the weighted average method, A2 and A3 are calculated. The output angle A3 is sent to the coach module. At this stage, the estimation precision of the robot's orientation can reach less than 3 degrees.

Fig. 11. 6050 IMUs

4   Experimental Result

4.1   Omnidirectional Wheel

Before 2012, a robot at least needed two to three sets of wheels. The cost was more than a thousand dollars. Robot failure caused by damage of the wheels happened frequently. This year, after we designed new wheels, all the wheels stayed intact. The new wheels also improved the time of the robot on duty. As the robot needs fewer wheels than before, the cost of the robot on wheels drops to four hundred dollars. Its reliability, robustness and durability is much better than the wheels we bought before. Moreover, these wheels can adapt to most environments, which enables us to use it in the real world and ordinary environment.

During the competition, we did not replace any wheels. Moreover, the new wheels provided larger grip force than before.

4.2   Ball Handling Mechanism

Through the test in 2013 RoboCup World Cup, the failure rates of dribbling fell, the average times of losing the ball per game are shown in Table 2. In fierce competition, the ball handling mechanism still can hold the ball well.

Table 2. Dribble Failure Times Per Game

Dribble condition	Previous Year	Current Year
Free dribble	4.3	1.7
Dribble in fierce competition	10.2	5.9

4.3   Mask

As Fig.12(Left) shows, we put two balls in the field. Ball1 is the target we wish to detect. Ball2 is an interference. Without the mask function, the image processing module binarizes both of the two balls in the image and fixes the center on Ball2 which does not belong to the match field. The solution is to use the mask function (Fig.12.Right) The mask function will pre-process the image before the image processing module takes over. The function excludes Ball2 which is outside the range of the mask. Then, the picture is sent to the image processing module which only needs to process Ball1. As the result, the center correctly is identified as Ball1.

During debugging, we always assigned our members to stand around the field to simulate interference to the robot. Therefore, we were able to test the effectiveness of the mask function (Table 3). The rates of fault recognition dropped by forty percent and the computation time cost of the image processing module decreased (Table 4).

Table 3. Wrong Recognitions Per Ten Tests

	Without Mask	With Mask
Ten balls around the Field	5	1
Ten people wearing green	7	2

Table 4. Image Processing Unit Time Per Iteration

	Without Mask	With Mask
Time	3ms	2ms

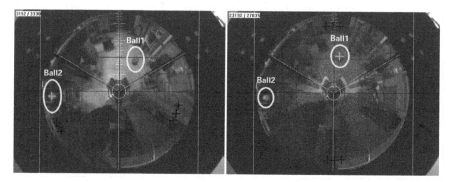

Fig. 12. Left: Not using template, red cross is on the Ball2. Right: Using template, red cross is on the Ball1.

In these pictures, the left ball is Ball2 and right is Ball1.The Red Cross is centered on the ball.

4.4   Localization

During a match, robots crash frequently while trying to get the ball. When a crash happens, the robots may be suddenly turned into a new direction. Therefore, the image received from the camera will suddenly change. Using the FLA which is just derived from the lines cannot be matched to the previous FLA due to the sudden change in physical location. It will result in the FLA being unable to match to the field template (Fig.13 Left). With the help of the IMUs (Fig.13 Right), the A1 corrects the FRA, so that the FLA can match with the field template and ensure the robot is able to determine its correct location, as shown in Fig. 13(Right).

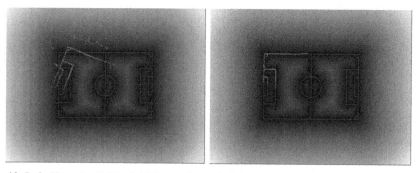

Fig. 13. Left: Not using IMUs, field data can't match field template. Right: Using IMUs, field data match field template after using IMUs data correction (red dots are field array data in the picture).

4.5   Competition

Via the test in this year's competition, all the changes we made this year resulted in a good effect. We did not change any omnidirectional wheels from preparing phrase to the end of competition. Moreover, the rubber wrapping around the rollers was almost always intact. Second, the failure of ball handling devices reduces. We just repaired a few times and replaced none. On software part, after introducing mask function, the robot could exclude the white interferences around the field. Furthermore, the mask function also decreased the time spent on image processing module from 5ms to 3ms. The localization also performed well. While collision with other robots, the IMUs could fix the wrong location information got from the suddenly changing image. It improved the precision of the robot's localization. Additionally, the data got from the IMUs reduced the computation time of the ICP method from 3ms to 2ms. In conclusion, all improvements made this year influenced the performance of our robots in a positive way. These improvements reduced the time we spent on repair and improved the time of the robot on duty. Finally, based on these improvements and the good performance of whole robots, we won the championship of 2013 RoboCup MSL.

5   Conclusion

Through these years' improvements, the collision among RoboCup MSL robots is more intense than before. Therefore, the robot requires more reliable mechanical design than before. Moreover, in order to hold and dribble the ball stably, the robot's ball handling device is improved, so that the robot can handle ball even violently impacted. The robot's software system is also updated. Thus, the robot can validly avoid the collision and possess the ball well.

How to command robots to cooperate efficiently and develop abundant tactics remains a very interesting research topic. Additionally, to test the efficiency between cooperation among five robots with different specific is also very challenging.

All of the above is what Team Water is going to focus on in the next year.

References

1. Zhang, W., Zhang, B., Zhu, M.: Intelligent Soccer Robot System. Tsinghua University Press, China (2010)
2. Ribeiro, F., Moutinho, I., Silva, P., Fraga, C., Pereira, N.: Controlling Omni-directional Wheels of a MSL RoboCup Autonomous Mobile Robot
3. de Best, J., van de Molengraft, R.: An Active Ball Handling Mechanism for RoboCup. In: 2008 10th Intl. Conf. on Control, Automation, Robotics and Vision, Hanoi, Vietnam, pp. 17–20 (December 2008)
4. Bradski, G.R., Kaehler, A.: Learning OpenCV: Computer Vision with the OpenCV Library. Reilly Media, Inc., USA (2008)

5. Lauer, M., Lange, S., Riedmiller, M.: Calculating the Perfect Match: An Efficient and Accurate Approach for Robot Self-Localisation. In: Bredenfeld, A., Jacoff, A., Noda, I., Takahashi, Y. (eds.) RoboCup 2005. LNCS (LNAI), vol. 4020, pp. 142–153. Springer, Heidelberg (2006)
6. Low, K.-L.: Linear Least-Squares Optimization for Point-to-Plane ICP Surface Registration
7. MPU-6000 and MPU-6050 Product Specification Revision 3.2. InvenSense Inc. (November 16, 2011)
8. Haykin, S.O.: Neural Networks and Learning Machines, 3rd edn. China Machine Press (2011)

RoboCup 2013
Humanoid Kidsize League Winner

Daniel D. Lee, Seung-Joon Yi, Stephen G. McGill, Yida Zhang,
Larry Vadakedathu, Samarth Brahmbhatt, Richa Agrawal,
and Vibhavari Dasagi

GRASP Lab, Engineering and Applied Science, Univ. of Pennsylvania, USA
{ddlee,yiseung,smcgill3,yida,vlarry,samarthb,richaagr,dvib}@seas.upenn.edu
http://www.seas.upenn.edu/~robocup

Abstract. The RoboCup Humanoid Kidsize League has rocketed into
an exceptionally competitive league, supported by recent introduction of
the open source hardware and software platforms. Teams must improve
substantially their robot teams every year to stay competitive. In addition to the important skills of walking speed, kick stability, and visual
acuity, the 2013 season introduced rule changes to test robot intelligence
– and team adaptability. In this paper, Team DARwIn, the winning team
of the Humanoid Kidsize League, presents its crucial improvements of its
robotic platform and soccer system to mitigate risks from same colored
goal posts, substitutions and noise in goalkeeper ball estimates.

1 Introduction

Team DARwIn from the University of Pennsylvania's GRASP lab has been competing in the Humanoid kidsize league since 2010 with a DARwIn series of robots,
and has been remaining champion since 2011. We also pushed forward to open
source both the hardware and software platforms: the latest generation of DAR-
wIn robots, the DARwIn-OP, is openly available in the form of blueprints and
as a commercial product. The modular, easy-to-use UPenn humanoid robotic
software platform has also been released as open source[1].

The open-source release of both software and hardware platforms made quite
a big impact on the humanoid kidsize league this year. Two thirds of the teams
used at least one DARwIn-OP robot or its variant, and more than half of the
teams used our software framework [1] to varying degrees. For our own team, this
competition was a time for focusing on localization to mitigate the complexity of
having same goal post color with no landmarks, adding more flexible walk and
kick transitions, and adapting our substitution execution. We will explain each
in detail in the following sections.

---

[1] http://seas.upenn.edu/~robocup

S. Behnke et al. (Eds.): RoboCup 2013, LNAI 8371, pp. 49–55, 2014.
© Springer-Verlag Berlin Heidelberg 2014

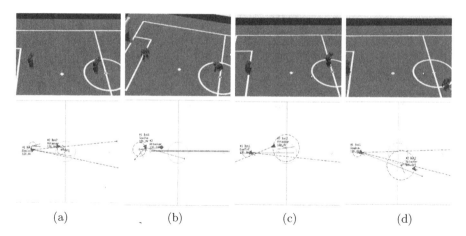

(a)   (b)   (c)   (d)

Fig. 1. The flip correction simulated in the Webots simulator. Each robot is represented by a blue triangle. The dotted circles show the error of the robot's localization relative to its actual location, the red lines show the robot's ball estimate while the dotted black lines show the robot's estimate of the goalposts. (a) Correct localization on both robot (b) Field player falls down (c) Field player got flipped (d) Flip corrected by the ball information.

2   Localization in a Symmetric Field

The biggest challenge for RoboCup 2013 was the adoption of a fully symmetric field to make gameplay more compatible with human soccer. The goalposts on each side of the field are now required to have the same color; additionally, the unique half-field localization landmarks have been removed. This means that no unique landmark exists on the field, and the robots must keep track of their localization status using ambiguous information.

We have been using a particle filter based localization strategy which can easily be generalized for fully ambiguous landmarks. However, with the field being fully symmetric, the robot has two possible pose candidate given the observation: the actual pose and the 'flipped' pose. A flipped pose occurs when the pose particles get attracted to an orientation that is mirrored about the center of the field. If this flipped pose is believed by the robot, then it will attack and score in our own goal. We have found that these flipped poses can occur when the robot gets pushed around near the center of field to accumulate lots of error in pose estimation.

To fix the flipped localization status, we used the localization information of the goal-keeper. With our strategy on the field, the goal-keeper should never approach the center of the field; thus it can act as a marshaling agent for robots that are confused in the middle of the field. The goal-keeper helps to reverse the pose flipping player by using the soccer ball as a common reference point for both robots – only one ball occupies the field at a given time. Our assumption is that the goalkeeper's ball estimation is fairly accurate. When the goal-keeper

Fig. 2. The ZMP preview control based kick during locomotion

estimates that the ball resides in its own half of the field and the confused robot can see the ball, the confused robot can properly disambiguate the goal posts.

However, we cannot use such a flip correction method if the ball is close to the center area. The goal keeper may easily misjudge the side of the field in which the ball resides, and the robot is able to score from the half field range. To prevent such mishaps, we assume that if the robot falls at the center field it instantly labels itself as confused. We do not allow the robot to kick directly to the goal until it is sure of its pose; instead, we make the robot move the ball to the side of the center field. In this way we can still keep the ball moving, while preventing the chance of scoring to our own goal. Additionally, it helps to disambiguate the localization based on the goalkeeper's observation.

3   Walk and Kick Controller Improvements

In spite of its relatively small size, the DARwIn-OP robot could walk with a speed up to 32cm/s with our ZMP-based walk controller. This was one of the fastest walk speed of the 2011 and 2012 competition, but this year many robots raised this bar and a few robots could walk with a speed exceeding 40cm/s. To remain a relatively speedy competitor, we increased the top walk speed of the robot to 40cm/s. Furthermore, we optimized the approach trajectory so that the robot walks in a curved trajectory to both minimize the traversal time and keep the robot stable. As a result, we could easily outrun larger and faster walking robots in spite of much shorter leg lengths.

Another improvement we have made this year is a new kick controller. Previously, we only used two kick controllers: a stationary one that is powerful but slow, and a dynamic one based on the locomotion engine that is very fast but relatively weak. This year, we debuted a new kick controller based on ZMP preview control. ZMP preview control is used by a few teams for locomotion and can yield a more powerful kick than the analytic ZMP algorithm we use. However, it has a big disadvantage of requiring a preview period that makes the robot less reactive to a changing environment. Thus, we only use the ZMP preview control for kick generation since kicks do not require reactive movements. The analytic ZMP algorithm continues to control locomotion. The transition between these two ZMP controllers is achieved by including the boundary conditions of the analytic ZMP controller as additional cost terms in the optimization process for the ZMP preview controller [2].

For the competition, we tuned the new kick controller to have the same speed as the previous dynamic walk kick, yet the new controller improved kicking distance by roughly 50%. Table 1 shows a comparison of our three different kicks. In game terms, the robot needed to successfully kick twice in succession to score from center field with the analytic walk kick, and now it can score with just one kick. This improvement greatly helped us this year.

Table 1. Comparison of average distances and execution times of kicks

	Stationary	Walk-Kick	ZMP Preview
Kick Distance (cm)	411	188	292
Start Time (sec)	2.60	0.70	0.70
Total Time (sec)	4.30	0.95	0.97

4   Goalkeeper Behavior

The main role of the goalkeeper in RoboCup is to defend the goal post from incoming shots; one way of blocking them is by diving. Unfortunately, the goalkeeper cannot lie down on the field indefinitely, as laying down for more than 5 seconds is against the rules and leads to the goalkeeper being ejected off the field. On-demand diving against an opponents' kick requires the precise determination of the position of, and time when, the ball will cross the goal line. Visually determining these crucial variables using the ball's velocity is difficult, as camera images are inherently noisy.

These images are captured at approximately 30 frames per second; and we use our standard ball detection algorithm to get the current position of the ball. The simplest way to determine the velocity is to find the displacement of ball in two consecutive image frames, with respect to the time difference between these frames – this is called an averaging filter. Unfortunately, this approach is subjective to noise due to false positive ball detections and the noise in each position reading.

Intuitively, we would want to train our model to use all previous information instead of relying on the last two frames alone. We implemented a two dimensional linear Kalman filter, similar to [3], maintaining a state consisting of four variables: $s = [x, y, \dot{x}, \dot{y}]$. x and y are the relative ball positions from the robot frame in their respective axes.

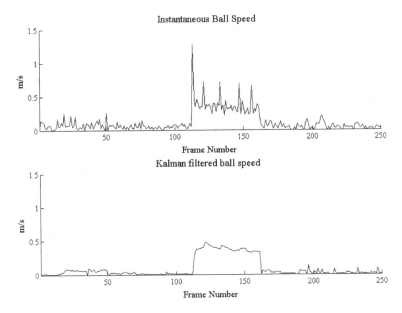

Fig. 3. When a ball is kicked, a standard frame by frame velocity filter is very noisy. In comparison, the Kalman filter eliminates much noise.

We assume a steady decrease in speed with the parameter α. α controls the velocity of the ball from a wrong estimation if the ball is occluded. If the velocity is not controlled, the Kalman filter would predict the next position from the previous velocity which might lead to erroneous predictions. No external input to the ball is modeled. Our Kalman process prediction is modeled with the following state transition formulation:

$$s_{t+1} = \begin{bmatrix} 1 & 0 & \Delta_t & 0 \\ 0 & 1 & 0 & \Delta_t \\ 0 & 0 & \alpha & 0 \\ 0 & 0 & 0 & \alpha \end{bmatrix} s_t \tag{1}$$

This Kalman estimation provides smoother, more reliable ball estimation. Figure 3 shows the comparison between the estimated ball velocities using averaging filter and the Kalman filter. From the ball state, we extrapolate Δt seconds into the future to see if the ball is in a dangerous zone where the goalkeeper must dive: $x_f = \dot{x}\Delta t + x$, and similarly for y. In the real matches, the new ball velocity filter worked very well, triggering almost no false positive during the whole match.

Fig. 4. The team table with three robots ready for play and two substitute robots waiting to get their chance

5 Rapid Substitutions

Using custom hardware, teams in the humanoid kidsize league inherently encounter more hardware issues than their standard platform league counterparts. With these issues, substitutions play a critical role in winning a match. The current rules allow for a maximum of two substitutions per match, and those substitutions can target either the goalkeeper or a field player. The problem is that we don't know which robot will require a substitute in advance. We may need to substitute two field players, or the goalkeeper twice. Booting up the robot manually and setting up the roles during a high pressure match leads to mistakes. On the other hand, taking too much time to switch could mean that less than three players occupy the field – a highly detrimental situation.

Our solution is to implement a script where the robot handler can set the role of the robot via a button press right before placing the robot on the field. The robot handler receives visual and auditory feedback from the robot to reduce the chance of a mistake. Another feature of the script is that malfunctioning robots taken off the field can be easily be placed into waiting state by button press. In this mode, the robot cannot affect teamplay by sending team messages. Figure 4 shows the typical team table setup right before the match, with 3 robots ready to play and two robots waiting in a waiting state for possible substitution.

6 Conclusions

The 2013 tournament was an extremely stiff competition, full of very tight, nail-biting matches. This RoboCup pushed the overall flexibility of our robots to the fore – being able to adjust unforeseen faulty localization, performing substitutions at the drop of the hat, and including more adaptable kicking engines. This flexibility required small but far reaching codebase adjustments, that hammer home the importance of modular design. Furthermore, with more avenues for failure, rigorous testing needed to be conducted on not just one robot at a time, but several – flip correction, for instance, requires both a goalie and field player for proper evaluation. With more changes upcoming in the RoboCup midsize league, the ability to adapt software very quickly will continue to be the number one priority. The same color goal post is just the first of many challenging changes to RoboCup software that must happen to achieve the 2050 goal of competing against the World Cup champions.

Acknowledgements. We acknowledge the support of the NSF PIRE program under contract OISE-0730206 and ONR SAFFIR program under contract N00014-11-1-0074. This work was also partially supported by the NRF grant of MEST (0421-20110032), the IT R&D program of MKE/KEIT (KI002138, MARS), and the ISTD program of MKE (10035348).

References

1. McGill, S.G., Brindza, J., Yi, S.-J., Lee, D.D.: Unified humanoid robotics software platform. In: The 5th Workshop on Humanoid Soccer Robots (2010)
2. Yi, S.-J., Hong, D., Lee, D.D.: A hybrid walk controller for resource-constrained humanoid robots. In: 2013 13th IEEE-RAS International Conference on Humanoid Robots (Humanoids) (October 2013)
3. Seekircher, A., Abeyruwan, S., Visser, U.: Accurate ball tracking with extended kalman filters as a prerequisite for a high-level behavior with reinforcement learning. In: The 6th Workshop on Humanoid Soccer Robots (2011)

Learning to Improve Capture Steps for Disturbance Rejection in Humanoid Soccer

Marcell Missura, Cedrick Münstermann, Philipp Allgeuer, Max Schwarz,
Julio Pastrana, Sebastian Schueller, Michael Schreiber, and Sven Behnke

Autonomous Intelligent Systems, Computer Science, Univ. of Bonn, Germany
{missura,schreiber}@ais.uni-bonn.de, behnke@cs.uni-bonn.de
http://ais.uni-bonn.de

Abstract. Over the past few years, soccer-playing humanoid robots
have advanced significantly. Elementary skills, such as bipedal walking,
visual perception, and collision avoidance have matured enough to allow
for dynamic and exciting games. When two robots are fighting for the
ball, they frequently push each other and balance recovery becomes crucial. In this paper, we report on insights we gained from systematic push
experiments performed on a bipedal model and outline an online learning
method we used to improve its push-recovery capabilities. In addition,
we describe how the localization ambiguity introduced by the uniform
goal color was resolved and report on the results of the RoboCup 2013
competition.

1 Introduction

In the RoboCup Humanoid League, robots with a human-like body plan compete
against each other in soccer games. The robots are largely self-constructed, and
are divided into three size classes: KidSize (<60 cm), TeenSize (90–120 cm), and
AdultSize (>130 cm). The TeenSize robots started to play 2 vs. 2 soccer games
in 2010 and moved to a larger soccer field of 9×6 m in the year 2011. In addition

Fig. 1. Left: Team NimbRo with robots Dynaped, Copedo, and NimbRo-OP. Right:
Team NimbRo vs. CIT-Brains in the RoboCup 2013 finals.

S. Behnke et al. (Eds.): RoboCup 2013, LNAI 8371, pp. 56–67, 2014.

to the soccer games, the robots face technical challenges, such as throwing the ball into the field from a side line.

For RoboCup 2013, the color coding of the goal posts was unified to yellow for both goals and the landmark poles at the ends of the center line were removed. Consequently, it was not possible anymore to determine the unambiguous position of a robot on the field based only on visual cues, which constitutes a problem for localization. However, most teams were able to implement suitable solutions and were able to reliably drive the ball towards the opponent goal. Our approach to disambiguate localization was to integrate a compass as an additional source of information. More details are given in Section 3.

Inspired by the success of the DARwIn-OP robot, we have constructed a Teen-Size open platform, the NimbRo-OP. Following the same spirit, the NimbRo-OP is a low-cost robot that is easy to construct, maintain, and extend. It is intended to provide access to a humanoid robot platform for research. The NimbRo-OP has matured enough to participate in the competitions. It participated in the Technical Challenges and scored its first official competition goal in the main event. More information about the NimbRo-OP is given in Section 5.

Bipedal walking is a crucial skill in robot soccer. It determines the success of a team to a substantial degree. Humanoid robots must be able to walk up to a ball and kick it, preferably without losing balance and falling to the ground. While most of the teams have mastered the skill of unperturbed walking on flat terrain, solutions to recover from strong disturbances, such as collisions with opponents, are not yet widespread. In ongoing research, team NimbRo has developed a stable bipedal gait control framework that has been designed to absorb strong perturbations. In Section 6, we report on the insights we gained from systematic push experiments, and introduce an online learning method that we used to improve push recovery capabilities. The learning controller is able to adjust the step size and recover balance quicker than the underlying simplified mathematical model.

2   Mechatronic Design of NimbRo TeenSize Robots

The mechatronic design of our robots is focused on robustness, weight reduction, and simplicity. All our robots are constructed from milled carbon fiber and aluminum parts that are assembled to rectangular shaped legs and flat arms. We use Dynamixel EX-106 and EX-106+ servos for the actuation of our classic robots Dynaped and Copedo. These robots are also equipped with spring-loaded protective joints that yield to mechanical stress and can snap back into place automatically. More information about the mechanical structure of the NimbRo classic robots can be found in [1] and [2]. The NimbRo-OP robot has a slightly different design with a reduced complexity. It is equipped with 6 DOF legs and 3 DOF arms that offer enough flexibility to walk, to kick, and to get up from the floor after falling. It is actuated by servos from the Dynamixel MX series. The mechatronic structure of the NimbRo-OP is best described in [3].

3 Perception

For visual perception of the game situation, we detect the ball, goal-posts, penalty markers, field lines, corners, T-junctions, X-crossings, obstacles, team mates, and opponents utilizing color, size and shape information. We estimate distance and angle to each detected object by removing radial lens distortion and by inverting the projective mapping from field to image plane.

For proprioception, we use the joint angle feedback of the servos and apply it to the kinematic robot model using forward kinematics. Before extracting the location and the velocity of the center of mass, we rotate the kinematic model around the current support foot such that the attitude of the trunk matches the angle we measured with the IMU. Temperatures and voltages are also monitored for notification of overheating or low batteries.

For localization, we track a three-dimensional robot pose (x, y, θ) on the field using a particle filter [4]. The particles are updated using a linear motion model. Its parameters are learned from motion capture data [5]. The weights of the particles are updated according to a probabilistic model of landmark observations (distance and angle) that accounts for measurement noise. To handle unknown data association of ambiguous landmarks, we sample the data association on a per-particle basis. The association of field line corner and T-junction observations is simplified using the orientation of these landmarks. Further details can be found in [6] and [2].

Integration of a compass: This year, we extended our sensory systems with a compass in order to help the particle filter to disambiguate the localization on the field. As starting from 2013 both goals have the same color and there are no landmarks that allow unambiguous localization based only on visual cues, it was necessary to add an additional source of information other than the objects detected by the computer vision. Using the compass output as observation of the global orientation in the particle filter greatly helps to reduce the number of hypothesis that can accumulate in the particle distribution. Figure 2 shows such an example. The robot observes a situation in the corner of the field, where field lines, L-shaped line crossings and a goal post have been successfully detected. Despite the high number of observations that the particles can be weighted with, two equally valid hypotheses form, as shown by the particle distribution

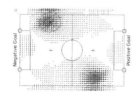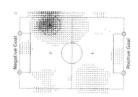

Fig. 2. Effect of the compass on localization confidence. The observed scene in the camera image (left) leads to two hypothesis peaks in the particle distribution of the particle filter (center). Adding the compass reading as an additional observation disambiguates the position estimation (right).

in the center. Adding the global heading as additional observation reduces the probability of particles that are facing in a wrong direction. Thereby one of the hypothesis in this example is invalidated (right). As an additional benefit of using a compass, we found that it not only improves localization, but also the effectiveness of our soccer behaviors. This is due to the fact that the rough direction of the opponent goal is always known. Thus, the ball is always moved in the right direction, even in cases where the particle filter reports a wrong pose.

4   Behavior Control

We control our robots using a layered framework that supports a hierarchy of reactive behaviors [7]. When moving up the hierarchy, the update frequency of sensors, behaviors, and actuators decreases, while the level of abstraction increases. Currently, our implementation consists of three layers. The lowest, fastest layer is responsible for generating motions, such as walking [8] —including capture steps [9], kicking, get-up motions [10], and the goalie dive [11]. At the next higher layer, we model the robot as a simple holonomic point mass that is controlled with the force field method to generate ball approach trajectories, ball dribbling sequences, and to implement obstacle avoidance. The topmost layer of our framework takes care of team behavior, game tactics and the implementation of the game states as commanded by the referee box. Please refer to [2] for further details.

5   NimbRo-OP TeenSize Robot

Our main innovation this year was the development of the NimbRo-OP robot along with a ROS framework based robot soccer software. The software contains many modules for basic functions required for playing soccer that we either started from scratch, or ported from our classic NimbRo system. In the now second release [12], the software package contains a compliant servo actuation module [13] and a visual motion editing component. Motions are replayed with a non-linear keyframe interpolation technique that allows to generate smooth and continuous motions while respecting configurable acceleration and velocity bounds. Kicking and get-up motions have been successfully implemented. For walking, we use a port of the same gait generator that we use for

Fig. 3. The NimbRo-OP

our classic robots [8]. For higher-level behavior control, we ported the NimbRo hierarchical reactive behavior architecture [7] [14] and the implementations of simple soccer behaviors within, such as searching for the ball, walking up to the ball and dribbling the ball. The vision processing module was rewritten from scratch as a ROS module along with accompanying tools for camera and color calibration. Utilizing a camera with higher resolution and more available processing power, we improved the quality of our object detection, which is described in [12] in more detail. A particle filter-based localization module is also provided. Apart from the core soccer software itself, graphical software components are available to maintain configuration parameters and to log the state of the system in great detail to support debugging and monitoring during games.

6   Online Learning of Lateral Balance

In recent years, team NimbRo has developed a gait control framework capable of recovering from pushes that are strong enough to force a bipedal walker to adjust step-timing and foot-placement. Only lateral balance mechanisms [9] have been used in competitions so far, but in simulation, the framework is now able to absorb pushes from any direction at any time during the gait cycle [15]. In a nutshell, the Capture Step Framework is based on an extremely simplified state representation in the form of a point mass that is assumed to behave like a linear inverted pendulum. A decomposition of the lateral and sagittal dimensions into independent entities, and a sequential computation of step-timing, zero-moment point and foot-placement control parameters facilitates the closed-form mathematical expression of our balance controller. Modeling, however, can only take one so far. Complex full-body dynamics, sensor noise, latency, imprecise actuation, and simplifying modeling assumptions will always result in errors that can limit the balancing capabilities of a humanoid robot. A good way to increase the efficiency of a model based approach are online learning techniques that can measure performance during walking and adjust the output of model-based push-recovery strategies.

Focusing on the simplified purely lateral setting, we have successfully implemented an online learning algorithm that learns the foot-placement error during disturbed walking on the spot and subtracts it from the model output in order to improve push recovery capabilities. In the following section, we briefly outline the concepts of lateral balance and introduce our evaluation method that can quantify and visualize the effects of isolated balance components. Subsequently, we describe the online learning algorithm we used, and show experimental results to verify the achieved improvement.

6.1   Lateral Gait Control

The pendulum-like dynamics of human walking has been long known to be a principle of energy-efficient locomotion [16]. Figure 4 shows stick diagrams of the idealized sagittal and lateral pendulum motions projected on the sagittal plane and the frontal plane. Interestingly, the sagittal and lateral motions exhibit strongly distinct behaviors. In the sagittal plane, the center of mass vaults over the pivot point in every gait cycle, while in the frontal plane, the center of mass oscillates between the support feet and never crosses the pendulum pivot point.

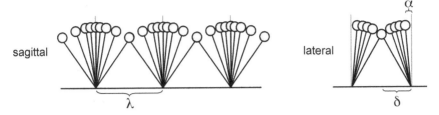

Fig. 4. Stick diagrams of idealized pendulum-like sagittal and lateral motion of a compass gait. In sagittal direction, the center of mass crosses the pendulum pivot point in every gait cycle, while in lateral direction it oscillates between the support feet. Parameter λ defines the stride length in the sagittal direction, parameter α denotes the characteristic lateral apex distance, and δ defines the support exchange location in the center of the step.

It is crucial not to tip over sideways, as the recovery from such an unstable state requires challenging motions that humanoid robots have difficulties performing.

The perpetual lateral oscillation of the center of mass appears to be the primary determinant of step timing. Disobeying the right timing can quickly destabilize the system after a disturbance, even if the disturbance itself would not have directly resulted in a fall [17]. Furthermore, we can identify two characteristic parameters in the lateral direction. We denote the minimal distance between the pivot point and the center of mass and that occurs at the apex of the step as α. The apex distance provides a certain margin for error. While during undisturbed walking the apex distance stays near α in every step, a push in the lateral direction can result in a smaller apex distance. As long as the apex distance is greater than zero, the center of mass will return and the walker will not tip over the support foot. Sooner or later, returning center of mass trajectories are guaranteed to reach the support exchange location that we denote as δ. While the support exchange location varies with increasing lateral walking velocity, for now we limit our setting to walking on the spot with zero velocity of locomotion and therefore we can assume δ to be a constant as well. To identify the model parameters α and δ for a real or a simulated biped, we induce the lateral oscillation by generating periodic, open-loop step motions using the walk algorithm described in [8]. Then, α and δ can be found by averaging the measured center of mass locations at the step apex and in the moment of the support exchange.

As a consequence of the principles described above, we can formulate the following control laws for our balance control computations:

- The timing of the step is determined by the moment when the center of mass reaches the nominal support exchange location δ.
- The lateral step size is chosen so that the center of mass will pass the following step apex with a nominal distance α with respect to the pivot point.

Formally, our balance controller is a function

$$(T, F) = \mathcal{B}(y, \dot{y}) \qquad (1)$$

that computes the step time T and the footstep location F as a function of the current state of the center of mass $(y, \dot{y})$. Here, y denotes the location of the center of mass along the lateral axis with respect to a right hand coordinate frame placed on the support foot, and $\dot{y}$ is the velocity of the center of mass. The step time T and the footstep location F are passed on to a motion generator that generates stepping motions with an appropriate frequency and leg swing amplitude. For the understanding of the experiments performed in this work, a conceptual insight of the lateral control laws presented above is sufficient. For more detailed information, we refer the reader to [15].

6.2 Experimental Setup

Using a physical simulation software, we performed a series of systematic push experiments on a simulated humanoid robot with a total body weight of 13.5 kg and a roughly human-like mass distribution. While the robot is walking on the spot, it is pushed in the lateral direction with an impulse targeted at the center of mass. After the impulse, the robot has some time to recover, before the next impulse is generated. If the robot falls, it is reset to a standing position and it is commanded to start walking again. The magnitude of the impulse is randomly sampled from the range $[-9.0, 9.0]$ Ns, where the sign of the impulse determines its direction (left or right). We generate 400 pushes for each of four balance controllers of increasing complexity:

- **No Feedback**: The controller ignores the pushes and does nothing. The robot executes an open-loop gait with a fixed frequency and step size.
- **Timing**: The controller adjusts only the timing of the step, but not the footstep location.
- **Timing + Step Size**: The controller adjusts the timing and the size of the steps using the mathematical model.
- **Timing + Step Size + Learning**: The controller responds to the disturbances using not only the model-based computation of the timing and the step size, but also a learned error that we subtract from the predicted step size. The error is learned online during the experiment.

The input space we use for learning is the lateral state space $\mathcal{S} = [y, \dot{y}] \in \mathbb{R}^2$ of the center of mass. When the support foot is the left foot, we flip the signs of y and $\dot{y}$ in order to exploit symmetry. During the experiment, the robot measures the efficiency of its steps and estimates an error that expresses a gradient, i.e. a desired scalar increase or decrease in the step size. The error is measured when the center of mass is at the step apex. It is given as simply the deviation from the nominal apex distance α. From the inverted pendulum model it follows intuitively that if the apex distance is greater than α, the step size was too large, and if the apex distance is smaller than α, the step size was too small. At the end of the step, we update the value of a function approximator for each of the states $(y, \dot{y})_{i \in I}$ that were encountered during the step. The update rule is

$$f((y, \dot{y})_i) = f((y, \dot{y})_i) + \eta(y_{i_a} - \alpha), i \in I, \tag{2}$$

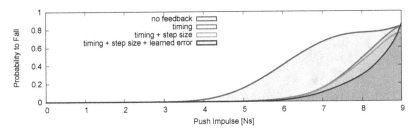

Fig. 5. Probability to fall versus the magnitude of the push impulse for four different controllers of increasing complexity

where $f((y, \dot{y})_i)$ is the value of the function approximator for the state $(y, \dot{y})_i$, y_{i_a} is the center of mass location that was measured at the step apex, and $\eta = 0.2$ is the learning rate. The function approximator is initialized with a value of 0 before learning. The step parameters that are passed on to the step motion generator are then

$$(T, F) = \mathcal{B}(y, \dot{y}) - (0, f(y, \dot{y})). \tag{3}$$

6.3   Evaluation of Results

Using the data we collected during the experiments, we can compare the efficiency of the four controllers. Figure 5 shows the probability to fall against the magnitude of the impulse and gives an impression of the push resistance of the controllers. Interestingly, the open-loop walk alone is able to handle pushes up to a strength of 3 Ns, in such a case returning slowly to a limit cycle. However, the three feedback controllers clearly increase the minimum impact required to make the robot fall and improve the ability to absorb an impact over the entire range of impulse strengths. The results of the three feedback controllers do not differ from each other significantly, leading to the conclusion that using the right step timing is already sufficient to predominantly stabilize returning center of mass trajectories. Why this effect can be achieved with step timing alone has a reasonable explanation. When the robot receives a push from the side, it typically first tilts towards the support leg and the center of mass approaches the outer edge of the support foot. If the robot was pushed in the direction away from the support leg, it will automatically tip onto the other leg in the center of the step, which leads to the same situation. Now, when the center of mass is moving towards the outer edge of the support foot, the robot may shorten the support leg if it does not adjust the motion timing, as internally the support leg is thought to be the swing leg at that time. This accelerates the center of mass additionally towards the support leg and reduces the lever arm, helping the robot to tip over the outer edge of the foot. Furthermore, the robot is likely to touch the floor with the other foot and can further accelerate itself in the wrong direction. And finally, if the center of mass returns, and it is moving away from the support leg, a badly timed extension of the support leg just before the support exchange adds energy to the lateral motion and increases the probability

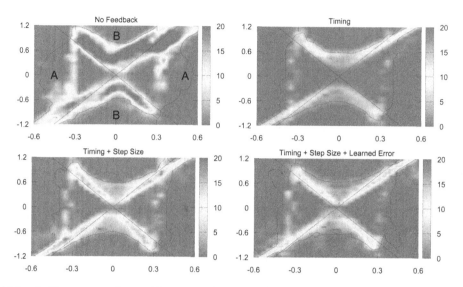

Fig. 6. Heat maps of unstable regions of the lateral phase space. Color coding marks the areas that have been crossed by falling trajectories. Thin black lines contour the cells that were visited at least ten times during the experiments. Straight zero-energy lines partition the phase space into stable regions of negative orbital energy (A), and unstable regions of positive orbital energy (B).

to tip over on the other side. Using adaptive timing, all of these undesired effects vanish. The adaptation of step timing prevents the robot from destabilizing itself due to badly timed leg motions in oblique poses and maximizes the minimal tip-over impulse to the value that can be passively absorbed. Using the torso as a reaction mass for active balancing could further increase the minimal tip-over impulse, but this is not in our scope at this time.

For a closer look, Figure 6 shows heat maps of the lateral phase space that were generated by backtracking from every fall to the first frame of a push and incrementing each grid cell that was touched by the center of mass on the way. The values of the cells are then used for color coding the unstable regions of the phase space for each controller. The thin black contours bound the regions of cells that were visited at least ten times during the experiments. The straight zero-energy lines are computed from the linear inverted pendulum model that is used to drive the feedback loops. The zero-energy lines partition the phase space into regions that we would expect to find based on model assumptions. The areas marked with the letter 'A' are regions of negative orbital energy. This is where all returning center of mass trajectories are located and stable lateral oscillations can take place. The sectors marked with the letter 'B' are of positive orbital energy and contain state trajectories that will inevitably cross the pivot point and tip over. The model is reflected by the experimental data, as the vast majority of the states encountered between a push and a fall are located in the unstable areas of the heat maps. The fall trajectories of all controllers must

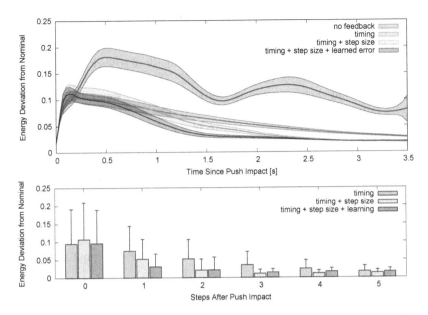

Fig. 7. Development of the lateral orbital energy after a push synchronized at the push impact (top), and at the individual steps after the push (bottom). While the "Timing" controller monotonically returns to a desired level of orbital energy, the adjustment of step size helps the robot to return to the nominal energy level much faster. The open-loop controller cannot be sensibly synchronized with the feedback controllers and thus it has been omitted from the bottom plot.

originate from the stable region, since the push is always applied in a stable state of the robot. The push changes the state trajectory abruptly and transfers it into the unstable section 'B'. It is evident that the heat map of the open-loop controller contains a much larger number of falls. The heat maps of the three feedback controllers look very similar with a strongly reduced number of falls in comparison with the "No Feedback" experiment. Again, we can conclude that step timing adaptation plays a pivotal role in preventing a fall.

In order to answer the question of how a bipedal walker can benefit from a well chosen step size, Figure 7 shows the development of the orbital energy after the disturbance in the cases where the robot did not fall. In the top half of the plot, the time series of the orbital energy deviation from a nominal value has been synchronized at the moment of the push impact. Since the open-loop controller has a tendency to amplify the push impulse, the peak energy shortly after the push is significantly higher. The wave-like form of the energy curve suggests that the open-loop controller occasionally disturbs itself. When using only timing feedback, the disturbance amplification and the self disturbances disappear and the orbital energy returns monotonically to a desired level. With the addition of a computed step size, the robot can absorb the orbital energy much faster. The controller with the learned step size error shows the best performance in terms of orbital energy dissipation. In the bottom half of the plot, the energy level with respect to the nominal value has been synchronized at the individual steps after the push. The fixed-frequency steps of the open-loop controller cannot be sensibly synchronized with the timed steps of the feedback controllers and thus have been omitted from the bottom plot. The first group of boxes show the energy deviation that has been measured during the step that was pushed. The second group of boxes at the index 1 represent the "capture step", the first step after the push. As in theory a full recovery is possible with one step, the efficiency of the capture step is of particular interest. The efficiency of a step can be computed as $1 - \frac{e_s}{e_{s-1}}$, where e_{s-1} and e_s are the excess energy levels before and after the step. The step efficiency of the step timing controller is 21%. Adding the step size modification improves the step efficiency to 51%, and learning further increases the energy absorption rate to 68%. Accelerating the return to a nominal, stable state has a positive effect on overall bipedal stability. The walker is ready to face the next disturbance in a shorter amount of time and thus not only the magnitude, but also the frequency of impulses that the robot can handle, is increased.

7   Conclusions

The TeenSize class experienced an uplift during the 2013 competition. Five teams were at the competition site and played games with more than one operational robot on the field from each team. Several technical challenges were completed. All teams were able to advance their software to cope with the new challenge of localization with symmetrical landmarks.

In the final, our robots met team CIT-Brains from Japan. In the beginning of the match, each team played with two players on the field. CIT-Brains played an offensive strategy with two strikers while team NimbRo designated one player as goal keeper. The CIT team managed to press onward towards the NimbRo goal, but the NimbRo robots defended against the attacks reliably. The obstacle avoidance feature of the CIT robots appeared to be a bit too aggressive and they approached the NimbRo robots too closely and often stepped on their toes, which made the CIT robots fall over. NimbRo striker Copedo used the opening gaps to score. Team NimbRo successfully demonstrated dynamic role assignment that temporarily assigned the goal keeper Dynaped the striker role when Copedo had to be taken out of the game. While in the second half, team CIT Brains had to reduce the number of players to one due to technical difficulties, team NimbRo managed to maintain two operational players throughout the game and scored reliably. Consequently, team NimbRo won the finals with a score of 4:0 and successfully defended its title for the fifth time in a row.

The stability of the gait of our robots and their robustness to disturbances was one of the key factors for our success. The online learning method outlined in this work will contribute to even faster stabilization of bipedal walking in future competitions.

Acknowledgment. This work is supported by Deutsche Forschungsgemeinschaft (German Research Foundation, DFG) under grants BE 2556/6 and BE 2556/10.

References

1. Missura, M., Münstermann, C., Mauelshagen, M., Schreiber, M., Behnke, S.: RoboCup 2012 Best Humanoid Award Winner NimbRo TeenSize. In: Chen, X., Stone, P., Sucar, L.E., van der Zant, T. (eds.) RoboCup 2012. LNCS (LNAI), vol. 7500, pp. 89–93. Springer, Heidelberg (2013)
2. Lee, D.D., et al.: RoboCup 2011 Humanoid League winners. In: Röfer, T., Mayer, N.M., Savage, J., Saranlı, U. (eds.) RoboCup 2011. LNCS, vol. 7416, pp. 37–50. Springer, Heidelberg (2012)
3. Schwarz, M., Schreiber, M., Schueller, S., Missura, M., Behnke, S.: NimbRo-OP Humanoid TeenSize Open Platform. In: Proceedings of 7th Workshop on Humanoid Soccer Robots, IEEE Int. Conf. on Humanoid Robots, Osaka, Japan (2012)
4. Thrun, S., Burgard, W., Fox, D.: Probabilistic Robotics. MIT Press (2001)
5. Schmitz, A., Missura, M., Behnke, S.: Learning footstep prediction from motion capture. In: Ruiz-del-Solar, J., Chown, E., Plöger, P.G. (eds.) RoboCup 2010. LNCS (LNAI), vol. 6556, pp. 97–108. Springer, Heidelberg (2010)
6. Schulz, H., Behnke, S.: Utilizing the structure of field lines for efficient soccer robot localization. Advanced Robotics 26, 1603–1621 (2012)
7. Behnke, S., Stückler, J.: Hierarchical reactive control for humanoid soccer robots. Int. Journal of Humanoid Robots (IJHR) 5, 375–396 (2008)
8. Missura, M., Behnke, S.: Self-stable Omnidirectional Walking with Compliant Joints. In: Proceedings of 8th Workshop on Humanoid Soccer Robots, IEEE Int. Conf. on Humanoid Robots, Atlanta, USA (2013)
9. Missura, M., Behnke, S.: Lateral capture steps for bipedal walking. In: Proceedings of IEEE Int. Conf. on Humanoid Robots (Humanoids) (2011)
10. Stückler, J., Schwenk, J., Behnke, S.: Getting back on two feet: Reliable standing-up routines for a humanoid robot. In: Proceedings of The 9th Int. Conf. on Intelligent Autonomous Systems, IAS-9 (2006)
11. Missura, M., Wilken, T., Behnke, S.: Designing effective humanoid soccer goalies. In: Ruiz-del-Solar, J., Chown, E., Plöger, P.G. (eds.) RoboCup 2010. LNCS (LNAI), vol. 6556, pp. 374–385. Springer, Heidelberg (2010)
12. Allgeuer, P., Schwarz, M., Pastrana, J., Schueller, S., Missura, M., Behnke, S.: A ROS-based software framework for the NimbRo-OP humanoid open platform. In: Proceedings of 8th Workshop on Humanoid Soccer Robots, IEEE Int. Conf. on Humanoid Robots, Atlanta, USA (2013)
13. Schwarz, M., Behnke, S.: Compliant robot behavior using servo actuator models identified by iterative learning control. In: 17th RoboCup Int. Symposium (2013)
14. Allgeuer, P., Behnke, S.: Hierarchical and state-based architectures for robot behavior planning and control. In: Proceedings of 8th Workshop on Humanoid Soccer Robots, IEEE Int. Conf. on Humanoid Robots, Atlanta, USA (2013)
15. Missura, M., Behnke, S.: Omnidirectional capture steps for bipedal walking. In: Proceedings of IEEE Int. Conf. on Humanoid Robots (Humanoids) (2013)
16. Kuo, A.D., Donelan, J.M., Ruina, A.: Energetic consequences of walking like an inverted pendulum: step-to-step transitions. Exercise and Sport Sciences Reviews 33(2), 88–97 (2005)
17. Missura, M., Behnke, S.: Dynaped demonstrates lateral capture steps, http://www.ais.uni-bonn.de/movies/DynapedLateralCaptureSteps.wmv

RoboCup 2013: Best Humanoid Award Winner JoiTech

Yuji Oshima[1], Dai Hirose[1], Syohei Toyoyama[1], Keisuke Kawano[1], Shibo Qin[1],
Tomoya Suzuki[2], Kazumasa Shibata[2], Takashi Takuma[3], and Minoru Asada[1]

[1] Dept. of Adaptive Machine Systems, Graduate School of Engineering,
Osaka University, Osaka, Japan
[2] Dept. of Mechanical Engineering, Faculty of Engineering,
Osaka Institute of Technology, Osaka, Japan
[3] Dept. of Mechanical Engineering,
Osaka Institute of Technology, Osaka, Japan
robocup@er.ams.eng.osaka-u.ac.jp

Abstract. This article presents the technical strategy employed by JoiTech in the
RoboCup 2013 humanoid league adult size championship. Two features focused
on by the team were the design of a versatile robot simulator and smart strategies
for the robot player JoiTech Messi. The input for the versatile robot simulator in
our system was data from the robot's camera and the output was a motor command given to the robot. Our system used real video data and a robot, as well as
virtual data and video data recorded from the real robot's camera. Thus, we could
select one of three inputs, i.e., real, virtual, and recorded, and one of two outputs,
i.e., real and virtual. This combination of data allowed us to debug the codes
used by Messi in an efficient manner. This reduced the number of real robot tests,
which minimized damage to the robot. In the design of the smart strategies for
the robot player JoiTech Messi, we developed a system that recognized opponent
robots using background subtraction, which improved the accuracy of the striker
and the goalkeeper. The detection of an opponent player was useful for finding the
space to shoot and to block the goal when an opponent player attempted to score.
These two features worked effectively during the competitions and our JoiTech
team won the championship, as well as the best humanoid award ("Louis Vuitton
Cup"), which demonstrated the success of our system.

1 Introduction

Team JoiTech participated in the soccer humanoid adult size league at RoboCup 2013.
The team comprised students from Osaka University and Osaka Institute of Technology.
We have participated in the RoboCup Japan Open and the RoboCup world competition
each year since 2010. Our team was originally derived from RoboCup team JEAP, who
participated in the humanoid league kid size competitions since 2006. Our team name,
JoiTech, is an acronym for "**J**EAP and **O**saka **I**nstitute of **Tech**nology," but it also means
"joint team with Osaka Inst. of Technology" and "enjoy technology." This was the third
humanoid adult size world RoboCup soccer competition. The adult size league has two
critical differences compared to the other leagues (kid size and teen size).

First, large and heavy robots cannot run for long periods because of the high load
on their motors. It is also expensive to prepare spare robots. Thus, it is preferable to

S. Behnke et al. (Eds.): RoboCup 2013, LNAI 8371, pp. 68–79, 2014.
© Springer-Verlag Berlin Heidelberg 2014

Table 1. Tichno-RN hardware specifications

Tichno-RN		
Height (mm)	1500	
Weight (kg)	25	
DOF	22	
Actuators	VS-SV410, VS-SV1150, VS-SV3310	
Camera Type	iBUFFALO BSW20KM11BK	
Controller	Main Controller	Sub Controller
CPU	Panasonic Let's Note Intel Corei5 2.6GHz	VS-RC003HV ARM7TDMI LPC2148
ROM	223 GB (SSD)	512 KB
RAM	8 GB	40 MB
OS	Windows 7	None

minimize the need for testing of large robots, which limits the development of code for these robots.

Second, adult size robots are required to strike the ball and block the goal. The rules of the adult size game differ from those of the other size leagues, because it is based on penalty kicks from human soccer. The adult size game involves an offensive player and a defending goalkeeper, and each side is limited to five penalty attempts. Shooting or blocking failures are not options in the championship. In recent years, the behavior of goalkeepers has become more important because many robots in the adult size league can score goals successfully. However, current adult size robots cannot move rapidly or drop down to block the goal. Previously almost all robot did not have any strategies in goalkeeper. Thus, we developed a practical strategy for the goalkeeper given the constraints on the robot's motions, which most teams had not considered.

We developed systems to solve the two critical problems encountered in the adult size league. This paper is organized as follows. We provide the hardware specifications in section 2. In section 3, we present an overview of the software and an object recognition system based on other software systems. We also describe the development environment used to solve the first problem, as well as the strategies used by the striker and the goalkeeper to address the second problem. The final section presents the results achieved by our team in the competitions and we summarize how our strategies performed during the games.

2 Robot Hardware

2.1 Hardware

In this section, we explain the hardware specifications for Tichno-RN, the mechanical structure of which was developed by Vstone Co., Ltd[7]. A front view and a schematic overview are shown in Fig. 1. The detailed specifications are given in Table 1.

Tichno-RN has 22 degrees of freedom (DoFs) as shown in Fig. 1(a). The legs and arms each have six DoFs and four DoFs, respectively. The structure and powerful

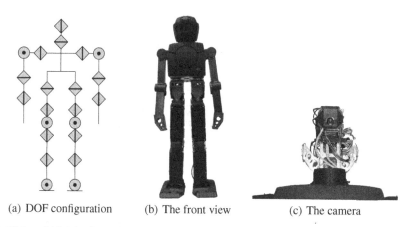

(a) DOF configuration   (b) The front view   (c) The camera

Fig. 1. Tichno-RN: (a) schematic overview of the actuators and the configuration of the body, (b) front view of the whole body, (c) camera mounted on the robot

electric motors generate a strong torque of about $32N \cdot m$, which allows our robot to squat down and turn rapidly.

Each actuator has a micro-controller with sensors that detect the angular position of joints, temperature, and speed, and which transmits the sensor information to a sub-controller. A camera is used as the eye of the robot and it has a wide angle view of 120 degrees. The camera position is adjusted to keep its own toes and the top of the goalposts within the visual field. Tichno-RN can squat down, stand up, hold a ball, and throw it, which are required to make a throw-in. Only the JoiTech robot succeeded in performing a throw-in during the technical challenge.

2.2 Control System

Tichno-RN has two controllers: a main controller and a sub-controller. The main controller has an advanced processor that allows object recognition and action decision making. This system uses a commercially available notebook with sufficient capacity for image processing. The main controller allowed us to develop a system without any special micro-controller programming skills. The sub-controller perceives information obtained from gyro and speed sensors in the actuators and it controls all of the actuators. The sub-controller stores the motor routines such as shooting and throw-in. The sequence of motor commands that comprise the motor routines are sent to the motors after the sub-controller receives an order from the main controller, and it subsequently receives sensor information such as the speed and angle in real time.

3 Software

Fig. 3 provides an overview of the software used by the main controller. This software was implemented using C++ and we utilized OpenCV [2] as the image processing library. The operating system was Ubuntu 12.04 LTS 64bit.

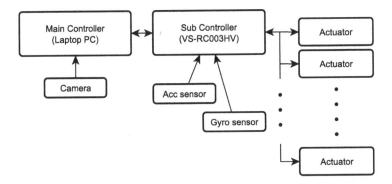

Fig. 2. Control system architecture: the main controller has an advanced processor for object recognition and action decisions. The sub-controller receives accelerator, gyro, and motor sensor information. The sub-controller sends a sequence of motor commands for one of the motor routines to the motors after it receives an order from the main controller.

The main program comprises three units. A strategy unit decides the robot's behavior based on sensory information. The motion module converts a behavior or motor routine into motor commands and initiates the motion. We define motions as motor routines such as walking or kicking. The vision module is an image processing unit. We also produced a simulation environment to facilitate efficient debugging.

The processing flow was as follows.

1. Recognize environmental information using the vision module.
2. Decide an action using the strategy.
3. Execute the motions using the motion module.

In the following subsections, we describe the image processing procedure, the strategies used by the attacker and the goalkeeper, a method for motion generation, and the test environment.

3.1 Method for Generating Motions

We use RobovieMaker2 to create the motions, which is a program developed by Vstone Co. Ltd. Fig. 4 shows the development environment of RobovieMaker2. Each value on the slider bars corresponds to the joint angles of the motors. The robot's poses or postures are generated using these slider bars. To create motions, we link a position to the next position and specify the time between them. Fig. 4 shows a motion flowchart on the right. RobovieMaker2 simplifies the process of producing motions. We produced 11 motions, including kicking and throw-ins. The walking motion we employed was developed by Vstone Co. Ltd.

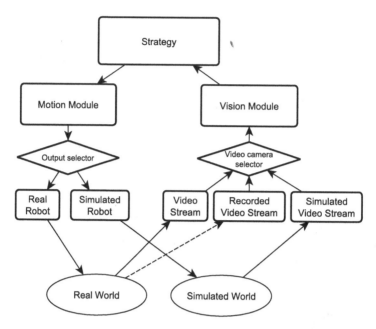

Fig. 3. System Architecture: The video stream and recorded video stream are produced from data captured by the real robot camera. The simulated video stream is produced from data generated by a simulator. The output selector selects a real or simulated robot. One output and one input can be selected. For example, it is possible to select recorded real world video data as the input and the simulated robot as the output.

3.2   Image Processing

The vision module is used for image processing and it detects the positions of key objects such as the ball and the goal. The vision module recognizes the field, the ball, lines, and obstacles.

An example result of the visual processing is shown in Fig. 5. The vision module extracts the biggest green region and considers it, filtered from noise, as the field area. The vision module assumes that objects such as the ball, goal, lines, and obstacles are in contact with the field area. The ball, lines, and obstacles are recognized based on the colors and shapes of the objects.

Goal recognition is important for shooting. Due to height of the robot The robot's camera captures a large area, including space outside the field area, which can lead to a false detection of objects. Therefore, we developed a strict goal recognition strategy. We specified the goal configuration as two vertical lines and a horizontal on the field area. However, this strict recognition method sometimes missed the goal because of image blurring when the robot was moving. To address this problem, we separated goal recognition into two phases: detection and tracking. First, the robot detects the goal lines using the Hough transform[4] and recognizes the goal as lines that meet the field. This recognition method prevents the false detection of objects outside the field. Second, the

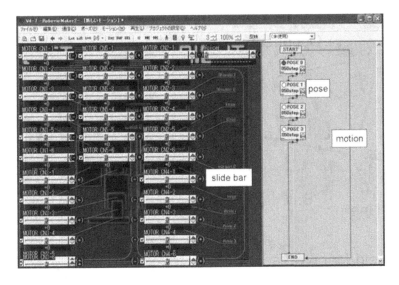

Fig. 4. RobovieMaker2: *left*, each slider corresponds to the joint angles of motors; *right*, flowchart showing an example of a motion

robot tracks the goal using a particle filter[6][3] based only on color until the goal is outside the camera image. This method allowed the robot to detect and track the goal in a stable manner, even while it was moving.

3.3 Test Environment

Any motions made by an adult size robot put a load on its motors and it is difficult to prepare spare robots or motors. Thus, it is desirable to minimize the number of trials when testing the software using a real robot.

To address this problem, we developed two test environments: a simulator and video recording/replay.

Combinations of these environments provided flexible and convenient systems for debugging the software. Next, we describe the simulator, the video recording and replay system, and combinations of these tools, which we refer to as the "versatile robot simulator."

Simulation. We created a simulation environment for software testing that does not require a real robot. Fig. 6 shows a snapshot captured by the simulator. This simulator was implemented using Open Dynamics Engine [8] and it reproduced the minimum set of game elements, i.e., the field, robots, lines, black obstacles, and a ball. We obtained the parameters for the simulated robot (e.g., moving speed, turning speed, and kicking strength) based on the movements of our real robot. We added noise to the parameters to test the robustness of the programs.

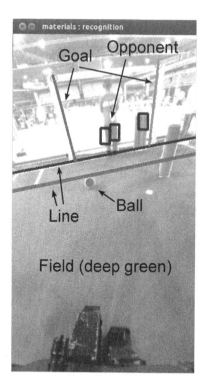

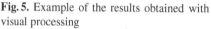

Fig. 5. Example of the results obtained with visual processing

Fig. 6. Snapshot of the simulation. The black pole in front of the goal represents the opponent goalkeeper.

Our simulator aimed to test the software used to generate the game strategies (and visual recognition), rather than improving the robot motions. Consequently, our environment did not simulate walking or other motions in detail.

Vision Recording and Replay. Our simulator was a convenient tool for software testing, but it lacked visual realism. Therefore, we created a visual testing system to record the camera data captured by a real robot during testing and we streamed the recorded data to the vision module. This system allowed us to test the vision software repeatedly using the real robot's camera data without operating the real robot. This system was more advantageous than real robot tests when debugging some errors in specific situations (e.g., the robot lost the goal in a particular position) because the tests utilized exactly the same data and were repeatable.

This system was used during the development phase and in the competition. For example, we analyzed the color information in the field over time and analyzed the recorded video data. The real competition movies were useful for analyzing the strategies used by opponents and making adjustments before the next game.

Versatile Robot Simulator. We could select the camera on the real robot, the simulated camera, or recorded video data as the input for the vision module, and the real robot or the simulated robot as the output for the motion module. Each combination (3 × 2) produced a different test environment for a specific purpose. The combinations are described below.

- Simulated camera + simulated robot:
  The strategies used by the robot could be tested in a completely simulated world. The real robot was never required. This test environment could be run in parallel with the software.

- Recorded video data + simulated robot:
  This combination allowed testing using previous real visual environments without damaging the robot. This was useful for fixing reproducible bugs because the recorded video could be used repeatedly. This test environment could be run in parallel with the software.

- Real camera + simulated robot:
  This was useful for detecting bugs because we could change the camera image by moving the video camera while watching the behavior of the robot in the simulator.

- Simulated camera + real robot:
  This is able to test in visual situation with real robot even impossible situation in real world. These environments were not valid for testing for competition.

- Recorded video data + real robot:
  It tests almost same parts of system as Recorded video data + simulated robot. These environments were not valid for testing.

- Real camera + real robot:
  This was the most basic test environment, which allowed testing in exactly the same environment as during the competition. We performed some tests in this environment, but they could damage the robot.

3.4   Strategies Used by the Attacker and the Goalkeeper

There are only limited opportunities for shooting in the adult size league. The probability of successful shooting and blocking will increased if the robot can dribble a ball and shoot it at the goal. Thus, we developed fundamental strategies for our robot, as follows.

Opponent Detection and Deciding the Shooting Direction. Recently, robots have been required to shoot accurately at the goal while avoiding the goalkeeper. In the 2013 competition, the robots also had to avoid a black obstacle placed in front of the goal. Next, we describe the strategy used by our robot, which allowed the robot to shoot at the goal accurately while avoiding the black obstacle and the goalkeeper.

1. Detect the location of the black obstacle using object recognition and the opponent robot by background subtraction (Fig. 7). Then the robot detects the positions of the goalposts by goal recognition.

The widest space

(a) Deciding the shooting direction.   (b) Image showing background subtraction, where the goalkeeper is highlighted.

Fig. 7. Goalkeeper recognition

2. Calculate the widest distances between the goalposts, the opponent robot, and the black obstacle.
3. Turn towards the center of the widest space and kicking the ball.

Using this method, the robot could kick the ball in the space with the highest likelihood of scoring a goal. Background subtraction allowed our robot to avoid the opponent robot, regardless of its color.

Goalkeeping Strategy. Previously, the strategy used by most adult size robot goalkeepers was standing upright in the goal area without moving. However, the blocking success has become important because of the increased number of successful shots in the championship. Thus, we developed the following strategy (Fig. 8).

1. Detect the opponent striker using background subtraction, and detect the approaching of the opponent striker by the distance between the striker and the ball.

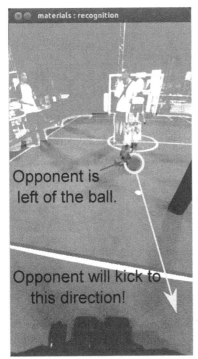

(a) Detection of the opponent robot and the shooting direction.

(b) Image with background subtraction, where the opponent is highlighted.

Fig. 8. Opponent recognition

2. Predict the direction of the ball that the opponent striker will kick by positional relationship between the striker and the ball. For example, if the opponent striker is located left of the ball, the opponent striker more likely to kick to right.
3. Move to the predicted direction.

Using this method, the goalkeeper was expected to make a rapid movement to make a block because the time between the prediction and approach to the ball was short. However, the walking speed of our robot was not sufficient to successfully make a block, depending on the speed of the opponent robot. To address this problem, we developed another goalkeeping strategy to decide the direction the goalkeeper moved in, which depended on the current position of the opponent striker and that of the black obstacle at the start of the game.

4   Results

Our results in the competitions are summarized in Table 2 [1].

The ratio of successful shots was only 24% because our vision system was not adjusted adequately in the round robin tournament. In the first trial of round robin 4,

Table 2. Results in the competitions

Match	Opponent	Result
Round Robin 1	Tsinghua Hephaestus	1-0
Round Robin 2	ROBIT	2-0
Round Robin 3	EDROM Adult Size	1-0
Round Robin 4	HeuroEvolution AD	1-4
Round Robin 5	Tech United Eindhoven	1-0
Semifinal	Tsinghua Hephaestus	4-1
Final	HeuroEvolution AD	4-3

Fig. 9. Shooting during the competition: the robot turned to the left because the opponent and the black obstacle were to the right of the goal

Fig. 10. Blocking during the competition: the robot moved to the left when the opponent was close to the right side of the ball

our robot successfully blocked the shot of the opponent HeuroEvolution AD. However, our robot failed to block any shots in the remaining trials, because the opponent team changed their attack strategy.

In the semifinal and the final, the ratio of successful shots improved to 80% (Fig. 9), because the robot adjustments were completed. Our robot made a successful block in the semifinal and the final (Fig. 10). Our ratio of successful blocks was only 27%, but HeuroEvolution AD's ratio was 0%. HeuroEvolution AD's ratio of successful shots throughout all of the competition was 80% that highlights the importance of successful blocks in our championship victory.

5   Conclusion

This study explained the technical strategy used by JoiTech to win the RoboCup 2013 humanoid league adult size championship. We had to solve specific problems to win the RoboCup soccer humanoid adult size championship. The first was a limitation of the number of times that testing could be performed, which was mainly due to hardware constraints. The second was that the number of shots was limited by the rules, which required successful shooting and blocking. We developed a versatile simulator to solve the first problem, which comprised six different input and output combinations. The virtual world with a recorded data stream allowed us to conduct testing in near-real environment without damaging the robot. This drastically reduced the number of tests on required the real robot. With respect to the second problem, predicting the direction in which the robot and the opponent robot kicked the ball improved the accuracy of shooting and blocking. These two improved the probability of successful shooting and blocking until the final match. Thus, we won the championship and the best humanoid award ("Louis Vuitton Cup").

The versatile robot simulator was useful for testing at the strategy level, but it could not be used for testing at the motion level (e.g., for throw-ins). It is difficult to simulate real robot motions because of the complex computations required, such as modeling contact with objects. In most cases, the robot's motions were not created by a simulator, but instead were refined by trial and error using real robots. Kawai et al.[5] proposed a method that can reduce the number of trials when adjusting robot motions. We could produce a system with a lower computational load to refine the robot motions by adding this method.

Recently, the quality of the adult size competition with simple rules has improved drastically. In the next step, the adult size robots will be required to be more dynamic and to use a flexible strategy like the other size league, so the rules will be closer to actual human soccer. Thus, it will be necessary to develop a more flexible and general system to ensure successful and accurate shooting and blocking in the future.

Acknowledgement. The authors gratefully acknowledge the contributions of the JoiTech team members. This work was supported by a Grant-in-Aid for Specially Promoted Research Number 24000012.

References

1. Robocup 2013 eindhoven (2013), http://www.robocup2013.org/ (retrieved August 30, 2013)
2. Bradski, G.: The OpenCV Library. Dr. Dobb's Journal of Software Tools (2000)
3. Gordon, N., Salmond, D.: Bayesian state estimation for tracking and guidance using the bootstrap filter. Journal of Guidance Control and Dynamics 18, 1434–1443 (1995)
4. Hough, P.V.C.: Machine Analysis of Bubble Chamber Pictures. Conf. Proc. 590914, 554–558 (1959)
5. Kawai, Y., et al.: Throwing Skill Optimization through Synchronization and Desynchronization of Degree of Freedom. In: Chen, X., Stone, P., Sucar, L.E., van der Zant, T. (eds.) RoboCup 2012. LNCS (LNAI), vol. 7500, pp. 178–189. Springer, Heidelberg (2013)
6. Kitagawa, G.: Monte Carlo Filter and Smoother for Non-Gaussian Nonlinear State Space Models. Journal of Computational and Graphical Statistics 5(1), 1–25 (1996)
7. Oshima, Y., Hirose, D., Park, J., Okuyama, Y., Sumita, R., Akutsu, D., Suzuki, T., Shibata, K., Takuma, T., Tanaka, K., Mori, H., Asada, M.: JoiTech Team Description. In: RoboCup 2013 Symposium Papers and Team Description Papers CD-ROM (2013)
8. Smith, R.: Open dynamics engine (2008), http://www.ode.org/

B-Human 2013:
Ensuring Stable Game Performance*

Thomas Röfer[1], Tim Laue[1], Arne Böckmann[2], Judith Müller[2],
and Alexis Tsogias[2]

[1] Deutsches Forschungszentrum für Künstliche Intelligenz,
Cyber-Physical Systems, Enrique-Schmidt-Str. 5, 28359 Bremen, Germany
[2] Universität Bremen, Fachbereich 3 – Mathematik und Informatik,
Postfach 330 440, 28334 Bremen, Germany

Abstract. The aim of a soccer game is to score more goals than the opponent does. The chances of doing so increase when all the own players are continuously present on the field. In the RoboCup Standard Platform League (SPL), the only reasons for players not to participate in a game are that they either have been penalized or they have not been ready for the match in the first place. This paper presents some of the methods the 2013 SPL world champion team B-Human uses to ensure a stable and continuous game performance. These include overcoming the weak obstacle avoidance of our 2012 system, reacting to a rule change by an improved detection of the field boundary, better support for analyzing games by logging images in real time, and quality management through procedures the human team members follow when participating in a competition.

1 Introduction

B-Human participates in the RoboCup Standard Platform League. It is a joint team of the Universität Bremen and the German Research Center for Artificial Intelligence (DFKI) and consists of numerous undergraduate students as well as three researchers. Since 2009, we won every world championship in the SPL except for 2012, where we became the runner-up. The main goal for 2013 was to win back the championship title. To do that, we had to analyze why we failed in 2012. Some of the reasons were external: the vibrating field construction and the bad wireless network introduced a high degree of noise into the games. They simply all looked bad, which made it hard to detect problems that were still present in our system. Thanks to the setup in Eindhoven, these problems were not present in 2013, at least not after UDP packet buffering was switched off in the wireless access points.

One of the major weaknesses we had in 2012 was the detection of other robots or obstacles in general [1] , which did not become apparent before the semifinal.

---

* The authors would like to thank all B-Human team members for providing the software base for this work.

S. Behnke et al. (Eds.): RoboCup 2013, LNAI 8371, pp. 80–91, 2014.
© Springer-Verlag Berlin Heidelberg 2014

The visual robot perception relied on the detection of the colored waistbands. Since they were rather small, it was limited to short ranges. In addition, our opponents in the semifinal and final played with arms behind the robot's back, which made the bands hard to detect when approaching such a robot from behind. The sonar-based obstacle detection struggled with the new ("silver") ultrasound sensors of the NAO V4 robots that we had received one week prior to the competition. It did not work well with the previously used method for acquiring the sonar measurements. As a result of both weaknesses, our robots committed an unusually high number of "player pushing" offenses. They sometimes also did not switch to their tackling behavior, which resulted in a number of "ball holding" penalties. As a result, our robots were often taken off the field in the last two games. Sometimes, only the goalkeeper was left. This was a huge advantage for our opponents.

For 2013, one of the main goals was to keep our robots on the field, i.e. to avoid receiving penalties. This goal was achieved while still scoring by far the most goals of all teams. According to the log files of the GameController [2], our team only had an average of 3.75 penalties per game. This was the lowest value in the whole competition. The average value of all teams was 11.06.

This paper focuses on a few examples of approaches, how this was achieved. To avoid "player pushing" and "ball holding" penalties, the detection of obstacles was significantly improved, using both ultrasound and vision. This is described in Sect. 2. In response to a rule change that allowed goals on neighboring fields to be visible, an improved detection of the field boundary was implemented, which is presented in Sect. 3. In the actual competition, it was more useful to avoid "leaving the field" penalties, because no neighboring goals were visible, but orange logos were placed next to the field that could have been mistaken for the ball in some situations. If a robot still misbehaves, it is often hard to determine what the actual reason for an error was. A new logging technique that includes both camera images of the NAO and runs in real time, simplifies debugging a lot because it gives the programmer the context needed to understand what happened. This method is described in Sect. 4. Some penalties such as "fallen robot" and "request for pickup" can be avoided by careful selection and preparation of the robots that play and by a general quality management. The organizational procedures that we use to ensure the best performance of our robots are outlined in Sect. 5.

2   Obstacle Detection

Obstacle detection is required for navigation on the field, but also for tackling. The NAO can perceive obstacles in various ways: visually [3,4,5], through ultrasound [6], with its foot bumpers, and with its arms [1]. We use all of these methods. However, the ultrasound detection method employed in 2012 did not work very well [1] and the previous approach of detecting robots visually relied on the colored waistbands, which only worked at short range and could not be applied anymore in 2013, because the waistbands were replaced by jerseys. For

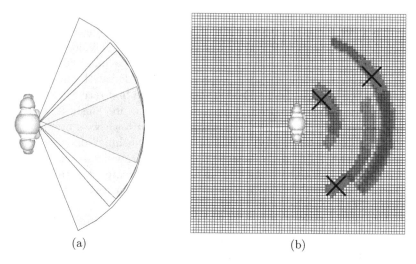

(a)   (b)

Fig. 1. Ultrasound obstacle detection. a) Overlapping measuring cones. b) Obstacle grid that shows that sometimes even three robots can be distinguished. Clustered cells are shown in red, resulting obstacle positions are depicted as crosses.

2013, the ultrasound-based obstacle detection was reimplemented and the previous visual robot detection was replaced by a visual obstacle detection. Since the new jerseys were not available before the competition, none of our methods is able to determine the team color of the objects perceived; they are simply treated as anonymous obstacles.

2.1   Ultrasound-Based

The NAO is equipped with two ultrasound transmitters (one on each side) and two receivers (again one on both sides), which results in four possible combinations of transmitter and receiver used for a measurement. We model the areas covered by these different measurement types as cones with an opening angle of 90° and an origin at the position between the transmitter and the receiver used (cf. Fig. 1a). Since NAO's torso is upright in all situations in which we rely on ultrasound measurements, the whole modeling process is done in 2-D. We also limit the distances measured to 1.2 m, because experiments indicated that larger measurements sometimes result from the floor. The environment of the robot is modeled as an approximately 2.7 m × 2.7 m grid of 80 × 80 cells. The robot is always centered in the grid, i. e. the contents of the grid are shifted when the robot moves. Instead of rotating the grid, the rotation of the robot relative to the grid is maintained. The measuring cones are not only very wide, they also largely overlap. To exploit the information that, e. g., one sensor sees a certain obstacle, but another sensor with a partially overlapping measuring area does not, the cells in the grid are ring buffers. These ring buffers store the last 16 measurements that concerned each particular cell, i. e. whether a cell was measured as free or as occupied. With this approach, the state of a cell always reflects recent measurements. Experiments have shown that the best results are achieved when cells are considered as part of an obstacle if at least 10 of the last 16 measurements have measured them as occupied. All cells above the threshold are clustered. For each cluster, an estimated obstacle position is calculated as the average position of all cells weighted by the number of times each cell was measured as occupied.

Since the ultrasound sensors have a minimum distance that they can measure, we distinguish between two kinds of measurements: close measurements and normal measurements. Close measurements are entered into the grid by strengthening all obstacles in the measuring cone that are closer than the measured distance, i. e. cells that are already above the obstacle threshold are considered to have been measured again as occupied. For normal measurements, the area up to the measured distance is entered into the grid as free. For both kinds of measurements, an arc with a thickness of 100 mm is marked as occupied in the distance measured. If the sensor received additional echoes, the area up to the next measurement is again assumed to be free. This sometimes allows narrowing down the position of an obstacle on one side, even if a second, closer obstacle is detected on the other side as well – or even on the same side (cf. Fig. 1b). Since the regular sensor updates only change cells that are currently in the measuring area of a sensor, an empty measurement is added to all cells of the grid every two seconds. Thereby, the robot forgets old obstacles. However, they stay long enough in the grid to allow the robot to surround them even when it cannot measure them anymore, because the sensors point away from them.

2.2   Vision-Based

The visual obstacle detection is based on the idea that obstacles are big non-green areas in the image. It is split into an initial detection stage and a refinement stage. The initial detection generates a list of spots in image coordinates that might be anywhere inside a possible obstacle. Afterwards, the refinement step filters these spots, using several sanity checks. In addition, it stitches obstacles together that span both camera images.

Initial Detection. The initial detection is based on the assumption that each non-green area that is bigger than a field line has to be an obstacle. The whole image is searched for such regions. To be able to search the image in a reasonable time, not every pixel is checked. Instead, a grid is utilized (cf. Fig. 2a). The vertical distance between two grid points is variable. It is calculated by projecting points to the image plane that are precisely 100 mm apart in the field plane. The horizontal distance between two grid points is fixed. To further reduce the runtime, pixels above the field boundary (cf. Sect. 3) are not checked.

For each grid point it is determined whether the point lies inside a non-green region or not. This is achieved by checking whether more than $k\%$ (currently $k = 93$) of the neighborhood around each point is not green. The neighborhood

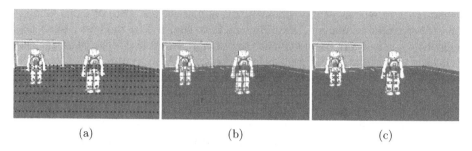

Fig. 2. For each blue spot in (a), it is checked whether it contains an obstacle or not. The highlighted pixels in (b) show the neighborhoods around each grid point. (c) shows the detected possible obstacle spots.

is a cross-shaped region that consists of the $2n$ neighboring pixels in vertical and horizontal direction (cf. Fig. 2b). n is the field line width at the given grid point, i. e. half the vertical distance between the grid point and its neighbor. The cross shape has been chosen due to its relative rotation invariance and ease of implementation.

Since the horizontal distance between two grid points is fixed, but the width of a neighborhood depends on the distance of the grid point, the neighborhoods of two adjacent grid points overlap. To avoid checking pixels for their greenness multiple times in the overlapping areas, a one-dimensional sum image is calculated for each grid row beforehand. The greenness of a certain region can then be calculated by subtracting the corresponding start and end pixels in the sum image. All grid points that lie inside non-green regions are considered to be possible obstacles.

Filtering. The possible obstacle spots show the general location of obstacle regions in the image. However, they usually do not show the foot point of an obstacle and neither do they show its size. In addition, they may contain false positives, in particular at field line junctions. Therefore, additional filtering is required. The filtering process consists of several steps:

Height check. The first step of the filtering process is to determine the height and the foot point of the obstacle at each possible spot. This is done by scanning the image vertically in both directions starting at each possible obstacle spot until either the horizon, i. e. the plane parallel to the field on the height of the camera, or a continuous patch of green is found. This way, the height can be determined well as long as the camera is more or less parallel to the ground. Since the NAO does not need to tilt its head and the body is kept upright while walking, this is usually the case.

If the height is less than a robot's width at the foot point, the obstacle spot will be discarded. This effectively removes most of the false positives. However, due to the fact that the environment around a RoboCup field is usually not green,

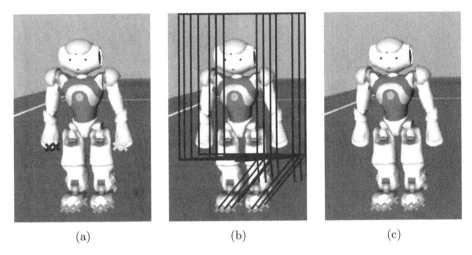

(a)   (b)   (c)

Fig. 3. Removal of false obstacle spots at the hands of other robots. The colored crosses in (a) are different obstacle spot clusters. The blue and the yellow clusters are false positives and should be removed. (b) shows the calculated ban regions and (c) shows the result.

false positives close to the field border cannot be removed using only a height criterion. Still, since robots are not allowed to leave the field, false positives at the field border do not matter.

Hand removal. Depending on their posture, the hands of other robots may be detected as obstacle spots. Technically, those spots are not false positives since they are part of another robot. However, the foot point of such obstacles is not on the field plane. Thus it is impossible to calculate the distance to the obstacle. These obstacles are removed using the assumption that the lowest (in image coordinates) obstacle spot belongs to the foot of a robot. Starting from the lowest spot, an area that should contain the robot's hands is calculated for each spot. All spots inside that area are removed. Figure 3 illustrates this process.

Obstacle Stitching. When obstacles are close, they span both the lower and upper camera image. The foot point of such obstacles is in the lower image; however, the lower image does usually not show enough of the obstacle to pass the height check. In this case, the visible height is calculated and the intersection of obstacle and image border is converted to relative field coordinates and buffered.

When the next image from the upper camera is processed, the buffered points are moved (still in relative field coordinates) according to the odometry and projected back into the upper camera image. Here, the height check that was started in the lower image is continued (cf. Fig. 4a, b).

Clustering. Finally, the remaining obstacle spots are clustered and the center of mass is calculated for each cluster.

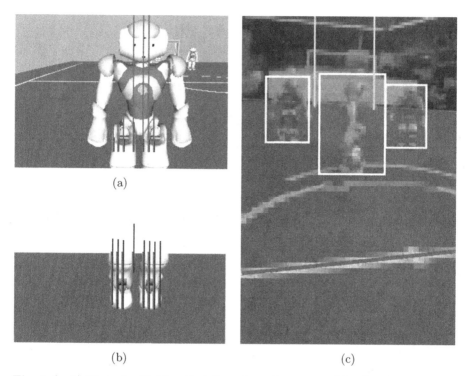

(a)

(b)   (c)

Fig. 4. (a, b) Obstacle stitching. Red lines show the measured height while blue lines indicate the minimal expected height. The image from the lower camera (b) does not contain enough of the obstacle to pass the height check; therefore, the height check is continued in the upper camera image (a). (c) Images of both cameras taken with the real time logger. The image quality is greatly reduced, but everything is still visible. The images are overlaid with perceptions that were also logged. See a video of the log file at http://www.youtube.com/watch?v=rv3KpBRSTVw.

3   Detecting the Field Boundary

The 2013 rules state that if fields are further away from each other than 3 m, a barrier between them can be omitted. This means that robots can see goals on other fields that look exactly like the goals on their own field. In addition, these goals can even be closer than a goal on their field. Therefore, it is very important to know where the own field ends and to ignore everything that is visible outside. Our old approach to determine the field boundary was to scan the image downwards to find the first green of sufficient size. This could now be easily on another field. Thus, a new approach was required. It searches vertical scanlines, starting from the bottom of the image going upwards. Simply using the first non-green pixel for the boundary is not an option, since pixels on field lines and all other objects on the field would comply with such a criterion. In addition, separating two or more fields would not be possible and noise could lead to false positives.

Fig. 5. Boundary spots (blue) and the estimated field boundary (orange) in a simulated environment with multiple fields

Our approach builds a score for each scanline while the pixels are scanned. For each green pixel a reward is added to the score. For each non-green pixel a penalty is subtracted from it. The pixel where the score is the highest is then selected as boundary spot for the corresponding scanline. The rewards and penalties are modified depending on the distance of the pixel from the robot when projected to the field. Field lines tend to have more green pixels above them in the image. As a result, they are easily skipped, but the gaps between the fields tend to be small in comparison to the next field when they are seen from a greater distance. So for non-green pixels with a distance greater than 3.5 meters, a higher penalty is used. This also helps with noise at the border of the own field. Figure 5 shows that most of the spots are placed on the actual boundary using this method.

Since other robots on the image also produce false positives below the actual boundary, the general idea is to calculate the upper convex hull from all spots. However, such a convex hull is prone to outliers in upward direction. Therefore, an approach similar to the one described by Thomas Reinhardt [7, section 3.5.4] is used. Several hulls are calculated successively from the spots by removing the point of the hull with the highest vertical distances to its direct neighbors. These hulls are then evaluated by comparing them to the spots. The best hull is the one with the most spots in near proximity. This one is selected as the field boundary. An example can be seen in Fig. 5.

4   Real-Time Logging with Images

Logging in real time during matches is, in general, nothing new to B-Human. However, until 2013, it was impossible to log images during actual games. Without images, log files are only of limited use for debugging, because the frame of reference for the programmer is missing. He or she can never be 100% certain about the situation or even whether looking at the correct situation at all.

The debugging of any perception-related mistake is close to impossible without images.

4.1   Logging Architecture

The main bottleneck when logging is the disk, i.e. a micro SD card in case of the NAO. Accessing the disk takes a very long time compared to everything else. The old logging implementation simply avoided this bottleneck by never writing to the disk during a game. Everything was buffered in memory until the game was over. However, when logging images, this approach is no longer feasible, because the robots would run out of memory very quickly. Therefore, we introduced a new non-real time process that continuously takes the oldest frames from the buffer and writes them to the disk. Due to the non-real time priority, data is only written when there is some spare time left. This way, the writing does not influence the real time property of the more important processes. In combination with a large buffer (600MB) and heavy image downscaling (cf. Sect. 4.2), this approach allows real time logging of images.

In addition, to further reduce the disk accesses, the data is compressed using Google's library Snappy [8] before it is written to the disk. Compressing the data and writing it to disk afterwards takes less time than to directly store the uncompressed data.

4.2   Image Compression

Compressing the images is done in two steps in order to get images that are as small as possible in minimal time.

At first, the images are scaled down by averaging blocks of 8×8 pixels. Since this is a very time consuming process, SSE instructions are used. In addition, it is very important not to process block after block since this impedes caching, resulting in great performance losses. Instead, we process line after line of the image buffering the results for eight lines until the average can be computed.

The second step reduces the number of representable colors and removes some unused space from the pixels. The cameras produce images in the YCbCr 4:2:2 format. This means that there are two luma channels for two chroma channels (the data has the form $\{Y_1, Cb, Y_2, Cr\}$). We only use one of the luma channels stored in each pixel since it would cost additional processing time to recreate the full resolution image. Thus one byte can be removed per pixel for the compressed image. Another byte can be gained by reducing the amount of bits each channel is allowed to use. In our case we use 6 bits for the luma and 5 bits for each chroma channel. Together, this results in another reduction of space by the factor of 2. This step might seem costly but it only takes 0.1 ms since it is done on a downscaled image.

The resulting compressed images are 128 times smaller than the original ones. Although a lot of detail is lost in the images, it is still possible to see and assess the situations the robot was in (see Fig. 4c).

5   Team Organization

The strict and careful organization of the tasks of the human team members is of significant importance for every match. This is definitely not a scientific issue but an inevitable necessity if one wants to make sure that all sophisticated algorithms and tactics are actually carried out effectively by the robots. There exist a number of mistakes that can be made and that ruin the performance of single robots or even the whole team and thereby render all scientific research useless. Common examples are using the wrong color calibration (making all robots blind or perceive false positives), mis-configured wireless settings (disabling communication among teammates), playing with buggy, untested code (perhaps causing crashes or deadlocks), and playing with damaged or insufficiently calibrated robots (letting the robot fall very often). In a perfect game, all five robots are well-configured, not damaged, and demonstrate their capabilities on the field for the whole time. Each robot that is not on the pitch – whether it is damaged, has not been prepared in time, or being penalized due to a rule violation such as Pushing (cf. Sect. 2) – weakens the overall team performance.

Therefore, our team has established a number of workflows that contribute to the goal of playing in this desired way. These procedures are the same in *all* matches, independent of the opponent team (no difference between The Invisibles and HTWK Leipzig) or the importance of the match (no difference between first round robin game and final). We consider this as very important as it trains all teammates and keeps concentration high.

5.1   Before a Game

To robustly walk at high speed and to perceive objects with high precision, continuously checking and calibrating all our robots is necessary. During a competition, multiple persons each spend many hours a day to make sure that we always have enough robots that can reliably play soccer. For this purpose, individual configurations are maintained, containing the maximum possible walk speeds and gait heights.

On the software side, we almost never start to play an official match in which we use software that has never played a match before. This could have been a test match, whether against another team or against an empty field, or the same code as in the official match before. This is necessary to avoid .any negatively surprising and unintended side-effects of code changes.

Almost exactly 30 minutes before a match starts, we always perform a *test kickoff* of the whole team. All robots walk to their kickoff positions, the test game is started, and the robots play until they score a goal. By performing this test, we ensure that software and configuration files have been deployed correctly and all robots are fully operational. In case of any problem, there is still enough time to solve it without the need to hurry. If everything works fine, all robots are turned off and remain at the field.

All robots are again turned on several minutes before the start of the game (or the half). Until the actual beginning, they remain connected to a computer that runs a monitoring tool. This enables us to react on unlikely but not impossible last minute problems such as a non-booting robot.

5.2   During a Game

During a game, the possible actions of the human team members are very limited: robots that have been picked-up can be fixed or replaced, a time-out or the half-time break can be used to change the configuration of the whole team.

In any way, these actions have to be carried out carefully and based on sound facts. As we are a huge team and many tasks are carried out by different persons, chaos needs to be avoided (situations such as somebody spontaneously shouting *"Time-Out!"* or disagreement about picking-up a certain robot). Therefore, we always have the same dedicated person that talks to the referee and one additional dedicated person that finally decides about all actions.

If the opponent team allows us to monitor our robots' communication, we always do. Our network packages include the so-called *robot health* that contains information about things such as battery power, joint temperatures, CPU temperature, or the frame rates of our two main processes. This allows us to identify potential problems before the robot starts to underperform and to discuss and schedule a replacement. During every game, a protocol is hand-written, including important facts for a later discussion but also statistics about which robot has fallen how many times. This allows us to watch out for potential hardware problems during the game and perhaps request a pick-up.

For every game, at least one already booted replacement robot is available at the field. This robot is always connected to a computer so that setting the player number and starting the B-Human software can be done immediately, making the robot ready to play in less than 30 seconds.

5.3   After a Game

After each game – test match or real one – the whole team discusses the match that was just played. The discussion is guided by the protocol that was written during the game. Tasks are assigned to individual team members. For analyzing certain mistakes, the log files (cf. Sect. 4) of the robots that misbehaved are watched to better understand the reason of an error.

6   Conclusions

At RoboCup 2013, B-Human's robots scored an average of 0.48 goals per minute of actual game time [2]. One reason for this success was that our robots were on the field most of the time. For instance, the new obstacle detection methods resulted in measurable improvements: While we do not know the number of "player pushing" penalties we received in 2012 (it certainly was high), we only received eight this year, in eight games. All other quarter-finalists committed more pushing offenses. Overall, B-Human was the team that received the least penalties per game, which surely helped our success. In our opinion, our workflows significantly contributed to the comparatively small number of inactive robots, fallen robots, and requests for pick-up. As a consequence of our team's reliability, we have been able to play most of the tournament with a full set of well-working robots that showed a convincing performance.

During fall 2013, a large part of the software that we used during RoboCup 2013 will be released along with a detailed description of all components, including those described in Sections 2–4.

References

1. Röfer, T., Laue, T., Müller, J., Bartsch, M., Batram, M.J., Böckmann, A., Lehmann, N., Maaß, F., Münder, T., Steinbeck, M., Stolpmann, A., Taddiken, S., Wieschendorf, R., Zitzmann, D.: B-Human team report and code release 2012 (2012), Only available online:
   http://www.b-human.de/wp-content/uploads/2012/11/CodeRelease2012.pdf
2. Röfer, T., Bartsch, M.: Statistics of GameController logs (2013),
   http://www.tzi.de/spl/pub/Website/Results2013/RoboCup2013Statistics.pdf
3. Engel, M.: Anwendung von maschinellem Lernen zur echtzeitfähigen und kalibrierungsfreien Erkennung von humanoiden Fußballrobotern. Bachlor's thesis, HTWK Leipzig (2012)
4. Metzler, S., Nieuwenhuisen, M., Behnke, S.: Learning visual obstacle detection using color histogram features. In: Röfer, T., Mayer, N.M., Savage, J., Saranlı, U. (eds.) RoboCup 2011. LNCS, vol. 7416, pp. 149–161. Springer, Heidelberg (2012)
5. Röfer, T., Laue, T., Müller, J., Fabisch, A., Feldpausch, F., Gillmann, K., Graf, C., de Haas, T.J., Härtl, A., Humann, A., Honsel, D., Kastner, P., Kastner, T., Könemann, C., Markowsky, B., Riemann, O.J.L., Wenk, F.: B-Human team report and code release 2011 (2011), Only available online:
   http://www.b-human.de/downloads/bhuman11_coderelease.pdf
6. Aquino, A., Akatsuka, C., Mushonga, T., Zhang, Y., He, Y., Lee, D.D.: The University of Pennsylvania RoboCup 2011 SPL Nao soccer team. In: Röfer, T., Meyer, N.M., Savage, J., Saranli, U. (eds.) RoboCup 2011: Robot Soccer World Cup XV Preproceedings, RoboCup Federation (2011)
7. Reinhardt, T.: Kalibrierungsfreie Bildverarbeitungsalgorithmen zur echtzeitfähigen Objekterkennung im Roboterfußball. Master's thesis, HTWK Leipzig (2011)
8. Google: Snappy – a fast compressor/decompressor (2013),
   http://code.google.com/p/snappy/

ZJUNlict:
RoboCup 2013 Small Size League Champion

Yue Zhao, Rong Xiong, Hangjun Tong, Chuan Li, and Li Fang

National Laboratory of Industrial Control Technology,
Zhejiang University, P.R. China
rxiong@iipc.zju.edu.cn
http://www.nlict.zju.edu.cn/ssl/WelcomePage.html

Abstract. The Small Size League is one of the most important events in RoboCup. The league is devoted to the advancing in mechanical design, artificial intelligence and multi-agent cooperation of mobile robots. ZJUNlict from Zhejiang University has participated in this league for ten years since 2004 and got the first place in RoboCup 2013. In this paper, we introduce the main ideas of the robots' hardware design of ZJUNlict, and emphasize the intelligent control system including the hierarchical architecture of strategy, the Lua script architecture, learning and selecting trajectories methods based on Dynamic Movement Primitives.

1   Introduction

Small Size League (SSL) is an important part of the RoboCup event. It is the fastest and most intense game in RoboCup's soccer competitions. The basic rules of SSL are based on the rules of a FIFA's soccer game, but each team consists of only six robots playing on field that is 6.05m long by 4.05m wide. There are two cameras mounted over the field to capture images, which are processed in real time by a shared vision system, SSL-Vision [1]. This vision system recognizes and locates the position and orientation of the robots and the position of the ball, then broadcasts the information package to each team via network. Of course, the objective of the game is to score more goals than the opponent. SSL emphasizes on the fast movement of robots in a complex, dynamic and competitive environment.

ZJUNlict from Zhejiang University has participated in this League for ten years since 2004. We won the regional competitions in RoboCup ChinaOpen 2006, 2007, 2008, 2011. In RoboCup world-wide competition, we run into the quarter finals in 2006, 2009, 2011, and got the semi finals in 2007 and 2008. Since 2012, we have made a great progress and received the second place in RoboCup2012, and won the championship in RoboCup2013.

The remainder of this paper is organized as follows. Section 2 briefly introduces the hardware design of ZJUNlict's robot. Section 3 introduces the intelligent control system including strategy selection and trajectory generation based on learning method. We present a play script writing in Lua and the results of the DMPs' execution in section 4, and section 5 concludes the paper.

S. Behnke et al. (Eds.): RoboCup 2013, LNAI 8371, pp. 92–103, 2014.
© Springer-Verlag Berlin Heidelberg 2014

2   Robot Design

2.1   Components of the Robot

The robot we developed is equipped with 4 omni-directional wheels driven by a 50 watt brushless Maxon motors. There are another three major mechanisms: a dribbling device, a shooting device and a chipping device. The robot and its mechanical components are shown in Fig.1.

Fig. 1. Mechanical design of the robot: (1)shooting device; (2)omni-directional wheel; (3)chipping device; (4)dribbling device

We employ the NiosII as the central processor module, which is a soft IP working in QuartusII and NiosII software programming environment. When powered on or reset, the robot initialize itself, then run into the main loop to execute the command sent from PC. The overview of our embedded software flowchart is show in Fig.2.

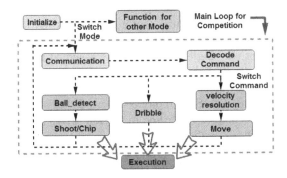

Fig. 2. Embedded software flowchart

3 Intelligent Control System

3.1 Hierarchical Architecture of Strategy

The software architecture of the intelligent control system is shown as Fig.3. It is the central module for planning and coordination among robots in both attack and defense modes. The whole system is composed of the World Model, the Decision Module and the Control Module [2].

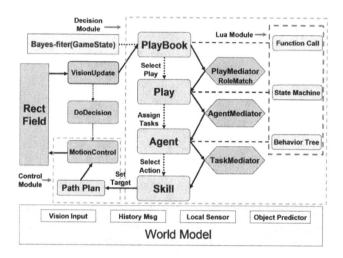

Fig. 3. Software architecture

The Decision Module is designed in a hierarchical structure, which is consist of the PlayBook module [3], the Play module, the Agent module and the Skill module. The PlayBook module selects the appropriate play by using a Bayesian filter to evaluate the game status, as described in section 3.1.1. The Play module focuses on coordination between teammates and is organized by a Finite State Machine (FSM), as described in section 3.1.2. The Agent module emphasizes on the planning skills of the single robot with the assigned tasks from the play. The Agent module selects its behaviors from Behavior Tree (BT), as described in section 3.1.3. The Skill module is a direct interface of the Decision Module with the Control Module. The Skill module generates target point and selects trajectory generation method, as described in section 3.1.4. The PlayBook, play-level and agent-level are all configured using the script programming language Lua, and will be detailed in Section 3.2.

The Control Module is responsible for the path planning and trajectory generation. It traditionally uses the Rapidly-exploring Random Trees (RRT's) algorithm [5] to find a feasible path and Bangbang-based algorithm [6] to solve two-boundary trajectory planning. We also adopt the idea of Dynamics Movement Primitives (DMPs) [9] to learn trajectories demonstrated by human, which will be shown in Section 3.3.

Finally, the World Model provides all the information of the match. The History Message records all the decisions from the Decision Module. Besides, the Vision Input means the original vision messages from two cameras, the Local Sensor Message is uploaded from the robots via wireless. The Object Predictor algorithm mainly focuses on our robots' real velocity information in a time-delayed system, and a Kalman Filter [10] method is appropriate for this situation. Thus, the closed-loop system adjusts all robots' behavior according to the change of the environment in real time.

Game State Evaluation. A basic problem in SSL is when to attack and when to defend, which means we should evaluate the game status based on observed information. But it's very complicated due to the complexity of game situation. If only the observation is taken into account, the evaluation result will be easily impacted by momentary sensing error, such as ball missing and robot misidentification. Imagining an evaluator only considering ball's position, then a wrong location of the ball would lead to a huge impact on the choice between attack and defense.

Thus, we propose a new method based on the Bayesian Theory [7], which evaluates the game state by combining the observed information and the historical strategy. The new evaluator of the game status is shown as Fig. 4,

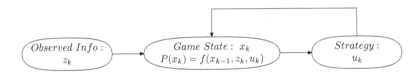

Fig. 4. Bayes-based evaluation method

where z_k is determined by the observation, u_k denotes the attribute of current play script, which can be discretized into values such as 0(attack), 1(stalemate), 2(defense), the game state x_k is calculated using a Bayesian filter method. Fig. 5 depicts the basic Bayesian filter algorithm in pseudo code.

Algorithm Bayesian filter $p(x_k, u_k, z_k)$
for all x_k **do**
    $\overline{p}(x_k) = \sum_{x_k} p(x_k | u_k, x_{k-1}) p(x_{k-1})$
    $p(x_k) = \eta p(z_k | x_k,) \overline{p}(x_k)$
end for

Fig. 5. General algorithm for Bayesian filter

Fig. 6 gives a practical application of Bayes-based filter in the system. We define three states for the game and there are corresponding plays for each state.

The selection and switch of the plays is up to the result of the filter. Furthermore, there also exists a score-evaluating mechanism between the plays with the same attribute, this mechanism is realized by the PlayBook module.

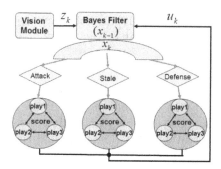

Fig. 6. Bayes Processing Flow Chart

The proposed method has two main features:

- **More stable:** Compared with the evaluation method that only considers observation, our evaluator gives a more appropriate analysis of the game state and helps to reduce the perturbation of the strategy and to strengthen the continuity of the strategy.
- **More flexible:** The values of prior probabilities $p(x_k|u_k, x_{k-1})$ and $p(z_k|x_k)$ is predefined in code according to different team's characteristics, making it more convenient and flexible to configure the attacking and defending strategy.

Play and Finite State Machine. Each play represents a fixed team plan which consists of many parts, such as applicable conditions, evaluating score, roles, finite state machine and role tasks, based on CMU's Skill, Tactics and Plays architecture [3]. The plays can be considered as a coach in the soccer game, who assigns different roles to robots at different states.

For a play script, the basic problem is "What should be done by Whom at What time". So we develop a novel FSM-based Role Match mechanism. Traditionally, a state node is described by the execution and switch condition [3]. We coupled a role match item in each state node to this basis. The new framework is shown as Fig.7, a role match item comprises several matching groups with priority, the priority determines the groups' execution order. Our goal is to find an optimal solution for every group by Munkres assignment algorithm [4], using the square of the distance between the current positions of the robots and expected roles' target positions as the cost function. In the implementation, we just consider five roles matching in the scripts, because the goalie corresponds a fixed robot by default.

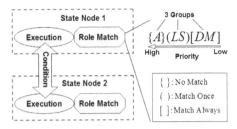

Fig. 7. FSM-based Role Match Framework. A role match item's syntax is described on the right: the letter in the string is short for the role's name in the state node(A-Assister, L-Leader, S-Special, D-Defender, M-Middle). We offer three modes for matching, *No Match* means to keep the role as same as last state, *Match Once* means match the role once stepping into this state, and *Match Always* means do role match every cycle. If a robot is missing, then ignoring the last role in the item and *No Match* group always has a higher priority.

Agent and Behavior Tree. Agent-level is responsible for the behavior planning such as manipulating ball to proper region, passing ball to a teammate and scrambling ball from the opponent, when a play-level decision is made. The main work of agent-level is to select a proper skill and the best target for the executor in each cycle. In order to make the behavior selection more intelligent and human-like, we build a behavior tree, which consists of a set of nodes including control nodes and behavior nodes.

– **Control Nodes** are utilized to set the logic of how different behavior nodes should be connected. In our agent behavior tree system, there are mainly three types of control nodes as shown in Fig.8. The first type is a sequence node that ensures all of its child nodes will be executed in a deterministic order if the precondition of the children node is satisfied. For example, as shown in Fig.8(a), when the precondition of the root node is satisfied, the step 1 node will be executed. After the step 1 node has been executed and the precondition of the step 2 node is satisfied, the step 2 node will be executed. This rule will last until the last child node is executed. The second type is a loop node which works like a counter. The execution of its child node depends on two conditions: one is that the precondition of this node is satisfied; the other is that the control node should have been executed for a predefined time period as illustrated in Fig.8(b). The third type is a priority selector node which executes its children nodes actions according to their priority. When the preconditions of a node with the higher priority have been satisfied, the existing node will be interrupted and the executing of the node with the higher priority will be executed, see Fig.8(c). All the nodes are associated with external preconditions that must be satisfied before the executing of the node. These preconditions should be updated before the update of the behavior tree.

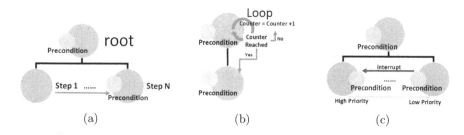

Fig. 8. Three types of control node: (a)sequence node; (b)loop node; (c)priority selector node

- **Behavior Nodes** are simply a set of atomic skills such as going to a specific position. They are all leaf nodes of a behavior tree and saved in an atom skill factory. They are also associated with preconditions that should be satisfied before executing.

Skill. Skill is a set of basic knowledge of actions different robots can perform, such as how to move to a point, how to get the ball and kick. Some skill will generate a next target point, which will be passed to the navigation module for path planning and trajectory generation. Some skill will generate the speed trajectory for some special behavior such as pulling the ball from an opponent front. Parameter tuning is always an important work for skill's performance in the competition.

3.2   Lua Script Architecture

We introduce a script language into our system to improve the flexibility and robustness of the system. Here we choose the script programming language Lua [8]. We have transplanted some repeated logic code to Lua such as positioning tactic, FSM configuration, Behavior Tree's generation, while left the complicated algorithms such as path planning, vision handling in C++ workspace. So the code is divided into two parts, as illustrated in Fig.9.

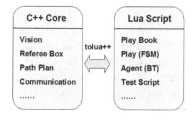

Fig. 9. Script architecture with Lua

Tolua++ is an extended version of tolua, which is a tool to integrate C/C++ code with Lua. At Lua side, we need to access some variables and functions written in C++. Tolua++ helps us deal with this using a package file. For more details about tolua++, please refer to its Reference Manual[1].

The design advantages of script architecture with Lua are:

- **Clear Logic:** Like other scripting language, Lua is easy to understand. We can pay more attention to the logic of the code rather than the syntax, and it's really easy for different people to express their tactics by Lua scripts even if the script's author has little knowledge on programming.
- **No Compiling:** In RoboCup Small Size League competition, each team has only four chances for time out, the total time is 10 minutes. Therefore, it is very important to rebuild the code as quickly as possible in the limited time. Usually, a modification in C/C++ code takes about 10-20 seconds, but the compiling takes 1 minute or more. Lua helps us to solve the problem, we can just modify our strategy in about 10 seconds, and then do a syntax check by Lua's own debug tool, which takes almost no time. So we can spend more time on modifying logic code rather than compiling and debugging.
- **Online Debugging:** A play script will be loaded every cycle in our code. So tuning some parameters or functions such as a FSM's switch condition or Behavior Tree's node action do not need to stop the whole program, the effects will be shown as soon as the modifications in a script file are saved, which enables easier and faster strategies adjustment.

3.3 Trajectory Generation with DMPs

We also introduce the Dynamic Movement Primitives (DMPs) framework [9] into our system to learn some special trajectories from human's demonstration. The DMPs presented below is just a basic model for learning a point-to-point trajectory without considering opponents and kinematic restrictions.

DMPs Framework. Generally speaking, a movement can be described by the following set of differential equations [11], which can be interpreted as a point mass attached to a spring and perturbed by an external force which is applied artificially in demonstration from an intuitive point of view:

$$\tau \dot{v} = K(g - x) - Dv - K(g - x_0)u + Kf(u) \tag{1}$$

$$\tau \dot{x} = v \tag{2}$$

$$\tau \dot{u} = -\alpha u \tag{3}$$

where

- x is the position for one DoF of the system,

---

[1] Tolua++ Reference Manual: `http://www.codenix.com/~tolua/tolua++.html`

- v is the velocity for one DoF of the system,
- x_0 is the start position,
- g is the goal position,
- K is the spring constant,
- D is the damping term,
- u is the phase corresponding to time which going from 1 towards zero,
- τ is temporal scaling factor,
- α is a pre-defined constant,
- f is a non-liner function which simulates as a extern force.

The equations (1) and (2) are referred as *transformation system*. And the differential (3) is called *canonical system*. Usually, K and D are chosen such that the transformation system are critically damped. Moreover, τ and α are chosen such that u is close to zero as time. These equations are time invariant and make the duration of a movement alter simply by changing τ. In addition, once the non-linear function f is defined, the shape of the resulting trajectory will be determined as well. Specifically, f can be defined as

$$f(u) = \frac{\sum_i^N w_i \psi_i(u) u}{\sum_i^N \psi_i(u)}, \qquad (4)$$

$$\psi_i(u) = exp(-h_i(u - c_i)^2) \qquad (5)$$

where ψ_i are Gaussian basis functions with center c_i and width h_i, N is the total number of Gaussian basis functions. And w_i are the weights need to be learned. All the equations written above are used for a one dimensional system. Our robot can be decoupled kinematically into three DoFs: X translational direction, Y translational direction, W rotational direction, which are shown as Fig.10.

Fig. 10. Three DoFs of robot in SSL

Learning and Executing Trajectory. An important advantage of DMPs is that a demonstration could be imitated by one-shot learning. In order to learn from the demonstrated trajectory, we should record the movement $x(t)$ first, then derivative $v(t)$ and $\dot{v}(t)$ from $x(t)$ in each time step, $t = 0, ..., T$. By combining the *transformation system* and *canonical system* together, $f_{target}(u)$ is obtained:

$$f_{target}(u) = \frac{\tau \dot{v} + Dv}{K} + (g - x_0)u - (g - x), \qquad (6)$$

where $x_0 = x(0)$ and $g = x(T)$. We can find the weights w_i in (4) by minimizing the error criterion $J = \sum(f_{target}(u) - f(u))^2$, which is a linear regression problem and can be solved efficiently [11]. Once w_i is determined for a demonstrated movement, we can generate trajectory with a new goal position by resetting $g = g_{new}$, $x_0 = x_{current}$ and $t = 0$. After this setup procedure, we could adjust τ to determine the duration of the movement.

4   Experiment Results

4.1   Play Script Writing in Lua

Fig.11 is an play script used for indirect kick in the frontcourt.

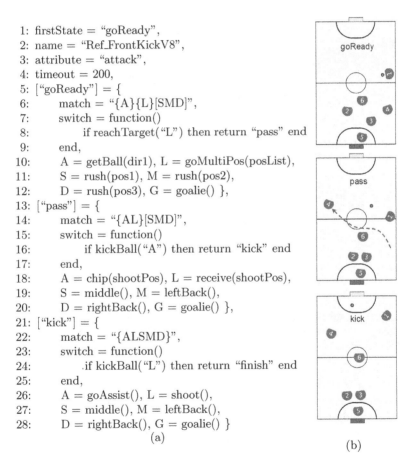

```
 1: firstState = "goReady",
 2: name = "Ref_FrontKickV8",
 3: attribute = "attack",
 4: timeout = 200,
 5: ["goReady"] = {
 6:     match = "{A}{L}[SMD]",
 7:     switch = function()
 8:         if reachTarget("L") then return "pass" end
 9:     end,
10:     A = getBall(dir1), L = goMultiPos(posList),
11:     S = rush(pos1), M = rush(pos2),
12:     D = rush(pos3), G = goalie() },
13: ["pass"] = {
14:     match = "{AL}[SMD]",
15:     switch = function()
16:         if kickBall("A") then return "kick" end
17:     end,
18:     A = chip(shootPos), L = receive(shootPos),
19:     S = middle(), M = leftBack(),
20:     D = rightBack(), G = goalie() },
21: ["kick"] = {
22:     match = "{ALSMD}",
23:     switch = function()
24:         .if kickBall("L") then return "finish" end
25:     end,
26:     A = goAssist(), L = shoot(),
27:     S = middle(), M = leftBack(),
28:     D = rightBack(), G = goalie() }
```
(a)

(b)

Fig. 11. An example of indirect free kick in frontcourt: (a)Lua script; (b)positions of robots in every state. In the script (a), the basic settings are from line 1 to line 4. Then we define the specific states in this play. For example, the state *goReady* comprises a role match item (line 6), switch condition (line 7-9) and execution (line 10-12).

4.2 Learning from Demonstration by DMPs

In the DMPs' experiment, we evaluated how well the DMPs' formulations generalize a quick turning and breaking through movement, as shown in Fig.12(a). First, we collected the movements in the three DoFs using the cameras mounted over the field. In the second step, we used DMPs method to train the data and got w_i for each DoF. Finally, we reset the DMPs formulas and adapted to a slightly nearer target, see Fig.12(b). In this experiment, the robot successfully moved to new positions which are at most $0.7m$ away from the original target, and the duration of the movement is about 1.5 second. Note that in the practical 3-DoFs motor system, every DoF sets the transformation system independently, but they share the canonical system as they are coordinated, the results of the motion reproduction in different DoF is shown as Fig.12(c) and Fig.12(d).

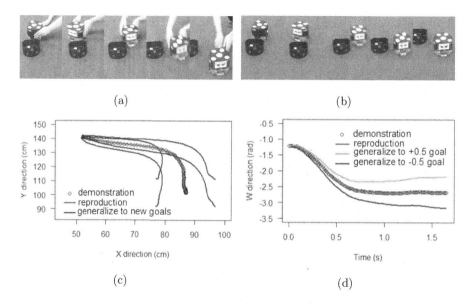

Fig. 12. Turning and breaking through example: (a)demonstration; (b)generalization to a slightly nearer target; (c)reproduction and generalization to new goals in X-Y panel; (d)reproduction and generalization to new goals in W rotational direction

In addition to the basic DMPs presented in section 3.3, we also try to add a dynamic potential field item to the DMPs' formula as in [12] for the purpose of obstacle avoidance. Moreover, we can change the target velocity of the movement while maintaining the overall duration and shape by developing several extensions and modifications to this approach as in [13].

5   Conclusion

In this year, we paid more attentions to the intelligent control system. First, Bayesian evaluation method and a role match framework based on STP [3] are proposed to make stable and flexible strategy decision. Second, we develop behavior tree for action connection. The FSM and BT both can be configured with Lua scripts. Finally, the DMPs are introduced to help generate human-like trajectory. These efforts make ZJUNlict score 43 goals and concede 2 goals in RoboCup2013. In next year, we will continue to focus on multi-agent cooperation and the trajectory generation model, also we hope to use 3D physics engine [14] to help simulate and predict common situations encountered in the game.

References

1. Zickler, S., Laue, T., Birbach, O., Wongphati, M., Veloso, M.: SSL-Vision: The Shared Vision System for the RoboCup Small Size League. In: Baltes, J., Lagoudakis, M.G., Naruse, T., Ghidary, S.S. (eds.) RoboCup 2009. LNCS (LNAI), vol. 5949, pp. 425–436. Springer, Heidelberg (2010)
2. Wu, Y., Yin, P., Zhao, Y., Mao, Y., Xiong, R.: ZJUNlict Team Description Paper for RoboCup2012. In: RoboCup 2012 (2012)
3. Browning, B., Bruce, J., Bowling, M., Veloso, M.: STP: Skills, tactics and plays for multi-robot control in adversarial environments. J. Syst. Control Eng. 219(I1), 33–52 (2005)
4. Munkres, J.: Algorithms for the assignment and transportation problems. Journal of the Society for Industrial & Applied Mathematics 5(1), 32–38 (1957)
5. Bruce, J., Veloso, M.: Real-time randomized path planning for robot navigation. In: IEEE/RSJ International Conference on Intelligent Robots and Systems 2002, vol. 3, pp. 2383–2388. IEEE (2002)
6. Sonneborn, L., Van Vleck, F.: The Bang-Bang Principle for Linear Control Systems. Journal of the Society for Industrial & Applied Mathematics, Series A: Control 2(2), 151–159 (1964)
7. Thrun, S., Burgard, W., Fox, D.: Probabilistic robotics, vol. 1. MIT Press, Cambridge (2005)
8. Ierusalimschy, R.: Programming in Lua. Lua.org, 2nd edn (2006)
9. Ijspeert, A.J., Nakanishi, J., Schaal, S.: Movement imitation with nonlinear dynamical systems in humanoid robots. In: International Conference on Robotics and Automation, pp. 1398–1403. IEEE, Washington, DC (2002)
10. Welch, G., Bishop, G.: An introduction to the Kalman filter (1995)
11. Pastor, P., Hoffmann, H., Asfour, T., Schaal, S.: Learning and generalization of motor skills by learning from demonstration. In: Proc. of the International Conference on Robotics and Automation (2009)
12. Park, D., Hoffmann, H., Pastor, P., Schaal, S.: Movement reproduction and obstacle avoidance with dynamic movement primitives and potential fields. In: IEEE-RAS International Conference on Humanoid Robotics (2008)
13. Kober, J., Mülling, K., Krömer, O., Lampert, C.H., Schölkopf, B., Peters, J.: Movement templates for learning of hitting and batting. In: IEEE International Conference on Robotics and Automation 2010, pp. 1–6 (2010)
14. Zickler, S., Veloso, M.: Playing Creative Soccer: Randomized Behavioral Kinodynamic Planning of Robot Tactics. In: Iocchi, L., Matsubara, H., Weitzenfeld, A., Zhou, C. (eds.) RoboCup 2008. LNCS (LNAI), vol. 5399, pp. 414–425. Springer, Heidelberg (2009)

The Walking Skill of Apollo3D – The Champion Team in the RoboCup2013 3D Soccer Simulation Competition

Juan Liu, Zhiwei Liang, Ping Shen, Yue Hao, and Hecheng Zhao

College of Automation, Nanjing University of Posts and Telecommunications,
Nanjing, 210046

Abstract. Quick and flexible walking is an indispensable skill for humanoid robots in the RoboCup soccer competition. So this paper mainly proposed a method to develop a flexible walking based on reinforcement learning for humanoid robots, which used Cerebellar Model Articulation Controller(CMAC) method and a linear inverted pendulum with a predictive control to generate a motion trajectory of the robots trunk in the premise of keeping dynamic balance of robots. Our team Apollo3D employed this walking skill, and won the championship in the RoboCup 2013 3D simulation competition.

Keywords: Cerebellar model articulation controller, preview control, linear inverted pendulum, trunk trajectory.

1 Introduction

How to make humanoid robots play soccer like human is a big challenge in the fields of artificial intelligence and intelligent robotics. It provides a benchmark platform for the state of the art in science and technology. Nowadays, the international competitions of robot soccer have become more and more common, the most influential of which is the Robot World Cup (RoboCup) and its ultimate goal is to develop a team of fully autonomous humanoid robots that can win against the human world soccer champion team in 2050. In order to complete the goal, different types of robot soccer league have been organized in the international and domestic, and the simulation game is the important one. The RoboCup 3D Simulated Soccer League allows external processes to control humanoid robots to compete against one another in a realistic simulation of the rules and physics of a soccer game. The simulation is executed in the RoboCup Simulated Soccer Server 3D (rcssserver3d) which runs on Linux, Windows and MacOS X. The underlying simulation engine is SimSpark.Since 2012, RoboCup3D simulation team is composed of eleven autonomous robots (NAO robots from Aldebaran) and each robot can communicate with other robots through the network.

In order to perform a fluent soccer match for humanoid robot teams, various basic actions must be implemented, including walking, kicking, rising from

S. Behnke et al. (Eds.): RoboCup 2013, LNAI 8371, pp. 104–113, 2014.

falling, and so on. Among these robotic skills, a steady dynamic walking is the most foundamental factor. Many studies have been proceeding in the walking of humanoid robot, with different degree of success. For example, Storm et al[1]. presented the concept of joint trajectory planning, which analyzed the smooth hip motion with the largest stability margin and derived the highly stable hip trajectory by iterative computation without calculating the desired ZMP (zero moment point) trajectory. This method is also suitable for different ground conditions by varying the values of the foot constraint parameters. However, the trajectory tracking is generated offline and computationally intensive. In order to make online feedback possible, Bavani et al. put forward a method called central pattern generator (CPG) [2-3], which is based on the neural network method. CPG is a circuit system, which can create a self-contained periodic signal and be initialized by the non-oscillation signal. The most difficulty in realizing this method is how to determine the appropriate weights of neural connections. Another approach by incorporating feedback into walking is the passive dynamic walking. Seung-Joon showed that a robot could walk down a slope without any actuator and control[4]. The operating forces need in walking lessened by gravity and swing in its natural frequency. This method can be applied to walk on a flat ground by relying on foot contact sensors.

Since the above methods only focus on stable biped movements in a particular period without high efficiency, they are not suitable for the soccer match environments with dynamic confrontions and limited space. To improve the walking skill of biped robots, it is important to realize a reinforcement learning walking method [5]. So this paper describes a robust reinforcement learning walking method which based on Cerebellar Model Articulation Controller(CMAC)[6-7].

2   CMAC Walking Method

The primary key to a successful robot team in the RoboCup3D simulation soccer competition was to realize the robots steady and robust gait, as well as to realize fully stable walking. The main advantage of this walking is that a robot can maintain moving forward, side and turning around in the premise of keeping stable velocities on approaching the destination.

In the process of competition, the changing external environment requires the robot to alter its orientation at any time, turn agilely and forward fast. The CMAC walking method is presented in the Fig.1. First we can get the feasible footholds and compute the ZMP values based on the foot-planning module. Subsequently, the trunk trajectory of robot can be attained based on a double linear inverted pendulum model (D-LIP) with a predictive control method[8]. As a result, we can plan the space trajectory of every two footholds in 3D space according to the cubic spline interpolation method[9]. Meanwhile, each joint angle can be calculated according to inverse kinematics method[10]. The pose of the robots trunk can easily computed by the gyro sensor of NAO[11]. Last but not least, we use CMAC optimization and correction algorithm to control the movement of the leg joint accurately. And the whole system is forming a feedback control with the D-LIP.

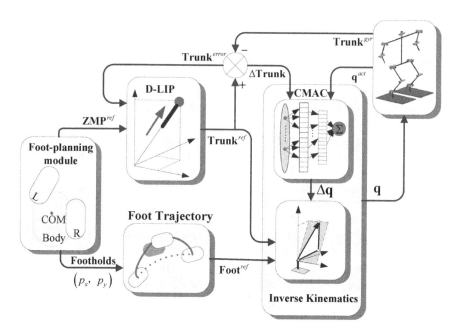

Fig. 1. Control block diagram of the CMAC walking

2.1   Foot-Planning Module

In the process of competition, players uses the designation of the footholds $(P_0, P_1, P_2, P_3 \cdots P_n)$ to plan each step of the robot. As shown in Fig.2, define $S_x^{(n)}$ as the step size in forward direction, $S_y^{(n)}$ as the step width in lateral direction, θ as the orientation of each step, superscript (n) as the nth step, and $(S_x^{(n)}, S_y^{(n)}, \theta)$ are called walking parameters. The nth foothold can be expressed as:

$$\begin{pmatrix} p_x^{(n)} \\ p_y^{(n)} \end{pmatrix} = \begin{pmatrix} p_x^{(n-1)} \\ p_y^{(n-1)} \end{pmatrix} + \begin{pmatrix} c_\theta^{(n)} & -s_\theta^{(n)} \\ s_\theta^{(n)} & c_\theta^{(n)} \end{pmatrix} \begin{pmatrix} s_x(n) \\ -(-1)^n s_y(n) \end{pmatrix} \quad (1)$$

where $c_\theta = cos\theta$, $s_\theta = sin\theta$, $(p_x^{(0)}\ p_y^{(0)})$ is the position of the first support leg and assume the left foot is the beginning foot (if not, then the $-(-1)^n$ will be turned into $(-1)^n$).

For the walking trajectory, every step of the 3D-LIPM track called walking unit is a hyperbolic in the x-y plane. Then the nth step walk unit $(\bar{x}^{(n)}, \bar{y}^{(n)})$ is determined by the following walking parameters:

$$\begin{pmatrix} \bar{x}^{(n)} \\ \bar{y}^{(n)} \end{pmatrix} = \begin{pmatrix} c_\theta^{(n+1)} & -s_\theta^{(n+1)} \\ s_\theta^{(n+1)} & c_\theta^{(n+1)} \end{pmatrix} \begin{pmatrix} s_x^{(n+1)}/2 \\ (-1)^n s_y^{(n+1)}/2 \end{pmatrix} \quad (2)$$

Given the supporting time of each step T_s, the altitude position of the center of mass $(COM)z_c$, the norm of the gravitational force g, the termination speed of the walking unit can be expressed as:

$$\begin{pmatrix} \bar{v}^{(n)} \\ \bar{v}^{(n)} \end{pmatrix} = \begin{pmatrix} c_\theta^{n+1} & -s_\theta^{n+1} \\ s_\theta^{n+1} & c_\theta^{n+1} \end{pmatrix} \tag{3}$$

where $T_c = \sqrt{Z_c/g}, C = cosh(T_s/T_c), S = sinh(T_s/T_c)$.

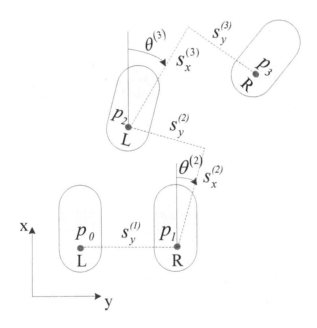

Fig. 2. The foothold planning

2.2   Based on the ZMP Predictive Control of D-LIP

The trajectory of the COM and the ZMP are decentralized by the sampling time t through a cubic polynomial. Apollo3D use a preview control method based on the trajectory of COM to predict the ZMP, and the N sample point values of the ZMP are need to calculate the current value of COM at the same time.

$$\mathbf{X}_{ZMP}(k+1) = \mathbf{A} \cdot \mathbf{X}_{COM}(k) + \mathbf{B} \cdot \dddot{\mathbf{X}}_{COM}(k) \tag{4}$$

$$\text{with } \mathbf{X}_{ZMP}(k+1) = \begin{pmatrix} x_{ZMP}(k+1) \\ \vdots \\ x_{ZMP}(k+N) \end{pmatrix}, \ddot{\mathbf{X}}_{COM}(k) = \begin{pmatrix} \ddot{x}_{COM}(k) \\ \vdots \\ \ddot{x}_{COM}(k+N-1) \end{pmatrix}$$

$$\mathbf{A} = \begin{pmatrix} 1 & t & t^2/2 - z_c/g \\ \vdots & \vdots & \vdots \\ 1 & Nt & N^2t^2/2 - z_c/g \end{pmatrix}, \mathbf{B} = \begin{pmatrix} b_0t^3/6 - tz_c/g & 0 & 0 \\ \vdots & \ddots & 0 \\ b_{N-1}t^3/6 - tz_c/g & \cdots & b_0t^3/6 - tz_c/g \end{pmatrix}$$

where z_{ZMP} denotes the horizontal displacement of the ZMP , $b_i = (1 + 3i + 3i^2)$ in the matrix B.

2.3   Foot Trajectory

The foot positions of every walk cycle are directly deduced from the foot-planning module. These footholds (p_x, p_y) are used to compute the continuous curve of foot movement. Due to a large amount of calculation, a violent oscillation and the poorly numerical stability about the high order function, Apollo3D employed the cubic spline interpolation method to compute the foot trajectory to solve all of the above problems.

2.4   CMAC Optimization and Correction Algorithm

CMAC is used as the walking leg controller of a robot to stabilize inverse kinematics model from position to joint space mapping. The trunk motion trajectory is defined as the given signal $Trunk_{t+1}^{ref}$ at time t+1, comparing with the real trunk position $Trunk_t^{gyr}$ at time t, we can obtain the position increment $\Delta\mathbf{Trunk}_{t+1}$, which combined with the leg angle $\mathbf{q}^{act}$ as the input of this control system. And the output of CMAC neural network is $\Delta\mathbf{q}_{t+1}$ in order to control the movement of the leg joint.

Due to the solution of leg inverse kinematics is not unique, we need a joint angle optimization to select a proper output angle. This optimization problem can be solved by gradient descent iteration, the optimization goal function can be written in vector form:

$$\mathbf{Y} = \frac{\alpha}{2}\mathbf{e}^T\mathbf{e} + \frac{\beta}{2}\Delta q^T \Delta q$$

Where $\mathbf{e}$ is the error vector, Δq is the joint angle increment.

The iterative algorithm about Δq can be expressed as follows:

$$\Delta q^{k+1} = \Delta q^k - \eta\frac{\partial \mathbf{Y}}{\partial \Delta q}, (\partial \mathbf{Y}/\partial \Delta q) = -\alpha\mathbf{e}^T\mathbf{J} + \beta\Delta q^T$$

where k indicate the number of iterations, η is learning step length, $\mathbf{J}$ is the leg Jacobin matrix.

Use the leg position deviation $\mathbf{e}_{t+1}$ in t+1 time to revise the weight of the controller ω_{ij}. Define the index function V and the correction algorithm about ω_{ij}:

$$\mathbf{V} = \mathbf{e}^T\mathbf{e}/2, \omega_{ij} = \omega_{ij} - \eta\frac{\partial\mathbf{V}}{\partial\omega_{ij}}, \partial\mathbf{V}/\partial\omega_{ij} = \mathbf{e}^T(-\mathbf{J}_i)$$

where i=1,2; j=1,2,$\cdots$ C indicate the logical address; η is the positive learning ratio.

2.5 Inverse Kinematics

The solution of the inverse kinematics equation is to compute all joint angle values when given a certain position of the object. The $\mathbf{Trunk}^{ref}$ and $\mathbf{Foot}^{ref}$ are respectively deduced from the double inverted pendulum model and the foot trajectory pattern, and correction of joint angle $\Delta\mathbf{q}$ that is deduced from CMAC module would be calculated in inverse kinematics.

2.6 Feedback Control Strategies

In the dynamic model of walking, the closed loop control is mainly to consider the error of the robot trunk position $\mathbf{Trunk}^{error}$. It is not easy to get the accurate value because of the unavoidable mechanical recoil force in each legs joints[10]. As is well known[8], the knees are always bending to avoid singular attitude in the process of walking. As long as the joints do not oscillate in the recoil force zone, it has assertion that the trunk error is aroused by this force. So the trunk error is considered in the double linear pendulum ($\mathbf{Trunk}^{error} = \mathbf{Trunk}^{ref} - \mathbf{Trunk}^{gyr}$) for the correction of D-LIPM original output value. We also used the double balance control mechanism to consolidate the stability of the robot, which is the center of mass balancing and gyroscope feedback-based balancing.

3 Experimental Results

This paper is based on a simulation model of the NAO robot, which has 22 degrees of freedom(DOF) to control the motion of its body and keep the center of mass of the robot above the area support by the feet. Our algorithm is applied in computer with configuration of Intel(R) Core(TM) i7-2600 CPU @ 3.40GHZ and simulation in the following charts.

Fig.4 shows the body rotation angle. The maximum value of the body rotation angle is 1.28 degrees, and this occurred when the support leg changed. The blue trajectories in Fig.3 and Fig.4 show the hip and the body rotation movement without using any reinforcement learning, and after the 19th learning that describe in red trajectories in Fig.4. From these two figures, the robot can walk and turn more stably.

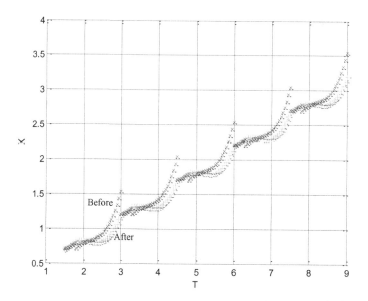

Fig. 3. Hip trajectory along the X axis

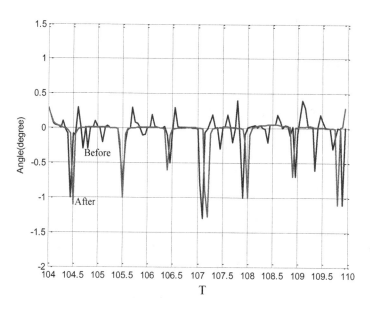

Fig. 4. Body rotation angle

Flexible turning walk for 22 steps, and simulate in the MATLAB environment is shown in Fig.5.

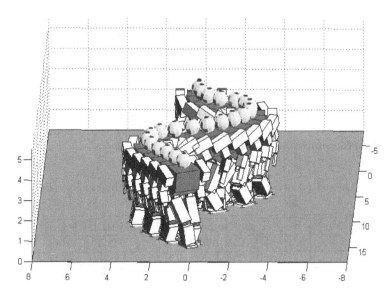

Fig. 5. Schematic diagram of CMAC walking

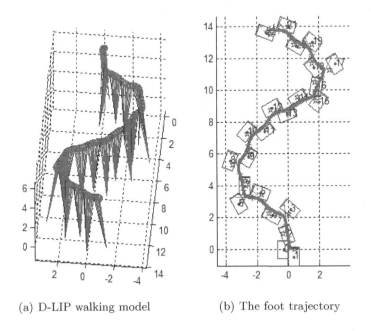

(a) D-LIP walking model   (b) The foot trajectory

Fig. 6. The trunk trajectory diagram

Fig.6 is a top view of Fig.5, the red lines in Fig.6a show the leg walking module based on D-LIP, and the black pentagons indicate the shifting process of the center of gravity between the two legs of robot. The Fig.6b shows the robots trunk trajectory (the red curve) and the foothold positions (the blue rectangles). The blue solid point indicates the actual foothold of each step and the pentagon means the foot position of the next time by using the preview control method. From these figures, it is known that the predicted values are basically coincidence with the actual foothold positions.

In addition, we perform some experiments to verify the validity of our algorithm. It can be seen from Table 1 is the average goals (twenty games) about our algorithm (Apollo3D) vs. Apollo3D__ Before (the only difference is without using any reinforcement learning walking) and the top eight teams in RoboCup 2013 3D soccer simulation competition.

Table 1. The data contrast results based on the omnidirectional walking pattern

Rank	Team Name	Goal Diff(error)
1-2	UT__Austinvilla	-0.95 (±0.06)
3	Apollo3D__Before	1.25 (±0.13)
4-5	Fcportugal	1.32 (±0.16)
4-5	SEU__Jolly	1.53 (±0.15)
6-7	Robocanes	1.74 (±0.20)
7-8	Boldhearts	2.51 (±0.14)
7-8	Magmaoffenburg	2.06 (±0.08)
9	Hfutengine	5.21 (±0.05)

4   Conclusion

Apollo3D won 53 goals during the RoboCup 2013 3D simulation competition and lose 5 goals. Although, the results of this competition not simply rely on walking skill, but agile and stable walking ability are the foundation and the most important factor in robotic soccer.

The establishment of a powerful humanoid robot soccer team not only need some good action skills, but also the upper decision module, and also involves a lot of optimization problems. Apollo3D takes various cooperation actions with a not optimal efficiency, so the next work is that how to optimize the parameters in the actions and the upper decision modules.

Acknowledgments. This paper was supported by the Open Foundation of Jiangsu Province Science Foundation (No.BK2012832) and the Natural Science Foundation of China (No.61104216 and No. 60805032).

References

[1] Strom, J., Slavov, G., Chown, E.: Omnidirectional walking using ZMP and preview control for the nao humanoid robot. In: Baltes, J., Lagoudakis, M.G., Naruse, T., Ghidary, S.S. (eds.) RoboCup 2009. LNCS (LNAI), vol. 5949, pp. 378–389. Springer, Heidelberg (2010)

[2] Bavani, M., Ahmadi, H., Nasrinpour, R.: A closed-loop Central Pattern Generator approach to control NAO humanoid robots' walking. In: Control, Instrumentation and Automation (ICCIA), pp. 1036–1041 (2011)

[3] Behnke, S.: Online trajectory generation for omnidirectional biped walking. In: Proc. of IEEE Int. Conf. on Robotics and Automation, USA, pp. 1597–1603 (2006)

[4] Seung-Joon, Y., Byoung-Tak, Z., Hong, D., Lee, D.: Online learning of a full body push recovery controller for omnidirectional walking. In: IEEE-RAS International Conference on Humanoid Robots (Humanoids), Bled, pp. 1–6 (2011)

[5] MacAlpine, P., Urieli, D., Barrett, S., Kalyanakrishnan, S., Barrera, F., Stone, P.U.: Austin Villa 2012 3D Simulation Team report (EB/OL). Technical Report AI11-10, The Univ. of Texas at Austin, Dept. of Computer Science, AI Laboratory (December 2012)

[6] Chih-Min, L., Chih-Hsuan, C., Chih-Min, L., et al.: CMAC-based dynamic-balancing design for humanoid robot. In: Proceedings of SICE Annual Conference 2010, August 18-21, pp. 1849–1854 (2010)

[7] Lee, J., Oh, J.-H.: Biped walking pattern generation using reinforcement learning. In: 2007 7th IEEE-RAS International Conference on Humanoid Robots, November 29-December 1, vol. 1, pp. 416–421 (2007)

[8] Gouaillier, D., Kilner, C.: Omni-directional closed-loop walk for NAO. In: IEEE-RAS International Conference on Humanoid Robots (Humanoids), Nashville, TN, pp. 448–454 (December 2010)

[9] Qiang, H., Kajita, S., Koyachi, N.: A High Stability, Smooth Walking Pattern for a Biped Robot. In: Proc. of IEEE Int. Conf. on Robotics and Automation, USA, pp. 65–71 (1999)

[10] Graf, C., Hartl, A., Rofer, T., Laue, T.: A robust closed-loop gait for the standard platform league humanoid. In: Proc. of the 4th Workshop on Humanoid Soccer Robots in Conjunction with IEEE-RAS Int. Conf. on Humanoid Robots, pp. 30–37 (2009)

[11] B-Human team. B-human:Team report and code release 2011, Chapter 5 (EB/OL) (Feburary 2012),
http://www.b-human.de/downloads/bhuman11coderelease.pdf (accessed)

The Decision-Making Framework of WrightEagle, the RoboCup 2013 Soccer Simulation 2D League Champion Team

Haochong Zhang and Xiaoping Chen

Department of Computer Science, University of Science and Technology of China
xpchen@ustc.edu.cn

Abstract. This paper presents the latest progress of WrightEagle, the champion of RoboCup 2D simulation league. We introduce a decision-making framework, an extension of MAXQ-OP framework using multiple heuristic functions and a reachable state checking method. The experimental results show that our approach improves the quality of solutions in complex situations.

1 Introduction

The most important characteristic of Robocup 2D simulation league is that the soccer matches only run in computers, where there is only software without any physical robots or other hardwares. Without physical layers, e.g., motion control and image processing, Robocup 2D simulation teams can avoid hardware limitations and focus on research topics such as high level decision-making, learning, and multi-agent collaboration. Therefore, the Robocup 2D simulation competition becomes an important test bed of large-scale decision-making, machine learning and multiagent system research. Each participating team will compete with many other teams, which may be developed in different approaches. For instance, at Robocup 2013, there were a total of 20 teams from 9 countries entering the 2D simulation competition[1].

Taking into account the scale of the 2D simulation competition, making good decisions against so many different opponents is a very challenging task and dealing with complex and varied situations is very difficult[3].

This paper introduces the RoboCup 2013 Soccer Simulation 2D League champion team, WrightEagle, that has won 4 champions and 5 runners-up in the past 9 years. Particularly, we introduce some latest innovations in WrightEagle2D Simulation Team. These innovations expand and implement MAXQ-OP framework and focus on how to extend the search depth in the decision-making procedure to deal with complex and diverse situations and improve the quality of solutions.

The remainder of this paper is organized as follows. Section 2 introduces some background knowledge and motivations. Section 3 presents the decision-making

---

[1] The detailed competition results can be found at:
http://www.oliverobst.eu/research/gliders2013-simulation-league-robocup-team-overview/robocup-2013

S. Behnke et al. (Eds.): RoboCup 2013, LNAI 8371, pp. 114–124, 2014.
© Springer-Verlag Berlin Heidelberg 2014

framework we developed. Section 4 and Section 5 describe more details about the decision-making framework including the reachable states checking method and the evaluation system. Section 6 presents the experimental results. We discuss some related work briefly in Section 7 and conclude the paper in Section 8.

2 Background

2.1 Robocup Soccer Simulation League 2D

In 2D simulation competition, each team is constructed of 12 agents, including 11 players and an online coach. Then, all the agents separately connect to the competition server which simulates the world including the dynamics and kinematics of the players and ball. The information sent to each player in every decision cycle by the server is highly uncertain and incomplete, which depends on the field's and the agent's own situation. Then, after receiving information, each agent has to send the actions back to server for execution in the simulated world. The whole process takes no more than 100 milliseconds (ms) which is one of the most difficult forces, so the agents must respond in a rapid way. In 2D simulator, actions are abstract commands such as turning the body and neck by a specified angle, dashing to a specified angle with a specified power, kicking to a specified angle with a specified power or tackling in a specified angle. The 2D simulator does not model the motions of any real robot but an abstracted player.

In Robocup 2D simulation league, we face up to two challenges. Firstly, the search space is extremely large. The simulator simulates the whole continuous 2D space on the standard $65m \times 105m$ soccer field. Moreover, the number of agents greatly influences the decision-making efficiency. Since each player has 10 teammates, 11 opponents and a coach on both sides, to build the teammate model and the opponent model is an extremely demanding job. All of the above factors increase the difficulty of designing an effective search algorithm for this problem. Secondly, the other challenge is the limited computation time in each decision cycle. In every cycle, which is 100 ms, considering the network delay and other necessary time cost, the agent has to make a decision within 70 ms. Due to the above restrictions, it is a great challenge to extend the search depth to a satisfactory level.

2.2 MAXQ Hierarchical Decomposition

The WrightEagle team is developed base on the Markov Decision Processes (MDPs) framework [1] with the MAXQ hierarchical structure [2] and heuristic approximate online planning techniques in the past years[3][4].

Generally, the MAXQ technique decomposes a given MDP into a set of sub-MDPs arranged over a hierarchical structure. Each sub-MDP is treated as a distinct subtask.

Given the hierarchical structure, a *hierarchical policy* π is defined as a set of policies for each subtask $\pi = \{\pi_0, \pi_1, \cdots, \pi_n\}$, where π_i is a mapping from active states to actions $\pi_i : S_i \to A_i$. The *projected value function* of policy π for a subtask M_i in state s, $V^\pi(i, s)$, is defined as the expected value after following policy π in a state s until the subtask M_i terminates in one of its terminal states in G_i. Similarly, $Q^\pi(i, s, a)$ is the expected value by firstly performing action M_a in state s, and then following policy π until the termination of M_i. It is worth nothing that $V^\pi(a, s) = R(s, a)$ if M_a is a primitive action ($a \in A$).

Dieterich [2] has shown that the value function of policy π can be expressed recursively as:

$$Q^\pi(i, s, a) = V^\pi(a, s) + C^\pi(i, s, a) \tag{1}$$

where

$$V^\pi(i, s) = \begin{cases} R(s, i) & \text{if } M_i \text{ is primitive} \\ Q^\pi(i, s, \pi(s)) & \text{otherwise} \end{cases} \tag{2}$$

$C^\pi(i, s, a)$ is the *completion function* that estimates the cumulative reward received with the execution of action M_a before completing the subtask M_i, as defined below:

$$C^\pi(i, s, a) = \sum_{s', N} \gamma^N P(s', N | s, a) V^\pi(i, s'), \tag{3}$$

where $P(s', N | s, a)$ is the probability that subtask M_a in s terminates in s' after N steps.

3   The Decision-Making Framework

This section presents a decision-making framework that has been implemented in WrightEagle 2D simulation team for for the macro action attack. This framework implements and extends the MAXQ-OP framework. We introduce more complete ideas of heuristic search and other techniques for transforming MAXQ-OP framework to adapt to the needs of the 2D simulation league. In past years, we have used MAXQ-OP framework to define a series of subtasks at different levels[4], the decision-making framework is operating at the level of shoot, dribble, and pass etc.

In order to maximally approximate the optimal $Q^\pi(i, s, a)$ values, we use a heuristic method to choose the direction of extension of the $Q^\pi(i, s, a)$.

Overall decision-making framework presented in Algorithm 1 is similar to the best-first search. The search process forms a search tree starting from the initial state and always extends the state which has the best evaluation. The framework generates next level states from a certain state and pushes these states into an ordered list sorted by evaluation. And we introduce the idea of anytime algorithm into the decision-making framework and limited search depth to an acceptable range.

In order to meet the time cost restrictions of Robocup 2D Simulation League meanwhile making good decision results, the framework has following features and components:

State and Action. The state in the framework records most of the information from the field, such as the positions of all players and ball. Except field information, we add the transition probability from an initial state to the current state. We will introduce more information of the transition probability in Section 4. We use macro actions, defined in MAXQ-OP framework[4], to get the successor states.

Evaluation System. In order to evaluate and choose an action, the MAXQ-OP framework compute the optimal *projected value function*. And to meet the needs of the Robocup 2D simulation league, we have to extend the MAXQ-OP framework. Then, we develop the evaluation system to evaluate all the actions and states generated by decision-making framework. More details are introduced in Section 4.

Reachable States Checking. To achieve better search results in limited time, we have to skip those unreachable states during the search process. We developed a reachable states checking method. We will presents more details in Section 5.

Algorithm 1. Decision(s)

Input: A state, s.
Output: Evaluation of the state.

1: $MaxEva \leftarrow -\infty$
2: $SearchNode \leftarrow 0$
3: Insert(SearchList,InitState)
4: **while** Empty(SearchList) = False **and** $SearchNode < MaxSearchNode$ **do**
5:     $State \leftarrow Top(SearchList)$
6:     Pop(SearchList);
7:     $SearchNode \leftarrow SearchNode + 1$
8:     **if** $GetDepth(State) \geq MaxSearchStep$ **then**
9:         **continue**
10:     **else**
11:         **for all** Teammate **do**
12:             $BehaviorList \leftarrow CalcAction(Teammate, State)$
13:             **for all** Behavior in BehaviorList **do**
14:                 $NextState \leftarrow GenerateNextState(Behavior)$
15:                 $NextEva \leftarrow Evaluate(NextState)$
16:                 $MaxEva \leftarrow Max(MaxEva, NextEva)$
17:                 Insert(SearchList,NextState)
18:                 SortByEvaluation(SearchList);
19:             **end for**
20:         **end for**
21:     **end if**
22: **end while**
23: **return** MaxEva

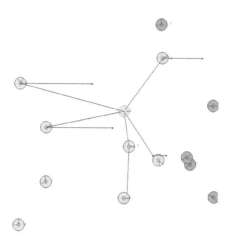

Fig. 1. Example of search process

4   The Evalution System

As we mentioned in Section 3, we hope that algorithm approximate the optimal $Q^\pi(i, s, a)$ value as soon as possible. We use heuristic method to choose the direction of extension of $Q^\pi(i, s, a)$. This heuristic method is natural to combine with the sparse property of the reward function of Robocup 2D simulation league.

From the perspective of soccer, when scoring, the soccer team get a high reward. Otherwise the reward is 0. The reward function has a sparse property which has the effect that the forward search process often terminates without any reward obtained. So we have to find an approach to help us to distinguish the states with a high possibility of scoring from all states to accelerate the search process and seize opportunities on the competition. An intuitive idea is using a heuristic function as the reward function.

We developed a heuristic evaluation system for long-term search algorithm. The input is a state and the output is the estimates of the possibility to score from this state. The evaluation system is composed of simple and intuitive heuristic functions such as the distance of the ball to the opponent's goal.

After combining MAXQ hierarchical decomposition and heuristic reward, we get a complete evaluation system as:

$$Q(i, s, a) = V(a, s) + C(i, s, a) \tag{4}$$

$$V(i, s) = \begin{cases} max_a Q(i, s, a) & \text{if i is composite} \\ R(s, i) & \text{otherwise} \end{cases} \tag{5}$$

and

$$C(i, s, a) = \sum_{s', N} \gamma^N T(s', N | s, a) V(i, s') \tag{6}$$

$$R(s,i) = \begin{cases} H(s,i) & \text{if s is a terminal state} \\ 0 & \text{otherwise} \end{cases} \tag{7}$$

Figure 2 and Figure 3 are examples of this Evaluation System.

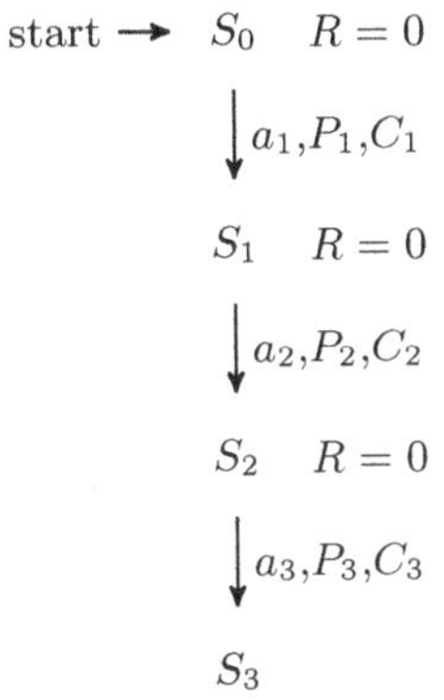

$$\text{start} \rightarrow \quad S_0 \quad R = 0$$

$$\downarrow a_1, P_1, C_1$$

$$S_1 \quad R = 0$$

$$\downarrow a_2, P_2, C_2$$

$$S_2 \quad R = 0$$

$$\downarrow a_3, P_3, C_3$$

$$S_3$$

$$HeuristicReward = H(S_3)$$

$$E_3 = P_1 * P_2 * P_3 * R(S_3) * \gamma^{(C_1 + C_2 + C_3)}$$
$$E(S_0, a_1) = max(E_1, E_2, E_3)$$

Fig. 2. An example of evaluation system. States S_0, S_1, S_2 are intermediate states and state S_3 is the terminal state. a_1, a_2, a_3 are actions. P_1, P_2, P_3 are transition probabilities. C is the number of steps of corresponding macro actions.

When the search precess begins, starting from the initial state, the evaluation system evaluates the initial state to get a heuristic reward, $H(s)$. Then the algorithm will extend this state to get the successor states. When a new state is found, the algorithm replaces the predecessor state's heuristic reward with the actual reward, which equals to 0 most of the time, and the evaluation system gives the new state a heuristic reward, $H(s')$.

In the search process, the terminal state will continue to be extended to become an intermediate state. This evaluation method not only depends on the current state but also the sequence of states that preceded it. However, if add a variable, the transition probability from the initial state to the state, into the state, then the evaluation system would not depend on the specific search history.

5   The Reachable States Checking

Through observation of the MAXQ-OP framework equation 8 and 2D simulation league, most of the transition probability, $P(s', N|s, a)$, between states is 0. So we only extend a small number of successor states in the decision-making process.

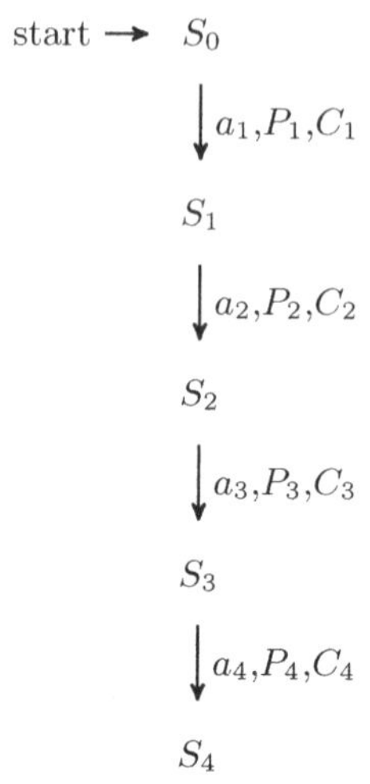

$$\text{start} \longrightarrow S_0$$

$$\downarrow a_1, P_1, C_1$$

$$S_1$$

$$\downarrow a_2, P_2, C_2$$

$$S_2$$

$$\downarrow a_3, P_3, C_3$$

$$S_3$$

$$\downarrow a_4, P_4, C_4$$

$$S_4$$

$$E_4 = P_1 * P_2 * P_3 * P_4 * R(S_4) * \gamma^{(C_1+C_2+C_3+C_4)}$$
$$E(S_0, a_1) = max(E_1, E_2, E_3, E_4)$$

Fig. 3. After extented state S_3, the evaluation system replace the heuristic reward of S_3 with 0 and give the new state S_4 a heuristic reward

And if we can greatly speed up the unreachable states detection speed, we can deepen the search depth.

$$C^\pi(i, s, a) = \sum_{s',N} \gamma^N P(s', N|s, a) V^\pi(i, s'), \qquad (8)$$

According to the transition function as given in [4], we simplify the state reachable problem as from an initial position after certain cycles whether a player is able to reach a particular position. So we developed a teammates control area marking system to check whether the state is reachable to speed up our search process.

In order to ensure the checking speed, we gave up the pursuit of absolute accuracy:

- We simplify player motion model, in the simplified motion model and player state, consider only dash, do not consider turn or other actions.
- We simplified player state, consider only distance, do not consider body angle or other information.
- We divide the continuous soccer field into the discrete space.

In fact, if we compute the minimum number of cycles that players from an initial state to reach another state, the result to use simplified state and motion model is always the lower bound of the result to use full state and motion model. It means if a state is unreachable in simplified state and motion model it must be unreachable in the normal model. But the other way around is non truth, if a state is reachable in simplified state and motion model, it may be reachable or unreachable in the normal model, such cases require a more precise computation.

As in the figure, Figure 4, we divided the field into a number of small areas. Each area has a controller, and in the example, different colors represent different controller. The longer idled player has a bigger control area. If a area is not being controlled by any player, it is an unreachable state.

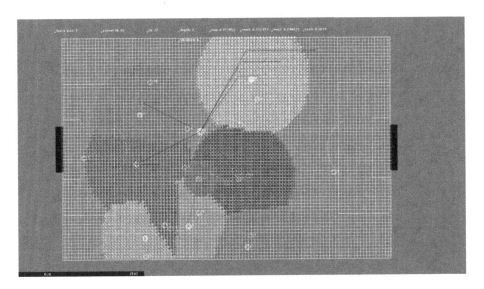

Fig. 4. Red lines mean the passes in history, black lines mean the considering passes. Different colors represent different teammates' control areas.

6   Experiments

In order to evaluate our framework, we use both time consumption and win rate to analyse the performance.

In the aspect of time consumption, worst case is most important, because time limitation is fixed. All we need to do is make sure the decision-making process in the worst case can be finished in time limitation.

To ensure the increase of time consumption is caused by the increase of search depth. We use tools[2], included in WrightEagleBase, DynamicDebug and Time-Test to finish this job. DynamicDebug allows us to reproduce the game scene and TimeTest allows us to analyse time consumption of specific function.

Table 1 shows the time consumption of different search depth with same DynamicDebug log files, which mean the same game and player role. Because of the search speed up methods, for example the reachable states checking, time

---

[2] Downloadable from `http://ai.ustc.edu.cn/en/robocup/`
`2D/releases/WrightEagleBASE-4.0.0.tar.gz`

consumption increases linearly. However, in practice, after taking into account other factors such as network communication delay and other behaviors, the max acceptable search depth is 3. When search depth reaches 4, the players begin to appear to miss the decision cycle.

Table 1. Time consumption

Search Depth	1	2	3	4
Sum Consumption (ms)	77.7	106.4	150.7	192.7
Max Consumption (ms)	14.3	26.2	39.7	54.9
Min Consumption (ms)	0.1	0.1	0.2	0.2

For each version, we tested the team in full games with Helios2012, the champion of Robocup 2012. We independently ran each version against the binary code of Helios2012 officially released by Robocup 2012 for 150 games on the same hardware and software environment.

Table 2. Winning Rate

Search Depth	Winning Rate
1	58.0%
2	62.0%
3	68.7%
4	59.2%

Table 2 shows the winning rates of different search depth of 150 games. The result shows the team performance is increased when search depth increases. However, in the case of search depth 4, the impact of computation timeout caused the decrease of winning rate.

The search depth 1 performs a greedy strategy, the greedy strategy is not bad in most of time. Players continue to play forward, and as long as we are lucky enough, the ball will be eventually scored the goal. Of course, we cannot be always lucky enough. In this situation, the players often ignore long-term rewards. The player only holds the ball and waits for opponents to scramble.

Then, in the search depth of 2 and 3, this problem is more and more improved. The players will avoid some of the obvious short-sighted behavior. For example, if the direction of attack stuck on, and the whole team will make several passes, thereby changing the direction of the attack. Otherwise in greedy strategy, the player will hold the ball and the match will fall into deadlock.

7 Related Work

Some teams in the 2D league using search techniques make decisions. The Helios team [5,6] have developed a framework for online multiagent planning. The framework includes ActionGenerator and FieldEvaluator. The ActionGenerator generates candidate action instances for a node in the search tree, a node from the root node to leaf node is an action sequence.The FieldEvaluator evaluates the value of the action-state pair instances that are generated by ActionGenerator. The value of an action sequence is the sum of the each state variables. The WrightEagle team used an approach based on MAXQ-OP to automate planning. It combines the benefits of a general hierarchical structure based on MAXQ value function decomposition with the power of heuristic and approximate techniques. In this paper, we further develop this work. We created a new decision-making framework based on the MAXQ-OP framework with heuristic functions and other methods to accelerate the search process.

More teams in the 2D league[7,8] tried to improve performance with machine learning (ML) techniques. The most preferred ML methods is Reinforcement learning (RL), based on the general idea that an autonomously acting agent obtains its behavior through repeated interaction with its environment on a trial and error basis, by processing and integrating the experience the agent has gathered into a value function that tells how much it may be worth entering some state by taking some specific action. For example, team AT Humboldt [7], focusing on the task of learning to dribble, used a specialized version of Watkins Q-learning algorithm and integrates it into a framework where a proper representation of the state and action spaces was learned using an evolutionary approach. The opposite problem, i.e. the task of attacking and disturbing a dribbling opponent player in order to scotch the opponent teams attack at an early stage and to gain ball possession, has been targeted by the Brainstormers team. Using a partial model of the soccer environment and utilizing a temporal difference RL approach in conjunction with neural net-based value function approximation, they learnt a behavior for the mentioned task which significantly boosted team performance [8].

8   Conclusion

This paper introduces the champion of the RoboCup 2013 2D simulation league, WrightEagle. We describe some recent efforts to deal with complex and diverse situations and improve the quality of the solutions.

First, we developed a decision-making framework which transformed and extended the MAXQ-OP framework into Robocup simulation league 2D. The decision-making framework combined the heuristic evaluation function and best-first search algorithm. Second, we used the heuristic function to extend the definition of reward function of MAXQ-OP framework, and discussed the properties and features of this method. Third, we introduced a reachable state checking method to speed up the search process. And we stated if the checking result of a state is unreachable, then it promised to be unreachable. Otherwise, it may or may not be reachable.

Combined above methods, we successfully extended the search depth and got better search results in limited decision-making time.

Acknowledgments. The authors would like to thank the other members of WrightEagle, Aijun Bai, Miao Jiang, Haibo Dai, Guanghui Lu, Xiao Li, Rongya Chen, for their contributions to WrightEagle2D simulation team.

This work is supported by the National Hi-Tech Project of China under grant 2008AA01Z150 and the Natural Science Foundation of China under grants 60745002 and 61175057, as well as the USTC 985 Project.

References

1. Puterman, M.L.: Markov decision processes: discrete stochastic dynamic programming, vol. 414. Wiley. com (2009)
2. Dietterich, T.G.: The maxq method for hierarchical reinforcement learning. In: ICML, pp. 118–126. Citeseer (1998)
3. Bai, A., Wu, F., Chen, X.: Online planning for large mdps with maxq decomposition. In: Proceedings of the Autonomous Robots and Multirobot Systems Workshop, at AAMAS 2012 (June 2012)
4. Bai, A., Wu, F., Chen, X.: Towards a principled solution to simulated robot soccer. In: Chen, X., Stone, P., Sucar, L.E., van der Zant, T. (eds.) RoboCup 2012. LNCS (LNAI), vol. 7500, pp. 141–153. Springer, Heidelberg (2013)
5. Akiyama, H., Nakashima, T.: HELIOS2012: RoboCup 2012 Soccer Simulation 2D League Champion. In: Chen, X., Stone, P., Sucar, L.E., van der Zant, T. (eds.) RoboCup 2012. LNCS (LNAI), vol. 7500, pp. 13–19. Springer, Heidelberg (2013)
6. Akiyama, H., Nakashima, T., Yamashita, K.: HELIOS2013 Team Description Paper. In: RoboCup 2012: Robot Soccer World Cup XVI, pp. 1–6 (2012)
7. Burkhard, H.-D., Hannebauer, M., Wendler, J.: At humboldt development, practice and theory. In: Kitano, H. (ed.) RoboCup 1997. LNCS, vol. 1395, pp. 357–372. Springer, Heidelberg (1998)
8. Gabel, T., Riedmiller, M., Trost, F.: A case study on improving defense behavior in soccer simulation 2d: The neurohassle approach. In: Iocchi, L., Matsubara, H., Weitzenfeld, A., Zhou, C. (eds.) RoboCup 2008. LNCS (LNAI), vol. 5399, pp. 61–72. Springer, Heidelberg (2009)
9. Stone, P.: Layered learning in multiagent systems: A winning approach to robotic soccer. The MIT Press (2000)
10. Stone, P., Sutton, R.S., Kuhlmann, G.: Reinforcement learning for robocup soccer keepaway. Adaptive Behavior 13(3), 165–188 (2005)

Clustering and Planning for Rescue Agent Simulation

Ahmed Abouraya, Dina Helal, Fadwa Sakr, Noha Khater, Salma Osama,
and Slim Abdennadher

German University in Cairo, Cairo, Egypt
{dina.helal,slim.abdennadher}@guc.edu.eg

Abstract. The paper describes the contribution of the GUC_ArtSapience team to the Rescue Agent Simulation competition in RoboCup in terms of the current research approach. The approach is divided into two parts: clustering and planning. Clustering is done through task allocation to divide the map among the agents. Planning is done after assigning the agents to parts of the map to determine how they should cooperate and coordinate together and how they should prioritize their tasks [2]. The agents can coordinate together using centers and communication if available or dynamically without the use of communication.

1 Introduction

Rescue planning and optimization is one of the emerging fields in Artificial Intelligence (AI) and Multi-Agent Systems. The RoboCup Rescue Agent Simulation provides an interesting test bench for many algorithms and techniques in this field. The simulation environment provides challenging problems that combine optimization (routing, planning, scheduling) and multi-agent systems (coordination, communication, noisy or missing communication) [3].

The Robotics and Multi-Agent Systems (RMAS) research group at the German University in Cairo (GUC) was established in September 2010. The goal of the research group is to study and develop AI algorithms to solve problems in robotics and simulation systems. These fields include computational intelligence, computer vision, multi-agent systems, and classical AI approaches.

The GUC_ArtSapience team made its third participation to the Rescue Agent Simulation in 2013 and won the first place. Our first participation (as RMAS ArtSapience) was in 2011 and the team ranked third in the final round. Our second participation was in 2012.

In this paper, we propose our approach which is divided into two parts: clustering and planning. Clustering is done using K-means to divide the map into small regions. The number of clusters is relative to the number of agents in the map. This is done in the preprocessing phase to set the initial positions of the agents in the map. This ensures that all regions in the map are covered by the three types of agents. As the simulation starts, each agent has a list of tasks to do which are prioritized. Agents communicate and coordinate together using communication channels and voice messages.

S. Behnke et al. (Eds.): RoboCup 2013, LNAI 8371, pp. 125–134, 2014.

This paper is organized as follows. Section 2 describes our clustering approach. Section 3 describes our planning technique regarding the agents' tasks and how they prioritize them. Section 4 describes our communication model. Finally, in Section 5 we give a summary of our results and performance in RoboCup 2013.

2   Clustering

In the rescue simulation environment, there is a lot of information that is obscure. For example, in disaster scenarios, the initial locations of fires, buried, injured civilians, and blockades are unknown. Moreover, tasks that are assigned to an agent do not specify where exactly this agent should carry out these tasks. Therefore, agents have no other choice but to traverse the whole map and search for their tasks. And since it is impractical for a single agent to cover the whole city map (all buildings and/or roads), our approach is to divide the map into smaller partitions.

Clustering is the method that we chose to divide the map. The clustering process is done in the preprocessing phase. Each type of agents cluster the map among themselves, so the maps is divided into clusters three times for Ambulance agents, Fire Brigades and Police agents. Ambulance agents and Fire Brigades clusters are groups of buildings while Police agents clusters are groups of roads close to each other. Each agent is assigned to a cluster, where it traverses all of the buildings/roads in it searching for events that require rescue actions. Figure 1 shows an example of Fire Brigade agents clustering the map in the preprocessing phase.

2.1   C-means

Initially, we used fuzzy c-means for clustering to generate overlapping clusters. C-means clustering is a method of clustering that allows one piece of data to belong to one, two or more clusters. The number of clusters was chosen to ensure that each cluster would have more than one agent. However, several tests showed that increasing the number of clusters to be equal to the number of agents produces better results. The problem with that was that increasing the number of clusters required extra computational time which exceeded the limited time allowed for preprocessing [6]. The advantages of using C-means clustering is overcoming the fact that there exist blockades all over the map which might prevent agents from reaching their targets. By increasing the number of agents heading for a specific target, the probability of this target being covered increases. However, time is wasted if two agents head for the same target.

2.2   K-means++

K-means++, as well as K-means clustering [5], aims to partition n observations into k clusters in which each observation belongs to the cluster with the nearest

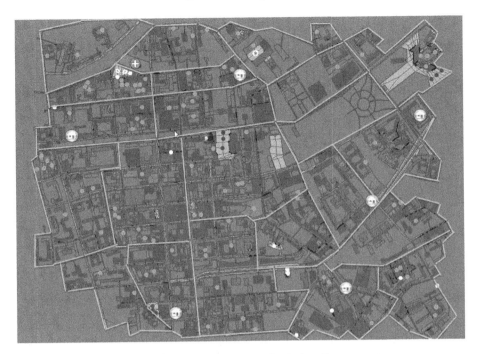

Fig. 1. Fire Brigade agents clustering the map

mean. To avoid sacrificing cluster quality, we chose to use K-means ++ clustering algorithm, which works similar to K-means, but uses a different heuristic for selecting the initial centroids which are selected from a uniform Gaussian distribution over the buildings in the map. K-means++ gives a much better start for the algorithm as well as guarantees a faster conversion to the optimal centroids. After dividing the map into partitions equal to the number of agents of a certain type in the map, each agent is assigned to the closest cluster to it.

Using K-means++ algorithm, we were able to overcome the time issue and we were able to compute clusters equal to the number of agents within the allowed time. Since the K-means++ algorithm is faster than the fuzzy C-means algorithm, we were able even to perform more computations during preprocessing such as calculating the shortest path distance between entities. In addition, through running the simulation on different maps, it was clear that there is a need for good strategies to find good starting values to reach the best possible solutions [4].

3   Planning

After assigning each agent to a certain region of map using the clustering technique, each agent has a set of tasks to perform as shown in figure 2. In this section we discuss the tasks of the Fire Brigades, Ambulance Team and Police Agents and how they are prioritized. Figure

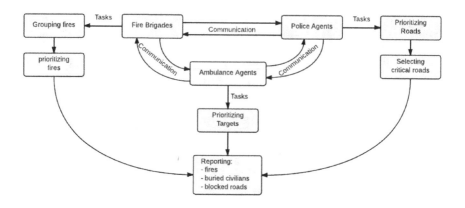

Fig. 2. Agents tasks after clustering

3.1 Fire Brigades

The main task of Fire Brigades is to extinguish fire in buildings. Thus, the Fire Brigades' targets in the map are buildings on fire. In the preprocessing phase, K-means++ algorithm is used to partition the map into clusters which are groups of buildings close to each other on the map. Each cluster is then assigned to the nearest Fire Brigade agent. Each agent keeps searching for fires inside its own cluster. Once it finds one, it starts extinguishing the fire and reports it to the other agents using radio communication.

3.1.1 Grouping Fires

Sometimes a single agent can not handle a fire inside its cluster by itself. Therefore, other agents respond to the fire report by going to the fire location and help extinguishing the fire. A problem might arise if all agents responded to the fire report leaving unprotected parts in the map where another fire can start with no one to extinguish it. Therefore not all agents should respond to the fire report, only the closest ones head to the fire. This is handled by grouping fires together where close fires are considered to be belonging to one group with a center c and radius r. Agents are distributed among different groups of fire such that an agent is assigned to the closest group. Thus, the agents can handle more than one fire simultaneously.

3.1.2 Prioritizing Fires

Warm buildings are reported by Fire Brigade agents and considered to be on fire until their temperature is below a certain threshold. Extinguishing a building if it is warm is easier than extinguishing them if they are on fire. Extinguishing warm buildings helps in containing the fire because warm buildings are usually at the outer part of a fire group. Based on that, the highest priority is given to the warm buildings, then to buildings with lowest degree of fieryness (a parameter to measure the degree of fire in a building) then to buildings of the highest degree of fieryness. This way, the outer part of the fire group is targeted first which increases the possibility of containing the fire.

3.2   Ambulance Teams

The Ambulance team's main task is to rescue buried civilians and carry them to refuges so their target is buried civilians. Each Ambulance agent is assigned to a cluster which is a group of buildings like the Fire brigades. An ambulance agents searches for buried civilians withing its cluster then moves on to other clusters.

3.2.1   Prioritizing Targets

Each Ambulance agent keeps track of the buried civilians it came across but couldn't rescue because of blockades or fires. Accordingly, Ambulance agents divide their tasks into three groups. The highest priority is assigned to the currently seen buried civilians followed by the ones it previously passed by but could not reach followed by the reported targets seen by other agents.

A problem that Ambulance agents face is deciding how many Ambulance agents should rescue a certain civilian. As the number of agents increase the rescuing process takes less time, also as the burriedness of a civilian increases, more time is required to rescue him. To tackle this problem first, we tried to calculate and estimate death time for the buried civilians using their buriedness and damage parameters retrieved from the simulator. The buriedness parameter measures how much a civilian is buried under a building. When its value increases, this means the civilian is buried deeper. However, this was rather hard to do because of inaccurate simulator parameters. Based on that, we limited the number of agents rescuing one civilian to a certain number which is a function of the total number of Ambulance agents in the map and the burriedness of the civilian. This ensures that not all agents would try to save one civilian leaving other civilians to die.

Another problem they face is deciding which civilians need to be rescued first and which civilians should not be rescued as they might not make it to the refuge. This is calculated using a threshold where civilians with damage less than a certain threshold are left behind as they most probably will die before reaching the refuge. If the damage is higher than the threshold, higher priority is given to the most damaged civilians followed by the less damaged.

A decision the Ambulance agents need to make is deciding which agent should load the civilian if more than one agent were rescuing him. Only one agent can load a civilian so we let the agent with the lowest id load the civilian. This is based on the fact that it won't matter which agent loads the civilian and takes him to the refuge so it is just a systematic decision.

3.3   Police Agents

The last type of agents we are going to discuss is the Police agents. The Police agents' main task is to clear block road so that other agents can move freely in the map and perform their tasks. Each Police agent is assigned to a certain cluster which is a group of roads unlike the previous two agents which clustered the map in groups of buildings. Police agents divide their tasks into two parts which are prioritizing their targets and determining the critical blockades.

3.3.1   Prioritizing Roads

Police agents clear blocked roads as they move. Their highest priority is clearing roads leading to buildings on fire so that Fire Brigades can reach the buildings and extinguish the fires. The second priority is for clearing roads with agents stuck in them so that they can start performing their tasks. The third priority is clearing roads leading to refuges where Fire Brigades need to go and refill water and civilians need to go to e rescued. The fourth priority is for roads leading to buried civilians. Finally the lowest priority is for the rest of the blocked roads.

3.3.2   Selecting Critical Roads

Critical blockades are blockades which blocking any of the roads or buildings entrances. If the distance two blockades on the same road is smaller than the agent's diameter, then this blockade is considered to be blocking the road. If the blockade passes this first check, we project the blockade's shape on each of the open road edges. Then, we check whether each of the edges is passable or not. If an edge is not passable, then this blockade needs to be cleared. These checks are performed on all the blockades within the Police agent's clearing range and that are on the path to its current target. This saved a lot of time that used to be wasted on clearing non-critical blockades on the roads where the agents could pass right by them.

3.4   Common Tasks

The common task between all types of agents is reporting what they see. All agents report anything in their line of sight while executing their tasks. They report blocked roads to the Police agents, buried civilians to the Ambulance agents and fires Fire Brigade agents. Communication is done using radio or voice communication which will be discussed in details in the following section.

4   Coordination and Communication

A very important part of the rescuing process is the ability of agents to communicate with each other. Some scenarios allow the agents to subscribe to communication channels and send broadcast messages to each other. In other scenarios, there is no communication, only voice messages that can be send from one agent to its closest neighbor agents. We adapt our algorithm to handle both kinds of communication. Communication channels do not always send accurate message, sometimes there is noise that distorts the sent messages.

4.1   Communication Strategy and Approach

In the scenarios with communication channels, the agents and centers subscribe to the communication channels according to available number of radio channels, maximum number of channels an agent can subscribe to, and the maximum number of channels a center can subscribe to [1].

The messages that are sent between the agents are prioritized according to their types. When agents send or receive messages, first they sort them in a priority queue based on the predetermined priorities. Such an approach allows the agents to utilize the resources and communicate efficiently, especially in the low communication or communication-less scenarios as the more important messages are given the higher priority.

In addition, a mapping technique that maps the IDs of various entities used in the communication process to IDs of smaller size has been devised. For instance, a road ID composed of a 6-digit number could be mapped to a unique ID which is a 2-digit number, thus reducing the message size by 4 bytes. Such an approach has proven to lead to better utilization of the bandwidth in case of radio channels and more information could be sent in a single voice message within the maximum allowed message size.

4.2   Noise Handling

Due to the heavy reliance of agents on communication, noise handling techniques are important to ensure that the messages are delivered and at the same to be able to make the communication process more efficient. In the communication channel, there are two types of noise: dropout noise and failure noise. Dropout noise is when the receiver receives a message without the content (i.e. content is lost), while failure noise is when the message disappears completely without notifying either the sender or the receiver. In order to overcome the noise, the messages are sent more than once. Each agent keeps track of the messages he/she sends per timestep. In the following timestep, the agent compares the messages heard with the messages that he/she sent through the channels during the previous timestep. During the comparison, the agent discards the duplicates and his/her own messages. The remaining messages which were sent by him/her and not received in the next timestep will be resent. This approach ensures that the cases of dropout or failure are handled without redundancy or resending of duplicates.

4.3   Voice Messages

Voice messages are important in scenarios with low communication or no communication at all. Each type of voice message has its own priority as explained in section 4.1. Furthermore, each voice message has its own time-to-live (ttl). The ttl basically limits the lifespan of the voice messages being propagated through the voice channels. Whenever a new voice message is created, it is set with the default ttl which is predetermined. Then as each agent receives a voice message, its ttl is decremented. If the ttl is greater than 0, then the agent rebroadcasts the message through the voice channel. And this process continues until the ttl of the voice message ultimately reaches 0. Then the message is discarded and will no longer be propagated. Such an approach provides a means of broadcasting the voice messages across a broader range and overcoming the limitations of the absence or unreliability of the radio communication channels.

In addition, the agents ensure that the entire allowed maximum size of the voice message is being used through grouping the information which needs to be reported according to a specific priority and sending them in a single message. In this way, different information and data could be sent in a single voice message per timestep, thus leading to more efficiency.

5   Results and Conclusion

In 2013, the enhancements in the agents' performance and in the communication model paid off in the competition. After the first day, we ranked fourth out of 13 teams. After the second day, we ranked third out of 13 teams. In the semi-finals, we ranked fourth out of 8 teams which led us to the final round. In the final round, we outperformed the other 3 teams and came in the first place. There were 6 maps in the final round as seen in table 1. After the 5th map, we were in the second place, however the last map (Istanbul) put us in first place. The last map (Fig. 3) showed a huge difference in the performance between our team and the other three finalists (Fig. 4). Our map is the only one that got partially burnt while the other three maps are completely burnt by the end of the simulation which resulted in a huge difference in the score between us and the other teams.

In this work, we have described our approach that outperformed all other teams during the RoboCup 2013 finals. Our approach is divided into two parts: clustering and planning. Clustering is done using K-means to divide the map into small regions. The number of clusters is relative to the number of agents in the map. This is done in the preprocessing phase to set the initial positions of the agents in the map. This ensures that all regions in the map are covered by the three types of agents. As the simulation starts, each agent has a list of tasks to do which are prioritized. Agents communicate and coordinate together using communication channels and voice messages. Using our approach, our scores greatly improved in 2013 as we improved our strategy in prioritizing the agents' tasks as well as the coordination between them.

As a future work, it would be interesting to use the centers in communication. In addition, further improvements could be done in the agents' tasks such as predicting how the fire spreads so that the Fire Brigades can start extinguishing the buildings more likely to spread fire to their neighboring buildings. Also, civilians' health could be better estimated so that the Ambulance agents can decide whether the civilians can be saved or not which could eliminate the time wasted on rescuing civilians that die on their way to the refuge.

Table 1. Rescue Simulation League 2013 Finals scores

Map	Kobe4	Eindhoven5	Mexico3	Berlin3	Paris4	Istanbul3
GUC_ArtSapience	169.57	169.22	145.42	79.01	106.25	66.97
MRL	170.51	165.92	156.16	66.14	84.64	12.90
Poseidon	165.62	155.64	127.37	85.50	136.54	10.27
S.O.S.	183.20	161.87	145.46	88.20	147.97	12.11

Fig. 3. GUC_ArtSapience Istanbul map in the final round

Fig. 4. Istanbul map of MRL, Poseidon, and S.O.S. in the final round

References

1. Guc artsapience team description paper (2012)
2. Multi-agent planning for the robocup rescue simulation: Applying clustering into task allocation and coordination. In: ICAART 2012 International Conference on Agents and Artificial Intelligence (2012)
3. Robocup rescue website, http://sourceforge.net/apps/mediawiki/roborescue/

4. Arthur, D., Vassilvitskii, S.: k-means++: The advantages of careful seeding. In: Proceedings of the Eighteenth Annual ACMSIAM Symposium on Discrete Algorithms, vol. 8(2006-13), pp. 1027–1035 (2007)
5. Kanungo, T., Mount, D.M., Netanyahu, N.S., Piatko, C.D., Silverman, R., Wu, A.Y.: An efficient k-means clustering algorithm: Analysis and implementation. IEEE Transactions on Pattern Analysis and Machine Intelligence 24, 881–892 (2002)
6. Winkler, R., Klawonn, F., Kruse, R.: Problems of fuzzy c-means clustering and similar algorithms with high dimensional data sets. In: Gaul, W.A., Geyer-Schulz, A., Schmidt-Thieme, L., Kunze, J. (eds.) Challenges at the Interface of Data Analysis, Computer Science, and Optimization. Studies in Classification, Data Analysis, and Knowledge Organization, pp. 79–87. Springer, Heidelberg (2012)

Increasing Flexibility of Mobile Manipulation and Intuitive Human-Robot Interaction in RoboCup@Home

Jörg Stückler, David Droeschel, Kathrin Gräve, Dirk Holz, Michael Schreiber, Angeliki Topalidou-Kyniazopoulou, Max Schwarz, and Sven Behnke

Rheinische Friedrich-Wilhelms-Universität Bonn,
Computer Science Institute VI: Autonomous Intelligent Systems,
Friedrich-Ebert-Allee 144, 53113 Bonn, Germany
{stueckler,droeschel,graeve,holz,schreiber,schwarz}@ais.uni-bonn.de,
{topalido,behnke}@cs.uni-bonn.de
http://www.NimbRo.net/@Home

Abstract. In this paper, we describe system and approaches of our team NimbRo@Home that won the RoboCup@Home competition 2013. We designed a multi-purpose gripper for grasping typical household objects in pick-and-place tasks and also for using tools. The tools are complementarily equipped with special handles that establish form closure with the gripper, which resists wrenches in any direction. We demonstrate tool use for opening a bottle and grasping sausages with a pair of tongs in a barbecue scenario. We also devised efficient deformable registration methods for the transfer of manipulation skills between objects of the same kind but with differing shape. Finally, we enhance human-robot interaction with a remote user interface for handheld PCs that enables a user to control capabilities of the robot. These capabilities have been demonstrated in the open challenges of the competition. We also explain our approaches to the predefined tests of the competition, and report on the performance of our robots at RoboCup 2013.

1   Introduction

The RoboCup@Home league [1,2] aims at fostering research in intelligent robots that perform tasks in the environments of our daily living. Since 2009, we participate with our team NimbRo@Home and won the competitions in 2011, 2012, and 2013. Our robot system joins lightweight but versatile hardware design with software skills for indoor navigation, mobile manipulation, and human-robot interaction.

At the 2013 RoboCup competition, our robots demonstrated novel abilities in object manipulation: tool use and skill transfer. We also continued the development of a remote user interface for handheld PCs. In this paper, we detail our robotic system and our approaches to perform the predefined tests of the competition as well as the aforementioned novelties.

S. Behnke et al. (Eds.): RoboCup 2013, LNAI 8371, pp. 135–146, 2014.
© Springer-Verlag Berlin Heidelberg 2014

2   The RoboCup@Home Competition 2013

In the RoboCup@Home competition [3], teams proceed through two stages to select the top five teams that participate in the final. The two stages consist of tests with predefined procedures and open demonstrations in which the teams can show the best of their own research. An additional technical challenge award can be achieved. This year's technical challenge tested the specific capability of recognizing and localizing furniture-type objects. The final is an open demonstration that is judged by an expert jury.

In the predefined tests, basic skills in mobile manipulation and human-robot interaction have to be demonstrated. The robots need to perform the tests at a scheduled time and within a specific time limit. The detailed arena setup is not known to the participants beforehand, but they have to take approaches that robustly work in typical everyday environments. The predefined tests allow for an objective score evaluation according to the achieved sub-tasks. Open demonstrations are evaluated by a jury that is composed of team leaders, or members from the league's executive committee, science, industry, or media.

The *Robot Inspection and Poster Session* is the first test in Stage 1. It tests navigation capabilities of the robots, its visual appearance and safety. In the *Follow Me* test, the robots must keep track of a previously unknown guide in an unknown (and crowded) environment. As in 2012, the robots have to keep track of the guide despite distractions, follow the guide in and out an elevator and find the guide behind a crowd. *Clean Up* tests object recognition and grasping capabilities of the robots. They have to retrieve as many objects as possible within the time limit, recognize their identity, and bring them to their designated locations. The *Cocktail Party* test is set in a butler scenario, where the robot gets called by three persons, learns their identity, and delivers drink orders to them. In the *Emergency Situation* test, a fire in the appartment is simulated. The robot has to detect the fire, and find persons that either stand, sit or lie on the ground and help them out of the appartment. The *Open Challenge* is the open demonstration of Stage 1. Teams can freely choose their 7 min demonstration.

The best 50% teams proceed to stage 2. The *Enduring General Purpose Service Robot* test has been changed from the 2012 test to last over an extended period of time (40 min). Three robots perform the test concurrently in the arena. The robots must understand and execute complex, incomplete, or erroneous speech commands given by an unknown speaker. The commands can be composed from actions, objects, and locations of the regular Stage 1 tests. In the *Restaurant* test, the robots are deployed in a previously unknown real restaurant, where a guide makes them familiar with drink, food, and table locations. Afterwards, the guide gives an order to deliver three objects. The *Demo Challenge* follows the theme "health care" and is an open demonstration.

3   Hardware Design

We designed our service robot Cosero [4] to cover a wide range of tasks in human indoor environments (see Fig. 1). It has been equipped with a flexible torso and

Fig. 1. The cognitive service robot *Cosero*. Left: Cosero receives a pair of tongs during the final of the RoboCup@Home competition 2013 in Eindhoven. Right: Cosero pushes a chair to its location during the Demo Challenge.

two anthropomorphic arms that provide human-like reach. A linear actuator moves the whole upper body up and down, allowing the robot to grasp objects from a wide range of heights—even from the floor. Its anthropomorphic upper body is mounted on a mobile base with narrow footprint and omnidirectional driving capabilities. By this, the robot can maneuver through narrow passages that are typically found in indoor environments, and it is not limited in its mobile manipulation capabilities by holonomic constraints. The human-like appearance of our robots also supports intuitive interaction of human users with the robot.

Cosero's grippers consist of two pairs of Festo FinGripper fingers on rotary joints (see Fig. 1). When the gripper is closed on an object, the bionic fin ray structure of the fingers adapts its shape to the object surface. By this, the contact surface between fingers and object increases significantly, compared to a rigid mechanical structure. A thin layer of anti-skidding material on the fingers establishes a robust grip on objects. By having two fingers on each side of the gripper, it supports stable grasps for torques in the direction of the fingers, and forces in the direction between opponent fingers.

For perceiving its environment, we equipped the robot with diverse sensors. Multiple laser scanners on the ground, on top of the mobile base, and in the torso measure objects, persons, or obstacles for navigation purposes. We use a Microsoft Kinect RGB-D camera in the head to perceive tabletop objects and persons.

4   Mobile Manipulation

Several regular tests in the RoboCup competition involve object handling, for which objects are usually placed separated on horizontal surfaces such as tables and shelf layers. The robot needs to drive to object locations, to perceive the objects, and to grasp them. We develop further manipulation capabilities, especially object manipulation skill transfer and tool use, that we demonstrate in open demonstrations.

4.1 Indoor Navigation

Our robots navigate in indoor environments on horizontal surfaces. Hence, we use the 2D laser scanner on the mobile base as the main sensor for navigation. We acquire 2D occupancy maps of the environment using simultaneous localization and mapping (gMapping, [5]). The robots localize in these 2D maps using Monte Carlo localization [6]. They navigate to goal poses by planning obstacle-free paths in the environment map, extracting waypoints, and following them.

Obstacle-free local driving commands are derived from paths that are planned towards the next waypoint in a local collision map. We incorporate 3D measurements of all distance sensors of our robots. The point measurements are maintained in an ego-centric 3D map and projected into a 2D occupancy grid map for efficient local planning.

4.2 Object Perception and Manipulation

Grasping objects from flat surfaces is a fundamental capability for which we developed efficient object detection and grasping methods [7]. Our object detection approach finds objects on planar segments and processes 160×120 range images at about $20\,\mathrm{Hz}$. It relies on fast normal estimation using integral images and efficient RANSAC plane estimation. Points above detected planes are clustered to objects. We consider two kinds of grasps on objects: top grasps that approach low objects from above and side grasps that are suitable for vertically elongated objects such as bottles or cans. We plan grasps by first computing grasp candidates on the raw object point cloud as perceived by the RGB-D camera. The grasp candidates are filtered for collisions during the execution of the grasping motion and are ranked to find the best grasp according to several convenience and stability criteria. The best grasp is finally executed using a parametrized motion primitive for either kind of grasp.

Our robots recognize objects by matching SURF interest points [8] in RGB images to an object model database and by enforcing spatial consistency between the features. In addition to the SURF feature descriptor, we store feature scale, feature orientation, relative location of the object center, and orientation and length of principal axes in the model. During recall, we efficiently match features between an image and the object database according to the descriptor using kd-trees. Each matched feature then casts a vote to the relative location, orientation, and size of the object. We consider the relation between the feature scales and orientation of the features to achieve scale- and rotation-invariant voting. When unlabeled object detections are available through planar RGB-D segmentation (see above), we project the detections into the image and determine the identity of the object in these regions of interest.

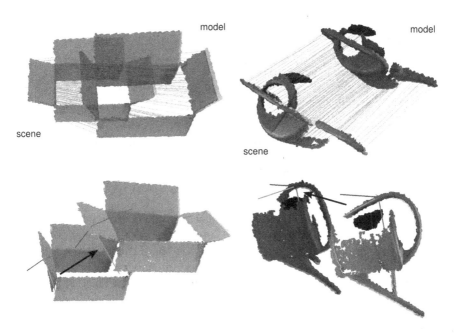

Fig. 2. Deformable registration for object manipulation skill transfer. Top: Deformable registration between two differently shaped watering cans and deformed instances of a box. Bottom: Local transformations estimated at selected points between the objects.

Our robots also support placing objects on planar surfaces and throwing objects into trash bins.

4.3   Object Manipulation Skill Transfer

In previous years, we showcased object perception and manipulation based on the tracked pose of the objects such as cooking plates, chairs, and watering cans. The manipulation skills were tailored to specific instances of the object. This year, we have shown how manipulation capabilities designed for a specific object instance can be transfered to other instances of the same class but with different shape. For this purpose, we developed efficient deformable registration between the shape of a perceived object with a model object (see Fig. 2).

We propose a multi-resolution extension to the coherent point drift (CPD) method [9] to deformably register RGB-D images efficiently. Instead of processing the dense point clouds of the RGB-D images directly, we utilize multi-resolution surfel maps (MRSMaps) [10] to perform deformable registration on an aggregated image representation. This image representation stores the joint color and shape statistics of points within 3D voxels (coined surfels) at multiple resolutions in an octree. The maximum resolution at a point is limited proportional to its squared depth according to the disparity-dependent noise of the RGB-D camera. In effect, the map exhibits a local multi-resolution structure which well reflects

  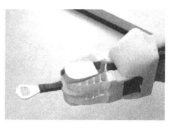

Fig. 3. Tool Adapters. We designed special adapters for tools to establish stable grasps that resist forces and torques in any direction. Left: Adapter attached to bottle opener. Center: Cosero's gripper design. Right: Grasp on the bottle opener adapter.

the accuracy of the measurements and compresses the image from 640×480 pixels into only a few thousand surfels. We also support the aggregation of multiple images into a single MRSMap, in effect overlaying the local multi-resolutions of the individual images.[1] We further improve the run-time performance of the CPD algorithm by aligning maps from coarse to fine resolutions. The registration on finer resolutions is initialized from the result on the coarser one. In addition to depth, we also utilize cues such as color and contours.

For the integrated skill transfer demonstration, we first segment the object of interest in the RGB-D image using techniques such as support-plane segmentation [7]. The RGB-D image segment is then transformed into a MRSMap and a reference object model MRSMap is aligned with the image. The grasp poses and motion trajectories are defined in terms of local coordinate frames relative to the object's reference frame. We assume that the poses and trajectories are close to the reference object's surface, and, hence, we find the local rigid transformation from the reference object towards the image segment. Finally, the motions are executed according to the transformed grasp and motion trajectories. Further details can be found here [11].

4.4   Tool Use

We implemented skills for using a bottle opener and a pair of tongs in a barbecue scenario. For a firm grip on the tool that can also resist torques along the finger direction, we augment the tools with specialized adapters (see Fig. 3). When the gripper is closed on the adapter, the fingers bend around the shape of the adapter and establish form closure. The ridge in the center of the adapter fits between the space of the fingers. It fixates the adapter for exerting torques in pitch direction. For some tools such as pairs of tongs, the opening of the gripper is also used to operate the tool. To create form closure with the fingers at various opening angles of the fingers, the adapters are equipped with flat springs for each finger.

---

[1] Our MRSMap implementation is available open-source from
http://code.google.com/p/mrsmap/

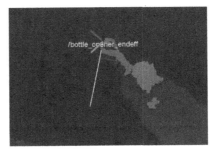

Fig. 4. Tool perception. For the opening of a bottle with a bottle opener tool, Cosero perceives the tip of the bottle and the tip of the tool.

Simple open-loop control is not feasible for tool use due to several sources of imprecisions. Firstly, an exact calibration between the robot's sensors and its end effector may not be known. Also, the pose of the tool in the gripper or the manipulated object cannot be assumed to be known precisely. We therefore implemented perception of the tips of the tool and the manipulated object using the head-mounted RGB-D camera (see Fig. 4). During manipulation, Cosero looks at the tool and the manipulated object, segments the objects from the surrounding using our efficient segmentation method (see Sec. 4.2), and detects the endings of the objects in the segments. For grasping sausages from a plate or a barbecue, we segment the sausages using plane segmentation and adapt the grasping motion to the position and orientation of the sausages. An adaptive motion then uses this information to perform the skill.

5   Human-Robot Interaction

A key prerequisite for a robot that engages in human-robot interaction is awareness of the whereabouts of people in its surrounding. We combine complementary information from laser scanners and vision to continuously detect and keep track of people. Using the VeriLook SDK, we implemented a face enrollment and identification system.

Our robots interact naturally with humans by means of speech, gestures, and body language. For speech, we use the Loquendo SDK. Its speech synthesis supports colorful intonation and sounds natural. Loquendo's speech recognition is speaker independent and is based on predefined grammars that we attribute with semantic tags for natural language understanding. The robots also support the interpretation [12] and synthesis of gestures.

Physical interaction between users and the robot occurs, for instance, when handing objects over, or when collaboratively working with objects. A key feature for this kind of interaction is compliant control of the arms [13,14]. When handing an object over to the user, the robot keeps the end effector compliant in the forward and upward direction such that the user can reach for the object and pull it. The robot detects the pulling and releases the object.

5.1 Specification of Tasks by Spoken Dialogue

Our robots support spoken dialogues for specifying complex commands that sequence multiple skills. The ability to understand complex speech commands, to execute them, to detect failures, and to plan alternative actions in case of failures is assessed in the *Enduring General Purpose Service Robot* test.

We decompose skill-level capabilities into a set of primitive skills that each are attributed only with a single object and/or location. The skill navigate_to_location, for example, depends on a goal location, while fetch_object_from_location both is determined through a target object and an object location.

The robot knows a set of specific objects with attached object labels which are used in spoken commands. Known specific objects are included in the visual object recognition database. It is also possible to define an unspecific object using labels such as "unknown", "some object", or the label of an object category (e.g., "tool"). If multiple skills with object references are chained, the reflexive pronoun "it" refers to the last object that occurred in the task command. Hence, objects are referred to by labels and may have the additional attributes of being specific, unspecific, and reflexive. Persons are handled in a similar way, but the notion of a person category is not included in our system. Our robots can enroll new persons and link their identity with their face appearance in the database of known persons.

Specific locations, location categories, unspecific locations, or location-specific adjectives (like "back") can be indicated by the user as well. We provide sets of navigation goal poses for specific locations as well as location categories. Different lists of poses are used for the purposes of object search, exploration for persons, or simple presence at a spot.

We utilize semantic tags in Loquendo's grammar specification to implement action, object, and location semantics in speech recognition. We appropriately designed the grammar such that recognition provides its semantic parse tree as a list of actions with attributed objects and locations.

Behavior control interprets the recognized semantic parse tree and sequences actions in a finite state machine. The robot executes this state machine and reports progress through speech synthesis. In case of a failure (e.g., desired object not found), the failure is recorded, the robot returns to the user, and reports the error through speech.

5.2 Convenient Remote User Interfaces

We develop handheld user interfaces to complement natural face-to-face interaction modalities [15,16]. Since the handheld devices display the capabilities and perceptions of the robot, they improve common ground between the user and the robot (see Fig. 5). They also extend the usability of the robot, since users can take over direct control for skills or tasks that are not yet implemented with autonomous behavior. Finally, such a user interface enables remote interaction with the robot, which is especially useful for immobile persons.

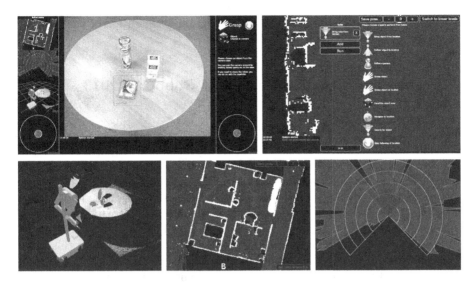

Fig. 5. Handheld user interface. Top left: Complete GUI with a view selection column on the left, a main view in the center, and a configuration column on the right. Top right: The user composes a task as several skills in a sequence by selecting skills from a list and configuring the involved objects and locations. Lower right: 3D external view. Lower middle: The navigation view displays the map, the estimated location, and the current path of the robot. Lower right: The sensor view displays laser scans and the field-of-view of the RGB-D camera in the robot's head.

The user interface supports remote control of the robot on three levels of autonomy. The user can directly control the drive and the gaze using joystick-like control UIs or touch gestures. The user interface also provides selection UIs for autonomous skills such as grasping objects or driving to locations. Finally, the user can configure high-level tasks such as fetch and delivery of specific objects.

The user interface is split into a main interactive view in its center and two configuration columns on the left and right side (see Fig. 5, top). One view displays live RGB-D camera images with object perception overlays (Fig. 5, top left). The user may change the gaze of the robot by sweep gestures, or select objects to grasp. A further view visualizes laser range scans and the field-of-view of the RGB-D camera (Fig. 5, bottom right). The navigation view shows the occupancy map of the environment, the pose of the robot, and its current path (Fig. 5, bottom center). Finally, we also render a 3D external view (Fig. 5, bottom left). The user can also choose to execute navigation or object manipulation skills. For fetching an object, for instance, the user either selects a specific object from a list, or chooses a detected object in the current sensor view.

High-level tasks such as fetch and delivery can be configured in a task spec-ification UI (Fig. 5, top right). The task-level teleoperation UI is intended to provide the high-level behavior capabilities of the robot. These behaviors se-quence multiple skills in a finite state machine. The user can compose actions, objects, and locations similar to a speech-based implementation of the parsing of complex speech commands as detailed in Sec. 5.1.

6   Results and Lessons Learned at RoboCup 2013

6.1   Competition Results

On the first competition day, our robots Cosero and Dynamaid [4] went through the inspection and registration test. Together with a poster presentation of the team, we achieved the highest score in this test. In the *Follow Me* test, Cosero lost track of the guide when a person was closely walking in front of the guide, such that we could not score. Cosero demonstrated that it could recognize the names of persons in the *Cocktail Party* test. In *Clean Up*, Cosero found two objects and brought one of it to its correct place. It got the highest score in this test, while only two out of 21 teams could score at all. The new *Emergency Situation* test was tackled by many teams. Here, Cosero achieved the third best score by finding a standing person, asking, if he requires help, and guiding him to the exit. Stage 1 was concluded by the *Open Challenge*, in which Cosero demonstrated the usage of a watering can, while we had predefined grasps and watering motion for another instance of cans. Cosero estimated the shape deformation between the two object instances, and transfered the manipulation skill to the new can. It grasped the watering can from a table, watered a plant, and placed the watering can back. Afterwards, Cosero was intended to push a chair to its location, but could not reach a good position for grasping it. The demonstration was well received by the jury of team members, and received 1361 points, only 11 points behind the team WrightEagle@Home from China. After Stage 1, we were closely behind team WrightEagle@Home which had about 2 % advantage in points.

Stage 2 began with the *Enduring General Purpose Service Robot* test. Cosero understood a complex speech command, fetched an object, and delivered it. A second command with missing information was also understood, for which it asked questions to retrieve the missing information. We achieved the best score in the test. In the *Restaurant* test, Cosero was shown the environment and the location of food and drinks, which it later found again. Cosero gained the second best score, behind WrightEagle@Home. The last test in Stage 2 was the *Demo Challenge*. We demonstrated a care scenario, in which the robot extended the mobility of a user with its mobile manipulation capabilities. Cosero moved a chair to its location and attempted to open a bottle. We also showed our teleoperation interface to the jury.

We reached the finals on the second place with 98% of the score of team WrightEagle@Home. In the final, Cosero demonstrated tool use. He used a pair of tongs to cook a sausage on a barbecue and a bottle opener to open a beer that he served to a person. This demonstration convinced the high-profile jury which awarded the highest number of points. Together with the results of Stage 1 and 2, the final normalized score was 99 points for NimbRo, followed by WrightEagle@Home (China, 86 points) and TU Eindhoven (Netherlands, 73 points).

6.2   Lessons Learned

Developing complex service robot systems as tested in the RoboCup@Home league is a challenge. Many individual skills must be integrated into a complete system. Implementing the procedures of the predefined tests is not solely programming a sequence of skills, but the unforeseeable setup of the arena and the unpredictable behavior of persons requires robust navigation, mobile manipulation, and human-robot interaction skills as well as robust high-level behavior control.

The competition tests the generality of the approaches by the teams. Typically, a test fails, if an assumption made during the development is not fulfilled in the actual competition scenario. For instance, in the *Follow Me* test, we did not foresee that the distracting person that crosses between the robot and the guide, would stand very close at the guide and both persons would stop. This made our person tracking approach fail which relies on tracking laser scan features. In future work, we plan to develop person detection and segmentation in RGB-D images, which still is a research challenge in the above situation.

In the *Cocktail Party* test, the place where the robot would get familiar with the persons in the scenario was not well chosen by us. After the persons introduced themselves, the implemented solution was expecting the persons to walk out of sight of the robot to trigger the fetching of the drinks. This problem could be resolved with a better design of the high-level behavior of the robot.

The most complex test is the *Enduring General Purpose Service Robot* test. Our performance demonstrates that we developed a complete system that integrates complex speech processing and person awareness with indoor navigation and mobile manipulation skills. It is a challenge to design dialogue engines and high-level behavior control that can react to all kinds of complex speech commands which involve the many possible actions, objects, locations, and speech expressions.

Our robots performed well in many tests. The fact that our robots won the competition for the third time in a row makes apparent that this is not a coincidence, but we developed a well-balanced and comparably robust system.

7   Conclusions

In this paper, we presented the contributions of our winning team NimBRo to the RoboCup@Home competition 2013 in Eindhoven. We advanced the mobile manipulation capabilities of our robots and demonstrated generalization of object manipulation skills and tool use. Our robots scored in many tests of the competition and came in second to the finals, only a few points behind the leading team WrightEagle@Home. In the final, our robot Cosero convinced the high profile jury with its tool-use skills demonstrated in a barbecue scenario, and won the competition. In future work, we aim at further increasing the robustness and generality of our approaches to navigation, object manipulation, and human-robot interaction.

Acknowledgments. We thank our student team members Marcel Brandt, Rainer Duppre, and Johann Heinrichs for their support.

References

1. van der Zant, T., Wisspeintner, T.: RoboCup X: A proposal for a new league where RoboCup goes real world. In: Bredenfeld, A., Jacoff, A., Noda, I., Takahashi, Y. (eds.) RoboCup 2005. LNCS (LNAI), vol. 4020, pp. 166–172. Springer, Heidelberg (2006)
2. Wisspeintner, T., van der Zant, T., Iocchi, L., Schiffer, S.: RoboCup@Home: Scientific competition and benchmarking for domestic service robots. Interaction Studies 10(3), 393–428 (2009)
3. Elara, M.R., Holz, D., Iocchi, L., Mahmoudi, F., del Solar, J.R., Stückler, J., Sugiura, K., Wachsmuth, S., Xie, J., van der Zant, T.: RoboCup@Home: Rules & regulations (2013), http://www.robocupathome.org/rules
4. Stückler, J., et al.: NimbRo@Home: Winning team of the RoboCup@Home competition 2012. In: Chen, X., Stone, P., Sucar, L.E., van der Zant, T. (eds.) RoboCup 2012. LNCS (LNAI), vol. 7500, pp. 94–105. Springer, Heidelberg (2013)
5. Grisetti, G., Stachniss, C., Burgard, W.: Improved techniques for grid mapping with Rao-Blackwellized particle filters. IEEE Trans. on Rob. 23(1), 34–46 (2007)
6. Fox, D.: KLD-sampling: Adaptive particle filters and mobile robot localization. In: Advances in Neural Information Processing Systems (NIPS), pp. 26–32 (2001)
7. Stückler, J., Steffens, R., Holz, D., Behnke, S.: Efficient 3D object perception and grasp planning for mobile manipulation in domestic environments. In: Robotics and Autonomous Systems (2012)
8. Bay, H., Ess, A., Tuytelaars, T., Van Gool, L.: Speeded-up robust features (SURF). Computer Vision and Image Understanding 110(3), 346–359 (2008)
9. Myronenko, A., Song, X.: Point set registration: Coherent point drift. IEEE Trans. on PAMI 32(12), 2262–2275 (2010)
10. Stückler, J., Behnke, S.: Multi-resolution surfel maps for efficient dense 3D modeling and tracking. Visual Communication and Image Representation (2013)
11. Stückler, J., Behnke, S.: Efficient deformable registration of multi-resolution surfel maps for object manipulation skill transfer. In: IEEE International Conference on Robotics and Automation, ICRA (2014)
12. Droeschel, D., Stückler, J., Behnke, S.: Learning to interpret pointing gestures with a time-of-flight camera. In: Proceedings of the 6th ACM International Conference on Human-Robot Interaction, HRI (2011)
13. Stückler, J., Behnke, S.: Following human guidance to cooperatively carry a large object. In: Proceedings of the 11th IEEE-RAS International Conference on Humanoid Robots (Humanoids), pp. 218–223 (2011)
14. Stückler, J., Behnke, S.: Compliant task-space control with back-drivable servo actuators. In: Röfer, T., Mayer, N.M., Savage, J., Saranlı, U. (eds.) RoboCup 2011. LNCS, vol. 7416, pp. 78–89. Springer, Heidelberg (2012)
15. Muszynski, S., Stückler, J., Behnke, S.: Adjustable autonomy for mobile teleoperation of personal service robots. In: Proc. of the IEEE International Symposium on Robot and Human Interactive Communication, RO-MAN (2012)
16. Schwarz, M., Stückler, J., Behnke, S.: Mobile teleoperation interfaces with adjustable autonomy for personal service robots. In: 9th ACM/IEEE International Conference on Human-Robot Interaction, HRI (2014)

How to Win RoboCup@Work?

The Swarmlab@Work Approach Revealed

Sjriek Alers[1], Daniel Claes[1], Joscha Fossel[1], Daniel Hennes[2], Karl Tuyls[1],
and Gerhard Weiss[1]

[1] Maastricht University, P.O. Box 616, 6200 MD Maastricht, The Netherlands
[2] European Space Agency, ESTEC, Keplerlaan 1, Noordwijk, The Netherlands

Abstract. In this paper we summarize how the Swarmlab@Work team
has won the 2013 world championship title in the RoboCup@Work league,
which aims to facilitate the use of autonomous robots in industry. The
various techniques that have been combined to win the competition come
from different computer science domains, entailing learning, (simultaneous) localization and mapping, navigation, object recognition and object manipulation. While the RoboCup@Work league is not a standard
platform league, all participants used a (customized) Kuka youBot. The
youBot is a ground based platform, capable of omnidirectional movement and equipped with a five degree of freedom arm featuring a parallel
gripper.

1 Introduction

RoboCup@Work is a recently launched competition which aims at flexible robotic
solutions in work-related scenarios. The leagues vision[1] is to "foster research
and development that enables use of innovative mobile robots equipped with advanced manipulators for current and future industrial applications, where robots
cooperate with human workers for complex tasks ranging from manufacturing,
automation, and parts handling up to general logistics".

In contrast to the well developed robotic solutions deployed in common mass-
production environments, RoboCup@Work targets smaller companies in which
flexible multi-purpose solutions are required, which are not yet available in industry. Example tasks are finding and acquiring parts, transportation to and
from dynamic locations, assembly of simple objects etc. From these industrial
goals various scientific challenges arise. For example, in perception, path planning, grasp planning, decision making, adaptivity and learning, as well as in
multi-robot and human-robot cooperation.

The competition is relative new and started in 2012. Due to the fact that the
it is still in a startup phase and not much of source code of last years event
was made public, no extensive resources regarding RoboCup@Work competition
where available to start from. From this year on all teams agreed to release their
codebase so that for next events the teams can improve their capabilities and
new teams are able to catch up with the competition.

---

[1] http://robocupatwork.org/

S. Behnke et al. (Eds.): RoboCup 2013, LNAI 8371, pp. 147–158, 2014.
© Springer-Verlag Berlin Heidelberg 2014

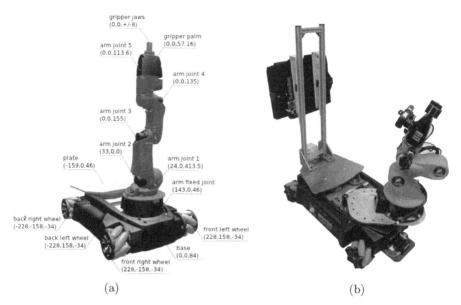

Fig. 1. (a) CAD model of a stock youBot. (b) Swarmlab@Work modified youBot.

2   Swarmlab@Work

Swarmlab[2] is the research laboratory of the Department of Knowledge Engineering at Maastricht University that focuses on designing Autonomous Systems. The general research goal and mission of the lab is to create adaptive systems and robots that are able to autonomously operate in complex and diverse environments. The Swarmlab@Work team [1] has been established in the beginning of 2013, consisting of 5 PhD candidates and 2 senior faculty members. Since then the team has won the @Work competitions of the 2013 RoboCup German Open, and the 2013 RoboCup world championship. The team's mission is (a) to apply Swarmlab research achievements in the area of industrial robotics and (b) identify new research challenges that connect to the core Swarmlab research areas: autonomous systems, reinforcement learning and multi-robot coordination.

In the remainder of this section we introduce the Swarmlab@Work robot platform that was used for the 2013 RoboCup@Work world championship. Especially, we describe the necessary hardware modifications that were made to adapt this standard platform to the winning configuration.

2.1   youBot Platform

The youBot is an omni-directional platform that has four mecanum [8] wheels, a 5 DoF manipulator and a two finger gripper. The platform is manufactured

---

[2] http://swarmlab.unimaas.nl

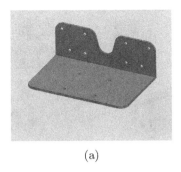

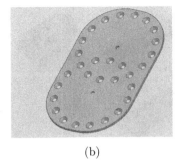

(a)   (b)

Fig. 2. (a) a CAD model of the laser mounts, (b) a CAD model of the arm extension plate

by KUKA[3], and is commercially available at the youBot-store[4]. It has been designed to work in industrial like environments and to perform various industrial tasks. With this open-source robot, KUKA is targeting educational and research markets. Figure 1a shows a model of the stock youBot.

The omnidirectional base is approximately 580x380 mm in size and it's height is $140mm$. It weighs 20 kg and it can carry a payload of 20 kg. Due to the fact that the platform has four mecanum wheels it has the possibility to move in every direction with a maximum speed of 0.8 m/s. The chassis is build up of $3mm$ thick stainless steel plates. Inside the base, a mini ITX Atom based PC-Board with 2 GB RAM is located together with a 32 GB SSD Flash drive. This computer enables onboard processing and control of the robot. Inside the base also the different motors and motor-driver boards are located as well as the power charging and power-distributing unit. The robot is powered by a 24Volt, 5 Ampere maintenance free rechargeable lead acid batteries. All hardware modules of the robot communicate internally over real-time EtherCat [9].

The youBot comes with a 5-degree-of-freedom arm that is made from casted magnesium, and has a 2-degree-of-freedom gripper. The arm is 655 mm high, weights 6.3 kg, and can handle a payload of up to 0.5 kg. The working envelope of the arm is 0.513 m^3, and is is connected over EtherCat with the internal computer, and has a power consumption limit of 80 Watts. The gripper has two detachable fingers that can be remounted in different configurations. The gripper has a stroke of 20mm and a reach of 50 mm, it opens and closes with an approximate speed of 1 cm/s.

In order to meet the requirements we demand from the youBot platform to tackle the challenges, we made a number of modifications to the robot. In this paragraph we describe which parts are modified and why these modifications are a necessity for our approach. Figure 1b shows the modified Swarmlab@Work youBot setup.

For perceiving the environment, two Hokuyo URG-04LX-UG01 light detection and ranging (LIDAR) sensors are mounted parallel to the floor on the front

---

[3] http://kuka.com
[4] http://youbot-store.com/

Fig. 3. Manipulation objects, from top left to bottom right: M20_100, R20, V20, M20, M30, F20_20_B, F20_20_G, S40_40_B, S40_40_G

and back of the robot. For mounting these LIDAR sensors a special laser mounting bracket was designed, see Figure 2a. In contrast to the thin aluminium stock brackets, these custom brackets are constructed from 4mm thick steel in order to prevent deformation, reduce vibration and to ensure constant horizontal alignment.

To extend the reach of the robot-arm in respect to the chassis, we designed an extension plate of 5 mm thick aluminium, the CAD model of this plate is shown in Figure 2b. This plate can extend the arm towards the bounds of the chassis, and is designed to be a multi-purpose extension for the youBot arm. This means that the plate can be mounted onto the chassis in angular steps of 22.5 degrees, placing the arm in the optimal positions for the given tasks. By using this plate we can gain an additional reach of the robot arm of 12.5 cm. A secondary advantage of this plate is that also the arm can be mounted onto this plate in steps of 22.5 degrees. This feature is very useful to optimize the turning position of the arm regarding its dead angle. While mounting this plate, we elevated the arm position approximately 2.5 cm, which also results in a reach advantage of about 2 cm. The maximum reach is achieved when the arm is stretched horizontally, i.e. the second joint is aligned such that the rest of the arm is parallel to the ground. To grasp an object from above, the fourth joint has to be perpendicular to the ground. In this configuration, the fingers extend further down than the mounting point of the arm by 7.5 cm. Our configuration mounts the arm at about 15.5 cm, i.e. the optimal picking height is at 8 cm, which is close to the 10 cm as specified as height of the service areas where the objects have to be picked up and placed.

In order to detect and recognize manipulation objects, an ASUS Xtion PRO LIVE RGBD camera is attached to the last arm joint. This camera is mounted, so that it faces away from the manipulator, as can be seen in Figure 7a. The main idea of this rather odd mounting position is that we want to use the RGB-D data of this camera, which is only available for distances larger than ∼0.5m. This is why we use a special pre-grip pose, in which we scan the area and try to look at the objects from above. A useful byproduct of this mounting technique is that grasped objects do not block the view of the camera, which typically happens with manipulator facing cameras. The downside is that it prevents the

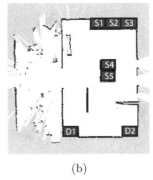

(a)   (b)

Fig. 4. (a) 2013 arena during BNT run with extra obstacles. (b) Map of the arena. Annotated with source (S1-S5) and destination (D1 and D2) areas.

use of visual servoing. However, we found that the arm can be steered precisely enough to make this disadvantage negligible.

The base computer is upgraded from the stock ATOM based architecture to an Intel i7 CPU and is powered by a dedicated 14.8V / 5 Ah Lithium Polymer battery pack that is charged and monitored by an OpenPSU power unit. For cooling we mounted an additional fan in the base of the robot. The base computer is supported by an i5 notebook, which is mounted on a rack at the backside of the robot.

For safety reasons the robot is equipped with an emergency stop button, that stops all robot movement, without affecting the processing units, so when the stop is released the robot can continue its movement without having to re-initialize it again.

In this section we described the hardware modifications needed to optimally execute the required RoboCup@Work tasks, in the next section these tasks will be describes in more detail, as well as the methods that we used to solve the specific tasks.

3   RoboCup@Work

In this section, we introduce the various tests of the 2013 RoboCup@Work world championship. Additionally, we sketch the different capabilities that we developed and explain briefly how these techniques are combined to tackle the different tests. Note that while the tests are slightly different for various levels of difficulty, we only give a description for the difficulty chosen by Swarmlab@Work. For a more detailed description we refer the reader to the RoboCup@Work rule book[5]. In the following the term service area refers to a small table from which typically objects have to be grasped, while destination area refers to a small table on which objects have to be placed. Figure 5a shows our robot performing a grasp in one of the service areas. A list of manipulation objects used in 2013

---

[5] http://robocupatwork.org/resources.html

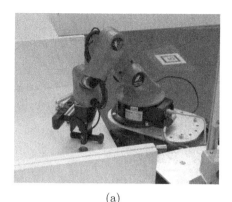
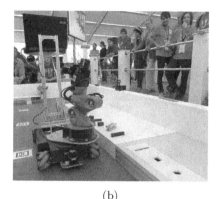

(a)   (b)

Fig. 5. (a) Grasping the M20_100 in a service area. (b) PPT setup.

is given in Figure 3. In the arena, there are five service areas and two destination areas as annotated in Figure 4b. A picture of the 2013 arena is shown in Figure 4a, in which two extra obstacles are placed.

Basic Navigation Test. The purpose of the Basic Navigation Test (BNT) is testing mapping, localization, path planning, obstacle avoidance and motion control. The robot has to autonomously enter the (known) arena, visit multiple way-points, and exit the arena again. Each way-point has to be accurately covered by the robot for a given duration. Additionally, obstacles may be spawned randomly in the arena, where a penalty is given if the robot collides with an obstacle.

Basic Manipulation Test. In the Basic Manipulation Test (BMT), the robot has to demonstrate basic manipulation capabilities such as grasping and placing an object. For this purpose three target and two decoy objects are placed on a service area, and the robot has to single out the three actual target objects and move them to any other area. Implicitly this also tests wether the object recognition works sufficiently well. Since navigation is not a part of this test, the robot may be placed at any location before the test starts.

Basic Transportation Test. The Basic Transportation Test (BTT) is a combination of the BNT and BMT. Here, certain objects have to be picked up from various service areas and transported to specific destination areas. Similar to BNT and BMT both decoy objects and obstacles are placed on the service areas and in the arena, respectively. The robot has to start and end the BTT outside of the arena. In addition to the challenges individually posed by BNT and BMT, the robot now has to approach multiple service and destination areas within a certain time window, hence for optimal performance determining the optimal route and payload configuration is necessary.

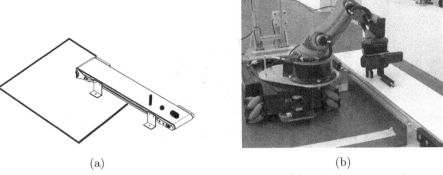

(a)   (b)

Fig. 6. (a) Schematic of the conveyor belt setup. (b) Our CBT approach.

Precision Placement Test. The Precision Placement Test (PPT) consist of grasping objects (see BMT), followed by placing them very accurately in object-specific cavities. Every object has one specific cavity for vertical, and one specific cavity for horizontal placement. Each cavity has the same outline as the object plus 10 percent tolerance. Figure 5b show the PPT destination area with some sample cavities.

Conveyor Belt Test. In the Conveyor Belt Test (CBT) the robot has to manipulate moving objects. For this purpose the robot has to autonomously approach the conveyor belt, grasp the object(s) which pass by, and place them on its rear platform. Due to the increased base difficulty, no decoy object have been used in the 2013 competition in this test. Figure 6a show the conveyor belt, which moves with approximately 3 cm/s.

4   Approach

In this section, the different techniques used to tackle the above mentioned tests are explained. We developed different modules for many different capabilities, e.g. basic global navigation, fine-positioning, object recognition, inverse kinematics of the arm, etc. We also explain our various recovery methods. By combining these capabilities in state-machines we are able to solve the tasks specified above.

Mapping and Localization. One of the most crucial capabilities of an autonomous agent is to localise itself efficiently in a known environment. To achieve this, we use gmapping [5] to build a map of the arena beforehand. The map of this years arena is shown in Figure 4b. After the map is recorded it can be used by AMCL [3] for efficient global localization.

Navigation. Another necessary capability of the robot is to navigate in the known environment without colliding with obstacles. The map created with

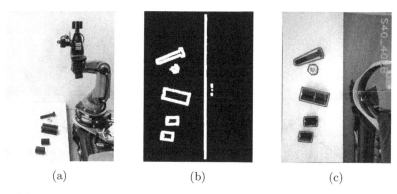

(a)   (b)   (c)

Fig. 7. (a) Pre-grip scan position. (b) Pre-processing of the image. (c) Detected objects, classification and grasp position of the closest object.

gmapping is used for the basic global navigation. The global path is computed by an A* algorithm and is then executed using a dynamic window approach [4] trajectory rollout for the local path planning. This planner samples different velocities and ranks them according to the distance to the goal and the distance to the path, while velocities that collide with the environment are excluded. We tuned the trajectory rollout in such a way that it incorporates the omni-directional capabilities of the robot and also included scores for the correct destination heading, when being in close distance to the goal.

Fine-Positioning. The navigation is very well suited to navigate between larger distances from different positions in the map. But the accuracy of the navigation and localization is not high enough to navigate with high reproducibility to previously known goals. Thus for aligning to those previously known locations another technique is needed, since we want to be as close as possible to the manipulation areas as possible. Also for the basic navigation test, the markers have to be covered exactly, which is not always the case when using only AMCL for localization. Thus, we implemented ICP based scan registration [7] for fine grain positioning. This techniques records a laser scan at a certain position. When the robot is close to this position, the differences between the current laser scan and the registered one are calculated into a direction which is used to steer the robot to the old position with very high accuracy. By using correspondence rejectors we make sure that slight changes in the environment, e.g. caused by humans standing around the robot, do not interfere with the scan registration accuracy.

Object/Hole Recognition. Besides all the navigation tasks, object detection and recognition is crucial to be able to interact with the environment, i.e. picking up objects and placing them in the correct target locations. We use the openCV-library[6] to detect the objects. An adaptive threshold filter is applied to the input

---

[6] http://opencv.org

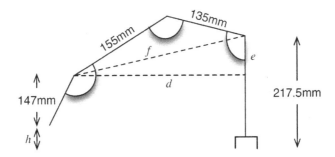

Fig. 8. Simple inverse kinematics: d and h are the grip distance and height, relative to the mount point of the arm. By always gripping from a top down position, e and f can be calculated and by that we can determine all angles for the joints.

image. Afterwards the image is converted into a black and white image and this is used to detect the contours of the objects as shown in Figure 7b. We use various features of the detected objects, e.g., length of principal axis, average intensity and area, and use a labeled data-set that we created on the location to train a J4.8 decision tree in weka [6] for the recognition of the objects. This decision tree is then used for the online classification of the objects. Figure 7c shows the detection in a service area.

Inverse Kinematics of the Arm. In order to manipulate the detected objects, the various joints of the arm have to be controlled such that the objects is grasped correctly. We implemented a simple inverse kinematics [10] module to calculate the joint values for any top-down gripping point that is in the reach of the robot. Since we are gripping from a top-down position, the inverse kinematics can be solved exactly, when we fix the first joint such that it is always pointing in the direction of the gripping point as shown in Figure 8. Then the remaining joints can be calculated in a straight forward manner, by solving the angles of a triangle with three known side lengths, since we know the distance of the grip and also the lengths of all the arm-segments. Since the position-reproducibility of the arm is in sub millimetre order, this proved to be sufficient for performing highly accurate grasp and place trajectories.

Recovery Methods. Of course, in robotics many things can go wrong, so we include various recovery methods. The navigation can easily get stuck, since the robot is tuned to drive very close to obstacles. Therefore, we implemented a force field [2] recovery. More specifically, the robot moves away from every obstacle that is picked up by the laser scan for a couple of seconds and the navigation is restarted. When the object recognition misclassifies an object, which happens especially with the two silver aluminium profiles, it can happen that the robot tries to pick up object S40_40_G. This is physically impossible, since the gripper cannot open sufficiently wide. Thus, we measure the voltages in the arm, and as

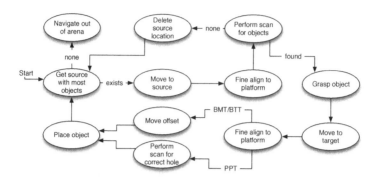

Fig. 9. The simplified global state-machine that is used for the BMT, BTT and PPT competitions, without recovery states

soon as a certain voltage is exceeded, we recover the arm to an upwards position and back up a couple of states.

4.1 State Machine

For the different tests, we combine the various capabilities explained in the previous section in different state machines to complete the tasks. For the BNT, we do not need any of the manipulation techniques, so we have a very small state machine that combines the navigation and the fine-positioning. More specifically, we use the navigation to get close to the position and then the fine-positioning moves the robot exactly to the asked position. If the navigation fails, the force-field recovery is activated. Afterwards it retries to move to the current location. When it fails more than a fixed number of times, the current goal is skipped.

In principle the BMT, BTT and PPT tasks do not differ very much. For instance, the BMT is basically a BTT, in which only one source and one target location are used and the PPT is a BMT in which the objects have to be placed in a specific location instead of anywhere on the platform. Thus for all these other tests (excluding CBT), we created a global state-machine that is shown in its simplified form in Figure 9. For sake of simplicity, the recovery connections are left out.

The general idea of the state machine is to move to the source location which has the most objects that we have to pick up. Then we perform a scanning motion along the platform, while looking for any object that we need. If we do not find anything we continue to the next source location or end, if we have visited all source locations specified. On the other hand, when an object is found, the robot grasps it and moves to the target location specified for this object. If it is a BMT or BTT, we move a certain offset to ensure that the objects are not stacked and place the object. If it is a PPT, we perform a scanning motion, however now looking for the matching hole for the object we are currently transporting. Afterwards, we again move to the current source location and repeat until everything

Table 1. Scores for the different tests

Place	Team	BNT	BMT	BTT	PPT	CBT	FINAL	Total
1	Swarmlab@Work	616	600	600	600	200	337.5	2953.5
2	LUHbots	490	225	500	112.5	75	400	1802.5
3	WF Wolves	211.25	0	37.5	131.25	0	206.25	586.25
4	RoboErectus@Work	243.75	37.5	0	150	0	37.5	468.75
5	b-it-bots	280	0	0	187.5	0	0	467.5

is transported. For almost every state we have a recovery state connected, that kicks in if the arm or the navigation fails. These recovery states are then usually connected to the previous state, which is re-initiated for a number of times and skipped as soon the number of retries are exceeded.

For the CBT, we did not have the facilities to prepare anything beforehand and there was only very little preparation time to solve the task. We came up with a straight forward solution. The robot aligns itself next to the conveyer belt and prepares the gripper such that it is hovering over the belt. An additional camera attached to the side of the arm searches for the object and as soon as the centre of the object is detected within a certain window the gripper starts to close. Figure 6b shows the approach.

5   Competition

Table 1 shows the final scores for the 2013 RoboCup championship, where "final" refers to a BTT with six objects instead of three.

While scoring highest in most tests, Swarmlab@Work did not perform outstandingly well in the finals. The reason for that is twofold: (a) We are the only team which did not use the back platform to store multiple objects, but instead transport only one object at a time. (b) Due to the lack of gripper feedback we always double check wether an object has indeed been gripped/transported. Both (a) and (b) resulted in insufficient time to finish the final runs.

Additionally, one can see that the average performance over all teams in the CBT is very low. This can be explained by the difficulty of the test, and also the lack of pre-competition preparation time since most teams do not have access to a conveyor belt outside the competition. Videos and further information can be found on our website[7].

6   Future Work

The Swarmlab@Work code will be released on Github[8]. For future competitions we plan on improving the object recognition by adding state of the art techniques, such as deep learning algorithms in order to decrease vulnerability to

---

[7] http://swarmlab.unimaas.nl/robocup_at_work/robocup-2013/

[8] http://github.com/swarmlab

lighting conditions. We also hope that this will allow for a more perspective stable object recognition. Furthermore, we are currently working on a learning by demonstration approach to allow for learning new arm trajectories on the fly. This might enable grasping various 'more difficult to grasp objects', e.g. glasses. It will also help to cope with future RoboCup@Work tests, which are yet to be introduced, e.g. assembly tasks. Combining the two previously mentioned approaches, we also hope to overcome the poor CBT performance.

Hardware-wise we plan to upgrade the gripper opening speed and distance in order to increase the range of graspable objects that the youBot can manipulate. We also plan on mounting the arm on a rotating extension plate, so that the arm can be turned towards the optimal grasp position. These modifications will unlock challenges with higher difficulty levels, which require amongst others the ability to grasp all available @Work objects - a capability we lacked in 2013. The use of the backplate to collect objects on, and thus enabling to transport more objects efficiently, will also be required in the next RoboCup challenge.

References

1. Alers, S., Bloembergen, D., Claes, D., Fossel, J., Hennes, D., Tuyls, K., Weiss, G.: Swarmlab@work team description paper. Robocup (2013)
2. Chong, N.Y., Kotoku, T., Ohba, K., Tanie, K.: Virtual repulsive force field guided coordination for multi-telerobot collaboration. In: Proceedings of the 2001 IEEE International Conference on Robotics and Automation (ICRA 2001), pp. 1013–1018 (2001)
3. Fox, D., Burgard, W., Frank, D., Thrun, S.: Monte carlo localization: Efficient position estimation for mobile robots. In: Proceedings of the Sixteenth National Conference on Artificial Intelligence, AAAI 1999 (1999)
4. Fox, D., Burgard, W., Thrun, S.: The dynamic window approach to collision avoidance. IEEE Robotics & Automation Magazine 4 (1997)
5. Grisetti, G., Stachniss, C., Burgard, W.: Improved techniques for grid mapping with rao-blackwellized particle filters. IEEE Transactions on Robotics 23, 43–46 (2007)
6. Hall, M., Eibe, F., Holmes, G., Pfahringer, B., Reutemann, P., Witten, I.H.: The weka data mining software: an update. SIGKDD Explor. Newsl. 11(1), 10–18 (2009)
7. Holz, D., Behnke, S.: Sancta simplicitas - on the efficiency and achievable results of slam using icp-based incremental registration. In: IEEE International Conference on Robotics and Automation, pp. 1380–1387 (2010)
8. Hon, B.E.: Wheels for a course stable selfpropelling vehicle movable in any desired direction on the ground or some other base, US Patent 3,876,255 (1975)
9. Jansen, D., Buttner, H.: Real-time ethernet: the ethercat solution. Computing and Control Engineering 15(1), 16–21 (2004)
10. McCarthy, J.M.: An introduction to theoretical kinematics. MIT Press, Cambridge (1990)

Analyzing and Learning an Opponent's Strategies in the RoboCup Small Size League

Kotaro Yasui, Kunikazu Kobayashi, Kazuhito Murakami, and Tadashi Naruse

Graduate School of Information Sciences and Technology,
Aichi Prefectural University, Japan
im121007@cis.aichi-pu.ac.jp,
{koboyashi,murakami,naruse}@ist.aichi-pu.ac.jp

Abstract. This paper proposes a dissimilarity function that is useful for analyzing and learning the opponent's strategies implemented in a RoboCup Soccer game. The dissimilarity function presented here identifies the differences between two instances of the opponent's deployment choices. An extension of this function was developed to further identify the differences between deployment choices over two separate time intervals. The dissimilarity function, which generates a dissimilarity matrix, is then exploited to analyze and classify the opponent's strategies using cluster analysis. The classification step was implemented by analyzing the opponent's strategies used in set plays captured in the logged data obtained from the Small Size League's games played during RoboCup 2012. The experimental results showed that the attacking strategies used in set plays may be effectively classified. A method for learning an opponent's attacking strategies and deploying teammates in advantageous positions on the fly in actual games is discussed.

1 Introduction

Robotic soccer in the RoboCup Small Size League (SSL) involves a competition between teams of at most six robots on a 6050 mm by 4050 mm field. Two cameras are positioned 4 m above the field to collect photographs of the field every 1/60th of a second. The photographs are sent to a vision computer dedicated to image processing. The vision computer calculates the positions of the robots and the ball on the field and sends these coordinates to each team's computer. Each team then uses this information to compute the next positions each robot will move to based on their strategy. These positions are then sent to the robots via radio communication. A referee box computer, which controls the progression of the game, sends messages such as 'start throw-in', 'start corner kick', etc. to each team's computer. The game advances automatically without the intervention of any person except for the referees and the person controlling the referee box.

The moving speeds of robots in the SSL have increased year-on-year. For instance, the champion team in the RoboCup 2012 moved its robots at a maximum speed of 3.5 m/s [1], and the champion team passed the ball at speeds exceeding 4 m/s. In this environment, predicting the behavior of an opponent is very important.

S. Behnke et al. (Eds.): RoboCup 2013, LNAI 8371, pp. 159–170, 2014.

In recent SSL games, each team has tended to decide on a strategy depending on the positions and velocities of the teammates, the opponents, and the ball. The referee box's signals are also used to select a strategy. By contrast, in human soccer, each player decides his/her behavior using, in addition to the above four factors, the history of play such as "how the opponents have played during the current game." It is reasonable to expect that a robotic soccer game may also play out by selecting an appropriate strategy based on an analysis of the opponent's strategies or behaviors during the previous game plays.

In an effort to analyze and learn from an opponent's strategies, a dissimilarity function that identified differences between an opponent's deployment choices at two different instances was developed. The dissimilarity function was then extended to identifying the differences between deployment choices made during two different time intervals. The opponent's strategies were then analyzed by using this dissimilarity function. The dissimilarity matrix generated from the dissimilarity function was used to perform a cluster analysis to classify the opponent's strategies. This method was applied to the data logged during SSL games played during the RoboCup 2012. This method was shown to be able to effectively classify the attacking strategies during set plays. A method for learning an opponent's attacking strategies and deploying teammates in advantageous positions on the fly in actual games is discussed.

2   Related Work

Bowling et al. [2] proposed a method for implementing an opponent-adaptive play selection. This method, which has been used in the SSL games, is formulated in the frame of an experts problem or a k-armed bandit problem. The method selects an effective play from a playbook based on the "regret" measure[1]. The selected play is then given an appropriate reward corresponding to the result of the play: success, failure, completion, or abort. The regret measure is updated based on the accumulated reward of each play.

Trevizan and Veloso [3] have proposed a method for comparing the strategies of teams played in the SSL. In their method, a team's strategy is represented as an episode matrix, the elements of which are the means and the standard deviations of variables over the time segment S in the time series T, where T represents a game. They selected 23 variables for analysis, including the distance between each robot and the ball. A similarity function $s(\cdot, \cdot)$ was then defined using the matrices to represent the closeness between two episodes. The method was applied during defense episodes, and they discussed the closeness of the defense strategies used by the two teams.

Visser and Weland [4] proposed a method for classifying an opponent's behavior based on a decision tree using the data logged by the RoboCup Simulation League. They classified, for instance, the behaviors of a goalkeeper into three

---

[1] The regret is the amount of additional reward that could have been received if an expert selection algorithm had known which expert would be the best and had chosen that expert at each decision point[2].

categories: the goalkeeper backs up, the goalkeeper stays in the goal, and the goalkeeper leaves the goal. They also analyzed the pass behaviors of the opponents in a similar way.

3 Comparison of Strategies

In the proposed dissimilarity method, an opponent's strategies were first classified using a dissimilarity function that quantifies the difference between two deployment decisions made by an opponent at times t_1 and t_2. The dissimilarity function was defined on a field coordinate system having an origin at the center of the field, with the x-axis pointing toward the center of the opponent's goal mouth and the y-axis representing the center line.

Let $R_i(t_k)$ be the coordinates of the opponent i at time t_k $(k = 1, 2)$. Assume that m opponents are present on the field at time t_1 and n opponents at time t_2. (Robots are numbered from 1 to m at time t_1 and from 1 to n at time t_2.) The dissimilarity function d is then defined as follows:

$$d(t_1, t_2) = \min_{U \in \{U_1, U_2\}} \left\{ \min_{\sigma \in S_6} \sqrt{\operatorname{trace}(F(U)P_\sigma)} \right\}, \tag{1}$$

$$U_1 = \begin{bmatrix} 1 & 0 \\ 0 & 1 \end{bmatrix}, \; U_2 = \begin{bmatrix} 1 & 0 \\ 0 & -1 \end{bmatrix},$$

$$F(U) = [f_{ij}],$$

$$f_{ij} = \begin{cases} \|U R_i(t_1) - R_j(t_2)\|^2 & (1 \leq i \leq m, \; 1 \leq j \leq n) \\ \Delta^2 & (\text{otherwise}) \end{cases},$$

where S_6 is the symmetric group of degree six, P_σ is the permutation matrix of a permutation σ, and U_2 is the matrix describing a reflection transformation through the x-axis used to change the sign of the y-coordinate. F is a 6×6 matrix, the element f_{ij} of which is the Euclidean distance between the position of the opponent i at time t_1 and the position of the opponent j at time t_2, if robots i and j are on the field. If one (or neither) robot is not on the field, this element is assigned the value Δ^2. Δ indicates a virtual distance and is assigned a constant value here. The selection of Δ requires a careful process that will be discussed elsewhere. d indicates the sum of the distances between all corresponding robots when the minimal distance mapping between robots i at t_1 and j at t_2, for all i and j, is achieved.

The opponent's strategy is assumed to rely on the type of skill attributed to each robot. A strategy is implemented based on the skill assigned to each robot, which can vary between t_1 and t_2. The dissimilarity function d is expected to function as designed, even in the event that the robots are permuted. In the context of robotic soccer, all strategies may be assumed to be symmetric along the x-axis[2], and the reflection U_2 given in Equation (1) can be considered. The ball and teammates' positions are excluded from Eq. (1) for the following reasons:

- The robots' positions depend on the ball so that relative positions between robots are sufficient to characterize the behavior on the field;
- The positions of the teammates can negatively affect the learning process because the behavior of the teammates are affected by the learning process.

Next, Eq. (1) was used to define a dissimilarity function $d_1(\cdot, \cdot, \cdot)$ between the deployment choice of an opponent at time t_1 and a set of deployment choices made during a time interval $[T_s, T_e]$, as follows:

$$d_1(t_1, T_s, T_e) = \min_{t \in [T_s, T_e]} d(t_1, t). \tag{2}$$

Equation (2) may be used to define a dissimilarity function $d_2(\cdot, \cdot, \cdot, \cdot)$ between two sets of deployments: one for the time interval $[T_s^{(i)}, T_e^{(i)}]$ and the other for $[T_s^{(j)}, T_e^{(j)}]$, as follows:

$$d_2(T_s^{(i)}, T_e^{(i)}, T_s^{(j)}, T_e^{(j)}) = \min_{t \in [T_s^{(i)}, T_e^{(i)}]} d_1(t, T_s^{(j)}, T_e^{(j)}). \tag{3}$$

Equation (3) shows the dissimilarity function between the two most similar deployment choices in the two distinct sets. The opponent's strategies may then be classified by applying Eq. (3) to the sets of deployments. The next section discusses the use of cluster analysis in classifying the opponent's strategies based on the dissimilarity function given by Eq. (3).

4   Cluster Analysis

Consider the N sets of deployment choices, each of which comprises a set of deployments made within a time interval. Equation (3) is then applied to any two sets in the N sets of deployments to calculate a dissimilarity matrix of size $N \times N$, the elements of which are the dissimilarities between the set plays. The matrix may be regarded as a proximity matrix, and a cluster analysis may be used to classify the opponent's strategies.

4.1   Clustering Methods

Many clustering methods are available for use in cluster analysis[8]. Typical methods include:

k-**means Algorithm.** The k-means algorithm is a cluster analysis method that aims to partition n observations into k clusters such that each observation belongs to the cluster characterized by the nearest mean. This creates a partitioning of the data space into Voronoi cells.

---

[2] In human soccer, a player may have his/her strong wing; however, in robotic soccer (especially in the SSL), none of the robots has a strong wing, in general. Therefore, it is generally acceptable to assume symmetric positions.

Ward's Method. Ward's method is a hierarchical cluster analysis method that shares many of the features of variance analysis. A linkage (of clusters) function, which specifies the distance between two clusters, is computed as an increase in the "error sum of squares" (ESS) after two clusters are fused into a single cluster. Ward's Method seeks to select successive clustering steps that minimize the increase in ESS at each step.

Group Average Clustering. The group average clustering method is a hierarchical method in which the distance between two clusters is calculated based on the average distance between all pairs of objects in the two different clusters.

Both the k-means algorithm and Ward's method are practical methods; however, they cannot be used to calculate the centroid of a cluster. In this problem, it is difficult to calculate the centroid because each element in a cluster denotes a continuous deployment choice. The group average clustering method, on the other hand, calculates the distance between objects (deployments in this case), and this distance is sufficient for implementing the clustering method. The group average clustering method was therefore chosen.

4.2   Estimating the Number of Clusters

The group average clustering method creates a hierarchical structure of clusters; however, the method does not estimate the number of clusters. To do so, the Davies–Bouldin index (DBI)[5] was used. The DBI for K clusters is defined as follows:

$$\text{DB}(K) = \frac{1}{K} \sum_{i=1}^{K} \max_{j \neq i} \frac{S_i + S_j}{M_{ij}}, \tag{4}$$

where M_{ij} is a measure of the **separation** between two clusters C_i and C_j, and S_i is a measure of **cohesion** within a cluster C_i. M_{ij} and S_i may be defined freely under some constraints[5]. The optimal number of clusters is given by K, which is selected to minimize $\text{DB}(K)$ over the range of K identified by some criterion, for example, using Sturges' formula . The definition of the DBI requires the judicious selection of an appropriate range for K because $\text{DB}(K)$ approaches 0 as the number of single object clusters increases.

Equation (3) is then used to define the separation M_{ij} and cohesion S_i as follows:

$$S_i = \frac{1}{|C_i|(|C_i| - 1)} \sum_{X_k \in C_i} \left\{ \sum_{X_l \in C_i, X_l \neq X_k} d_2(T_s^{(k)}, T_e^{(k)}, T_s^{(l)}, T_e^{(l)}) \right\},$$

$$M_{ij} = \frac{1}{|C_i||C_j|} \sum_{X_k \in C_i} \sum_{X_l \in C_j} d_2(T_s^{(k)}, T_e^{(k)}, T_s^{(l)}, T_e^{(l)}),$$

where S_i is the mean distance between any two objects in C_i and M_{ij} is the distance between C_i and C_j, computed using the group average clustering method. The values of S_i and M_{ij} satisfy the constraints given in [5].

5   Experimental Results

Many goals in the SSL games are scored based on the set plays, such as the identification of a throw-in, corner kick, or goal kick. Each team implements a variety of strategies in the set plays. The method proposed in the previous sections was applied here to the set plays executed by the teams participating in the RoboCup 2012 games in an effort to analyze and classify the opponents' strategies.

Let X_i be a set of deployment choices made during the i-th set play ($1 \leq i \leq N$). The time interval surrounding X_i is $[T_r^{(i)}, T_e^{(i)}]$, where $T_r^{(i)}$ is the time the team receives a start command for the set play from the referee box and $T_e^{(i)}$ is the time at which a kick is carried out in the set play.

$T_s^{(i)}$ can be defined as

$$T_s^{(i)} = \max(T_e^{(i)} - T_{behavior}, T_r^{(i)}), \tag{5}$$

where $T_{behavior}$ is a constant that specifies the time at which the robot takes an action. Small $T_{behavior}$ values are sufficient for an expert team because their robots move fast. Somewhat larger $T_{behavior}$ values absorb the dispersion of the set plays based on a strategy. For unfamiliar teams, the assumption of a small $T_{behavior}$ value is recommended. In this paper, $T_{behavior}$ is set to 1.0 sec.

Equation (3) was used to compute the dissimilarity functions

$$d_2(T_s^{(i)}, T_e^{(i)}, T_s^{(j)}, T_e^{(j)}), \ (1 \leq i \leq N, \ 1 \leq j \leq N). \tag{6}$$

A dissimilarity matrix[3] was then generated, and a dendrogram was calculated based on the group average clustering. Finally, the number of clusters present was estimated using Equation (4). The opponents' strategies used in the set plays among the K strategies were then classified. The following Sturges' formula[6] was used to estimate the range of K:

$$1 \leq K \leq \lceil \log_2 N + 1 \rceil, \tag{7}$$

where $\lceil x \rceil$ is the ceiling function of x.

5.1   Classifying RoboDragons' Strategies

First, the RoboDragons' strategies (our team's strategies) were classified using the proposed method. RoboDragons used four attacking strategies during the set plays that took place in the RoboCup 2012 world championship. These strategies were denoted $A_i, (1 \leq i \leq 4)$. The RoboDragons' simulation system[4] was used

---

[3] In the experiments described in this section, Δ in Eq. (1) was not used because six robots on each team were always on the field.

[4] The proposed method was first applied to the real games played during the RoboCup 2012. These games were easy to classify because the parameter values used to implement the strategies were fixed. The simulation system was then used to execute set plays in which the values of the parameters were varied.

to execute a pseudo-game involving RoboDragons (Blue) vs. RoboDragons (Yellow)[5]. Twenty-four set plays in total were executed. One strategy was used in each set play.

Each strategy was used six times over the course of the 24 games. According to the rules of the SSL, the ball was placed at the (x, y) coordinates at the start of a set play, where the y-coordinate is 1915 mm and the x-coordinate is randomly selected from within a range of values that permits execution of the set play.

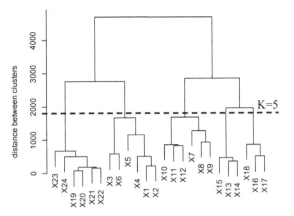

Fig. 1. Dendrogram (RoboDragons)

Figure 1 shows a dendrogram of the experimental results obtained[6], and Figure 2 shows the Davies–Bouldin index. The number of clusters fell within the range $1 \leq K \leq 6$, according to Equation (7). Figure 2 shows that an estimate of $K = 5$ in this case was reasonable. The dendrogram shown in Fig. 1 was cut

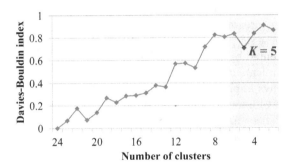

Fig. 2. Davies–Bouldin index (RoboDragons)

off at the height at which the number of clusters reached five to obtain the following five clusters.

$$C_1 = \{X_1, X_2, X_3, X_4, X_5, X_6\}$$
$$C_2 = \{X_7, X_8, X_9, X_{10}, X_{11}, X_{12}\}$$
$$C_3 = \{X_{13}, X_{14}, X_{15}\}$$
$$C_4 = \{X_{16}, X_{17}, X_{18}\}$$
$$C_5 = \{X_{19}, X_{20}, X_{21}, X_{22}, X_{23}, X_{24}\}$$

The set plays implemented in C_1 used the A_1 strategy, C_2 used A_2, and C_5 used A_4. The set plays characterized by the strategy A_3 could be classified as belonging to one of two clusters, C_3 and C_4. If $K = 4$, however, then C_3 and C_4

---

[5] Blue and Yellow are the colors used to identify the teams in the SSL.
[6] The dendrograms in this paper were drawn using the R statistical software package (http://www.r-project.org/).

were unified, and the clusters correctly separated the strategies. These results revealed the utility of the proposed method. Figure 1 showed that the strategy A_4 was easier to identify than the other strategies because the height of C_5 was the smallest. (Strategy A_3 was characterized as the height of $(C_3 \cup C_4)$.)

5.2   Classifying the Opposing Teams' Strategies

This section discusses attempts to classify the opposing teams' strategies. The data logged during the final game of the RoboCup 2012, in which Skuba (Blue) opposed ZJUNlict (Yellow), were analyzed here. In the game, 62 set plays were restarted from 10 cm inside of the touch boundary crossed by the ball. Of these set plays, 37 were implemented by Skuba and 25 were

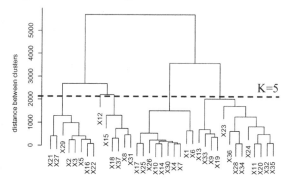

Fig. 3. Dendrogram (Skuba)

implemented by ZJUNlict. Figures 3 and 4 show the dendrograms of the set plays implemented by Skuba and ZJUNlict, respectively. Equation (4) was used to estimate the number of clusters: $K = 5$ for Skuba, and $K = 6$ for ZJUNlict.

The analysis of Skuba's strategies will be discussed first. Figure 5 shows the classified results. Each image illustrates a deployment choice made immediately after the kick was taken during a set play. In these images, generated by our logged data review system, the sizes of the ball and robots are enlarged, and the number in each robot is the robot's ID. Note that the attacking direction changed from set play X_{17} because the teams changed sides after half time.

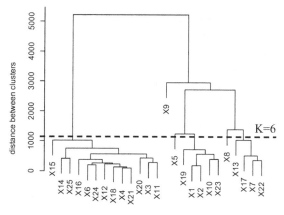

Fig. 4. Dendrogram (ZJUNlict)

An analysis of the clusters readily identified the strategies used. The strategies were found to be characterized as follows:

- C_1 encompassed a strategy in which the ball was kicked directly toward the goal without passing between robots.
- C_2 encompassed a strategy in which the ball was passed to a teammate at the far side of the opponent's goal area after a corner kick had been taken.

- C_3 encompassed a strategy in which the ball was passed to the teammate at the near side of the center line after a throw-in had been taken.
- C_4 encompassed a strategy in which the ball was passed to the teammate at the far side of the field. This strategy resembled the strategy C_1.
- C_5 encompassed a strategy in which the ball was passed to the teammate at the center of the field.

The analysis of ZJUNlict's strategies will be discussed next. The following classification results were obtained from this analysis. (The corresponding images are omitted due to limitations on the article space.)

$$C_1 = \{X_1, X_2, X_{10}, X_{19}, X_{23}\}$$
$$C_2 = \{X_3, X_4, X_6, X_{11}, X_{12}, X_{14}, X_{15}, X_{16}, X_{18}, X_{20}, X_{21}, X_{24}, X_{25}\}$$
$$C_3 = \{X_5\}$$
$$C_4 = \{X_7, X_{13}, X_{17}, X_{22}\}$$
$$C_5 = \{X_8\}$$
$$C_6 = \{X_9\}$$

The strategies could be characterized as follows:

- C_1 encompassed a strategy in which the ball was passed to the teammate at the far side of the opponent's goal area.
- C_2 encompassed a strategy in which short passes were made to the teammate located along the direction of the goal. The team used this strategy many times.
- C_3 was similar to strategy C_1.
- C_4 encompassed a strategy in which the ball was passed to the teammate at the far side of the field after a throw-in had been taken on the opponent's side.
- C_5 was similar to strategy C_4.
- C_6 appeared to be similar to strategy C_4, although the placement of two robots on opposite sides of the ball may have led to the use of another strategy.

The results of this experiment revealed that the classification of an opponent's strategies was possible.

6   Application: On-line Learning

Section 5 demonstrated that an opponent's strategies could be classified using the method proposed here. This method was applied to an on-line learning algorithm to assist in selecting an advantageous action during the opponent's $(N + 1)$th

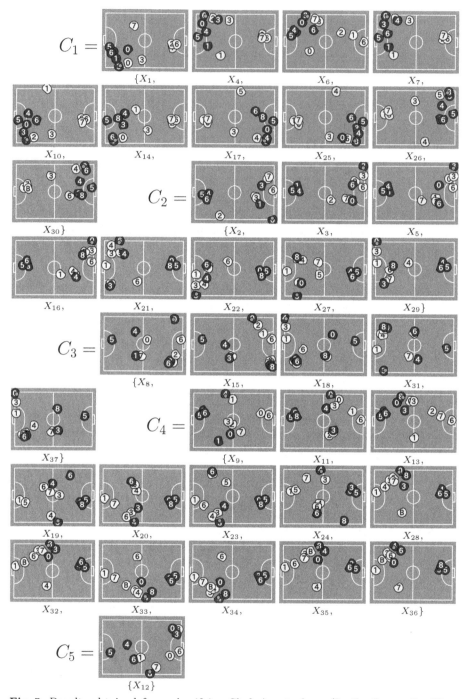

Fig. 5. Results obtained from classifying Skuba's set plays. (In the figure, the Blue team's robots are shown as black circles and the Yellow team's robots are shown in white. Figure 6 uses the same color convention.)

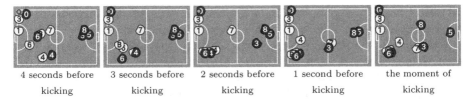

| 4 seconds before | 3 seconds before | 2 seconds before | 1 second before | the moment of |
| kicking | kicking | kicking | kicking | kicking |

Fig. 6. Positions of robots, every second in set play X_{37}

behavior, based on the classification results obtained from the N behaviors executed previously. As an example, consider the process of learning during a set play in which the game is restarted by placing the ball near the touch boundary.

A dissimilarity function d_3 is defined between the deployment choice X_j in C_i and the current deployment choice X as follows:

$$d_3(t, C_i) = \frac{1}{|C_i|} \sum_{X_j \in C_i} d_1(t, T_r^{(j)}, T_e^{(j)}). \tag{8}$$

This equation gives the mean value of the dissimilarity between X and X_j for each X_j in C_i. For all clusters obtained thus far, Equation (8) is computed to estimate the most likely deployment strategy selected by the opponent. The real-time computation is easy. Equation (8) used $T_r^{(j)}$ instead of $T_s^{(j)}$, as defined in Equation (5) because the action immediately prior to the kick as well as the precursors to the action taken in the strategy were of interest.

The data logged during the final game of the 2012 RoboCup competition were used to compute Eq. (8) for the 37th set play X_{37} of Skuba, assuming that the set plays $X_1...X_{36}$ could be classified as described in Section 5.2 (excluding X_{37} from C_3). Figure 7 shows the obtained results, and Figure 6 shows the positions of the robots every second during the set play X_{37}.

Because the value of $d_3(t, C_i)$ was high for C_1 and C_4 4 seconds prior to kicking, the strategy executed here was not selected from among these strategies. Clusters C_2 and C_3 had low values of $d_3(t, C_i)$ between 4 seconds and 2 seconds prior to kicking; however, the value for C_2 began to increase 2 seconds prior to kicking. Skuba's ID3 robot dashed out at this time. One second prior to kicking, the ID8 robot dashed out. Only C_3 had a low value of $d_3(t, C_i)$ after receiving the signal from

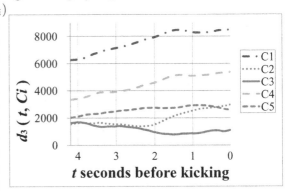

Fig. 7. The values obtained from Eq. (8). Duration: from the receipt of a start signal for a set play (from the referee box) to the completion of the kick.

the referee box. These observation, in conjunction with Eq. (8), suggested that the strategy corresponding to C_3 would be executed once again.

According to the strategy corresponding to C_3, the ball was passed to the robot that dashed out from the defense area in all cases. Therefore, in X_{37}, the ID8 robot was expected to shoot, and the ID3 robot, which dashed out 2 seconds prior to kicking, would be a bait robot. From this analysis, marking the ID8 robot 2 or 4 seconds prior to kicking will break the opponent's strategy.

7 Concluding Remarks

A dissimilarity function d that measures the differences between two different deployment choices in an SSL game has been designed. Additionally, a method for classifying and analyzing an opponent's strategies based on a clustering algorithm has been proposed. This method was applied to simulated games to demonstrate the utility of the method. Finally, the method was applied to the final game of the 2012 RoboCup SSL to show that an unknown strategy could be accurately classified. It has been demonstrated that the method could be used to an identify opponent's strategies during a game on the fly and in real time. Future work will further improve the method described here.

Acknowledgements. This work was supported by funding from the Aichi Prefectural University.

References

1. Panyapiang, T., Chaiso, K., Sukvichai, K., Lertariyasakchai, P.: Skuba 2012 Extended Team Description (2012), http://robocupssl.cpe.ku.ac.th/robocup2012:teams (accessed September 15, 2013)
2. Bowling, M., Browning, B., Veloso, M.M.: Plays as Effective Multiagent Plans Enabling Opponent-Adaptive Play Selection. In: Proceedings of International Conference on Automated Planning and Scheduling 2004, pp. 376–383 (2004)
3. Trevizan, F.W., Veloso, M.M.: Learning Opponent's Strategies In the RoboCup Small Size League. In: Proceedings of AAMAS 2010 Workshop on Agents in Real-time and Dynamic Environments (2010)
4. Visser, U., Weland, H.-G.: Using Online Learning to Analyze the Opponent's Behavior. In: Kaminka, G.A., Lima, P.U., Rojas, R. (eds.) RoboCup 2002. LNCS (LNAI), vol. 2752, pp. 78–93. Springer, Heidelberg (2003)
5. Davies, D.L., Bouldin, D.W.: A Cluster Separation Measure. IEEE Transactions on Pattern Analysis and Machine Intelligence 1(2), 224–227 (1979)
6. Sturges, H.A.: The Choice of a Class Interval. Journal of the American Statistical Association 21, 65–66 (1926)
7. SSL Technical Committee. Laws of the RoboCup Small Size League 2012 (2012), http://robocupssl.cpe.ku.ac.th/_media/rules:ssl-rules-2012.pdf (accessed September 15, 2013)
8. Everitt, B.S., Landau, S., Leese, M., Stahl, D.: Cluster Analysis, 5th edn. Wiley (2011)

An Entertainment Robot for Playing Interactive Ball Games

Tim Laue[1], Oliver Birbach[1], Tobias Hammer[2], and Udo Frese[1]

[1] Deutsches Forschungszentrum für Künstliche Intelligenz,
Cyber-Physical Systems, Bremen, Germany
{Tim.Laue,Oliver.Birbach,Udo.Frese}@dfki.de
[2] Institute of Robotics and Mechatronics, DLR, Germany
Tobias.Hammer@dlr.de

Abstract. This paper presents a minimalistic robot for playing interactive ball games with human players. It is designed with a realistic entertainment application in mind, being safe, flexible, reasonably cheap, and reactive. This is achieved by a clever, minimalistic robot design with a 2 DOF roll tilt unit that moves a bat with a spherical head. The robot perceives its environment through a stereo camera system using a circle detector and a multiple hypothesis tracker. The vision system does not require a specific ball color or background structure. The paper motivates the proposed robot design with respect to the above mentioned requirements, describes our solution to the tracking, calibration, and control issues involved and presents indoor and outdoor experiments where the robot bats balls tossed by different players.

1 Introduction

RoboCup Soccer has been founded as a basic research endeavour, as "an attempt to foster AI and intelligent robotics research by providing a standard problem where a wide range of technologies can be integrated and examined" [15]. However, unlike other basic research questions robot soccer is easily understood by the general public making the RoboCup competitions both a scientific and a public event. As Kitano said, a "publicly appealing but formidable challenge". This unique combination also motivates other "sport robotics" research activities, such as ball catching [5]. Now, being a basic research program, RoboCup soccer and other sport robotics activities are far from actual applications, they only contribute indirectly, e.g. by stimulating household robotics research. This paper is an attempt to identify a direct, commercially realistic application of sport robotics technology in the entertainment industry.

Our proposed system (cf. Fig. 1) is a minimalistic ball-playing robot serving at events, such as office parties, fairs or open house presentations. The robot is stationary and if a human throws a ball towards the robot it is supposed to hit it back, engaging the human in a robot-human ball game (at the moment it can technically only intercept not hit back). It is intended not as a long-term game, but rather as a short exciting experience being driven by the fascination

S. Behnke et al. (Eds.): RoboCup 2013, LNAI 8371, pp. 171–182, 2014.
© Springer-Verlag Berlin Heidelberg 2014

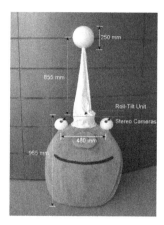

Fig. 1. The developed robot system, its major components, and dimensions. The computer and the motor's power supply are inside the robot's body.

of the unusual combination of sport, technology, and interactivity. Consequently, it is not an end-consumer product, rather than a device bought and operated by professional agencies organizing an event. The robot is specifically designed to meet the following requirements for this kind of operation:

1. *Safety.* For interacting with humans, the system has – of course – to be safe. It must be guaranteed that no person gets injured by the robot.
2. *Flexibility and easy appliance.* Similar to existing entertainment devices, the hardware and software must operate under various conditions. Amongst others, this concerns elements in the perceivable environment, lighting conditions, or the behavior of humans participating in the game. The system must be transportable and a non-expert must be able to set it up.
3. *Reasonably low costs.* The costs for the construction and maintenance of many current robot systems are not economical for any commercial activities. A successful entertainment robot should have a prize comparable to current non-robot entertainment devices.
4. *Reactivity.* A robot that interacts with people in a game (e. g. some kind of ball game) must have a level of reactivity comparable to that of humans. A significantly lower performance would result in a boring game that does not challenge the humans.
5. *Throughput.* Often events have many visitors and a large number of people should interact quickly with the system.

The contribution of this paper is the proposed design of a minimalistic ball-playing robot serving at events. We have developed a working, low-cost robot system that has been evaluated in various environments regarding different aspects. Although being complete at a technical level, the current module can only play a minimalistic game up to now. Thus it is a first step towards a commercial product.

This paper is organized as follows: Section 2 describes related work in the areas of sport robotics and tracking in sport environments. The developed event module and the underlying design considerations are presented in Sect. 3. Section 4 and Sect. 5 describe the necessary subcomponents for ball tracking and their calibration, followed by the motion control approach in Sect. 6. The robot's overall behavior and recent applications and experiments are presented in Sect. 7 and Sect. 8 respectively.

2   Related Work

Up to now, entertainment robots have been quite rare and are mostly settled in the context of artificial pets or toys. Prominent examples are the *Sony AIBO* [11] and the *RoboSapiens* [21] respectively. A few recent systems actually play ball games with humans. An example is Segway-soccer [2]. In this game, autonomous Segways play together and against humans standing on Segways. As a result, the actuation disadvantages of the robots are compensated. The same holds for the table soccer robot *KiRo* [22], where the competing humans only have access via rotary handles. This machine plays on a remarkable level and is even commercially available. Two robots that play simpler but more direct games are the *RoboKeeper* [9], a robot goalkeeper that parries penalty shots by professional soccer players, and DLR's *Rollin' Justin*, a humanoid torso that catches balls thrown by humans [5]. The former is already a successful product that can be hired for events, the latter is a research prototype actually built for service robotics not sports. Both systems share the same requirement: The ball has to be detected accurately and tracked robustly, so an exact prediction of upcoming trajectory is available in time and the robot can act accordingly.

Most of the previously presented work on perception systems with subsequent robotic actuation is based in the lab or controlled environments, usually extracting the ball due to its distinct color [18,19,3,9] or from the difference to a reference image [1,10]. A more flexible way would be to detect the ball as a circle in the image [24]. Due to the large measurement space, the native use of the popular Hough-Transform [14] is precluded. In [4], we proposed an efficient alternative that is also used here. Detected balls are usually tracked in 3-D either in a global or frame-by-frame based way. Simple global approaches fit a parabola to non-ambiguous ball measurements ([13,18,3]). Sophisticated global approaches also handling clutter exist (e.g. [23]), but their global operation forbids real-time usage. In contrast, frame-by-frame approaches such as Extended Kalman Filter (EKF) ([10,19]) are considered to be the optimal single-target tracking solution. They are generalized by Multiple Hypothesis Tracking [7] handling false-alarm measurement, starting and ending of ball flights in a sound probabilistic way [4].

Detecting and tracking balls is also of interest in broadcast television, with prominent examples for tennis and cricket [16], and for baseball [12]. The former is even used in the adjudication process after controversial calls by umpires, actually influencing the game's outcome.

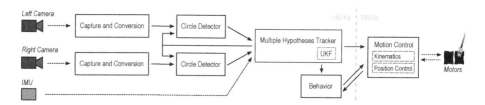

Fig. 2. The major software components of the presented robot system. Each box represents a single process (implemented as a ROS node). The arrows depict the data flow with solid arrows representing ROS communication links and dashed arrows representing direct communication with hardware devices.

3   System Overview

The contribution of the system presented in this paper is a combination of a minimalistic, low-cost hardware design and elaborated software which together provide the efficiency and performance needed for an interactive ball game.

3.1   Conception

The key idea in our robot design is to use as few degrees of freedom (DOF) as possible. The design of conventional 6 DOF robots is dominated by the fact that lower motors have to support and accelerate the weight of higher motors. Avoiding this setup dramatically reduces weight and costs and increases safety and reactivity.

How many DOF are needed to play a ball game? Surprisingly, two DOF suffice. First, we use a bat with spherical head to hit the ball, so we do not need to control the bat's orientation (cf. Fig. 1). This avoids three motors that need to be carried by other motors in conventional robots. Second, we attach the sphere with a rod to the first two robot axes leaving the elbow axis out. This reduces the robot's workspace to (roughly) a sphere, the missing DOF is contributed through the ball's motion by choosing when to hit the ball. With this design, the robot becomes inherently safe, as all moving parts are lightweight (335g) and soft with plastic foam for the sphere and a cushioned rod. For the same reason, we discarded mobility, because a mobile robot with battery, motors, and computing power would have a weight that makes inherent safety unrealistic at velocities needed for sports. A lower limit to the bat's weight comes from the fact that elasticity between motor and bat is needed to protect the motor from the ball's impact. Hence the impact mostly works as if bat and ball collide in free space, requiring the bat to be much heavier (230g) than the ball (53 − 60g).

An interactive game requires the robot to pass the ball back to a human player, raising the following question:

How much control has a 2 DOF robot over the ball? Surprisingly much: If the spherical bat hits the ball centrally, i. e. such that sphere and ball center would collide, the ball is reflected back into the direction it came from (apart from friction of course). With an offset in the two dimensions of the spherical workspace, the direction in which the ball is reflected can be controlled. Furthermore, the chosen bat velocity gives two additional DOFs on the reflected ball and also allows to add energy, if the ball is hit in motion. This leads to 4-DOF overall available to control the ball. Ball control is actually redundant as the motion of a ball reflected at a given point has only 3 DOF, namely velocity. We conclude, that two motors, at least in principle, suffice to intercept and control a flying ball in a minimalistic, hence cheap, and intrinsically safe setup.

Which kind of games is appropriate for such a robot? Realistically, with low-cost motors and intrinsic safety, the robot will not be competitive with a human player. In comparison, the RoboKeeper is competitive even with soccer professionals, but has a single expensive motor and needs a safety barrier and safety equipment. Our idea is to implement cooperative games that require passing between the players and the robot.

3.2 Hardware

The spherical head of the bat is made of foam-filled polystyrene, the rod is made of carbon and cushioned with rubber foam. Having a total height of 1.945m (cf. Fig. 1), the robot's workspace roughly resembles the one of a child stretching its arms. The bat is moved by two *Dynamixel EX-106+* servo motors that are paired to a roll tilt unit. These motors provide nominally 10.49Nm drive torque, well above the 2.9Nm needed to hold the bat. The nominal speed without load is $546°/s$, effectively we operate at $180°/s$ with 70% torque to reduce gear load. This still allows sufficiently dynamic actuation (cf. Sect. 8). We use different, unmodified toy rubber balls ($53-60$g weight, $90-121$mm diameter) that are small and light enough to be played by the robot and heavy enough to be conveniently thrown by a human.

The robot's main sensors are two synchronized *AVT Marlin F-046* FireWire cameras. To be able to track a ball during its whole flight (cf. Fig. 3), the cameras are equipped with lenses of 4.8mm focal length ($\approx 57°$ horizontal field-of-view) and are mounted on the robot in an angle of about $35°$. We have also tried Microsoft's Kinect sensor, but found that it cannot detect the ball due to motion blur. Furthermore, an IMU is used to provide gravity information to the ball flight tracker as well as to determine impacts on the robot. All computations are performed by a modern personal computer (*Intel Core i7 860* 4×2.80GHz). The overall hardware costs of the robot as presented in this paper are less than 5000€ (including: 1000€ computer, 2500€ sensors, 1000€ actuators), fitting the requirement of reasonable low costs. The whole system is self-contained, i.e. it does not require any further constructions or modifications in the environment, such as external sensors or special lighting.

The current robot hull is designed as a plushy blue pig to provide a pleasant and funny counterpart to human players. However, the system is not technically bound to any specific character and might have a completely different appearance in the future.

Fig. 3. A ball tracked indoors. Circles (red) are detected in the image by a contrast-normalized gradient criterion and passed to a Multi Hypothesis Tracker (MHT) that handles clutter and missing detections and uses an Unscented Kalman Filter (UKF) to estimate position and velocity that are used to predict future states (green circles).

3.3   Software

The software consists of a number of components depicted in Fig. 2. The whole system is embedded in the Robot Operating System (ROS) [17] framework, running on Ubuntu Linux. Each component is a separate process, so the computationally demanding image processing and ball tracking components utilize the available CPU cores. The overall architecture is quite straightforward: Images (780×580 8 Bit grayscale) are captured with about 50 Hz. On these images, a generic search for circles is performed. The circles from both camera images are fused within the tracking component (cf. Sect. 4). The most likely ball tracks are sent back to the circle detection processes, allowing a refined search for circles around the predicted position. Finally, the robot's motors are controlled to steer the bat to the first intersection point of the ball's trajectory with the robot's workspace (cf. Sect. 6). In case of no tracks, some other actions are carried out by the robot (cf. Sect. 7). Position control runs with 100 Hz on the PC, as the servo's built-in controller is not able to handle the large inertia in our setup.

4   Ball Tracking

The subcomponent for tracking balls is the two-staged bottom-up approach introduced in [4], which is also in use for DLR's ball-catching humanoid *Rollin' Justin* [5]. In the first stage, balls are detected as circles (cf. Fig. 3). By computing the average fraction of image energy that is a radial gradient and evaluating this gradient response along the contour of all possible circle positions and radii, the N best circles in the image are determined. To achieve real-time operation, evaluation of all circle responses is done in a multiple-scale fashion. The detector is illumination invariant. Also, in contrast to other ball detection methods that detect the ball by its color ([3,6,18,19]), this method needs no calibration, facilitating robust and easy appliance.

In the second stage, the detected circles are passed to a multi-target filter, namely the Multiple Hypothesis Tracker (MHT) [7], to estimate the cardinality and the individual states, i. e. position and velocity, of all flying balls. MHT seeks for the data association hypothesis with the highest posterior probability. On arrival of a new set of measurements, each existing hypothesis from the previous time step is expanded to a set of new hypotheses by considering all possible

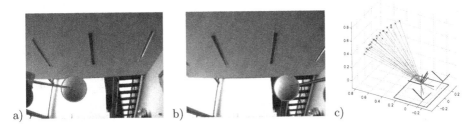

Fig. 4. a) and b) Left and right camera image showing the head of the bat and its projection as a circle and a line obtained from the calibration given the measured joint angles. c) Positions of the head's center (marked as a cross) and of the bat (line) used for calibration and visualization of all frames involved in the calibration procedure.

associations of measurements to tracks. Also, handling of track initiation and termination, as well as the case of having clutter measurements, is considered while generating hypotheses. To increase robustness, prior information encoding typical positions from where the ball is thrown and that the ball is thrown towards the robot is used when starting tracks. This helps generating only the tracks actually demanding the robot to react and discards any that might accidentally look like possibly flying ball (e. g. a moving head). The states of the involved tracks are estimated using an Unscented Kalman Filter (UKF) [20].

5   Calibration

Although being mechanically simple, a calibration of the setup has to be performed prior to playing. Fortunately, sensors and actuators can be *factory calibrated*, since all components are rigidly mounted and not expected to change.

During the robot's normal operation, the tracks estimated by the MHT must be converted from the camera frame into the base frame of the kinematic chain. We calibrate this transformation, the joint's scale factors and offsets, the camera's intrinsic parameters and stereo calibration jointly by using a checkerboard and by observing how the spherical head of the bat moves in both cameras when joint angles change (cf. Fig. 4). Given both joint angles $\theta_{roll}, \theta_{tilt}$, the forward kinematics, transforming from the frame of the tilting servo into the base frame, reads as a chain of transformations

$$M_{\theta_{roll}, \theta_{tilt}} = M(\theta_{roll}) \cdot T \cdot M(\theta_{tilt}) \tag{1}$$

with

$$M(\theta) = \begin{bmatrix} \cos\theta & -\sin\theta & 0 & 0 \\ \sin\theta & \cos\theta & 0 & 0 \\ 0 & 0 & 1 & 0 \\ 0 & 0 & 0 & 1 \end{bmatrix}, T = \begin{bmatrix} \cos\frac{\pi}{2} & 0 & \sin\frac{\pi}{2} & 0 \\ 0 & 1 & 0 & l_{off} \\ -\sin\frac{\pi}{2} & 0 & \cos\frac{\pi}{2} & 0 \\ 0 & 0 & 0 & 1 \end{bmatrix}.$$

$M(\theta)$ represents a rotation along the joint axis and T describes the fixed rotational and translational displacement between the joints. Here, l_{off} is -35 mm.

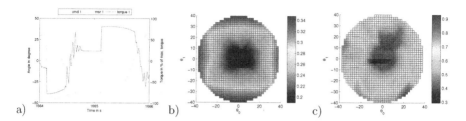

Fig. 5. a) Position controller's behavior: commanded (green) and measured (dashed green) angle and motor torque (blue). **b)** Required time [s] to reach a joint angle configuration (θ_0, θ_1) when starting from $(\theta_0 = 0, \theta_1 = 0)$ with $\leq 2°$ error. **c)** Required time [s] to reach $\leq 0.5°$ error. The robot reached the whole workspace reasonably quickly (< 0.34s), but needs significant time (> 0.4s) to damp oscillations. This problem is mainly caused by backlash and has to be addressed in the future. The steady state error was always $< 0.4°$. Overall, the system appears fast enough for a ball game.

Calibration data is acquired in two consecutive steps. First, observations of a checkerboard from different views define camera parameters and their relation (stereo). In the second step, a grid of different bat poses (cf. Fig. 4, right) is commanded and corresponding joint angles and camera images are saved. Measuring the spherical head as a circle and the connection to the tilt frame as a point on the rod, the transformation between the kinematic base frame and the cameras as well as the servo's scale factors and offsets are defined.

From 9 checkerboard images and 25 bat poses, this procedure results in a residual rms $(0.62, 0.88)$ px for checkerboard corner positions and $(3.8120, 2.2419)$ px and 0.7307 px for head positions and radii, respectively.

6 Motion Control

The motion control component has the task to move the bat to a location where it hits the ball in midair. Precisely, the trajectory of the ball's center should hit the bat's center. Hitting the ball non-centrally or in motion, as outlined in Sect. 3.1, is future work. The approach is straightforward: First, angles for both servo motors are computed by inverse kinematics. Afterwards, the movement is executed by a textbook PD controller.

The predicted ball tracks from the MHT are converted into robot-base coordinates and the one closest to a position reachable by the robot is determined. For this task, an analytical inverse of (1) is used:

$$\theta_{roll} = \arctan \frac{x}{z} \quad \text{for } z > 0 \tag{2}$$

$$\theta_{tilt} = -\arctan \frac{y}{\sqrt{x^2 + z^2} - l_{off}} \tag{3}$$

where x, y, z are in robot base coordinates and l_{off} is the displacement between the two servo axes. Feeding the calculated angles back into (1) allows to determine the distance to a reachable position and therefore the closest position where the ball is intercepted. The resulting angles are passed to the position controller that actuates the servo motors based on the commanded and measured servo angles. In addition, it enforces position limits to avoid hardware damage. The core of this component is a PD-Controller for each servo with angle-position error e_t as process variable and percentage of maximum torque as output.

$$\tau_{out} = k_p e_t + k_d d_t + \tau_{comp} \tag{4}$$

where d_t are the derivatives of e_t, low-pass filtered by

$$d_t = e^{-2\pi f \Delta t} \cdot d_{t-1} + \left(1 - e^{-2\pi f \Delta t}\right) \cdot \frac{e_t - e_{t-1}}{\Delta t} \tag{5}$$

with f as the cutoff frequency defined later. To compensate the steady-state errors induced by the bat's weight, a torque τ_{comp} is added, assuming a weight F and the center of gravity being in the sphere center.

$$\tau_{comp_{roll}} = -(l \cos \theta_{tilt} + l_{off}) F \sin \theta_{roll} \tag{6}$$

$$\tau_{comp_{tilt}} = -lF \cos \theta_{roll} \sin \theta_{tilt} \tag{7}$$

The controller has to cope with two main sources of inaccuracy. The first one are elasticities of the outer hardware, i.e. the post and connectors which allow oscillation of the end effector on acceleration. The second is large backlash (ca. 3°), presumably inside the servo gear. This is a common problem of low-cost servos. It causes, together with latencies of the control loop, massive oscillations of ca. 24Hz if k_d is increased. To reduce this effect and allow for higher k_d and k_p the cutoff frequency of the low-pass filter on the derivations is chosen as $f = 20$Hz. The controller's typical operation is shown in Fig. 5a.

7   Robot Behaviors

The sole ability to return balls is – independent of its quality – not satisfying for a robot that is made to entertain people. Therefore, a couple of interactive behaviors have been added to the system to give users some kind of appealing feedback as well as the impression of an intelligent robot. Even with only two 2 DOF, some meaningful gestures are already possible.

The overall behavior is realized as a simple finite state machine (cf. Fig. 6). After booting the system, the robot remains in the *sleeping* state until its bat has been lifted by a user. This haptic state transition allows an operation without any additional external controls. In its *waiting state*, the robot does not perform any actions and awaits a flying ball. If no ball is thrown for a certain amount of time, the robot starts to *cheer* the user by moving its bat and playing some sound. A similar gesture is performed in case of *complaining* about a badly thrown ball that did not reach the robot's workspace. In case of a thrown ball, the cheering and complaining states can both be interrupted at any time to avoid any mode confusion, i.e. to avoid the user's impression that the robot ignores the ball. In addition to these states, two simple reactive behaviors, which are

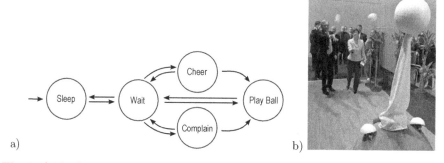

Fig. 6. a) The finite-state machine that controls the robot's behavior. b) Demonstration at the DFKI's booth during the CeBIT 2012. Visitors throw balls towards the robot.

always active, have been added. Whenever the robot is unexpectedly hit by a ball or pushed by a person, the event can be detected by the built-in IMU and a loud "Ouch!" sound is triggered. Furthermore, the bat's waiting pose is dynamically adjusted to always face the player, providing an impression of the robot's attention. The according person tracking is currently realized using a *Microsoft Kinect* sensor. To save this additional sensor and to be able to track persons outdoors, an alternative image-based solution using the *Histogram of Oriented Gradients* approach [8] is currently under development. Most of these behaviors are shown in the supplementary video[1].

8   Experimental Results and Applications

In a first experiment, the reachability of the workspace given a limited amount of time was determined. The results that are depicted in Fig. 5 indicate a sufficient speed and reactivity for an interactive game.

Given a technically working system, the most obvious experiment is to throw balls towards the robot and count how often it really plays the ball. We conducted this inside the entrance hall of our research building as well as on the lawn next to the building. The latter experiment was conducted during the afternoon, having a sunny sky with several clouds. In both experiments, the ball was thrown from about 5-6 meters which we consider as a reasonable distance in a real game. Different persons have thrown balls. As presented in detail in Tab. 1, the robot was able to play 84% of all balls that reached its workspace, even 95.1% outdoors. However, there is a remarkable difference why the robot fails to play the ball if it does. Indoors, the major problem was ball tracking, 14.6% of the balls crossing the robots workspace have not been tracked, outdoors only 4.9%. At a first glance, this is surprising since computer vision under natural lighting conditions is considered to be more error-prone. In fact, the applied approach is linearly illumination independent but affected by circular clutter. The latter occurs indoors (cf. Fig. 3) but not in the sky. Instead, the major

---

[1] http://www.informatik.uni-bremen.de/ timlaue/video/piggy_demonstration.mp4

Table 1. Results of the conducted hitting experiments

	Throws	Hits total	Workspace missed	Hits in Workspace	Not tracked	Missed
Indoor	196	145 (74.0%)	18 (9.2%)	81.5%	14.6%	3.9%
Outdoor	56	39 (69.6%)	15 (26.8%)	95.1%	4.9%	0%
Total	252	184 (73.0%)	33 (13.1%)	84.0%	12.8%	3.2%

problem outdoors was the throwers missing the robot's workspace, probably caused by wind. In both experiments, the robot nearly always hit a tracked ball.

The system has already been presented to the public on several occasions. One particular highlight was the CeBIT 2012 computer fair in Hanover, Germany, at which the robot has continuously played with arbitrary visitors for five whole days, nine hours a day. At this demonstration, the robot proved to be already applicable under realistic conditions (cf. Fig. 6). In particular, it is well suited for active entertainment of a large number of people: We simply continuously handed balls to the crowd who threw them towards the robot without any instructions necessary.

Regarding a commercial perspective, discussions with an entertainment professional were very positive: Even when assuming a final robot price of € 15000, written-off in 5 years, plus € 1000/year maintenance, yearly costs would be € 4000. With a realistic price of € 950 per use minus 2×€ 200 personnel costs and 30 uses per year, we would have € 12500 (or 83%) return of invest per year.

9 Conclusions and Future Work

In this paper, we presented a commercially realistic entertainment robotics application beyond a lab demo scenario. The current version of the proposed robot system already fulfills important requirements for an interactive entertainment robot, such as safety, low costs, and flexibility regarding setup and application. In different experiments, we demonstrated the reactivity needed for playing a minimalistic first ball game. However, for more sophisticated ball games, preferably including multiple persons, several issues still have to be addressed in the future. To allow tracking in a larger area around the robot, the cameras need to be moved. This could be realized almost cost-neutral by replacing the robot's roll tilt unit by a pan tilt construction and mounting the cameras above the pan joint. Furthermore, the precision of the system needs to be increased to achieve multi-directional reflections reliably.

References

1. Andersson, R.L.: Dynamic sensing in a ping-pong playing robot. IEEE Trans.on Robotics and Automation 5(6) (1989)
2. Argall, B., Browning, B., Gu, Y., Veloso, M.: The first segway soccer experience: Towards peer-to-peer human-robot teams. In: Proc. Conf. on Human-Robot Interaction (2006)
3. Bätz, G., Yaqub, A., Wu, H., Kühnlenz, K., Wollherr, D., Buss, M.: Dynamic manipulation: Nonprehensile ball catching. In: Proc. IEEE Mediterranean Conf. on Control and Automation (2010)

4. Birbach, O., Frese, U.: A multiple hypothesis approach for a ball tracking system. In: Fritz, M., Schiele, B., Piater, J.H. (eds.) ICVS 2009. LNCS, vol. 5815, pp. 435–444. Springer, Heidelberg (2009)
5. Birbach, O., Frese, U., Bäuml, B.: Realtime perception for catching a flying ball with a mobile humanoid. In: Proc. IEEE Int. Conf. on Robotics and Automation (2011)
6. Bruce, J., Balch, T., Veloso, M.: Fast and inexpensive color image segmentation for interactive robots. In: Proc. IEEE/RSJ Int. Conf. on Intelligent Robots and Systems, pp. 2061–2066 (2000)
7. Cox, I.J., Hingorani, S.L.: An efficient implementation of reid's multiple hypothesis tracking algorithm and its evaluation for the purpose of visual tracking. IEEE Trans. on Pattern Analysis and Machine Intelligence 18(2), 138–150 (1996)
8. Dalal, N., Triggs, B.: Histograms of oriented gradients for human detection. In: Proceedings of the 2005 IEEE Computer Society Conf. on Computer Vision and Pattern Recognition, pp. 886–893 (2005)
9. Fraunhofer, I.M.L.: 4attention GmbH & Co. KG: Robokeeper web site (2011), http://www.robokeeper.com
10. Frese, U., Bäuml, B., Haidacher, S., Schreiber, G., Schaefer, I., Hähnle, M., Hirzinger, G.: Off-the-shelf vision for a robotic ball catcher. In: Proc. IEEE/RSJ Int. Conf. on Intelligent Robots and Systems (2001)
11. Fujita, M., Kitano, H.: Development of an Autonomous Quadruped Robot for Robot Entertainment. Autonomous Robots 5(1), 7–18 (1998)
12. Guéziec, A.: Tracking pitches for broadcast television. Computer 35(3), 38–43 (2002)
13. Hove, B., Slotine, J.: Experiments in robotic catching. In: Proc. of the American Control Conf., pp. 380–385 (1991)
14. Kimme, C., Ballard, D., Sklansky, J.: Finding circles by an array of accumulators. Comm. of the ACM 18(2) (1975)
15. Kitano, H., Asada, M., Kuniyoshi, Y., Noda, I., Osawa, E., Matsubara, H.: RoboCup: A challenge problem for AI. AI Magazine 18(1), 73–85 (1997)
16. Owens, N., Harris, C., Stennett, C.: Hawk-eye tennis system. In: Int. Conf. on Visual Information Engineering, pp. 182–185 (2003)
17. Quigley, M., Conley, K., Gerkey, B.P., Faust, J., Foote, T., Leibs, J., Wheeler, R., Ng, A.Y.: Ros: an open-source robot operating system. In: Proceedings of the Open-Source Software workshop of the International Conference on Robotics and Automation (ICRA), Kobe, Japan (2009)
18. Riley, M., Atkeson, C.G.: Robot catching: Towards engaging human-humanoid interaction. Autonomous Robots 12(1), 119–128 (2002)
19. Smith, C., Christensen, H.I.: Using COTS to construct a high performance robot arm. In: Proc. IEEE Intern. Conf. on Robotics and Automation (2007)
20. Thrun, S., Burgard, W., Fox, D.: Probabilistic Robotics. MIT Press, Cambridge (2005)
21. Tilden, M.W.: Neuromorphic robot humanoid to step into the market. The Neuromorphic Engineer 1(1), 12 (2004)
22. Weigel, T., Nebel, B.: KiRo - an autonomous table soccer player. In: Kaminka, G.A., Lima, P.U., Rojas, R. (eds.) RoboCup 2002. LNCS (LNAI), vol. 2752, pp. 384–392. Springer, Heidelberg (2003)
23. Yan, F., Kostin, A., Christmas, W., Kittler, J.: A novel data association algorithm for object tracking in clutter with application to tennis video analysis. In: Proc. IEEE Int. Conf. on Computer Vision and Pattern Recognition (2006)
24. Yuen, H., Princen, J., Illingworth, J., Kittler, J.: A comparative study of hough transform methods for circle finding. In: Alvey Vision Conf., pp. 169–174 (1989)

Self-calibration of Colormetric Parameters in Vision Systems for Autonomous Soccer Robots

António J.R. Neves, Alina Trifan, and Bernardo Cunha

ATRI, IEETA / DETI
University of Aveiro, 3810–193 Aveiro, Portugal
{an,alina.trifan}@ua.pt, mbc@det.ua.pt

Abstract. Vision is an extremely important sense for both humans and robots, providing detailed information about the environment. In the past few years, the use of digital cameras in robotic applications has been significantly increasing. The use of digital cameras as the main sensor allows the robot to capture the relevant information of the surrounding environment and take decisions. A robust vision system should be able to reliably detect objects and present an accurate representation of the world to higher-level processes, not only under ideal light conditions, but also under changing lighting intensity and color balance. To extract information from the acquired image, shapes or colors, the configuration of the colormetric camera parameters, such as exposure, gain, brightness or white-balance, among others, is very important. In this paper, we propose an algorithm for the self-calibration of the most important parameters of digital cameras for robotic applications. The algorithms extract some statistical information from the acquired images, namely the intensity histogram, saturation histogram and information from a black and a white area of the image, to then estimate the colormetric parameters of the camera. We present experimental results with two robotic platforms, a wheeled robot and a humanoid soccer robot, in challenging environments: soccer fields, both indoor and outdoor, that show the effectiveness of our algorithms. The images acquired after calibration show good properties for further processing, independently of the initial configuration of the camera and the type and amount of light of the environment, both indoor and outdoor.

1 Introduction

Nowadays, digital cameras are used as the main sensor on robots and allow them to acquire relevant information of the surrounding environment and then take decisions about the actions to take. We can point out some areas of application of these robots, as the case of the industry, military, surveillance, service robots and more recently, vehicles for assisted driving.

To extract information from the acquired image, such as shapes or colors, the camera calibration procedure is very important. If the parameters of the camera are wrongly calibrated, the image details are lost and it may become

S. Behnke et al. (Eds.): RoboCup 2013, LNAI 8371, pp. 183–194, 2014.

Fig. 1. Images acquired with wrong parameters. From left to right, wrong value of gamma (high), exposure (low), saturation (high), saturation (low), white-balance (high both in Blue and Red gains).

almost impossible to recognize anything based on shape or color (see for example Fig. 1).

Our experience, as well as the experience of all the researchers that work in the field of computer vision, show that digital cameras working in automode fail regarding the quality of the images acquired under certain situations, even considering the most recent cameras. Algorithms developed for calibration of digital cameras assume some standard scenes under some type of light, which fails in certain environments. We did not find scientific references for these algorithms.

As far as we can understand from the published work regarding other teams of RoboCup Soccer, as well as according to our personal contact with them in the competitions, most of them there do not use any algorithm running on the robots to self-calibrate the colormetric parameters of the digital cameras, which means that their cameras are only adjusted manually before the robot starts its operation. Some of the teams have tried to solve the problem by developing algorithms for run-time color calibration (see for example [1,2]).

In this work, we show that the problem can be solved by adjusting the colormetric parameters of the camera in order to guarantee the correct colors of the objects, allowing the use of the same color classification independently of the light conditions. This allows also a correct processing of the image if other features have to be extracted. We think that this paper presents an important contribution to the RoboCup community.

We propose an algorithm to configure the most important colormetric parameters of the cameras, namely gain, exposure, gamma, white-balance, brightness, sharpness and saturation, without human interaction, depending on the availability of these parameters in the digital camera that is being used. This approach differs from the well known problem of photometric camera calibration (a survey can be found in [3]), since we are not interested in obtaining the camera response values but only to configure its parameters according to some measures obtained from the acquired images in robotic applications. The self-calibration process for a single robot requires only few seconds, including the time necessary to interact with the application, which is considered fast in comparison to the several minutes needed for manual calibration by an expert user. Moreover, the developed algorithms can be used the real-time by the robots while they are operating.

The work that we present in this paper was tested and is being used by two robotic projects developed in the University of Aveiro, in the challenging environment of robotic soccer (where the main goal is the development of multi-agent robotic teams), both with wheeled and humanoid robots. These robots are presented in Fig. 2. In these applications, the robots have to adjust, in a constrained amount of time, its camera parameters according to the lighting conditions [4].

Fig. 2. Robotic platforms used in this work. From left to right, CAMBADA (RoboCup Middle Size league (MSL) robotic soccer) and NAO (RoboCup Standard Platform League (SPL) soccer robot).

This paper is structured in five sections, the first of them being this introduction. Section 2 provides an overview on the most important colormetric parameters of digital cameras and the properties of the image that are related to each one. Section 3 describes the proposed algorithm. In Section 4 the results and their discussion are presented. Finally, Section 5 concludes the paper.

2   Configuration of the Camera Parameters

The configuration of the parameters of digital cameras is crucial for object detection and has to be performed when environmental conditions change. The calibration procedure should be effective and fast. The proposed calibration algorithm processes the image acquired by the camera and computes several measurements that allow the calibration of the most important colormetric parameters of a digital camera, as presented in Fig. 3. Besides the referred parameters, the hardware related parameters gain and exposure are also taken into consideration.

Starting by some definitions, luminance is normally defined as a measurement of the photometric luminous intensity per unit area of light travelling in a given

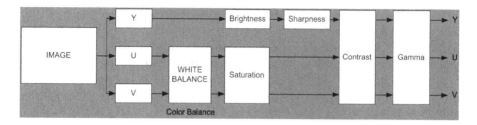

Fig. 3. A typical image processing pipeline (inside the image device) for a tri-stimulus system. This processing can be performed on the YUV or RGB components depending on the system. This should be understood as a mere example.

direction. Therefore, it is used to describe the amount of light that goes through, or is emitted from, a particular area, and falls within a given solid angle.

Chrominance is a numeral that describes the way a certain amount of light is distributed among the visible spectrum. Chrominance has no luminance information but is used together with it to describe a colored image defined, for instance, by an RGB triplet. Any RGB triplet in which the value of R=G=B has no chrominance information.

Gain, exposure, gamma and contrast are related and we use the information of the luminance of the image to calibrate them. The priority is to keep gamma out and the exposure to the minimum possible value to reduce noise in the image and the effect of the moving objects in the image. If the light conditions are very hard, the algorithm calibrates the gamma and exposure time.

Gain is a constant factor that is applied to all the pixels in the image when the image is acquired. Exposure time is the time that the image sensor (CCD or CMOS) is exposed to the light. Gamma correction is the name of a nonlinear operation used to code and decode luminance or TGB tristimulus values. One of the most used definition of contrast is the difference in luminance along the 2D space that makes an object distinguishable. To calibrate all these parameters, the image histogram of luminance is calculated and a statistical measure to balance the histogram of the acquired image is used, as presented next.

The histogram of the luminance of an image is a representation of the number of times that each intensity value appears in the image. Image histograms can indicate if the image is underexposed or overexposed. For a camera correctly calibrated, the distribution of the luminance histogram should be centered around 127 (for a 8 bits per pixel image). An underexposed image will have the histogram leaning to the left, while an overexposed image will have the histogram leaning to the right (for an example see the Fig. 8 a)).

Statistical measures can be extracted from digital images to quantify the image quality [5,6]. A number of typical measures used in the literature can be computed from the image gray level histogram. Based on the experiments presented in [7], in this work we used the mean sample value (MSV):

$$MSV = \frac{\sum_{j=0}^{4}(j+1)x_j}{\sum_{j=0}^{4} x_j},$$

where x_j is the sum of the gray values in region j of the histogram (in the proposed approach we divided the histogram into five regions). When the histogram values of an image are uniformly distributed in the possible values, then $MSV \approx 2.5$.

The brightness parameter is basically a constant (or offset) that can be added (subtracted) from the luminance component of the image. It represents a measure of the average amount of light that is integrated over the image during the exposure time. If the brightness it too high, overexposure may occur which will white the saturated part or the totality of the image. The proposed algorithm considers a black area in the image as reference to calibrate this parameter. The concept is that the black area should be black – in the RGB color space, this means that the average values of R, G and B should be close to zero in this region.

White balance is the global adjustment of the intensities of the colors (typically red, green, and blue primary colors). An important goal of this adjustment is to render specific colors particularly neutral colors correctly; hence, the general method is sometimes called gray balance, neutral balance, or white balance. This balance is required because of different color spectrum energy distribution depending on the illumination source. The proposed algorithm uses a white area as reference to calibrate this parameter. The idea is that the white region should be white – in the YUV color space this means that the average value of U and V should be 127.

The black and white regions are defined manually beforehand and corresponds to regions of the robots in the image. In the case of the NAO robots, the robot stops and look to a defined position of its own body where these regions are. In the case of the CAMBADA robots, due to the use of an omnidirectional vision system [8], the own body is seen in the image and a white and black area was placed close to the mirror.

The saturation of a color is determined by a combination of light intensity that is acquired by a pixel and how much this light is distributed across the spectrum of different wavelengths. Saturation is sometimes also defined as the amount of white you have blended into a pure color. In the proposed algorithm, we consider the histogram of the Saturation (obtained in the HSV color space) and we force the MSV value of this histogram to 2.5, following the explanation above about the use of the MSV measure regarding the histogram of intensities, calibrating this parameter.

Sharpness is a measure of the energy frequency spatial distribution over the image. It basically allows the control of the cut-off frequency of a low pass spatial filter. This may be very useful if the image is afterward intended to be decimated, since it allows to prevent spatial aliases artifacts. We do not consider this parameter in the proposed calibration algorithm as in the referred applications of the robots we work with the resolution of the images acquired by the camera.

A graphical representation of the statistical measures extracted from the image acquired by the camera and its relation to the parameters to be calibrated is presented in Fig. 4.

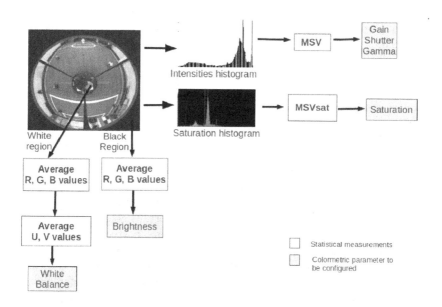

Fig. 4. A graphical representation of the statistical measures extracted from the image acquired by the camera and their relation to the parameters to be calibrated on the camera

3   Proposed Algorithm

The algorithm configures the most important parameters, as referred above. For each one of these parameters, and that are available on the camera, a PI controller was implemented. PI controllers are used instead of proportional controllers as they result in better control, having no stationary error. The constants of the controller have been obtained experimentally for both cameras, guaranteeing the stability of the system and an acceptable time to reach the desired reference.

The algorithm presented next starts by the configuration of the parameters related to the luminance on the image, namely gain, exposure and gamma, by this order if necessary. To improve the quality of the image, i. e. to have as less noise as possible, exposure should be as high as possible. On the other hand, the gamma should be the one that gives the best dynamic range for the intensity and we only want to change it if the gain and exposure alone cannot deliver good results.

When the image acquired has enough quality in terms of luminance, considering that the MSV for the histogram of intensities is between 2 and 3, the algorithm starts calibrating the other parameters, namely white-balance, saturation and brightness, according to the ideas expressed in the previous section.

The algorithm stops when all the parameters have converged. This procedure solves the problem of the correlation that exists between the parameters.

```
do
  acquire image
  calculate the histogram of Luminance
  calculate the MSV value for Luminance
  if MSV != 2.5
    if exposure and gain are in the limits
      apply the PI controller to adjust gamma
    else if gain is in the limit
      apply the PI controller to adjust exposure
    else
      apply the PI controller to adjust gain
    end
    set the camera with new gamma, exposure and gain values
  end
  if MSV > 2 && MSV < 3
    calculate the histogram of saturation
    calculate the MSV value for saturation
    calculate average U and V values of a white area
    calculate average R, G and B values of a black area
    if MSVsat != 2.5
      apply the PI controller to adjust saturation
      set the camera with new saturation value
    end
    if U != 127 || V != 127
      apply the PI controller to adjust WB_BLUE
      apply the PI controller to adjust WB_RED
      set the camera with new white-balance parameters
    end
    if R != 0 || G != 0 || B != 0
      apply the PI controller to adjust brightness
      set the camera with new brightness value
    end
  end
while any parameter changed
```

4   Experimental Results

To measure the performance of the proposed self calibration algorithm, experiments have been made on two robotic platforms, namely the CAMBADA robots (RoboCup Middle Size league robotic soccer robots) and NAO robots (RoboCup Standard Platform League soccer robot), in indoor and outdoor scenarios. In all the cases, the images acquired after the proposed autonomous colormetric calibration have good properties, both in terms of objective and subjective analysis.

A video was submitted together with this paper, showing the experiments described in this paper.

More details on the CAMBADA vision system can be found in [8] and more details about the vision system of the NAO robots used by the Portuguese Team can be found in [9].

The experiment that follows has been conducted using the cameras of the robots with different initial configurations inside a laboratory with both artificial and natural light sources. In Fig. 5, the experimental results are presented both when the algorithm starts with the parameters of the camera set to lower value as well as when set to the higher values. As it can be seen, the configuration obtained after using the proposed algorithm is approximately the same, independently of the initial configuration of the camera.

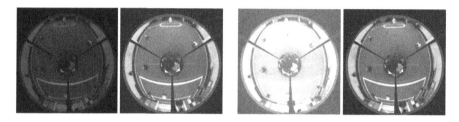

Fig. 5. Some experiments using the proposed automated calibration procedure. From left to right, an image captured with some of the parameters of the camera set to lower values, the corresponding image obtained after applying the proposed calibration procedure, an image captured with some of the parameters of the camera set to higher values and the corresponding image obtained after applying the proposed calibration procedure.

In Fig. 6 is presented the variation of the camera parameters related to the experiment described above. As we can see, the convergence of the parameters is fast. It took less than 100 cycles for the camera to converge to the correct parameters in order to obtain the images presented in Fig. 5. In this experiment, the camera was working at 30 fps that means a calibration time below 3 seconds. These are the worst case scenarios in calibration. Most of times, in practical use, the camera can start in auto mode and the algorithm will be applied after that. An example of this situation is presented in the video, where the lamps of the laboratory are switched on and off after the camera has been calibrated. In these situations, the camera converges in a reduced number of cycles.

In the NAO robot, we work with the cameras at 15 fps due to the limitations on the processing capabilities, which leads to times close to 10 seconds. However, due the fact that the NAO robots do not have graphical interface with the user, the proposed algorithm is very useful to calibrate the two cameras of the robots.

In Fig. 7 we present an image acquired with the camera in auto mode. The results obtained using the camera with the parameters in auto mode are

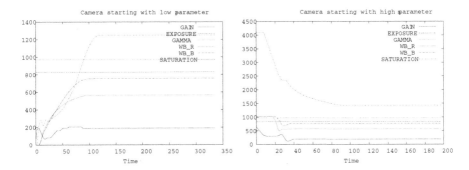

Fig. 6. On the left, a graphic showing the variation of the parameters of the camera when it was started with higher values. On the right, a graphic showing the variation of the parameters of the camera when it was started with higher values. In both experiments, we can see a fast convergence of the parameters.

overexposed and the white balance is not correctly configured, both for the CAMBADA and NAO robots. The algorithms used by the digital cameras that are on the robots (we tested also some more models and the results are similar) is due to several reasons explained next. First, the cameras are typically expecting a natural image and, in most robotic applications, as the example of robotic soccer, shows that the scenario has specific characteristics (omnidirection vision systems, green carpets, black robots, ...). Moreover, as the case of the CAMBADA omnidirectional vision system, the camera analyzes the entire image and, as can be seen in Fig. 5, there are large black regions corresponding to the robot itself. Moreover, and due to the changes in the environment around the robot as it moves, leaving the camera in auto mode leads to undesirable changes in the parameters of the camera, causing problems to the correct feature extraction for object detection.

Fig. 7. From left to right, an example of an image acquired with the camera of CAMBADA in auto mode, an image in the same robot after the proposed algorithm, the image with the camera of NAO in auto mode and an image in the same robot after the proposed algorithm

The good results of the automated calibration procedure can also be confirmed by the histograms presented in Fig. 8. The histogram of the image obtained after applying the proposed automated calibration procedure (Fig. 8b) is centered near the intensity 127, which is a desirable property, as visually confirmed in Fig. 5. The histogram of the image acquired using the camera in auto mode (Fig. 8a) shows that the image is overexposed, leading to the majority of the pixels to have saturated values.

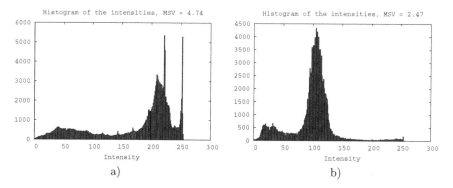

a)   b)

Fig. 8. The histogram of the intensities of the two images presented in Fig. 5. a) shows the histogram of the image obtained with the camera parameters in automode. b) shows the histogram of the image obtained after applying the automated calibration procedure.

As far as we know about the work of other teams that participate in RoboCup Soccer (MSL, SPL and humanoids), as well as according to our personal experience, most of the robots use their digital cameras in manual mode in what concerns colormetric parameters, even with advanced and recent cameras. This statement is easily supported and verified in practice.

In a near future, it is expected that the MSL robots will have to play under natural lighting conditions and in outdoor fields. This introduces new challenges. In outdoor fields, the illumination may change slowly during the day, due to the movement of the sun, but also may change quickly in short periods of time due to a partial and temporally varying covering of the sun by clouds. In this case, the robots have to adjust, in real-time, the camera parameters, in order to adapt to new lighting conditions.

The proposed algorithm was also tested outdoors, under natural light. Figure 9 shows that the algorithm works well even with different light conditions. It confirms that the algorithm can be used in non-controlled lighting conditions and under different environments.

Besides the experimental results presented in this paper, the proposed algorithm have been used with success during 2012 in the competitions where the referred robots participate, namely during RoboCup 2012 both by CAMBADA team in MSL and Portuguese Team in SPL, DutchOpen 2012 by CAMBADA team in MSL and Portuguese Robotics Open 2013 by CAMBADA team in MSL.

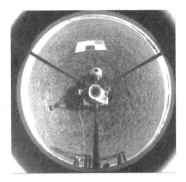

Fig. 9. On the left, an image acquired outdoors using the camera in auto mode. As it is possible to observe, the colors are washed out. That happens because the camera's auto-exposure algorithm tries to compensate the black around the mirror. On the right, the same image with the camera calibrated using the implemented algorithm. As can be seen, the colors and their contours are much more defined.

5   Conclusions

We propose an algorithm to autonomously configure the most important colormetric parameters of a digital camera. This procedure requires a few seconds to calibrate the colormetric parameters of the digital cameras of the robots, independently of the initial parameters of the cameras, as well as the light conditions where the robots are being used. This is much faster than the manual calibration performed by an expert user, that even having as feedback the statistical measures extracted from the digital images that we propose in this paper needs several minutes to perform this operation.

The experimental results presented in this paper show that the algorithm converges independently of the initial configuration of the camera. These results allow the use of the same color ranges for each object of interest independently of the lighting conditions, improving the efficiency of the object detection algorithms.

The calibration algorithm proposed in this paper is also used in run-time in order to adjust the camera parameters during the use of the robots, accommodating some changes that can happen in the environment or in the light, without affecting the performance of the vision algorithms.

As a future work, we would like to extend this work to more robotic applications in non-controlled environments, as well as to present more detailed results regarding its use in more models of digital cameras with different type of lens, that also affects the calibration of these parameters.

Aknowledgements. This work was developed in the Institute of Electronic and Telematic Engineering of University of Aveiro and was partially funded by FEDER through the Operational Program Competitiveness Factors - COMPETE and by National Funds through FCT - Foundation for Science and

Technology in the context of the project FCOMP-01-0124-FEDER-022682 (FCT reference PEst-C/EEI/UI0127/2011).

References

1. Heinemann, P., Sehnke, F., Streichert, F., Zell, A.: Towards a calibration-free robot: The act algorithm for automatic online color training. In: Lakemeyer, G., Sklar, E., Sorrenti, D.G., Takahashi, T. (eds.) RoboCup 2006. LNCS (LNAI), vol. 4434, pp. 363–370. Springer, Heidelberg (2007)
2. Takahashi, Y., Nowak, W., Wisspeintner, T.: Adaptive recognition of color-coded objects in indoor and outdoor environments. In: Visser, U., Ribeiro, F., Ohashi, T., Dellaert, F. (eds.) RoboCup 2007. LNCS (LNAI), vol. 5001, pp. 65–76. Springer, Heidelberg (2008)
3. Krawczyk, G., Goesele, M., Seidel, H.: Photometric calibration of high dynamic range cameras. Research Report MPI-I-2005-4-005, Max-Planck-Institut für Informatik, Stuhlsatzenhausweg 85, 66123 Saarbrücken, Germany (April 2005)
4. Mayer, G., Utz, H., Kraetzschmar, G.K.: Playing robot soccer under natural light: A case study. In: Polani, D., Browning, B., Bonarini, A., Yoshida, K. (eds.) RoboCup 2003. LNCS (LNAI), vol. 3020, pp. 238–249. Springer, Heidelberg (2004)
5. Shirvaikar, M.V.: An optimal measure for camera focus and exposure. In: Proc. of the IEEE Southeastern Symposium on System Theory, Atlanta, USA (March 2004)
6. Nourani-Vatani, N., Roberts, J.: Automatic camera exposure control. In: Proc. of the 2007 Australasian Conference on Robotics and Automation, Brisbane, Australia (December 2007)
7. Neves, A.J.R., Cunha, B., Pinho, A.J.P., Pinheiro, I.: Autonomous configuration of parameters in robotic digital cameras. In: Araujo, H., Mendonça, A.M., Pinho, A.J., Torres, M.I. (eds.) IbPRIA 2009. LNCS, vol. 5524, pp. 80–87. Springer, Heidelberg (2009)
8. Neves, A.J.R., Pinho, A.J., Martins, D.A., Cunha, B.: An efficient omnidirectional vision system for soccer robots: from calibration to object detection. Mechatronics 21(2), 399–410 (2011)
9. Trifan, A., Neves, A.J.R., Cunha, B., Lau, N.: A modular real-time vision system for humanoid robots. In: Proceedings of SPIE IS&T Electronic Imaging 2012 (January 2012)

BRISK-Based Visual Feature Extraction for Resource Constrained Robots

Daniel Jaymin Mankowitz and Subramanian Ramamoorthy

School of Informatics, University of Edinburgh, Edinburgh, EH8 9AB
`daniel@mankowitz.co.za`, `s.ramamoorthy@ed.ac.uk`

Abstract. We address the problem of devising vision-based feature extraction for the purpose of localisation on resource constrained robots that nonetheless require reasonably agile visual processing. We present modifications to a state-of-the-art Feature Extraction Algorithm (FEA) called Binary Robust Invariant Scalable Keypoints (BRISK) [8]. A key aspect of our contribution is the combined use of BRISK0 and U-BRISK as the FEA detector-descriptor pair for the purpose of localisation. We present a novel scoring function to find optimal parameters for this FEA. Also, we present two novel geometric matching constraints that serve to remove invalid interest point matches, which is key to keeping computations tractable. This work is evaluated using images captured on the Nao humanoid robot. In experiments, we show that the proposed procedure outperforms a previously implemented state-of-the-art vision-based FEA called 1D SURF (developed by the rUNSWift RoboCup SPL team), on the basis of accuracy and generalisation performance. Our experiments include data from indoor and outdoor environments, including a comparison to datasets such as based on Google Streetview.

Keywords: BRISK, BRISK0 - U-BRISK, feature extraction, localisation, resource constrained robot, *Nao* Humanoid Robot.

1 Introduction

The emergence of field robots that must persistently operate in dynamic environments brings with it the need for localisation based on features that may not have been explicitly engineered with the robot in mind. The issue is particularly problematic for resource constrained robots that must adopt a low complexity approach to computation. Generally robust localisation needs rich features such as is available from Feature Extraction Algorithms (FEAs) which form a crucial part of vision systems. FEAs are utilised in vision systems in order to detect *landmarks* (also known as *interest points*) and match them between corresponding images. FEAs can therefore be used for tasks such as localisation and are commonly used in systems such as automated-driving and underwater exploration [6,13]. Scale Invariant Feature Transform (SIFT) and Speeded-Up Robust Features (SURF) are examples of FEAs that can be used for this task [3,5]. However, these algorithms have significant processing requirements and therefore are not applicable to a wide variety of resource-constrained systems.

S. Behnke et al. (Eds.): RoboCup 2013, LNAI 8371, pp. 195–206, 2014.

Many current vision-based localisation techniques utilise stereo-vision in order to identify interest points [6, 12]. Many robot platforms, such as the Nao humanoid robot used in the Standard Platform League of RoboCup [10], do not allow for the possibility of utilising stereo vision.

In the RoboCup domain, a visual feature extraction technique termed 1D SURF, has been developed by the *rUNSWift* team for the purpose of localising Nao robots on the football pitch [2]. This method is computationally efficient but could suffer from limited accuracy when generalising to different environments, as will be illustrated in the sections to follow. To address these issues, feature extraction algorithms are required that can both generate rich features and at the same time be accurate and computationally efficient. We also focus on the single camera case, which is the current setup for the Nao humanoid robot [1].

We present a computationally efficient and accurate FEA called BRISK0 - U-BRISK which is a variant of the Binary Robust Invariant Scalable Keypoints (BRISK) FEA [7]. We show that this FEA can detect features which can be utilised for localising resource constrained mobile robots. We developed two novel matching constraints as well as a novel scoring function which is used to find the optimal parameters for FEAs. We present experiments verifying these developments and also highlight the potential for outdoor navigation using these techniques. We also outline a localisation routine incorporating the FEA as a concluding remark.

2   Algorithm Overview

Our proposed architecture of a vision-based feature extraction algorithm, as used in a localisation application, is shown in Figure 1. An image is captured by the robot's vision system (in this case, a Nao humanoid robot) and this forms the input to our FEA. Our FEA then tries to match this image to a set of stored images in an image bank. The stored images, along with their corresponding location in the environment, are manually captured by the robot prior to executing the algorithm. Descriptors for each stored image are computed and stored by the robot. The stored image whose descriptors generate the largest matching score with the input image descriptors is flagged as a match. The matching score and the stored image's corresponding geographic coordinates are passed to a localisation module which then updates the robot's position on a map. The robot will always assume a stationary position when performing this algorithm; this is analogous to a person gathering their bearings when lost in an environment.

Throughout this paper a match between two images will be referred to as an *image match*. A match between corresponding interest points between a pair of images will be referred to as an *interest point match*.

3   BRISK0 - U-BRISK

Binary Robust Invariant Scalable Keypoints (BRISK) has been recently developed by Leutenegger et al. [7]. Interest points are detected by computing a

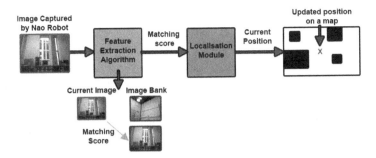

Fig. 1. The FEA incorporated with a proposed localisation module to be used by a resource constrained robot for localisation [9]

Features from Accelerated Segment Test (FAST) [11] score for each pixel in the image. If the pixel is above the pre-defined score threshold, then it is detected as an interest point. A binary feature vector of length 512 bits is then generated from some simple brightness comparison tests. Interest points are then matched between image pairs by computing the Hamming distance between the feature descriptors. This algorithm yields computational performance that betters SURF by an order of magnitude in various domains [7].

Our contribution is a modification to this original BRISK method that is more directly aimed at resource constrained robot systems with limited computation. This detector-descriptor variation is called BRISK0 - U-BRISK. BRISK0 - U-BRISK is based on SU-BRISK developed by Leutenegger et al. [7] and includes a modification to the image processing routine. This modification involves processing only a subset, such as the upper 300 pixel rows, of the captured image as this section contains more repeatable, static features such as ceiling lighting. The lower and more dynamic, less repeatable section of the image is discarded. Of course, this restriction may lose information. We compensate for this by introducing additional tests for consistency, discussed in Section 8.

3.1 BRISK0 Detector

The detector module of the BRISK0 - U-BRISK FEA is responsible for detecting the interest points. The interest points in the original BRISK implementation are invariant to both scale and rotation. To achieve scale invariance, BRISK utilises a scale-space consisting of an image pyramid whereby the lowest layer of the pyramid is the original image and the higher layers of the pyramid are down-sampled versions of the original image.

We discard the scale-space pyramid and only detect interest points on the first octave corresponding to the original image, creating a detector termed BRISK0. This is computationally efficient since down-sampled versions of the original image are discarded as well as the continuous scale refinement procedure [7]. Once all of the interest points are detected, a 512 bit descriptor vector needs to be computed for each of the interest points. This is achieved using the U-BRISK descriptor.

3.2 U-BRISK Descriptor

In the original BRISK implementation, the descriptor vector is calculated by initially generating a pre-defined sampling pattern to sample the neighborhood surrounding the detected interest point k. The samples p_i, are equally-spaced in concentric circles surrounding the detected interest point [7]. The pattern is then rotated based on an angle α that is generated from a set of gradient calculations [7]. This is repeated for each interest point. A set of brightness comparison tests are then used to generate the 512 bit descriptor vector which is used in the matching procedure.

We do not rotate the sampling pattern, as in the original BRISK implementation, to create the U-BRISK descriptor. This prevents the FEA from being rotation invariant but improves the FEA's computational efficiency. It has been shown that the above-mentioned detector-descriptor pair is robust to rotations of up to 10° as well as scale changes of 10% or less [7]. This creates a more computationally efficient FEA which is crucial for resource constrained robots.

3.3 Image Processing

In addition to utilising a computationally efficient FEA, we implemented further optimisations. This includes optimising the 640 × 480 YUV image captured by the Nao robot [1]. The image is first converted to gray scale. Only interest points in the upper 300 pixels of the gray scale image are detected. These correspond to interest points near the ceiling which are less prone to changing over time. This implies that the lower portion of the captured image does not need to be processed. This results in a significant increase in computational performance.

4 Matching Feature Descriptors

Once the descriptors are computed for a pair of images, it needs to be determined whether or not the images overlap one another. We achieve this by utilising an interest point descriptor matching technique. Two techniques are utilised which include 2-NN Matching and Radius Matching.

2-NN Matching and Radius Matching both compute the Hamming or Euclidean distance between a pair of interest point descriptors in feature space for corresponding images, in order to determine whether or not the images overlap one another. The Hamming distance is computed for BRISK-based descriptors, whereas the Euclidean distance is computed for SURF-based descriptors.

As shown in Figure 1, interest points will be computed for the current image and will be matched against interest points corresponding to each image in the image bank. 2-NN Matching will compute the two closest interest point matches, ip_{match1} and ip_{match2} respectively, from an image in the image bank to an interest point $ip_{current}$ in the current image. Only the closest matching interest point ip_{match1} is paired with $ip_{current}$. However, both ip_{match1} and ip_{match2} are required to remove invalid matches by using the well-known $2 - NN$ Ratio constraint. Radius Matching will only assign an interest point match between a pair of images if the distance between the interest point descriptors is below a pre-defined threshold. An example of matched correspondences (indicated by green lines) between two images is shown in Figure 2.

Fig. 2. Matching correspondences between two images

4.1 Matching Score

Once interest point matches are assigned to the current image, it needs to be determined whether the current image and the image in the image bank overlap one another. We achieve this by using a Matching Score (MS). The MS for a pair of interest point descriptors is calculated by taking the inverse Hamming or Euclidean distance between the descriptors [2, 4] as shown in Equation 1. For both 2-NN and Radius Matching, the inverse distance between the interest point descriptor $ip1$ in the current image and its closest match $ip2$ in an image from the image bank are used to compute the MS. Thus, interest points that are very similar will have a small distance corresponding to a high MS and vice versa. The individual MS values for every interest point is summed to produce the Image Matching Score (IMS) between a pair of images as shown in Equation 2. If the IMS is above a pre-defined threshold, then the images overlap and are flagged as an image match.

$$MS_{ip1,ip2} = \frac{1}{D_{ip1,ip2}} \tag{1}$$

$$IMS_{i1,i2} = \sum_{ip1,ip2} MS_{ip1,ip2} \tag{2}$$

For example, in Figure 2, the MS for each corresponding interest point descriptor pair (connected by a green line) is calculated. In the figure, there are 20 individual MS. These scores are summed together to produce the IMS between these images. Since these images do indeed overlap one another, the matching score between the images should be above the pre-defined IMS threshold, indicating an image match.

5   Novel Matching Constraints

For both 2-NN and Radius matching, it is important to determine whether interest point matches are indeed valid matches. Thus, we developed two novel geometric matching constraints in an attempt to remove invalid interest point matches. These constraints are termed the Angle and Distance constraints respectively.

In order to determine whether an interest point match is valid, we present an algorithm that initially places the images, containing each of the relevant interest points, next to one another as shown in Figure 3 and Figure 4. The angle constraint calculates the angle from the interest point in the left image to the interest point in the right image. Through visual analysis, it has been determined that the angle should be less than $10°$ in order for two interest points to match one another[1]. As can be seen in Figure 3, α is larger than the pre-defined angle threshold and therefore the interest points are invalid. β however, is smaller than the threshold and therefore a valid match (shown in green) is found.

The distance constraint states that the number of image columns separating two interest points in corresponding images should be the width of the image plus a pre-defined threshold. We have assumed small changes in rotation and scale which would again be applicable to the RoboCup domain. Through visual analysis, a pre-defined threshold of 200 pixels has been chosen. As can be seen in Figure 4, interest points that are within this threshold are flagged as valid matches (shown in green).

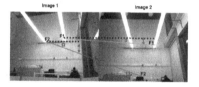

Fig. 3. The angle constraint that has been developed to remove invalid matches

Fig. 4. The distance constraint that has been developed to remove invalid matches

6   Experimental Setup

We tested BRISK0 - U-BRISK using images captured from the upper camera of a Nao humanoid robot [1]. The Nao Robot uses an ATOM Z530 1.6 GHz processor [1]. In addition to this, the Nao's camera can process 30 frames per second. Therefore, in order to process every frame, the FEA must be able to run in under $33ms$ on the Nao's processor. All experimental image matching procedures were evaluated on a computer with an Intel Core 2 Duo T6400 2.00GHz processor[2].

---

[1] This is very useful for the RoboCup domain as only small rotations are expected whilst the robot is localising itself.

[2] It must be noted that the computer's processor is approximately 2.5 times faster than the Nao's processor. Therefore the computational times presented in this section are faster than the real-time performance on the Nao.

In order to ensure that the BRISK0 - U-BRISK FEA exhibits the best over-all performance, we tested it against the four FEAs mentioned in Section 7. The three environments utilised for this testing procedure include a standard RoboCup environment, a large hall and an office. Examples of images captured in each of these environments are shown in Figure 5.

Robocup   Large Hall   Office

Fig. 5. Examples images from each of the environments used for the experiments

We generated in each environment a dataset of overlapping images and a dataset of non-overlapping images for the experiments to follow. A pair of images are considered to overlap one another if at least 50% of the first image visually overlaps the second image. This has been manually performed to ensure that the images do indeed overlap. Pairs of images with no overlap are referred to as non-overlapping images. In total 200 images were captured over all three datasets at scales varying over a range of 3 meters and orientations varying over the Nao head's full yaw range. This generated in total 2540 overlapping image pairs and 4290 non-overlapping image pairs.

7   Parameter Optimisation

In order to verify that BRISK0 - U-BRISK is a computationally efficient and ac-curate FEA, we compared BRISK0 - U-BRISK to variations of well known FEAs. The detector-descriptor FEAs compared include BRISK-BRISK, BRISK0-BRISK, BRISK0-SURF2D and 1D SURF as implemented by the *rUNSWift* team.

Each of these FEAs has a set of parameters that need to be pre-defined in order to accurately and efficiently match images. The parameters include the Minimum Interest Point Detection Threshold (MIPDT) which is the threshold above which a pixel is detected as an interest point; the Maximum Accepted Hamming Distance (MAHD) which is the maximum number of bits that can differ between interest point descriptors for the interests points to be flagged as a match; the Maximum Accepted Euclidean Distance (MAED) is the maximum Euclidean distance below which two descriptors are flagged as a match.

The optimal parameters $MIPDT^*$, $MAHD^*$ and $MAED^*$ for each FEA are found using a grid search and a novel scoring function that we developed for this purpose. 1D SURF uses the optimal parameter settings recommended by the *rUNSWift* team [2] and therefore is left out in this experiment. To determine the optimal parameters, 108 training images were captured at different scales and orientations, generating a total of 1421 and 2808 overlapping and non-overlapping image pairs respectively. These images are captured by a Nao Robot that is placed on a RoboCup football pitch in a standard RoboCup environment.

The novel scoring function consists of a number of sub-functions which are now detailed. The Single Image Score (SIS) function, shown in Equation 3, represents how well a pair of images i_1, i_2 overlap one another in a particular dataset d for a particular Feature Extraction Algorithm FE and parameter values $\mathbf{p}$.

$$SIS_{(i_1,i_2),d}^{FE,\mathbf{p}} = \alpha f(t_{i_1,i_2}) + (1-\alpha)g(NVM_{i_1,i_2}) \quad 0 \le \alpha \le 1 \tag{3}$$

$f(t_{i_1,i_2})$, shown in Equation 4, represents the timing score for a pair of images based on the overall time (in milliseconds) taken to perform image processing, detection, extraction and matching. t_{max} is defined as the largest time tabulated for the current dataset d in milliseconds and normalises the score between 0 and 1. A large time is undesired and will result in a low matching score.

$$f(t_{i_1,i_2}) = \mid log_{10}(\frac{0.9t_{i_1,i_2}}{t_{max}^{FE}} + 0.1) \mid \quad f(t_{i_1,i_2}) \in [0,1] \tag{4}$$

$g(NVM_{i_1,i_2})$, shown in Equation 5, rewards a pair of images if they have a large Number of Valid interest point Matches (NVM). Here, M_{total} is the total number of interest point matches between a pair of images. ϵ can be set to any value above 0 and has been pre-defined with a value of 0.1. The parameter α is a weighting parameter and takes values between 0 and 1.

$$g(NVM_{i_1,i_2}) = \frac{NVM_{i_1,i_2}}{M_{total,(i_1,i_2)} + \epsilon} \quad g(NVM_{i_1,i_2}) \in [0,1] \tag{5}$$

Once the SIS score has been calculated for each pair of images i_1, i_2, these scores are then summed together for a particular set of parameter values $\mathbf{p}$ using a specific FE in a particular dataset d. The resulting score is called the Multi-Image Score (MIS) and is shown in Equation 6.

$$MIS_{\mathbf{p},d}^{FE} = \beta\tau + (1-\beta)h(IZM) \quad 0 \le \beta \le 1 \tag{6}$$

The first term of the MIS score is the computation of the mean of all SIS scores for a particular set of parameter values $\mathbf{p}$, using a specific feature extraction algorithm FE, in a particular dataset d. This value is referred to as τ and is shown in Equation 7. β is a weighting parameter between 0 and 1.

$$\tau = \frac{\sum_{i_1,i_2=1,i_1 \ne i_2}^{N} SIS_{(i_1,i_2),d}^{FE,\mathbf{p}}}{N} \tag{7}$$

$h(IZM)$, defined in Equation 8, is a scoring function that accounts for the number of Image Zero Matches (IZM) for a particular set of parameter values $\mathbf{p}$ in a particular dataset d using a specific FEA. An IZM is defined as a pair of images containing no valid interest point matches. The denominator IZM_{max}^{FE} represents the maximum number of IZMs found for a particular parameter setting in a particular dataset using a particular FEA. This function penalises parameter settings that result in a large number of IZMs since IZMs should not be present in a dataset of overlapping images.

$$h(IZM)_{\mathbf{p},d}^{FE} = | log_{10}(\frac{0.9 IZM_{\mathbf{p}}^{FE}}{IZM_{max}^{FE}} + 0.1) | \quad h(IZM_p^{FE}) \in [0,1] \quad (8)$$

The maximum $MIS_{\mathbf{p},d}^{FE}$ is then found across all datasets and the corresponding parameters $\mathbf{p}$ for each FEA are then chosen as the optimal parameters. Using this parameter optimisation procedure, we show that BRISK0 - U-BRISK produces the best overall image matching performance.

8   Experimental Results

8.1   Comparative Performance

We found that BRISK0 - U-BRISK produced the best overall performance whilst utilising the Radius Matching technique as shown in Table 1. 1D SURF utilises the RANSAC Matching technique as developed by the $rUNSWift$ team [2]. As seen in Table 1, in the worst case BRISK0 - U-BRISK can match an image pair in 12.82 ms. This is within the 33 ms time constraint required to process a pair of images for every image frame[3]. It should be noted that the times tabulated for 1D SURF utilise a sub-optimal image processing routine, which is not utilised on the $rUNSWift$ robots. The times for 1D SURF can therefore be improved upon [2].

Table 1. The comparative performance for each of the FEAs in three different environments

	RoboCup		Large Hall		Office	
FEA Algorithm	AUC (%)	Time (ms)	AUC (%)	Time (ms)	AUC (%)	Time (ms)
BRISK0-BRISK	92.455	9.734	96.74	14.64	94.85	11.71
BRISK-BRISK	83.290	14.627	95.52	19.89	96.23	17.20
BRISK0-SURF2D	96.033	13.027	98.80	20.35	96.05	16.63
BRISK0-UBRISK	**97.242**	**8.805**	**93.15**	**12.82**	**96.15**	**10.45**
SURF 1D	74.039	13.301	90.84	14.03	92.33	14.14

We generated a ROC curve in the standard RoboCup environment which is shown in Figure 7. The ROC curve has been generated by utilising the IMS as the threshold which is varied from the maximum IMS in the dataset to 0.

---

[3] The $12.82ms$ generated by the computer has been converted to the approximate time expected on the Nao's processor. This time is still within the $33ms$ time constraint.

All overlapping image pairs with a threshold above IMS are classified as a True Positive (TP) match. False Positive (FP) image pairs are generated by non-overlapping images that are above the IMS threshold. In total we generated 1740 overlapping image pairs and 3480 non-overlapping image pairs in order to calculate the ROC curve.

The percentage Area Under the ROC Curve (AUC) for BRISK0 - U-BRISK is comparable with the other FEA methods as seen in the table. In addition, it is desirable to have a FP rate of 0 in order to prevent the robot from generating incorrect image matches. Therefore we had to determine the maximum TP rate that can be attained with a FP rate of 0, $TP_{FP=0}^{max}$, for BRISK0 - U-BRISK. As can be seen in Figure 6 BRISK0 - U-BRISK performs well in all environments attaining a minimum $TP_{FP=0}^{max}$ value of 67% in the RoboCup environment. In addition to this, it out-performs the 1D SURF routine indicating that BRISK0 - U-BRISK has superior performance in each tested environment.

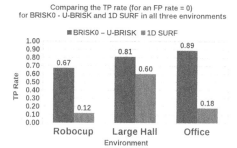

Fig. 6. The $TP_{FP=0}^{max}$ values for each of the three environments for BRISK0 - U-BRISK

Fig. 7. A comparison of the ROC curves for the RoboCup dataset using Radius Matching

8.2   Varying Lighting Conditions

Since BRISK0 - U-BRISK has been chosen as the best overall FEA, we needed to test it under varying conditions to determine its robustness. One such test is under varying illumination. In total 5040 overlapping image pairs and 10008 non-overlapping image pairs were generated whilst switching off combinations of electrical lights in a standard indoor RoboCup environment. The image scale is fixed for this experiment. We found that the $TP_{FP=0}^{max}$ value decreased in poor lighting conditions. A minimum $TP_{FP=0}^{max}$ value of 40% resulted in the poorest lighting conditions; this implies that it is still possible to match image pairs under varying illumination albeit with poorer matching capabilities.

8.3   Outdoor Navigation

A possible application of the BRISK0 - U-BRISK FEA is utilising it on a resource constrained robot to perform navigation in an outdoor environment. Since the algorithm is sensitive to rotation and scaling, this algorithm would be better suited to fairly stable outdoor environments. Examples include a home robot wandering around a yard, navigating around a housing complex or small suburb. In order to test this application, a dataset has been generated using Google Street View (GSV) images. If a resource-constrained robot can match its captured image I_{robot} to images that I_{robot} overlaps in the GSV dataset, I_{google}, then in principle the robot can localise itself.

For the experiment, 30 GSV images were downloaded at the same resolution as that of a Nao robot. The images were captured at latitude: 55.94474 and longitude: -3.18779. The GSV images were combined with images captured by the Nao robot at the same location. This produced 210 overlapping image pairs and 420 non-overlapping image pairs.

After running BRISK0 - U-BRISK FEA, it was found that the Nao's images can be matched to the corresponding GSV images that the Nao's images overlap, albeit with a significant decrease in matching performance. An example of interest point matches between I_{robot} and I_{google} can be seen in Figure 8. The ROC curve generated an AUC value of 77.57%. The $TP_{FP=0}^{max}$ value is 39% which implies that it is possible to match overlapping images whilst never matching non-overlapping images. One of the main reasons for the decrease in performance was due to buildings with many similar windows which caused matching ambiguities.

Fig. 8. Matching correspondences, shown in green, between an image captured by the Nao's camera (left image) and the Google StreetView image (right image)

9   Conclusions

We have presented BRISK0 - U-BRISK which is a unique detector-descriptor variation of the BRISK FEA and has not previously been utilised on a resource constrained robot. We found that BRISK0 - U-BRISK produced the best overall performance in terms of accuracy and generalisation to different environments. It performs significantly better than 1D SURF on these metrics. BRISK0 - U-BRISK is sensitive to changes in scale and rotation in return for computational efficiency. This may limit the applications in which the algorithm can be utilised. In harsh environments, the full power of the original BRISK implementation may be required. This FEA can still match images under varying lighting conditions albeit at a slight decrease in performance. It can potentially be used for some forms of outdoor navigation as seen in the Google Street View experiment. We are well aware that many sophisticated techniques already exist for outdoor navigation and we do not expect our method to outdo others in terms of absolute performance. However, many scenarios involving resource constrained robots that operate in mixed environments, e.g., in the patio of a home or office, could benefit from this capability. In current work, we are looking at implementing this method as part of a particle filter localisation module shown in Figure 1.

Acknowledgements. This work has taken place in the Robust Autonomy and Decisions group, School of Informatics, University of Edinburgh. Research of the RAD Group is supported by the UK Engineering and Physical Sciences Research Council (grant number EP/H012338/1) and the European Commission (TOMSY Grant Agreement 270436, under FP7-ICT-2009.2.1 Call 6).

References

1. Aldebaran Robotics, Nao Hardware (2013), http://www.aldebaran-robotics.com/documentation/family/robots/video_robot.html
2. Anderson, P., Yusmanthia, Y., Hengst, B., Sowmya, A.: Robot Localisation Using Natural Landmarks. In: Chen, X., Stone, P., Sucar, L.E., van der Zant, T. (eds.) RoboCup 2012. LNCS (LNAI), vol. 7500, pp. 118–129. Springer, Heidelberg (2013)
3. Bay, H., Ess, A., Tuytelaars, T., Van Gool, L.: Speeded-up robust features (SURF). Computer Vision and Image Understanding 110(3), 346–359 (2008)
4. Briggs, A., Yunpeng, L., Scharstein, D.: Feature matching across 1D panoramas. In: Proc. IEEE Workshop on Omnidirectional Vision and Camera Networks (2005)
5. Juan, L., Gwun, O.: A comparison of sift, pca-sift and surf. International Journal of Image Processing (IJIP) 3(4), 143–152 (2009)
6. Lategahn, H., Geiger, A., Kitt, B.: Visual SLAM for autonomous ground vehicles. In: 2011 IEEE International Conference on Robotics and Automation (ICRA), pp. 1732–1737. IEEE (2011)
7. Leutenegger, S., Chli, M., Siegwart, R.: BRISK: Binary robust invariant scalable keypoints. In: Proc. International Conference on Computer Vision (ICCV), pp. 2548–2555. IEEE (2011)
8. Mair, E., Hager, G.D., Burschka, D., Suppa, M., Hirzinger, G.: Adaptive and Generic Corner Detection Based on the Accelerated Segment Test. In: Daniilidis, K., Maragos, P., Paragios, N. (eds.) ECCV 2010, Part II. LNCS, vol. 6312, pp. 183–196. Springer, Heidelberg (2010)
9. Mankowitz, D.J.: BRISK-based Visual Landmark Localisation using Nao Humanoid Robots. MSc. Thesis, MSc. in Artificial Intelligence, University of Edinburgh (2012)
10. RoboCup, Standard Platform League (2012), http://www.tzi.de/spl/bin/view/Website/WebHome
11. Rosten, E., Drummond, T.W.: Machine learning for high-speed corner detection. In: Leonardis, A., Bischof, H., Pinz, A. (eds.) ECCV 2006, Part I. LNCS, vol. 3951, pp. 430–443. Springer, Heidelberg (2006)
12. Se, S., Lowe, D., Little, J.: Vision-based Mobile Robot Localization and Mapping using Scale-Invariant Features. In: Proc. International Conference on Robotics and Automation (ICRA), pp. 2051–2058 (2001)
13. Thomas, S., Salvi, J., Petillot, Y.: Real-time Stereo Visual SLAM. MSc. Thesis, MSc. Erasmus Mundus in Vision and Robotics (2008)

Compliant Robot Behavior Using Servo Actuator Models Identified by Iterative Learning Control

Max Schwarz and Sven Behnke

Rheinische Friedrich-Wilhelms-Universität Bonn
Computer Science Institute VI: Autonomous Intelligent Systems
Friedrich-Ebert-Allee 144, 53113 Bonn, Germany
`max.schwarz@uni-bonn.de, behnke@cs.uni-bonn.de`

Abstract. System parameter identification is a necessary prerequisite for model-based control. In this paper, we propose an approach to estimate model parameters of robot servo actuators that does not require special testing equipment. We use Iterative Learning Control to determine the motor commands needed to follow a reference trajectory. To identify parameters, we fit a model for DC motors and friction in geared transmissions to this data using a least-squares method. We adapt the learning method for existing position-controlled servos with proportional controllers via a simple substitution. To achieve compliant position control, we apply the learned actuator models to our humanoid soccer robot NimbRo-OP. The experimental evaluation shows benefits of the proposed approach in terms of accuracy, energy efficiency, and even gait stability.

1 Introduction

DC servo motors are popular actuators in the field of robotics because of their ease of use and low cost. Traditional control methods often ignore the dynamics of the motor, in particular friction forces, and compensate the loss of knowledge about the system through sensory feedback. While it is possible to reach very small position errors with this method, high-gain position control often results in undesirable behavior like stiffness and oscillations (limit cycles).

The research field of humanoid soccer robots places unique demands on the performance of robot joints. Having the ability to perform highly dynamic motions is more important than accurate setpoint tracking at low speeds. These motions require considerable torques and moments and can be dangerous to the joint—in particular to the gear—if not executed properly.

The use of motor and friction models enables the controller to demand exactly the torque needed to follow a position trajectory. In addition to minimizing energy consumption, this leads to a more compliant motion of the robot in the face of unexpected obstacles or perturbations—a feature that is important in many robot applications.

Determining motor model coefficients can be a daunting task, since specific test runs in controlled situations are needed. Often the motor cannot remain in the robot for parameter measurement. Additionally, special test setups might

S. Behnke et al. (Eds.): RoboCup 2013, LNAI 8371, pp. 207–218, 2014.

be needed to produce fixed load conditions. These problems call for a simpler identification method, which is the main objective of this work.

In this paper, we define models for DC motors and friction in gear transmissions. We apply Iterative Learning Control [1] to follow a reference trajectory. From the resulting motor commands, we identify the model parameters. We evaluate this learning and identification process on our NimbRo-OP humanoid soccer robot [2]. Finally, we apply the model for feed-forward control during full-body walking motions.

2   Related Work

Friction effects in robot joints have been thoroughly investigated and characterized. Waiboer et al. [3] successfully modeled friction forces in robotic joints as friction between lubricated discs in a rolling-sliding contact. For parameter estimation, a four step least-squares fitting involving the hand-tuning of Stribeck parameters is needed.

A smaller version *(AX-12+)* of the actuator used for evaluation *(MX-106)* has been modeled before [4] with a similar friction model. Special test setups were used to determine viscous and static friction constants. The Stribeck parameters were also hand-tuned.

A few approaches for online learning of friction compensation torques exist. Kim et al. [5] apply reinforcement learning with a neural network to control a 1-DOF system with changing friction parameters.

Iterative Learning Control has been proposed as a method for friction compensation before. Liu [6] used a PD type iterative control for learning torque commands to overcome friction effects on a fixed trajectory. However, generalization to other trajectories or operating conditions was not attempted.

In contrast to the existing work, the proposed system identification method does not require special test benches or isolation of the actuator. The parameters are estimated using a trajectory that is relevant to the robot's general operation. This ensures good results in important position, velocity, and acceleration ranges. While the Stribeck curve parameters still might need to be tuned by hand, all other parameters are calculated using a single linear optimization.

Of particular interest here is the direct modeling of DC motors, which has not been at the center of scientific attention since most robot actuators are controlled by position or torque, hiding the internal workings of the servo motor behind a controller interface.

3   Modeling Robot Joints

3.1   Motor Model

The motor model we are considering here describes a simplified ideal DC motor. This assumption is a good one for most robotics actuators, since they contain high-quality DC motors.

A simple model for an ideal DC motor can be derived by considering the power balance present in the motor at a constant voltage U:

$$P_{el} = P_{mech} + P_J, \tag{1}$$

where P_{el} is the electrical power consumed, P_{mech} denotes the mechanical output power and P_J corresponds to the Joule heating losses in the motor winding. Further substitution yields

$$UI = \omega\tau + RI^2 \tag{2}$$

$$\Leftrightarrow U = \omega\frac{\tau}{I} + RI, \tag{3}$$

where U is the voltage applied to the motor, I is the current flowing through the winding, ω and τ are the present angular velocity and torque, respectively, and R denotes the motor winding resistance.

The torque constant k_τ describes the relationship of τ and I:

$$\tau = k_\tau I, \tag{4}$$

which gives

$$U = \omega k_\tau + \frac{R}{k_\tau}\tau. \tag{5}$$

This equation determines the required motor voltage at a given angular velocity to produce a target torque, which can be calculated using the inverse dynamics as discussed below.

3.2 Friction Model

The above motor model describes an ideal DC motor. Real robot joints suffer from friction effects not only inside the motor itself but also in connected transmissions. The focus of our research is on gear transmissions, as they are the most common type of transmission used. We will not consider gear inertia, but discuss a way how they could be included.

The torque τ generated by the DC motor is converted into the output torque τ_o and the friction torque τ_f. Motor axis angular velocity ω and joint velocity $\dot{q}$ are tightly coupled by the gear constant:

$$\tau = \tau_o + \tau_f, \tag{6}$$

$$\omega = k_{gear}\dot{q}. \tag{7}$$

The dominant friction forces result from static and Coulomb friction in the gears and bearings. The transition between these friction states is known as the Stribeck curve [7]. A common choice for this transition is an exponential decrease from static to Coulomb friction developed by Bo and Pavelescu [8] and refined by

Armstrong-Helouvry [9] with an additional viscous friction term for lubricated gear teeth:

$$\tau_f = \text{sgn}(\dot{q})\left((1-\beta)\tau_c + \beta\tau_s\right) + c^{(v)}\dot{q}, \tag{8}$$

$$\beta = \exp\left(-\left|\frac{\dot{q}}{\dot{q}^{(s)}}\right|^{\delta}\right), \tag{9}$$

where q is the joint position (angular or linear), τ_c and τ_s describe Coulomb and static friction torques, and $c^{(v)}$ is the viscous friction constant. The Stribeck parameter $\dot{q}^{(s)}$ determines the transition velocity between static and Coulomb friction. The empirical exponent δ depends on the material surfaces and ranges from 0.5 to 1.0 [3].

The combination of motor and friction models results in the equation

$$U = \omega k_\tau + \frac{R}{k_\tau}\left(\tau_o + \text{sgn}(\dot{q})\left((1-\beta)\tau_c + \beta\tau_s\right) + c^{(v)}\dot{q}\right) \tag{10}$$

$$= \dot{q}k_{gear}k_\tau + \frac{R}{k_\tau}\tau_o + \tau_c\,\text{sgn}(\dot{q})\frac{R}{k_\tau}(1-\beta) + \tau_s\,\text{sgn}(\dot{q})\frac{R}{k_\tau}\beta + \frac{R}{k_\tau}c^{(v)}\dot{q} \tag{11}$$

$$= \frac{R}{k_\tau}\tau_o + (k_{gear}k_\tau + \frac{R}{k_\tau}c^{(v)})\dot{q} + \tau_c\frac{R}{k_\tau}\,\text{sgn}(\dot{q})(1-\beta) + \tau_s\frac{R}{k_\tau}\,\text{sgn}(\dot{q})\beta, \tag{12}$$

which can be simplified to

$$U = \alpha_0\tau_o + \alpha_1\dot{q} + \alpha_2\,\text{sgn}(\dot{q})(1-\beta) + \alpha_3\,\text{sgn}(\dot{q})\beta. \tag{13}$$

As exact values for the physical motor and friction constants are not required, it suffices to determine the α_i to obtain a motor model usable for control. The α_i are linear coefficients and can be estimated from experimental data using regression techniques.

One should note that β depends on the Stribeck parameters δ and $\dot{q}^{(s)}$, which cannot be estimated using linear optimization. Non-linear optimization techniques may be employed to solve this problem, but have their own shortcomings, as they might find local optima which generate physically incorrect solutions [10].

As the influence of the Stribeck parameters is limited to very low velocities, it is satisfactory in most cases to use a reasonable fixed transition speed and set the exponent δ to 1 in order to simplify the calculations. If low velocities are an important aspect of the robot's operation, hand-tuning of the Stribeck parameters might be necessary.

If explicit consideration of gear inertia is needed, a fifth term needs to be added, as the torque generated by gear inertia is directly proportional to the angular acceleration $\ddot{q}$ of the joint axis:

$$U = \alpha_0\tau_o + \alpha_1\dot{q} + \alpha_2\,\text{sgn}(\dot{q})(1-\beta) + \alpha_3\,\text{sgn}(\dot{q})\beta + \alpha_4\ddot{q}. \tag{14}$$

In our experiments, consideration of gear inertia was not required for good model performance.

A reaction time t_d needs to be included in the model equation if there are significant time delays in the system:

$$U(t - t_d) = \alpha_0 \tau_o(t) + \alpha_1 \dot{q}(t) + \alpha_2 \operatorname{sgn}(\dot{q}(t))(1 - \beta(t)) + \alpha_3 \operatorname{sgn}(\dot{q}(t))\beta(t). \quad (15)$$

3.3   Adaption to Position-Controlled Actuators

Most of the available intelligent actuators for robotics are position-controlled servo motors. Since the applied motor voltage cannot be directly influenced, a relationship between the command input of the actuator and the applied motor voltage needs to be found. The actuators usually include a complete PID position controller with configurable gains. In most cases, however, only the proportional part P is used, which results in:

$$U = U_{bat} k_{ctrl} k_P (q_d - q), \quad (16)$$

where U_{bat} is the supply voltage, k_P is the configurable P controller gain and k_{ctrl} maps the controller output to motor voltage U. The current and desired servo positions are described by q and q_d, respectively. Since q_d is the command variable, we solve for it:

$$q_d = \frac{1}{k_{ctrl} k_P U_{bat}} U + q \quad (17)$$

$$= \frac{1}{k_{ctrl} k_P U_{bat}} (\alpha_0 \tau_o + \alpha_1 \dot{q} + \alpha_2 (1 - \beta) + \alpha_3 \beta) + q. \quad (18)$$

As k_{ctrl} is unknown, it is combined with the model coefficients α_i:

$$q_d = \frac{1}{k_P U_{bat}} (\hat{\alpha}_0 \tau_o + \hat{\alpha}_1 \dot{q} + \hat{\alpha}_2 (1 - \beta) + \hat{\alpha}_3 \beta) + q. \quad (19)$$

The proposed learning and identification procedure can then be performed as presented below. Care should be taken to select a small enough k_P. If k_P is big, the model influence will be small, resulting in bad model fit precision.

The explicit consideration of the supply voltage U_{bat} compensates drops in voltage due to battery draining and ensures good model match over wide voltage ranges.

One should note that the feedback characteristic of the P controller is not modified as long as the model is correct. This means positional errors lead to proportional responses in voltage, just offset by the voltage required to generate τ_o.

4   System Identification

4.1   Reference Trajectory and Learning Process

During the learning process, the robot repeatedly executes a fixed reference trajectory function $q_{ref}(t)$ on a single joint, which in turn defines the goal servo position over time. The reference trajectory should cover a wide range of servo motions, especially those relevant in later operation. A reference trajectory for a soccer robot should, for instance, include walking and kicking motions.

The first and second derivative of the reference trajectory are needed, since both appear in the motor and friction models and/or dynamics equations which are required to calculate τ_o.

We apply the classic Iterative Learning Control (ILC) algorithm introduced by Arimoto et al. [1] to determine the command inputs from one iteration k to the next iteration $k + 1$:

$$U^{(0)}(t) = 0, \tag{20}$$

$$U^{(k+1)}(t) = U^{(k)}(t) + \lambda(e^{(k)}(t + t_d)), \tag{21}$$

where λ is the learning feedback coefficient and $e^{(k)}$ denotes the trajectory error in the k-th iteration. The first command curve $U^{(0)}(t)$ can also be initialized with a guessed command curve (e.g. using the output of an existing controller) to make the algorithm converge faster.

4.2   Parameter Estimation

When the trajectory error after K ILC iterations is sufficiently small, the combined motor and friction model can be fitted against the command curve $U^{(K)}$ generated with ILC. A discrete sampling with N samples is used, while simultaneously compensating the system reaction time:

$$\hat{U}(n) = U^{(K)}(n\Delta t), \tag{22}$$

$$\hat{\tau}_o(n) = \tau_o(n\Delta t + t_d), \tag{23}$$

$$\hat{q}(n) = q(n\Delta t + t_d), \tag{24}$$

$$\hat{\beta}(n) = \beta(n\Delta t + t_d). \tag{25}$$

The parameter identification can then be modeled as a least-squares linear optimization problem:

$$\begin{pmatrix} \alpha_0 \\ \alpha_1 \\ \alpha_2 \\ \alpha_3 \end{pmatrix} = \arg\min \left\| A \begin{pmatrix} \alpha_0 \\ \alpha_1 \\ \alpha_2 \\ \alpha_3 \end{pmatrix} - \begin{pmatrix} \hat{U}(0) \\ \vdots \\ \hat{U}(N-1) \end{pmatrix} \right\|^2 \tag{26}$$

with

$$A = \begin{pmatrix} \hat{\tau}_o(0) & \dot{\hat{q}}(0) & \mathrm{sgn}(\dot{\hat{q}}(0))(1 - \hat{\beta}(0)) & \mathrm{sgn}(\dot{\hat{q}}(0))\hat{\beta}(0) \\ \vdots & \vdots & \vdots & \vdots \\ \hat{\tau}_o(N-1) & \dot{\hat{q}}(0) & \mathrm{sgn}(\dot{\hat{q}}(N-1))(1 - \hat{\beta}(N-1)) & \mathrm{sgn}(\dot{\hat{q}}(N-1))\hat{\beta}(N-1) \end{pmatrix}. \tag{27}$$

Traditional methods including SVD or QR factorization can be used to solve for α_i.

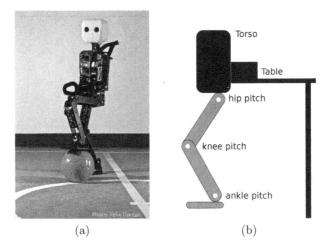

(a)   (b)

Fig. 1. Experimental setup: (a) NimbRo-OP prototype on which the experiments were carried out; (b) Setup for single joint evaluations. Fixed robot parts are drawn in *black*, while the *orange* leg can move freely.

5 Experimental Results

To validate our approach, we performed experiments on the NimbRo-OP Humanoid TeenSize Open Platform robot [2] (see Fig. 1(a)). NimbRo-OP primarily uses Dynamixel MX-106 intelligent actuators from Robotis Inc. in its legs.

We used a newly introduced inverse dynamics module based on the Recursive Newton-Euler Algorithm [11] contained in the open-source Rigid Body Dynamics Library (RBDL, see [12]) to calculate the torques needed for achieving desired joint trajectories. For these calculations, we created a full kinematic model of our robot, including inertial information. We estimate the direction of gravity using the built-in inertial measurement unit of NimbRo-OP.

5.1 Single Joint Evaluation

For evaluation on the single joint level, we fixed the NimbRo-OP torso to a table and used a hip pitch actuator of the robot (see Fig. 1(b)). The other actuators of the leg were commanded to hold position. Given that the servo is position controlled, Equation (19) models the system. We choose a reference trajectory that contained elements relevant to the robot application (robot soccer), as shown in Fig. 2. We derived velocities and accelerations from the reference trajectory and calculated torques using the inverse dynamics module.

We fixed the Stribeck parameters $\dot{q}^{(s)}$ and δ to initial assumptions and estimated the system reaction time t_d from the time delay between command and reaction under P control (see Fig. 5). We increased the ILC learning coefficient λ slowly until a sufficient convergence speed was reached. The final parameter values are summarized in Table 1(a).

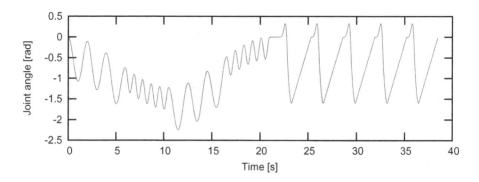

Fig. 2. The reference trajectory used in the experiments. The first section contains sinusoidal oscillations as they occur during walking motions. The second section is composed of fast sinusoidal and linear parts often used for kicking motions.

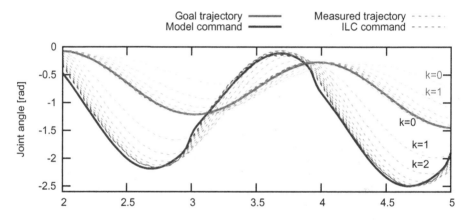

Fig. 3. Learning process applied on the reference trajectory with 12 iterations of ILC. Note that the first ILC iteration ($k = 0$) starts at the goal trajectory, with further iterations converging towards the model command curve fitted in the 12th iteration.

The learning process is visualized in Fig. 3. During the learning process, the trajectory deviation was measured in a windowed fashion to allow for latencies:

$$q_{max} = \max_{t} \min_{-D<a<D} |q_{ref}(t) - q_m(t+a)|, \qquad (28)$$

where $q_m(t)$ denotes the measured trajectory.

A total of 12 ILC iterations were needed to produce an acceptable maximum trajectory deviation of approx. 0.02 rad for a window size $D = 25$ ms. Note that the learned commands sent to the motor often significantly deviate from the reference trajectory by more than 1 rad. The model was then fitted using least squares. The fitted model produced a maximum trajectory deviation of 0.1 rad on the reference trajectory. Fig. 4 shows the detailed development of the trajectory error. The identified parameters are summarized in Table 1(b).

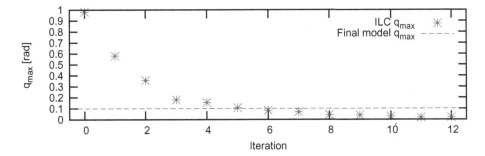

Fig. 4. Maximum windowed trajectory error q_{max} (see Eq. 28) during the learning process

Table 1. (a) Chosen parameters of the learning process; (b) identified model parameters for ROBOTIS Dynamixel MX-106

$\dot{q}^{(s)}$	δ	t_d	λ	$\hat{\alpha}_0$	$\hat{\alpha}_1$	$\hat{\alpha}_2$	$\hat{\alpha}_3$
0.1 rad/s	1.0	0.03 s	0.8	0.19817	0.49586	0.13729	0.03006
	(a)					(b)	

5.2   Integration into a Walking Motion

We integrated the resulting motor model into the emerging NimbRo-OP soccer software framework [13]. We used the model for feed-forward control only and did not incorporate feedback mechanisms in order to avoid oscillations, which could be caused by latencies, and to keep the robot compliant to outside disturbances.

Since our robot currently does not provide a way of measuring foot contact forces, we move the origin of the inverse dynamics calculation to the support foot under the assumption that it is essentially fixed to the ground (i.e. does not slide or tilt). In this way, the calculation does not depend on the contact forces. During support transition, when both feet are on the ground, the estimated torque is faded in a linear fashion between the computed results of the inverse dynamics for both feet.

The actuators of previous NimbRo robots [14] had to be driven with relatively high proportional gains to meet trajectory requirements. The new feed-forward control based on the learned motor model allows very low P gains to be used while still following the target trajectory (see Fig. 5). In comparison to hard, high-gain P control, the model-based control uses less power: Energy consumption was measured with the internal current sensor of the actuator. After 40 full steps, the left knee joint had consumed 189 J. The model-based control resulted in 140 J of consumed energy. Two factors contribute to this reduction of energy usage. Current spikes are avoided by predicting higher loads, as can be seen in Fig. 5. Steady-state current is also reduced. Less current consumption leads to longer battery life and less heat in the actuators—both positive effects.

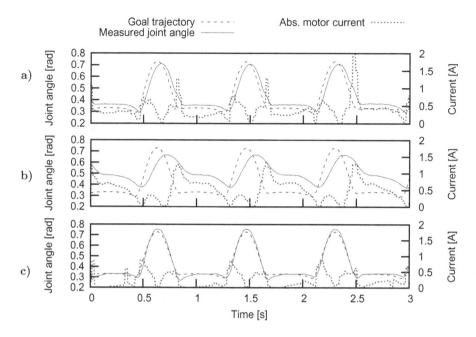

Fig. 5. Evaluation of feed-forward control with the learned joint model on a full-body walking motion. Shown are the joint targets and measured angles of the left knee. The leg contracts with a positive knee angle. **a)** Trajectory tracking using high-gain P control ($k_p = 1.0$). **b)** Trajectory tracking using low-gain P control ($k_p = 0.35$). **c)** Trajectory tracking using low-gain P control ($k_p = 0.35$) with model-based feed-forward control.

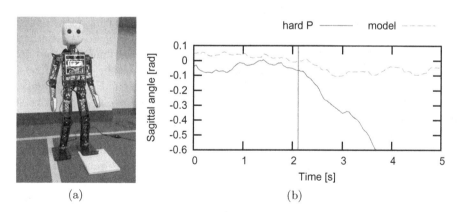

(a)   (b)

Fig. 6. Obstacle experiment. (a) Experiment setup. The obstacle has a height of 15 mm; (b) Development of sagittal trunk angle during the experiment. The *purple* line denotes the time of impact of the first step on the obstacle. The applied torque by the hard P controller leads to the robot falling backwards, while the model-based controller reacts softly enough to maintain stability.

As current is directly coupled with torque produced by the motor (see Eq. 4), potentially damaging torque spikes are also avoided.

The incorporated knowledge about the system also results in less latency than the high-gain controller. Furthermore, the explicit modeling of battery voltage in Eq. 19 results in equal performance over the whole battery voltage range.

In the walking experiments, trajectory errors were mainly caused by erroneous foot contact estimation. For example, undesired behavior could be observed when the robot mistakenly thought that one foot is firmly on the ground and moved the base of the dynamics calculations to this foot. If the foot is still in the air, the applied torques caused strange behavior. This process can be seen in the overshoot seen in Fig. 5 on the downward slope. However, since a low-gain P controller is used, the legs are very compliant and move back into proper position as soon as ground contact is established as can be seen in Fig. 5. Hence, walking performance is not adversely affected.

On the contrary, the higher compliance helps NimbRo-OP to walk over small obstacles (see Fig. 6(a)) without stabilizing algorithms. We tested rectangular obstacles with a height of up to 15 mm and commanded the robot to walk forward with a very slow velocity. Fig. 6(a) illustrates the robot behavior when executing the first step onto the obstacle. One can observe that the hard P controller reacts with too much force and causes the robot to fall backwards while the model-based controller is more compliant and just leads to a slight backward leaning. An additional stabilizing factor in this situations is the fact that calculated joint torques reflect the orientation of the robot, as the direction of gravity is estimated by the IMU.

6   Conclusion

In this paper, we demonstrated that simple motor and friction models can be learned using Iterative Learning Control in a fast and straightforward way, avoiding dedicated test runs for single parameters but instead fitting all parameters at once. The learned model performs better than traditional P controllers on real walking motions. Even intelligent servo actuators like the Dynamixel MX series can profit from an accurate motor model for feed-forward control, resulting in less power consumption and more compliant robot movement. Due to the generality of the used DC motor model, and the single assumption of a low-gain position-based P controller, our approach is applicable to a wide range of robotic actuators.

Some of the simplifications made when modeling motor and friction (e.g. unmodelled gear inertia) may be improved in future work. Since friction is the dominant problem in robot joint control, it may for example be prudent to investigate more complex friction models like the more general robot joint friction model proposed by Waiboer [10].

If the model is used in feed-forward control of position-controlled servos, wrongly estimated model parameters or an incorrect dynamics model can result in steady-state trajectory errors which are not corrected. Further work may include the careful application of feedback mechanisms (e.g. online model refinement) to reduce these errors.

Acknowledgment. This work has been partially funded by grant BE 2556/6 of German Research Foundation (DFG).

References

1. Arimoto, S., Kawamura, S., Miyazaki, F.: Bettering operation of robots by learning. Journal of Robotic Systems 1(2), 123–140 (1984)
2. Schwarz, M., Schreiber, M., Schueller, S., Missura, M., Behnke, S.: NimbRo-OP humanoid TeenSize open platform. In: 7th Workshop on Humanoid Soccer Robots, IEEE-RAS International Conference on Humanoid Robots, Osaka (2012)
3. Waiboer, R., Aarts, R., Jonker, B.: Velocity dependence of joint friction in robotic manipulators with gear transmissions. In: ECCOMAS Thematic Conference Multibody Dynamics (2005)
4. Mensink, A.: Characterization and modeling of a dynamixel servo. Technical report, University of Twente, Individual research assignment (2008)
5. Kim, Y.H., Lewis, F.L.: Reinforcement adaptive learning neural-net-based friction compensation control for high speed and precision. IEEE Transactions on Control Systems Technology 8(1), 118–126 (2000)
6. Liu, J.-S.: Joint stick-slip friction compensation for robotic manipulators by iterative learning. In: IEEE/RSJ International Conference on Intelligent Robots and Systems (IROS), vol. 1, pp. 502–509 (1994)
7. Stribeck, R., Schröter, M.: Die wesentlichen Eigenschaften der Gleit- und Rollenlager: Untersuchung einer Tandem-Verbundmaschine von 1000 PS. Springer (1903)
8. Chun Bo, L., Pavelescu, D.: The friction-speed relation and its influence on the critical velocity of stick-slip motion. Wear 82(3), 277–289 (1982)
9. Armstrong-Helouvry, B.: Control of Machines with Friction. International Series in Engineering and Computer Science, vol. 128. Springer (1991)
10. Waiboer, R.R.: Dynamic Modelling, Identification and Simulation of Industrial Robots – for off-line programming of robotised laser welding. University of Twente (2007)
11. Featherstone, R.: Rigid Body Dynamics Algorithms. Springer, Berlin (2008)
12. Rigid Body Dynamics Library, https://bitbucket.org/rbdl/rbdl/
13. Pastrana, J., Schwarz, M., Schreiber, M., Schueller, S., Missura, M., Behnke, S.: Humanoid TeenSize open platform NimbRo-OP. In: Proceedings of 17th International RoboCup Symposium (2013)
14. Missura, M., Münstermann, C., Mauelshagen, M., Schreiber, M., Behnke, S.: RoboCup 2012 Best Humanoid Award winner NimbRo TeenSize. In: Chen, X., Stone, P., Sucar, L.E., van der Zant, T. (eds.) RoboCup 2012. LNCS (LNAI), vol. 7500, pp. 89–93. Springer, Heidelberg (2013)

Unexpected Situations in Service Robot Environment: Classification and Reasoning Using Naive Physics

Anastassia Küstenmacher[1], Naveed Akhtar[1], Paul G. Plöger[1],
and Gerhard Lakemeyer[2]

[1] Department of Computer Science, Bonn-Rhein-Sieg University of Apply Science,
Sankt Augustin, Germany
{anastassia.kuestenmacher,naveed.akhtar,paul.ploeger}@h-brs.de,
[2] Knowledge-Based Systems Group, RWTH Aachen University, Aachen, Germany
gerhard@kbsg.rwth-aachen.de

Abstract. Despite perfect functioning of its internal components, a robot can be unsuccessful in performing its tasks because of unforeseen situations. Mostly these situations arise from the interaction of a robot with its ever-changing environment. In this paper we refer to these unsuccessful operations as external unknown faults. We reason along the most frequent failures in typical scenarios which we observed during real-world demonstrations and competitions using our Care-O-bot III robot. These events take place in an apartment-like environment.

We create four different - for now adhoc - fault classes, which refer to faults caused by a) disturbances, b) imperfect perception, c) inadequate planning or d) chaining of action sequences. These four fault classes can then be mapped to a handful of partly known, partly extended fault handling techniques.

In addition to existing techniques we propose an approach that uses naive physics concepts to find information about these kinds of situations. Here the naive physics knowledge is represented by the physical properties of objects which are formalized in a logical framework. The proposed approach applies a qualitative version of physical laws to these properties to reason about the fault. By interpreting the results the robot finds the information about the situations which can cause the fault. We apply this approach to scenarios in which a robot performs manipulation tasks (pick and place). The results show that naive physics hold great promises for reasoning about unknown external faults in the field of robotics.

Keywords: faults in robotics, unexpected situations, naive physics.

1 Introduction

Robots, operating in the open world, outside of lab conditions, may face situations where they are not able to perform their task successfully. These situations occur instantaneously, are sporadic in nature and are caused by interaction with the environment.

In the RoboCup@Home League the main focus is to enable a service robot to perform household chores. Clear or set the table, clean up the apartment, fetch and deliver objects: these are typical tasks for a service robot. To perform them in a desired manner in an uncertain environment, the robot needs the ability to handle unforeseen situations.

S. Behnke et al. (Eds.): RoboCup 2013, LNAI 8371, pp. 219–230, 2014.

(a) Scenario 1   (b) Scenario 2   (c) Scenario 3   (d) Scenario 4

Fig. 1. Examples of unexpected situations

To illustrate the range of these situations we will consider the following scenarios:

Scenario 1: Collision with unknown obstacle

A service robot needs to build a map of a domestic environment using its sensors (e.g. cameras and laser rangefinders). How should it react when colliding with an unexpected obstacle like a glass door or a desk of a table which may not be detected by any of its perception devices (Fig. 1. (a))?

Scenario 2: Impact of inaccurate action performance

A robot has to transfer a die from the floor to a table (Fig. 1. (b)). While bending down, the robot kicks the die accidentally and as a consequence grabs the die wrongly (i.e. at two opposite corners instead of at two parallel faces). The wrong grip causes the die to fall from the table while being placed. How can the real cause of the fault, namely kicking the die, be identified?

Scenario 3: Pushed/held by an external actor

A robot has to travel from one place to another. During its movement it is pushed or held by an external actor (Fig. 1. (c)). Hence the robot can not reach its desired location. How can we enable the robot to identify the external fault so that it can reach the desired location?

Scenario 4: Imprecise knowledge about properties of an object

In this scenario a robot has to place a bottle on the table. It should stand stable in the final state, but it does not and falls. It is possible that the robot was not holding the bottle properly before releasing it on the table. The robot's gripper with the bottle inside could be slightly inclined. How may the robot learn from this example and avoid this fault in the future (Fig. 1. (d))?

Note that in the described situations all of the robot's internal components are working perfectly fine. Here the causes of failing the task originate from outside and cannot be explained by the robot's knowledge. Therefore, these situations are termed as exogenous causes of failure or external unknown faults. In our scenarios, external unknown faults refer to the existence of an external impact, partially known properties of objects and unmodeled consequence of executed actions.

The majority of robots utilize the traditional fault handling approaches which work well under the assumptions of known a priori faults. However, their performance substantially drops if they are subject to unexpected events. Unfortunately, we cannot expect that all possible fault sources have already been encountered in practice.

Based on our research and on the studies conducted by other researchers we developed a method to handle some of these situations, classified them and assigned appropriate techniques to each class individually. For now we select four fault classes and enumerate a set of techniques for solving them.

Our reasoning approach uses naive physics knowledge and a qualitative version of physical laws to find reasons of the occurrence of the detected unexpected situations, namely external unknown faults. These situations occur when the behavior of the objects in the robot environment deviates from its expected values.

2   Classification of Unexpected Situations

Traditional model-based fault diagnosis techniques focus on improving the reliability of robots in certain well-known fault situations. The widely known autonomous robot Shakey [1] was already equipped with model-based fault detection capabilities. Shakey's execution monitoring system PLANEX monitored a sequence of actions, provided by the STRIPS planner for accomplishing a required task. After executing each action the monitor checked for inconsistencies between observations and the robots internal representation. These inconsistencies indicated a faulty plan but could not identify the cause of fault.

Despite successfully implementing the model-based algorithms, they do not diagnose exogenous causes of failure. Nevertheless, model-based techniques allow a robot to *detect* unexpected situations if its nominal behavior is modeled. Pettersson [2] presents a model-free approach to execution monitoring. He assumes that instead of having a model with predefined faulty states, only the nominal behavior of a robot is given. Therefore the detection of faults is performed based on the comparison between observed and expected behaviour.

Steinbauer [3] notes that most of the faults that can be observed during a RoboCup may be avoided by a more careful design of hardware devices and software components. He based his observations on fault-reports from the RoboCup competitions, especially when systems operate in conditions similar to lab environments and chances for unexpected situations are reduced to a minimum. But he also states that unexpected situations could not be dealt so easily.

The researches dealing with unexpected situations refer to them with keywords like *errant expectations* [5], *unexpected events* [6], *exogenous events* [4] and *external unknown faults*. In the scenarios mentioned in section 1 and in many other similar situations, a successful completion of robot's tasks suffers from many exogenous factors. Such kind of faults cannot be solved by any of the existing fault handling techniques.

At the same time each of the described scenarios may on its own be handled using particular algorithms. For instance, Ueda [7] proposes to re-plan the robotic task in the case of the robot being pushed by an external actor as in scenario 3. Mendoza [8] utilizes a *hidden Markov model*-based model for detecting motion interference events in mobile robots. He successfully tests his approach for three types of motion interference: collision with undetected obstacles (scenario 1), held by a human (scenario 3) and stuck wheels. In his research Sundvall [6] uses the other estimation-based method *extended Kalman filter* (especially successful for the navigation problems as in scenario 1) for detecting unexpected events, such as a collision with an unexpected obstacle. Gspandl [4] uses a belief-based management system to reason about the sequence of past executed actions of the robot and their results. This algorithm can be suitable for solving the problem of scenario 2.

Although none of these algorithms can be a common solution for the problem of handling all different unexpected situations together, individually these situations may be solved. Therefore, it makes sense to evaluate these algorithms and enumerate assumptions for them as well. Based on these assumptions we suggest to classify various exogenous events into different classes, and then associate each class with one or more appropriate fault handling algorithms. Based on existing methods and observations from the given scenarios we propose to group the unexpected situations into four major classes:

1. *Inadequate description of planner operator* (e.g. unsuccessful completion of action because of its insufficient description)
2. *External disturbance* (e.g. not reaching expected location because of being pushed/held by external actor)
3. *Imprecision of perception devices* (e.g. delusion of robot due to glass door)
4. *Consequence of previous inaccuracy* (e.g. incorrect release of die because of inaccurate grasping)

Our contributions in this field have been made in reasoning about faults from the *inadequate description of planner operator* class. We mainly focussed on actions which deal with the *release* of objects (like scenario 2) and do not consider the presence of external agents in the environment of robots. We assume natural physical phenomena as the only cause of external faults and propose to utilize *naive physics* for reasoning about these situations.

3   Naive Physics

In [9], Hayes proposes to formalize everyday knowledge of the physical world into a declarative symbolic theory. In principle, such a theory represents the *naive physics* knowledge which can be used for commonsense reasoning about everyday physical phenomena. This proposal of Hayes is one of the widely acclaimed ideas in AI, but it has never been truly followed [10]. For commonsense reasoning, the literature in AI is mostly concerned with what is more appropriately called *qualitative physics* [11]. In contrast to *naive physics*, *qualitative physics* concentrates on reasoning programs rather than knowledge representation. Furthermore, *qualitative physics* emphasizes on restricted systems rather than daily life physical phenomena.

One approach of representing commonsense physical knowledge, is the methodology of *microworlds* [10]. In his work [10] Davis proposes to structure the knowledge in *microworlds*, where *"a microworld is an abstraction of a small part of physical interactions, sufficient to support some interesting collection of inferences"*. Component-based electronics [11], rigid object kinematics and dynamics are few examples of *microworlds*. Such structuring of the knowledge makes its formalization easier. In comparison to Hayes' proposal, any knowledge formalized using the *microworlds* approach is more inferencing oriented. It also implies that keeping in view the *collection of inferences* to be supported by a *microworld*, we can limit the scope of the knowledge to be formalized for that *microworld*. This aspect of the *microworlds* approach makes it more practical in comparison to the original proposal of Hayes.

Reasoning about the unknown external faults in robotics requires the ability to reason about physical phenomena encountered in our *daily life*. As noted in [12], approaches in *qualitative physics* (QSIM [13] and qualitative process theory [14] etc.) are inadequate for such reasoning. On the other hand, it is also true that in current AI literature there does not exist any large scale formalization of daily life physical phenomena (formalized in a *naive physics* theory or structured into *microworlds*) which can be used directly for unknown external faults. Therefore, we use insights from the works referenced in this section to formalize a small body of *naive physics* knowledge for reasoning about unknown external faults in robotics. Our proposed approach illustrates how *naive physics* knowledge can be formalized in a way that we can benefit from it for reasoning about unknown external faults.

4   Naive Physics Approach for Unknown External Faults

When a fault is detected by a robot because of unforeseen situations in its environment, it needs to find out the information about the cause of the fault to avoid it in the future.In the proposed approach, this information is produced in the form of hypotheses, which describe the possible situations that could have caused the detected fault. Each of these hypotheses is generated as a state of the object, whose parameters are the properties of that object. These properties are representative of naive concepts behind them which are defined over extended time. Thus, the states produced by the reasoning process characterize the behavior of the object over an extended time. This allows the robot to achieve the goal of finding the cause behind a a detected fault with minimal information and computation.

In the reasoning scheme (Fig. 2.), the hypotheses only represent those situations which could result in the observed behavior of the object under the influence of physical phenomena e.g. gravity and friction. In a detected fault, the physical phenomena cause the object to achieve a *final state* that is not the *goal state* for the robot. The reasoning scheme uses this *final state* of the object and the physical phenomena to trace the possible situations that could have caused the detected fault.

4.1   Proposed Scheme

The name of each module in Fig. 2. represents its function in the reasoning scheme. We describe each of the modules with a running example of scenario 2 in section 1. The proposed scheme focuses on finding the reasons for unknown faults which are associated with the current action of the robot (i.e. releasing the die). [1]

Fault Categorizer. In our example the robot detects an unknown fault when it finds that its action was unsuccessful. This fact becomes known to the robot when it detects that a relation on(table,die) is unsatisfied in the *effects* of the action performed

---

[1] The description of modules uses Prolog syntax to show the relations and logical sentences used for the scenario.

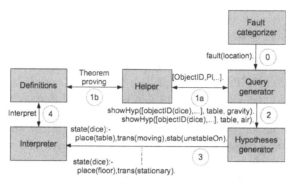

Fig. 2. Scheme of reasoning

by the robot. Based on this unsatisfied relation, the *fault categorizer* categorizes the detected fault. The categories of the detectable faults are defined a priori (see section 4.2). In the running example, the category of the fault is chosen to be location. In this work we assume that the detection of the fault has already been made by the robot and the *fault categorizer* has the ability to categorize any of the unsatisfied relations in the *effects* of robot's actions. The transmission of the signal from the *fault categorizer* is shown as step 0 in the scheme of Fig. 2.

Query Generator. The signal of the fault occurrence is passed on the *query generator* along with the category of the fault. In Fig. 2., fault(location) shows this signal. The *query generator* uses this signal to automatically generate relevant queries for the *hypotheses generator*.

Helper. The process of the generating of queries requires the knowledge of the *final state* of the object. That is, the state of the die being on the floor. This knowledge is made available to the *query generator* by a *helper*. The *helper* helps in defining the *final state* of the object by filling in the values of a list of properties, which is transmitted to it by the *query generator* (step 1a in the Fig. 2.). The *helper* also provides the *query generator* with a certain value of a property of the object that is involved in the detection of the fault. In our example, the constant symbol table represents this value. This value is decided from the failed relation on(table, die), which indicates that the desired *place* 'table' was not achieved by the object as an *effect* of the robot's action.

Definitions. The *helper* uses the observations of the robot and the definitions of properties stored in the module named as *definitions*, to define the *final state* of the object. The definitions of the properties are formalized in the form of a logical framework (see section 4.3). In the work presented here, the job of the *helper* is performed manually by using logical theorem proving to define the *final state* of the object. Therefore, step 1b in Fig. 2. is indicated as theorem proving.

Hypotheses Generator. In step 2, the *query generator* transmits the queries to another module which is called *hypotheses generator*. This module generates hypotheses about the situations that could have caused the detected fault. These hypotheses are represented as states of the concerned object. Each of these states is a conjunction of possible values of relevant properties of the object. These values are constraint by the qualitative version of physical laws used by the *hypotheses generator*. The physical laws let the module generate only those states of the object which can result in its observed *final state* under the phenomenon represented by the physical law. We define the physical laws by simple `Prolog` clauses which can generate lists of values of properties for the hypotheses. For our example, the clause shown in Fig. 3. (a) serves the purpose of the gravitational law for the *hypotheses generator*. Here, when a relevant query is posed to the *hypotheses generator* it automatically uses this clause to generate hypotheses. Another clause (not shown here) used by the *hypotheses generator* in this scenario is regarding the phenomenon of motion due to air. Both the clauses are used by the module to generate only relevant hypotheses for the occurred fault.

Interpreter. The generated hypotheses are received by an *interpreter* (step 3 in the Fig. 2.). In addition to receiving the hypotheses, this module performs the function of interpreting the hypotheses using the definitions of the properties in the *definitions* module. This interpretation (i.e. step 4) results in generating the information that can be used by the robot to avoid the occurrence of the detected fault in the future. In the work presented here, the function of this module is performed manually and the interpretation only represents the meanings of the hypotheses in natural language. This is done in order to show that meaningful information about the occurred situation can be extracted from *naive physics* knowledge using the proposed approach.

4.2 Ontology

In the scheme shown in Fig. 2., the *query generator* and the *hypotheses generator* require the knowledge of relevant properties of the object and relevant physical laws to perform their functions effectively. This knowledge is embedded in these modules in the form of an ontology, shown in Fig. 3. (b). This ontology is mainly derived from a list of properties of physical objects (made of substances) proposed in substance schema [15]. These properties represent the concepts used by naive reasoners to reason about daily life physical phenomena.

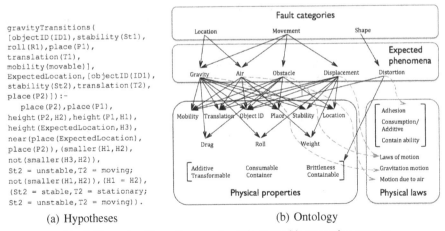

(a) Hypotheses   (b) Ontology

Fig. 3. Example of hypotheses and ontology used in reasoning process

The *fault categories* shown in the ontology are formed by considering the tasks which can be performed by the robots for which the ontology is being created. In this work, we restrict these broad *categories* only to those shown in the Fig. 3. (b), because we only consider the robots which can perform simple manipulation tasks (e.g. picking and dropping/placing objects). In such tasks, faults can be detected when the robot is unable to achieve the desired `location` of the object. Or, when the object behaves unexpectedly because of some `movement`, or when its `shape` gets distorted. We do not claim here that these *categories* are enough for all unknown faults for the considered robots. However, it is easy to see that these *categories* cover a large range of situations that can result in detectable faults for such robots.

The ontology associates different *categories* of faults to different *expected phenomena* which can cause these types of faults. These phenomena provide links between the *physical laws* and the *physical properties* of objects which are relevant for the *expected phenomena*. In our scheme the list of relevant properties is used by both the *query generator* and the *hypotheses generator*. However, the *physical laws* are only utilized by the *hypotheses generator* to generate possible hypotheses.

4.3   Framework

We formalize the definitions of properties of objects in the form of a logical framework. This framework is meant to illustrate how the properties of objects can be defined such that the definitions are useful for reasoning about unknown external faults. The mention framework only deals with the *fault category* `location`. We propose formalization of different frameworks for different *fault categories*. Such structuring of naive physics knowledge enables easy identification of physical interactions which are sufficient to support inferencing for their relevant *fault categories*. A detailed account on the developed framework can be found in [16], here we only summarize the framework by illustrating its important aspects.

The developed framework is meant only for solid objects. It considers the geometry to be $\mathbf{R}^3$, a subset of which is occupied by any object in the domain. Also, the shape of any object is equal to the closure of its interior and the object itself is considered as a primitive entity. All these notions are the same as those used in the framework proposed in [12]. However, our work differs greatly from Davis's work in dealing with time and physics. Our framework captures the notion of time by the concepts of *intervals* and *instants*, where an *interval* is just a finite set of *instants*. The framework formalizes *naive physics* concepts in the form of definitions of *predicates* and *functions* in first-order logic. Those concepts which depend on time, have explicit mentioning of respective temporal notions in the definitions. Definitions of atemporal concepts are time independent. These concepts are mostly related to describing the geometry of the objects.

$\forall_{object} \forall_t$ Unstable(object,t) $\Leftrightarrow$
On(object,object1,t) $\land \exists$ i $\in$ t [Coordinates(v,CG-F(object),i) $\land$
Parallel(Make-line-F(v,v1),Up) $\land$ v1 $\notin$ xx = Top-F(object1,t)]

The sentence shown above in first-order logic gives the definition of an `object` being *unstable* on its place in an interval t. In this sentence, the relations with their names ending with -F represent *functions*, whereas the other relations represent *predicates*. Each relation and constant symbol in the sentence is also defined in the framework. Below are the interpretations of these relations and symbols.

- On(object,object1,t): is true when object is *on* object1 during the interval t.
- Coordinates(v,centerOfGravity,i): is true when v is the vector representing the *coordinates* of some object's centerOfGravity at an instant i.
- CG-F(object): returns the vector of *center of gravity* of an object.
- Parallel(l,ll): is true when the line l is *parallel* to the line ll.
- Make-line-F(v,v1): returns a *line* passing through the points represented by coordinates v and v1.
- Top-F(object,t): returns a set of points (xx) representing the *top* of an object in the interval t.
- Constant symbol Up: represents an imaginary line in the straight upward direction normal to the floor.

As a result of the interpretations given above, the predicate Unstable(object,t) is true if and only if the projection from the object's center of gravity in the straight downward direction leaves the top of the other object (on which it is places) in some instant of the interval t.

Here, some characteristics of the definition of Unstable should be noticed. Firstly, the definition is only describing a *naive* concept of an object being *unstable*. This definition does not represent the actual concept (as in statics or dynamics) of *stability* of objects. Secondly, there is a bias in the definition towards *fault category* location. This means that according to the definition if the object is *unstable* on another object, it can change its *location* at its own (because of gravity). Such a bias also enables easy identification of the information (regarding the naive concepts) that needs to be formalized in the form of a definition. It can also be seen that the definition of Unstable describes the concept regarding the *stability* of an object by considering its behavior over an extended time. And, this description is independent of any particular object. All the characteristics of the definition of Unstable mentioned above also hold for the other definitions formalized in the framework.

The framework only uses very basic concepts from geometry and mechanics (e.g. center of gravity and lines) to define the naive concepts behind the properties. This keeps the evaluation of these properties arithmetically simple and computationally effective. In the framework, the definitions of properties *do* represent some redundant information. For instance, according to the definitions, an *unstable* object is also an object that is *moving*. However, both these concepts are defined separately. Such redundancies are unavoidable because these are found in the naive concepts represented by the definitions. However, the redundancies have been kept from creating any inconsistency in the framework.

5 Results

We apply our scheme to our running example in which the die falls on the floor instead of staying on the table. In this case, after the completion of step 0 and 1(a & b) of Fig. 2. the *query generator* generates two queries for the *hypotheses generator*. One of these queries is shown below.

```
showHyp([objectID(die),place(floor),stability(stable),
mobility(movable), translation(stationary),roll(rollable)],
table,gravity).
```

This query consists of following three components.

1. A list of predicates, in which each predicate represents a property of the `die`. The arguments of the predicates show the values of respective properties of the `die` in its *final state*.
2. A value of the *place* (i.e. `table`) that was required to be achieved by the robot in its action.
3. The *expected phenomena* that can cause the detected fault, i.e. `gravity`.

A theoretical proof that as per definitions the *final state* of the `die` is represented by the values of the properties used in the query, can be found in [16]. The other query generated by the *query generator* only replaces the *physical phenomenon* of `gravity` with `air` (see step '2' in Fig. 2.). It should be noticed that the *query generator* knows which *physical phenomena* should be used in the queries because it uses the ontology shown in Fig. 3. The *hypotheses generator* generates two separate lists of hypotheses for the phenomena of `gravity` (a) and `air` (b) in response to the posed queries.

```
1. State(die):-
   place(table),stability(unstable),translation(moving).
2. State(die):-
   place(floor),stability(unstable),translation(moving).
3. State(die):-
   place(floor),stability(stable),translation(stationary).
```

```
1. State(die):-
   translation(moving),
   place(floor).
2. State(die):-
   translation(stationary),
   place(floor).
```

(a) Hypotheses for 'gravity'

(b) Hypotheses for 'air'

As stated earlier, each of the generated hypotheses represents a situation that can be the cause of the detected fault. As human beings we can immediately see that hypothesis 1 for `gravity`, is the most relevant one for the considered scenario. Interpretation of this hypothesis gives the reason of the detected fault. According to the *definitions* module this interpretation reads (in natural language) as following. *There existed some interval (after the release of the die) in which the die was on the table and its center of gravity was not at a fixed point in space and the projection from its center of gravity, in straight downward direction, left the top of the table in at least one of the instants of that interval.* This interpretation represents the occurred situation correctly. The information obtained from the interpretation can be used by the robot to avoid the same situation in the future. In our work we assume that the *interpreter* interprets the hypotheses by simply selecting them one by one, however it can be seen that it is not very difficult to prune the list of the hypotheses using some heuristics before the interpretation.

We also apply our approach to a different scenario. In this scenario when the robot picks a `bottle`, the `bottle` slips from its `gripper` and falls on the `table`. Again, the detected fault is categorized under the type `location`. Therefore, the *query generator* also generates very similar queries. The only difference in the queries is that the value `table` is replaced by `gripper`.

Again, the hypothesis 1 for `gravity` (c) is the most relevant one for the described scenario. However, it should be noticed that if the interpretation of this hypothesis is done by using the definition of `Unstable` shown in section 4.3, then this hypothesis would not represent the occurred situation correctly. Therefore, the proposed framework (discussed in section 4.3) also allows multiple definitions for a single property

```
1. State(bottle):-
   place(gripper),stability(unstable),translation(moving).
2. State(bottle):-
   place(table),stability(unstable),translation(moving).
3. State(bottle):-
   place(table),stability(stable),translation(stationary).
```

(c) Hypotheses for 'gravity'

```
1. State(bottle):-
   translation(moving),
   place(table).
2. State(bottle):-
   translation(stationary),
   place(table).
```

(d) Hypotheses for 'air'

of objects. Although, the current framework (reported in [16]) contains only one definition for `Unstable`, however another definition for the same concept can easily be formed by using the insights from our work. The relations representing these different definitions can be distinguished by their names or arity. There is another important observation that can be made about the results of both the scenarios. That is, all the situations referred by the generated hypotheses *can* result in the respective *final states* of the objects in each scenario. This means that in principle all the generated hypotheses are useful for the robot. The work [16] presents the other experiments where the current approach was applied.

6   Conclusions

This paper is directed towards developing a strategy for service robots to successfully handle unexpected situations that occur during interactions of the robot with its dynamic environment. The idea is to utilize the existing algorithms to deal with unknown external faults in the appropriate manner. We defined four classes of these faults. In addition we proposed an approach that uses *naive physics* knowledge to find information regarding the situations which result in the occurrence of faults from *inadequate description of planner operator* class.

Application of the proposed approach to a simple pick and place scenario shows that using *naive physics* for unknown external fault reasoning is indeed beneficial. By using *naive physics* it is possible to gain useful information about the occurred fault without detailed modeling and complex computations. We find that the use of *naive physics* knowledge is not very complicated, because we can distribute this knowledge under different categories of faults. This distribution eases the process of formalization of the knowledge by limiting its scope. The formalization is guided by the category of the fault for which the knowledge is being formalized. By knowing the category a priori it is possible to formalize only those aspects of the knowledge which are relevant for a particular category of faults. Therefore, we propose to formalize the knowledge into different frameworks for different fault categories. We also propose that the categories of faults should be decided based on the tasks which can be performed by the robot. This means that the overall knowledge for fault reasoning also has its scope limited to a particular (type of) robot (e.g. mobile manipulators).

The future direction of the *naive physics* approach is to extend the ontology used in this work for other categories of faults and then formalizing more knowledge of naive physics reasoner for these categories. Another interesting issue which has been left open in this work is related to pruning the list of hypotheses generated by the *hypotheses generator*. Our approach can greatly benefit by developing effective methods for selecting more relevant hypotheses for interpretation.

Lastly we may specify an enlarged set of classes with corresponding handling techniques. The new classes can originate from a design of the particular robot as well as from new realistic scenarios. As one source of scenarios we can chose the RoboCup@ Home competition where a robot has to operate in the apartment environment and perform everyday tasks. After this has been done successfully, the complete approach will contribute to the improvement of the robot's robustness towards exogenous causes of failure. These skills will make robots more adaptive and compliant with real world environments.

Acknowledgments. This work was sponsored by the B-IT foundation and the Strukturfond des Landes Nordrhein-Westfalen for female PhD students.

References

1. Nilsson, N.J.: Shakey the robot. Technical Report 323, AI Center, SRI International, 333 Ravenswood Ave, Menlo Park, CA 94025 (April 1984)
2. Pettersson, O., Karlsson, L., Saffiotti, A.: Model-free execution monitoring in behavior-based robotics. IEEE Trans. on Systems, Man and Cybernetics, Part B 37(4), 890–901 (2007)
3. Steinbauer, G.: Survey on faults of robots used in robocup (2012), http://www.ist.tugraz.at/rfs/
4. Gspandl, S., Pill, I., Reip, M., Steinbauer, G., Ferrein, A.: Belief management for high-level robot programs. In: 22nd International Joint Conference on Artificial Intelligence (2011)
5. Karg, M., Sachenbacher, M., Kirsch, A.: Towards expectation-based failure recognition for human robot interaction. In: 22nd International Workshop on Principles of Diagnosis (2011)
6. Sundvall, P., Jensfelt, P.: Fault detection for mobile robots using redundant positioning systems. In: IEEE International Conference on Robotics and Automation, ICRA 2006 (2006)
7. Ueda, R., Kakiuchi, Y., Nozawa, S., Okada, K., Inaba, M.: Anytime error recovery by integrating local and global feedback with monitoring task states. In: IEEE, 15th International Conference on Advanced Robotics ICAR, pp. 298–303 (2011)
8. Mendoza, J., Veloso, M., Simmons, R.: Motion interference detection in mobile robots. In: IEEE/RSJ International Conference on Intelligent Robots and Systems, IROS (2012)
9. Hayes, P.J.: The naive physics manifesto. In: Michie, D. (ed.) Expert Systems in the Micro Electronic Age, pp. 242–270. Edinburgh University Press (1979)
10. Davis, E.: The naive physics perplex. AI Magazine 19, 51–79 (1998)
11. Kleer, J.D., Brown, J.S.: A qualitative physics based on confluences, pp. 83–126. Morgan Kaufmann Publishers Inc., San Francisco (1990)
12. Davis, E.: A logical framework for commonsense predictions of solid object behaviour. Artificial Intelligence in Engineering 3(3), 125–140 (1988)
13. Kuipers, B.: Qualitative simulation. Artificial Intelligence 29, 289–338 (2001)
14. Forbus, K.D.: Qualitative process theory. Artificial Intelligence 24, 85–168 (1984)
15. Reiner, M., Slotta, J., Chi, M., Resnick, L.: Naive physics reasoning: A commitment to substance-based conceptions. Cognition and Instruction (2000)
16. Akhtar, N.: Fault reasoning based on naive physics. Technical report, Hochschule Bonn-Rhein-Sieg, Sankt Augustin, Germany (April 2011)

Person Following by Mobile Robots: Analysis of Visual and Range Tracking Methods and Technologies

Wilma Pairo[1,2], Javier Ruiz-del-Solar[1,2], Rodrigo Verschae[1], Mauricio Correa[1], and Patricio Loncomilla[1]

[1] Advanced Mining Technology Center, Universidad de Chile
[2] Department of Electrical Engineering, Universidad de Chile
{wpairo,jruizd,mcorrea,ploncomi}@ing.uchile.cl,
rodrigo@verschae.org

Abstract. Person following by mobile robots in unconstrained environments has not been yet successfully solved, and new approaches to tackle this problem need to be developed. The main goal of this article is to analyze the use of state-of-the-art computer vision methods for human detection and tracking when a robot is trying to follow a person. The methods were selected taking into account their accuracy in previous studies as well as being real-time or near real-time. Thus, *tracking based on a HOG person detector*, *tracking-by-detection with Kernels* and *compressive tracking* were analyzed and compared to methods based on the use of Kinect and laser sensors using a database built specifically for this purpose. The database was captured using a service robot, and it considers real-world conditions. The results show that the vision-based methods are much more robust for tracking purposes than standard range-based methods used by the robotics community, although being slower.

Keywords: Person following, RoboCup, depth image, laser, benchmark.

1   Introduction

The detection and tracking of humans by robots are key abilities of service and social robots. These basic abilities are required to implement several applications, among them person following. If robots will be part of our daily life in a near future, then they need to be able to follow us, as well as to walk with us and navigate among us, in daily life environments. However, person following by mobile robots in unconstrained environments has not been yet successfully solved. For instance, in the 2012 RoboCup@Home competition[1], 14 of 17 service robots were not able to follow a human in the "Follow Me" test, competition which was ran in an indoor setup under dynamic conditions (non-controlled illumination, audience near the robots, etc.) [6][7][8].

In the RoboCup@Home community most of the research teams address the person following problem by using active infrared-based depth sensors such as the Kinect and lasers. But, as it will be shown in this article, the use of this kind of sensors is not

---

[1] Robocup@Home competition website: http://www.robocupathome.org/

S. Behnke et al. (Eds.): RoboCup 2013, LNAI 8371, pp. 231–243, 2014.

always reliable, because their sensing capabilities depend on the reflecting material (e.g. in many cases black materials do not reflect enough infrared light), and on the amount of solar light received by the sensor in outdoors or through windows (sunlight can saturate the IR sensors or the projected IR patterns).

We think that the use of computer vision algorithms applied to visible-spectrum images needs to be considered and further analyzed as an alternative for solving the person following problem. Computer vision approaches can eventually replace or at least complement systems based on depth sensors. In this context, the main goal of this article is to analyze the use of state-of-the-art computer vision methods for human detection and tracking when a robot is trying to follow a person using visual (RGB) cameras. The methods to be analyzed were selected taking into account their accuracy in previous studies as well as being real-time or near real-time. The methods are: Histograms of Gradients[4], Tracking-by-Detection with Kernels [2], and Visual tracking using Compressing Tracking [1].

Human following systems based on these methods and on the use of the Kinect and laser sensors will be compared using a database built specifically for this purpose. Thus, RGB video sequences, Kinect sequences and laser sequences, taken using a tele-operated service robot that followed a person in different setups, were captured. They include realistic conditions such as dynamic backgrounds, multiple persons, occlusions, and uncontrolled illumination. This dataset will be made public for future studies. The article is organized as follows. In Section 2 problems associated to the use of infrared-based depth sensors are described. In Section 3, the human detection and tracking methods under comparison are presented. In Section 4 the evaluation of the different methods is described. Finally, conclusions are given in Section 5.

2   Operation Problems of IR-Based Range Sensors

Infrared (IR) based range sensors such as the Kinect or lasers, suffer some operation problems that depends on environmental conditions of daily life environments, such as the color and texture of the reflecting material, and the amount of solar light perceived by the sensor in outdoors or through windows. These factors affect the robustness of perception systems built based on the use of these sensors, and in particular the performance of person following systems.

We conducted a series of experiments in order to show that these problems do exist. Our goal is not to quantify these problems, but to show their existence. In the experiments, a person or an object was placed in front of a Kinect sensor, a Hokuyo laser and a standard RGB camera, and the sensor response was recorded and analyzed. Some of the experimental results are shown under different visualizations in Figure 1. In the case of the Kinect images (second column in graphs shown in Fig. 1), grey scale values indicate depth (distances), with higher values (e.g. bright) indicating larger distances and low values (dark) indicating closer distances. The black color indicates that no object was detected. In the case of the laser sensor (third column in graphs shown in Fig. 1)), blue and green areas correspond to segmented object candidates. The blue ones correspond to objects that are discarded based on their geometry (size, shape, distance, etc.), while the green ones correspond to person candidates. Two green hues are used to facilitate the visualization of person candidates. The laser has a 120° of field of view. In the image the upper-central part corresponds to the location were the subject is located in front of the robot.

The first set of experiments analyzes the response of the sensors to black clothes. Persons wearing different black clothes were placed at 2 meters in front of the sensors: 18 different black clothes were analyzed and in 4 cases (22.2%) the cloth was not detected properly by the Kinect and by the Hokuyo laser. The Figure 1 (a) shows two of these four cases (cloth: cotton & leather).

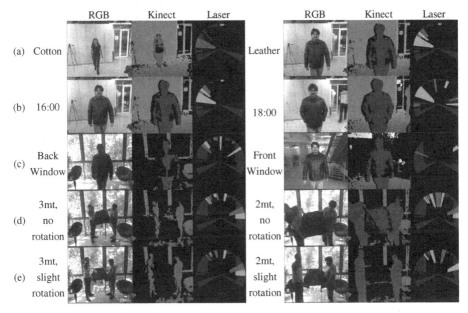

Fig. 1. Responde of the sensors under different conditions: (a) cloth, (b) day time, (c) light direction, (d)-(e) rotation. See the main text for details.

The second set of experiments addresses the sensor's dependence on illumination. Figures 1 (b) and (c) present the response of the different sensors at different day times and with different illumination conditions determined by nearby window locations, respectively. The cloth used in this experiment corresponds to the one that presented worst results among the 18 clothes that were considered in experiment 1. In Figure 1 (b) it can be noticed that the time at which the images were taken affects the detection: 16:00 (left), 18:00 (right). At times when sunlight is stronger (e.g. closer to midday) the performance of the Kinect is affected more. Also, the location of the light source (in this case a nearby closed window) with respect to the camera may affect the detection as shown in Figure 1 (c). In Figure 1 (c) it can be observed that the Kinect is more affected when the light source is in front of the subject (right images) --saturating the pattern projected by the Kinect--, while the laser seems to be more affected when the light source is in front of the laser sensor (left images) -- saturating the sensor. Thus, clearly both sensors are affected by illumination conditions. Recall that these experiments were performed indoors, in an environment were all nearby windows were closed. The detected problems are much more important outdoors.

The third set of experiments addresses the sensors' dependence on the angle of the reflecting surface of the object to be detected. Figures 1 (d) and (e) present the output at different distances and rotations for the same cloth (black leather jacket) considered in Figures 1 (b-c). In Figures 1 (d) no rotation is applied to the cloth, case where the laser and the depth camera do not detect the subject at 3mts (the cloth appears as black in the image). Moreover, in Figures 1 (e) a slight rotation (pitch) is applied to the fabric and the laser does not detect the object at 1 meter or more, while the depth camera does not detect the fabric at 2mts or more. This clearly shows that active depth sensors are not robust enough for detecting some materials under real-world – but not difficult-- conditions.

3   Person Following Methods under Comparison

Person following can be modeled as a human tracking problem plus a basic control of the robot pose. It can be tackled using RGB or range cameras. In these two cases bounding boxes are tracked in the image space, and the tracking is implemented by using tracking algorithms or detection algorithms. In the case of detection algorithms, the tracking is performed by integrating detections over time.

In order to analyze the performance of visual-based human tracking and detection algorithms for person following, we have selected three state-of-the-art methods: Histograms of Gradients (HOG) [4] applied to human detection, Tracking-by-Detection with Kernels [2], and Visual tracking using Compressing Tracking [1]. The use of these algorithms is compared with the Kinect sensor (together with the OpenNI[2] framework) and the use of a laser-based approach. In this last case the method does not track explicitly a human, but "the object" in front of the robot.

Based on these basic tracking algorithms, 10 different tracking methods are implemented and compared. Table 1 presents a summary of the methods. Note that the processing times in the table correspond only to the running of the algorithm and do not consider the sampling frequency of the sensor. The visual methods are evaluated using the implementation of the original authors. We briefly describe each of the methods:

HOG: Visual tracking using a HOG person detector [4][3]. This detector performs an exhaustive search over different scales and locations of the image using a sliding window approach: each window of the image is analyzed and classified either as a person or a non-person. It uses HOG features and a part-based analysis that makes use of SVMs classifiers. We used only one cluster (training samples can be clustered based on their aspect ratio), and only one model for the person class was built. The training was done using the original database of the VOC challenge 2007 --obtained from its web page-- but after removing the images of subjects riding bicycles.

HOG1/2: Visual tracking using a HOG torso detector based on [4]. This is the same as **HOG**, but only the upper part (torso) of the subjects is used for the training

---

[2] OpenNI: http://www.openni.org/
[3] Implementation used: http://people.cs.uchicago.edu/~rbg/latent/, Version 5

of the detector. This allows the system to be able to detect subjects standing [4] close to the robot --when only the torso is visible--, but also in cases when the subject is far away from the camera.

TDK: Visual tracking using Tracking-by-Detection with Kernels [2][4]. During operation the method (re)-trains a classifier at every frame of the video. During the training it uses samples from the previous frame of the object being tracked (thus it is called Tracking-by-Detection). The main difference with existing similar methods is that the training and the classification is done in the Fourier-space thanks to the use of a circulant matrix representation that allows to perform the training and the detection at a very high frame rate. For including temporal information, the method performs a temporal averaging of the classifier's parameters. See [2] for details.

CT: Visual tracking using Compressing Tracking [1][5]. This method extracts Haar-like features from the image and projects them using a sparse matrix to obtain the features to be used by a classifier. The sparse matrix is selected using a procedure inspired in Compress Sensing. The used classifier is a naïve Bayes classifier that is updated online over time. As in the case of TDK, a classifier is trained using positive and negative samples obtained from the previous frame.

Kinect: A depth image for detecting and tracking people is processed by using the OpenNI framework (using UserGenerator.GetUsers(aUsers, nUsers) function available in the NiUserTracker subproject, OpenNI v1.5.2.23). This method applies a segmentation algorithm to the depth image to detect people, and it uses motion information to initialize the tracking.

Laser: A laser model URG04LX is used to capture scan-lines, which are transformed into segments. At the initialization stage, a person stands in front of the robot and the corresponding segment is initialized as the one corresponding to the person to be tracked. Afterwards the segment is tracked across the frames [5].

For all visual methods we implemented a version that re-initializes the tracking when the detection is lost. We refer to these variants with a +**RE** postfix. Given the fact that these methods are based on statistical classifiers, in case the classifier gives a very low confidence output, the classifier is re-initialized and re-trained (TDK and CT) or the temporal tracking is re-initialized (HOG and HOG1/2). The corresponding variants are: **HOG+RE:** Visual tracking using a HOG person detector with re-initialization, **HOG1/2+RE:** Visual tracking using a HOG torso detector with re-initialization, **TDK+RE:** Visual tracking using TDK with re-initialization, and **CT+RE:** Visual tracking using CT with re-initialization.

Temporal Tracking, Initialization and Re-initialization

The similarity between detection windows is measured following the criteria used at the Visual Object Challenge (VOC) [3]. It corresponds to the ratio of the intersection and the union of a pair of bounding boxes. Given two bounding boxes, B_p, B_q:

$$a_0(B_p, B_q) = \frac{\text{area}(B_p \cap B_q)}{\text{area}(B_p \cup B_q)} \tag{1}$$

---

[4] Implementation used:
http://www2.isr.uc.pt/~henriques/circulant/index.html
[5] Implementation used:
http://www4.comp.polyu.edu.hk/~cslzhang/CT/CT.htm

If two bounding boxes are identical in size and they are located in the same position, this measure will be 1, and it will diminish as the boxes are farther away from each other, or if the sizes of the bounding boxes differ. This measure is used in this work for the following purposes:

- Maintaining temporal coherence for HOG and HOG1/2 methods. HOG-based methods may detect more than one person in a scene. To select the one being tracked, all detections of the current frame will be compared with the detections of the previous frame, and the one with maximum a_0 score will be selected as the detection of the current frame. If no detection is found in a frame, the prediction from the previous frame is kept as the current prediction.
- Comparing a detection B_p with the ground truth B_{gt} during the evaluation of the methods. A detection B_p will be considered correct iff $a_0(B_p, B_q) > 0.5$.
- Selecting the bounding box corresponding to the object to initialize or re-initialize the methods: during tracking initialization, given that in all videos the person to be tracked is centered in the image and standing in front of the robot at the first frame, we used a HOG1/2 torso detection as the bounding box (window) corresponding to the subject at the first frame. All methods are initialized using this window. In the cases where HOG1/2 detected more than one person in the first frame, the initial detection was selected using the detection that was closest to a window of size 340x400 with the following coordinates: top-left (150,80), and bottom-right (500,480) – we will call this window W_0.

In the case a detection is not found, a **+RE** detector is re-initialized completely using the same criteria used for initializing the tracking: a HOG1/2 detector is applied, and the HOG1/2 detection closest to W_0 is selected as the new detection. If HOG1/2 did not detect any subject, the detection window from the previous frame of the corresponding method is used instead of HOG1/2. Note that in the case of TDK+RE and CT+RE, when a re-initialization is performed, the classifier is trained again from scratch without using information from previous frames.

Table 1. Overview of methods under comparison. Processing times (fps) do not include the time required to capture the data.

Name	Sensor	Methodology	Processing Time (fps)**			Reference
			VD1*	VD2*	VD3*	
Kinect	Kinect	Person detection	79.14	75.58	77.83	OpenNI
Laser	Laser	Laser Scanning	1316.41	1575.92	1631.39	[5]
HOG	Visual	Person detection, HOG Features	0.40	0.40	0.39	Based on [4]
HOG1/2	Visual	Torso detection, HOG Features	0.41	0.41	0.41	Based on [4]
CT	Visual	Compressive Tracking	31.15	30.87	31.35	[1]
TDK	Visual	Tracking by detection using Kernels	18.33	16.39	8.47	[2]
HOG+RE	Visual	HOG + re-initialization	0.40	0.40	0.39	Based on [4]
HOG1/2+RE	Visual	HOG1/2 + re-initialization	0.41	0.41	0.41	Based on [4]
CT+RE	Visual	CT + re-initialization	18.27	12.08	18.98	Based on [1]
TDK+RE	Visual	TDK + re-initialization	18.56	1.53	8.08	Based on [2]

* A description of videos VD1, VD2 and VD3 is given in the next section.

** This time does not consider the sampling time of the sensor, but only the processing time.

4   Evaluation

Database Description

The UCHFollow database was created in order to carry out the evaluation of the methods described in the previous section. The database is divided into three sets: *Visual Set (V set), Visual-Occlusion Set (VO set), Visual & Depth Set (VD set).* The data was captured using a tele-operated service robot that followed a person in different setups. The data sequences were acquired using the 8-bit VGA resolution camera (640×480 pixels) available on the Kinect (we will refer to this camera as the RGB camera), a Kinect range sensor, and a Hokuyo laser sensor. In the case of the V and VO sets, only the RGB camera was used to capture the data, while in the VD case the RGB camera, the Kinect's range sensor and a laser sensor were used. The data sequences include realistic conditions such as dynamic backgrounds, occlusions, uncontrolled illumination, and multiple persons. In the VD and V sets, the person being followed walks at about 4 [km/h]. The database is available for research purposes and can be downloaded at http://vision.die.uchile.cl/personfollow/.

The V set consists of 8 videos, V1 to V8, obtained under different conditions (natural variations in illumination, indoor setup, pose, clothes, occlusions, and background). The VO set consists of 4 videos, VO1 to VO4. In these videos the followed person remains at a fixed distance and is occluded by another person for 1, 2, 4, 8, 16 and 32 seconds. The VD set consists of 3 videos (VD1 to VD3), obtained under different conditions (natural variations in illumination, indoor setup, pose, clothes, occlusions, and background) captured using the three sensors.

Details of each video are available in Table 2. For each video the ground truth was generated using the following procedure: for every 10th frame of the video the bounding box of the subject to be tracked was annotated, and the bounding boxes for the remaining frames were obtained by interpolation. Special attention was given to

Table 2. Database Summary: Visual Set (V Set), Visual Occlusion Set (VO Set) and Visual & Depth Set (VD Set)

AvgOL: Average Occlusion Length. OC: Occlusions. PBG: people in the background

Video	Duration Frames	Time	# Subjects	# OC	AvgOL [frames]	Type of Cloth	Black cloth	Description
V1	3444	4:23	1	0	0	Texture	No	Good illumination
V2	2786	3:18	1	0	0	Uniform	No	Good illumination
V3	3101	3:57	1	0	0	Texture	No	Good illumination
V4	3341	4:29	>5	4	5.25	Texture	No	Good illumination, interaction w/ people, OC
V5	1731	2:10	>20	1	6	Texture	No	Complex scenario, PBG, interaction w/ people, OC
V6	5001	5:57	>20	19	9.47	Uniform	Yes	Complex scenario, PBG, interaction w/ people, OC
V7	3011	3:44	>20	8	9.25	Uniform	Yes	Complex scenario, PBG, 2 persons walking side by side, interaction w/ people, OC
V8	4431	5:21	2	2	388	Uniform	No	Complex scenario: subject enters *elevator*, OC
VO1	2141	2:37	2	9	97.78	Uniform	No	Subject remains at a fixed distance, OC
VO2	1561	1:52	2	6	155.83	Uniform	No	Subject remains at a fixed distance, OC
VO3	1681	1:54	2	6	155	Texture	No	Subject remains at a fixed distance, OC
VO4	1851	2:41	2	8	120.37	Texture	No	Subject remains at a fixed distance, OC
VD1	1151	1:27	>20	0	0	Texture	No	Complex scenario, PBG
VD2	1131	1:24	>20	7	9.57	Texture	No	Complex scenario, PBG, OC
VD3	1901	2:19	>20	1	10	Texture	No	Complex scenario, PBG, interaction of followed person w/ people, OC

occlusions, situation in which frames just before and after the occlusion were annotated. A subject was considered as not occluded (i.e. visible) if at least three of the following four parts of the body were visible: left arm, right arm, head, and torso.

Evaluation Methodology

Three sets of experiments were carried out. The first two make use of the V and VO sets, respectively, to compare the different methods that use RGB images. The third experiment uses the VD set to compare visual-based and depth-based methods.

For every frame, each method gives as output a bounding box of the region that contains the object of interest; the only exception is the laser tracking, which gives as output a line segment containing the object of interest. (This segment is first mapped to a segment (of length d) in the image plane. The laser's mapped bounding box's bottom side is set to the mapped segment, and its height is *1.6*d.*) For each frame the bounding box is compared to the ground truth (eq. (1)), and if the similarity between both is larger than 0.5, the detection is considered correct, otherwise incorrect. If the person is occluded, the ground truth does not exist for that frame, and thus that frame is not considered during the evaluation. For the whole video the detection rate (DR) is measured as the number of correct detections (counted over the frames with ground truth annotations) divided by the number of frames where the subject is visible. All methods give one and only one prediction per frame, thus the DR summarizes the performance of the method (false positives do not need to be evaluated).

Evaluation Results

In Table 3, we can see the detection results of the methods evaluated in the V set. HOG1/2+RE shows the best mean performance, followed by HOG1/2 that is able to obtain up to 100% in a difficult video (see description of video V5 in Table 2). It can also be observed that the full body HOG method does not have a good performance. This behavior is expected because in this video most of the time the whole body of the human being tracked is not completely visible. Video V8 is difficult because in this situation the robot must follow a person that enters and exits from an elevator, and therefore the robot must enter and exit the elevator too. When the person is in the elevator, the tracked person is very near to the robot and only partially visible, thus the ground truth is not defined at these frames. In this video, HOG1/2+RE obtains a 95.13% of correct detections. It can be also observed that the performance of the TDK and CT tracking methods is low, but these results improve largely if the methods are re-initialized. For example, the detection accuracy of CT increases from 29% to 75% in video V1, and increases from 36% to 84% in video V5. The best performance in terms of detection rate was obtained by HOG1/2+RE, followed by CT+RE. In terms of processing times, as presented in Table 1, CT is the fastest method with almost 30fps, while CT+RE – which uses HOG for re-initialization-- performs at 4 to 10 fps depending on the video. On the contrary HOG1/2 is the best performing one, but at the same time is rather slow, running only at 0.4 fps.

Figure 2 shows the results as accumulated detections over time. In all subfigures the dashed red line indicates the performance of a perfect tracking (the red dashed line is not visible in some images because some of the methods had perfect performance), and vertical red lines are used to indicate occlusions of the subject being followed. Recall that frames where an occlusion happened were not evaluated (no person to track!), and thus all methods have a flat performance during that time. In Figure 2 it can be observed that occlusion is one of the greatest problems affecting the performance of the methods (e.g. Figure 2, Video V6).

Table 4 and Figure 3 present results for the experiments with occlusions using the VO set. In these experiments, the subject was static and completely visible (torso and legs), and a second subject occluded the one being tracked for increasing periods of time, going from 2 to 32 seconds. The HOG+RE method has the best performance, followed by HOG1/2+RE, and HOG. TDK+RE has a clearly better performance than CT+RE. In these videos, unlike in the V set, the full body of the person is visible, which favors HOG variants over HOG1/2 ones. It is important to note that the recovering time of the methods to occlusions is critical for their performance: if the robot loses a subject for too long time, it may be impossible to catch up him/her again (the robot may be far from the subject). Also a tracking method that looses a subject after even short occlusions is not robust enough for robust robot applications.

Table 3. Performance summary DR (%) on V set. Bold letters indicate best working methods for variants with and without re-initialization. #RE = Number of re-initializations

	V1	V2	V3	V4	V5	V6	V7	V8	Avg	Std	#RE
HOG	45.27	79.11	13.25	49.40	43.77	33.19	7.35	46.84	39.77	22.47	-
HOG1/2	**98.58**	**99.43**	**86.42**	71.05	**100.0**	89.67	55.57	94.83	86.94	15.89	-
TDK	36.50	62.20	70.24	43.40	14.61	24.50	10.90	7.85	33.77	23.56	-
CT	29.76	66.58	58.69	47.26	36.99	8.57	0.00	8.89	32.09	24.72	-
HOG+RE	28.66	65.69	13.51	40.18	71.94	39.10	48.92	47.91	44.48	18.87	1129.6
HOG1/2+RE	**97.91**	**99.43**	**86.36**	95.45	**99.19**	90.91	71.33	95.13	91.96	9.44	104.3
TDK+RE	36.50	64.07	72.17	43.43	16.46	41.78	36.26	23.01	41.71	18.80	352.7
CT+RE	75.64	55.13	64.43	44.43	84.35	40.51	17.23	51.08	54.10	21.16	72.5

Table 4. Performance summary on VO set. Bold letters indicate best working methods for variants with and without re-initialization.

MRT := Maximum recover time, #RE = Number of re-initializations

	VO1		VO2		VO3		VO4		Average		#RE
	MRT	DR %	MRT	DR %	MRT	DR %	MRT	DR %	DR %	Std	
HOG	**2s**	**32.99**	**32s**	**100**	**32s**	**100**	2s*	14.19	**61.79**	44.78	-
HOG1/2	2s*	12.05	0s	11.66	0s	19.44	**8s**	**71.39**	28.63	28.73	-
TDK	0s	7.45	0s	11.98	0s	19.44	1s	30.07	17.23	9.88	-
CT	0s	7.45	0s	11.66	0s	19.44	1s	3.83	10.59	6.71	-
HOG+RE	**8s**	**90.25**	**32s**	**100**	**32s**	**100**	**32s**	81.19	**92.86**	9.04	771.8
HOG1/2+RE	1s	45.36	**32s**	89.29	**32s**	82.02	**32s**	76.24	73.23	19.33	275.8
TDK+RE	0s	54.64	0s	11.66	0s	19.44	**32s**	**93.24**	44.75	37.35	202
CT+RE	0s	12.13	0s	11.66	0s	19.44	8s*	30.40	18.41	8.75	94

* The method lost the subject and recovered it before the last recovered occlusion.

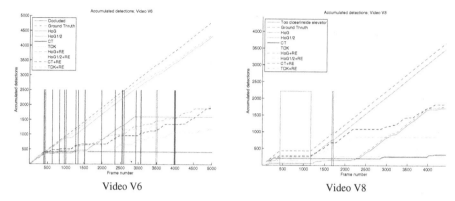

Video V6   Video V8

Fig. 2. Results on the V Set (accumulated detection rates). The step signal (red) indicates occlusions. The red, dashed signal represents a perfect tracking.

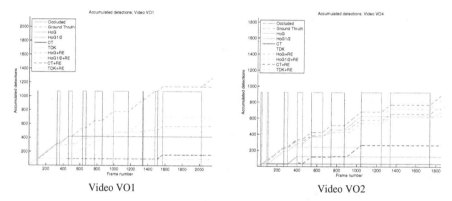

Video VO1   Video VO2

Fig. 3. Results on the VO set (accumulated detection rates). The step signal (red) indicates occlusions. The Red, dashed signal represents a perfect tracking.

In Table 5 and Figure 3, we can see the results of the methods evaluated in the VD set. HOG1/2+RE shows again the best performance in all videos and it is closely followed by HOG1/2. The performance of the TDK and CT tracking methods is low again. However their results improve if they are re-initialized. Kinect-based method performs better than all versions of the TDK and CT tracking methods, but worse than the HOG-based ones. The laser tracking method has a good performance in the first two videos, but in the video VD3 it has the worst performance, maybe because in this video a lot of people stay besides the subject being followed, and the information used by the method (tracked objects are represented as segments) is not robust enough.

In average the Kinect has a better performance (45.84%) than pure tracking methods (TDK and CT), both with and without re-initialization, but the Kinect performance is not even close to the HOG based methods. This happens although the Kinect uses depth information which one would expect to simplify the segmentation and tracking of the subject. Note that in these videos the cloth wore by the subject is

Table 5. Performance (Detection rate) summary on VD set. Bold letters indicate best working methods for variants with and without re-initialization. #RE = Number of re-initializations

	VD1 [DR%]	VD2 [DR%]	VD3 [DR%]	Avg [DR%]	Std	#Re
Laser	73.50	68.42	9.10	50.34	35.81	-
Kinect	42.05	58.08	37.39	45.84	10.85	-
HOG	60.90	71.71	26.18	52.93	23.79	-
HOG1/2	**100**	**93.51**	**98.94**	**97.48**	3.48	-
TDK	24.50	16.82	23.32	21.55	4.14	-
CT	16.07	17.67	27.18	20.31	6.01	-
HOG+RE	60.64	71.43	53.20	61.76	9.17	892.3
HOG1/2+RE	**98.78**	**98.31**	**95.82**	**97.64**	1.59	56.6
TDK+RE	31.79	38.25	21.36	30.47	8.52	146
CT+RE	48.48	41.73	13.59	34.60	18.51	12.7

detected by the Kinect and the laser (it is an easy cloth for those sensors). In a case with "difficult" clothes (e.g. back), one would expect that the performance of the Kinect and laser based methods would decrease. We believe that this happens because a more robust segmentation and/or better features for classification need to be developed for depth data (the amount of research that has been done using visual data is much higher that the one devoted to depth data).

HOG-based methods are much more robust than all other methods, but at the same time they are rather slow (e.g. they are 100 times slower that the Kinect). Fast methods using depth-sensors could be re-initialized using HOG-based ones in order to make them robust to occlusions and give them a better initialization (e.g. the Kinect requires the subject to move to initialize the tracking). In summary, we think that visual methods are robust enough, and could be used as to complement or even replace depth-based methods, as depth-based methods cannot be used in outdoors, and may present problems detecting some kinds of cloth, among other problems.

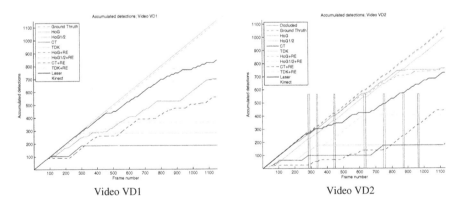

Video VD1   Video VD2

Fig. 4. Results on VD set (accumulated detection rate). The step signal (red) indicates occlusions. The Red, dashed signal represents a perfect tracking.

5   Discussion and Conclusions

In this work, a preliminary analysis of the reliability of the Kinect and laser sensors for measuring depth under different environmental conditions was performed. A set of images of subjects wearing different clothes under various conditions was captured using those sensors. From these captured images it was observed that the range sensors did not correctly detected 4 out of 18 clothes. The distance between the sensor and the cloth/object to be captured, the amount of illumination on the cloth and on the sensor, and the viewpoint angle are all factors that increase the probability that the depth of the cloth cannot be measured properly. The detections are affected by sun light, even in indoors, and this effect is dependent on the time of the day.

Three databases of videos were built for evaluating and comparing the tracking methods. Results from the V Set database show that the HOG1/2+RE method has the highest detection rate (91.96%), followed by HOG1/2 (86.94%) and then CT+RE (54.10%). Re-initialization proves to be useful as it improves the results around 10%. The results from the VO Set show that the best method is HOG+RE (92.86%) followed by HOG1/2+RE (76.24%) and then by HOG (61.79%). In this database people are far from the robot and the whole body is seen in the images; this is the reason of HOG getting the best performance instead of HOG1/2. These results show that HOG based methods can be robust under occlusions (temporal coherence is required), that tracking methods such as CT and TDK are not robust enough, and that re-initialization helped when occlusions occurred. Results from the VD set show that best results are obtained by HOG1/2+RE (97.64%), followed by HOG1/2 (97.48)%, HOG+RE (61.76%), the laser tracker (50.34%) and by the Kinect tracker (45.84%). Results from this last database show that image-based methods derived from HOG outperform depth-based methods even in cases where the cloth color did not affect the depth-based methods. The laser has good performance in two videos, but in a third video, where the subject approaches a crowd, its performance decreased, thus lacking robustness in this situation. It is important to notice that the performance of HOG1/2+RE is 30% higher than the one of the Kinect based method!

The runtime of the methods is critical for tracking targets that can realize quick movements. Using the VD set, runtimes were compared, measured as frame per second (fps), but not including the sensing time. Results show that the laser-based method has the highest frame rate (1,507fps) followed by the Kinect tracking (77fps), by CT (31 fps), and by TDK (14fps). HOG-based methods have very low frame rates (0.39fps), which make them unsuitable when agile tracking of people is required. In this case, tracking methods CT+RE and TDK+RE can be useful because they run fast, and re-initialization using HOG only occurs when the tracker loss the target. Future work includes developing a system that is able to integrate visual, Kinect and laser information in an optimal way for obtaining a highly reliable tracker.

Acknowledgments. This work was partially financed by FONDECYT grants 3120218 and 1130153.

References

[1] Zhang, K., Zhang, L., Yang, M.-H.: Real-time compressive tracking. In: Fitzgibbon, A., Lazebnik, S., Perona, P., Sato, Y., Schmid, C. (eds.) ECCV 2012, Part III. LNCS, vol. 7574, pp. 864–877. Springer, Heidelberg (2012)

[2] Henriques, J.F., Caseiro, R., Martins, P., Batista, J.: Exploiting the circulant structure of tracking-by-detection with kernels. In: Fitzgibbon, A., Lazebnik, S., Perona, P., Sato, Y., Schmid, C. (eds.) ECCV 2012, Part IV. LNCS, vol. 7575, pp. 702–715. Springer, Heidelberg (2012)

[3] Everingham, M., Van Gool, L., Williams, C.K.I., Winn, J., Zisserman, A.: The PASCAL Visual Object Classes (VOC) Challenge. International Journal of Computer Vision 88(2), 303–338 (2010)

[4] Felzenszwalb, P.F., Girshick, R.B., McAllester, D., Ramanan, D.: Object Detection with Discriminatively Trained Part Based Models. IEEE Transactions on Pattern Analysis and Machine Intelligence 32(9) (September 2010)

[5] Chen, Y.J., Liao, C.T., Tsai, A.C.: Mobile robot based human detection and tracking using range and intensity data fusion. In: IEEE Workshop on Advanced Robotics and Its Social Impacts, ARSO 2007 (2007)

[6] Stückler, J., Droeschel, D., Gräve, K., Holz, D., Schreiber, M., Behnke, S.: NimbRo@Home 2012 Team Description. In: RoboCup 2012 (2012)

[7] Seib, V., Luing, V., Sarnecki, L., Grun, T., Knauf, M., Barthen, S., Paulus, D.: RoboCup 2012 - homer@UniKoblenz. In: RoboCup 2012 (2012)

[8] Ruiz-del-Solar, J., Correa, M., Smith, F., Loncomilla, P., Pairo, W., Verschae, R.: UChile HomeBreakers 2012 Team Description Paper. In: Robocup 2012 (2012)

Fast Monocular Visual Compass
for a Computationally Limited Robot

Peter Anderson and Bernhard Hengst

School of Computer Science and Engineering,
University of New South Wales, Australia

Abstract. This paper introduces an extremely computationally inexpensive method for estimating monocular, feature-based, heading-only visual odometry - a visual compass. The method is shown to reduce the odometric uncertainty of an uncalibrated humanoid robot by 73%, while remaining robust to the presence of independently moving objects. High efficiency is achieved by exploiting the planar motion assumption in both the feature extraction process and in the pose estimation problem. On the relatively low powered Intel Atom processor this visual compass takes only 6.5ms per camera frame and was used effectively to assist localisation in the UNSW Standard Platform League entry in RoboCup 2012.

1   Introduction

Bipedal robots need to walk on a variety of floor surfaces that can cause them to slip to varying degrees while walking. They can also be bumped or impeded by undetected obstacles. Accurate navigation therefore requires an autonomous robot to estimate its own odometry, rather than assuming that motion commands will be executed perfectly.

If a robot is fitted with one or more cameras, visual odometry algorithms can be used to estimate the relative motion of the robot between subsequent camera images. However, these methods are typically computationally expensive for resource constrained robots operating in real time. To overcome this problem, this paper exploits the planar motion assumption in both the image feature extraction process, and in the calculation of relative pose estimates. The result is an extremely computationally inexpensive method for estimating heading-only visual odometry using a monocular camera and 1-dimensional (1D) SURF features.

This technique was developed and used in the RoboCup Standard Platform League (SPL) robot soccer competition, and implemented on an Aldebaran Nao v4 humanoid robot. The Nao is equipped with a 1.6 GHz Intel Atom processor and two 30 fps cameras. These cameras have minimal field of view overlap, necessitating the use of monocular visual odometry methods. As the Nao is not fitted with a vertical-axis gyroscope, heading odometry can only be estimated using visual methods.

S. Behnke et al. (Eds.): RoboCup 2013, LNAI 8371, pp. 244–255, 2014.

On the soccer field, the combination of fast motion and collisions with other robots exacerbates the difficulties of biped navigation, making robot soccer an excellent test domain for this technique. However, the technique itself in no way relies on the characteristics of the robot soccer application domain.

Although the use of heading-only visual odometry has some limitations, practical experience suggests that when a bipedal robot manoeuvres, the greatest odometric inaccuracy is observed in the robot's heading, which can change very quickly when the robot slips, rather than in the forward or sideways components of the robot's motion. The visual heading odometry is therefore used in conjunction with the odometry generated by assuming perfect execution of motion commands (which we will refer to as command odometry). Results suggest that the combination of this technique with command odometry results in a dramatic improvement in the accuracy of odometry information provided to localisation filters. A further benefit of the visual odometry module is the ability to detect collisions and initiate appropriate avoidance behaviour.

The remainder of this paper is organised as follows: Section 2 outlines related work, Section 2.1 provides a brief introduction to 1D SURF image features, Section 3 describes calculation of heading odometry using 1D SURF features, and Section 4 presents experimental results.

2   Background

If robots are fitted with multiple cameras with overlapping fields of view, stereo visual odometry algorithms can be used to recover the relative 6 degree of freedom (DoF) motion of the robot. This is typically achieved by measuring the 3D position of detected image features in each frame using triangulation [1]. A robot fitted with a single camera must use monocular visual odometry methods, which cannot solve the general 6 DoF relative motion estimation problem. Generally, in this case only heading information can be accurately obtained as the effect of translation is small compared to that of rotation, while the absolute scale of motion is unobservable.

To simplify the 6 DoF visual odometry problem for monocular cameras, a number of previous papers have used constraints on the motion of the agent. Approaches that use the planar motion assumption, such as [2], [3], typically rely on the detection of repeatable local image features that can be tracked over multiple frames. These features are then used to recover the relative robot pose using epipolar geometry; the assumption of planar motion on a flat surface makes relative pose estimation simpler and more efficient. Even greater efficiencies in pose estimation have been demonstrated when the planar motion assumption is coupled with the nonholonomic constraint of a wheeled vehicle [4] [5].

The planar motion assumption has also been used to develop monocular methods to determine the scale of motion [6]. These methods use the surface context approach developed by [7] to split a single image into three geometric regions: the ground, sky, and other vertical regions. Using this technique, scale can be determined based on the motion of features on the ground plane.

A unifying feature of these previous works is that although the planar motion constraint has been applied to the pose estimation problem, it has not been applied to the feature detection and extraction process. This process is computationally expensive, for example, using the OpenCV implementation, [8] demonstrated that SIFT and SURF feature extraction took on the order of 50 - 100 ms on a desktop PC for relatively small images. As a result, the most efficient pose estimation processes such as [4] have found that the overall framerate is limited by the feature extraction process, in this case using SIFT, Harris, and KLT image features. The computational cost of these feature extraction methods is prohibitively expensive for resource constrained robots such as the Nao.

Although impressive results have been reported using visual odometry in many application areas, the use of visual odometry in dynamic environments containing many independently moving objects remains challenging. Typical approaches to feature-based relative motion estimation are sensitive to wrong feature matches or feature matches on moving objects, even with the use of RANSAC based outlier rejection schemes [9]. To overcome these issues in dynamic environments, authors such as [9] have used image patch classification to improve rejection of independently moving objects such as other cars.

2.1 1D Surf Features

1D SURF is an optimised feature detector designed to exploit the planar motion constraint. We implemented the algorithm developed by authors in [10] for fast mobile robot scene recognition. It consists of a modified one dimensional variant of the SURF [11] algorithm. As shown in Figure 1, the algorithm processes a single row of grey-scale pixels captured from a 30 pixel horizontal band at the robot's camera level (the robot's horizon). The horizon band is chosen for feature extraction because, for a robot moving on a planar surface, the identified features cannot rotate or move vertically. For a humanoid robot, the position of the horizon in the image is determined by reading the robot's limb position sensors and calculating the forward kinematic chain from the foot to the camera, or by using an IMU.

As shown in [10], the use of a 1D horizon image and other optimisations dramatically reduces the computational expense of SURF feature extraction, exploits the planar nature of the robot's movement, and still provides acceptable repeatability of the features. Consistent with the original SURF algorithm, the extracted features are robust to lighting changes, scale changes, and small changes in viewing angle or to the scene itself. On a 2.4GHz Core 2 Duo laptop 1D SURF runs more than one thousand times faster than SURF, achieving sub-millisecond performance. This makes the method suitable for visual navigation of resource constrained mobile robots; on the Nao v4 we find the mean extraction time of 1D SURF features is 2 ms.

Fig. 1. Image captured by the Nao robot showing superimposed 30 pixel horizon band in red, and the extracted grey-scale horizon pixels used for 1D SURF feature extraction at the top of the image

3   Relative Pose Estimation

The 1D SURF visual odometry algorithm is of the monocular, feature-based, heading-only variety. Although it is possible to estimate the scale of camera motion using a monocular system as in [6], 1D SURF features are found only on the robot's horizon, not on the ground plane, and therefore do not lend themselves to these techniques. The error in the heading component of the robot's command odometry has a much greater effect on localisation accuracy than the errors in the forward and sideways components.

The soccer field is a dynamic environment with multiple independently moving objects in the form of other robots, referees, and spectators in the background, as illustrated in Figure 2. To help prevent the movement of other robots on the field from influencing the visual odometry, in this application domain a visual robot detection system is used to discard features that are part of other robots. This system uses region-building techniques on a colour classified image to detect other robots. It is similar in spirit to the feature classification approach used by [9]. However, the system is also robust to the movement of undetected robots and referees in front of the camera, as described further below.

To estimate the relative heading motion between two subsequent camera frames, the horizon features in each image are matched using nearest neighbour matching with a distance ratio cutoff as outlined in [10]. For each corresponding feature pair, the horizontal displacement in pixels of the feature between frames is calculated. Using this data, only a single parameter needs to be estimated: the robot heading change between the two frames. Since this can be estimated using only one feature correspondence, it is not necessary to use RANSAC for robust model estimation. Similarly to [4], histogram voting can be used, which is more efficient than RANSAC.

Fig. 2. A typical camera frame captured by the Nao during a soccer match could include both other robots and the referee. In this case the robot is detected and it's horizon features (indicated as blue and red blobs) are discarded. The referee cannot be detected, constituting an independently moving object.

Using the histogram voting approach, the robot's heading change between two frames is estimated by the mode of all feature displacements in a feature displacement histogram, as illustrated in Figure 3. Knowing the resolution and horizontal field of view of the camera, it is trivial to convert the robot's heading change from pixels to degrees or radians, and to adjust for the movement of the robot's neck joint between frames.

Provided the stationary background is always the largest contributor of features in the image, the histogram mode will remain unaffected by the introduction of independently moving objects. If there are many or large moving objects in the frame, and the identification of the static background is uncertain, the distribution of feature displacements will be strongly multi-modal. Multi-modality enables this scenario to be easily detected, in which localisation filters can fall back to using command odometry only. In contrast, when the visual odometry is reliable, the distribution of feature displacements will be approximately uni-modal, and localisation filters can use visual odometry in preference to command odometry for heading. Using this approach makes the system relatively robust to the movement of undetected robots and referees, as shown in Figure 3. At all times the forward, sideway, and turn components of odometry used for localisation are generated by command odometry.

On the soccer field, robots are frequently bumped or impeded by other robots and obstacles that are not visible. It is advantageous to detect these collisions to prevent robot falls. This can be done by differencing visual heading odometry with command odometry. When an exponential moving average of this quantity breaches certain positive and negative bounds, it indicates that the robot is slipping with a rotation to the left or right respectively. In the robot soccer domain, we have found that reducing the stiffness of the Nao's arms at this point is sufficient to avoid a significant number of falls.

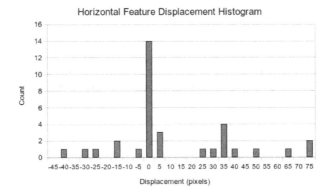

Fig. 3. The distribution of feature displacements over subsequent camera frames indicates two modes: one representing the viewing robot's heading change, measured against static background features, and the other representing the independent motion of the referee. In this case, the larger mode can be easily identified and visual heading odometry determined using this mode. If the two modes are similar in size the algorithm is able to fail gracefully by reverting to command odometry.

3.1   Reducing Drift

Computing visual odometry by integrating over all adjacent frames of a video sequence leads to an accumulation of frame-to-frame motion errors, or drift. To minimise this drift, many visual odometry techniques make periodic optimisations over a number of local frames, known as sliding window bundle adjustments. In this paper, adjustments to the estimated robot trajectory are made at each step by choosing to potentially discard some frames from the image sequence. This allows the system to remain robust in the presence of single frames corrupted by horizon location error, blur, or feature occlusion.

To implement this adjustment, whenever a new frame is obtained the heading change between the new frame and each of the three previous frames is calculated. The current heading odometry is then calculated relative to the 'best' of these three prior frames. The notion of the 'best' prior frame takes into account two factors. The first is the confidence level of the heading change estimate between the prior frame and the new frame. The second is the reliability of the prior frame's own odometry estimate (itself a recursive function of confidence levels). More formally, at time t the robot's heading odometry is calculated relative to prior frame at time $t - b_t$, with $b_t \in \{1, 2, 3\}$ given by:

$$b_t = \arg\max_{k \in \{1,2,3\}}\{\min\{reliability_{t-k}, confidence_{t-k,t}\}\}$$

where the *reliability* of the odometry at time t is determined recursively by the *reliability* of the best prior frame odometry and the *confidence* of the heading change estimate between $t - b_t$ and t:

$$reliability_t = \min\{reliability_{t-b_t}, confidence_{t-b_t,t}\}$$

$$reliability_0 = \infty$$

The measure of the confidence of the heading change estimate between two frames could be calculated in several different ways. It should always reflect higher confidence when the distribution of feature displacements is more uni-modal, and lower confidence when the distribution is more multi-modal (indicating difficulty in resolving independently moving objects from the stationary background).

Our approach was to calculate confidence based on the difference in the count of the first and second modes of the feature displacement histogram. Overall, the choice to consider three previous frames in the odometry calculation represents a trade off between computational cost and drift reduction. If the robot changes heading quickly there is little to be gained from increasing the size of this sliding window.

4   Results

In order to evaluate the performance and robustness of 1D SURF visual odometry in comparison to naive command odometry, a quantitative benchmark test is required. In this paper the University of Michigan Benchmark test (UMBmark) is used [12]. In addition, tests are undertaken that include obstacle collisions that disrupt the natural motion of the Nao, and repeated observations of independently moving objects at close range.

By way of background, UMBmark is a procedure for quantifying the odometric accuracy of a mobile robot. In the test, the robot is pre-programmed to move in a bi-directional square path, in both the clockwise and anti-clockwise directions, and the accuracy of the return position is assessed. Although the test was designed for assessing wheel odometry error in differential-drive robots, several results of the paper are also relevant for bipedal robots. In particular, the paper illustrates that a uni-directional square path test is unsuitable for evaluating robot odometric accuracy due to the possibility of compensating turn and forward motion errors. Using the bi-directional square path test, however, these errors are revealed when the robot is run in the opposite direction.

To assess the performance of visual odometry on the Nao, the UMBmark square path procedure was repeated five times in the clockwise direction, and five times in the counter-clockwise direction. In recognition of the greater inaccuracy of bipedal robots compared to wheeled robots, the side lengths of the square path were reduced from 4 m to 2 m. Prior to this test, the command odometry was calibrated to provide reasonable performance using the same settings across five robots, but it was not calibrated to suit this particular robot.

The test was conducted on the SPL field, and the robot was controlled to ensure that in each test it walked as near to a perfect square path as possible, at a speed of approximately 15 to 20 cm/s, including turning on the spot at each corner, and returning to the starting position. The position of the robot

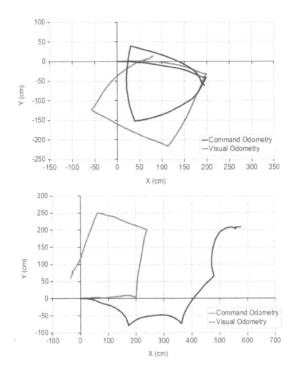

Fig. 4. Clockwise (top) and Anti-clockwise (Bottom) odometry track of an uncalibrated Nao robot walking in a 2m x 2m square path

calculated using command odometry dead-reckoning was then compared with the position of the robot using visual odometry dead-reckoning.

The estimated odometry tracks for the first trial in each direction are shown in Figures 4, illustrating a substantial deviation from the true square path in the case of the command odometry, and a much smaller deviation when the 1D SURF visual odometry was used. It is evident from these diagrams that the robot used for the test has a systematic left turn bias. In order to walk around the square field, continuous right turn corrections were required, which can be observed in the paths generated by odometry dead reckoning. The use of visual odometry has compensated for a significant proportion, but not all, of this systematic bias.

Figure 5 illustrates the final positioning error using both odometry methods at the end of five trials in each direction. Using the centre of gravity approach outlined in [12], the command odometry has a measure of odometric accuracy for systematic errors of 306 cm, relative to the visual odometry approach with an accuracy of 84 cm; a reduction of 73%. These results suggest that the 1D SURF visual odometry technique can compensate for a significant proportion of the systematic odometry error of an uncalibrated robot. The standard deviation for the walk-engine return positions is 550 cm, relative to 160 cm for the visual

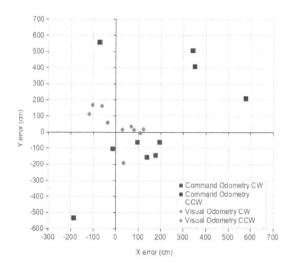

Fig. 5. UMBmark results for an uncalibrated Nao robot walking in a 2 m x 2 m square path in both clockwise (CW) and counter-clockwise (CCW) directions. Results indicate a significant increase in accuracy using visual odometry relative to naive walk-engine generated odometry.

odometry return positions. This indicates that the visual odometry technique also compensates for non-systematic odometry errors.

In the next test, the robot performed five trials of a simple out-and-back manoeuvre along a 2 m long straight line. On both legs of the path, a block of wood was placed in front of one shoulder of the robot to cause a collision. Figure 6 illustrates an odometry track for a trial during which the robot experienced a gentle collision on the way out, and a more significant collision on the way back to the starting position that resulted in an uncommanded turn. Over five trials of this test, the command odometry had a measure of odometric accuracy of 140 cm with standard deviation of 102 cm, versus the combined visual odometry approach with an accuracy of 43 cm and standard deviation of 53 cm.

The final test of 1D SURF visual odometry was designed to illustrate the performance of the system with multiple visible moving objects. For this test, the Nao was programmed to maintain position in the centre of the SPL field while walking on the spot, while a number of people wearing blue jeans (simulating referees as in the RoboCup soccer application domain) were instructed to walk in front of the Nao at a range of around 50 cm. The first five people walked in the same direction, from right to left. Figure 8 illustrates that at this range the referees jeans will typically fill approximately half of the robot's horizontal field of view.

During this test, the robot's visual odometry was monitored to see if the movement of the objects in front of the camera could trigger a false positive heading change. As illustrated in Figure 9, during this process there is no

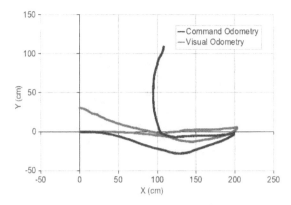

Fig. 6. Odometry track from an uncalibrated Nao on a single 2 m out and back trial with collisions in both directions

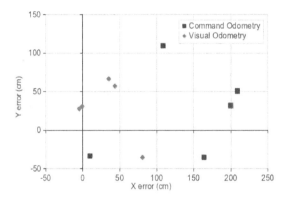

Fig. 7. Final positioning error over five out and back trials with collisions. Results again indicate a significant increase in accuracy using visual odometry relative to naive command odometry.

apparent drift in the robot's visual heading odometry in over one thousand frames (approximately 33 seconds). The oscillating pattern of the heading is attributed to the robot's constant adjustments to maintain position in the centre of the field. During these experiments, the mean execution time of the visual odometry module on the Nao v4 (including the execution time required to extract 1D SURF features from the camera image) was 6.5 ms. This equates to a theoretical maximum frame rate of around 150 fps (although in practise the Nao camera is limited to 30 fps and the remaining computational resources are used for other tasks).

The visual compass was used by the UNSW team in the 2012 Standard Platform League with great effect to help localisation accuracy [13].

Fig. 8. Moving object tests as they appear to the Nao robot. Approximately half of the robot's horizontal field of view was obscured by the moving object. Although the person's jeans appear very dark, features are still detected in this area as indicated by the red and blue blobs.

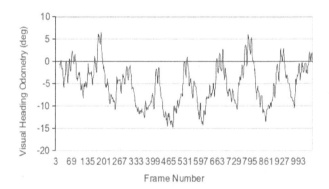

Fig. 9. Visual heading odometry with the Nao walking on the spot during the moving object test. There is no evidence of large spikes which would indicate false positive heading changes, or heading drift, during this 33 second period. An oscillating pattern can be seen which reflects the robot's actual movement adjustments to maintain position while walking on the spot.

5   Conclusions

We have presented a robust and extremely computationally inexpensive method for estimating monocular, feature-based, heading-only visual odometry using 1D SURF features. Further work is required to investigate the feasibility of full planar visual odometry (including both heading and translation) using the same features.

References

1. Scaramuzza, D., Fraundorfer, F.: Visual odometry: Part I - the first 30 years and fundamentals. IEEE Robotics and Automation Magazine 18 (2011)
2. Nister, D., Bergen, O.N., Visual, J.: odometry for ground vehicle applications. Journal of Field Robotics 23 (2006)
3. Liang, B., Pears, N.: Visual navigation using planar homographies. In: IEEE International Conference on Robotics and Automation (ICRA 2002), pp. 205–210 (2002)
4. Scaramuzza, D., Fraundorfer, F., Siegwart, R.: Real-time monocular visual odometry for on-road vehicles with 1-point ransac. In: Proceedings of IEEE International Conference on Robotics and Automation (ICRA) (2009)
5. Scaramuzza, D.: 1-point-RANSAC structure from motion for vehicle-mounted cameras by exploiting non-holonomic constraints. International Journal of Computer Vision 95, 75–85 (2011)
6. Choi, S., Joung, J.H., Yu, W., Cho, J.I.: What does ground tell us? monocular visual odometry under planar motion constraint. In: 11th International Conference on Control, Automation and Systems (ICCAS), pp. 1480–1485 (2011)
7. Hoiem, D., Efros, A.A., Hebert, M.: Recovering surface layout from an image. International Journal of Computer Vision 75 (2007)
8. Juan, L., Gwun, O.: A comparison of SIFT, PCA-SIFT and SURF. International Journal of Image Processing (IJIP) 3(4), 143–152 (2009)
9. Kitt, B., Moosmann, F., Stiller, C.: Moving on to dynamic environments: Visual odometry using feature classification. In: IEEE International Conference on Intelligent Robots and Systems, IROS (2010)
10. Anderson, P., Yusmanthia, Y., Hengst, B., Sowmya, A.: Robot localisation using natural landmarks. In: Chen, X., Stone, P., Sucar, L.E., van der Zant, T. (eds.) RoboCup 2012. LNCS (LNAI), vol. 7500, pp. 118–129. Springer, Heidelberg (2013)
11. Bay, H., Ess, A., Tuytelaars, T., Van Gool, L.: Speeded-up robust features (SURF). Computer Vision and Image Understanding 110(3), 346–359 (2008)
12. Borenstein, J., Feng, L.: UMBmark: A benchmark test for measuring odometry errors in mobile robots. In: SPIE Conference on Mobile Robots (1995)
13. Harris, S., Anderson, P., Teh, B., Hunter, Y., Liu, R., Hengst, B., Roy, R., Li, S., Chatfield, C.: Robocup standard platform league - rUNSWift 2012 innovations. In: Proceedings of the 2012 Australasian Conference on Robotics and Automation, ACRA 2012 (2012)

Reusing Risk-Aware Stochastic Abstract Policies in Robotic Navigation Learning*

Valdinei Freire da Silva[1], Marcelo Li Koga[2],
Fábio Gagliardi Cozman[2], and Anna Helena Reali Costa[2]

[1] Escola de Artes, Ciências e Humanidades,
Universidade de São Paulo, São Paulo, Brazil
[2] Escola Politécnica, Universidade de São Paulo, São Paulo, Brazil
{valdinei.freire,mlk,fcozman,anna.reali}@usp.br

Abstract. In this paper we improve learning performance of a risk-aware robot facing navigation tasks by employing transfer learning; that is, we use information from a previously solved task to accelerate learning in a new task. To do so, we transfer risk-aware memoryless stochastic abstract policies into a new task. We show how to incorporate risk-awareness into robotic navigation tasks, in particular when tasks are modeled as stochastic shortest path problems. We then show how to use a modified policy iteration algorithm, called AbsProb-PI, to obtain risk-neutral and risk-prone memoryless stochastic abstract policies. Finally, we propose a method that combines abstract policies, and show how to use the combined policy in a new navigation task. Experiments validate our proposals and show that one can find effective abstract policies that can improve robot behavior in navigation problems.

Keywords: Risk-Awareness, Memoryless Stochastic Abstract Policies, Transfer Learning.

1 Introduction

Consider an agent that displays risk-awareness in the sense that she has preferences amongst risk-prone, risk-neutral and risk-averse attitudes. This agent may face a series of tasks; for instance a robot may face navigation problems in a variety of rooms, each one in a different day. It is reasonable to suppose that insights obtained in the solution of the first task will be useful in the solution of the second task, and so on. Hence the agent may be interested in learning not only the solution of the first task, but also an *abstract* description of the solution that may be used when solving the second task, and so on. How can this risk-aware agent find a suitable abstract policy? How can the agent reuse and transfer into new problems this abstract policy, and how useful can it be?

* This research was partly sponsored by FAPESP – Fundação de Amparo à Pesquisa do Estado de São Paulo (Procs. 11/19280-8, 12/02190-9, and 12/19627-0) and CNPq – Conselho Nacional de Desenvolvimento Científico e Tecnológico (Procs. 311058/2011-6 and 305395/2010-6).

S. Behnke et al. (Eds.): RoboCup 2013, LNAI 8371, pp. 256–267, 2014.

Substantial previous work can be found on transfer learning methods. One must decide *what* to transfer: value functions [12,19], features extracted from the value functions [1,8], heuristics and policies [3,5]. We focus on policy transfer; all transferred information is encoded by policies. An advantage of policy-based transfer is that it requires only a mapping between the states/actions of the *source* task and the states/actions of the *target* task; this amounts to less information than required by methods that transfer value functions [5]. Policy-search is in fact quite appropriate when coupled with abstraction [9]. Our approach is to transfer learning by exploiting abstract policies encoded through *relational* representations that compactly capture domains in terms of relations and objects [14]. Each abstract state aggregates a set of concrete states; given a single abstract state, one cannot know which concrete state obtains, but each concrete state is mapped to a unique abstract state. We consider abstract policies that are *memoryless* and *stochastic*. We use the AbsProb-PI algorithm [17] to construct such abstract policies in such a way that abstraction captures the main features of the source task. A memoryless policy chooses actions only according to the current observation of the system. No matter how smart the agent may be in constructing the abstract policy from the source problems, a direct application of an abstract policy in a new problem will almost certainly generate sub-optimal behavior, so one must always mix actions prescribed by an abstract policy with actions learned in the new, concrete task.

Our main contribution in this paper is to incorporate risk-awareness into the abstract policy built from a source task, so that the agent can modulate her prior experience by her preferences over risk, when learning features of a new target task. Introducing explicit preferences over risk requires us to measure subjective attitudes towards risk and to evaluate policies as risky or conservative [6,15]. A decision maker can have three distinct attitudes towards risk: averse, neutral or prone [6]. Relatively few activity in AI and Robotics addresses risk-awareness in policy construction, despite the practical importance of this issue, possibly due the difficulty in optimizing risk-aware measures. Liu and Koenig [11] consider probabilistic planning with nonlinear utility functions. Delage and Mannor [4] consider a relaxed version of minimax by defining probabilistic constraints on minimal performance. Mannor and Tsitsiklis [13] consider a trade-off between expected and variance of accumulated rewards. In all of these approaches, optimal policies are non-stationary in general.

We focus on robotic navigation problems modeled as Stochastic Shortest Path (SSP) problems.

We show here that an SSP problem, modeled as an MDP using the infinite-horizon discounted-reward criterion, can incorporate risk-awareness by varying the discount factor [15]. We then apply the AbsProb-PI algorithm [17] in the construction of memoryless stochastic abstract policies showing different attitudes towards risk, as a consequence of different discount factors. We also propose a strategy that balances between risk-neutral and risk-prone attitudes when reusing abstract policies in a new target task. Our experiments show that at the beginning of a new task, better results are obtained by using a risk-prone attitude, and as the execution of the task progresses, it is better to adopt a risk-neutral attitude. Both risk-prone and risk-neutral attitudes are dominated by the *non-stationary* policy that combines both attitudes in the new target task.

The remainder of this paper is structured as follows. Section 2 presents our proposals for tuning the discount factor in order to incorporate risk-awareness. Section 3 explains how to use relational representations in MDPs, describes the AbsProb-PI algorithm and presents our approach to combine different risk-aware policies. We present experimental results and analysis in the robotic navigation domain in Section 4. Finally, in Section 5 we draw our final conclusions.

2   Modeling Risk Sensitivity in SSPs with Negative Constant Rewards

In this section we formalize Markov Decision Processes (MDPs) and the Stochastic Shortest Path (SSP) problem we use when modeling the robotic navigation task. We will show that risk-awareness in this setting can be equated with the choice of the discount factor used in evaluating the performance of an agent in an MDP.

2.1   Markov Decision Process

For our purposes an MDP is a tuple $\langle \mathcal{S}, \mathcal{A}, T_a(\cdot), \beta, r(\cdot) \rangle$, where $\mathcal{S}$ is a finite set of process states s; $\mathcal{A}$ is a finite set of possible actions a; $T_a : \mathcal{S} \times \mathcal{S} \to [0,1]$ is a conditional probability distribution over $\mathcal{S}$ given each state in $\mathcal{S}$; $\beta : \mathcal{S} \to [0,1]$ is an initial probability distribution; and $r : \mathcal{S} \times \mathcal{A} \to \mathbb{R}$ is a reward function [16]. An MDP starts at a state $s_0 \in \mathcal{S}$, selected with probability $\beta(s_0)$. If the process is in state i at time t and action a is applied, then the next state is j with probability $T_a(i,j)$ and a reward $r(i,a)$ is received. In an MDP, a *stationary stochastic policy* $\pi : \mathcal{S} \times \mathcal{A} \to [0,1]$ specifies the probability $\pi(s,a)$ of executing action a in state s. We denote by $\mathcal{A}^s$ the set of allowable actions in state s; that is the support of $\pi(s,\cdot)$.

In this paper we wish to consider risk-aware decision profiles; to do so, we introduce a utility function that encodes preferences over the value of histories. We denote by $u(R(h))$ the utility of a history. The quality of a policy is measured by its expected utility:

$$E[u(R(h))] = \sum_h P(h|\pi)u(R(h)). \tag{1}$$

Clearly if $u(x) = \lambda x$ for some $\lambda > 0$, we return to the usual $E[R(h)] = \sum_h P(h|\pi)R(h)$. If instead $u(x) = e^{-\lambda x}$ (exponential utility), then:

$$E[u(R(h))] = \sum_h P(h|\pi)e^{-\lambda R(h)}. \tag{2}$$

An optimal policy π^* is a policy that yields the highest expected utility [16].

2.2 Stochastic Shortest Path Problem

A special case of MDPs is the Stochastic Shortest Path (SSP) problem with a unique goal state $s_G \in \mathcal{S}$ that is an absorbing state, i.e. if $r(s_G, a) = 0 \; \forall a \in \mathcal{A}$ and $T_a(s_G, s_G) = 1 \; \forall a \in \mathcal{A}_{s_G}$. Additionally, we assume that at each state except s_G the reward function is negative and constant, and fixed at -1. Suppose the decision maker wishes to minimize the number of steps to reach s_G. In this case each history h is summarized by its number of transitions $|h|$, because

$$R(h) = -(1 + \gamma + \gamma^2 + \cdots + \gamma^{|h|-1}). \tag{3}$$

Focus for a moment on $\gamma < 1$. Then

$$R(h) = -\frac{1 - \gamma^{|h|}}{1 - \gamma} = \alpha e^{\eta|h|} - \alpha, \tag{4}$$

where $\alpha = 1/(1 - \gamma)$ and $\eta = \ln \gamma$. Note that $\eta < 0$, so this expression for $R(h)$ holds even if history h is infinitely long (we obtain $R(h) = -\alpha$). In fact in this case $-\alpha \le R(h) \le 0$, hence $E[R(h)]$ exists for every policy and is finite, a well-known result.

Additionally, if $\gamma = 1$, then $R(h) = -|h|$; in this case a proof that $E[R(h)]$ is finite depends on further conditions [2]. In particular, say that a stationary policy is *proper* if the probability that it reaches s_G starting from any state s goes to 1 as t goes to infinity: If the SSP is such that there exists a proper policy, and any policy that is not proper yields infinite expected value over histories starting at some state, then a finite $E[R(h)]$ exists for some stationary policy.

In fact, with an additional assumption we can even consider values of γ larger than one. For each fixed policy π, write down the Markov chain that is obtained by fixing transition probabilities at π. Now construct an auxiliary Markov chain M_π by removing all arcs from s_G into itself, and adding arcs from s_G to each one of the other states, associating each arc from s_G to s with $\beta(s)$. We say the SSP problem is *recurrent* if each M_π (for each π) defines a recurrent Markov chain [20]. Recall that a finite Markov chain is recurrent if it is irreducible (any state can reach any state with positive probability); and if a finite Markov chain is recurrent, it is positive recurrent (the expected number of transitions to return to a state is finite). If the SSP problem is positive recurrent, every policy is proper. Moreover, Expression (4) holds for values of γ larger than one, and a finite $E[R(h)]$ exists for some stationary policy.

2.3 Risk-Awareness as γ Selection

Suppose our SSP problem has a proper policy, and every non-proper policy yields infinite expected value over histories starting at some state.

We can interpret $E[R(h)]$ for our SSP problem as the expected utility of *another* SSP problem, with the same states, actions and transition probabilities, but where $\gamma = 1$, and an associated utility function $u(\cdot)$. That is, the original SSP problem is not risk-aware and runs with some $\gamma < 1$, while the associated

SSP problem is risk-aware and runs with $\gamma = 1$. We can do this because when $\gamma = 1$, we have $R(h) = -|h|$, and we can then take the utility function:

$$u(x) = \alpha e^{-\eta x} - \alpha.$$

This device produces an expression analogue to Expression (4). So, the risk-aware analysis with $\gamma = 1$ and utility function $u(\cdot)$ yields the same results as our original SSP without risk-awareness.

Consequently, we can interpret γ in the original problem as an expression of a risk-aware judgement on the part of the agent [15]. We note that there exist other interpretations of γ; for instance, γ may reflect the fact that the agent focuses more intensely in close rewards; another interpretation is that γ is the probability of surviving one more time step [21].

Obviously, if the original problem already has $\gamma = 1$, then the agent already has a risk-neutral attitude. But if $\gamma < 1$ in the original problem, we can say that the agent has a risk-prone attitude.

Now suppose our SSP is recurrent. In this case the analyzes in the previous paragraph holds even for values of γ larger than one in the original problem. In this case we obtain $\eta > 0$ and the utility function reflects a risk-averse behavior.

3   Risk-Sensitive Abstract Policies

We can now use the proposed interpretation of the discount factor γ to model different attitudes toward risk of an agent aimed at solving an SSP. As stated earlier, SSP is a special case of MDP; so we can use any algorithm that solves MDPs to solve our SSP problem. For each solution, we can tune the agent's attitude towards risk by varying γ. However, our interest is not only in tuning the agent's attitude toward risk, but also in abstracting the solution so that this abstract solution can be reused in other new problems.

3.1   Relational Representations and Abstraction

Let us introduce some definitions first. C is a set of *constants*, P_S is a set of *state predicates* and P_A is a set of *action predicates*. If $t_1, \ldots, t_n$ are *terms*, each one being a constant (represented with lower-case letters) or a variable (represented with capital letters), and if p/n is a predicate symbol, $p \in P_S$ (or $p \in P_A$), with arity $n \geq 0$, then $p(t_1, \ldots, t_n)$ is an *atom*. If an atom does not contain any variable, it is called a *ground atom*. We call a set of atoms a *conjunction*. The state set $\mathcal{S}$ of a relational MDP is defined as the set of possible ground conjunctions over P_S and C, and the action set $\mathcal{A}$ is the set of possible ground atoms over P_A and C (for more details, please refer to [14,7]).

When $\mathcal{S}$ and $\mathcal{A}$ in a relational MDP are ground sets, we call this a *concrete* MDP. Note that a relational representation enables us to aggregate states and actions by using variables instead of constants in the predicate terms. A *substitution* θ is a set $\{X_1/t_1, \ldots, X_n/t_n\}$, binding each variable X_i to a term t_i. We call *abstract state* σ (and *abstract action* α) the representation in which all the constants of a ground state s (and a ground action a) are replaced by variables. In this work we consider only the level of abstraction that preserves every predicade in the conjunction without any constants in the terms. We denote by $\mathcal{S}_\sigma$ the set of ground states covered by an abstract state σ. Similarly we define $\mathcal{A}_\alpha$ as the set of all ground actions covered by an abstract action α. We also define $\mathcal{S}_{ab}$ and $\mathcal{A}_{ab}$ as the set of all abstract states and the set of all abstract actions in a relational representation of an MDP, respectively. Consider an abstract state $\sigma = \{p_1(X_1), p_2(X_1, X_2)\}$, $\sigma \in \mathcal{S}_{ab}$; a ground state $s_1 \in \mathcal{S}$, $s_1 = \{p_1(t_1), p_2(t_1, t_2)\}$ is abstracted by σ with $\theta = \{X_1/t_1, X_2/t_2\}$ and a ground state $s_2 \in \mathcal{S}$, $s_2 = \{p_1(t_3), p_2(t_3, t_4)\}$ is abstracted by σ with $\theta = \{X_1/t_3, X_2/t_4\}$. In this case σ represents an abstraction of both s_1 and s_2, and $s_1, s_2 \in \mathcal{S}_\sigma$.

An abstract policy π_{ab} specifies a mapping from abstract states into abstract actions: $\pi_{ab} : \mathcal{S}_{ab} \times \mathcal{A}_{ab} \to [0, 1]$. The challenge is to apply an abstract policy in a concrete problem: we must provide a way to translate from the concrete to the abstract level, and vice-versa. Problems that can use the same predicates to describe its states and actions are in the same *domain class*; that is, problems that have the same spaces of abstract states $\mathcal{S}_{ab}$ and abstract actions $\mathcal{A}_{ab}$ are in the same domain class. When the set of objects and a transition function are added, we specifie a *domain* $\mathcal{D}$. It describes the dynamics of the world and also the number of states and actions are now fixed. Finally, a *task* Ω is a tuple $\langle \mathcal{D}, r, \mathcal{G}, \beta \rangle$, where $\mathcal{D}$ is a domain, r is the reward function, $\mathcal{G}$ is the set of goal states, that indicates which states of the domain are desirable and β is the probability distribution over initial states. A task fully describes an MDP. Furthermore, in this work we focus on the transfer learning among different tasks within the same domain class.

Now we can explain how to use an abstract policy in a concrete MDP. Assume a stochastic abstract policy π_{ab} is given. For a ground state s we find the corresponding abstract state σ so that $s \in \mathcal{S}_\sigma$, i.e., $\xi_s(s) = \sigma$. The current stochastic abstract policy π_{ab} defines a probability distribution over abstract actions $\mathcal{A}_{ab}$ to be applied in the abstract state $\sigma \in \mathcal{S}_{ab}$. We then choose probabilistically the abstract action $\alpha \in \mathcal{A}_{ab}$ to be executed in σ. Note that an abstract action α defines $\mathcal{A}_\alpha$; thus, applying the abstract policy π_{ab} results in the mapping from an abstract state into a set of ground actions. To produce a particular ground action, we define $\xi_a : \mathcal{A}_{ab} \to \mathcal{A}$. Here we define ξ_a as a function that randomly selects (with uniform probability) a concrete action a from the set of concrete actions $\mathcal{A}_\alpha \cap \mathcal{A}^s$, which combines $\mathcal{A}_\alpha$ with the set $\mathcal{A}^s$ of allowable actions in state $s \in \mathcal{S}$. Obviously, other schemes may be used to produce a concrete action a. This whole process yields the grounding of a stochastic abstract policy $\pi_{ab}(\xi_s(s), \alpha)$, denoted by $a = \text{grounding}_{\pi_{ab}}(\alpha, s)$.

3.2   Solving MDPs with AbsProb-PI

The task of the agent in an MDP is to find a policy. In a concrete fully-observable MDP, the set of deterministic memoryless policies, $\pi : \mathcal{S} \to \mathcal{A}$, contains an optimal policy [16,10]. When considering abstract states and actions, stochastic policies are appropriate, because as they are more flexible by offering more than one choice of action per state, they can be arbitrarily better than deterministic policies [18,17]. We define a *memoryless stochastic abstract policy* as $\pi_{ab} : \mathcal{S}_{ab} \times \mathcal{A}_{ab} \to [0,1]$, with $P(\alpha|\sigma) = \pi_{ab}(\sigma, \alpha)$, $\sigma \in \mathcal{S}_{ab}$, $\alpha \in \mathcal{A}_{ab}$.

AbsProb-PI [17] is an algorithm that, given an MDP, uses cumulative discounted reward evaluation to build an abstract policy. It is a model-based algorithm, i.e., it requires prior knowledge of the whole model, including transition and reward functions. AbsProb-PI is a Policy Iteration algorithm, designed to perform in an abstract level. At each iteration, a gradient function G is used to determine the improvement direction of the current policy. As the abstract state space $\mathcal{S}_{ab}$ does not necessarily hold the Markov property, AbsProb-PI considers the initial state distribution $\beta(s_0)$. Two parameters must be defined: $\epsilon > 0$ which guarantees that the policy converges at most to an ϵ-greedy policy; and $\rho(i)$, the step size used in the gradient descent method for each iteration i. We refer to [17] for a thorough presentation of AbsProb-PI.

The use of a relational representation enables us to generalize experiences and to define abstract policies, making it easier to transfer knowledge between tasks in different domains. Abstract policies can effectively be used in new similar problems, or even be used to accelerate reinforcement learning [14,7], as we show in Section 4.

3.3   Transferring Risk-Aware Abstract Policies

To achieve transfer learning, we first need to decide which policy will be transferred. Policies found using different risk attitudes result in different performances (i.e., history size) of the agent. Figure 1 shows the probability as a function of history size $|h|$ (number of steps to reach goal) given discount factors $\gamma = 1.0$ and $\gamma = 0.9$ for several tasks (i.e., several goal states) on the navigation environment shown in Figure 2-left (see section 4 for the environment description). This figure shows that: (i) if γ is small (risk-prone attitude) there is a high probability for short histories, but also high probabilities for very long histories (histories with $|h| > 200$ steps accumulate almost half of the histories); and (ii) if γ is large (risk-neutral attitude) there is a medium probability for short histories, but also low probability for very long histories (histories with $|h| > 200$ steps happen with probability lower than 0.02).

We have options: we can transfer a risk-prone policy π_{RP} (found with discount factor γ_{RP}), or a risk-neutral policy π_{RN} (found with γ_{RN}, with $\gamma_{RP} < \gamma_{RN}$). Under π_{RP} there are higher probabilities for short histories, whereas under π_{RN} there are lower probabilities for longer histories. Our strategy is to combine both. It starts by applying π_{RP}, and then changes to π_{RN} as time goes on. The transition between policies is done linearly with time, i.e., let π_{NS}^t be the abstract policy applied at time step t, then the non-stationary policy π_{NS}^t is defined by:

$$\pi_{NS}^t(\sigma, \alpha) = (1 - \mu(t))\pi_{RP}(\sigma, \alpha) + \mu(t)\pi_{RN}(\sigma, \alpha) \ \forall \sigma \in \mathcal{S}_{ab}, \forall \alpha \in \mathcal{A}_{ab},$$

with $\mu(t) = \min\left\{ \frac{t}{thr}, 1 \right\}$.

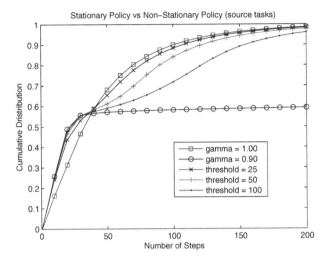

Fig. 1. Probability of success versus history size of stationary policies found with different γ values and non-stationary policies with different threshold values

The parameter thr is such that π_{RN} is fully applied for $t > thr$. Figure 1 also shows the performance of such strategy under different values of thr; our strategy is applied on source tasks, in which the policies π_{RP} and π_{RN} were found. If threshold parameter thr is well tuned, the non-stationary policy π_{NS} shows a better balance between short histories and long histories.

Now that we have a policy to transfer, we have to reuse this policy when learning a new task. This policy is used preferably to guide the exploration of the space state. Obviously, other approaches can be used.

4   Experiments

We conducted experiments to evaluate our proposal; that is, to find and use a risk-aware abstract policy in inter-task transfer applications. All experiments were made on a simulated robotic navigation domain, described below.

Navigation Domain Class

We use a robotic navigation domain class, in which the task of the agent is to navigate through an environment to reach a specific location. The space is divided in several cells of equal size, each representing a single state. Besides, the results of actions are probabilistic: if the agent performs a movement action, there is a probability p (we use $p = 0.9$) that it succeeds (and thus changes state), while it does not move with probability $1 - p$. Figure 2 shows instances of this domain.

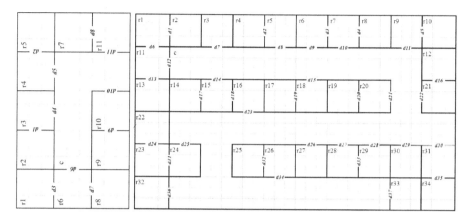

Fig. 2. The robotic navigation domain class. Left: an environment with 11 rooms (*source domain*). Right: an environment with 34 rooms (*target domain*).

There are four types of objects the agent can recognize: rooms, doors, markers and corridors. Additionally, it has a positioning system as well, being able to tell if it is near or far from the goal location. Therefore, we describe each ground state and action using the relational alphabet $C \cup P_S \cup P_A$. C is the set of objects. P_S describes predicates related to observations, $P_S = \{$inRoom(R), inCorridor(C), seeDoorFar(D), seeAdjRoom(D), seeAdjCorridor(D), seeEmptySpace(M), nearGoal, farGoal, appGoal(X), awayGoal(X)$\}$, where R is a room, C is a corridor, M is a marker, D is a door and X is any object. The range of vision of the robot is of one cell for markers, but it can perceive doors that are two cells distant. The predicate seeDoorFar means that the robot can see a door, but it is at least two cells distant, whereas seeAdjRoom(D) and seeAdjCorridor(D) indicate that the robot is so close to that door (one cell distant) that it can even see what lies through it (if it is a room or a corridor). seeEmptySpace(M) means that the robot sees a free space with a marker, where it could move to. nearGoal is true if the robot is at a close distance to the goal, i.e., at a Manhattan distance to the goal smaller than 5 and farGoal is true if the robot is at a far distance from the goal, i.e., at a Manhattan distance to the goal bigger than 8. There are indications of two possible directions: appGoal(X) (approaching goal) or awayGoal(X) (moving away from the goal) and these predicates indicates that the object X is closer or farther than the goal in relation to the agent.

All actions the agent can take moves the agent toward an object, with a step of one cell (unless it stays in the same state due to environment dynamics). Agent also has to consider whether it is approaching the goal or moving away from it. That being said, the predicates for actions are:

$$P_A = \{\ \text{goToDoorAppGoal}(di), \text{goToDoorAwayGoal}(di),$$
$$\text{goToRoomAppGoal}(ri), \text{goToRoomAwayGoal}(ri),$$
$$\text{goToCorridorAppGoal}(ci), \text{goToCorridorAwayGoal}(ci),$$
$$\text{goToEmptyAppGoal}(mi), \text{goToEmptyAwayGoal}(mi)\}.$$

Each task has a unique goal state $(\mathcal{G} = \{s_G\})$, which in our experiments is always the cell in the center of a room. The experiments conditions and their results are discussed in the next section.

Results

We use two domains in our experiments, both of which are represented in Figure 2. The first one, the source domain, contains 11 rooms and the center of all of them is used as the goal state for a task, resulting in a total of 11 tasks. The second one, the target domain, contains 34 rooms, and similarly, 34 tasks. Considering all 11 tasks together in the source domain, the agent groups them as a single big task and then uses AbsProb-PI with $\gamma = 0.9$ and $\gamma = 1.0$ to find, respectively, abstract policies π_{RP} and π_{RN}. Parameters used in AbsProb-PI were: $\rho(i) = \frac{1}{1+0.5i}$, $\epsilon = 0.05$, initial state distribution is uniform for all states, and stop criterion after 500 iterations. With these two learned policies at hand, a non-stationary policy π_{NS} is also created after setting $thr = 50$. To assess if the non-stationary policy performs better than policies π_{RP} and π_{RN} when applying in different tasks (transfer), we execute them in each task of the target domain, one at a time. Figure 3-left shows the probability of success versus histories size $|h|$ when applying these three policies. We also compare against random policy π_{rand} as a reference point.

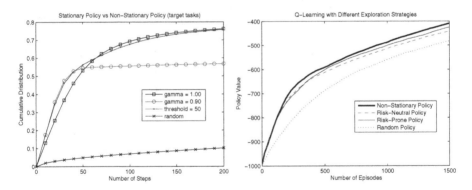

Fig. 3. Left: Cumulative probability distribution over history size of π_{RP}, π_{RN} and π_{NS}. Right: Transfer learning with the combined policy π_{NS} compared to learning reusing π_{RP}, π_{RN} and π_{random}.

One can notice that the non-stationary policy indeed explores the best of the two stationary policies and all of them are substantially better than the random policy.

We now evaluate the effectiveness of transfer learning when using risk-prone, risk-neutral, non-stationary, and random policies. On each of the 34 tasks in the target domain, an agent learns a policy using the Q-Learning algorithm with an

ϵ-greedy exploitation/exploration strategy. We always use the transferred policy in the exploration phase. We used the random policy in the exploration phase as traditionally is done in the ϵ-greedy strategy and compared the results when replacing it by a past abstract policy (π_{RP}, π_{RN} or π_{NS}). It is worth mentioning that the abstract policies are also stochastic and all actions have a minimum probability of ϵ, hence exploration is guaranteed. We set learning rate to 0.05 for the Q-Learning algorithm, $\epsilon = 0.1$ and $\gamma = 0.999$. Each episode has maximum number of steps 1,000 and learning was done for 15,000 episodes; each task in the target domain was learned 5 times.

Results for transfer learning are shown on Figure 3-right that plots the current policy value versus the number of steps. We observe that the non-stationary policy outperforms all of them.

5 Conclusions

In this paper we have proposed algorithms for a risk-aware robot that must transfer abstract policies from already solved tasks to new tasks. Our abstract policies employ relational representations, and our policies are memoryless, hence compact, and stochastic. We have derived the necessary algorithms, and our experiments demonstrate that our methods are effective in navigation problems.

We have presented a formulation of risk-aware SSP problems (with constant negative reward and unique goal state) that reduces risk-awareness to γ-tuning; our analysis is novel in that it justifies even values of γ that are larger than one. We then used these insights to combine abstract policies learned by the AbsProb-PI algorithm. This algorithm generates policies under the infinite-horizon discounted criterion; by appropriately changing γ, we obtain risk-aware behavior out of AbsProb-PI. We have proposed a method that combines linearly the various abstract policies generated by AbsProb-PI, emphasizing a risk-prone behavior in the first stages of learning, and a risk-neutral behavior in latter stages. The resulting combined policy is not only stochastic, but also non-stationary. Our experiments have focused on a robot in indoor navigation, and have shown that a robot that mixes the combined policy with learning in a new task does have an advantage over a robot learning from scratch a new policy in the new task.

Our contributions are both an extended interpretations for γ-tuning as risk-awareness, and a method that employs risk-awareness in transfer learning. It is necessary still to better understand the risk-averse behavior that obtains when $\gamma > 1$ is employed; we plan to focus on this development in the future.

References

1. Banerjee, B., Stone, P.: General game learning using knowledge transfer. In: Proc. of the Twentieth Int. Jt. Conf. on Artif. Intell., pp. 672–677. AAAI Press (2007)
2. Bertsekas, D.P., Tsitsiklis, J.N.: An analysis of stochastic shortest path problems. Math. of Oper. Res. 16(3), 580–595 (1991)

3. Bianchi, R., Ribeiro, C., Costa, A.: Accelerating autonomous learning by using heuristic selection of actions. J. of Heuristics 14, 135–168 (2008)
4. Delage, E., Mannor, S.: Percentile optimization for markov decision processes with parameter uncertainty. Oper. Res. 58(1), 203–213 (2010)
5. Fernández, F., García, J., Veloso, M.: Probabilistic Policy Reuse for inter-task transfer learning. Robotics and Auton. Syst. 58(7), 866–871 (2010)
6. Howard, R.A., Matheson, J.E.: Risk-sensitive markov decision processes. Management Science 18(7), 356–369 (1972)
7. Koga, M.L., Silva, V.F., Costa, A.H.R.: Speeding-up reinforcement learning tasks through abstraction and transfer learning. In: Proc. of the Twelfth Int. Jt. Conf. on Auton. Agents and Multiagent Syst. (AAMAS 2013), pp. 119–126 (2013)
8. Konidaris, G., Scheidwasser, I., Barto, A.: Transfer in reinforcement learning via shared features. J. of Mach. Learn. Res. 13, 1333–1371 (2012)
9. Li, L., Walsh, T.J., Littman, M.L.: Towards a Unified Theory of State Abstraction for MDPs. In: Proc. of the Ninth Int. Sympos. on Artif. Intell. and Math., pp. 531–539. ISAIM (2006)
10. Littman, M.L.: Memoryless policies: theoretical limitations and practical results. In: Proc. of the Third Int. Conf. on Simul. of Adapt. Behav.: from Animals to Animats 3, pp. 238–245. MIT Press, Brighton (1994)
11. Liu, Y., Koenig, S.: Probabilistic planning with nonlinear utility functions. In: ICAPS, pp. 410–413 (2006)
12. Liu, Y., Stone, P.: Value-function-based transfer for reinforcement learning using structure mapping. In: Proc. of the Twenty-First Natl. Conf. on Artif. Intell., pp. 415–420. AAAI Press (2006)
13. Mannor, S., Tsitsiklis, J.: Mean-variance optimization in markov decision processes. In: Proc. of the Twenty-Eighth Intl. Conf. on Mach. Learn. (ICML 2011), pp. 177–184. ACM (2011)
14. Matos, T., Bergamo, Y.P., Silva, V.F., Cozman, F.G., Costa, A.H.R.: Simultaneous Abstract and Concrete Reinforcement Learning. In: Proc. of the Ninth Symp. of Abstr., Reformul., and Approx (SARA 2011), pp. 82–89. AAAI Press (2011)
15. Minami, R., da Silva, V.F.: Shortest stochastic path with risk sensitive evaluation. In: Batyrshin, I., González Mendoza, M. (eds.) MICAI 2012, Part I. LNCS (LNAI), vol. 7629, pp. 371–382. Springer, Heidelberg (2013)
16. Puterman, M.: Markov Decision Processes: Discrete Stochastic Dynamic Programming. John Wiley & Sons, Inc. (1994)
17. da Silva, V.F., Pereira, F.A., Costa, A.H.R.: Finding memoryless probabilistic relational policies for inter-task reuse. In: Greco, S., Bouchon-Meunier, B., Coletti, G., Fedrizzi, M., Matarazzo, B., Yager, R.R. (eds.) IPMU 2012, Part II. CCIS, vol. 298, pp. 107–116. Springer, Heidelberg (2012)
18. Singh, S.P., Jaakkola, T., Jordan, M.I.: Learning without state-estimation in partially observable markovian decision processes. In: Proc. of the Eleventh Int. Conf. on Mach. Learn. (ICML 1994), vol. 31, p. 37. Morgan Kaufmann (1994)
19. Taylor, M.E., Stone, P., Liu, Y.: Transfer learning via inter-task mappings for temporal difference learning. J. of Mach. Learn. Res. 8(1), 2125–2167 (2007)
20. Wasserman, L.: All of Statistics: A Concise Course in Statistical Inference. Springer (2003)
21. Whittle, P.: Why discount? the rationale of discounting in optimisation problems. In: Heyde, C., Prohorov, Y., Pyke, R., Rachev, S. (eds.) Athens Conference on Applied Probability and Time Series Analysis. Lecture Notes in Statistics, vol. 114, pp. 354–360. Springer, New York (1996)

Motivated Reinforcement Learning for Improved Head Actuation of Humanoid Robots

Jake Fountain, Josiah Walker, David Budden, Alexandre Mendes, and Stephan K. Chalup

The University of Newcastle, Australia
{jake.fountain,josiah.walker,david.budden}@uon.edu.au,
{alexandre.mendes,stephan.chalup}@newcastle.edu.au

Abstract. The ability of an autonomous agent to self-localise within its environment is critically dependent on its ability to make accurate observations of static, salient features. This notion has driven considerable research into the development and improvement of feature extraction and object recognition algorithms, both within RoboCup and the robotics community at large. Instead, this paper focuses on the rarely-considered issue imposed by the limited field of view of humanoid robots; namely, determining an optimal policy for actuating a robot's head, to ensure it observes regions of the environment that will maximise the positional information provided. The complexity of this task is magnified by a number of common computational issues; specifically high dimensional state spaces and noisy environmental observations. This paper details the application of motivated reinforcement learning to partially overcome these issues, leading to an 11% improvement (relative to the null case of uniformly distributed actuation policies) in self-localisation and ball-localisation for an agent trained online for less than one hour. The method is demonstrated as a viable method for improving self-localisation in robotics, without the need for further optimisation of object recognition or tuning of probabilistic filters.

Keywords: motivated reinforcement learning, localisation, Fourier basis, head actuation, simulated curiosity.

1 Introduction

Effective localisation is essential to solving tasks such as the Robocup soccer competition. The behaviour of a robot is a function strongly dependent on the robot's localisation model. Inaccurate localisation leads to ineffective behaviour and poor performance. In a game of humanoid robot soccer, successful localisation of the robot and field objects relies on the vision system to measure relative field object locations. Moreover, the behaviour of the head determines the efficacy of the vision system; limited field of view implies that all field objects cannot be measured at once. The head actuation problem involves choosing neck motor positions to optimise localisation. If the robot is completely unlocalised,

S. Behnke et al. (Eds.): RoboCup 2013, LNAI 8371, pp. 268–279, 2014.
© Springer-Verlag Berlin Heidelberg 2014

panning the head can be used to localise at least partially. The problem of interest involves achieving and maintaining a high level of localisation assuming that initially the robot is partially localised. Since there are a finite number of useful field objects, this problem can be abstracted, with inverse kinematics, to that of sequentially choosing field objects on which to focus gaze to minimise localisation uncertainty while playing soccer.

Wong, et al. [1], used humanoid robots to model human gaze behaviour in urban environments. They utilised a gaze direction model based on visually salient objects placed in a model urban environment. If a robot saw a salient object it would slow down to view it for a period of time before scanning for other objects as it continued to walk. This behaviour was then used to benchmark the performance of gaze vector computer vision techniques; that is, measuring where the robot is looking with a wide field of view camera. We postulate that this anthropomorphic behaviour is desirable in making head actuation decisions to localise humanoid robots effectively. Therefore, we utilised motivated reinforcement learning techniques [2] to implement 'curious', anthropomorphic behaviours with the goal of optimising localisation in the RoboCup KidSize soccer league [3].

There are two types of field objects important to localisation of a humanoid robot while playing soccer: *landmarks* and *objects*. Landmarks are used to localise the robot while objects must be localised in the world model by measuring position relative to a localised robot, and then transforming the information into global coordinates [4]. In our approach, the model of the soccer field world is a collection of Unscented Kalman Filters [5]; each robot maintains a filter for its own position and a filter for the ball's position. Each filter is a Gaussian probability distribution for the possible field locations of an object. This probabilistic model allows uncertainty to be managed quantitatively for each robot and object, and this allows the robot to assess its localisation performance without external reference.

The head actuation problem was phrased as a Markov decision process and solved using Q-learning [6] with a motivated agent. A motivated agent uses the concept of novelty to seek out events which are similar-yet-different to previous experiences; this *curious* behaviour was useful in partitioning head actuation decisions between measuring landmarks and objects.

2   Motivated Reinforcement Learning

Reinforcement learning is a form of machine learning used to solve problems involving a series of decisions made by an *agent* based on perceptions of its *environment*, with a metric indicating performance after every decision [7]. This type of problem is called a Markov decision process, and is described fully by: a state space S; a set of actions A and a function $\Lambda : S \to 2^A$ where $\Lambda(s)$ is the subset of actions available in state $s \in S$; a transition function $T : S \times A \times S \to [0, 1]$ describing the probability of state transitions; and a reward function $R : S \times A \to \mathbb{R}$ [6]. Here $2^A = \{U : U \subseteq A\}$ is the power set of A, or the set of all subsets of A. We additionally simplify the problem for the purpose of head actuation by including the assumption that $\Lambda(s)$ is finite and discrete for each $s \in S$. Q-learning is a method of learning the optimal value function Q^* and can be performed as an *online* learning method; that is, the model of Q^* is updated between actions to reflect experiences. Actions are chosen from successive states to affect the environment and thus explore the state action space $S \times A$. After each action, a function $Q : S \times A \to \mathbb{R}$ is updated to approximate Q^* using the rule

$$Q(s,a) \leftarrow Q(s,a) + \alpha[R(s,a) - Q(s,a) + \gamma \max_{a' \in \Lambda(s')} Q(s',a')]$$

where s' is sampled according to the transition function T, and α and γ are constants known as the learning rate and the discount factor. It has been shown that this method converges to the optimal value function Q^* provided the agent explores all states and each action from each state sufficiently [6]. A further assumption for convergence is that the process is *Markov*. A process satisfies the Markov property if the reward and transition functions depend only on the current state, and not the history of the system. The state space in the head actuation problem was constructed with with the aim of satisfying this property (Sec. 4). Once Q^* is known, the agent can make optimal decisions given each state s in an effort to maximise long term reward by choosing a to maximise $Q(s,a)$.

Motivation theory attempts to describe the behaviour of biological intelligent agents by giving motivational reasons for behaviour. This includes describing the behaviour of agents removed from stimuli. When reward is a constant function in some region of the state action space, the choice of actions becomes ambiguous as the environment no longer discriminates. Typically, an animal given no environmental stimuli will seek out novel experiences, rather than taking no action [8]. Reinforcement learning infrastructure has been used to model motivated behaviour for application in generating complex, exploratory behaviours in unsupervised intelligent agents for non-player characters in online multi-player games [2]. Such motivated reinforcement learning agents differ from standard agents in that they generate their own reward, independent of the environmental reward, based on state perceptions and their own actions. It has been established that natural agents will seek out a middle ground in terms of novelty of sensation, resulting in an aversion to experiences too familiar or too unfamiliar. Saunders and Gero [9] implemented motivated reinforcement learning agents to study the progression of architectural designs, with successive designs similar-yet-different to previous designs. Merrick, et al. [10], have used motivated reinforcement learning agents to create game content procedurally, conforming to the similar-but-different concept of motivation, to simulate creativity. Further research by Merrick [11] involves agents which generate goals based on motivation. The agent then seeks to learn these goals with standard reinforcement learning techniques.

Given the novelty $N = N(s,a)$ of a state action pair, the Wundt curve is used to model the motivation reward function $M : S \times A \to \mathbb{R}$

$$M(s,a) = M_0 + \frac{M_1}{(1 + e^{-\rho_1(N-N_1)})} - \frac{M_2}{(1 + e^{-\rho_2(N-N_2)})}$$

where $\rho_1, \rho_2, N_1, N_2, M_0, M_1$ and M_2 are real parameters which define the function's behaviour; Fig.1 illustrates the Wundt curve utilised in solving the head actuation problem. The novelty $N(s,a)$ can be calculated using methods such as a Habituated Self-Organising Map [9]. In contrast, the novelty detection method utilised for the head actuation agents involved maintaining a model $T' : S \times A \to S$ of the expected transition function $E\{T\} : S \times A \to S$. $T'(s,a)$ modeled the most likely next state after taking action a from state s. After each action a from state s, the expected next state, $T'(s,a)$, is compared to the actual next state, s', using the Euclidean norm on $S \subseteq \mathbb{R}^m$. That is, the novelty is given by

$$N(s,a) = \|T'(s,a) - s'\|^2 \tag{1}$$

After each novelty calculation, the function T' was updated to agree more closely with $E\{T\}(s,a) = s'$, simulating habituation, with novelty declining over multiple similar experiences. This method measures the agent's ability to predict environmental reaction and ascribes high prediction accuracy with low novelty values. The magnitude of the novelty depends on numerous unpredictable factors, and so the motivation function required extensive tuning to achieve desirable behaviour (see Sec. 5 for more detail).

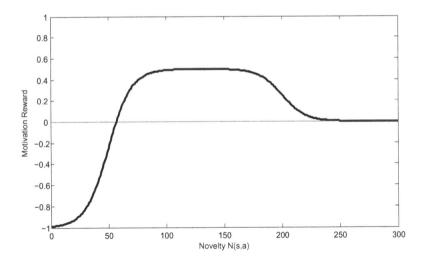

Fig. 1. The Wundt function [8] used for calculating the motivation reward $M(s,a)$ is visualised. A reinforcement learning agent interprets motivation as reward and this drives exploration of the state action space.

3   Approximating Continuous Value Functions

A method for storing a continuous value function, Q, in reinforcement learning involves a weighted sum of basis functions and learning of a set of scalar weights using gradient descent. Based on the Fourier series expansion of periodic functions, a Fourier basis can be used to approximate the value functions in a given domain [12]. The Fourier basis linear approximator is given by the cosine part of a truncated Fourier series and is updated with a sampled point, gradient descent update rule. By using only the cosine terms of the series, the number of required terms for a given accuracy is halved. This comes at the cost of the approximator being even; we overcome this limitation by restricting the state space to a non-periodic domain of the function, namely $S = [0, \tau]^m$ for some $\tau > 0$. In the domain $[0, \tau]^m$ the function is neither periodic nor even, and thus arbitrary continuous functions $f : [0, \tau]^m \to \mathbb{R}$ can be approximated. The performance of the Fourier basis linear approximator at value function approximation has been shown to compete with other leading methods such as radial basis estimation and popular learned basis approximation architectures [12]. For an approximator $F : [0, \tau]^m \to \mathbb{R}$ of order $k \in \mathbb{N}$, the value of the approximation at $\mathbf{x} \in [0, \tau]^m$ is given by

$$F(\mathbf{x}) = \langle \mathbf{w}, \overrightarrow{\phi}(\mathbf{x}) \rangle = \sum_{\mathbf{c} \in C} w_{\mathbf{c}} \cos \left[\frac{\pi}{\tau} \langle \mathbf{c}, \mathbf{x} \rangle \right]$$

where C is a subset of $(\mathbb{Z}_{k+1})^m$. We say $\phi_{\mathbf{c}}(\mathbf{x}) = \cos(\frac{\pi}{\tau}\langle \mathbf{c}, \mathbf{x} \rangle)$ is the basis function corresponding to $\mathbf{c} \in C$ and $w_{\mathbf{c}} \in \mathbb{R}$ is the weight of the basis function corresponding to $\mathbf{c} \in C$. $\mathbb{Z}_j$ denotes the set of integers modulo j, thus C is a collection of $m-$dimensional vectors of integers less than or equal to k. $\overrightarrow{\phi}(\mathbf{x})$ is the vector of basis functions. If $f : \mathbb{R}^m \to \mathbb{R}$ is to be approximated by F with the form above, then sampling f at $\mathbf{x} \in \mathbb{R}^m$ allows F to be updated according to a gradient descent update rule to shift the value of $F(\mathbf{x})$ to agree more closely with $f(\mathbf{x})$. To approximate a function $g : \mathbb{R}^m \to \mathbb{R}^n$, n Fourier basis approximators are used, one for each component of the output.

4   Experiment and Results

The head actuation problem required a solution which accounted for the subtleties involved with maintaining field models with high levels of noise in measurements. This involves making appropriate decisions based on what is most likely to be seen given the estimated locations of objects, which objects give the most information and several other factors. Thus, the problem of choosing head actuation decisions was framed as a Markov decision process and solved with online reinforcement learning methods.

The Markov decision process was constructed by sampling data from the robot infrastructure. The states were constructed as vectors of useful data. Each entry in the state vector was scaled using sigmoid $(f(x) = \frac{1}{1+e^{-a(x-b)}})$ or decaying growth $(f(x) = 1 - e^{-ax})$ functions to confine the ranges within $[0, \tau]$ to ensure convergence of the Fourier basis approximator which was used to store the value function Q. The action space was the collection of landmarks and objects which were on the field; namely the four goal posts and the ball. For a given action, the robot scans the region in which the object is likely to be found, given by the Kalman filter, for a set period of time, or until it finds the object. An unlocalised robot would thus scan its full field of view, giving a good initial level of localisation. The state space had 10 dimensions and the action space contained 5 actions. The state variables were selected to be approximately independent for inclusion in an uncoupled Fourier basis linear approximator. The state variables were chosen to be the distances to each goal, the total uncertainty in the robot's location filter, the distance to the ball from the robot, the total uncertainty in the ball's location filter, and five *object priorities* corresponding to the action space. The priority of an object is given by the time since the object was last seen, t, and the head movement cost to look at the object, measured in radians, c. If $c \neq 0$ then the priority is given by

$$\tau(1 - e^{-c_p \frac{t}{c}})$$

where c_p is a constant which accounts for the different units and ranges of t and c. If $c = 0$ then the priority is zero, corresponding to when the object is in the field of view of the robot. Objects which have not been seen for longer times or which require little movement to view have a high priority. Thus it can be predicted that high priority objects should be chosen more frequently. This combination of two state variables with complementary expected optimal policies minimises the dimensionality of the state space and reduces computation times. Distances were calculated based on data stored on the robot or measured by the robot. Measured location was used when the object was in the field of view of the robot, and filtered location was used otherwise. The robot's (x, y) location was not included in the state variables because sufficient information is encoded in the distance to the goals. Two reflectively symmetric possibilities exist for a given pair of goal distances. This information is sufficient for head movement decisions as the best action will not depend on which wing of the field the robot is positioned.

The action function $\Lambda(s)$ was defined as the set of objects the robot could look at in state s without exceeding the restrictions on head rotation imposed by the KidSize League RoboCup rules ($-135^o \leq \theta \leq 135^o$, $-90^o \leq \varphi \leq 90^o$; where θ and φ are the angles measured from forward facing direction in the horizontal and vertical planes respectively [13]). The environmental reward function was constructed as a decreasing function of total localisation variance of the robot model and field objects model. The reward was calculated according to

$$R(s,a) = \frac{1}{2}[e^{-c_b(\sigma_{bx}^2 + \sigma_{by}^2)} + e^{-c_r(\sigma_{rx}^2 + \sigma_{ry}^2) - c_\phi(\sigma_{r\phi}^2)}] \tag{2}$$

where $\sigma_{b\zeta}^2$ and $\sigma_{r\zeta}^2$ are the variances in the variable ζ in the next state s' for the ball and the robot respectively. The variable ϕ denotes the robot's heading on the field and the terms of the form c_ζ are constants which were adjusted to weight each variance equally, accounting for differing units and ranges. The reward is normalised to the interval $[0,1]$, with perfect localisation indicated by $R = 1$. The reward function is a function of data also directly included in the state vector, namely the total ball error, $c_b(\sigma_{bx}^2 + \sigma_{by}^2)$, and the total self localisation error, $c_r(\sigma_{rx}^2 + \sigma_{ry}^2) + c_\phi(\sigma_{r\phi}^2)$. Thus, the reward depended on the current state only and did not compromise the Markov property. All data was sourced from the world model of the robot, or from the robot's measurements. Data could have been generated by overhead cameras and other external sensors, particularly the absolute localisation error. However, by restricting data to that stored and computed on the robot, the process could be solved online, even during a RoboCup match. Online learning is necessary for the motivated reinforcement learning agent, as the behaviour relies on the changing novelty and thus changing reward function. Standard reinforcement learning agents should also benefit from the adaptability of online learning. Additionally, by using the localisation uncertainty for the reward, and not the absolute error, the performance of the agents can be measured with noise from other systems, like vision, excluded.

Three agents used Q-learning to solve the Markov decision process with reward functions differing between agents. Training sessions were performed with off-policy (RL Agent) or on-policy (others) action selection. During games all agents operate by making decisions on-policy and learning simultaneously. One agent was trained to maximise the reward given by the environment, $R(s,a)$;

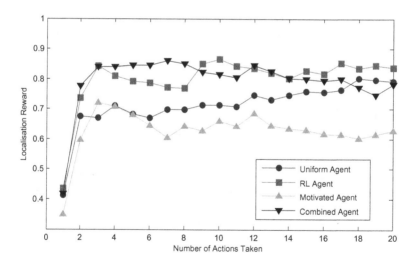

Fig. 2. Results of on-policy, kidnapped robot testing of the head actuation agents. The performance of each agent was assessed by placing the robot in fifteen field positions and allowing the head actuation agent to make twenty decisions while learning. The localisation reward, $R(s,a)$, was recorded after each action was taken and averaged over the 15 positions to provide a metric of performance over the whole field. A localisation reward of 1 corresponds to complete certainty in localisation and the relative localisation uncertainty is given by simply inverting this graph.

this agent will be denoted the *RL Agent*. The second agent was trained to ignore the reward given by the environment and instead use a motivation reward, M(s,a); this agent will be denoted the *Motivated Agent*. A third agent, the *Combined Agent*, used a sum of the environmental and the motivation reward, biased additively by -0.5 (Table 1). The Wundt parameters used for motivation were $N_1 = 50, N_2 = 200, M_0 = -1, M_1 = 1.5, M_2 = 0.5$ and $\rho_1 = \rho_2 = 0.1$ (Fig. 1). These parameters were chosen to suit the range of the novelty by experimental tuning, and are specific to the novelty calculation method and desired behaviour. It was found that, for the purpose of the head actuation, it was more effective to penalise highly novel states minimally, hence the parameters chosen. This choice of parameters gave a motivation reward function $M(s, a)$ ranging from -1 to 0.5 (Fig. 1). The combination of environmental and motivation rewards for the Combined Agent was selected by tuning to balance the importance of exploration and exploitation. The agents utilised Fourier basis linear approximators of order $k = 30$ with gradient descent update rule to store and learn the value function Q and the expected transition function T'. The set $C \subseteq (\mathbb{Z}_{k+1})^m$ for the approximators was chosen to be $C = \{\mathbf{c} \in (\mathbb{Z}_{k+1})^m : c_i \neq 0$ for at most one $i \in \mathbb{N}$ s.t. $1 \leq i \leq m\}$. This is called a *decoupled* Fourier basis, as only single variable cosines form the basis functions. The state variables were chosen carefully because of this; variables which may have been correlated were re-formulated to be independent variables where possible. For example, the robot's position was not included in the state variables, as the goal distances and position would not be independent.

Table 1. The reward schemes for learning agents are summarised

Agent	Reward
RL Agent	$R(s, a)$
Motivated Agent	$M(s, a)$
Combined Agent	$R(s, a) + M(s, a) - 0.5$

The agents were trained for approximately 30 minutes each. Fifteen minutes while playing soccer and fifteen minutes of *fixed position training*. Fixed position training involves placing the robot on the field in a series of positions and allowing the agent to make decisions and learn while the robot remains stationary. Fixed position training was used to guide the robot to explore states that rarely occur during soccer playing but which are still important. During training, off-policy soft-max action selection was used to train the RL Agent, whereas the motivated agents made all decisions on-policy. Off-policy training was required only for the RL Agent as the others had inbuilt explorative tendencies in the form of motivation reward. Another technique for off-policy training, ϵ-greedy, was also tested, but found to produce no better results. The purpose of this training was to provide the agents with a base value function and policy upon which they can build during game play by learning online.

Obtaining quantitative results from a soccer playing robot proved too difficult due to the stochastic elements introduced by the interaction of several other robot systems, so the agents were assessed by measuring on-policy performance during fixed position, *kidnapped robot* localisation problems [14]. That is, each agent was moved to a random location on the field and was allowed to localise by making twenty head movement decisions, on-policy, while learning. This simulates the scenario where the robot loses localisation, perhaps due to falling down, and must re-localise before re-engaging with the game. This was done for 15 positions for each agent and the reward after a given number of actions was averaged over the 15 positions. The environmental reward for each action was used to measure performance as it directly indicates the relative localisation accuracy of the robot (Equation 2). A third agent was used as a control: the *Uniform Agent*. The Uniform Agent simply chose head movement with uniform probability from the available objects, as indicated by Λ. The Uniform Agent performed well and, as it was simple to implement, the Uniform Agent became the baseline to which the other agents were compared.

The results of the experiment are shown in Fig. 2 and Table 2. The mean environmental localisation reward over the 15 field positions was used as a metric for agent performance after each action. The environmental reward after a given number of actions varied on average 23% from the displayed mean value. Although this seems large, it must be recognised that between different field locations, during kidnapped robot testing, the bounds on the localisation uncertainty vary due to the differing amounts of information available. For example, the magnitudes of σ_{bx}^2 and σ_{by}^2 will be bounded below by some value determined by the distance from the ball to the robot and by the parameters involved with the Kalman filter. Therefore a spread of values for the localisation uncertainty is expected. Learning was fast enough to be performed online without affecting the other robot systems. However, this result was sensitive with respect to the size of the Fourier linear approximators. This is why a decoupled approximator was a necessity.

5   Discussion

The Uniform Agent performed better than previous methods, such as the simple hard coded logic statement method which was used in previous competitions, with a steady increase in mean reward over the twenty actions after the initial localisation. The Uniform Agent performed well because it was obtaining information regardless of its choices. The RL Agent generally performed better than the Uniform Agent at playing soccer. It was able to balance looking at the goals and the ball in most situations, and this enabled goal scoring to increase and defending to be more effective. Moreover, the Combined Agent performed better than the RL Agent for the first 8 actions due to the interaction of *boredom* with the environmental reward. Boredom is the state where, due to low novelty, the motivation reward is low. By choosing to penalise boredom heavily (Fig. 1) relative to the environmental reward in the Combined Agent, a curious but focused

Table 2. The localisation performance of reinforcement leaning agents during kidnapped robot testing is summarised (corresponding to Fig. 2). The RL Agent and Combined Agent outperformed the Uniform Agent both at localising quickly and in long-term performance. The expression 'localised over x' means achieving a localisation reward greater than x.

Agent Type	Uniform	Motivated	RL	Combined
Number of cases(out of 15) localised over 0.8 within 4 actions	10	10	13	13
Actions to achieve over 0.8 mean localisation reward	18	>20	3	3
Average % improvement over Uniform Agent over 20 actions	-	-11%	12%	11%

head actuation agent was trained. Overall the Combined Agent was able to make head actuation decisions to localise robustly. The worsening performance of the Combined Agent for more than 8 actions is likely due to the agent becoming bored after being in the one field position for extended periods and revisiting the same subset of the state space. This would cause worse decisions due to the motivation reward dominating, and this is reflected in the reward curve trending similarly to the Motivated Agent's curve after 12 actions. This would not seriously affect soccer playing as the field location changes quickly during a soccer match, so the robot will likely never make more than a few decisions from one field position at a time. The motivated agent performed poorly as it could not distinguish which actions were better and it did not become bored quickly enough due to the Wundt function. A better agent may have used a *hyperactive* Wundt function, with high novelty required for positive reward, causing the agent to choose the next action based on the least frequently chosen action. However, the Wundt function shown in Fig. 1 was effective for the Combined Agent, so it was also examined independently.

The Combined Agent is only random in its initial state, before training. After the first learning iteration, it uses no random variables to make decisions but rather a complex motivation system that balances exploration and exploitation. It should be noted that the Combined Agent requires no off-policy action selection during training as the motivation reward induces a natural exploration. This exploration may provide a viable alternative to using off-policy action selection techniques, such as soft-max or ϵ-greedy, during training. Due to its balanced exploration, motivation may prove more effective than these techniques; more research would be required to measure its effectiveness.

As with any machine learning task, finding an acceptable set of training parameters was difficult. The key parameters for the Fourier basis approximators was the domain size τ, order k and the basis function set C; k and C must be chosen to balance computation time and resolution of the function. The dimensionality costs of choosing $C = (\mathbb{Z}_{k+1})^m$ were too large for a high dimensional state space (m=10) as $|(\mathbb{Z}_{k+1})^m| = (k+1)^m$. The use of the uncoupled basis function set, with size $|C| = m(k+1)$, allowed for liberal choice for the order k at the cost of requiring uncoupled state variables. The calculation of the novelty proved effective for the purposes of motivating the head actuation agents to explore new actions and states. It was found that the expected transition function T' does not have to approximate $E\{T\}$ with arbitrary accuracy. However, due to the way in which the novelty was calculated with a norm on $\mathbb{R}^m$, it was impossible to predict the range of values which the novelty would take or where the Wundt function should be most sensitive. An upper bound for the novelty can be estimated to be $m\tau^2$, based on the Equation 1 and the restriction of the range of the state vectors to $[0, \tau]$. However, T' is not bounded by these limits and this value says little about the useful range of novelty once the function approximator has been trained. For example, this experiment had a state space dimension of $m = 10$ and a function approximator range $\tau = 10$ giving a theoretical maximum novelty of 1000. However, the most useful set of Wundt parameters only distinguished between novelties in the range of 50 to 300 (Fig. 1). These Wundt parameters were obtained by, on the first instance of training an agent on a given state space, re-adjusting the parameters between training sets. Training sets involved between 20 and 100 actions with full training involving about 500 actions over about an hour. This training was not automated; specifically the fixed position training required manual repositioning the robot and ball after each training set. After one agent was trained, and the Wundt function parameters found, other agents could be trained without adjustment of the parameters provided the state space and function approximator parameters did not change.

6   Conclusion

A motivated reinforcement learning framework was successfully developed and implemented for the optimisation of head actuation policies. Self-localisation was demonstrated as improving by over 11% relative to the null case of uniformly distributed actuation policies, for agents trained online for no more than one hour. Within 4 actions, the reinforcement learning agents were able to localise accurately for 13 out of 15 cases. In contrast, the uniform agent localised accurately in 10 from 15 cases within 4 actions. It was observed that, by exhibiting some level of artificial 'boredom' and 'curiosity', the motivated reinforcement learning agent is able to modestly improve its self-localisation by observing its environment in a more intelligent manner; a viable method for improving localisation performance without the need for improved object recognition algorithms or the tuning of probabilistic filters.

Acknowledgements. This project was supported by the Australian Mathematical Sciences Institute (AMSI) summer research scholarship program. Thanks to the University of Newcastle RoboCup team, the NUbots, for providing hardware resources and support.

References

1. Wong, A.S.W., Chalup, S.K., Bhatia, S., Jalalian, A., Kulk, J., Nicklin, S., Ostwald, M.J.: Visual gaze analysis of robotic pedestrians moving in urban space. Architectural Science Review 55(3), 213–223 (2012)
2. Merrick, E.K., Maher, M.L.: Motivated Reinforcement Learning: Curious Characters for Multiuser Games. Springer, Dordrecht (2009)
3. Kitano, H., Asada, M., Kuniyoshi, Y., Noda, I., Osawa, E., Matsubara, H.: Robocup: A challenge problem for ai. AI Magazine 18(1) (1991)
4. Budden, D., Fenn, S., Walker, J., Mendes, A.: A novel approach to ball detection for humanoid robot soccer. In: Thielscher, M., Zhang, D. (eds.) AI 2012. LNCS, vol. 7691, pp. 827–838. Springer, Heidelberg (2012)
5. Wan, E., van der Merwe, R.: The unscented kalman filter for nonlinear estimation. In: Adaptive Systems for Signal Processing, Communications, and Control Symposium, AS-SPCC 2000, pp. 153–158. The IEEE (2000)
6. Watkins, C.: Learning from Delayed Rewards. PhD thesis, Cambridge University (1989)
7. Sutton, R.S., Barto, A.G.: Reinforcement Learning: An Introduction. MIT Press, Cambridge (1998)
8. Wundt, W.: Principles of Physiology and Psychology. Macmillan, New York (1910)
9. Saunders, R., Gero, J.S.: Designing for interest and novelty - motivating design agents. In: de Vries, B., van Leeuwen, J., Achten, H. (eds.) Proceedings of the Ninth International Conference on Computer Aided Architectural Design Futures, pp. 725–738. Kluwer Academic Publishers (2001)
10. Merrick, K.E., Isaacs, A., Barlow, M., Gu, N.: A shape grammar approach to computational creativity and procedural content generation in massively multiplayer online role playing games. Entertainment Computing 4(2), 115–130 (2013)
11. Merrick, K.: Intrinsic motivation and introspection in reinforcement learning. IEEE Transactions on Autonomous Mental Development 4(4), 315–329 (2012)
12. Konidaris, G., Osentoski, S., Thomas, P.S.: Value function approximation in reinforcement learning using the Fourier basis. In: Burgard, W., Roth, D. (eds.) Proceedings of the Twenty-Fifth AAAI Conference on Artificial Intelligence, AAAI 2011, pp. 380–385. AAAI Press, San Francisco (2011)
13. The RoboCup Institution: RoboCup Soccer Humanoid League Rules and Setup for the 2013 Competition in Eindhoven, DRAFT (2012), http://www.tzi.de/humanoid/bin/view/Website/Downloads
14. Majdik, A., Popa, M., Tamas, L., Szoke, I., Lazea, G.: New approach in solving the kidnapped robot problem. In: Robotics (ISR), 2010 41st International Symposium on and 2010 6th German Conference on Robotics (ROBOTIK), pp. 1–6 (2010)

Efficient Distributed Communications
for Multi-robot Systems

João C.G. Reis[1], Pedro U. Lima[1], and João Garcia[2]

[1] Instituto de Sistemas e Robótica, Instituto Superior Técnico, Portugal
{jreis,pal}@isr.ist.utl.pt
[2] INESC-ID Lisboa, Instituto Superior Técnico, Portugal
jog@gsd.inesc-id.pt

Abstract. Wireless communications are one of the technical problems that must be addressed by cooperative robot teams. The wireless medium often becomes heavily loaded and the robots may take too long to successfully transmit information, resulting in outdated shared data or failures in cooperative behaviors that require synchronization among teammates. This paper introduces a novel solution to enable the immediate transmission of synchronization data in a way designed to reduce and better tolerate packet loss. It does so by categorizing the communications in multi-robot systems in two classes, robot state diffusion and synchronization messages. For the former, an existing adaptive transmission method (RA-TDMA) is used, and for the latter a novel solution was developed. Experiments show an important delay reduction when sending synchronization messages over a loaded network.

1 Introduction

One of the challenges in developing multi-robot systems lies with the communications among them. Wireless networks used without special structure or restrictions provide adequate communication facilities. However, in the face of restrictions or high load, it is necessary to use the network carefully to better exploit its capabilities without overloading it.

An example scenario of multi-robot communication is a robot soccer game, where this work was applied. The *Robot World Cup (RoboCup)*[5] was proposed in 1997 as an attempt to foster AI and intelligent robotics research. The *RoboCup Middle Size League (MSL)* is a senior competition where two teams of five robots play a soccer game. The rules used are a subset of the official *FIFA* Laws with added constraints on the robots and environment[8]. During games, robots communicate using only *IEEE 802.11a* or *IEEE 802.11b* network modes. Unicast and multicast communication modes are allowed and broadcast is forbidden. The maximum transmission bit rate allowed per team is 20% of *IEEE 802.11b* (2.2 Mbps). However, the rules are mostly not enforced in practice and several problems occur frequently, degrading communications quality.

Efficient communications are a key factor for the success of teams during competitions but most teams simply schedule their robots to transmit information periodically, without any synchronization among robots. Thus, in the worst

S. Behnke et al. (Eds.): RoboCup 2013, LNAI 8371, pp. 280–291, 2014.
© Springer-Verlag Berlin Heidelberg 2014

case situation, all robots might try to transmit at the same time. Under such a situation, some packets would be transmitted before others, causing communication delays. Furthermore, collisions are very likely to occur, further delaying communication.

Some works show how identifying communication patterns is an important step in developing reusable software[12]. It is possible to distinguish two categories regarding communications in cooperative multi-robot systems. On the one hand *Robot State Diffusion* has to do with the robot's perception of the surrounding world using its own sensors (own position, the ball position the positions of obstacles). The robots continuously exchange this data to improve each robot's knowledge about the world. On the other hand, *Synchronization Messages* are related to the robots need to communicate to agree and keep relational behaviors[3,6] synchronized.

In this paper, a solution for multi-robot wireless communication is presented. This solution takes advantage of the communication patterns observed in cooperative multi-robot systems to better exploit the medium capabilities, thus enhancing a previous solution for periodic data transmission[10] with a mechanism to enable agents to quickly and reliably communicate with each other or in groups. This work is currently integrated into the *Robot Operating System (ROS)* middleware[9] and published as open source software[1].

Some good solutions for communications in multi-robot systems can be found in the literature. However, little attention is given to the types of communication required, as described above. The *Cooperative Autonomous Mobile Robots with Advanced Distributed Architecture (CAMBADA)* team[2] of the University of Aveiro, Portugal, uses a middleware infrastructure specifically developed for their team. One of the components implements a *Reconfigurable and Adaptive Time Division Multiple Access (RA-TDMA)* communication protocol[11] with self-configuration capabilities to dynamically adapt to the number of active team members[10]. It works in a fully distributed way and has proved to be effective in game situations. There is an implementation available as free software[2]. This protocol, in which the solution presented in this paper is based, is adequate for robot state diffusion, but lacks explicit support for synchronization messages, for which our work provides a novel solution.

A communication system centered around a *message dispatcher*[13] is used by the RoboCup MSL *1. RFC Stuttgart* team[4] from the University of Stuttgart, Germany. In this solution, all agents establish a TCP connection to a central message dispatcher, a special entity in the system that filters messages according to time and other defined constraints. Messages with high priority are always sent; other messages are deleted if they become older than a specified threshold. In this solution, robot state is transmitted together with synchronization messages and the medium access can be affected by transmissions of outdated information. The priority mechanism transmits synchronization messages first, but messages must be transmitted to the dispatcher before this mechanism is

---

[1] SocRob Multicast, `http://www.ros.org/wiki/socrob_multicast`
[2] rtdb - Real-Time DataBase Middleware, `http://code.google.com/p/rtdb/`

used, adding an extra level of indirection. The message dispatcher also transmits all messages directly to all destinations, instead of using the multicast channel. This consumes bandwidth that could be used to transmit robot state information more frequently, thus keeping it more updated.

ROS[9] is a robotic middleware solution that has become widely used because of its quality of design and implementation. A *ROS* system is organized as a collection of nodes that communicate using messages. Nodes transmit and receive messages through two procedures: a publish-subscribe mechanism, named *topics*, where multiple nodes can publish messages and multiple nodes can subscribe to receive messages, and a direct node to node communication mechanism named *services*. However, there is no organized multicast solution that takes advantage of the characteristics of multi-robot systems to enhance communication. Furthermore, bandwidth limitation is not possible.

Transmission of real-time traffic in wireless networks is an interesting problem for many domains, e.g., factory automation. In [7], the authors propose a solution that can be implemented in *IEEE 802.11a/b/g* networks. However, it requires that some parameters of the network card are adjusted to give real-time stations priority over other stations. *E-MAC*[1] is a similar solution that avoids starvation of non real-time traffic. These solutions are not applicable to *RoboCup MSL* because of the need for modifying network card parameters, which although it is not explicitly forbidden in the rules, it would give an unfair advantage against teams not using it. Additionally if two teams would use this kind of solutions, conflicts would occur, unless the two teams cooperate somehow to arbitrate medium access. This is currently not allowed by the rules.

The rest of this document is organized as follows. Section 2 presents a detailed description of the *RA-TDMA* protocol. Section 3 describes in detail the proposed solution. Section 4 shows how the solution is integrated in the system. Section 5 presents the evaluation conducted on the solution. The conclusions are presented in section 6.

2   Reconfigurable and Adaptive TDMA

The RA-TDMA protocol[11] tries to disperse transmissions of all team members in time to avoid collisions within the team as much as possible, since the remaining network load cannot be controlled. Time is divided in slots of duration T_{tup} (team update period) in which all team members transmit once. T_{tup} is a configuration parameter set prior to execution and determines the global responsiveness of the system. Each of these slots is equally subdivided in slots for each active team member of duration T_{xwin}. Agents transmit at the beginning of their respective slots, thus spacing the transmissions as much as possible. Each agent uses only a fraction of its slot, the remaining time is used to accommodate delays in transmission and give the other team a chance to transmit.

This protocol does not need clock synchronization. After its own transmission slot, each agent keeps registering the exact time of arrival of its teammates' packets. The reception delay (δ) is calculated with respect to the expected time

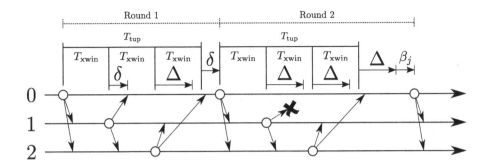

Fig. 1. Time diagram of two example rounds by a team of three agents

of arrival, which corresponds to the beginning of the transmitting agent own slot. The current round period is enlarged by the greatest of these delays. Only delays up to Δ are considered, with Δ being a global configuration parameter. Therefore, the effective round period will vary in the interval $[T_{tup}, T_{tup} + \Delta]$. An example situation is depicted in the first round of Fig. 1. At agent 0, the packet from agent 1 was expected after T_{xwin} but is received with a slight delay of δ. However, the packet from agent 2 is received with a delay greater than Δ, thus is ignored. The next packet transmitted by agent 0 is delayed by δ, because it was the only delay shorter than Δ.

When a robot does not receive any packet with a delay below Δ in a round, the next transmission will be delayed by a further β_j, different for every robot j. In this situation, the effective round period will be $T_{tup} + \Delta + \beta_j$. This is used to prevent situations in which the robots all keep transmitting but are unsynchronized. Having different round times in this situation will force the robots to resynchronize, since after a few rounds the robots will again receive a packet with a delay below Δ. An example is presented in the second round of Fig. 1, where the packet from agent 1 is lost and the packet from agent 2 is received with a delay greater than Δ.

Dynamic Reconfiguration. While in operation, the robots divide the *TDMA* round period by the number of active robots[10]. Since the robots can come in and out of play and malfunction during games, the number of active robots must be determined dynamically. Consequently, the transmissions are always separated as much as possible leaving no unused slots. Agents have two identifications: static *IDs* (SIDs) are given to each agent prior to execution within a pre-defined and known interval, and are used for agent identification. Dynamic *IDs* (DIDs) are used to identify only active teammates locally at each agent and are never transmitted. DIDs are assigned to each active agent, sorted by their SIDs. Each agent maintains a membership vector for all possible agents, indexed by their SIDs. Each agent may be in one of four states: *not running*, *insert*, *running* and *remove*. This vector is shared with the team by adding it to every transmission.

When an agent starts communicating, it sets its own state to *insert*. Agents that receive these initial packets set their state for the new agent as *insert*. When an agent X that has agent Y in *insert* state and detects that all agents in state *running* have agent Y in *insert* or *running* states, it updates the state of agent Y to *running*. The *DID* of agent Y is calculated and it is considered part of the *TDMA* round. Removing an agents follows a dual process. If nothing was received from that agent in the last rounds (the number of rounds is a configuration parameter), its state is changed to *remove*. When all agents have a given agent in *remove* or *not running* states, its state is changed to *not running*. The *DIDs* are reassigned and the round slots adjusted accordingly.

3   Solution Architecture

The *RA-TDMA* solution provides a suitable approach to dispersing robot state. However, it does not provide any special mechanism for transmission of synchronization messages. This is a problem for two reasons:

- Synchronization messages have to wait to be transmitted. This is a minimal delay for one single message but will accumulate in situations of especially bad network conditions where many packets fail to arrive correctly. This can be even worse if several rounds of communication are needed to synchronize the robots.
- There is no reception feedback. If the client software needs to be sure of correct reception, some mechanism must be implemented on top of the communication protocol. Again, this may take almost a full round time if no packets are lost, or much more with bad network conditions.

The solution proposed here can be seen as a protocol with two different modes of operation, each concerning one of the two robot communication patterns.

3.1   Robot State Diffusion

Robot state is transmitted using rounds of *Adaptive-TDMA* with a long period, divided by the number of active agents. To differentiate it from the transmission of synchronization messages, this mechanism is called *long rounds*. The long rounds closely are implemented using RA-TDMA, as described in section 2.

3.2   Synchronization Messages

When an agent needs to transmit synchronization information that is urgent or requires an answer as soon as possible, it initiates a new question round. Each question round is composed by the *question*, which is the initial data transmitted by the starting agent, and *answers*, that can be complex information or a simple acknowledgment. The mechanism to transmit synchronization messages is called *short rounds*, and might handle several question rounds simultaneously. In order to easily distinguish short and long round packets, a different multicast socket is used. This socket uses the same multicast *IP* address that is used in the long rounds, but with a different port. This way, long and short round packets can be distinguished without increasing the packet size.

Basic Scenario. Each question round is identified by the SID of the agent that started it along with a *question identifier*, a number that uniquely identifies each question from a given agent. Every packet contains five elements:

- The SID of the starting agent (SSID);
- The *question identifier* (QID);
- The *question* itself;
- A list of SIDs of agents that are *required* to answer;
- A list of *answers* along with the SIDs of the agents that produced each answer.

The initial packet transmitted by the starting agent contains the *question* and the list of SIDs of agents that are *required* to answer it. When this packet is received by another agent, it first verifies if its SID is in the *required* list. If this is the case, the agent will transmit a packet with its answer and its SID removed from the *required* list. Therefore, in a given packet regarding a specific question, each agent might be in one of three states:

- *Required*: The agent is required to answer this question and its SID is in the *required* list.
- *Answer*: An answer is present in the *answers* list of this packet.
- *Complete*: The answer was correctly received by the stating agent or no answer is required.

The question is only transmitted on packets that have any agent in the *required* state. Agents in the *answer* or *complete* states do not need the question, so if all agents are in either of these states the question field is left blank. When the starting agent has all the answers it requires, it will transmit a terminator packet with all agents in the *complete* state. This packet contains the two identifiers, but all other fields are left blank. With this last packet the question round is successfully completed. A simple example can be found in the first round of Fig. 2, where all packets are successfully received. The table shows the contents of the packets.

The proposed solution is flexible enough to accommodate various possibilities, an agent might want to transmit something to only one or to multiple agents. In any case, it is only necessary to add the proper SIDs to the *required* list. The proposed solution makes sure that all those agents receive the *question* and all their *answers* are returned to the starting agent, unless they are not reachable. Furthermore, what is important in a question round might be the initial data transmitted (*question*), the data that is transmitted by the agents in the *required* list, or both.

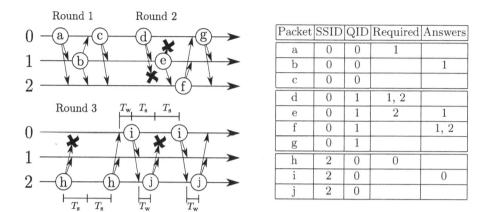

Fig. 2. Time diagram of three example short rounds by a team of three agents

All agents begin inactive. They become active as soon as a question round is started locally or a packet with a question round arrives. However, an agent will not become active if its SID is not in the *required* list of the received question. Note that several question rounds might be active at once. An agent might have started some, be required to answer some, and still there might be some others with which the agent is completely uninvolved. While an agent is active, it will transmit all it knows about all question rounds, even if they do not involve it. Only question rounds that involve it will keep the agent active.

Tolerance to Packet Loss. The protocol was designed to reduce and tolerate packet loss using as much redundancy as possible. When a packet is received with a question round started by some other agent, the agent will react to it if its SID is in the *required* list. It does not matter if the packet is the initial packet transmitted by the starting agent, or a packet already containing answers transmitted by some other agent. This way, if some agent does not receive the initial packet, it will likely get all the needed information from the next packet regarding that question round. Each agent also keeps saving all the answers from other agents. These answers will all be transmitted along with its own answer. This way, if some packet with an answer fails to reach the starting agent, the next packet will likely contain a copy of that answer. If an agent does not receive the final packet signaling completion, it will transmit its answer again. When the starting agent receives this, it will resend the terminator packet. The second round of Fig. 2 shows a situation where two packets are lost yet the round completes without any added delay.

Transmission Timing. Transmission of the initial packet determines the start of a question round. Agents transmit at the beginning of their own slots, which have a fixed duration of T_s. When an agent successfully receives a packet from some other agent, it recalculates its next transmission time based on how many agents are active in the short rounds. Packet reception can only cause this time to be anticipated, to avoid situations where the transmission time is over delayed because of the reception of delayed packets.

When the received packet belongs to the agent that should have transmitted right before, the agent will ignore the slot duration and transmit much sooner. Transmission could happen immediately, but this would put an undesirable load on the medium since all agents might transmit in sequence without any interval. To avoid this, a short waiting time T_w is used between transmissions of successive agents, to create a window in which the medium is available to the other team. The slot duration and the short waiting time are configuration parameters. In the last round of Fig. 2 two packets were lost. Since only two agents are active, agents schedule retransmission with a period of $2 \times T_s$. When a packet is received from the other agent, the next packet is transmitted after T_w.

Active Agents Estimation. For one agent to know if it is its turn to transmit, it must keep a record of which agents are active in the short rounds. It is not possible to know this for sure since packets might be lost or delayed, but it is possible to have an estimate that will be accurate in situations where all packets are promptly delivered. When this estimate fails, two or more agents might try to transmit simultaneously. This will increase the probability of packet loss. However, the waiting time between packets gives a better chance for these transmissions to succeed. The estimation is made based on the information that is kept about all active question rounds. Starter agents are always considered active, since the question round will only end after the terminator packet. All agents whose SID is in the *required* list of some active question round are also considered active.

3.3 Shared Concerns

Long and short rounds operate almost independently. Still, there is some information that must be shared between the two. Long rounds keep track of which agents are active at a given moment. This information is used by the short rounds to avoid waiting for transmissions from an agent that is not active. During operation, short rounds keep verifying if agents in the *required* list are in any state different from *not running*. This is done by the starting agent to ensure that the answers are delivered as soon as all the agents in the *required* list have answered or changed to the state *not running*. It is also done by all agents to keep the estimation of active agents as accurate as possible.

Short rounds are expected to be occasional and resolved quickly. Therefore, when the time comes to transmit in a long round, the agent will first check if there is a short round active. In this case, it will simply not transmit. Because short rounds may extend in time, this is only done once. After that, the agent will transmit anyway. This is necessary to resolve the case where an agent becomes inactive during a short round, and to guarantee that robot state keeps being diffused even in the presence of a heavy load of short rounds.

Bandwidth Management. It is easy to estimate how much bandwidth is needed by the long rounds based on the team update period T_{tup} and the size of the information that is transmitted. However, for the short rounds bandwidth usage is much harder to estimate because it depends on events in the game that will trigger question rounds and on packet loss because it uses retransmission. Packet sizes and the T_{tup} must be kept reasonable. Some attention must also be spent on ensuring that there are no situations that trigger an excess of question rounds. Still, a mechanism to ensure that the allowed bandwidth is not exceeded is necessary.

The bandwidth manager keeps track of a moment in time that represents the end of the last authorized transmission at the maximum allowed bandwidth. When an authorization request is received, if this moment is in the future, the request is immediately denied. If this moment is in the past, the request is accepted and the moment is then updated with the greatest of two moments: the current time or the sum of the previous moment and the duration of the requested transmission at maximum allowed bandwidth.

4   Integration with ROS

This solution is implemented as a library called *SocRob Multicast*. This library depends only on the *Boost C++* Libraries[3] for socket programming, threads and time control and on ROS[9] for logging and serialization of messages. To make the bridge between this library and the rest of the system, a ROS node must be created with a structure loosely following the one represented in Fig. 3. This node is responsible to create and process long round messages and knows what information must be shared, and also converts *ROS* services into short rounds. The correspondence between robots and SIDs is done by this node and can not be automated since it completely depends on the domain. In *RoboCup MSL*, the SIDs are the robot number minus one, because robots are numbered starting in 1 and SIDs start in 0. Although it is designed to comply with *RoboCup MSL* rules, this library is domain independent, enabling a *ROS* system to use one master per agent. If the connection between agents is severed, the agents will remain functional.

5   Evaluation

The evaluation of the proposed solution is centered on how fast information can travel from one robot to another, under different network conditions. Since the protocol presented is divided in two modes, it makes sense to evaluate these modes separately. The tests were conducted using one laptop per agent, with a team of five agents, in order to simulate real game conditions. The T_{tup} used was 100 milliseconds, T_{s} was 5 milliseconds and T_{w} was 1 millisecond. The laptops were connected to a *IEEE 802.11a* network. Each test was run for about

---

[3] Boost C++ Libraries, http://www.boost.org/

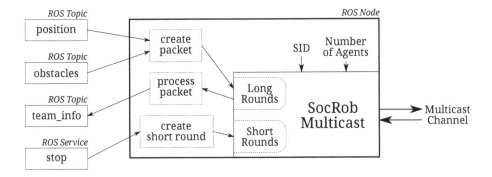

Fig. 3. Integration with ROS

one minute. The agent with SID 0 played the special role of initiating all question rounds and saving test results. The laptop clocks were synchronized using *chrony*[4]. During the tests, the greatest time difference reported by *chrony* was almost 2 milliseconds. Therefore, all results that depend on clock synchronization can be affected by this clock skew. All tests were run with and without load on the network. It is not possible to guarantee that the wireless medium is completely clear, thus when analyzing results with no network load it is necessary to consider that some interference is still possible.

5.1   Evaluation of Robot State Diffusion

CAMBADA team already proved that using *Adaptive-TDMA* is a better solution than a simple periodic transmission[11]. To prove that the same happens with our implementation of the RA-TDMA protocol, we created a periodic transmission solution that tries to transmit 10 times per second at a well defined moment. Thus, depending on the accuracy of time synchronization between computers, the agents will all try to transmit very close in time, simulating the worst case for periodic transmissions. Without network load, this degraded solution loses 5.629% of the packets, while our implementation of RA-TDMA loses almost no packet: only 0.038% of the total. In a loaded network, the degraded solution loses 7.981% of the packets, while RA-TDMA looses only 5.465%.

5.2   Evaluation of Synchronization Messages

Synchronization messages should be evaluated by the time it takes for all the receivers to get the question (question delay), and how long it takes for the initial agent to receive all the answers (answer delay). The accuracy of the question delay values depends on clock synchronization, since timestamps are taken in different computers.

---

[4] Chrony Home, http://chrony.tuxfamily.org/

Table 1. Synchronization messages delay. Time values are in milliseconds.

Network load:		Without		With	
Transmit using:		Long Rounds	Short Rounds	Long Rounds	Short Rounds
Question Delay	Minimum	100.487	1.098	101.934	1.374
	Average	116.488	2.407	136.935	22.227
	Maximum	431.995	51.470	948.631	358.010
Answer Delay	Minimum	145.564	6.976	177.353	13.368
	Average	214.030	12.044	247.494	69.924
	Maximum	509.755	102.012	1256.729	396.034

Four scenarios were considered: sending synchronization messages together with robot state using long rounds and using short rounds, both with and without network load. For each scenario, two experiments were made. In one experiment, agent with SID 0 sends a message to all other agents simultaneously. The results for this experiment can be found in Table 1. In another experiment, messages are sent to a single agent in turn. These results are not shown as they show similar, yet slightly smaller values. To simulate the worst case scenario when transmitting in the long rounds, the synchronization messages were started only in the moment after a robot state message was transmitted. Therefore, they had to wait a full round to be transmitted, hence the difference to the best case scenario is about 100 milliseconds in every value. The times achieved using short rounds are much smaller both with and without network load. The short rounds are particularly effective in avoiding extreme values, as the maximum value observed during the experiments was much smaller.

The experimental results are as expected, the short rounds bring a great advantage. The average delays are greatly reduced in all scenarios. Furthermore, the short rounds are effective in keeping the maximum values much lower.

6   Conclusions

The solution presented in this paper greatly improves the transmission of synchronization messages in the described scenarios. Transmission of robot state can be efficiently done using *Adaptive-TDMA*. However, waiting for the *Adaptive-TDMA* transmission slot can cause a great delay. This delay can have a great impact in robot performance, especially in situations where many packets are lost or more that one round of communication is needed to reach a decision. Middleware for robotics has seen a great evolution in recent years, with stabilization of good solutions like *ROS* that are now used in many robotic applications with evident advantages. However, the better known solutions lack support of advanced communication protocols like the one presented here. The proposed solution is a step further in the communication capabilities of multi-robot systems. It can even be used in many scenarios, including the *RoboCup MSL*, or if a team of robots needs to be deployed in a situation where a public wireless network must be used.

Acknowledgments. The authors would like to thank José Carlos Castillo Montoya for his comments and careful review of this paper.

References

1. Aad, I., Hofmann, P., Loyola, L., Riaz, F., Widmer, J.: E-MAC: Self-Organizing 802.11-Compatible MAC with Elastic Real-time Scheduling. In: IEEE Internatonal Conference on Mobile Adhoc and Sensor Systems 2007, pp. 1–10. IEEE (2007)
2. Dias, R., Neves, A.J.R., Azevedo, J.L., Cunha, B., Cunha, J., Dias, P., Domingos, A., Ferreira, L., Fonseca, P., Lau, N., Pedrosa, E., Pereira, A., Serra, R., Silva, J., Soares, P., Trifan, A.: CAMBADA 2013: Team Description Paper (2013)
3. Drogoul, A., Collinot, A.: Applying an Agent-Oriented Methodology to the Design of Artificial Organizations: A Case Study in Robotic Soccer. Autonomous Agents and Multi-Agent Systems Journal 1(1), 113–129 (1998)
4. Käppeler, U.P., Zweigle, O., Rajaie, H., Häussermann, K., Tamke, A., Koch, A., Eckstein, B., Aichele, F., DiMarco, D., Berthelot, A., Walter, T., Levi, P.: 1. RFC Stuttgart Team Description 2010 (2010)
5. Kitano, H., Asada, M., Kuniyoshi, Y., Noda, I., Osawa, E.: Robocup: The robot world cup initiative. In: Proceedings of the First International Conference on Autonomous Agents, pp. 340–347. ACM (1997)
6. Lima, P., Custódio, L.: Artificial Intelligence and Systems Theory Applied to Cooperative Robots. International Journal of Advanced Robotic Systems (3) (September 2004)
7. Moraes, R., Vasques, F., Portugal, P.: A TDMA-based mechanism to enforce real-time behavior in WiFi networks. In: IEEE International Workshop on Factory Communication Systems, WFCS 2008, pp. 109–112. IEEE (2008)
8. MSL Technical Committee: Middle Size Robot League Rules and Regulations for 2013, version - 16.1 20121208 (January 2013)
9. Quigley, M., Gerkey, B., Conley, K., Faust, J., Foote, T., Leibs, J., Berger, E., Wheeler, R., Ng, A.: ROS: an open-source Robot Operating System. In: ICRA Workshop on Open Source Software, vol. 3 (2009)
10. Santos, F., Almeida, L., Lopes, L.: Self-configuration of an Adaptive TDMA wireless communication protocol for teams of mobile robots. In: IEEE International Conference on Emerging Technologies and Factory Automation, ETFA 2008, pp. 1197–1204. IEEE (2008)
11. Santos, F., Almeida, L., Pedreiras, P., Lopes, L., Facchinetti, T.: An Adaptive TDMA Protocol for Soft Real-Time Wireless Communication among Mobile Autonomous Agents. In: Proc. of the Int. Workshop on Architecture for Cooperative Embedded Real-Time Systems, WACERTS, vol. 2004, pp. 657–665 (2004)
12. Schlegel, C.: Communication patterns as key towards component-based robotics. International Journal of Advanced Robotic Systems 3(1), 49–54 (2006)
13. Zweigle, O., Kappeler, U., Haussermann, K., Levi, P.: Event based distributed real-time communication architecture for multi-agent systems. In: 2010 5th International Conference on Computer Sciences and Convergence Information Technology (ICCIT), pp. 503–510. IEEE (2010)

Distributed Formation Control of Heterogeneous Robots with Limited Information

Michael de Denus, John Anderson, and Jacky Baltes

Autonomous Agents Laboratory,
Dept. of Computer Science, University of Manitoba,
Winnipeg, Manitoba, Canada R3T2N2
{mdedenus,andersj,jacky}@cs.umanitoba.ca
http://aalab.cs.umanitoba.ca

Abstract. In many multi-robot tasks, it is advantageous for robots to assemble into formations. In many of these applications, it is useful for the robots to have differing capabilities (i.e., be heterogeneous) in terms of perception and locomotion abilities. In real world settings, groups of robots may also have only imperfect or partially-known information about one another as well. Together, heterogeneity and imperfect knowledge provide significant challenges to creating and maintaining formations. This paper describes a method for formation control that allows heterogeneous robots with limited information (no known population size, shared coordinates, or predefined relationships) to dynamically assemble into formation, merge smaller formations together, and correct errors that may arise in the formation. Using a simulation, we have shown our approach to be scalable and robust against robot failure.

1 Introduction

Formations are desirable in collections of robots for many reasons, including maximizing area coverage, sensor coverage, and minimizing contact while moving quickly as a group. Heterogeneity in robots is similarly advantageous in many settings: parsimony in control, budgetary considerations, and specialization of tasks all make simpler robots with divergent capabilities desirable. Knowledge may be inconsistent between robots, making for more diversity, and commonly-assumed global knowledge may not be available. For example, in a rescue scenario, robots may be lost due to failure, and agencies may arrive with new equipment over time, preventing total population from being known with certainty. Communication may be noisy or temporarily disrupted, leading to inconsistency in information across members. All of these factors conspire to present a difficult challenge for the creation of formations and their maintenance over time.

In this paper, we present a distributed approach to formation control in multi-robot teams that deals with such challenging scenarios: robots may be heterogeneous in terms of movement abilities, sensing, and other equipment; knowledge of others, including identity and the overall size of the team, is assumed to be

S. Behnke et al. (Eds.): RoboCup 2013, LNAI 8371, pp. 292–303, 2014.
© Springer-Verlag Berlin Heidelberg 2014

imperfect; and communication is assumed to be unreliable. Our approach is also intended to assume that failure of robot hardware is a strong likelihood over time, and that other robots may be available to replace them by physically sensing or encountering the ongoing formation in the environment.

Our approach deals with these challenges by limiting the number of robots that each formation member interacts with directly, while not putting any restriction on the size of the team itself. This theme has been successful in the past [1,2]. Existing methods, however, do not address addition and removal of robots from the formation, and are not particularly tolerant of robot failure. We have created a mechanism for new robots to gain entry to the formation, and for existing robots to adapt given the failure of others. The combination of these two properties addresses the fact that we do not know how many other robots are in the environment, and that the robot population can change at any time. This builds on our previous work [3], which assumed only that communication had to be local (no broadcast) and where heterogeneity in sensing was limited to the inability to identify other robots.

2   Related Work

There are many existing approaches to formation control. One common theme is to employ elements of central processing. Some approaches restrict processing to a single robot [4], while others pre-compute portions of the paths centrally [5]. Any central processing introduces a single point of failure, and should be avoided in cases when possible. Under our approach, the responsibility for formation control is distributed, and no single robot is essential.

Another common theme is to use leader and follower robots. One way to accomplish this is through hierarchical leadership [2,6]. These methods define multiple levels of leadership, but still suffer from the fact that the failure of any of the leaders would disable some portion of the formation. Our approach would be affected by the loss of a robot, but has mechanisms to recover from such a loss. Techniques involving chains of robots can also be considered to fall under this category. Such techniques can either have a predefined ordering of robots [1] or can determine this ordering at runtime [3]. Our technique employs the concept of following a target, but does not have any predefined relationships.

Another large class of formation control techniques that use local rules to maintain formations in a somewhat decentralized way. For example, Hadaegh et al's approach [7] allows for the dynamic addition and removal of formation members, but does not scale well, as all members need to communicate with one another. Leonard and Fiorelli [8] demonstrate a method that employs virtual robots, whose purpose is to direct and organize the formation. This approach is robust against robot failure, but requires robots to have a shared frame of reference. Other approaches of this type [9,10] are able to maintain formations, but are very domain specific and do not allow large numbers of potential formations. Our approach allows for a large number of potential formations, and requires communication only with a robot's immediate neighbors.

3   Approach

Our approach to formation control employs several strategies to deal with heterogeneity and limited information. To create a formation in a distributed manner,
formations must be built up as robots encounter one another in the environment. Procedures must be in place for smaller sub-formations to merge, and for
the formation to continue to emerge even as robots fail. As groups of robots
join into larger formations, the formation itself must be continually balanced, so
that a globally coherent shape emerges (e.g. if forming a V, smaller formations
either cannot all join one side, or must be repositioned if they do). Each of these
operations is detailed in the following subsections, following some preliminary
discussion to explain common elements of our approach. Because the approach
is distributed, it must also be possible for *any* robot in the formation to perform any required operation as needed. Under our approach, robots can initiate
direct communication with other robots that have been perceived or have been
previously identified, but can not perform broadcast communication.

3.1   Preliminaries

Our approach falls into the *neighbor referenced* category of Balch and Arkin's
categorization [11]. Each robot in the formation has one associated target robot
that it maintains a position with respect to. This target robot is determined
dynamically, and is used to position the robot within the formation. It is also
possible for a robot in a formation to follow no one in some specifications (e.g.
the point of a V).

A formation in our approach is specified as a collection of interconnected
segments, each consisting of chains of robots formed over time. Each segment
has an associated angle and distance, specifying the desired angle and distance at
which each robot in the segment should keep its target. These may be specified
as functions, allowing segments to be lines or curves. Each segment also has
a relative length, which defines proportion in the formation by specifying the
desired length of the segment relative to the length of the shortest segment in the
formation's definition. Overall, this allows a broad range of specified formations,
including open shapes (lines, Vs) and closed shapes (squares, rectangles) as well
as formations with curved segments. The only current restriction on shape in
our implementation is that an agent can have only one target to follow.

Any number of segments may connect together, and these connection points
are termed *entry points*. The robots occupying these positions in the formation
are *entry point robots*. These robots are responsible for handling joining and
merging operations, as well as estimating population size. These entry point
robots are not pre-determined: any robot becomes one if it happens to be located
at the point where a segment ends. A robot knows this by examining the segment
membership of its neighbours, and the number of neighbouring robots.

Each possible formation is common knowledge among all participating robots,
as is the desired current formation out of the many the robots may possibly know,
and a goal direction in which to move.

To deal with an unknown population size, each entry point robot maintains an estimated count of each segment it belongs to. These in turn can be used to estimate the size of an entire formation. To maintain a segment population count, entry point robots send out *counting* messages to each of their neighbours. Upon receiving a counting message, if the receiver occupies an entry point, it increments the count in the message by 1, and sends it back to the sender. A non-entry point robot will have at most two neighbours, and so a non-entry point-robot receiving a counting message increments the count by one forwards it on to the neighbor who did not send it. If messages are lost, or the count becomes stale, it will be corrected by a future message.

To obtain an estimated count of the entire formation, a robot examines all segments to which it belongs. It sums up the relative lengths of these segments, and calculates the percentage of the total formation in these segments. It then uses this information to extrapolate the number of robots in the complete formation. This must be considered a rough estimate, since it is making the assumption that a robot's current neighborhood is representative of the formation as a whole. However, it strikes a balance between accuracy and the extensive communication beyond immediate neighbours that forcing entry point robots to exchange population counts would entail.

3.2   Joining a Formation

There are two ways in which a robot can join a formation: creating a new formation with another robot, or joining an existing formation. In our approach, both of these options are initiated in the same way. When not participating in a formation, robots periodically send out *join* requests to those they encounter. A robot receiving such a request (the *responding* robot) may be in three different situations. First, the responding robot may not yet be in a formation, in which case it would reply with a *success* message, along with instructions to the sender for joining the new formation. Second, the responding robot may be in a formation but not at an entry point, in which case the responding robot would reply with a *rejection* message. If the responding robot has a target, it will identify that target to the requesting robot as an alternative. Finally, in the case where this is a valid entry point, the responding robot can reply with a success message and integrate the joining robot. To perform this integration, the entry point robot will choose the destination segment that is most out of balance and adopt the requestor as a neighbor there (this is a common element to balancing an existing formation, and is explained in Sec. 3.4). An example of successfully joining a formation is illustrated in Fig. 1. Loss of communication during joining will cause the joining robot to maintain its heading and speed while trying to locate another place to join.

3.3   Merging Formations

In our approach, merging two distinct formations forces robots in one formation to sequentially join the other, and thus makes use of the techniques described in

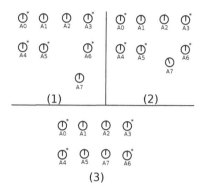

Fig. 1. The joining process. The formation pictured is a rectangular box formation. Entry points are marked with '*' (1) Robot A7 detects a nearby formation, initiating communication with A6. (2) A6, being an entry point, directs A7 towards its left side. (3) A7 Joins the formation. A5 recognizes that it is now part of a longer segment and drops entry point status.

Sec. 3.2. When any robot perceives another, it queries that robot for formation membership, and if it is part of a formation, requests a size estimate. At the same time it performs a size estimate of its own formation (Sec. 3.1). The smaller of the two formations is asked to transform into a line formation (this involves a chain of adopting one's nearest neighbour, rejecting anyone further, until everyone has only a single neighbour), to facilitate joining the larger formation one at a time. The robot at the front of the new line has an entry point to merge to, located in the other formation. This causes a chain of formation joining, as each robot in the line recognizes it has no target and attempts to join the formation next to it.

3.4   Balancing a Formation

Because robots may join at any entry point in the formation, balance needs to be maintained between segments and ultimately across the formation. To accomplish this, we employ a balancing technique that allows the entry point robots to move other robots between segments. The entry point robot first computes the ideal number of robots in each of the segments in which it participates. This is done by estimating the total number of robots in these segments (Sec. 3.1), and then performing a vector multiplication according to the fractions that should be in each segment according to the desired proportion for each segment, specified in the formation description. The resulting values are called the *ideal* counts for these segments. The segment whose actual count is furthest below its ideal count is selected as the destination segment, and the segment which is furthest above its ideal count is selected as the source segment. Assuming these differences are at least one, the entry point agent instructs its neighbor in the source segment to become its neighbour in the destination segment, and its neighbour in the destination segment to become the neighbour to the shifted agent (Fig. 2). While this moves only one robot at a time, it is important to view this occurring in a distributed manner: if a distant part of the formation is short on robots, multiple local shifts will eventually cause them to move there.

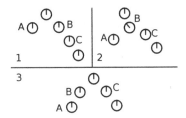

Fig. 2. In panel 1, the V formation is unbalanced. In panel 2, the entry point robot has instructed robot B to change segments. In panel 3, we see the result of the rearrangement of robots into the new balanced formation.

3.5   Communication Interruptions

All the above operations require ongoing communication, but only across small local distances. Still, these may be interrupted at times. In general, the worst case for communication interruption during any of these operations results in a breaking up of a formation (a heartbeat not heard after ten seconds in our implementation leads one robot to think its neighbour has failed). In that worst case, the robots are still close to one another and can pick one another up again as communication resumes, through the operations defined in the previous subsections.

Inconsistencies are also possible because of communication problems. For example, because it is possible for communications to break down in the middle of a balance, it is possible for an entry point robot to ask its neighbor to shift to a different segment, and yet have the neighbor occupying that position to remain where it is, resulting in two agents occupying the same position in the formation. This is repaired by periodically checking the intended segment ID for each of an entry robot's neighbours with that actually reported by the neighbour, and managing conflicts by requesting one robot to move, essentially restarting the rebalance. There are a number of other analogous measures for dealing with potential inconsistencies; all of these may be found in [12].

4   Implementation

We implemented this approach using the well-known player/stage [13] simulation package, using Pioneer2 DX robots equipped with laser and fiducial scanners, with both of these sensors operating in a forward direction. We used three different sensing ranges, allowing for heterogeneity in sensing. We also used three separate maximum speeds, giving heterogeneity in locomotion ability.

Our robot controller is behaviour-based [14], and consists of three behaviours. These behaviors are the same as those used in our previous work [3]: *goal seeking*, *formation keeping* and *neighbor avoidance*. Goal seeking causes all robots to move towards a known common goal location. This is used to simulate localization and goal selection. Formation keeping ensures that robots stay at the desired distance and angle from their target robot. Finally, neighbor avoidance ensures that robots don't get too close together or hit each other.

Because some robots will have short sensing distances, we allow them to be assisted by those with better sensing abilities, by querying their immediate neighbors for the visibility of other robots. This allows robots to get a reasonable idea of whether a specific robot is in front of or behind them. In practice, this type of message is usually sent to a robot's target. The target then returns whether or not it can sense the sending robot. If a robot is visible to its target while the target faces forward, this indicates it has passed the target and should be behind it. In such a case, the following robot slows down and allows the target to resume its appropriate position.

Heterogeneity in locomotion ability is addressed by having each robot send out messages to its neighbors. These messages indicate the minimum of the robot's top speed, and the top speed of its neighbors. Eventually, this will propagate the speed of the slowest member to all formation members. This allows all formation members to properly maintain formation without leaving members behind.

Each robot has a common representation of the formation in terms of relationships between segments. Segment membership is stored as a bit vector in each robot.

5   Evaluation

Because the main thrust of this approach is its support of heterogeneity, we examined the effects of heterogeneity on scalability and tolerance of failure. Since other approaches do not support the range of conditions for which ours is intended, a direct comparison was not possible.

As a basis for performance, we defined a measure of formation error. Based on the total number of robots in the formation, we calculate the ideal number of robots per segment. We then calculate the absolute difference between the number of robots in the segment, and this ideal. The formation error is the sum of these absolute differences from the ideal segment counts. Other metrics employed include the maximum number of robots in formation, and the number of trials that result in a single formation. Further metrics and additional experimentation can be found in [12].

The trials were run in an obstacle-free simulated environment, which was approximately 40 metres wide by 1000 metres long. Robots assembled into a V formation with two segments of equal length. Robots were required to maintain a 45 degree angle, and maintain a separation of 2 metres. Robots were initially distributed across random grid points within a 6 metre x 6 metre area within the environment. Further dispersion of robots from this is observed initially, because of obstacle avoidance and desire to move toward the goal.

We defined three different levels of perceptual and locomotion ability, and we refer to these levels as *poor, moderate* and *good*. For sensing, we used a range of 6 metres for poor, 8 metres for moderate, and 10 metres for good. For locomotion ability, we used 1 m/sec for poor, 2 m/sec for moderate, and 3 m/sec for good. We established 7 profiles, each describing the percentage of robots in the simulation that have each level of ability along each of these dimension. These profiles are described in Table 1. In the case when the number of robots in a trial is not evenly divisible into the specified groups, remaining robots were granted moderate sensing or locomotion.

Table 1. Sensing and locomotion profiles

Profile Number	Poor	Moderate	Good
0	0%	100%	0%
1	25%	75%	0%
2	50%	50%	0%
3	0%	75%	25%
4	0%	50%	50%
5	25%	50%	25%
6	50%	0%	50%

To examine scalability, we varied the number of robots in the simulation. We used populations of 5, 10 and 15 robots. For population, we tested all 7 locomotion and all 7 sensing profiles. To examine robustness against failure, we used 10 robots, and selected 3 chances of failure, each paired with a maximum number of failed robots. Every 30 seconds, our logging server tells a robot to fail with a given probability. The probabilities of failure were 0, 0.1 and 0.4. These were paired with a maximum number of 0, 1 and 2 robot failures respectively. As with the scalability trials, we tested all 7 sensing and locomotion profiles within each category of failure.

We ran our trials on Amazon EC2, using 50 c1.medium instances, performing 50 repetitions for each trial. A trial was considered complete when either 5 minutes had elapsed, or all robots were in the same formation with no robot changing segments for 10 seconds.

6   Results

This section provides an overview of our findings. A more detailed description, along with many additional results, can be found in [12]

Our results showed that neither heterogeneity in sensing, nor heterogeneity in locomotion ability had a large impact on robots' ability to create and maintain formations using our approach. In Figs. 3 and 4, we see that the maximum number of robots in formation at a given time is high for all sensing profiles. This indicates that formations are being created, and maintained in different areas in the environment.

Despite a high percentage of robots in formation, ending a trial by achieving a single overall formation was not common ($\sim 5\%$ of trials). Part of the reason for this is the distance between robots given the distributed nature of this approach. Once robots have coalesced into only a few sub-formations, it can become difficult for robots from different formations to encounter one another and merge further. While a single formation may not always be the end goal, we intend to experiment further with an additional behaviour attracting robots to others perceived at a distance, and other related behaviours that may influence the likelihood of a single end formation.

The results in Fig. 5 and 6 show that the formations that are being created are generally correct. The large error bars are present because this statistic is sampled at the end of the trial. If a correct formation is not achieved, the formations may be in the middle of a joining or balancing operation. This potential for increased error means that this is actually quite a conservative metric.

The results referenced in this section are for the variation of sensing profiles only. Variation of locomotion profiles had a similar effect. Full results can be found in [12].

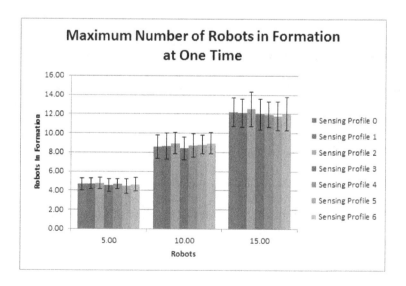

Fig. 3. Maximum number of robots in formation as population size changes

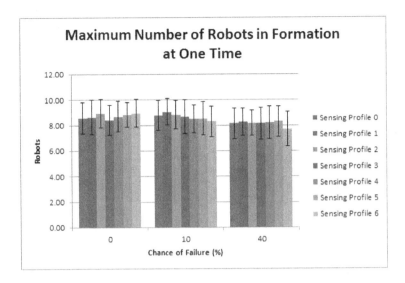

Fig. 4. Maximum number of robots in formation as chance of failure changes

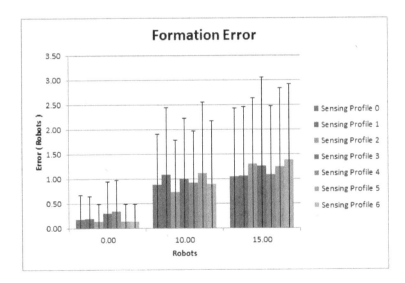

Fig. 5. Formation error as population size changes

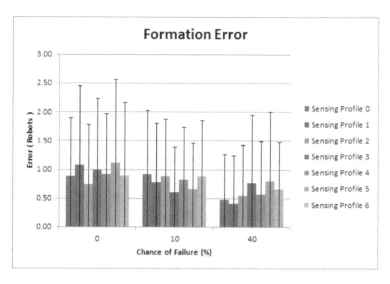

Fig. 6. Formation error as chance of robot failure changes

7 Discussion and Future Work

We have described a distributed approach to formation control that is independent of the number of robots and supports heterogeneity in sensing and locomotion. Using a simulation, we have shown that our approach is robust against robot failure, and that it is scalable across a number of different population sizes.

These are positive results given the significant challenges of the conditions under which this operates. We demonstrated that neither heterogeneity in sensing nor heterogeneity in locomotion ability seriously impact the performance of this approach. The establishment of formations also indicates that our approach adequately compensated for limited knowledge of others and the inability to assume a given perceptual or locomotion ability.

There are a number of immediate avenues of future work. First, we would like to experiment with even larger populations of robots, and extend heterogeneity to elements such as robot size and terrain that may not be traversable by all robots. In addition, we would like to experiment with more complex formations. We also plan to explore metrics for evaluating formations that better reflect decentralized approaches such as this. We would like to implement several existing approaches to formation control, and use these metrics to get a better idea of how our approach compares to others.

We are currently moving to replicate these experiments using Citizen microrobots in a mixed reality environment using global vision [15], after the prior RoboCup Mixed Reality League. The small size of the microrobots allows a reasonable population on a small physical area, while the virtual elements employed in mixed reality can allow a large surface by having terrain that scrolls under the robots, as well as obstacles represented virtually. Since catastrophic obstacles (e.g. fire, pits) can be simulated, would also allow a rationale for robot failure to replace the statistical failure used in the simulation studies.

References

1. Fredslund, J., Mataric, M.: A general algorithm for robot formations using local sensing and minimal communication. IEEE Transactions on Robotics and Automation 18(5), 837–846 (2002)
2. Kwon, J.W., Chwa, D.: Hierarchical formation control based on a vector field method for wheeled mobile robots. IEEE Transactions on Robotics 28(6), 1335–1345 (2012)
3. de Denus, M., Anderson, J., Baltes, J.: Heuristic formation control in multi-robot systems using local communication and limited identification. In: Baltes, J., Lagoudakis, M.G., Naruse, T., Ghidary, S.S. (eds.) RoboCup 2009. LNCS (LNAI), vol. 5949, pp. 437–448. Springer, Heidelberg (2010)
4. Hattenberger, G., Lacroix, S., Alami, R.: Formation flight: Evaluation of autonomous configuration control algorithms. In: Proceedings of the 2007 IEEE/RSJ International Conference on Intelligent Robots and Systems, San Jose, USA, pp. 2628–2633 (October 2007)
5. Rampinelli, V., Brandao, A., Martins, F., Sarcinelli-Filho, M., Carelli, R.: A multi-layer control scheme for multi-robot formations with obstacle avoidance. In: International Conference on Advanced Robotics, ICAR 2009, pp. 1–6 (June 2009)
6. Consolini, L., Morbidi, F., Prattichizzo, D., Tosques, M.: Stabilization of a hierarchical formation of unicycle robots with velocity and curvature constraints. IEEE Transactions on Robotics 25(5), 1176–1184 (2009)
7. Hadaegh, R., Kang, B., Scharf, D.: Rule-based formation estimation of distributed spacecraft. In: The 10th IEEE International Conference on Fuzzy Systems 2001, vol. 3, pp. 1331–1336 (2001)
8. Leonard, N., Fiorelli, E.: Virtual leaders, artificial potentials and coordinated control of groups. In: Proceedings of the 40th IEEE Conference on Decision and Control 2001, vol. 3, pp. 2968–2973 (2001)
9. Lee, G., Chong, N.Y.: Adaptive self-configurable robot swarms based on local interactions. In: Proceedings of the 2007 IEEE/RSJ International Conference on Intelligent Robots and Systems, San Jose, USA, pp. 4182–4187 (October 2007)
10. Yamaguchi, H.: Adaptive formation control for distributed autonomous mobile robot groups. In: Proceedings of the 1997 IEEE International Conference on Robotics and Automation, vol. 3, pp. 2300–2305 (April 1997)
11. Balch, T., Arkin, R.: Behavior-based formation control for multirobot teams. IEEE Transactions on Robotics and Automation 14, 926–939 (1998)
12. de Denus, M.: Adaptive formation control for heterogeneous robots with limited information. Master's thesis, University of Manitoba (2013)
13. Gerkey, B., Vaughan, R., Stoy, K., Howard, A., Sukhatme, G., Mataric, M.: Most valuable player: a robot device server for distributed control. In: Proceedings of the 2001 IEEE/RSJ International Conference on Intelligent Robots and Systems, vol. 3, pp. 1226–1231 (November 2001)
14. Arkin, R.C.: Behavior-based Robotics. MIT Press, Cambridge (1998)
15. Anderson, J., Baltes, J., Tu, K.Y.: Improving robotics competitions for real-world evaluation of AI. In: Proceedings of the AAAI Spring Symposium on Experimental Design for Real-World Systems, Stanford, CA, pp. 1–8 (March 2009)

RoSHA: A Multi-robot Self-healing Architecture*

Dominik Kirchner, Stefan Niemczyk, and Kurt Geihs

University Kassel, Distributed Systems Group,
Wilhelmshoeher Allee 73, 34121 Kassel, Germany
{kirchner,niemczyk,geihs}@vs.uni-kassel.de
http://www.vs.uni-kassel.de

Abstract. Reliability is one of the key challenges in multi-robot systems to increase practicable applicability and hence the commercial usage. This paper presents RoSHA, a self-healing architecture for multi-robot systems. RoSHA is based on the established robot middleware ROS and provides components for application independent analysis and repair. A plug-in architecture enables the developer to simply add new components for repair and analysis. Bayesian networks are used to diagnose failures and their root causes. ALICA, a domain specific language for multi-robot systems, is applied to coordinate recovery plans in multi-robot systems.

Keywords: self-healing, multi-robot system, system monitoring, failure diagnosis, system recovery.

1   Introduction

Multi-robot systems become more and more important for many application domains, like search and rescue, warehouse management, or urban exploration. Often the complexity of tasks and domains demands autonomous operation in unknown and dynamic environments. These systems are confronted with complex tasks, like planning, localization, or coordination. The dynamic environment and the intricate interaction between these tasks result in a failure-prone setting [1,2]. Accordingly, an architecture of multi-robot systems often consists of connected components. A failure in a single component can lead to a degeneration or even a crash of the total system. For example, in a single robot task, the localization of the robot relies on the fusion of multiple sensor information, like a direction, determined by a compass module and a distance scan, extracted from a camera image. If an error occurs in only one of the involved sensors, the robot could be delocalized. Due to the coordination and interactions of single robots in a multi-robot task, the resulting susceptibility to failures is expected to become even worse.

---

* The project IMPERA is funded by the German Space Agency (DLR, Grant number: 50RA1112) with federal funds of the Federal Ministry of Economics and Technology (BMWi) in accordance with the parliamentary resolution of the German Parliament.

S. Behnke et al. (Eds.): RoboCup 2013, LNAI 8371, pp. 304–315, 2014.

In order to foster real-world applications and to increase the commercial use, reliability is a key design challenge. Carefully tested system design is one step in this direction. However, many errors result from the dynamic operation conditions, thus automatic failure detection and self-healing is required at runtime. To achieve self-healing capabilities, the system needs to identify root causes of a failure and execute a recovery policy. A more advanced concept than static failure handling is required. In this paper we present a generic and abstract robot self-healing architecture, called RoSHA. This architecture is based on ROS (Robot Operating System) diagnostics[1] and is inspired by the MAPE-K cycle [3]. It contains the five blocks monitoring, diagnostic, recovery planning, repair execution, and a knowledge base. We use an abstract system model and Bayesian networks as knowledge base to determine the root causes for detected failures, like a component crash, a deadlock, etc..

The rest of this paper is organized as follows. Section 2 identifies the requirements for a self-healing architecture. In Section 3 we introduce ROS and the existing diagnostic stack. The architecture and basic concepts of RoSHA are presented in Section 4. In Section 5 we name and discuss existing solutions. Finally, in Section 6 we conclude the paper and propose future work.

2   Requirements

Robustness of multi-robot systems is a key challenge that only little research has been directed to. In order to contribute to this challenge the proposed architecture offers self-healing capabilities. Often the architecture of a multi-robot system is characterized by a high degree of complexity. Therefore, the interaction between the robot system and the self-healing add-on should be carefully designed and limited to few well defined interfaces. The integration effort for robot developers should be minimized. Moreover, the self-healing add-on should not be application nor domain specific, rather provide its services in a general way. Most robot systems have limited resources and need to perform their tasks in soft real-time. Therefore, the architecture of the self-healing add-on should be resource-efficient to prevent indirect interferences. Scalability is another important requirement. The self-healing add-on should be independent from size and distribution of a multi-robot system.

Beside these envisioned features of a self-healing architecture, humans should be still able to oversee and control the system. If executed wrongly, repair actions can disturb the operational performance of the robots, e.g. in the case of an unnecessary restart of a core component. To cope with these risks, human operators should be able to intervene at any time. Furthermore, logged human intervention actions provide a valuable source of labeled training data for improvements. In summery we identified the following requirements of a self-healing architecture:

Ease of integration: To support existing robot systems without extensive integration efforts, well-defined interfaces between the robot system and the add-on have to be claimed.

---

[1] http://www.ros.org/wiki/diagnostics

Resource-efficient: The runtime support must not decrease the robot's performance. Therefore, the self-healing add-on has to be implemented in a light-weighted resource-efficient way.

High degree of configurability: The application and domain independent components of the self-healing add-on require a high degree of configurability. This is necessary to adjust these components to special needs of an existing robot system.

Human controllability: An unnecessary executed recovery action can decrease the performance of the whole robot system. Therefore, human intervention and control should be supported.

Extensibility and modularity: To extend the self-healing add-on to domain and application specific requirements a flexible design is required. Further application specific components, e.g. a new specialized control interface, should be simple to develop and integrate.

Multi-robot support: To improve the recovery, inter-robot interactions should be considered in the recovery execution. Therefore, multi-robot coordination for complex recovery processes should be supported.

3   Background

In order to realize the requirements stated in Section 2, we use ROS (Robot Operating System) [4], which is introduced in the remainder of this section.

3.1   Robot Operating System

ROS is a software framework for single and multi-robot system development. It provides standard operating system services such as hardware abstraction, low-level device control, implementation of commonly-used functionality, message-passing between components, and package management. The middleware is based on a graph architecture where processing takes place in nodes. These nodes receive and send messages using a publish/subscribe mechanism. This multi-component structure directly supports the required ease of integration and extensibility. The middleware services, like connecting different nodes, are dynamically organized from a central arbiter, the roscore. Communication in this system is not limited to local inter process communication (IPC) and could be distributed over a network as well. ROS contains repositories for additional packages. Users can contribute packages to the community that implement common functionalities for reuse, such as simultaneous localization and mapping, planning, perception, simulation, and so on.

3.2   ROS Diagnostics

In ROS the task of analyzing and intuitive reporting the system state is provided by the diagnostics stack. It consists of development support for collecting and publishing information, an analysis and aggregation node, and visualization tools. This tool-chain is built around standardized interfaces, namely the *diagnostic* topic for monitoring information and the *diagnostic_agg* topic for the analyzed and aggregated results.

To make relevant information available, the diagnostic stack provides support for integration of a monitoring publisher in a ROS nodes. This allows the nodes to publish diagnostic data, the node status, or to monitor processing timings. Gathered data are published continuously on the *diagnostic* topic. Additional, there exists support for triggered node self-tests. The *diagnostic_aggregator* node is responsible for analyzing and aggregating reported data at runtime. The aggregation is used to categorize the monitoring data to an information item that summarizes one aspect of the system, e.g. status information of one component. The analysis is performed using a plug-in model. Each plug-in provides one analyzer that analyzes and aggregates information on a system's aspect. This aggregated information is sent on the *diagnostic_agg* topic for notification or visualization. A standard tool for visualization is the *robot_monitor* node that highlights the overview. Detailed information can be easily accessed for all aspects. Due to the defined interface the integration of application specific reporting is facilitated.

In summary the ROS diagnostic stack provides the required features for multi-robot support, extensibility and modularity, and human control. The ROS middleware offers local and network communication and hence supports multi-robot systems. The multi-component support in combination with the defined interfaces in the ROS diagnostic stack facilitates extensions in a convenient way.

Beside the discussed strength, there exist some shortcomings as well. The information gathering has to be integrated in the source code of the components. Moreover, the robot system must be developed with ROS to communicate the diagnostic data from the nodes. Therefore, the claimed ease of integration to an existing system is not fulfilled. However, the main shortcoming is the lack of autonomous failure recovery. For system recovery ROS diagnostics strictly relies on human operators to manually repair the system.

4   RoSHA Design and Architecture

In this section, we describe our approach to realize a self-healing add-on that fulfills the requirements identified in Section 2. Apart from the discussed shortcomings, ROS diagnostics already provide some of these requirements discussed in the previous section. Thus, we decided to build on ROS and the ROS diagnostics stack to develop RoSHA. In the development, we explicitly consider the special needs of multi-robot recovery, as described later.

As depicted in Figure 1, the structure of the overall system consists of two basic parts, which outline the local and the distributed aspect of the system. Here, the distribution of a multi-robot system is abstracted as interactions of a local robot with its team. Details of our proposed self-healing architecture are presented in the robot part. This part consists of four basic blocks (presented as gray areas). The ROS middleware and the operation system serve as the fundamental hardware abstraction and execution layer that provides basic system services. The robot system realizes functionalities for the intended operation of the robot. The last block is RoSHA, which supports the reliability of the robot system and the user, who operates the system. The internal structure of RoSHA is built on the concept of MAPE-K [3] as the underlying decision cycle. This cycle is composed of five basic blocks, monitoring, analyzing, planning, execution, and a knowledge base. Following the requirement of modularity, we realized each block as an independent component.

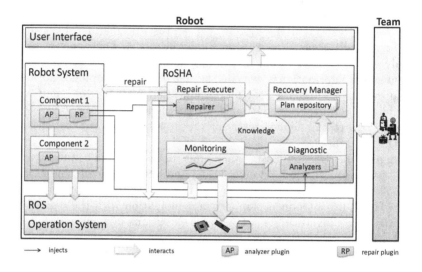

Fig. 1. Overview of the overall robot system architecture

The monitoring component collects information about the current system state. Therefore, two different aspects of the system are monitored. The first one is the knowledge provided by the operation system, like the current resource usage of a component (CPU load, memory usage, thread count, etc.). The second aspect is information provided by the components themself. If used, each component can directly send status information on the ROS monitoring topic, which is received by the monitoring component of RoSHA.

The diagnostic component uses the collected information to identify failures and their root causes. A set of basic plug-ins exist that supports generic and component independent analysis. In our opinion generic analyzing is not enough to identify all failures. The developer of the component is best capable of providing component specific analysis. Each component can provide its own analyzing plug-in (AP). This is injected into the diagnostic component at runtime. A system model is used to determine the root causes for each detected fault.

Detected faults are reported to the recovery manager. This component selects a recovery plan from a set of predefined policies to recover from the failure. The repair execution component performs the repair actions included in the selected recovery policy. Similar to the diagnostic component the execution component provides a set of generic repair actions. This set can be extended with component specific actions, which can be offered as repair plug-ins (RP).

4.1   System Model

In accordance to the decision cycle, we use a single model to represent the accumulated knowledge. This model describes the current configuration of the multi-robot system. Additionally, configuration information of the self-healing add-on are included, e.g. the configuration details of a robot specific monitoring.

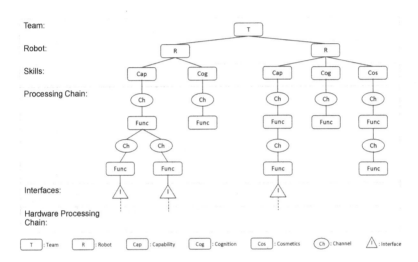

Fig. 2. Layered visualization of an exemplary system model in a multi-robot setting

We define a representation of the system according to the information flow between the components (see Figure 2). Therefore, we transformed a data flow diagram (graph representation) in a tree structure to express sequence and parallel component configuration. Furthermore, we include additional semantic elements on the upper levels of the model. The top-level element is the *team* node that represents the total multi-robot system. A team consists of *robot* nodes, which are placed on the second layer. All configuration-related to the overall robot system is specified here. On the next level, the model is decomposed in the robot's basic skills, like the ability to move, detect objects, localize, or plan. The set of skills is divided in *capabilites, cognition,* and *cosmetics. Capabilities* are used for sensor and actor skills, *cognition,* for cognitive skills like the planning and decision making of an autonomous systems, and *cosmetics,* for pure supportive skills, like a graphical user control interface. Each skill consists of *functionalities* and *channels* to represent the processing flow. Functionalities correlate with the system components of the robot system and channels represent the communication links between them. Components that are present in multiple skills are presented separately in each skill and linked together. This processing chain ends with an *interface* element that marks the end of the software domain layer and the beginning of the hardware layer. Modeling the information flow on the hardware layer is not yet included, but part of the future work.

4.2   Monitoring

The monitoring is responsible to observe the current system state. The system, as described in Section 4.1, is composed of connected components. Therefore, capturing the entire robot system's state means to collect information of each component. However, properly working components do not directly induce a working system, the interactions between these components have to be considered as well. Therefore, we define two categories: the *component monitoring*, to capture components' state information and the *flow monitoring*, to supervise the communication flow between components.

The subject of the monitoring is one aspect of the information collection, the provider is the other. A common way to do this is to extend the components of the robot system to send current state information. An example for this is a heart beat signal. The component itself continuously sends a message to show its liveliness. We call that type of information collection *active monitoring*, because the components of the robot system have to provide their state information by their own. The self-healing architecture supports this monitoring, however we focus on a more flexible information collection. The proposed monitoring process focuses on the collection of general characteristic information of a component, like its resource usage. No direct support of the component is needed. This is referred as *passive monitoring* in the context of our work.

Passive monitoring can be very resource-inefficient. Therefore, we developed an adaptive monitoring, which is based on system specific aspects defined in the system model. For each component a set of characteristics are defined and monitored due to a set of properties. For example one characteristic could be the CPU usage of a component. The property describes the expected range from 15% to 25% with a typical value of 17%. The distance between the typical value and current value in respect to the borders is used as the distance metric to adapt the monitoring process. If the CPU usage is near the typical value, the monitoring level will decrease and the monitoring interval will increase and vice versa. The adaptive monitoring reduces the system load in error free situations. Furthermore, it reduces the amount of data to analyze as well. In a typical robot system components are not equally important. Therefore, the system model supports the configuration of initial monitoring levels for each functionality and channel.

4.3   Diagnostics

After monitoring the state of the system, the interpretation has to be done. This is addressed by the diagnostic component. The perceived state information can be seen as symptoms of the overall system health. Therefore, the goal of the diagnostic component is to aggregate related symptoms, calculate an estimation of the health state, and identify possible root causes.

As described in Section 4.1, we defined a model of the component dependencies. This structure is composed of component items and communication items. Both are subjects of our monitoring and hence, state information are continuously updated. In our work we apply Bayesian networks for the diagnosis [5]. Bayesian inference needs a structured knowledge representation, a Bayesian network, to perform the diagnostic analysis. In our self-healing add-on this knowledge representation models the correlation of symptoms, root causes, and the resulting failure probability (see Figure 3). The resulting model can be structured in three layers, the symptom layer: the root cause layer, and the component failure layer. In the context of the Bayesian formalism, the monitored information is regarded as continuously updated evidences of the symptoms. Therefore, ongoing inference of the model is needed. In order to address temporal characteristics, we extend the models to dynamical Bayesian networks by including temporal dependencies to capture time series properties. With this extension we are able to model and analyze trends in symptoms to improve the failure analysis, e.g. an increasing memory usage in the case of a memory leak.

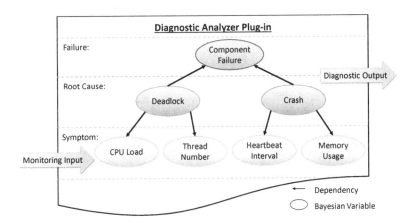

Fig. 3. Bayesian network for the diagnostic task

Besides a similar layered structure of the failure models, the included knowledge, like a-priori probabilities and dependencies, is specific for each component. This enables the distribution of the models to the location of the corresponding component to support an alignment of failure modeling. In order to process these distributed models, we apply the plug-in model to loosely connect the failure models with the diagnostic component of the self-healing add-on.

With these distributed models, inference of failure probabilities and root cause probabilities can be inferred separately. In order to infer a reliability estimate for the entire robot system, or some subsystem of it, the interference of components' failure probabilities has to be considered. This combination can be calculated using reliability theory [6]. Therefore, we apply the formalism of reliability block diagrams (RBD) to compute the (sub-) system reliability.

4.4   Recovery Manager

After diagnosis, detailed information of failure probabilities of the robot system are available. In addition, the diagnostic component provides probabilities of root causes to identify the most likely reason for a potential failure. Based on these information, the responsibility of this component is to select a fitting recovery policy to restore the system. As mentioned before, monitoring information, and hence the diagnostic results, change in real-time. Dynamic planning for each situation is a time consuming task that requires much resources [7]. This is in conflict with the requirement of recourse efficiency. Therefore, we decided to provide a set of predefined, but proven recovery policies. These policies are stored in the plan repository of the component.

For plan selection and team recovery coordination we use the multi-agent coordination language ALICA [8]. ALICA inherently fulfills the requirement of multi-agent support. The policies, called plans in ALICA, set up a strategy of repair and assessment actions for a team of agents. In such a plan we are able to model complex repair and check procedures with multiple robots involved.

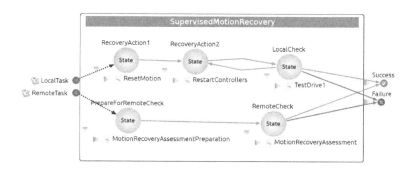

Fig. 4. Recovery plan in ALICA

Figure 4 illustrates an example recovery policy that is designed to recover a robot's motion process. It performs two recovery actions and locally verifies the success by a test drive. Additionally, an independent assessment of the recovery is done by supportive robots that confirm the success through their independent observation of a matching movement. Therefore, explicit knowledge exchange between the robot system and the recovery manager is used. The recovery policies are case specific and modeled manually. The needed environment knowledge for the recovery assessment has to be communicated to the self-healing add-on by the robot system. The communication can rely on IPC or the plug-in model. Without the robot system's support, the recovery manager is limited to local repair coordination based on monitoring and diagnostic information.

4.5 Repair Execution

The recovery manager selects an appropriate recovery policy and coordinates the involved repair actions. These actions are communicated to the repair executor node. Two types of repair actions are distinguished. Generic repair actions that provide repair services independent of the target component and specific repair actions that are designed for one specific target node. Both types are realized as plug-ins, similar to the analyzers of the diagnostic component. The recovery impact of generic repair, like a restart of a component, is limited due to their general scope. Specialized component knowledge is needed to implement more specific repair actions, e.g. a runtime reconfiguration of the component. Only the component's developers possess this detailed knowledge and seems best fitted to implement these specific repairs. To support the extensibility of the self-healing add-on, the specific repair plug-ins are located in the components. A third, more sophisticated way to restore a system is an architectural recovery. The goal of this recovery policy is to change the architecture of the robot system in order to restore the operational mode. Thereby, the architectural changes comprise adding, removal, and connection of system components. In this policy a sequence of architectural repair actions are performed. These repairs are not limited to local actions, but involve the total system. In a multi-robot system we need to perform distributed architectural actions. After an unsuccessful local recovery, distributed repairs allow to compensate this failure through relocation of the failed component to another robot system. If local and distributed recovery failed, the robot looses some of its capabilities and is seen as degenerated. However, such a robot could still contribute to the global task. In order to continue this task as good as possible, the degree of degeneration has to be considered and appropriate adaptation actions should be performed, e.g. a global task reallocation. These repair actions are ongoing work in RoSHA .

5 Related Work

The general problem of self-healing is addressed in several projects. In the following we like to present some of the related work and discuss their strength and shortcomings.

The SHAGE (Self-Healing, Adaptive, and Growing Software) framework [9] consists of two parts that enable the self-management of a robot. The first part comprises several components to manage the system, while the second part contains internal and external repositories to store architectural reconfiguration descriptions and components. These two parts work together to observe the situation of the environment and to trigger appropriate architectural reconfigurations. Furthermore, SHAGE includes a learning component to improve the reconfiguration due to previous experiences. This reconfiguration aims to adapt the behavior of the robot in exceptional situations. However, the challenge how to identify these situations is kept open. Recovery of these situations is done by changing the behavior of the robot through architectural adaptation while component failure recovery is not addressed. The framework includes interfaces to external repositories to share knowledge, but does not support coordination of multiple robots.

The Rainbow project [10,11] proposes an architecture-based self-adaptation approach. It provides reusable infrastructure together with mechanisms to tailor these to the domain's needs. These specializations allow the developer to define aspects of the system, like monitoring targets or adaptation conditions and actions. The adaptation is done through statically associated sets of action rules for each identified adaption cause. The information collection and the execution of adaptation actions rely on direct support of the target system. Therefore, integration in an already existing system seems difficult. Furthermore, the architecture is inherently centralized and thus sensitive to single-point of failures and of limited use for distributed systems, like multi-robot systems.

LeaF (learning-based fault diagnosis) [12] from Parker and Kannen is based on an adaptive causal model for fault diagnosis and recovery. The approach focuses on multi-robot systems. The causal model represents expected faults and is initially created by the developers at design time. A case based reasoning is used to handle unexpected faults during runtime. The system extracts possible recovery options from the existing model and adds the new fault and a recovery method afterwards. Furthermore, the model can represent faults that only occur during the interaction between robots. However, this approach does not cover hardware related monitoring and fault detection. The authors make no statement how to integrate LeaF in an already existing robot system or how to change or replace existing components.

Often, a robot system is built without considerations of system management, including self-healing. The main concern is to create a working system for a given task. Developers tend to neglect reliability in this design phase. That could lead to the situation in which it is necessary to integrate self-healing abilities to an already existing system. The discussed projects propose remarkable results in the field of self-healing through architectural adaptation. However, they do not address questions of practicable usability, like the integration in an existing system, or resource efficiency. We argue that these properties are central design decisions and should be reflected as architectural requirements. We argue as well that coordination and assessment of multi-robot recovery actions are central requirements for a domain and task independent self-healing architecture. In comparison with the discussed projects, RoSHA considers these requirements.

6   Conclusion and Future Work

Due to the system complexity of modern robot systems, reliability cannot be ensured in design time. Hence, runtime failure recovery in a self-healing add-on is needed. The integration of the self-healing add-on in an already existing multi-robot systems is essential in the sense of practicable usage. Therefore, increased usability and multi-robot support is required. In this paper, we presented a robot self-healing architecture to address these challenges. We propose a framework that is tailored for generic use in multi-robot scenarios. The design of the framework addresses the need for resource efficiency through the usage of an adaptive monitoring. Plug-in support with existing generic repair and analysis plug-ins enables an ease of integration. The coordination of multi-robot recovery policies is given by the language ALICA .

As part of our ongoing research we plan to integrate distributed failure learning methods to reduce the dependency of expert knowledge. A comprehensive real-world evaluation of the self-healing add-on will be done at the RoboCup[2] world championships 2013. In this dynamic setting the suitability for multi-robot teams will be evaluated. Furthermore, we plan to compare the recovery performance of a fully supported self-healing add-on with results from limited system support, e.g. with restricted support to generic monitoring and recovery actions.

References

1. Carlson, J., Member, S., Murphy, R.: How UGVs Physically Fail in the Field. IEEE Transactions on Robotics 21(3), 423–437 (2005)
2. Carlson, J., Murphy, R.: Reliability analysis of mobile robots. In: International Conference on Robotics and Automation, vol. 1, pp. 274–281. IEEE (2003)
3. Huebscher, M., McCann, J.: A survey of autonomic computing degrees, models, and applications. ACM Computing Surveys 40(3), 1–28 (2008)
4. Quigley, M., Conley, K., Brian, G., Josh, F., Tully, F., Jeremy, L., Rob, W., Andrew, N.: ROS: an open-source Robot Operating System. In: ICRA Workshop on Open Source Software. Number Figure 1 (2009)
5. Pearl, J.: Probabilistic Reasoning in Intelligent Systems: Networks of Plausible Inference. Morgan Kaufman Publ. Inc. (1997)
6. Shooman, M.: Reliability of Computer Systems and Networks: Fault Tolerance, Analysis, and Design. John Wiley & Sons, Inc. (2002)
7. Ghallab, M., Isi, C.K., Penberthy, S., Smith, D.E., Sun, Y., Weld, D.: PDDL - The Planning Domain Definition Language. Technical report, CVC TR-98-003/DCS TR-1165, Yale Center for Computational Vision and Control (1998)
8. Skubch, H.: Modelling and Controlling Behaviour of Cooperative Autonomous Mobile Robots. Phd thesis, University of Kassel (2012)
9. Kim, D., Park, S., Jin, Y., Chang, H.: SHAGE: a framework for self-managed robot software. In: Proceedings of the International Workshop on Self-adaptation and Self-managing Systems, pp. 79–85. ACM Press, Shanghai (2006)
10. Garlan, D., Cheng, S.W., Schmerl, B., Steenkiste, P.: Rainbow: Architecture- Based Self-Adaptation with Reusable. Computer 37(10), 46–54 (2004)
11. Cheng, S.W., Garlan, D., Schmerl, B.: Evaluating the effectiveness of the Rainbow self-adaptive system. In: 2009 ICSE Workshop on Software Engineering for Adaptive and Self-Managing Systems, pp. 132–141 (May 2009)
12. Parker, L., Kannan, B.: Adaptive Causal Models for Fault Diagnosis and Recovery in Multi-Robot Teams. In: Intelligent Robots and Systems, pp. 2703–2710 (2006)

---

[2] http://www.robocup2013.org

Routing with Dijkstra in Mobile Ad-Hoc Networks

Khudaydad Mahmoodi, Muhammet Balcılar, M. Fatih Amasyalı, Sırma Yavuz,
Yücel Uzun, and Feruz Davletov

Yıldız Technical University, Computer Science Department, Istanbul
{khudaydad.mahmoodi,fdavletov}@gmail.com,
{muhammet,mfatih,sirma}@ce.yildiz.edu.tr,
yuceluzun@windowslive.com

Abstract. It is important that robot teams have an effective communication infrastructure, especially for robots making rescue operations in debris areas. The robots making rescue operation in a large area of disaster are not always directly connected with central operator. In such large areas robots can move around without losing communication with each other only by passing messages from one to another up to the central operator. Routing methods determine from which node to which node the messages are conveyed. In this work blind flooding and table-based routing methods are tested for three different scenarios to measure their effectiveness using the simulation environment USARSIM and its wireless simulation server WSS. Message delay times and maximum data packet streaming rates are considered for measuring the effectiveness. Although it has some deficiencies, it was observed that table-based approach is more advantageous than blind flooding.

Keywords: Mobile ad-hoc networks, Routing protocols, USARSim, WSS.

1 Introduction

In recent years, the importance of the teleoperation of mobile robots and teams of mobile robots increased. Recently, more and more mobile robots are developed which are capable of operating in impassable or hazardous environments with little or no communication infrastructure [2]. Along with technological advances robots became much more intelligent and much more capable. It means that they must be developed to possess the capability of constructing a network and performing cooperative works [1].

A key driving force in the development of cooperative mobile robotic systems are their potential for reducing the need for human presence in dangerous applications. Such applications as the disposal of toxic waste, nuclear power processing, firefighting, civilian search and rescue missions, planetary exploration, security, surveillance and reconnaissance tasks have elements of danger. In these cases, wireless communication provides the low-cost for mobile robot networks to cooperate efficiently [1].

There is increasing demand for connectivity in places where there is no base station or infrastructure available. This is where ad-hoc networks came into existence.

S. Behnke et al. (Eds.): RoboCup 2013, LNAI 8371, pp. 316–325, 2014.

Wireless networks can be classified into infrastructure networks and infrastructure-less networks or mobile ad-hoc networks (MANETs) [3].

MANETs are autonomously self-organized and self-configuring networks without infrastructure support. To create a temporary network there is no need for any centralized administration or infrastructure. In such networks, due to the absence of dedicated routers, each member node is also responsible for routing messages to other nodes. If mobility is very high, then the network may experience frequent and unpredictable topology changes [3], [5], [8].

Recently, mobile ad-hoc networks became a hot research topic among researchers due to their flexibility and independence of network infrastructures such as base stations. The infrastructure-less and the dynamic nature of these networks demand a new set of networking strategies to be implemented in order to provide efficient end-to-end communication. MANETs can be deployed quickly at a very low-cost and can be easily managed [3].

There have been a number of ad-hoc routing protocols developed for MANETs, each with benefits relating to specific usage scenarios. The majority of routing protocols for MANET try to reduce bandwidth usage, minimum energy consumption, throughput, packet delay time, etc. Different routing protocols use different measures to determine the optimal route between the sender and receiver. Each protocol has its own advantages and disadvantages. In this application, we try to reduce packet delay time by minimizing hop count for better and faster communication between robots.

In this study, we implement Dijkstra's algorithm with minimum hop in order to minimize packet transmission time and tested our method on USARSim simulation software. This paper proceeds as follows. Section 2 shows the classification of mobile ad-hoc routing protocols. USARSim and WSS are briefly introduced in section 3. Section 4 explains implemented the Dijkstra's algorithm. Experimental results of algorithms are analyzed in section 5. Finally, Section 6 concludes the paper.

2   Routing Protocols

Routing protocols in MANETs can be classified into two categories based on routing strategies and network structure [5]:

1. Proactive Protocols (Table-Driven)
2. Reactive Protocols (On-Demand)

2.1   Proactive Routing Protocols

Proactive or table-driven routing protocols maintain the routing information even before it is needed. Each and every node in the network maintains routing information to every other node in the network. Routes information is generally kept in the routing tables and is periodically updated as the network topology changes. The proactive protocols are not suitable for larger networks, as they need to maintain node entries for each and every node in the routing table of every node. This causes more overhead in the routing table leading to consumption of more bandwidth [6], [7].

2.2   Reactive Routing Protocols

Reactive or on-demand routing protocols create routes only when required by a node. If a node requires a path to a destination, it starts a route discovery process in the network. It can be either source initiated or destination-initiated. Once a route has been established, the route discovery process ends and the route will be valid until it breaks down or is no longer desired [4]. Packets are sent through this route. If there is no communication between two nodes then it is not necessary to maintain routing information at these nodes [3]. Such protocols often perform better than proactive protocols when implemented in a large network due to a smaller overhead. However a large amount of network traffic can cause the performance to deteriorate sharply as most such protocols flood the network while looking for a route, and this can lead to clogging of links. Another major disadvantage is the delay required to find a route which in some applications might be unacceptable [5], [7].

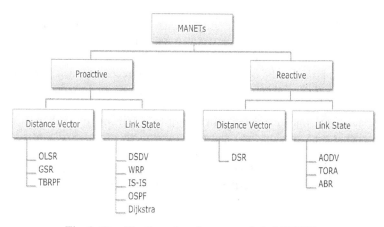

Fig. 1. Classification of routing protocols in MANETs

3   USARSim and WSS

The Urban Search and Rescue Simulator (USARSim) is built on top of the Unreal Engine, and uses the Unreal Engine to simulate the environment and the robots [5]. It is designed as a simulation companion to the National Institute of Standards (NIST) reference test facility for autonomous mobile robots for urban search and rescue [9].

Creating robots and preparing test environments are very costly and difficult tasks in real world. USARSim helped us to create robots easily in desired positions and prepare test environment with minimal cost. Maps in RoboCup competition contain specific disaster environment, victims, routes with different types of obstacles, etc. Robots have some certain missions due to competition rules. The main goal of robots are accomplishing mission as a desired manner in a given disaster environment.

The Wireless Simulation Server (WSS) is developed by Jacobs University which is used to simulate wireless communications between all robots and the operator that are created in USARSim simulation program. According to RoboCup virtual robot competitions rules, all communications between robots should be through WSS software. WSS allows us to get the signal strength of any pairs of robots whenever it is required. In order to know status of link between two nodes we have to compare the signal strength value with predefined threshold value. If the signal strength value is equal or bigger than threshold value (-93 dBm) then connection available, otherwise connection between nodes breaks down [10].

Registering is required to a specific port of WSS before starting communication. Once the connection has been opened, the WSS allows sending of messages and closing the connection. Each time a message is sent, the path loss between the end points is checked. If it is better than the threshold then the message is forwarded, otherwise message will be discarded and connection will be closed [5].

4   Routing with Blind Flooding Algorithm

The blind flooding approach is most widely employed strategy to perform and distribute messages to all robots in the networks [11]. Implementation of this algorithm is very simple. Because of exploring every possible nodes in the network, it increases the possibility of delivering packet to destination. On the other hand, flooding algorithm has some drawbacks like implosion, overlap and resource blindness. For instance, many unneeded packets on the network leads to use more bandwidth. Flooding exhibits a desirable behavior when adopted in wired networks.

Algorithm: Routing with Blind Flooding Algorithm

Inputs: $r_{dest}, r_{cur}, m, S_{i,j}, th$
when packet m is received **do**
    if $r_{dest} = r_{cur}$
        Take packet and process it
    else
        if $S_{r_{cur}, r_{dest}} \geq th$
            Send the packet m directly to r_{dest}
        else
            Send the packet to all neighbors of r_{cur} except
            itself and the robot which packet came from.
        endif
    endif
endDo

---

5   Routing with Dijkstra Algorithm

Dijkstra's algorithm was created in 1959 by Dutch computer scientist Edsger Dijkstra. Dijkstra's Algorithm is a graph search algorithm that solves the single-source shortest path problem for a graph with non-negative edge path costs, producing a shortest path tree. This algorithm is often used in routing and other network related protocols [12].

Dijkstra's algorithm finds the shortest path from a given starting node to all of the other nodes in the graph. It requires that the weights of all edges are non-negative. It operates by maintaining a set of visited nodes and continually updating the tentative distance to all of the unvisited nodes. At each iteration, the closest unvisited node is added to the visited set and the distances to its unvisited neighbors are updated.

Algorithm: Finding optimum route from comstation to other robots

---

`Inputs:` $N = \{1,2,..\#Robots\}$, $S_{i,j}$ $i = 1..\#Robots$, $j = 1..\#Robots$
`Output:` $Path(n)$ $n = \{2..\#Robots\}$
$R = \{1\}$ `// Currently only robot1 (comstation) can be reach`
`from robot1`
Calculate $Cost(i,j)$ `using with equation (1)`
`for each` robot n **`in`** $N - \{1\}$ **`do`** `//initialization step`
    $TCost(n) = Cost(1,n)$
    $Path(n) = \{n\}$
`endfor`
`while` $R \neq N$ **`do`** `//` actual algorithm steps
    Find the robot m in $\{N - R\}$ that has minimum value for
    $TCost(m)$
    $R = R \cup \{m\}$
    `for each` robot n **`in`** $\{N - R\}$ **`do`**
        $TCost(1,n) = min(TCost(n), TCost(m) + Cost(m,n))$
        `if` $TCost(m) + Cost(m,n) < TCost(n)$
        $Path(n) = Path(m) \cup n$
        `endif`
    `endfor`
`endwhile`

Dijkstra's algorithm works for graphs with non-negative edges, but in our application robot signal strengths take zero or negative values. According to these values, if the signal strength is between 0 and -93, then the robot is in the coverage area. If the signal strength is below -93, then the robot is out of coverage area. In order to apply Dijkstra's algorithm in our application we normalized the robot signal strengths as in equation (1).

$$Cost(i,j) = \begin{cases} -93 \leq S_{i,j} \leq 0 & 1 - (S_{i,j}/1000) \\ S_{i,j} < -93 & \infty \end{cases} \tag{1}$$

In the above mathematical equation, $S_{i,j}$ is used to represent the signal strength value between ith robot and jth robot. $Cost(i,j)$ is used to represent link cost value between ith robot and jth robot which is used by the algorithm.

Let's consider $\#Robots$ is number of total robots, R is set of reachable robots, $Path(n)$ is routing path between comstation and nth robot. $TCost(n)$ is total cost of routing path between comstation and nth robot. According to these definitions, the steps in the algorithm are as follows [13].

In our application optimum route calculation is performed once by comstation in every 5 second. Then comstation sends updated path information to all robots periodically. Robots are using these paths for sending messages. Table-based algorithm which is used by robots in order to send messages in the network is as follow.

Algorithm: Routing with Table-Based Approach

```
Inputs: r_dest, r_cur, m, Path
when packet m is received do
    if r_dest = r_cur
        Take packet and process it
    else
        Get Path from packet m
        Get next robot node from Path
        Forward the packet m to the next robot node
    endif
endDo
```

6 Experimental Results

In this section, implemented routing algorithms are tested on two different propagation models of WSS. For all tests USARSIM and WSS software run on a PC that used as a server, test code ran on another PC that was used as a client. Client connected to the server through a router which has maximum 100Mps link speed.

6.1 Noop Propagation Model

For an increasing number of robots, the average message delays are tested for the situation that all the robots are directly connected with comstation, before testing the rates the routing algorithms transmitted messages to destination. By doing this the average message delays could be seen for different number of robots in the most ideal situation, independent of routing approaches. For test scenarios, except from comstation respectively 1 robot and 2, 4, 8, 15 robots are created. All the robots are keep in touch with comstation. Each robot continuously sends a 2048 byte message to

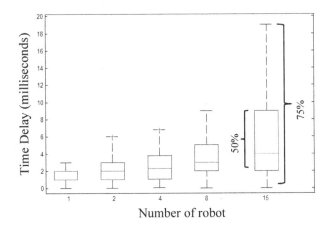

Fig. 2. Boxplot of message delay versus robots number in Noop Propagation Model

the comstation and once it sends a message, robot logs its id along with sending time. In addition, comstation logs the id of the message and its receiving time. Test continues till each robot sends 20000 messages to the comstation.

Fig.2 shows the boxplot diagram indicating the densities of the distribution of message delay times for varying the number of robots between 50% and 75% confidence interval, according to test results. When the test results are analyzed it can be said that in the situation that no routing method is needed, mainly when the Noop Propagation model is used, message delays are increased linearly with the number of robots. However, even with 15 robots no bad condition occurs while controlling the robots because the message delays have 11.04 average and 22.4 ms standard deviation at this situation.

6.2   Distance and Obstacle Propagation Model

In the Distance and Obstacle Propagation Model, robots are not in communication if their signal strength value is smaller than the threshold value. In this case, the robots' sensor information and comstation's controls commands are sent via other robots. Route selection process is very important for sending message packets to destination. In Blind Flooding method, each node sends the message packet received from other nodes to its neighbor. This method is known as baseline method in the literature. It is compared with table-based method which finds shortest path between robots dynamically and sends it to other robots periodically. At runtime, robots can be anywhere on the map and robots communication graph changes dynamically. It's obvious that whenever the graph structure changes, test results changes as well. In order to do a fair test 4, 8 and 15 robots are created except comstation and positioned on map as Fig.3. Graphs in Fig.3 are bidirectional and colored nodes represent the comstation. This test is done by calculating the number of packets sent by robots to the comstation per unit of time and packet delay. During the test comstation does not send any message to robots in blind flooding method. On the other hand, the table-based method, the comstation sends dynamically calculated route information to robots at specified intervals.

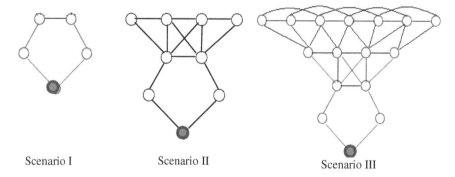

Scenario I   Scenario II   Scenario III

Fig. 3. Test scenarios

In a disaster area, researcher robot teams must send sensor information to the central operator as fast as possible. In this test optimum data transmission (packet number) is researched to prevent systematic network delay and packet losses caused by buffer overflows for both routing algorithm. Different delay time is set between sequential message packets to examine buffers in the network. For each of the three scenarios, packet success rates versus data transmission rates obtained with blind flooding algorithm are shown in Fig.4.

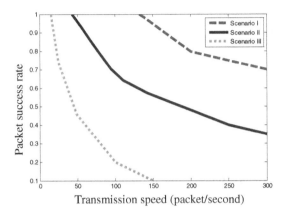

Fig. 4. Packet success rate versus packet/second rate for blind flooding algorithm

For all three scenarios even with maximum packet rates the system allows, any systematic delay or packet loss does not occur in the table-based routing method. Each tests carried on until each robot sends 5000 messages to the comstation. Total packet rate and packet rate per robot is shown as below.

Table 1. Results of maximum packet transmission speed

Scenario	Method	Total Packet Speed (pck/sec)	Packet Speed for each robot (pck/sec)
I	BF	132.45	33.11
I	TB	398.32*	99.58
II	BF	43.52	5.44
II	TB	312.76*	39.09
III	BF	15.16	1.01
III	TB	229.11*	15.27

In the table above, (*) represents the maximum packet rate allowed by the system. When table-based method is used neither packet losses nor systematic delay is observed. Therefore, rates with (*) signs are not actual rates; these rates may increase if faster data generated. The high number of unnecessary internal messaging in the network causes unnecessary bandwidth usage in Blind Flooding method. When Blind Flooding is used for routing, to avoid network buffer overflow each robot must generate no more than 1.01 packets/sec. This number is approximately 15 times greater in Table-Based routing.

In order to measure the delays of packets sent from robots to the comstation for each scenario, it's ensured that robots generate packets at rates obtained from previous analysis. Each test is carried on until each robot sends 5000 messages to the comstation and the delay of each message packet is calculated.

Table 2. Results of packet delay

Scenario	Method	Worst Robot mean delay (ms)	Best Robot mean Delay(ms)	Mean Delay (ms)	Standard Deviation of delay (ms)
I	BF	29.75	10.03	18.46	16.80
I	TB	10.10	5.27	7.67	6.58
II	BF	91.39	11.07	52.41	42.45
II	TB	20.56	7.27	15.32	10.46
III	BF	361.87	15.25	218.74	155.77
III	TB	42.10	13.16	30.40	17.36

Test results are shown in Table 2. According to the table it's seen that blind flooding (BF) gives similar results to table-based routing (TB) for lower number of nodes. But for bigger network structures (the more connection between nodes) the delay of BF method is increased exponentially. However for TB method this increase occurs linearly. The histogram of packet delays occurred with blind flooding and table-based method for Scenario III is as in Fig.5.

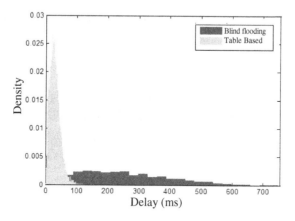

Fig. 5. Histogram of packet delay in scenario III using with Blind flooding and Table-based method

7   Conclusion

Ad hoc mobile robot communications are a promising networking technology for multi-robot communication in debris areas. Yet ad hoc robot communication represents a relatively underdeveloped application field. Blind Flooding approach can be preferred at situations with relatively few nodes, due to the simplicity of its application, its ensuring packets to reach target in any architecture (if there is no overflow in buffer). One of the advantage of this approach is does not need extra configuration messages. However, it must be cautioned that message delay increases with the square of number of connections and the slowdown in packet transmission speed. Message delay increase linearly with the number of connections between nodes and high packet transmission speeds are important advantages of table-based routing. However, as a requirement of its method, the messages including dynamic routing tables having to be sent to other nodes periodically consume extra bandwidth. When the frequency of these messages are decreased, the packet loss is increase because the robots find out routing table late or even does not. A good ad-hoc network routing algorithm can be developed with a table based approach by balancing this value in real or simulation debris areas.

References

1. Wang, Z., Zhou, M., Ansari, N.: Ad-hoc robot wireless communication. In: IEEE International Conference on Systems, Man and Cybernetics 2003, vol. 4. IEEE (2003)
2. Zeiger, F., Kraemer, N., Schilling, K.: Commanding mobile robots via wireless ad-hoc networks—A comparison of four ad-hoc routing protocol implementations. In: IEEE International Conference on Robotics and Automation, ICRA 2008. IEEE (2008)
3. Wahi, C., Sonbhadra, S.K.: Mobile Ad Hoc Network Routing Protocols: A Comparative Study. International Journal of Ad Hoc, Sensor & Ubiquitous Computing 3(2) (2012)
4. Wang, Z., Liu, L., Zhou, M.: Protocols and applications of ad-hoc robot wireless communication networks: An overview. Future 10, 20 (2005)
5. Nevatia, Y.: Ad-Hoc Routing for USARSim (2007)
6. Gorantala, K.: Routing protocols in mobile ad-hoc Networks. Master's Thesis in Computing Science (June 15, 2006)
7. de Morais Cordeiro, C., Agrawal, D.P.: Mobile ad hoc networking. Center for Distributed and Mobile Computing, ECECS. University of Cincinnati (2002)
8. Shrivastava, A., et al.: Overview of Routing Protocols in MANET's and Enhancements in Reactive Protocols (2005)
9. Carpin, S., et al.: USARSim: a robot simulator for research and education. In: 2007 IEEE International Conference on Robotics and Automation. IEEE (2007)
10. Pfingsthorn, M.: RoboCup Rescue Virtual Robots: Wireless Simulation Server Documentation, pp. 1–8 (October 2008)
11. Giudici, F.: Broadcasting in Opportunistic Networks, Universita Degli Studi di Milano, thesis (2008)
12. Dijkstra, E.: Dijkstra's algorithm. Dutch scientist Dr. Edsger Dijkstra network algorithm (1959), http://en.wikipedia.org/wiki/Dijkstra's_algorithm
13. Dijkstra's Algorithm (2013), http://www.mathcs.emory.edu/~cheung/Courses/455/Syllabus/5a-routing/dijkstra.html (accessed May 21, 2013)

Shape Based Round Object Detection Using Edge Orientation Histogram

Hamid Mobalegh, Lovísa Irpa Helgadóttir, and Raúl Rojas

Institut für Informatik, Freie Universität Berlin

Abstract. In this paper we introduce a shape based method to globally detect the ball in a RoboCup soccer scenario. The method can be used for any round object with detectable edges. The concept of integral images presented in Viola & Jones 2001, is used, however the integration is applied to a vector representation of the gradient orientation histogram of each pixel. The method takes advantage from the fact that large areas of the image can be filtered out, as these are only covered by straight edges. An overlapped binary search quickly reduces the search area and locates ball candidates in the image. The candidates are finally selected using an outlier elimination technique.

1 Introduction

Shape based object detection methods are often more reliable than methods that rely on color information. However, this is generally achieved at the cost of more processing power. The computational effort can be strongly reduced if the generality of the scenario is limited.

RoboCup Soccer is such an example. According to the RoboCup 2050 mission, robots shall be capable of tolerating outdoor lighting conditions and play with a standard FIFA ball. To reach this goal, the rules are shifted year by year to motive researches.

The work presented here is inspired by the previous work on *Histogram of oriented Gradients*[1] introduced by Dalal and Triggs[4]. Originally, the HOG algorithm focused on the problem of pedestrian detection in static images. Today it has expanded to other objects such as animals, vehicles and other media such as video streams. The method is of great importance as it is only based on the shape data of the image rather than other more environment-dependent information like brightness and color. The method we present here extends this idea by using integral images and an overlapped binary search. To accelerate the algorithm we were also inspired by intensity integral images presented in Viola et al. [12].

Ball detection without color information has been focused on in RoboCup competitions since its early years, as the RoboCup 2050 goal includes the use of a standard FIFA ball. The use of an arbitrary ball in RoboCup competitions is still limited to the "technical challenges" in many leagues. The biggest improvements

---

[1] HOG

S. Behnke et al. (Eds.): RoboCup 2013, LNAI 8371, pp. 326–335, 2014.
© Springer-Verlag Berlin Heidelberg 2014

have occurred in the middle size league. A descriptive survey on the developments in this league is given in [6]. Researchers have different approaches to tackle this problem. Hough transformation is used by [9],[1] and [7]. It is computationally expensive and cannot be used for a global search. Arc and Model fitting are used in [5] and [3]. Both solutions are iterative and need a prior knowledge of the ball position. Partial occlusion and clutter is well tolerated by these methods. Another group of methods focuses on Adaboost classifiers (see e.g.[11] and [8]). These methods have been shown to have a low false positive rate.

Many of the algorithms addressed above fail to perform globally, due to their computational complexity. A local ball search is however more relevant in middle size and small size leagues as the camera does not have fast motion and the position of the ball can be predicted fairly well. In contrast, in the humanoid leagues the camera moves very fast and a prediction is not possible due to high amount of measurement error. Therefore more focus is needed on global search methods.

The rest of this paper is organized as follows: First we show how a gradient image is calculated. A description of how a histogram of the edge orientation is constructed follows. Then we explain how a intensity integral images can be used to accelerate our algorithm and represent the original image. Next we describe the overlapped binary searching algorithm that find the possible ball candidates and the statistical test we use to eliminate outliers. Finally we show the results of the algorithm tested on images recorded on a RoboCup humanoid field.

2   Structure of Method

The method is structured as presented in figure 1. It includes five stages starting with the calculation of the gradient vector. This is common in many shape based object detection techniques. Using non-maxima-suppression, the edges are thinned and normalized. A so-called *"Histogram Integral Image"*[2] is then constructed based on the orientation of the gradient vector in each pixel. This representation of the original image helps to accelerate the search algorithm, especially, as it is possible to implement this part in hardware. An overlapped binary search recursively scans the pyramid down and finds the best-fitting box around the object using edge orientation statistics from the HII. The final result is then once more filtered using further statistical criteria.

2.1   Gradient Vector Calculation and Thresholding

The gradient vector can be calculated using one of the standard methods. For a better performance we use Robert's cross operator[10]. Strong edges are selected using a simple thresholding. Non-maxima-suppression is then performed along the gradient direction similar to the method used in canny edge detector [2]. This reduces the edge thickness and provides normalized results for the procedure introduced in section 2.5. Figure 2 shows the intermediate results.

---

[2] HII

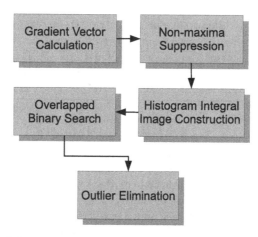

Fig. 1. Structure of the Ball Detection Method: First the gradient image is calculated. Then it is non-maxima-suppressed. The data is used to calculate the HII. An overlapped binary search locates the ball candidates. Finally outlier canidates are filtered based on statistical criteria.

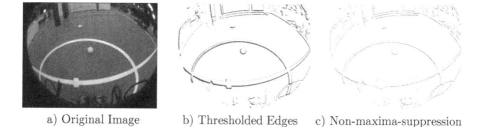

a) Original Image   b) Thresholded Edges   c) Non-maxima-suppression

Fig. 2. Edge Points Selected for Ball Detection Algorithm: Gradient of the input image is calculated. Strong edges are selected upon thresholding. Edges are thinned using non-maxima suppression.

2.2   Histogram of Edge Orientations

A histogram of edge orientations is an important measure used in shape-based object detection algorithms. In our method, the histogram presents the distribution of edge directions. The direction of the gradient vector is extracted in a 360° range, which is quantized into 18 groups of 20 degrees each. The detection algorithm relies indirectly on the fact that an ideal round object has a uniform distribution of edge orientations. However, this feature gets lost rapidly as the window size grows and other contents are added to the image. Additional contents of the image always add positive values to the histogram.

A direct result gained from this is that large regions of the image can be entirely rejected if the orientation histogram contains zeros in more than a given number of directions. Theoretically, an edge orientation histogram belonging to a window containing the round object must be zero free. This, however, cannot always be fulfilled in real test conditions due to shadows, partial occlusions or some other effects. Therefore a certain number of zeros are allowed in the histogram, which can be adjusted in accordance to the image condition.

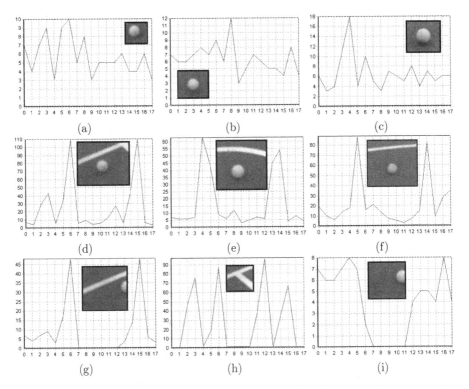

Fig. 3. Histogram of Edge Orientations: Examples of windows containing an entire ball (a-f), a partial view of the ball (g,i) or no ball (h). Only the examples without a complete ball (g-i) have bins containing zero pixels.

To visualize what kind of information an edge orientation histogram provides and how strong this measure is, several example histograms of the images captured from the robots are presented in figure 3. In figure 3a, b and c, the histograms of windows containing only a ball are presented. The histograms show more or less a uniform distribution over all directions. In figure 3c, d and e, the window size is enlarged and some other content is added. The histograms have lost the uniformness, however all components have remained non-zero. In figure 3g and i, the ball is partially visible which has lead to a significant number of zero components and finally the histogram of a ball-free window is shown in figure 3h. As it can be seen, a round object placed on the border of a window cannot be detected. A solution to this problem is discussed in section 2.4.

2.3   Histogram Integral Image

The idea of intensity integral images is introduced in [12]. In this work we extend
this idea to the histogram of edge orientation.

Assume an image with 18 channels, i.e. each pixel value is an 18 dimensional
vector. The vector describes the histogram of edge orientations according to a
rectangle, stretched from the origin of the source image to the given coordinates
as presented in figure 4a. This is described in equation 1.

$$\mathbf{I}(x,y) = \sum_{\substack{x\prime < x \\ y\prime < y}} \mathbf{H}(x\prime, y\prime) \tag{1}$$

$\mathbf{H}$ is a vector filled with zeros except for the component corresponding to the
direction of the gradient vector at $(x\prime, y\prime)$, which is filled with 1 if the pixel is
identified as an edge pixel.

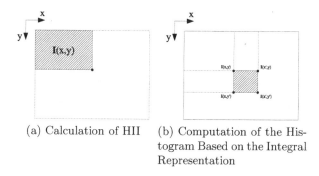

(a) Calculation of HII   (b) Computation of the His-
                           togram Based on the Integral
                           Representation

Fig. 4. Computation of Histogram Integral Image: Each entry in the HII is a vector
representing the edge orientation histogram of a rectangle from the origin of the image
to the indexed pixel. Based on this image a calculation of the histogram of an arbitrary
window is simply done using 4 vector additions.

According to [12], an integral image can be computed in a single scan by using
the following recurrence relation:

$$\mathbf{S}(x,y) = \mathbf{S}(x-1, y) + \mathbf{H}(x,y) \tag{2}$$

$$\mathbf{I}(x,y) = \mathbf{I}(x, y-1) + \mathbf{S}(x,y) \tag{3}$$

where $\mathbf{S}$ is a temporary vector holding a histogram of the current line of the image. It is enough to store $\mathbf{S}$ as a single accumulator because there is no reference
to its history.

Integral representation reduces the computation of the histogram for any given
window to two additions and one subtraction as follows:

$$\mathbf{Hist}(x,y,x\prime,y\prime) = \mathbf{I}(x,y) + \mathbf{I}(x\prime, y\prime) - (\mathbf{I}(x\prime, y) + \mathbf{I}(x, y\prime)) \tag{4}$$

This accelerates the operation to a great extent.

Algorithm 1. Overlapped Binary Search

Boolean search_ball(window, level)

begin

      if   window is already scanned then

            return   true

      window   <- scanned
      calculate_histogram(window)
      if   histogram has at least one zero component then

            return   false

      if   (level>5) then

            return   false

      b   <- false
      for   all sub windows

            b   <- b or search_ball(sub window, level+1)

      if   not b then

            push_ball_candidate(window)

end

2.4 Overlapped Binary Search

So far we have defined a method to determine the areas in the image that do
not cover an entire ball. However, this does not imply that an area contains a
ball if it is not rejected using this measure. Furthermore, it does not guarantee
that if the ball exists in the area, it is the only object surrounded by the area.
In this section a recursive function is suggested to find the best ball candidates
using the edge orientation histogram measure.

    The function is presented in algorithm 1.. It examines the given window using
the above described method. The function first verifies if the window could
contain a ball by scanning the histogram. If the window is not rejected it is
divided into several overlapping sub-windows, which are recursively processed

by the algorithm 1.. The base cases of the recursion are windows, which either have more than a certain number of zero components in their histogram or are smaller than the ball is expected to be. This is demonstrated in figure 5a.

Thanks to the overlapped search algorithm, no problem occurs when the ball lies on the border of neighboring sub-windows. The object can be entirely covered by at least one sub-window. As demonstrated in figure 5b, the sub-windows are a quarter of the area of the parent window and are distributed both horizontally and vertically, each one fourth of the parent window edge length. A window is thus divided into 9 sub-windows.

The overlapped searching can reference a window more than once. We there-fore suggests a look up table with an element for each possible window, down to the desired depth storing the search result for that window. A repeated reference can be detected at the beginning of the function and replied with the pre-stored search result.

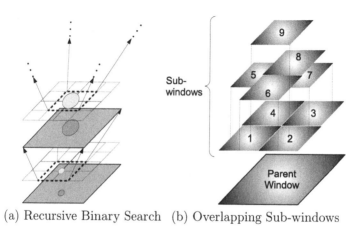

(a) Recursive Binary Search   (b) Overlapping Sub-windows

Fig. 5. Overlapped Binary Search: (a) The search recursion focuses on windows con-taining possible ball candidates and goes deeper until the size of the widow becomes smaller than the size of the ball. (b) A window is divided to 9 overlapping sub-windows.

2.5   Outlier Elimination

In accordance with algorithm 1., a ball candidate is found if all sub-windows of an accepted parent window are rejected to contain an entire ball. However the results are still subject to false positives. It is therefore required to further filter the output of the algorithm using a geometrical criterion. Two measures are suggested, both of which can be obtained from the histogram so that no further reference to the image is needed.

The first measure computes the standard deviation and the average of the histogram. The following condition verifies how uniform the distribution is.

$$\sigma < \alpha\mu \tag{5}$$

where σ is the standard deviation and μ is the average of the histogram. α is a constant, which determines the accepted uniformity of the distribution. The higher α becomes, the more candidates are accepted as balls. We set $\alpha = 2.0$ for our scenario.

The second measure verifies whether the number of edge points the window contains matches the circumference of the window. Assuming d to be the edge length of the window and n the number of edge pixels found in the window, the following is the condition to find balls.

$$\frac{\pi}{2}d(1 - \beta) < n < 4d(1 + \beta) \tag{6}$$

As window size is halved in each level, the window can be up to double the ball size. This sets the lower bound of the pixel count. The upper bound is set to the circumference of the window. β is an adjustable tolerance added to the condition. As a prerequisite to this condition the edge thickness should be reduced to one pixel using non-maxima-suppression.

2.6 Results

The algorithm was tested off-line using images recorded from RoboCup humanoid field. A data base of 3500 images from the recordings during RoboCup 2010 is used for the evaluation of the algorithm. The dataset includes images without balls, images including one ball as well as images containing more than one ball. Table 1 summarizes the evaluation results. As a performance measure, the number of references to the recursive function is counted. Comparing this to the number of pixels of the image (in this case 512x512) this shows a promising optimization of the search method. However, the initial image scan to calculate the *HII* should also be considered but authors intend to implement this stage into the hardware. Some example results are presented in figure 6.The results are overlaid with windows that the algorithm has accessed, which shows how the method approaches the image.

Table 1. Evaluation of the Ball Detection Algorithm

Property	Value	Unit
Number of Evaluation Images	3500	-
Total Number of Balls in Ground Truth (N)	2872	-
Detection Rate (TP/N)	88	%
False Positive Rate (FP/N)	9	%
False Negative Rate (FN/N)	12	%
Average recursive function calls	673	Calls/Frame

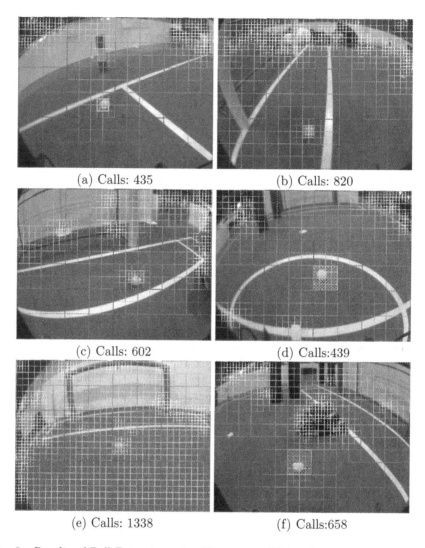

(a) Calls: 435   (b) Calls: 820

(c) Calls: 602   (d) Calls:439

(e) Calls: 1338   (f) Calls:658

Fig. 6. Results of Ball Detection using Histogram of Edge Orientations. Images are overlaid with the windows checked recursively by the algorithm. Detected ball candidates are presented as colored windows.

3 Conclusion

In this paper, we have presented a method for shape based round object detection using a edge orientation histogram. It can be used to globally detect the ball in the RoboCup soccer scenario, based only on the shape of the ball and is completely independent on factors like color and brightness. The image is represented using a histogram of the orientation of gradient vectors in each pixel. The representation can be implemented in hardware and accelerates the search algorithm considerably. A recursive search algorithm is then used to detect the ball in the image in relatively few function calls, compared to the number of pixels in the image. A statistical measure is used to eliminate most outliers. This method is fast due to few references to the image, and robust, as it has very few false positives.

References

[1] Bonarini, A., Furlan, A., Malago, L., Marzorati, D., Matteucci, M., Migliore, D., Restelli, M., Sorrenti, D.: Milan Robocup Team 2009. In: RoboCup 2009 Graz, Graz, Austria (2009)

[2] Canny, J.: A computational approach to edge detection. IEEE Transactions on Pattern Analysis and Machine Intelligence PAMI-8(6), 679–698 (1986)

[3] Coath, G., Musumeci, P.: Adaptive arc fitting for ball detection in robocup. In: Proceedings of APRS Workshop on Digital Image Analysing, Brisbane, Australia, pp. 63–68 (2003)

[4] Dalal, N., Triggs, B.: Histograms of oriented gradients for human detection. In: Proc. IEEE Computer Society Conf. Computer Vision and Pattern Recognition, CVPR 2005, vol. 1, pp. 886–893 (2005)

[5] Hanek, R., Beetz, M.: The contracting curve density algorithm: Fitting parametric curve models to images using local self-adapting separation criteria. International Journal of Computer Vision 59, 233–258 (2004)

[6] Li, X., Lu, H., Xiong, D., Zhang, H., Zheng, Z.: A Survey on Visual Perception for RoboCup MSL Soccer Robots. International Journal of Advanced Robotic Systems 1 (2013)

[7] Martins, D., Neves, A., Pinho, A.: Real-time generic ball recognition in RoboCup domain. In: Proc. of the Ibero- . . . , 11th edn. (2008)

[8] Mitri, S., Frintrop, S., Pervolz, K., Surmann, H., Nuchter, A.: Robust object detection at regions of interest with an application in ball recognition. In: Proceedings of the 2005 IEEE International Conference on Robotics and Automation, ICRA 2005, pp. 125–130 (April 2005)

[9] Neves, A.J., Pinho, A.J., Martins, D.A., Cunha, B.: An efficient omnidirectional vision system for soccer robots: From calibration to object detection. Mechatronics 21(2), 399–410 (2011)

[10] Roberts, L.G.: Machine perception of three-dimensional solids. Technical report, DTIC Document (1963)

[11] Treptow, A., Zell, A.: Real-time object tracking for soccer-robots without color information. Robotics and Autonomous Systems 48(1), 41–48 (2004)

[12] Viola, P., Jones, M.: Rapid object detection using a boosted cascade of simple features. In: Proc. IEEE Computer Society Conf. Computer Vision and Pattern Recognition, CVPR 2001, vol. 1 (2001)

RoboCup Logistics League Sponsored by Festo: A Competitive Factory Automation Testbed

Tim Niemueller[1], Daniel Ewert[2], Sebastian Reuter[2],
Alexander Ferrein[3], Sabina Jeschke[2], and Gerhard Lakemeyer[1]

[1] Knowledge-based Systems Group, RWTH Aachen University, Germany
{niemueller,gerhard}@kbsg.rwth-aachen.de
[2] Institute Cluster IMA/ZLW & IfU, RWTH Aachen University, Germany
{firstname.lastname}@ima-zlw-ifu.rwth-aachen.de
[3] Electrical Engineering Department, Aachen Univ. of Appl. Sc., Germany
ferrein@fh-aachen.de

Abstract. A new trend in automation is to deploy so-called cyber-physical systems (CPS) which combine computation with physical processes. The novel RoboCup Logistics League Sponsored by Festo (LLSF) aims at such CPS logistic scenarios in an automation setting. A team of robots has to produce products from a number of semi-finished products which they have to machine during the game. Different production plans are possible and the robots need to recycle scrap byproducts. This way, the LLSF is a very interesting league offering a number of challenging research questions for planning, coordination, or communication in an application-driven scenario. In this paper, we outline the objectives of the LLSF and present steps for developing the league further towards a benchmark for logistics scenarios for CPS. As a major milestone we present the new automated referee system which helps in governing the game play as well as keeping track of the scored points in a very complex factory scenario.

1 Introduction

A new trend in automation is to deploy so-called cyber-physical systems (CPS) to larger extents. These systems combine computation with physical processes. They include embedded computers and networks which monitor and control the physical processes and have a wide range of applications in assisted living, advanced automotive systems, energy conservation, environmental and critical infrastructure control, or manufacturing [1]. One application area of CPS are logistics scenarios in automation settings. As production is going to move away from mass production towards customized products, the challenges for the automation process will increase. This will open the floor to mobile robots in order to help with the manufacturing process.

In particular, mobile robots will be deployed for transportation tasks, where they have to get semi-finished products in place to be machined in time. This is right where the novel *RoboCup Logistics League Sponsored by Festo* (LLSF)

S. Behnke et al. (Eds.): RoboCup 2013, LNAI 8371, pp. 336–347, 2014.
© Springer-Verlag Berlin Heidelberg 2014

Fig. 1. A LLSF competition during the RoboCup 2012 in Mexico City

starts. Teams of robots have to transport semi-finished products from machine to machine in order to produce some final product according to some production plan. Machines can break down, products may have inferior quality, additional important orders come in and need to be machined at a higher priority. Due to increasing demands for flexibility production facilities will become dynamic environments, where shop floor layouts and the number, location and type of the engaged machinery change constantly. The robots therefore need to be able to identify these machines either visually or by direct communication. For the LLSF, a team consisting of up to three robots starts in the game area of about 5.6 m × 5.6 m. A number of semi-finished products is represented by RFID-tagged *pucks*. Each is in a particular state, from raw material through intermediate steps to being a final product. The state cannot be read by the robot but must be tracked and communicated among the robots of a team. On the playing field are *machines*, RFID devices with a signal light indicating their processing status. When placed on a proper machine type, a puck changes its state according to the machine specification which is communicated via broadcast messages. The outcome and machine state is indicated by particular light signals. During the game a number of different semi-finished products need to be produced with ten machines on the playing field. Orders are posted to the robots requesting particular final products to be delivered to specific delivery gates and in specified time slots. All teams use the same robot base, a Festo Robotino which may be equipped with additional sensor devices and computing power, but only within certain limits.

The LLSF has a number of very interesting research questions to be addressed. The robots need to be autonomous, detect the pucks, detect and identify the light signals from the machines, know where they are and where to deliver the final product to. These are the basic robotics problems such as localization, navigation, collision avoidance, computer vision for the pucks and the light signals. Of course, all these modules have to be integrated into an overall robust software architecture. On top of that, the teams need to schedule their production plan. This opens the field to various concepts from full supply chain optimization approaches to simpler job market approaches where a master robot gives sub-tasks to other robots. If the robots are able to prioritize the production of certain goods they can maximize the points they can achieve. In order to avoid resource conflicts (there can only be one robot at a machine at a time), to update other robots about the current states of machines and pucks, and to delegate (partial) jobs of the production process the robots must communicate. The only means allowed is via a wireless network. Since at RoboCup events there are so many wireless networks, connectivity cannot be guaranteed and is even likely to be interrupted every now and then. Hence robots must have both, useful cooperative behavior, but also sensible single-robot fallbacks. Robots that can recognize each other can have a decisive advantage. In the first phase exploration and roaming concepts must be implemented. A lot of these tasks present cutting edge research problems for the fields of factory automation and CPS, especially in the logistics sector. We intend to emphasize these problems in future tournaments to improve the attractiveness of the league as a benchmark for research and development in these fields.

The LLSF competition started in 2012 for the first time awarding a first RoboCup champion. In order to develop the league further, a number of rule changes have been proposed and implemented such as allowing more computing power on- and off-field, more diverse production plans or the introduction of a Technical Challenge. Another major novelty is the introduction of the referee box which, as in the other leagues, governs the automated game play. A particular benefit in the LLSF is the fact that the game is easy to observe automatically, and hence a large autonomy of the game play can be achieved. To foster the acceptance and implementation of the connection by the teams—while still being backward compatible for this year—certain aspects of the rules have been added and modified to grant bonus points for robots that can communicate with the refbox. For example, during the exploration phase, teams must announce their findings to the refbox, only then they can score.

This paper presents the LLSF as a competitive RoboCup league and details the objectives as well as the ensuing research questions. We outline the leagues' road-map for the next years which we intend to gradually increase in complexity, presenting the automated referee system as a major milestone. The rest of the paper is organized as follows. In the next section, we review cyber-physical systems and their connection to logistics problems arising in automation processes. In Section 3, we discuss the current version of the rules and detail the game objectives. In Section 4, we present the automated referee box (refbox). We first sketch the CLIPS rule base which the refbox is based on, before we describe the tasks and the implementation of the refbox in detail. After proposing possible future advancements of the league in Sect. 5 we conclude in Sect. 6.

2   CPS and Logistics Challenges

Digital devices have infused and changed the private and industrial world. Almost every process is supported, controlled or monitored by a digital device. Taking this a step further leads to the development of so-called cyber physical systems (CPS) [1]. Here, every entity of a system is equipped with suitable hardware and software for carrying relevant information and for communicating autonomously with its environment to exchange these information [2]. In this way, the entity is transformed from a passive element into an acting agent, which directly interacts with and manipulates its surrounding. This allows for decentralized control as well as increased flexibility and robustness of the affected processes. Examples for possible applications include a.o. the fields of advanced automotive systems, process control, manufacturing, and distributed robotics [3]. Especially logistic scenarios will benefit from CPS by promoting transparency and robustness due to decentralized control. Shipping goods or material can carry address information, handling instructions, or even directly instruct operating facilities. Additionally, due to throughout interconnections and built-in sensors, real-time tracking and tracing becomes possible in so far unrivaled accuracy and quality. However, to allow for large-scaled, reliable, and flexible CPS, research must still find answers to a number of challenges. According to [4] these include finding solutions for

- situation recognition,
- planning and anticipatory, partially or completely autonomous behavior,
- cooperation and negotiation,
- strategies of self-organization and adaptation.

For the application of CPS in manufacturing systems, the requirements for communication within the smart factory are detailed in [5]: CPS must allow for horizontal and vertical communication within the automation pyramid. *Vertical communication* refers to information exchange between the different levels of manufacturing management and control. Nowadays, these are associated to organization-wide enterprise resource planning systems (ERP), manufacturing execution systems (MES) for detailed production planning across machines and the underlying supervisory control and data acquisition (SCADA). While the boundaries between the levels are not fixed, in the long run CPS are expected to completely obliterate them. CPS must therefore be able to exchange information on different levels of abstraction. Thus, non-hierarchical communication structures will evolve which shift the production from a centralized production control to a production controlled by the products which are to be manufactured [6]. *Horizontal communication* refers to data exchange on the same level along the production chain. Thereby, information is handed to downstream production processes which will adapt their production parameters to the individual characteristics of the currently manufactured product. As we will show in the following sections, these challenges are similar to the ones faced in the LLSF.

3   Game Rules and Objectives

As presented in Sect. 2, the deployment of CPS in the logistics of production sites opens a new field of challenges for factory automation. To recommend the LLSF as a suitable testbed for research and education regarding CPS in logistics, in Sect. 3 we will present the ideas of the LLSF in more detail.

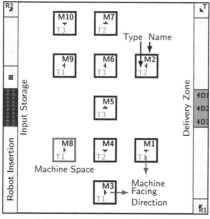

(a) LLSF Competition Area

(b) Machine with light signal, RFID device, and game puck

Fig. 2. LLSF Competition Area and field machine

In 2012 the Logistics League Sponsored by Festo (LLSF) was officially founded. The general intention is to create a simplified and abstracted *factory automation scenario* with an emphasis on logistics applications. Teams of up to three robots operate in a fenced area of about 5.6 m × 5.6 m as shown in Fig. 2(a). Fig. 3 shows the original Festo Robotino and a modified version of the Carologistics RoboCup team[1]. The task is to complete a production chain by carrying a (semi-finished) product (a puck in the game) along different machines (signal lights on the playing field as shown in Fig. 2(b)). Points are awarded for intermediate and completed products.

On the field two margin areas on opposite sides contain the puck input storage, delivery zone, and several machines that deal as delivery gates for the final products or as recycling stations. Each puck has a programmable radio frequency identification (RFID) chip with which the different product states S_0, S_1, S_2, and P_1, P_2, P_3 are distinguished. Initially, all pucks are in state S_0. In the enclosed inner field, ten signals equipped with an RFID device mounted on its front represent production machines. Each machine is assigned a random but defined type out of the types T_1–T_5, which is initially unknown to the robots. The type determines the input and output of a machine. Pucks transition through their states by being processed through machines. The complete production tree is shown in Fig. 4. Circular nodes indicate a puck's state and rectangular nodes show the respective machine type. For example, the T_1 machine in the upper branch takes an S_0 puck as input with an S_1 puck as output. If a machine, like T_2, requires multiple inputs, these can be presented to the machine in any order. However, until the machine cycle completes, all involved pucks must remain in

---
[1] http://www.carologistics.org

(a) Standard Festo Robotino   (b) Modified Robotino of the Carologis-
                                     tics RoboCup team

Fig. 3. Standard Festo and modified Carologistics Robotino

the machine space. The last input puck will be converted to the output puck, all others become junk and must be recycled at a recycling station.

The machines indicate their state after processing a puck using light signals. A green signal means that the particular machine production cycle has been completed, i.e., all required input products have been presented one after another, and now the puck has been transformed to the machine's respective output; for instance, after a T_1 machine transformed a puck from state S_0 to S_1, an orange light indicates partial success (more pucks are required).

Besides typical robotics tasks such as motion planning or self-localization, the robot needs to plan an efficient sequence of actions to produce as many products as possible in a fixed amount of time. Moreover, the robot has to deal with incomplete knowledge as it is not known in advance what machine has which type. Thus, the robots need to combine *sensing and reasoning* to incrementally update their belief about the world. Based on the knowledge gained, it has to find a strategy to maximize its production output, ideally minimizing costs such as travel distance.

4   Autonomous Referee Box

The LLSF game comprises a rather constrained task for now. Yet overseeing the game requires tracking of more than 20 pucks and their respective states, watching machine areas of 10 machines to detect pucks that are moved out of bounds, placing late order pucks for visual triggering at specified places at certain times, and overseeing completion of the production chain awarding points and keeping a score. This can easily overwhelm a human referee and make the competition hard to understand for a visitor. In fact, in 2012 we needed to review a camera recording of a game to award points in hindsight because a situation was overseen by the two human referees. Therefore, we strive for the implementation

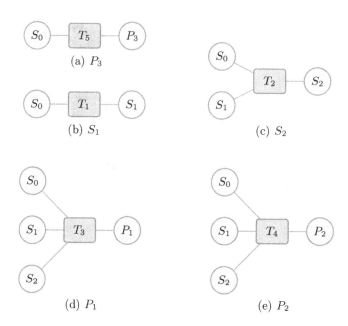

Fig. 4. Production Chain Diagrams showing the machines and inputs relative to their outputs

of a (semi-)autonomous referee box (refbox). The refbox shall control, monitor, and score the overall game. The human referees instruct the refbox (e.g., to start or pause the game), which then communicates with the robots on the field. The humans act as a second tier referee correcting for misjudgment of yet unforeseen situations. Additionally, full autonomy cannot be achieved, yet, because there is no sensor to detect pucks moved out of a machine area.

In the following we will first introduce CLIPS, a logic-based system used to implement the core functionality of the refbox and to formally describe the competition rules. We then give details about the implementation of the refbox.

4.1   The CLIPS Rule-Based Production System

CLIPS is a rule-based production system using forward chaining inference based on the Rete algorithm [7]. The CLIPS rule engine [8] has been developed and used since 1985 and is thus mature and stable. It was designed to integrate well with the C programming language[2], which specifically helps to integrate with the refbox. Its syntax is based on LISP as it was conceived to replace LISP-based systems. The Carologistics team has already used CLIPS to implement the task coordination component [9]. Therefore it appeared suitable for the refbox implementation.

---

[2] And C++ using clipsmm, see http://clipsmm.sf.net

CLIPS has three building blocks [10]: a fact base or working memory, the knowledge base, and an inference engine. *Facts* are basic forms representing pieces of information which have been placed in the fact base. They are the fundamental unit of data used by rules. Facts can adhere to a specified template. It is established with a certain set of slots, properties with a specified name which can hold information. The *knowledge base* comprises heuristic knowledge in the form of condition-action rules, and procedural knowledge in the form of functions. *Rules* are a core part of the production system. They are composed of an antecedent and consequent. The antecedent is a set of conditions, typically patterns which are a set of restrictions that determine which facts satisfy the condition. If all conditions are satisfied based on the existence, non-existence, or content of facts in the fact base the rule is activated and added to the agenda. The consequent is a series of actions which are executed for the currently selected rule on the agenda, for example to modify the fact base. *Functions* carry procedural knowledge and may have side effects. They can also be implemented in C++. In our framework, we use them to utilize the underlying robot software, for instance to communicate with the reactive behavior layer described below. CLIPS' *inference engine* combines working memory and knowledge base. Fact updates, rule activation, and agenda execution are performed until stability is reached and no more rules are activated. Modifications of the fact base are evaluated if they activate (or deactivate) rules from the knowledge base. Activated rules are put onto the agenda. As there might be multiple active rules at a time, a conflict resolution strategy is required to decide which rule's actions to execute first. In our case, we order rules by their salience, a numeric value where higher value means higher priority. If rules with the same salience are active at a time, they are executed in the order of their activation, and thus in the order of their specification. The execution of the selected rule might itself trigger changes to the working memory, causing a repetition of the cycle.

4.2   Tasks of Referee Box

The refbox has several tasks it must fulfill. It must control the game, communicate with the robots, represent the current state of the game, and interface with the devices on the playing field. We will now briefly explain the four tasks before we describe some aspects of the implementation.

- *Control.* The refbox must oversee the game implementing the rules defined in the rule-book[3]. For this very purpose we use CLIPS. This part is also responsible for awarding points if the robots accomplished a (partial) task.
- *Communication.* It must communicate with the robots on the field to provide information, send orders, and receive reports.
- *Representation.* A textual or graphical application is required to visualize the current state of the game and to receive command input from the human referees. A simplified visualization can be used to explain the game to visitors.

---

[3] The current rules can be found at http://www.robocup-logistics.org/rules

– *Interfacing.* The referee box needs to communicate with the programmable logic controller (PLC) which is used to set the light signals and read the RFID sensors on the pucks.

4.3   Implementation

The referee box has been implemented by members of the LLSF TC. Its infrastructure is written in C++ and the game controller core in CLIPS. It uses Boost for some of its internals, for example asynchronous I/O and signal propagation.

The base program creates the environment for the CLIPS core, in which the actual game controller is implemented. This core is a knowledge-based system. The facts in the working memory are used to keep track of the state of the game and to communicate within the core. Rules trigger on specific conditions and events, for example the reception of a message, or the completion of a production cycle of a machine. A time fact is periodically asserted (currently at 25 Hz) to allow for time-based triggering, such as in the case of the production completion. This allows us to specify durative actions.

There are currently two interfaces to represent the game state and to accept commands. A textual shell which uses the ncurses library is used for quick operation by the human referee. It shows the most important information and accepts commands. A graphical user interface (GUI) has been implemented using the Gtkmm library. It features a visual display of the playing field and will be focused on visualization and explanation of the game to the audience in 2013.

There are two ways of communication: client-server stream connections using TCP, and peer-to-peer broadcast communication using UDP. This might be changed to multicast at a later point in time. The stream connections are used to connect the refbox tools like the shell or GUI. Broadcast communication is used to communicate with the robots. Both protocols use Google protocol buffers (protobuf) for message specification and serialization. A small framing protocol allows for transmitting messages of different types over the same connection. This is particularly important in RoboCup, where network resources are scarce and combining messages and reducing the number of connection handshakes is beneficial. Using protobuf for message specification gives a very efficient serialization in terms of message size, and allows for forward compatibility, i.e. older clients can still read the messages as new fields are added, they are simply ignored. It also allows for optional fields, further reducing the amount of data required to be sent.

The refbox has been released as Open Source software and can be downloaded at `http://www.robocup-logistics.org/refbox`. The page also contains links to the documentation of the refbox.

5   Gradual Advancement of the Logistic Scenarios

In this section we will propose possible advancements of the LLSF to keep the league challenging for existing teams while preserving a sufficiently low entry barrier for new teams. The basic idea for developing the league further is to gradually increase the complexity of the game. The objective is to create a benchmark for logistics tasks scenarios addressing the challenges mentioned in Sect. 2. Therefore, we need to push the boundaries towards more realistic scenarios in the following years. We envision challenges in the following areas.

Basic Robotic Problems. For now, only one team is in the arena at a time. Mostly it is sufficient to navigate with odometry or correct the pose estimate by the unambiguous positions of the observable landmarks. At some point, an opposing team could be in the arena at the same time. It could even try to obstruct machines and hinder the opposing team (within certain limits). With opponents, the whole setup becomes much more complicated and teams need to have solutions for increased situation awareness, collision avoidance, navigation, and self-localization.

Mobile Manipulation. A realistic logistics robot will most probably be equipped with a handling device. In the next few years, the pucks will not only be pushed around by the robots, but also have to be lifted up and put into the machine or a conveyor belt. The possibilities to extend the concept of a "machine" as we have it right now are manifold. They would need to be in such a way that a manipulator arm or a fork lift mechanism on the robots needs to be installed. The interaction with machines must not stay limited to one interacting robot. Also cooperative scenarios, where one robot moves the puck, while a second simultaneously activates the machine in a given way or tasks which have to be executed nearly in parallel on different machines are imaginable.

Multi-robot Coordination. At this time multiple robots can be advantageous if they split the serial task and complete parts of it with each robot. In the future, we could require multiple robots to cooperate to be able to complete the task at all. For example, two goods may be required to be delivered to two machines in a small time window. Also cooperation of multiple teams would be very interesting, as it would require to agree on a common communication protocol. For instance, one team could produce an item that another has to deliver, and only completing both steps by robots of different teams would score.

Logistics Management. The game setup allows for numerous interesting scenarios for logistics management. While the current task still is fixed in many aspects, this should change in the future. Production tasks, machine and product specifications could be changed dynamically, as well as a shift of the optimization goals from a purely market driven maximization of the production output to a cost driven maximization of machine utilization. Also the playfield structure should allow for changes by rotating machines as a first step towards a change of the factory layout, e.g., from a workshop-based layout to a flow-production system. While a part of these changes is communicated by the refbox (see Sect. 4), other changes have to be recognized by the robots. Competitive teams should have to build more autonomous robotic agents that can cope with such an environment are able to exchange relevant information and react as a team.

Robot Hardware Restrictions. Currently, the Robotino robot by Festo is the only legitimate robot platform that can be used for the game. Step-wise, hardware restrictions could be dropped to move towards more realistic scenarios. One way for the league could be to proceed towards a number of autonomous fork-lift systems in a real factory scenario.

Referee Box Development. The current referee box is the first of its kind in the RoboCup domain which has achieved such a high level of autonomy. For the future we want to further increase this autonomy. By adding an overhead camera over the field, for instance, we could automatically track the pucks and robots on the field. This would put the refbox in a position to automatically declare foul play if a puck is moved out of a machine area while it was still needed.

The rules should be implemented in a *rule tic-toc* way. This means that major changes to the rules only become implemented every two years. Major changes will be announced in one year, but only the year after this rule will come into effect. This will give all the teams enough time to tackle the new problem and work towards solutions that are required by the rule change.

Another way to push boundaries further is the introduction of Technical Challenges. In the 2013 competition, the TC announced three technical challenges where teams need to show their technical skills in the field of robotic navigation, localization and computer vision. As in other RoboCup leagues, participation in the Technical Challenge will not be mandatory, but teams will very likely use this possibility to showcase their latest development there. This way, after one particular Technical Challenge has been solved by most of the teams, it can be replaced by a more complicated task. Additionally an open challenge, where teams can show a task of their choosing, can serve as a showcase to propose ideas for future development of the league and its rules. This concept has already been applied successfully to other RoboCup leagues with a strong Technical Committee which is pushing boundaries from year to year.

6   Conclusion

In this paper we gave an overview of the new RoboCup Logistics League Sponsored by Festo (LLSF). The idea is to simulate a factory automation scenario with mobile robots. A team of robots has to "machine" different semi-finished products in order to produce some end-product and then deliver it at a certain delivery point. The team needs to identify machines, derive feasible production plans and realize them in due time. Additionally, they need to cope with failures in the production process that may occur, products need to be recycled and high priority orders may be placed arbitrarily. All this must be dealt with to optimize the outcome of the factory (and earn as many points as possible). With this league, exactly the challenges that arise in logistics setting of cyber-physical systems (CPS) are covered. The requirements of the systems are that they enjoy among others planning abilities and autonomous behavior, self-organization strategies or cooperation facilities. Therefore, the LLSF is very well suited to become a benchmark for logistics scenarios of CPS.

We outlined the rules of the league and presented a road-map which will develop the league further towards this eligibility as a benchmark. Requirements towards cooperation and communication, both within the team and with external entities will play an increasing role. A number of rule changes have already been proposed such as allowing more computing power and fostering autonomous behavior with several technical challenges. A major improvement, which we described is the introduction of the referee box, an autonomous referee for the LLSF. Different than in other leagues where referee boxes are mainly used to control the flow of the game, in LLSF the referee box keeps track of the machines types and the produced items and awards the points to the teams as the factory scenario is very complex for a human referee to follow. In the near future, we also want to extend the referee box with an automated robot and product tracking system to take another step towards the automated factory.

Acknowledgments.   We thank the anonymous reviewers for their helpful comments.

References

1. Lee, E.: Cyber Physical Systems: Design Challenges. In: 11th IEEE International Symposium on Object Oriented Real-Time Distributed Computing (ISORC 2008), pp. 363–369 (2008)
2. Marwedel, P.: Embedded System Design. Springer (2006)
3. Lee, E.A.: Computing foundations and practice for cyber-physical systems: A preliminary report. University of California, Berkeley. Tech. Rep. UCB/EECS-2007-72 (2007)
4. Broy, M., Cengarle, M.V., Geisberger, E.: Cyber-physical systems: Imminent challenges. In: Calinescu, R., Garlan, D. (eds.) Monterey Workshop 2012. LNCS, vol. 7539, pp. 1–28. Springer, Heidelberg (2012)
5. Vogel-Heuser, B., Witsch, M. (eds.): Increased Availability and Transparent Production. Embedded Systems - I Tagungen und Berichte, vol. 2. Kassel University Press, Kassel (2011) (in German)
6. Hensel, R.: Industry 4.0 revolutionizes production. VDI Nachrichten (2012) (in German)
7. Forgy, C.L.: Rete: A fast algorithm for the many pattern/many object pattern match problem. Artificial Intelligence 19(1) (1982)
8. Wygant, R.M.: CLIPS: A powerful development and delivery expert system tool. Computers & Industrial Engineering 17(1-4) (1989)
9. Niemueller, T., Lakemeyer, G., Ferrein, A.: Incremental task-level reasoning in a competitive factory automation scenario. In: Designing Intelligent Robots: Reintegrating AI II, Papers from the 2013 AAAI Spring Symposium. AAAI Technical Report, vol. SS-13-04. AAAI (2013)
10. Giarratano, J.C.: CLIPS Reference Manuals (2007), http://clipsrules.sf.net/OnlineDocs.html

Parallel Computation Using GPGPU to Simulate Crowd Evacuation Behaviors: Planning Effective Evacuation Guidance at Emergencies

Toshinori Niwa, Masaru Okaya, and Tomoichi Takahashi

Meijo University, Aichi, Japan
{e0930064,m0930007}@ccalumni.meijo-u.ac.jp, ttaka@meijo-u.ac.jp

Abstract. We propose parallel computing to simulate crowd evacuation behavior. It allows evacuation of ten thousands of agents to be simulated faster than does the existing system. Our prototype system consists of a new traffic simulator and scenario generator. The traffic simulation system uses a general purpose graphics processing unit (GPGPU) and simulates the agents' movements in a three-dimensional map. Our proposal enables realistic evacuation simulations and provides a platform that widens the applications of RoboCup Rescue Simulation to, for example, crowd evacuation from buildings. The evacuation simulations help security offices to prepare manuals for emergencies.

1 Introduction

The RoboCup Rescue Simulation (RCRS) system was designed in 1998 and the Rescue Simulation league started using its version 0 system [1]. The objectives of RoboCup Rescue Project are

- to apply agent technology to social problems and contribute human welfare,
- to provide a practical problem for development of research fields,
- to promote international research collaboration via the project.

RCRS aims to simulate rescue operations and human behaviors in disaster situations over an area of a few km^2 considering the number of people and facilities that exist in the area. The simulation results will be put to practical use during disasters and even before a disasters occurs.

It has been recognized that there are issues related to achieving the objectives that were set at the beginning of the RoboCup Rescue project [2]; that is it cannot simulate the behavior of a large number of agents or behaviors inside buildings. They are key issues for rescue simulation systems to model realistic disasters. Daidaitoku project tried to solve these issues [3]. The project aimed to simulate the evacuation of 10,000 people and operations to rescue them from a real 4km^2 area. Their system divided the large area into small areas. The size

S. Behnke et al. (Eds.): RoboCup 2013, LNAI 8371, pp. 348–359, 2014.

of the divided area was the order of RCRS. The project revealed that communication between agents in different areas impeded the efficient simulation of 4km^2.

In the RoboCup Rescue community, various approaches have been proposed and implemented to increase the scale of simulation: the number of agents, the size of area, etc. [4][5]. In 2009, at the Infrastructure competition of the Rescue Simulation League, a Java based kernel and a new traffic simulator were proposed. They were adopted as the RCRS version 1 system. However, the basic architecture of the RCRS version 1 remained the same as that of version 0. The number of agents remains at the order of a few hundred and it takes a lot of computing resources, time and memory to simulate disaster situations in which a large number of agents and evacuations from three dimensional (3D) buildings were involved.

We propose a system that can simulate situations where more than thousand agents start their evacuations by announcements from emergency broadcast system and share information on the emergency by communications. The system uses a general purpose graphics processing unit (GPGPU) and allows rescue and evacuation simulation to be performed faster than does using only CPU when the number of agents is large. In addition, it represents the agents' simulated movements on a 3D rather than a two-dimensional (2D) map. Section 2 overviews RCRS's architecture and lessons from past cases. Our GPGPU-based system, and parallel computations of the agents' behaviors are described in Section 3. In Section 4, the results of two simulations are shown. The possibility that the evacuation simulations can help security offices prepare manuals for emergencies is discuss in Section 5.

2   Lessons from Past Evacuation Cases and Overview of RCRS's Architecture

2.1   Lessons from Past Evacuation Cases

At the beginning, the RCRS was designed to simulate rescue operations at disaster situations that were modeled after the Hanshi-Awaji earthquakes at 1995 [2]. The importance of evacuations has been reaffirmed by reports on following disasters: the September 11, 2001 attacks and the Great East Japan Earthquake that occurred on March 11, 2011, along with the ensuing tsunami, took many lives and caused serious injuries [6][7].

Evacuation announcements greatly influence human behavior when they start evacuation and during their egress from buildings. Human behavior at emergencies are different. According to the reports on the September 11 and the tsunami 2011; some people evacuated immediately when the disasters occurred, while others did not evacuate, even when they heard emergency alarms from authorities. In the case of the two World Trade Center (WTC) buildings, WTC1 and WTC2 had similar layouts and size. Similar numbers of people were in the buildings during the attacks. However, the evacuation behaviors in each building were different. People in both buildings started evacuation when WTC1 was

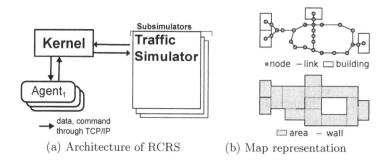

(a) Architecture of RCRS   (b) Map representation

Fig. 1. Architecture and map representation of RoboCup Rescue simulation. (In version 0, network representation was used; in version1, open area representation was adopted.)

attacked. After 17 minutes, when WTC2 was attacked, about 83% of survivors from WTC1 remained inside, while 60% of survivors remained inside WTC2.

It is important to egress safely from buildings and to do rescue operations quickly during emergencies. Evacuation drills have been conducted at schools and shopping malls with intent to smooth evacuations and to conduct rescue operations properly. The evacuation drills are used to estimate the required safe egress time and to improve prevention plans for predictable emergent situations

2.2   Overview of RCRS's Architecure

While the RCRS was expected to simulate situations where 10,000 agents were involved, the number of agents remains a few hundred. Figure 1 (a) shows the basic architecture of RCRS, which consists of a kernel, sub-simulators and agents. The kernel, sub-simulator and agents are connected using the TCP/IP protocol, and therefore they can be executed on one computer or on several computers that are connected to via a network. This architecture allows RCRS to be run on distributed computers when simulations require significantly large computer resources. In fact, at the beginning of the Rescue Simulation league, several computers were used to run a simulation of RCRS. However, the execution of one sub-simulator is confined in one computer, which makes it difficult to simulate disasters at bigger area and with a lot of agents involved.

Figure 1 (b) shows two kinds of map representations. The upper one is a network representation and the lower one is an open area map representation. The network representation was used in RCRS version 0 to simulate the movements of agents from a building to refuges. The nodes of the net representation are intersections and buildings. The links represents roads which has the number of traffic lanes as its property. In the open area representation, roads and floors inside buildings are represented as polygons. The edges of the polygons represent walls or boundaries to adjoining areas. The traffic simulator of version 1 is one of sub-simulators and calculate the positions of the agent in one area and the translation from one are to the other area.

The kernel exchanges data and commands between agents and sub-simulators via kernel at every simulation step (Δt). The simulation step is the time of an agent's sense-decision-action cycle and its default values is one minute in the RCRS. The traffic simulator simulates the agents' movements in one simulation step by calculating their locations in the area at $\Delta \tau$ step, which is finer than a Δt step, $\Delta t = N_{step} \times \Delta \tau$. In the default setting, N_{step} is 6,000.

3   GPGPU Based Simulation System

3.1   Parallel Computation in Calculating Agents' Positions

Our idea is to use GPGPU in calculation of agents' positions, $\mathbf{r}_i$, in a Δt. The changes of agent's position are calculated at evey micro step $\Delta \tau$ according to the Helbing social force model [8]. A social force model is a particle model in which human movements are controlled with Newton dynamics:

$$m_i \frac{d\mathbf{v}_i}{dt} = m_i \frac{v_i^0(t)\mathbf{e}_i^0(t) - \mathbf{v}_i(t)}{\tau_i} + \sum_{j(\neq i)} \mathbf{f}_{ij} + \sum_W \mathbf{f}_{iW} \tag{1}$$

where the first term is the intention-driven force that moves the agent to their desired positions. The agent's intentions determine positions that they want to go and $\mathbf{e}_i^0$ is a unit vector to the target places. m_i and v_i^0 are the weight and the speed of walking of agents i, respectively. The speed is set according to the age and sex of the agent and becomes faster when the agent is afraid and zero when it arrives at its destination. $\mathbf{v}_i(t)$ is a walking vector at t. $\mathbf{f}_{ij}$ and $\mathbf{f}_{iW}$ are repulsion forces to avoid collision with other agents or walls, respectively. The resultant force is the change of velocity, $\mathbf{v}_i(t) = \frac{d\mathbf{r}_i}{dt}$, and the change is used to calculate the next position of the agent after Δt.

Algorithm 1 shows a typical of traffic simulator code. The computations of the agents' position are performed for every agents and are iterated N_{step} times in one Δt. Simulations involving a large number of agents requires computations of $O(n^2)$ where n is the number of agents, because there are interaction terms between one agent and other agents that are around. More than 1,000 people are involved in real situations, and therefore many computer resources are required: CPU power, memory, and communications among the agent. Algorithm 2 shows a code that uses parallel computation. Codes from line 5 to 12 are computed in parallel on multi-threads. Line 1, 4 and 13 are data conversion and read/write to parallel computations.

3.2   Prototype System of GPGPU-Based Traffic Simulator

Figure 2 shows the architecture of our system, which is a subset of RCRS and specializes in crowd evacuation in emergencies. Our prototype system consists of three components: scenario generator, 3D traffic simulator, and agents:

Algorithm 1. Computation with one thread:

1: call micro-step function N_{step} times.
2: **function** MICROSTEP
3:     n is the number of agents.
4:     **for** $i = 0, \cdots, n$ **do**
5:         calculate own intention-driven force and list walls near agent$_i$
6:         **for** agent$_j$ near agent$_i$ **do**
7:             calculate $\mathbf{f}_{ij}$
8:         **end for**
9:         calculate $\mathbf{f}_{iW}$
10:        calculate next candidate position of agent$_i$.
11:    **end for**
12:    calculate next position of all agents and return.
13: **end function**

Algorithm 2. Parallel computation with multiple threads:

1: call initialization data conversion function for OpenCL.
2: call micro-step function N_{step} times.
3: **function** MICROSTEP
4:     transfer data to device and calculate following on parallel M threads.
5:     for $k(0, \cdots, M-1)$ thread,
6:     **for** $i = k \times n/M, k \times n/M + 1, \cdots, k \times n/M + n/M - 1$ **do**
7:         calculate own intention-driven force and list walls near agent$_i$
8:         **for** agent$_j$ near agent$_i$ **do**
9:             calculate $\mathbf{f}_{ij}$
10:        **end for**
11:        calculate $\mathbf{f}_{iW}$ and next position of agent$_i$.
12:    **end for**
13:    read data from device and calculate next position of all agents
14:    wait termination of all thread and return
15: **end function**

---

scenario generator: It plays the same role as the kernel of RCRS and controls the related 3D data and commands; for example, oral communication and vision are limited on the same story of the building. The initial positions of agents are also set as three dimensional coordinates by this scenario generator.

3D traffic simulator: It is based on the traffic simulator of version 1 which we developed further. The new features of the traffic simulator are (1) it can handle agents' movement in three dimensional buildings, and (2) it can use GPGPU to compute the calculations of the social forces of agents and the behaviors of mass agents in parallel at micro-steps.

agent: It is the same as the one used in RCRS version 1. It receives sensor data and decides its action based on their Belief-Desire-Intention (BDI) model at every simulation step (Δt). The BDI model features the evacuation behaviors of the agents. [9].

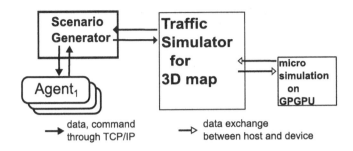

data, command
through TCP/IP

data exchange
between host and device

Fig. 2. Architecture of prototype evacuation system

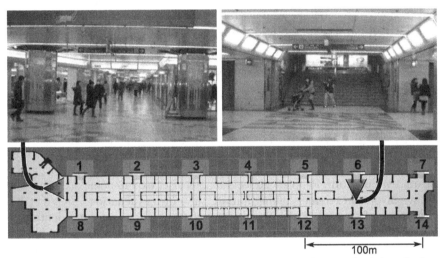

(a) Layout of subterranean shopping mall: (Left photo is taken at the left end of one street and right photo is at exit 6. The numbers indicate exits to ground level.)

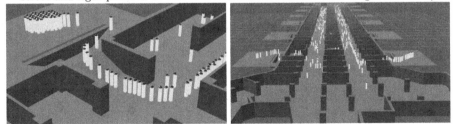

(b) Snapshot of evacuation simulation from a subterranean shopping mall. (left: a stair way to ground level from street in the mall. right: a birds-eye view of the mall.)

Fig. 3. Simulations of evacuations at subterranean shopping mall

4   Evacuation Simulation Experiments

4.1   Evacuation Scenario

Simulations of crowd's evacuations are useful to make a good prevention plan for emergencies. Figure 3 (a) shows the layout of the subterranean shopping mall in our city, Nagoya. The length and width are about 300m and 30m respectively, and there are about 90 shops are in three rows; two main streets pass between the rows. Exits to the ground are located at intervals of 50m.

At big earthquakes, the city office assumes that many pepole stay at this area. For security offices, it is important to guide people in the mall promptly to the ground at emergencies. Figure 3 (b) is a snapshot of the simulation of evacuating 1,000 agents to ground level from a subterranean shopping mall. We show two simulation experiments: one demonstrates an usage of simulation and the other is the effect of parallel computation using GPGPU.

4.2   Experiment 1: Effect of Guidance to Evacuation

This simulation shows how announcements from the security office effect the time of evacuations. The simulation scenario was as follows. A thousand agents are randomly located in the mall. The behaviors of the agents are modeled according to the report of the Great East Japan Earthquake that shows three types of human reactions when alarms are given.

instant evacuation: People who feel anxious after experiencing accidents that have involved extreme shaking initiate their own evacuation.
evacuation after tasks: People who do not feel anxious after accidents evacuate after completing their current activity. They do, however, feel anxious when they hear the guidance information announcement.
emergent evacuation: People who do not feel anxious and do not evacuate after completing their current activity and after hearing the evacuation guidance information initiate evacuation when they become extremely anxious after receiving new information from others.

The percentages of the types are 57%, 31%, and 12%, respectively. When evacuation guidances are announced, the agents start to evacuate according to their behavior types.

Following two different evacuation messages are announced.

– Message 1 gives instructions to evacuate by using two exits, numbers 1 and 8.
– Message 2 gives instructions to evacuate using the nearest exit except two exits, number 2 and 10.

Figure 4 (a) shows snapshots of the evacuation. The snapshots in the left and right column show evacuation guided by message 1 and message 2, respectively. In the snapshots, it can be seen that message 2 results in a faster evacuation than with guidance 1. Figure 4 (b) shows time sequence of evacuations and the rate of

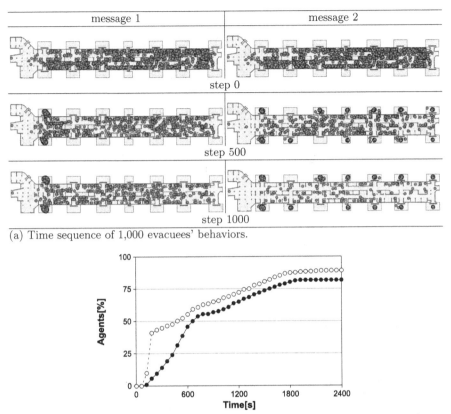

(a) Time sequence of 1,000 evacuees' behaviors.

(b) Rates of evacuation from shopping mall (Black and white points correspond to evacuation guided with messages 1 and 2.)

Fig. 4. Simulations of evacuations at subterranean shopping mall

evacuees over simulation step. The evacuation rate of white points corresponding to message 2 is faster at start time, and the evacuation rates for both messages are similar for both guidances after congestion started at exits, The evacuation rate did not reach 100%, because all the agents are not instant evacuation type.

This experiment shows that the guidances given to evacuees influences their behaviors and, furthermore, that evacuation simulations can facilitate planning manuals for emergencies.

4.3 Experiment 2: Parallel Computation in on Simulation Step

The mall is connected to other subterranean malls and buildings. At emergencies, people pass through the mall and evacuate to safe places in the ground level. The number of people at emergencies become larger than at normal times. Table 1 shows the computation time of one evacuation simulation (Δt). The first row

Table 1. Execution time of one simulation step (ms)

Algorithm & coding	number of agents										
	1,000	2,000	3,000	4,000	5,000	6,000	7,000	8,000	9,000	10,000	11,000
1 A	4.7	8.9	13.7	19.7	26.5	36.9	43.9	55.5	65.0	100.5	116.6
2 B_{cpu}	11.1	16.1	21.5	29.0	34.8	52.1	67.0	79.2	96.4	111.4	149.6
device	(2.5)	(5.3)	(7.5)	(11.8)	(13.9)	(19.2)	(26.2)	(33.2)	(41.4)	(53.6)	(74.1)
host	(0.7)	(1.4)	(2.2)	(3.6)	(4.7)	(5.7)	(7.0)	(9.1)	(14.6)	(12.0)	(16.5)
B_{gpgpu}	10.8	13.6	16.9	21.4	24.3	38.8	42.7	52.7	62.6	80.7	90.3
device	(1.4)	(2.9)	(3.4)	(4.3)	(5.3)	(7.5)	(7.8)	(10.3)	(12.7)	(14.2)	(22.9)
host	(0.7)	(1.1)	(1.7)	(3.4)	(3.4)	(5.2)	(5.6)	(9.7)	(12.1)	(16.3)	(16.8)
C_{cpu}	3.4	8.3	12.1	20.8	28.8	40.5	55.4	59.3	83.3	114.0	117.4
device	(1.9)	(5.6)	(8.1)	(15.3)	(22.2)	(31.9)	(46.4)	(48.9)	(69.8)	(98.3)	(97.9)
host	(1.1)	(2.1)	(3.0)	(4.3)	(5.1)	(6.8)	(7.0)	(8.2)	(10.7)	(12.6)	(16.1)
C_{gpgpu}	6.1	10.8	15.6	17.7	22.2	29.1	35.4	42.9	48.5	58.8	80.6
device	(4.6)	(8.1)	(11.8)	(12.4)	(16.0)	(21.2)	(26.0)	(31.8)	(37.1)	(46.0)	(60.3)
host	(0.9)	(1.7)	(2.5)	(3.6)	(4.3)	(5.6)	(6.8)	(8.1)	(8.3)	(9.2)	(15.2)

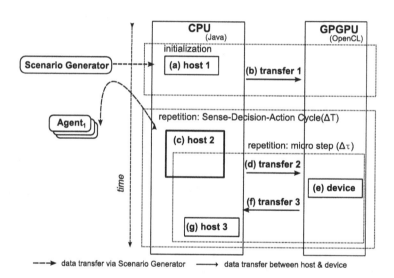

Fig. 5. Task assignments and data flow between CPU and GPGPU

of the table, labeled as algorithm 1 and coding A, shows the time and the time increases as the number of agents varies from 1,000 to 11,000. The simulations are executed on one PC with Fedora 14, and the simulation program is written by Java. Its CPU was Intel Core I7-3960X 3 (30GHz, 6core, 64GB memory (48GB heap)). The second and third rows of Table 1 are execution times of a simulation based on Algorithm 2. Coding B differs from coding C in data exchange timing between host and device. Coding B exchanges data at every $\Delta\tau$ and coding C at every Δt. The values in parentheses: labeled as device and host of Table 1 are time executed in GPGPU and CPU.

Figure 5 shows task assignments between CPU and GPGPU. Agents send their goals to the traffic simulator every Δt. The traffic simulator is consisted of two modules; one is executed in the host (CPU) and the other is executed in parallel in the device (GPGPU) at every $\Delta\tau$. The codes in the host are written in Java and the codes in the device are written in Open Computing Language (OpenCL). OpenCL is a framework for programming on GPGPU hardware. GPGPU is a utilization of a GPU; it provides a parallel computing platform. We use OpenCL because codes written in OpenCL can be executed on PCs with or without GPGPU hardware.

- Routine (a) of Figure 5 corresponds to lines 1 and 2 in Algorithm 2 and is called once. It converts the structure of data and transfers them between codes written by Java and OpenCL in routine (b).
- Routines from (c) to (g) are executed every RCRS step. At the first step, the destinations where agents want to go are sent from the modules of agents in routine (c).

In code B, routines (c) to (g) are repeated N_{step} times at $\Delta\tau$ cycle. The destinations, and the present polygon and position in the polygon are transferred to GPU. The next positions of agents are calculated according to equation in routine (e). The positions are returned to host. Codes in routine (g) check overlaps between the positions and whether the agents move to other polygons. And it is repeated from routine (c). In code C, the repetition of N_{step} times are calculated in routine (e) on GPGPU. The same programs are executed on GPGPU, so the checks of overlapping and transfer to other polygons are executed for all agents

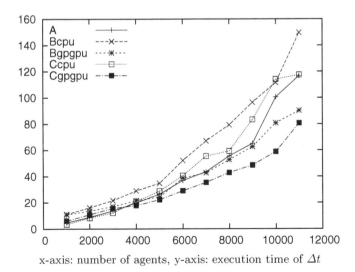

x-axis: number of agents, y-axis: execution time of Δt

Fig. 6. Execution time of vacuation simulation at shopping mall per step

whether it is necessary for not. Routine (c), (d), (f) and (g) are executed once in Δt.

The performance is trade-off cost of data transfer between CPU and GPCPU and the parallel programming in GPGPU. The subscriptions, $_{cpu}$ and $_{gpgpu}$, represents the computations with and without GPGPU. B_{cpu} stands for the code B running only CPU and C_{gpgpu} for the code C running on CPU and GPGPU. The GPU was NVIDIA Tesla C2075 (448core, 1.15GHz, 6GB memory).

Figure 6 shows the execution time in Table 1. At first, the overhead of data transfer negates the benefit of parallel computation using GPGPU. Over 4,000, the simulation coded using OpenCL with GPGPU (C_{gpgpu}) outperforms the others and the execution time remains quite low at ten thousands of agents.

5   Discussion and Summary

For achieving the objectives of the RCRS, the scale of simulation and the behaviors of agents in disaster situations are of key issues. While proposals and implementations have been made to apply distributed computing or parallel computing to RCRS, the interactions among agents over the area: communication among them and their movements impede the distributed computations. RCRS has not been sufficiently scaled up to simulate situations at the level of that the simulation results are put to practical use.

In this paper, we proposed a parallel computation in a traffic simulator. The traffic simulation system uses a GPGPU, which allows us to simulate evacuation of thousands of agents faster than the present system. Further, it can simulate the movements of agents inside buildings and guide the agents at their sense-decision-action cycle. These makes evacuation simulation realistic one, and the results show that evacuation behaviors of a large number evacuees can be simulated without an excessive increase in the required computation resources.

Our proposed system provides a platform that simulates ten thousands of crowd evacuation inside buildings and the evacuation simulations help security offices to prepare manuals for emergencies. This widens the applications of RCRS to make plans for decrease dameges from disasters.

References

1. Kitano, H., Tadokoro, S., Noda, I., Matsubara, H., Takahashi, T., Shinjou, A., Shimada, S.: Robocup rescue: Search and rescue in large-scale disasters as a domain for autonomous agents research. In: IEEE International Conference on System, Man, and Cybernetics (1999)
2. Takahashi, T., Tadokoro, S., Ohta, M., Ito, N.: Agent based approach in disaster rescue simulation - from test-bed of multiagent system to practical application. In: Birk, A., Coradeschi, S., Tadokoro, S. (eds.) RoboCup 2001. LNCS (LNAI), vol. 2377, pp. 102–111. Springer, Heidelberg (2002)
3. Takeuchi, I., Kakumoto, S., Goto, Y.: Towards an integrated earthquake disaster simulation system. In: First International Workshop on Synthetic Simulation and Robotics to Mitigate Earthquake Disaster (2003)

4. Skinner, C., Barley, M.: Robocup rescue simulation competition: Status report. In: Bredenfeld, A., Jacoff, A., Noda, I., Takahashi, Y. (eds.) RoboCup 2005. LNCS (LNAI), vol. 4020, pp. 632–639. Springer, Heidelberg (2006)
5. Takahashi, T.: http://sakura.meijo-u.ac.jp/ttakaHP/MyPresentation/ agentCompetition09.pdf
6. Averill, J.D., Mileti, D.S., Peacock, R.D., Kuligowski, E.D., Groner, N.E.: Occupant behavior, egress, and emergency communications (NIST NCSTAR 1-7). Technical report. National Institute of Standards and Technology, Gaitherburg (2005)
7. C. O. G. of Japan. Prevention Disaster Conference, the Great West Japan Earthquake and Tsunami. Report on evacuation behavior of people (in Japanese), http://www.bousai.go.jp/jishin/chubou/higashinihon/7/1.pdf (date: February 9, 2012)
8. Helbing, D., Farkas, I., Vicsek, T.: Simulating dynamical features of escape panic. NATURE 407, 487–490 (2000)
9. Okaya, M., Takahashi, T.: Evacuation simulation with guidance for anti-disaster planning. In: Chen, X., Stone, P., Sucar, L.E., van der Zant, T. (eds.) RoboCup 2012. LNCS (LNAI), vol. 7500, pp. 202–212. Springer, Heidelberg (2013)

Vision Based Referee Sign Language Recognition System for the RoboCup MSL League

Paulo Trigueiros[1], Fernando Ribeiro[1], and Luís Paulo Reis[2]

[1] Departamento de Electrónica Industrial da Universidade do Minho,
Campus de Azurém 4800-058, Guimarães, Portugal
[2] EEUM – Escola de Engenharia da Universidade do Minho – DSI,
Campus de Azurém 4800-058, Guimarães, Portugal
ptrigueiros@gmail.com, fernando@dei.uminho.pt,
lpreis@dsi.uminho.pt

Abstract. In RoboCup Middle Size league (MSL) the main referee uses assisting technology, controlled by a second referee, to support him, in particular for conveying referee decisions for robot players with the help of a wireless communication system. In this paper a vision-based system is introduced, able to interpret dynamic and static gestures of the referee, thus eliminating the need for a second one. The referee's gestures are interpreted by the system and sent directly to the Referee Box, which sends the proper commands to the robots. The system is divided into four modules: a real time hand tracking and feature extraction, a SVM (Support Vector Machine) for static hand posture identification, an HMM (Hidden Markov Model) for dynamic unistroke hand gesture recognition, and a FSM (Finite State Machine) to control the various system states transitions. The experimental results showed that the system works very reliably, being able to recognize the combination of gestures and hand postures in real-time. For the hand posture recognition, with the SVM model trained with the selected features, an accuracy of 98,2% was achieved. Also, the system has many advantages over the current implemented one, like avoiding the necessity of a second referee, working on noisy environments, working on wireless jammed situations. This system is easy to implement and train and may be an inexpensive solution.

Keywords: Hand gesture recognition, Support Vector Machine (SVM), Hidden Markov Model (HMM), Finite State Machine (FSM), RoboCup MSL.

1 Introduction

RoboCup is an international joint project to promote Artificial Intelligence (AI), robotics, and related fields since 1996 [1]. RoboCup project has been held every year, and the progress that has been seen in applied technology is amazing, especially in the Soccer Middle Size League (MSL) [2]. RoboCup MSL is one of the most challenging, using real non humanoid robot teams to play with an ordinary soccer ball in an autonomously way. Since the main goal consists of a soccer game in 2050 between the

S. Behnke et al. (Eds.): RoboCup 2013, LNAI 8371, pp. 360–372, 2014.

humans world champions team against the robots world champions team, this paper presents a new system, able to interpret dynamic and static gestures of the referee, thus eliminating the need for a second person. In RoboCup MSL the main referee uses assisting technology, controlled by a second referee, to support him, in particular for conveying referee decisions for robot players with the help of a wireless communication system. On this system, the referee's gestures are interpreted by a vision-based system and sent to the Referee Box, which then sends the commands to the robots. The system uses a depth image acquired with a Kinect camera for hand feature extraction (centroid distance signature [3]) and an SVM model to classify hand postures (Fig. 1). For dynamic gestures, each gesture to be recognized is scored against 11 different HMMs that consists our defined language set (Fig. 2). The model with the highest score is selected as the recognized gesture. A finite state machine controls the transition between each system state namely: DYNAMIC, STATIC and PAUSE, to build the final referee command (Table 2).

The main advantages of using such a system identified to date are:

- The RoboCup gaming environment is quite noisy, so a vision-based system would improve in the identification of commands;
- The communication between referees is sometimes difficult due to ambient noise, and lack of visibility caused by enthusiastic public;
- The response time can be improved, since the command identified by the vision system is automatically transmitted to the robot;
- Reducing the number of referees during a game;
- Reducing the number of indecisions that happen many times during a game;

The system also has disadvantages and one major limitation is currently the restriction on the referee movements, which can reduce the visibility of the game, limitation that can be improved if the referee is placed at a higher level like in tennis games.

Fig. 1. The defined hand postures for the referee static commands

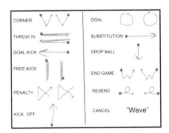

Fig. 2. The defined dynamic gestures used in the Referee CommLang

The rest of the paper is as follows. Firstly, the related work is review in section 2. Section 3 introduces the actual data pre-processing stage, feature extraction and models creation. The *Referee CommLang* is introduced in section 4. Experimental methodology and results are explained in section 5. Conclusions and future work are drawn in section 6.

2   Related Work

Hand gestures, either static or dynamic, for human computer interaction in real time systems is an area of active research and with many possible applications. However, vision-based hand gesture interfaces for real-time applications require fast and extremely robust hand detection, and gesture recognition. Support Vector Machines (SVM's) is a technique based on statistical learning theory, which works very well with high-dimensional data. The objective of this algorithm is to find the optimal separating hyper plane between two classes by maximizing the margin between them [4]. Faria et al. [5, 6] used it to classify robotic soccer formations and the classification of facial expressions, Ke et al. [7] used it in the implementation of a real-time hand gesture recognition system for human robot interaction, Maldonado-Báscon [8] used it for the recognition of road-signs and Masaki et al. [9] used it in conjunction with SOM (Self-Organizing Map) for the automatic learning of a gesture recognition mode. He first applies the SOM to divide the sample into phases and construct a state machine, and then he applies the SVM to learn the transition conditions between nodes. Almeida et al. [10] proposed a classification approach to identify the team's formation in the robotic soccer domain for the two dimensional (2D) simulation league employing Data Mining classification techniques. Trigueiros et al. [11] have made a comparative study of four machine learning algorithms applied to two hand features datasets. In their study the datasets had a mixture of hand features. Hidden Markov Models (HMMs) have been widely used in a successfully way in speech recognition and hand writing recognition [12], in various fields of engineering and also applied quite successfully to gesture recognition. Oka et al. [13] developed a gesture recognition system based on measured finger trajectories for an augmented desk interface system. They have used a Kalman filter for the prediction of multiple finger locations and used an HMM for gesture recognition. Perrin et al. [14] described a finger tracking gesture recognition system based on a laser tracking mechanism which can be used in hand-held devices. They have used HMM for their gesture recognition system with an accuracy of 95% for a set of 5 gestures. Nguyen et al. [15] described a hand gesture recognition system using a real-time tracking method with pseudo two-dimensional Hidden Markov Models. Chen et al. [16] used it in combination with Fourier descriptors for hand gesture recognition using a real-time tracking method. Kelly et al. [17] implemented an extension to the standard HMM model to develop a gesture threshold HMM (GT-HMM) framework which is specifically designed to identify inter gesture transition. Zafrulla et al. [18] have investigated the potential of the Kinect depth-mapping camera for sign language recognition and verification for educational games for deaf children. They used 4-state HMMs to train each of the 19 signs defined in their study. Cooper et al. [19] implemented an isolated sign recognition system using a 1st order Markov chain. In their model, signs are broken down in visemes (equivalent to phonemes in speech) and a bank of Markov chains are used to recognize the visemes as they are produced. In the field of feature selection and statistical techniques, Ribeiro et al. [20] proposed a method that consisted in using a histogram of white-green transitions for robot orientation. This technique does not take much computational time and it proved to be very reliable.

3   Pre-processing and Feature Extraction

3.1   Static Hand Gesture Classification

For static gesture classification, hand segmentation and feature extraction is a crucial step in computer vision applications for hand gesture recognition. The pre-processing stage prepares the input image and extracts features used later with classification algorithms [3]. In our system we used feature vectors composed of centroid distance values to train a SVM (Support Vector Machine) for hand posture classification.

3.1.1   Centroid Distance

The centroid distance signature is a type of shape signature [3]. The centroid distance function is expressed by the distance of the hand contour boundary points, from the centroid (x_c, y_c) of the shape calculated in the following manner:

$$d(i) = \sqrt{[x_i - x_c]^2 + [y_i - y_c]^2}, \qquad i = 0, \dots, N - 1 \qquad (1)$$

where $d(i)$, is the calculated distance, x_i and y_i are the coordinates of contour points and N is the number of contour points used in the calculation. This way, we obtain a one-dimensional function that represents the hand shape signature as shown in Fig. 3. In our system we have used 32 equally spaced points on the contour to build the hand signature.

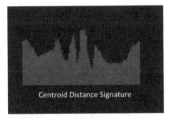

Fig. 3. Hand Centroid Distance Signature

Due to the subtraction of centroid, which represents the hand position, from boundary coordinates, the centroid distance representation is invariant to translation as Rayi Yanu Tara et al. [21]   have demonstrated in their paper.

3.1.2   SVM

A SVM is used to learn the pre-set hand postures (Fig. 1). The SVM is a pattern recognition technique in the area of supervised machine learning, which works very well with high-dimensional data. When more than two classes are present, there are several approaches that evolve around the 2-class case [22]. The one used in our system is the one-against-all, where c classifiers have to be designed. Each one of them is designed to separate one class from the rest. One drawback with this kind of approach is that after the training phase there are regions in the space, where no training data lie, for which more than one hyperplane gives a positive value or all of them result in negative values [23].

3.2   HMM Dynamic Gesture Classification

Dynamic hand gestures are time-varying processes, which show statistical variations, making HMMs a plausible choice for modeling the processes [24]. So, for the recognition of dynamic gestures a HMM (Hidden Markov Model) model was trained for each possible gesture. A Markov Model is a typical model for a stochastic (i.e. random) sequence of a finite number of states [25]. A human gesture can be understood as a Hidden Markov Model where the true states of the model are hidden in the sense that they cannot be directly observed. HMMs have been widely used in a successfully way in speech recognition and hand writing recognition [12]. In our system the 2D motion hand trajectory points are labeled according to the distance to the nearest centroid based on Euclidean distance, resulting in a discrete feature vector like the one shown in Fig. 4, and translated to origin. In our case the 2D features are sufficient for hand gesture recognition.

The feature vectors thus obtained are used to train the different HMMs and learn the model parameters: the initial state probability vector (π), the state-transition probability matrix ($A = [a_{ij}]$) and the observable symbol probability matrix ($B = [b_j(m)]$). In the recognition phase an output score for the sample gesture is calculated for each model, given the probability that the corresponding model generated the underlying gesture. The model with the highest output score represents the recognized gesture. In the proposed system, a left-right HMM (Fig. 5) was used [26, 27]. This kind of HMM has the states ordered in time so that as time increases, the state index increases or stays de same. This topology has been chosen, since it is perfectly suitable to model the kind of temporal gestures present in the system.

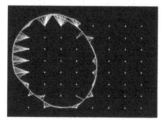

[1 2 2 2 3 3 4 4 4 5 6 13 13 14 14 14 14 14 14 14 13 13
20 19 19 18 18 17 24 23 22 22 22 22 22 29 29 29 29
29 29 29]

Fig. 4. Gesture path with respective feature vector

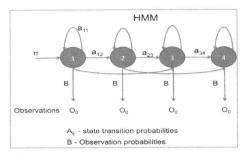

Fig. 5. Example of a left-right HMM

The model with the highest output score represents the recognized gesture. In the proposed system, a left-right HMM (Fig. 5) was used [26, 27]. This kind of HMM has the states ordered in time so that as time increases, the state index increases or stays de same. This topology has been chosen, since it is perfectly suitable to model the kind of temporal gestures present in the system.

4   The Referee CommLang.

The Referee CommLang is a new and formal definition of all the commands that the system accepts. As in [28], the language must represent all the possible gesture combinations (static and dynamic) and at the same time be simple in its syntax. The language was defined with BNF – BakusNaur form [29], and has three types of commands: *Team commands*, *Player commands* and *Game commands*.

This way, a language is defined to be a set of commands that can be a TEAM_COMMAND, a GAME_COMMAND or a PLAYER_COMMAND. The TEAM_COMMAND is composed of the following ones: KICK_OFF, CORNER, THROW_IN, GOAL_KICK, FREE_KICK, PENALTY, GOAL or DROP_BALL. A GAME_COMMAND can be the START or STOP of the game, a command to end the game (END_GAME), cancel the just defined command (CANCEL) or resend the last command (RESEND). For the END_GAME command, we have to define the PART_ID (identification of the part of the game ending) with one of four commands – 1ST, 2ND, EXTRA or PEN (penalties).

```
<LANGUAGE> ::= {<COMMAND>}
<COMMAND> ::= <TEAM_COMMAND> | <GAME_COMMAND> | <PLAYER_COMMAND>
<TEAM_COMMAND> ::= <KICK_OFF> | <CORNER> | <THROW_IN> | <GOAL_KICK> |
<FREE_KICK> | <PENALTY> | <GOAL> | <DROP_BALL>
<GAME_COMMAND> ::= <START> | <STOP> | <END_GAME> | <CANCEL> | <RESEND>
<PLAYER_COMMAND>::=<SUBSTITUTION> | <PLAYER_IN> | <PLAYER_OUT> |
<YELLOW_CARD> | <RED_CARD>
```

For the TEAM_COMMANDS we have several options: KICK_OFF, CORNER, THROW_IN, GOAL_KICK, FREE_KICK, PENALTY and GOAL that need a TEAM_ID (team identification) command, that can be one of two values – CYAN or MAGENTA, and finally the DROP_BALL command.

```
<KICK_OFF> ::= KICK_OFF <TEAM_ID>
<CORNER> ::= CORNER <TEAM_ID>
<THROW_IN> ::= THROW_IN <TEAM_ID>
<GOAL_KICK> ::= GOAL_KICK <TEAM_ID>
<FREE_KICK> ::= FREE_KICK <TEAM_ID>
<PENALTY> ::= PENALTY <TEAM_ID>
<GOAL> ::= GOAL <TEAM_ID>
<DROP_BALL> ::= DROP_BALL
```

For the PLAYER_COMMAND, first we can have a SUBSTITUTION command with the identification of the player that leaves (PLAYER_OUT) and the one that enters (PLAYER_IN) the game with the PLAYER_ID command. The PLAYER_ID can take one of seven values (PL1, PL2, PL3, PL4, PL5, PL6, PL7). For the remaining commands, PLAYER_IN, PLAYER_OUT, YELLOW_CARD or RED_CARD there exists the need to define the TEAM_ID as explained above, and the PLAYER_ID.

```
<SUBSTITUTION> ::= SUBSTITUTION <PLAYER_IN> <PLAYER_OUT>
<PLAYER_IN> ::= PLAYER_IN <TEAM_ID> <PLAYER_ID>
<PLAYER_OUT> ::= PLAYER_OUT <TEAM_ID> <PLAYER_ID>
<YELLOW_CARD> ::= YELLOW_CARD <TEAM_ID> <PLAYER_ID>
<RED_CARD> ::= RED_CARD <TEAM_ID> <PLAYER_ID>
<START> ::= START
<STOP> ::= STOP
<END_GAME> ::= END_GAME <PART_ID>
<CANCEL> ::= CANCEL
<RESEND> ::= RESEND
<TEAM_ID > ::= CYAN | MAGENTA
<PLAYER_ID> := PL1 | PL2 | PL3 | PL4 | PL5 | PL6 | PL7
<PART_ID> ::= 1ST | 2ND | EXTRA | PEN
```

5   Experimental Methodology and Results

The experimental methodology was divided into three parts, each one with its own application built and tested in an Intel Core i7 (2,8 GHz) Mac OSX computer with 4GB DDR3. First, a feature extraction phase and SVM model training of the defined static hand postures. Second, acquisition of hand motion sequences for each of the defined gestures, and HMM model training. The final one, the Referee CommLang implementation, is the system with posture identification and gesture recognition, controlled by a event driven Finite State Machine that controls the transition between the three possible states: DYNAMIC, STATIC and PAUSE, and builds the final command that is transmitted to the robots. All the datasets were built with data collected from four people making the defined postures and gestures in front of the Kinect. For feature extraction and SVM training, the application built in C++ (Fig. 7) uses openFrameworks (http://www.openframeworks.cc/), OpenCV [30] (computer vision library), OpenNI [31] (Open Natural Interface) that enables real-time hand tracking and skeleton tracking, and the machine-learning library dlib (http://dlib.net/). The dataset was built from features extracted from four users performing the set of commands in front of the camera, and was analyzed with the help of RapidMiner [32], in order to find the best parameters for the learner. The experiments were performed with a *10-fold cross validation*, and we were able to achieve an accuracy of

98,2% for the given features, with an RBF (radial basis function) kernel with a C value equal to 6. The obtained confusion matrix can be seen in Table 1 where can verify the existence of some high values between command number four and three, between number five and four and between number six and seven, and that contributed to the 1,8% of false positives. For dynamic gesture learning and training, we built another application (Fig. 8) using openFrameworks with OpenNI [31] and a personal HMM class library for gesture learning and classification. In the case of the HMM the number of centroids used to label the hand path points, is a predefined value, equally spaced in a 2D grid as was shown in Fig. 4. For each gesture, a HMM is trained and the model parameters (the initial state probability vector, the state-transition probability matrix and the observable symbol probability matrix) are learned. The number of observation symbols (alphabet) was defined to be 64 with 4 hidden states. Several values for the number of observations and hidden states were tried out, without significant improvements for values greater than the defined ones.

Table 1. Centroid distance confusion matrix

		Actual class						
		1	2	3	4	5	6	7
Predicted class	1	**588**	0	0	0	0	2	0
	2	2	**706**	0	0	1	0	1
	3	0	1	**578**	1	0	0	0
	4	0	0	12	**715**	3	0	0
	5	0	1	1	13	**536**	1	3
	6	1	8	0	1	5	**693**	12
	7	2	0	0	2	6	9	**751**

The Referee CommLang prototype (Fig. 9), uses the SVM model and the HMM models controlled by a FSM (Finite State Machine) to recognize the defined command language.

5.1   Finite State Machine

Since the Referee CommLang uses a combination of dynamic and static gestures, we needed to model the command semantics. A Finite State Machine is a usually employed technique to handle this situation [33, 34]. Our system uses a FSM, as shown in Fig. 6, to control the transition between the three possible states: DYNAMIC, STATIC and PAUSE. By using a rest or pause state it is possible to identify changes in gestures and somehow eliminate all unintentional actions between dynamic-static or static-static gestures. The combination of a dynamic gesture and a static or set of static gestures gives us a command as defined in the Referee CommLang and listed in Table 2.

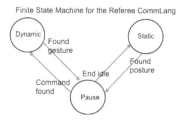

Fig. 6. The Referee CommLang Finite State Machine

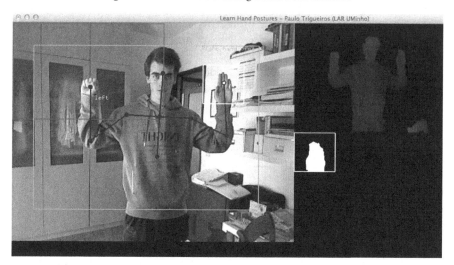

Fig. 7. Application interface for the posture learning and training system

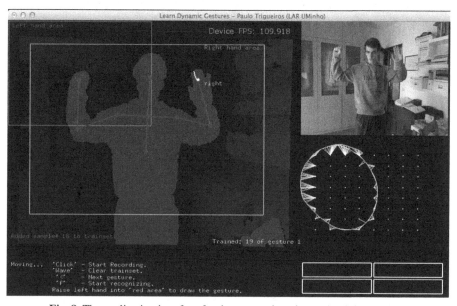

Fig. 8. The application interface for the gesture learning and training system

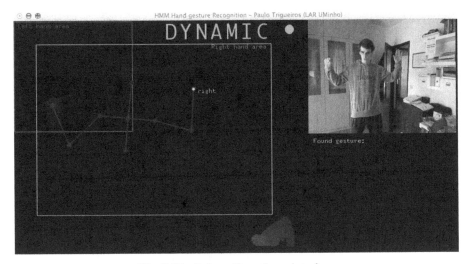

Fig. 9. The Referee CommLang interface

6 Conclusions and Future Work

This paper presents a system able to interpret dynamic and static gestures of the refer-ee, thus eliminating the need for a second person. The referee's gestures are interpret-ed by the vision-based system and sent to the Referee Box, which sends then the proper commands to the robots. The set of dynamic gestures defined is a first ap-proach to the problem, and future adjustments or definition of new ones is always possible. For static posture recognition an SVM model was learned from centroid distance features and a good recognition rate was achieved. For temporal gesture recognition one HMM was trained for each gesture and it was also possible to achieve a good recognition rate, proving that this kind of models, as was already seen in other references, works very well for this type of problem. A vision-based sign language was defined – *the Referee CommLang* – that is able to represent all the possible ges-ture combinations (static and dynamic) and at the same time be simple in its syntax. The experimental results showed that the system works very reliably, being able to recognize the combination of gestures and hand postures in real-time, although with some limitations due to the nature of the camera used. Despite these limitations, for the hand posture recognition, with the SVM model trained with the selected features, an accuracy of 98,2% was achieved. The system has many advantages over the cur-rent implemented one, where there exists the need for two referees, and is an easy and inexpensive solution to implement and train. As further work it is intended to keep improving this system in order to eliminate some of the limitations, like the referee movements restrictions. The current system only allows 2D gesture paths, although as further work it is thought to test and include not only the possibility of 3D gestures but also to work with several cameras to thereby obtain a full 3D environment and achieve view-independent recognition. Also, the existence of different new cameras, with improved depth resolution and with a higher frame frequency, now available on the market will be tested, to improve some of the limitations of the Kinect, thus leading to better static gesture recognition rates.

Table 2. Command set definition with associated codes and text description

Nr.	Command	1st Gesture	2nd Gesture	3rd Gesture	4th Gesture	Code (TEXT)
1	CORNER		TEAM			11 – CORNER, TEAM 1 12 – CORNER, TEAM 2
2	THROW_IN		TEAM			21 – THROW_IN, TEAM 1 22 – THROW_IN, TEAM 2
3	GOAL_KICK		TEAM			31 – GOAL_KICK, TEAM 1 32 – GOAL_KICK, TEAM 2
4	FREE_KICK		TEAM			41 – FREE_KICK, TEAM 1 42 – FREE_KICK, TEAM 2
5	PENALTY		TEAM			51 – PENALTY, TEAM 1 52 – PENALTY, TEAM 2
6	KICK_OFF		TEAM			61 – KICK_OFF, TEAM 1 62 – KICK_OFF, TEAM 2
7	GOAL		TEAM	PLAYER		71(1-7) – GOAL, TEAM1, PLAYER(1-7) 72(1-7) – GOAL, TEAM2, PLAYER(1-7)
8	SUBSTITUTION		TEAM	PLAYER_IN	PLAYER_OUT	81(1-7)(1-7) – SUBSTITUTION, TEAM 1, PLAYER_IN(1-7), PLAYER_OUT(1-7) 82(1-7)(1-7) – SUBSTITUTION, TEAM 2, PLAYER_IN(1-7), PLAYER_OUT(1-7)
9	DROP_BALL					9 - DROP_BALL
10	END_GAME					101 – END_GAME, PART 1 102 – END_GAME, PART 2
11	RESEND					11 - CANCEL
12	RESEND	"Wave"				12 - RESEND

Acknowledgment. The authors wish to thank all members of the Laboratório de Automação e Robótica (LAR), at University of Minho, Guimarães. We would like to thank IPP (Instituto Politécnico do Porto) for the contributions that made all this work possible. Also a special thanks to all others who contributed to making this work possible.

References

[1] Kitazumi, Y., Ishii, K.: Survey of cooperative algorithm in robocup middle size league. In: World Automation Congress, WAC 2010, pp. 1–6 (2010)

[2] M. T. Committe, Middle Size Robot League Rules and Regulations for 2013 (2013), http://wiki.robocup.org/wiki/Middle_Size_League-Rules

[3] Trigueiros, P., Ribeiro, F., Reis, L.P.: A Comparative Study of different image features for hand gesture machine learning. In: 5th International Conference on Agents and Artificial Intelligence, Barcelona, Spain (2013)

[4] Ben-Hur, A., Weston, J.: A User's Guide to Support Vector Machines. In: Data Mining Techniques for the Life Sciences, vol. 609, pp. 223–239. Humana Press (2008)

[5] Faria, B.M., Lau, N., Reis, L.P.: Classification of Facial Expressions Using Data Mining and machine Learning Algorithms. In: 4ª Conferência Ibérica de Sistemas e Tecnologias de Informação, Póvoa de Varim, Portugal, pp. 197–206 (2009)

[6] Faria, B.M., Reis, L.P., Lau, N., Castillo, G.: Machine Learning Algorithms applied to the Classification of Robotic Soccer Formations ans Opponent Teams. presented at the IEEE Conference on Cybernetics and Intelligent Systems (CIS), Singapore (2010)

[7] Ke, W., Li, W., Ruifeng, L., Lijun, Z.: Real-Time Hand Gesture Recognition for Service Robot, pp. 976–979 (2010)

[8] Maldonado-Báscon, S., Lafuente-Arroyo, S., Gil-Jiménez, P., Gómez-Moreno, H.: Road-Sign detection and Recognition Based on Support Vector Machines. IEEE Transactions on Intelligent Transportation Systems, 264–278 (June 2007)

[9] Oshita, M., Matsunaga, T.: Automatic learning of gesture recognition model using SOM and SVM. In: Bebis, G., et al. (eds.) ISVC 2010, Part I. LNCS, vol. 6453, pp. 751–759. Springer, Heidelberg (2010)

[10] Almeida, R., Reis, L.P., Jorge, A.M.: Analysis and Forecast of Team Formation in the Simulated Robotic Soccer Domain. In: Lopes, L.S., Lau, N., Mariano, P., Rocha, L.M. (eds.) EPIA 2009. LNCS (LNAI), vol. 5816, pp. 239–250. Springer, Heidelberg (2009)

[11] Trigueiros, P., Ribeiro, F., Reis, L.P.: A comparison of machine learning algorithms applied to hand gesture recognition. In: 7th Iberian Conference on Information Systems and Technologies, Madrid, Spain, pp. 41–46 (2012)

[12] Rabiner, L.R.: A tutorial on hidden Markov models and selected applications in speech recognition. Proceedings of the IEEE 77, 257–286 (1989)

[13] Oka, K., Sato, Y., Koike, H.: Real-time fingertip tracking and gesture recognition. IEEE Computer Graphics and Applications 22, 64–71 (2002)

[14] Perrin, S., Cassinelli, A., Ishikawa, M.: Gesture recognition using laser-based tracking system. In: Sixth IEEE International Conference on Automatic Face and Gesture Recognition, Seoul, South Korea, pp. 541–546 (2004)

[15] Binh, N.D., Shuichi, E., Ejima, T.: Real-Time Hand Tracking and Gesture Recognition System. In: Proceedings of International Conference on Graphics, Vision and Image, Cairo, Egypt, pp. 362–368 (2005)

[16] Chen, F.-S., Fu, C.-M., Huang, C.-L.: Hand gesture recognition using a real-time tracking method and hidden Markov models. Image and Vision Computing 21, 745–758 (2003)

[17] Kelly, D., McDonald, J., Markham, C.: Recognition of Spatiotemporal Gestures in Sign Language Using Gesture Threshold HMMs. In: Wang, L., Zhao, G., Cheng, L., Pietikäinen, M. (eds.) Machine Learning for Vision-Based Motion Analysis, pp. 307–348. Springer, London (2011)

[18] Zafrulla, Z., Brashear, H., Starner, T., Hamilton, H., Presti, P.: American sign language recognition with the kinect. presented at the 13th International Conference on Multimodal Interfaces, Alicante, Spain (2011)

[19] Cooper, H., Bowden, R.: Large lexicon detection of sign language. In: Lew, M., Sebe, N., Huang, T.S., Bakker, E.M. (eds.) HCI 2007. LNCS, vol. 4796, pp. 88–97. Springer, Heidelberg (2007)

[20] Ribeiro, F., et al.: Robot Orientation with Histograms on MSL. In: Röfer, T., Mayer, N.M., Savage, J., Saranlı, U. (eds.) RoboCup 2011. LNCS, vol. 7416, pp. 507–514. Springer, Heidelberg (2012)

[21] Tara, R.Y., Santosa, P.I., Adji, T.B.: Sign Language Recognition in Robot Teleoperation using Centroid Distance Fourier Descriptors. International Journal of Computer Applications 48 (June 2012)

[22] Theodoridis, S., Koutroumbas, K.: An Introduction to Pattern Recognition: A Matlab Approach: Academic Press (2010)

[23] Theodoridis, S., Koutroumbas, K.: Pattern Recognition, 4th edn. Elsevier (2009)

[24] Wu, Y., Huang, T.S.: Vision-Based Gesture Recognition: A Review. In: Braffort, A., Gibet, S., Teil, D., Gherbi, R., Richardson, J. (eds.) GW 1999. LNCS (LNAI), vol. 1739, pp. 103–115. Springer, Heidelberg (2000)

[25] Fink, G.A.: Markov Models for Pattern recognition - From Theory to Applications. Springer (2008)

[26] Camastra, F., Vinciarelli, A.: Machine Learning for Audio, Image and Video Analysis. Springer (2008)

[27] Alpaydin, E.: Introduction to Machine Learning. MIT Press (2004)

[28] Reis, L.P., Lau, N.: COACH UNILANG - A Standard Language for Coaching a (Robo) Soccer Team. In: Birk, A., Coradeschi, S., Tadokoro, S. (eds.) RoboCup 2001. LNCS (LNAI), vol. 2377, pp. 183–192. Springer, Heidelberg (2002)

[29] Backus, J.W., Bauer, F.L., Green, J., Katz, C., McCarthy, J., Perlis, A.J., et al.: Revised Report on the Algorithmic Language ALGOL 60. Communications of the ACM, 1–17 (1960, January 1963), http://doi.acm.org/10.1145/366193.366201

[30] Bradski, G., Kaehler, A.: Learning OpenCV: Computer Vision with the OpenCV Library, 1st edn. O'Reilly Media (2008)

[31] OpenNI, The standard framework for 3D sensing (2013), http://www.openni.org/

[32] Miner, R.: RapidMiner: Report the Future (December 2011), http://rapid-i.com/content/view/181/196/

[33] Buckland, M.: Programming Game AI by Example. Wordware Publishing, Inc. (2005)

[34] Millington, I., Funge, J.: Artificial Intelligence for Games, 2nd edn. Elsevier (2009)

Unsupervised Recognition of Salient Colour for Real-Time Image Processing

David Budden[1,2] and Alexandre Mendes[3]

[1] National ICT Australia (NICTA), Victoria Research Lab
[2] Department of Electrical and Electronic Engineering,
The University of Melbourne, Parkville, VIC 3010, Australia
david.budden@unimelb.edu.au
[3] School of Electrical Engineering and Computer Science,
The University of Newcastle, Callaghan, NSW 2308, Australia
alexandre.mendes@newcastle.edu.au

Abstract. Humans have the subconscious ability to create simple abstractions from observations of their physical environment. The ability to consider the colour of an object in terms of "red" or "blue", rather than spatial distributions of reflected light wavelengths, is vital in processing and communicating information about important features within our local environment. The real-time identification of such features in image processing necessitates the software implementation of such a process; segmenting an image into regions of salient colour, and in doing so reducing the information stored and processed from 3-dimensional pixel values to a simple colour class label. This paper details a method by which colour segmentation may be performed offline and stored in a static look-up table, allowing for constant time dimensionality reduction in an arbitrary environment of coloured features. The machine learning framework requires no human supervision, and its performance is evaluated in terms of feature classification performance within a RoboCup robot soccer environment. The developed system is demonstrated to yield an 8% improvement over slower traditional methods of manual colour mapping.

Keywords: Computer vision, colour vision, robotics, RoboCup, LUT generation.

1 Introduction

Szeliski describes image segmentation simply as the task of finding groups of pixels that "go together" [9]. This is an abstract notion that corresponds with an inherently subconscious human process: the ability to look at an image and identify salient features, such as a person, landmark or household item. Hundreds of algorithms exist for image segmentation, with most relying on the assignment of a *feature vector* (containing spatial and/or colour information) to each pixel. For real-time applications, the dimensionality of this feature vector is commonly reduced to contain only colour information, with each pixel assigned a colour *class label* corresponding with higher level notions of colour (such as "red" or

S. Behnke et al. (Eds.): RoboCup 2013, LNAI 8371, pp. 373–384, 2014.

Fig. 1. An example of poor (left) versus good (right) quality colour image segmentation of a common RoboCup scenario. Segmentation was accomplished using the k-means algorithm ($k = 5$) over RGB (left) and YC_bC_r (right) colour space representations [2].

"yellow"). This reduction has two primary advantages: the reduction of computational complexity, by processing over a lower dimensional search space; and the removal of spatial class dependencies, allowing for the pixel-label mapping to be calculated offline and stored in a static data structure for constant time access. An example of colour segmentation within a RoboCup environment is presented in Fig. 1.

This work focuses on this reduced task of *colour segmentation*, within an environment where salient features exhibit some significant degree of colour-coding. RoboCup robot soccer [6] was chosen as such as scenario for experimentation, where field lines, goals and the ball are each assigned unique colours, and where maintaining real-time processing performance is critical for robot responsiveness. A system for generating and storing a pixel-label mapping without human supervision was developed for the Robotis DARwIn-OP humanoid robot, with the mapping quality evaluated in terms of feature classification performance. Finally, the developed method was quantitatively compared to mappings generated manually by a human expert, both in terms of the aforementioned performance and required time for mapping generation.

2   Colour Vision Methodology

2.1   Colour Spaces

Past research has demonstrated the YC_bC_r colour space as optimal (compared to RGB, HSV and $CIE\ L^*a^*b$) for unsupervised colour segmentation, in terms of both *internal* and *external* validation techniques [2]. As YC_bC_r is also the native colour space for many cameras, including the Logitech C905 fitted to the utilised DARwIn-OP robot platform, all pixel values throughout this work are processed and represented in YC_bC_r format.

2.2 Colour Look-Up Tables

In computer vision, a mapping from an arbitrary 3-component colour space C to a set of colours M assigns a class label $m_i \in M$ to every point $c_j \in C$ [3]. If each channel is represented by an n-bit value and $k = |M|$ represents the number of defined class labels, then

$$C \to M,$$

where

$$C = \{0, 1, \ldots, 2^n - 1\}^3 \quad \text{and} \quad M = \{m_0, m_1, \ldots, m_{k-1}\}.$$

Concretely, in a colour space C, each pixel in an image is represented by a triplet with each value representing the contribution of each component to the overall colour of that pixel. Projecting the pixel values into the colour space constructed by the orthogonal component axes results in a projected colour space distribution of the original image. If computational resources are limited, the colour segmentation process is performed offline, with the resultant mapping represented in the form of a $2^n \times 2^n \times 2^n$ look-up table (LUT). This LUT can then be used for efficient, real-time colour classification; as such, the focus of this work reduces to the task of generating effective LUTs without human supervision.

3 Mean Shift and Mode Finding Algorithms

As described in Sec. 1, image segmentation algorithms typically require the assignment of a feature vector to each pixel in an image. Where this feature vector contains only colour (i.e. not spatial) information, the segmentation task is reduced to that of *colour segmentation*. Although many algorithms exist for the task of colour segmentation (and image segmentation is general), this work considers only *mean shift* and *mode finding* techniques. Such techniques involve the assumption that each pixel's feature vector is sampled from some unknown underlying probability distribution, and attempts to locate *clusters* (modes) within this distribution [9].

Specifically, three algorithms were considered: k-means clustering [5], which parameterises the underlying distribution as a superposition of hyperspherical distributions; *expectation maximisation* [1], which generalises k-means by assuming a mixture of Gaussians; and *mean shift* [4], which generates a non-parametric model of the entire distribution by convolving all feature vectors by some *kernel function*.

3.1 k-Means Clustering

Given a set of m data points (corresponding with feature vectors in the scenario of colour segmentation) $P = \{x^{(1)}, \ldots, x^{(N)}\}$ $(x^{(i)} \in \mathbb{R}^m)$, k-means clustering attempts to partition P into K sets (known as *clusters*) $\mathbf{S} = \{S_1, \ldots, S_K\}$ such that the following objective function J is minimised

$$J(c^{(1)}, \ldots, c^{(m)}, \mu_1, \ldots, \mu_K) = \frac{1}{m} \sum_{i=1}^{m} \|x^{(i)} - \mu_{c^{(i)}}\|^2,$$

where c^i is the index of the cluster $(1, \ldots, K)$ to which data point $x^{(i)}$ is currently assigned, μ_k is the *cluster centroid* of S_k ($\mu_k \in \mathbb{R}^n$), and therefore $\mu_{c^{(i)}}$ is the centroid of the cluster to which $x^{(i)}$ has been assigned [5,9]. This is accomplished by repeating the following two-step algorithm until convergence:

Assignment Step:

$$S_i^{(t)} = \{x^{(p)} : \|x^{(p)} - \mu_i^{(t)}\| \leq \|x^{(p)} - \mu_j^{(t)}\| \; \forall \, 1 \leq j \leq k\}.$$

Update Step:

$$\mu_i^{(t+1)} = \frac{1}{|S_i^{(t)}|} \sum_{x^{(j)} \in S_i^{(t)}} x^{(j)}.$$

3.2 Expectation Maximisation

As opposed to k-means, which associated each data point to a respective label via the *nearest neighbours* method (i.e. minimum Euclidean distance), *expectation maximisation* (EM) utilises a *Mahalanobis distance* [7,9]

$$d(x^{(i)}, \mu_k; \Sigma_k) = (x^{(i)} - \mu_k)^T \Sigma_k^{-1}(x^{(i)} - \mu_k),$$

where $x^{(i)}$ and μ_k are data points and cluster centroids respectively ($x^{(i)}, \mu_k \in \mathbb{R}^n$), and Σ_k are their corresponding covariance estimates. As opposed to the hard assignment result of k-means, EM then allows for each data point to be probabilistically associated with several clusters. This is performed by iteratively re-estimating the parameters for a mixture of Gaussians density function [1,9]

$$p(x|\pi_k, \mu_k, \Sigma_k) = \sum_k \pi_k \, \mathcal{N}(x|\mu_k, \Sigma_k),$$

where π_k is the mixing coefficient, μ_k and Σ_k are the Gaussian mean and covariance respectively, and $\mathcal{N}(x|\mu_k, \Sigma_k)$ is the multivariate Gaussian distribution [1,9]

$$\mathcal{N}(x|\mu_k, \Sigma_k) = \frac{1}{|\Sigma_k|} e^{-d(x, \mu_k; \Sigma_k)}.$$

This process is accomplished by repeating the following two-step algorithm until convergence:

Expectation Step: Estimates the likelihood of a data point $x^{(i)}$ having been generated from the kth Gaussian cluster

$$z_{ik} = \frac{1}{Z_i} \pi_k \, \mathcal{N}(x|\mu_k, \Sigma_k), \quad \sum_k z_{ik} = 1.$$

Maximisation Step: Updates the parameter values

$$\mu_k = \frac{1}{N_k} \sum_i z_{ik} x^{(i)},$$

$$\Sigma_k = \frac{1}{N_k} \sum_i z_{ik} (x^{(i)} - \mu_k)(x^{(i)} - \mu_k)^T$$

$$\pi_k = \frac{N_k}{N},$$

where N_k is an approximation of the number of data points belonging to each cluster

$$N_k = \sum_i z_{ik}.$$

Where hard cluster assignment is required in the generalised case (including the traditional colour segmentation scenario), each data point $x^{(i)}$ may be mapped simply to the cluster k that maximises z_{ik}.

3.3   Mean Shift

As opposed to k-means and EM, which rely on explicitly *parametric* mixture models of the underlying distribution, the mean shift algorithm estimates the density function $f(x)$ by convolving the sparse set of data points (corresponding with pixel feature vectors) by some kernel function $k(r)$ [4,9]

$$f(x) = \sum_i K(x - x^{(i)}) = \sum_i k\left(\frac{\|x - x^{(i)}\|^2}{h^2}\right),$$

where h is the kernel width and $x^{(i)}$ are the data points ($x^{(i)} \in \mathbb{R}^n$). Unfortunately for a very large number of data points (potentially millions of pixels in the case of an entire image), calculating the density function over the entire search space can become too computationally expensive. Instead, a method known as *multiple restart gradient descent* [9] can be utilised, which iteratively estimates the gradient vector at some point y_t and takes an "uphill" step in that direction. The gradient of the probability density function with *derivative kernel* $G(x)$ is estimated as

$$\nabla f(x) = \sum_i (x^{(i)} - x) G(x - x^{(i)}) = \left(\sum_i G(x - x^{(i)})\right) m(x),$$

where $g(r) = -k'(r)$ is the first derivative of $k(r)$, and m(x) (the *mean shift*) is calculated as

$$m(x) = \frac{\sum_i x^{(i)} G(x - x^{(i)})}{\sum_i G(x - x^{(i)})} - x.$$

Finally, the procedure updates the point y_t (at iteration t) by its locally weighted mean ($y_{t+1} = y_t + m(y_t)$) [9]. Although a non-parametric method of clustering by definition, the estimated probability density function $f(x)$ is implicitly parameterised by the selection of kernel function $k(r)$. Two such kernels were considered for colour segmentation: the *Gaussian kernel function*

$$k_N(r) = e^{-\frac{1}{2}r},$$

and *flat kernel function*

$$k_F(r) = \begin{cases} 1 & \text{if } |r| \leq 1 \\ 0 & \text{otherwise} \end{cases}.$$

The latter was chosen due to the simplified implementation and improved computational efficiency. As long as the kernel $k(r)$ is chosen to be monotonically decreasing, the mean shift algorithm is proven to converge [4].

4   System Implementation

4.1   Image Stream Generation

A MATLAB utility was developed to allow for the creation of DARwIn-OP compatible *image streams* from a set of selected images. Specifically, this work utilises a library of 100 images captured by the DARwIn-OP for LUT generation, with each salient RoboCup feature (goals, lines, ball, etc.) equally represented. This library is partitioned into equal sized training and evaluation sets, with each partition maintaining this equal representation[1].

Although this library of images was captured manually, a known initial placement of the DARwIn-OP allows for simple development of a behaviour to automatically capture equivalent representative images.

4.2   Look-Up Table Generation

Given an input training library of 50 images, the process of LUT generation involves the following four steps:

1. **Selection of training images:** It was hypothesised that the classification performance of an autonomously generated LUT would be some function of the number of training images considered during its generation. Too few images would provide insufficient information to allow robust classification, whereas too many would result in the over-classification of colours, introducing substantial background noise. As such, for each proposed method of LUT generation, LUTs were generated for different sized subsets of the training set of 50 images. For each method, the number of training images ranged from 5 to 50, increasing in multiples of 5, with each image selected at random from the initial training set while ensuring the preservation of feature distribution.

---

[1] Available at: `http://davidbudden.com/experimental_data/testImages0912.zip`

2. **Cluster generation:** Each of the three clustering algorithms described in Sec. 3 (k-means, expectation maximisation and mean shift) were applied separately to each selected training image within the 3-dimensional YC_bC_r colour space. The pixel-cluster mapping for each pixel is stored.

3. **Cluster merging:** Although cluster validation assigns a label to each pixel within the original image, there are two fundamental problems. Firstly, the number of clusters is in no way representative of the number of salient colours actually present within the image, but rather restricted by some algorithm parameter. Secondly, there is no direct relation between cluster labels (an arbitrary enumeration) and actual high-level colours (red, blue, etc.) These issues are solved by *cluster merging*, which requires a human user to initially define a number of YC_bC_r (or RGB equivalent) triplets for potential colours of interest[2]. Each pixel is assigned its final colour class label by determining which optimal triplet is the nearest neighbour of that pixel's assigned cluster centroid.

4. **LUT generation:** After each clustering is performed, the mapping between every YC_bC_r triplet and respective colour class label is stored. However, there are often clashes between mappings across a set of images; for example, a triplet may map to the yellow class label in one image and the orange class label in another, due to different colour distributions affecting the nearest neighbour of the cluster centroid. To deal with this, a voting system is implemented, where the class label most frequently associated with a given triplet defines the value eventually stored in the LUT. The LUT is simply a data file storing a flattened $256 \times 256 \times 256$ matrix of colour class labels (0 for unclassified), indexed by YC_bC_r value and generated in a raw format (no headers or formatting bytes) understood by the remainder of the DARwIn-OP vision system.

4.3   Look-Up Table Generalisation

Previous research has indicated that the quality of a manually generated LUT may be improved via the process of *generalisation*; filling gaps and removing outliers by *support vector machine* (SVM) post-processing [8]. For each colour class label, a single one-class SVM was created. The purpose of each SVM was to determine the likelihood (between -1 and 1) of each of the 256^3 data points (i.e. YC_bC_r pixel coordinates, also scaled between -1 and 1 for each axis) belonging to that respective class. Specifically, a MATLAB implementation of the LIBSVM one-class SVM implementation was used [10].

Despite the fact that previous research has indicated the implementation of such a generalisation process improves LUT performance [8], a course-grain grid search of SVM parameters yielded consistent reduction in performance. This discrepancy may be influenced by a number of factors: the improved quality of input LUTs, the smoother nature of input class regions (resulting from clustering

---

[2] In the RoboCup environment, all salient features are encoded with colours corresponding with the corners of the RGB cube.

rather than some manual process) and the selection of a different colour space (YC_bC_r rather than HSV).

4.4   Performance Evaluation

As several thousand LUTs were generated for this work, it is infeasible to qualitatively assess the performance of each by "how good" the corresponding classified images appear (the standard approach the evaluating a manually generated LUT within the NUview utility). Furthermore, although previous research has demonstrated a correlation between *internal* and *external cluster validation techniques* in the context of colour classification [2], a more explicit method of evaluating LUT performance is necessary.

As described in Sec. 4.1, a library of 50 images was used for the purpose of LUT performance evaluation. A feature vector of binary features was manually generated for each image, which each bit corresponding with the presence of some salient RoboCup feature (goals, lines, ball, etc.) within that frame. By generating a similar feature vector by processing each LUT-classified image with the DARwIn-OP robot vision system, classification performance can be calculated directly. Furthermore, by maintaining the same vision system version and same set of 50 images between experiments, the colour segmentation represented by the LUT becomes the only experimental variable. In this case, it follows that the following metrics are representative of the performance of a LUT:

- **Sensitivity:** Measures the probability of classifying some feature, given that the feature is present

$$\text{sensitivity} = \frac{\text{number of true positives}}{\text{number of true positives} + \text{number of false negatives}}.$$

- **Specificity:** Measures the probability of not classifying some feature, given that the feature is not present

$$\text{specificity} = \frac{\text{number of true negatives}}{\text{number of true negatives} + \text{number of false positives}}.$$

5   Results

As described in Section 4, look-up tables (LUTs) were generated for differing sized subsets of the initial training set of 50 images, based on the presumption that an optimal quantity of training data should be evident. As such, for each clustering technique, 10 LUTs were generated to assess a range of training set sizes from 5 to 50 images. Furthermore, to ensure the statistical significance of the experimental results, each of these 10 LUTs were generated 120 times, with the presented results representing the mean performance. This step was necessary due to two stochastic elements of the generation process: the random selection of image subsets from the original training set; and the random initial placement of cluster centroids, resulting in final clusters of varying density and separation.

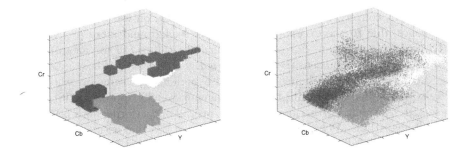

Fig. 2. Two examples of colour look-up tables generated for the training set of 50 images: a manually tuned LUT (left), created within the NUview utility using a point-and-click approach over the period of 1 hour; and an autonomously generated LUT (right), utilising k-means clustering over a subset of 30 training images and generated in less than 10 seconds.

5.1 Classification Performance

Fig. 3 presents the mean sensitivity and specificity values for LUTs produced using the k-means (blue), expectation maximisation (red) and mean shift (magenta) algorithms. These plots represent data collected from 3600 LUTs, each tested over 50 images, resulting in the binary classification of more than of 1.2 million features.

Some interesting observations follow from these results. Firstly, the specificity curve for all methods follow the expected trend, decreasing as the number of training images increases as the result of over-classification of colour. Secondly, the sensitivity curve for mean shift and expectation maximisation confirmed the

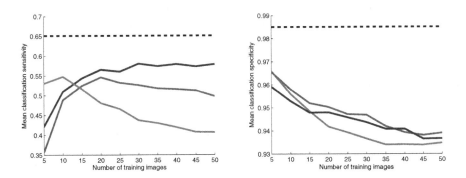

Fig. 3. Comparison of colour look-up table feature classification sensitivity (left) and specificity (right) as a function of the number of training images, for k-means (blue), expectation maximisation (red) and mean shift, utilising a flat kernel (magenta). The performance of a manually tuned LUT is indicated by the dashed lines. It is evident that k-means clustering yields the best classification sensitivity for most training set sizes, with comparatively little difference evident between the specificities of each method.

initial hypothesis: poor classification for too few images due to a lack of training data. Similarly, convergence to relatively poor classification was evident for too many images, due to the excess classification of background noise preventing effective feature extraction. For some intermediate values (10 for mean shift, and 20 for EM), optimum sensitivity is achieved.

Fig. 3 demonstrates that k-means does not experience the same overshoot in sensitivity as EM and mean shift, but rather asymptotically approaches its optimum sensitivity value. Furthermore, k-means overall outperforms both EM and mean shift algorithms. As k-means is a specific case of EM that assumes underlying hyperspherical distributions of data, it can be inferred that human-perceived salient colours may form approximately spherical distributions within the YC_bC_r colour space, therefore allowing more optimal partitions to be formed within the algorithm's internal iteration limit. This observation is consistent with earlier research suggesting that clusters of maximum density and separation occur within the YC_bC_r space, and correspond with qualitatively accurate segmentation of an image into regions of salient colour [2].

5.2   Optimal Method Selection and Evaluation

From the previous experimentation, it has been demonstrated that the best performing colour look-up tables are produced by k-means clustering. Given this information, the remaining task is to determine the optimal set of internal clustering parameters. Specifically, the MATLAB implementation of k-means clustering is parameterised by two values of interest: k, the initial number of parameters (later reduced to the number of colour class labels by the LUT generation implementation); and the *iteration count*[3]. For an iteration count greater than 1 (the default value used previously in experimentation), each clustering is repeated an equal number of times, with only the clusters yielding the best spatial properties (lowest average distance from centroids to corresponding points) being retained. Although it is intuitive that clusters of improved spatial properties may yield LUTs which exhibit improved classification performance, this correlation has not been experimentally verified.

The experimental range for k was set to be from 5 to 14 (with the previous default as 10), and that for the iteration count set from 1 to 10. For each of the 100 resulting pairs of parameter values, 120 LUTs were generated, resulting final autonomous creation of 15,600 LUTs and respective evaluation of more than 5 million binary features. The average sensitivity and specificity for each parameter pair are presented in Fig. 4.

Fig. 4 demonstrates that increasing both the number of clusters and number of iterations results in an overall diminishing performance increase, in terms of both sensitivity and specificity. The optimal pair of parameters was experimentally determined to be: number of clusters $k = 11$, and number of iterations $= 10$.

---

[3] Note that "number of iterations" refers to the MATLAB parameter (i.e. complete reruns of the clustering process), with each allowed to run until convergence. The cluster assignments retained are those that maximise cluster density and separation.

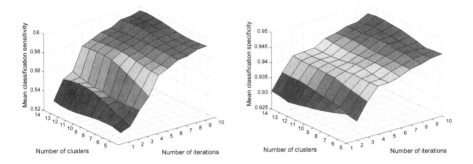

Fig. 4. Mean classification performance for colour look-up tables generated by k-means clustering, parameterised by both the number of clusters and clustering iteration count, presented in terms of sensitivity (left) and specificity (right). See Sec. 4.4 for definitions.

The mean performance of the 120 LUTs generated for this parameter pair is presented in Table 1, along with the performances of the manually generated benchmark LUT, and the overall best performing LUT from the set of 120.

Table 1. Salient feature classification performance over 50 images, for 120 LUTs generated using k-means clustering ($k = 11$, number of iterations $= 10$). The best generated LUT yields an 8% improvement sensitivity for zero tradeoff in specificity.

	Benchmark	Mean	Best	Performance
Sensitivity	0.6596	0.5964	0.7163	1.08
Specificity	0.9768	0.9465	0.9747	1.00

It is evident that, although the mean performance of the 120 generated LUTs is overall slightly worse than the manually generated benchmark, the best performing LUTs yield an 8% increase in classification sensitivity, for zero tradeoff in specificity.

6   Conclusion

It has been demonstrated that a system can be developed to solve a seemingly subjective problem (the mapping of millions of pixel values to a small set of colour class labels) without human supervision, despite the inherent similarity to subconscious human abstraction. Moreover, it has been demonstrated that, in addition to solving the problem in orders of magnitude less time than is required by a human expert (in the order of seconds, rather than hours), the autonomous colour look-up table (LUT) generation system is able to produce LUTs that yield up to an 8% improvement in classication performance over those manually tuned within the NUview utility. Although demonstrated solely within a RoboCup robot soccer environment, this development has strong potential in any image processing or machine vision system requiring real-time recognition of coloured features.

Acknowledgement. NICTA is funded by the Australian Government as represented by the Department of Broadband, Communications and the Digital Economy and the Australian Research Council through the ICT Centre of Excellence program.

References

1. Bishop, C.: Pattern recognition and machine learning, vol. 4. Springer (2006)
2. Budden, D., Fenn, S., Mendes, A., Chalup, S.: Evaluation of colour models for computer vision using cluster validation techniques. In: Chen, X., Stone, P., Sucar, L.E., van der Zant, T. (eds.) RoboCup 2012. LNCS (LNAI), vol. 7500, pp. 261–272. Springer, Heidelberg (2013)
3. Budden, D., Fenn, S., Walker, J., Mendes, A.: A novel approach to ball detection for humanoid robot soccer. In: Thielscher, M., Zhang, D. (eds.) AI 2012. LNCS, vol. 7691, pp. 827–838. Springer, Heidelberg (2012)
4. Comaniciu, D., Meer, P.: Mean shift: A robust approach toward feature space analysis. IEEE Transactions on Pattern Analysis and Machine Intelligence 24(5), 603–619 (2002)
5. Hartigan, J., Wong, M.: Algorithm AS 136: A k-means clustering algorithm. Journal of the Royal Statistical Society. Series C (Applied Statistics) 28(1), 100–108 (1979)
6. Kitano, H., Asada, M., Kuniyoshi, Y., Noda, I., Osawa, E.: Robocup: The robot world cup initiative. In: Proceedings of the First International Conference on Autonomous Agents, pp. 340–347. ACM (1997)
7. Mahalanobis, P.C.: On the generalized distance in statistics. In: Proceedings of the National Institute of Sciences of India, New Delhi, vol. 2, pp. 49–55 (1936)
8. Quinlan, M.J., Chalup, S.K., Middleton, R.H.: Application of svms for colour classification and collision detection with aibo robots. In: Advances in Neural Information Processing Systems, vol. 16 (2003)
9. Szeliski, R.: Computer vision: algorithms and applications. Springer (2010)
10. Witten, I.H., Frank, E.: Data Mining: Practical machine learning tools and techniques. Morgan Kaufmann (2005)

Improved Particle Filtering for Pseudo-Uniform Belief Distributions in Robot Localisation

David Budden and Mikhail Prokopenko

Information and Communications Technologies Centre, Adaptive Systems
Commonwealth Scientific and Industrial Research Organisation (CSIRO)
PO Box 76, Epping, NSW 2121, Australia
{david.budden,mikhail.prokopenko}@csiro.au

Abstract. Self-localisation, or the process of an autonomous agent determining its own position and orientation within some local environment, is a critical task in modern robotics. Although this task may be formally defined as a simple transformation between local and global coordinate systems, the process of accurately and efficiently determining this transformation is a complex task. This is particularly the case in an environment where localisation must be inferred entirely from noisy visual data, such as the RoboCup robot soccer competitions. Although many effective probabilistic filters exist for solving this task in its general form, pseudo-uniform belief distributions (such as those arising from course-grain observations) exhibit properties allowing for further performance improvement. This paper explores the RoboCup 2D Simulation League as one such scenario, approximating the artificially constrained noise models as uniform to derive an improved particle filter for self-localisation. The developed system is demonstrated to yield from 38.2 to 201.3% reduction in localisation error, which is further shown as corresponding with a 6.4% improvement in goal difference across approximately 750 games.

Keywords: Robotics, localisation, particle filter, robot soccer.

1 Introduction

The RoboCup 2D Simulation League incorporates a number of critical challenges in the areas of artificial intelligence, machine learning and distributed computing [6,9]. These include, but are not limited to:

- Distributed client/server model, introducing the challenges of fragmented, localised and imprecise information (both in terms of noise and latency) about the environment [8].
- Asynchronous perception-action activity, and a limited window of opportunity to perform a desired action [5].
- No centralised controllers or central world model, resulting in a lack of global vision or localisation information [10,11].

S. Behnke et al. (Eds.): RoboCup 2013, LNAI 8371, pp. 385–395, 2014.

This paper focuses on the final point; specifically, how to infer localisation information from noisy observations of the environment.

In its simplest form, the task of localisation (particularly self-localisation of an agent) can be viewed as a problem of coordinate transformation [13]; specifically, determining the transformation between the agent's local coordinate system and the environment's global coordinate system. Knowledge of this transformation allows the agent to consider global features (such as unique markers and goalposts) with reference to its own coordinate frame, facilitating navigation and execution of more complex behaviours.

Knowing the position and orientation of an agent is both sufficient and necessary for determining this coordinate transformation. As the RoboCup 2D Simulation League does not provide each agent with global vision information, the position and orientation must be inferred from observing unique and stationary *markers* within the environment. In a traditional robotics scenario, the agent employs physical sensors (such as cameras or range-finders) and computer vision techniques [3,4] to infer the relative position and orientation of such markers. These observations typically correspond with a well-defined Gaussian error, allowing for the effective applications of Kalman and particle filtering techniques [12,13].

Although the need for image processing is removed, the RoboCup 2D Simulation League introduces a number of unique challenges to the task of self-localisation. Particularly, the artificial observational noise introduced by the server is non-Gaussian (and approximately uniform within well-defined radial regions). The remainder of this paper presents a formal definition of the approximate positional error region given an arbitrary number of observations, and describes a modified particle filter that employs knowledge of the noise model to increase self-localisation accuracy. A linear approximation step is introduced to reduce the algorithm's time complexity, with the final performance results compared to the well-known agent2D implementation developed by Akiyama *et. al.* [1].

2   Particle Filter

The *particle filter* is a nonparametric implementation of the *Bayes filter*, where the posterior distribution $bel(x_t)$ is approximated by a set of random state samples (*particles*) drawn from this posterior [13]. Concretely, the set $\mathcal{X}_t$ of M particles are denoted

$$\mathcal{X}_t = \left\{ x_t^{[1]}, x_t^{[2]}, \ldots, x_t^{[M]} \right\},$$

where each particle $x_t^{[m]}$ $(1 \leq m \leq M)$ is an instantiation of the state at time t. Such particles are proportional to the Bayes filter posterior distribution

$$x_t^{[m]} \sim bel(x_t) = p(x_t | z_{1:t}, u_{1:t}),$$

where $z_{1:t}$ is the set of all previous *observations* and $u_{1:t}$ the set of all previous *actions*. Conceptually, the particle filter consists of the following steps [13]

Algorithm 1. `particleFilter(`$\mathcal{X}_{t-1}, u_t, z_t$`)`

$\overline{\mathcal{X}}_t = \mathcal{X}_t = \emptyset$

for $m = 1 \ldots M$ **do**

  Sample $x_t^{[m]} \sim p(x_t | u_t, x_{t-1}^{[m]})$

  $w_t^{[m]} = p(z_t | x_t^{[m]})$

  $\overline{\mathcal{X}}_t = \overline{\mathcal{X}}_t + \langle x_t^{[m]}, w_t^{[m]} \rangle$

end for

for $m = 1 \ldots M$ **do**

  draw i with probability $\propto w_t^{[i]}$

  add $x_t^{[i]}$ to $\mathcal{X}_t$

end for

return $\mathcal{X}_t$

---

1. **Action update:** Sample M particles $\overline{\mathcal{X}}_t = \left\{ x_t^{[1]}, x_t^{[2]}, \ldots, x_t^{[M]} \right\}$ at random from the state transition distribution $p(x_t | z_{1:t}, u_{1:t})$. These particles form a nonparametric approximation of $\overline{bel}(x_t)$, where $\overline{bel}(x_t) = p(x_t | z_{1:t-1}, u_{1:t})$ is the *prediction* posterior before considering observation z_t.
2. **Calculate weights:** Calculate an *importance factor* (*weight*) $w_t^{[m]} = p(z_t | x_t^{[m]})$ for each particle $x_t^{[m]}$ given observation z_t. Add each weighted particle $\langle x_t^{[m]}, w_t^{[m]} \rangle$ to the temporary particle set $\overline{\mathcal{X}}_t$.
3. **Resample:** Draw M particles (with replacement) from $\overline{\mathcal{X}}_t$ and add to the particle set $\mathcal{X}_t$. The probability of drawing the particle $x_t^{[m]}$ is proportional to the corresponding weight $w_t^{[m]}$, and often results in the inclusion of duplicate particles (and consequent exclusion of particles with low importance weight).

Similarly to the Bayes filter, the particle filter recursively calculates the posterior $bel(x_t) \sim \mathcal{X}_t$ from the prior $bel(x_{t-1}) \sim \mathcal{X}_{t-1}$ by considering the observation z_t and action u_t at time t. A basic implementation `particleFilter(`$\mathcal{X}_{t-1}, u_t, z_t$`)` of such a particle filter is defined in Algorithm 1.

3 Problem Model

As described in Sec 1, a lack of centralised controllers or a central world model within the RoboCup 2D simulation league results in a lack of global vision or localisation information. Agents are therefore required to self-localise by observing the positions of a number of uniquely identifiable markers, as illustrated in Fig. 1. The soccer server then introduces artificial noise to each observation. This section provides a concrete definition of this error model, as well as deriving the approximated self-localisation error that results from N marker observations.

3.1 Non-gaussian Observation Noise

As described in Sec. 1, an agent r in a RoboCup 2D Simulation League scenario self-localises by observing a number of fixed markers $m^{[n]}$, with the global

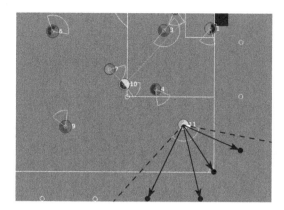

Fig. 1. Visualisation of an agent r making observations $r_m^{[n]}$ of markers $m^{[n]}$. The soccer server [6] introduces noise (as defined in (3.1)) to each observation, resulting in the self-localisation error model defined in Sec. 3.2.

position of each marker known a priori. If r were able to make noise-free measurements of the environment, a single observation of any such marker would result in zero localisation error. Instead, the soccer server [6] introduces artificial noise to each observation, such that the noise magnitude is proportional to the actual distance between agent and marker.

In a traditional localisation problem, such noise could be accurately modeled by a Gaussian distribution $\mathcal{N}(\mathbf{r}_m, \sigma)$ centered about the observed marker location $\mathbf{r}_m$, allowing for the implementation of well-known Kalman or particle filtering techniques [12,13]. The soccer server instead implements a more complex noise model $\mathcal{N}_s(d, \Delta_1, \Delta_2)$, involving the quantisation of exponential terms [6].

Concretely, Given the actual distance d of an agent from an observed marker, the noisy distance d' communicated by the server to the agent is defined as

$$\mathcal{N}_s : d \to d' = Q\left(e^{Q(\log(d), \Delta_1)}, \Delta_2\right),$$

where Δ_1 and Δ_2 represent the *quantisation step sizes* (set to 0.01 and 0.1 respectively), and the *quantisation function* Q is defined as

$$Q(x, y) = y \left\lceil \frac{x}{y} \right\rceil.$$

Although the nonparametric nature of a particle filter is suitable for such a distribution (by approximating the posterior distribution $bel(x_t)$ as a set of particles $\mathcal{X}_t$), further approximating this complex distribution as uniform within a radial region (corresponding with the agent's radial field of view[1]) allows for a simplified filter implementation. The remainder of this section derives a formal definition of the resultant self-localisation noise distribution.

---

[1] The field of view of an agent r is chosen to be either π, $2\pi/3$ or $\pi/3$ radians, depending on the observed distance of the ball from r.

3.2 Self-localisation Error Model

One Observed Marker. The uncertainty in the position of a marker observed at $\mathbf{r}_m = (x_m, y_m)^\top$ resulting from observation $\mathbf{r}$ by an agent r at $\mathbf{r}_r = (x_r, y_r)^\top$ may be approximated as an annulus sector D of uniform probability, as illustrated in Fig. 2a. D may be concretely defined as

$$D = \left\{ \mathbf{p} \in \mathbb{R}^2 : \left| \|\mathbf{p} - \mathbf{r}_r\| - \|\mathbf{r}_r - \mathbf{r}_m\| \right| \leq \Delta_r, \quad |\phi_p| \leq \Delta_\theta \right\},$$

where Δ_θ and Δ_r parameterise the maximum angular and distance errors (as demonstrated in Fig. 2), and $\phi_p = \arccos\left(\frac{(\mathbf{r}_m - \mathbf{r}_r) \cdot (\mathbf{p} - \mathbf{r}_r)}{\|\mathbf{r}_m - \mathbf{r}_r\| \|\mathbf{p} - \mathbf{r}_r\|} \right)$. The uniform distribution D only has physical meaning when considering coordinates relative to the robot (as the exact position of the marker is known a priori). For the purpose of self-localisation, we are interested instead in the (closed) region E: the set of all possible locations $\mathbf{q}$ of the agent r resulting from observation $\mathbf{r}$.

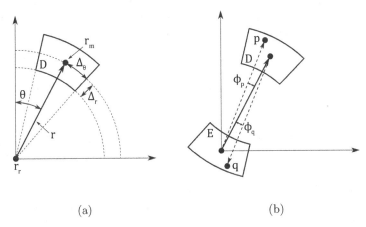

(a)   (b)

Fig. 2. Visualisation of the region D defining all possible locations of a marker m, given that an agent r at $\mathbf{r}_r$ observes m at $\mathbf{r}_m$ (a). As the position of m is known a priori, this observation results in the region E of all possible locations of the agent (b).

Fig. 2b illustrates the construction of the region E. As the position of the marker is known to be $\mathbf{r}_m$, it follows that a observation of the marker residing at $\mathbf{p} = (p_x, p_y)^\top \in D$ corresponds with an error of $\mathbf{p} - \mathbf{r}_m$. As the agent knows the absolute position of the marker a priori, it follows that positional self-belief may be defined as $\mathbf{q} = (q_x, q_y)^\top = \mathbf{r}_r - (\mathbf{p} - \mathbf{r}_m)$. It may be demonstrated that the equivalent mapping

$$T : \mathbf{p} \mapsto \mathbf{q} = (\mathbf{r}_r + \mathbf{r}_m) - \mathbf{p} \tag{1}$$

results in the image

$$E = \left\{ \mathbf{q} \in \mathbb{R}^2 : \mathbf{q}' = \mathcal{T}(\mathbf{r}_r)\mathcal{R}(\pi)\mathcal{T}(-\mathbf{r}_m)\mathbf{p}' \right\},$$

where $\mathbf{q'} = (q_x, q_y, 1)^\top$ and $\mathbf{p'} = (p_x, p_y, 1)^\top$ are the homogeneous representations of $\mathbf{p} \in D$ and $\mathbf{q} \in E$ respectively, $\mathcal{T}(\mathbf{v})_{3,3}$ is the homogenous translation matrix by 2-dimensional vector $\mathbf{v}$, and $\mathcal{R}(\theta)_{3,3}$ is the homogenous rotation matrix by angle θ[2]. Concretely,

$$\mathbf{q'} = \begin{bmatrix} \mathbf{q} \\ \hline 1 \end{bmatrix} = \mathcal{T}(\mathbf{r}_r)\mathcal{R}(\pi)\mathcal{T}(-\mathbf{r}_m)\mathbf{p'}$$

$$= \begin{bmatrix} \mathbf{I}_2 & \mathbf{r}_r \\ \hline \mathbf{0} & 1 \end{bmatrix} \begin{bmatrix} -\mathbf{I}_2 & \mathbf{0} \\ \hline \mathbf{0} & 1 \end{bmatrix} \begin{bmatrix} \mathbf{I}_2 & -\mathbf{r}_m \\ \hline \mathbf{0} & 1 \end{bmatrix} \begin{bmatrix} \mathbf{p} \\ \hline 1 \end{bmatrix}$$

$$= \begin{bmatrix} \mathbf{r}_r + \mathbf{r}_m - \mathbf{p} \\ \hline 1 \end{bmatrix}$$

$$\therefore \mathbf{q} = (\mathbf{r}_r + \mathbf{r}_m) - \mathbf{p},$$

where $\mathbf{I}_2$ is the 2-dimensional identity matrix. This result is equivalent to the mapping T defined in (1), confirming that E is the image of D under T. The region E, representing all possible locations of an agent r observing a single marker at $\mathbf{r}_m$ (as illustrated in Fig. 2), may finally be represented as

$$E = \left\{ \mathbf{q} \in \mathbb{R}^2 : \left| \|\mathbf{q} - \mathbf{r}_m\| - \|\mathbf{r}_r - \mathbf{r}_m\| \right| \leq \Delta_r, \quad |\phi_q| \leq \Delta_\theta \right\}. \tag{2}$$

This represents the intuitive definition of the region E; the region D rotated 180 degrees about the midpoint of $\mathbf{r}_r$ and $\mathbf{r}_m$.

Multiple Observed Markers. In a RoboCup 2D simulation league scenario, it is far more commonplace for an agent to observe not one, but N markers $m^{[n]}$ ($n \in [1, \cdots, N]$), each at a separate position known a priori. This results in N regions of uniform observation belief $D^{[1]}, \cdots, D^{[N]}$, which map to N corresponding regions of positional self-belief $E^{[1]}, \cdots, E^{[N]}$ as per Sec. 3.2. As the probability $p(\mathbf{q} \notin E^{[n]}) = 0$ ($\forall n \in [1, \cdots, N]$), it follows that the final region of self-belief E resulting from N such observations may be expressed as

$$E = \bigcap_{n=1}^{N} E^{[n]}, \quad E \neq \emptyset \tag{3}$$

where $E^{[n]}$ is the region representing all possible locations of an agent r given an observation of marker $m^{[n]}$, defined as per (1).

4   Implementation

This section details the implementation of the Gliders2013 localisation system, based on the particle filter framework detailed in Sec. 2 and problem model of Sec. 3.

---

[2] Homogeneous representation of an $(n - 1)$-dimensional transformation as an n-dimensional matrix allows for any affine transformation to be decomposed into the product of elementary shear, scaling, rotation and translation matrices [7].

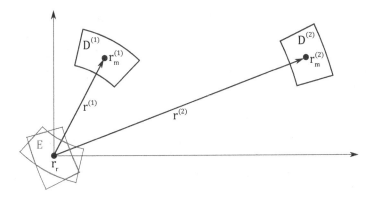

Fig. 3. Visualisation of the region E defining all possible locations of an agent r at $\mathbf{r}_r$, given that r observes markers $m^{[1]}$ and $m^{[2]}$ at $\mathbf{r}_m^{[1]}$ and $\mathbf{r}_m^{[2]}$ respectively

4.1   Action Update

As described in Sec. 2, the particle filter *action update* step involves sampling M particles $\overline{\mathcal{X}}_t = \left\{ x_t^{[1]}, x_t^{[2]}, \ldots, x_t^{[M]} \right\}$ at random from the state transition distribution $p(x_t | z_{1:t}, u_{1:t})$. In the Gliders2013 implementation, this nonparametric approximation of $\overline{bel}(x_t)$ is defined as

$$\overline{\mathcal{X}}_t = \left\{ x_t^{[m]} \in \mathbb{R}^2 : x_t^{[m]} = x_{t-1}^{[m]} + \mathbf{d}_{t-1} \quad \forall\, m \in [1, \cdots, M] \right\},$$

where $\mathbf{d}_{t-1}$ is the displacement resulting from the agent's action at time $t - 1$, and the initial set of particles $\overline{\mathcal{X}}_0$ are chosen as M random positions satisfying a single observation of the nearest marker $m^{[n]}$.

4.2   Calculate Weights

As described in Sec. 2, the particle filter *calculate weights* step involves calculating an *importance factor* (*weight*) $w_t^{[m]} = p(z_t | x_t^{[m]})$ for each particle $x_t^{[m]}$ given observation z_t. As the problem model described in Sec. 3 involves only uniform probability distributions, these weights are simply defined as

$$w_t^{[m]} = \begin{cases} 1 & \text{if } x_t^{[m]} \in E_t \\ 0 & \text{otherwise} \end{cases}, \tag{4}$$

where E_t is the region E defined in (3), for observations of markers $m^{[n]}$ $(1 \leq m \leq M)$ at time t.

4.3   Resample

As described in Sec. 2, the particle filter *resample* step involves drawing M particles from $\overline{\mathcal{X}}_t$ to add to the particle set $\mathcal{X}_t$, with the probability of drawing the particle $x_t^{[m]}$ proportional to the corresponding weight $w_t^{[m]}$. As the particle weights defined in (4) take only binary values, it follows that any particle $x_t^{[m]} \notin E_t$ will never be drawn. Similarly, any particle $x_t^{[m]} \in E_t$ will be drawn with equal probability. Following this observation, the Gliders2013 implementation of the posterior $\mathcal{X}_t \sim bel(x_t)$ is defined as

$$\mathcal{X}_t = \overline{\mathcal{X}}_t \cap E_t \cup \mathcal{Y}_t, \tag{5}$$

where $\mathcal{Y}_t$ is a set of $|\overline{\mathcal{X}}_t - \mathcal{X}_t|$ points drawn at random from the uniform distribution E_t. This maintains a set of M particles between iterations, where each particle is a valid positional self-belief given all observations at time t^3.

5   Results

The performance of the presented Gliders2013 localisation system was evaluated against two metrics: the effect on self-localisation error, explicitly measurable by calculating the difference between server and self-belief player positions; and the effect on overall team performance, implicitly measured through the goal statistics of a statistically significant number of games.

Fig. 4 demonstrates the results for self-localisation error. Specifically, average self-localisation error (for all players) is plotted as a function of ball distance, for the previous agent2D [1] localisation system (blue) and introduced Gliders2013 particle filter system (red). Mean error across 10 games (60,000 cycles, or approximately 750,000 measurements) is presented as a solid line, with the corresponding standard deviation presented as a dashed line. As the x-axis represents the maximum distance from considered players to the ball, values approaching the right of the graph indicate the total mean, whereas values toward the left indicate the performance of players in critical positions (i.e. very close to the ball).

It can be seen from Fig. 4 that the Gliders2013 localisation system yields a mean self-localisation error reduction of 1.7 cm; an overall improvement of 38.2%. More significant is the improvement for players in close proximity to the ball, where localisation performance is especially critical. Concretely, self-localisation error decreases by 6.1 cm for players within 1 meter of the ball; an improvement of 201.3%.

Although these results appear significant as a percentage, a reduction of localisation error by centimeters may seem insignificant in the context of a full-sized soccer field. It is therefore critical that these results be demonstrated to correspond with a statistically significant improvement in game performance. To establish the extent of the overall team improvement brought upon by the increased localisation accuracy, multiple iterative experiments were carried out matching Gliders2012 up against teams not based on agent2D [1]. Among the

---

^{3} Although this definition of the posterior $\mathcal{X}_t$ does not map exactly to the formal particle filter framework presented in Sec. 2, it is better suited to the scenario of uniform distributions of positional self-belief.

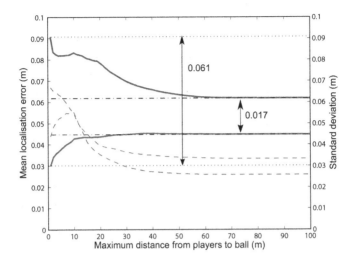

Fig. 4. Self-localisation error as a function of ball distance, for the previous agent2D [1] localisation system (blue), and introduced Gliders2013 particle filter system (red). Mean error is presented as a solid line, with standard deviation presented as a dashed line. An overall localisation improvement of 38.2% (or 1.7 cm) is evident, and a dramatic improvement of 201.3% (6.1 cm) for players within 1 meter of the ball (when accurate performance is most critical). Results presented are for 10 games (60,000 cycles, or ∼ 750,000 measurements) between Gliders2013 and WrightEagle.

latter class, WrightEagle [2] (the runner-up of RoboCup 2012) was selected. Approximately 750 games were conducted against this benchmark team in total, both for the baseline Gliders2012 team, and Gliders2012 incorporating the improved Gliders2013 localisation system (such that this system is the only variable between teams). These results are presented in Table 1.

Table 1. Game statistics between the Gliders2012 baseline team and WrightEagle [2] (the runner-up of RoboCup 2012), for both the old (agent2D [1]) and new (Gliders2013) localisation systems. As these results have been generated over approximately 750 games and the localisation system is the only variable between Gliders2012 teams, it can be confidently inferred that the 6.4% improvement in goal difference results directly from the 6.1 cm reduction in self-localisation error (for critically-positioned players; 1.7 cm overall).

	Old Localisation	New Localisation	Improvement
Goals Scored	0.36	0.41	14%
Goals Conceded	2.19	2.13	2.8%
Goal Difference	-1.83	-1.72	6.4%

6   Conclusion

Self-localisation is a complex yet critical task, both within the RoboCup 2D Simulation League, and the multi-billion dollar robotics industry at large. It allows an agent to infer its own position within some local environment, thus facilitating navigation around or interaction with obstacles, features and other agents. This paper has described a modified particle filter algorithm for non-Gaussian belief distributions, and demonstrated how approximating these distributions as uniform (given constrained noise models, such as those resulting from course-grain observations or artificial environments) facilitates reduction in self-localisation error. The proposed system yielded an improvement of 38.2% for agents within a RoboCup 2D Simulation League environment, which increased to 201.3% as agents approach performance-critical regions of the field. It was further demonstrated that this improvement results in a 14% increase in goals scored and 2.8% reduction in goals conceded for the Gliders2012 baseline team, evaluated across approximately 750 games against the WrightEagle team [2] (RoboCup 2012 runners-up).

Future research will focus on applying similar methodologies to ball and player localisation systems (including the incorporation and optimisation of team communication strategies), in an attempt to approach the performance leveraged by enabling full game state information for all player agents.

Acknowledgements. The authors would like to thank the University of Newcastle's NUbots RoboCup team in assisting with the preparation of this manuscript.

References

1. Akiyama, H.: Agent2D Base Code (2010), http://www.rctools.sourceforge.jp
2. Bai, A., Zhang, H., Lu, G., Jiang, M., Chen, X.: Gliders2012 wrighteagle 2d soccer simulation team description 2012. In: RoboCup 2012 Symposium and Competitions: Team Description Papers, Mexico City, Mexico (June 2012)
3. Budden, D., Fenn, S., Mendes, A., Chalup, S.: Evaluation of colour models for computer vision using cluster validation techniques. In: Chen, X., Stone, P., Sucar, L.E., van der Zant, T. (eds.) RoboCup 2012. LNCS (LNAI), vol. 7500, pp. 261–272. Springer, Heidelberg (2013)
4. Budden, D., Fenn, S., Walker, J., Mendes, A.: A novel approach to ball detection for humanoid robot soccer. In: Thielscher, M., Zhang, D. (eds.) AI 2012. LNCS, vol. 7691, pp. 827–838. Springer, Heidelberg (2012)
5. Butler, M., Prokopenko, M., Howard, T.: Flexible synchronisation within RoboCup environment: A comparative analysis. In: Stone, P., Balch, T., Kraetzschmar, G. (eds.) RoboCup 2000. LNCS (LNAI), vol. 2019, pp. 119–128. Springer, Heidelberg (2001)
6. Chen, M., Dorer, K., Foroughi, E., Heintz, F., Huang, Z., Kapetanakis, S., Kostiadis, K., Kummeneje, J., Murray, J., Noda, I., et al.: Robocup soccer server, Manual for Soccer Server Version 7 (2003)

7. Hill, F., Kelley, S.: Computer graphics: using openGL. Prentice Hall, Upper Saddle River (2001)
8. Noda, I., Stone, P.: The RoboCup Soccer Server and CMUnited Clients: Implemented Infrastructure for MAS Research. Autonomous Agents and Multi-Agent Systems 7(1-2), 101–120 (July-September)
9. Prokopenko, M., Obst, O., Wang, P., Held, J.: Gliders2012: Tactics with action-dependent evaluation functions (2012)
10. Prokopenko, M., Wang, P.: Evaluating team performance at the edge of chaos. In: Polani, D., Browning, B., Bonarini, A., Yoshida, K. (eds.) RoboCup 2003. LNCS (LNAI), vol. 3020, pp. 89–101. Springer, Heidelberg (2004)
11. Prokopenko, M., Wang, P.: Relating the entropy of joint beliefs to multi-agent coordination. In: Kaminka, G.A., Lima, P.U., Rojas, R. (eds.) RoboCup 2002. LNCS (LNAI), vol. 2752, pp. 367–374. Springer, Heidelberg (2003)
12. Russell, S., Norvig, P., Canny, J., Malik, J., Edwards, D.: Artificial intelligence: a modern approach, vol. 2. Prentice Hall, Englewood Cliffs (1995)
13. Thrun, S., Burgard, W., Fox, D.: Probabilistic robotics, vol. 1. MIT Press, Cambridge (2005)

Robust and Efficient Object Recognition for a Humanoid Soccer Robot

Alexander Härtl[1], Ubbo Visser[1], and Thomas Röfer[2]

[1] University of Miami, Department of Computer Science,
1365 Memorial Drive, Coral Gables, FL, 33146 USA
{a.haertl,visser}@cs.miami.edu
[2] Deutsches Forschungszentrum für Künstliche Intelligenz,
Cyber-Physical Systems, Enrique-Schmidt-Str. 5, 28359 Bremen, Germany
thomas.roefer@dfki.de

Abstract. Static color classification as a first processing step of an object recognition system is still the de facto standard in the RoboCup Standard Platform League (SPL). Despite its efficiency, this approach lacks robustness with regard to changing illumination. We propose a new object recognition system where objects are found based on *color similarities*. Our experiments with line, goal, and ball recognition show that the new system is real-time capable on a contemporary NAO (version 3.2 and above). We show that the detection rate is comparable to color-table-based object recognition under static lighting conditions and substantially better under changing illumination.

1 Introduction

Color-based recognition of geometrically simple objects is a well known problem that has been extensively studied for over two decades [18,5]. Prominent examples for successful image processing methods are edge detection [2,10], region-growth algorithms [21,8], and histogram-based algorithms [12,19,14]. A popular approach is based on edge detection and subsequent Hough transformation [4] in which the authors show a method for line detection that can be used for more general curve fitting. Although the literature shows a broad range of variations and implementations of the mentioned approaches, only a few can be used for embedded systems that are constrained by limited resources such as time, memory, and/or CPU power.

The RoboCup Soccer environment demands efficient real-time object recognition. Many systems are still based on fixed color tables that are similar or based on the *CMVision* system [1]. It is well suited for static lighting conditions but lacks robustness when illumination varies. Röfer [15] improved the robustness by introducing ambiguous color classes and delaying hard decisions to a later processing stage. Reinhardt [14] uses different heuristics applied to color histograms to cope with variations in illumination.

We propose a new object recognition system where objects are found based on *color similarities*. As a first step, a subsampling is created considering the

S. Behnke et al. (Eds.): RoboCup 2013, LNAI 8371, pp. 396–407, 2014.

perspective projection of objects on the field plane (Section 2.1). Based on the subsampling, the image is segmented line-wise considering color similarities of neighboring pixels, similar to region growing (Section 2.2). The actual object recognition (Section 3) is based on the result of this segmentation. Line and ball detection are based on region growing, considering the shape of the objects, whereas the goal detection is histogram-based.

We have conducted our experiments for line, goal, and ball recognition (Section 4) and our results show that it is real-time capable on a contemporary NAO (version 3.2 and above).

2 Preprocessing

Processing the image in its full resolution (YUV422 640×480) is computationally expensive and would not allow for real-time operation. Therefore, the effective resolution has to be reduced. The naïve approach would be to reduce the resolution for the actual processing globally. This, however, does not account for the perspective projection of objects on the field plane: closer objects appear larger in the image. In order to rectify this effect the image is locally subsampled with different resolutions.

2.1 Subsampling

The image is split into two areas: a) the field plane and b) the area above this plane, for which a static subsampling is applied with a high horizontal and low vertical resolution suitable for goal post detection. The basic idea for a) is to project a homogeneous grid on the field plane into the image [11]. However, the exact projection of a grid is approximated by two means to increase the efficiency of the following segmentation process. First, the rotation about the image axis is ignored because a humanoid robot operates mostly upright and therefore the camera possesses a small rotation about that axis. This has been verified experimentally, which showed that the average deviation while walking and moving the head is only 1.15 degrees [6]. The second approximation limits the sampling points within a line to a fixed interval, i.e., the distance between the sampling points is limited to a fixed number of pixels.

Both approximations decrease the complexity and therefore increase the efficiency of the subsequent segmentation process. The resulting subsampling is reduced to rectangular blocks of homogeneous grids of sampling points, as shown in Figure 1. The computation of the grid is thus reduced to finding the coordinates of the borders of those blocks of different resolutions. Note that the computation of the subsampling requires the camera pose to be known; here it is computed based on sensor fusion of forward kinematics and the IMU (cf. [16]). First, the border between the upper and lower area is found by projecting a point in the distance of the field diagonal into the image plane. This is motivated by the fact that the farthest object to be recognized on the field plane has a

(a)   (b)

Fig. 1. Visualization of the subsampling using greyed out pixels (a) and successive supersampling (b). The resolution within the field plane incorporates the perspective projection, whereas the area above the field plane (marked by the red line) is subsampled statically. It can be seen that the shape of all objects remains sound despite the subsampling, regardless of their distance.

distance of the field diagonal. Second, the borders between the blocks of different granularity within the lower area are found using the following formula:

$$y_\Delta = \frac{\Delta_f\, h + \Delta\sqrt{\mathbf{R}_{11}^2 + \mathbf{R}_{12}^2}}{\Delta\left(\mathbf{R}_{13}\, f - \mathbf{c}_y\,\sqrt{\mathbf{R}_{11}^2 + \mathbf{R}_{12}^2}\right)} \tag{1}$$

with Δ being the desired pixel spacing in the image plane, Δ_f the mesh size in the field plane, f the effective focal length of the camera, $\mathbf{R}$ the rotation matrix of the camera, $\mathbf{c}$ the optical center, and h the height of the camera above the field plane. To compute the blocks, y_Δ is computed for increasing pixel spacings Δ, beginning with 1, which delivers the vertical positions of the borders of the blocks in the image plane.

The mesh size of the grid in the field plane has a significant influence on both the processing time and the recognition accuracy. A fine mesh size results in a large processing time, whereas a coarse mesh size results in an impaired detection rate and accuracy. Therefore, a good compromise has to be found. Having in mind that the resulting system is supposed to be real-time capable, it is desirable to pursue a constant processing time. Assuming that the processing time is proportional to the number of sampling points, the mesh size should be chosen in a way that a certain constant number of sampling points is generated. We use the secant method [13] as a numerical approximation algorithm to find an appropriate value for Δ_f, which usually converges after approx. 5 iterations.

2.2   Segmentation

After the subsampling is determined, it is used to segment the camera image. The general idea of the segmentation process is based on the popular *region growing* [9], which has been modified to reduce the processing time. It combines

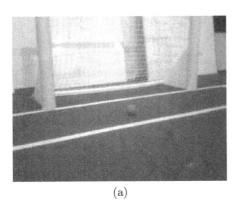 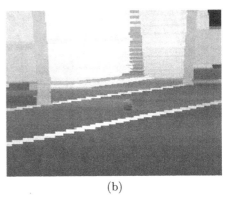

(a)   (b)

Fig. 2. Original camera image (a) and the result of the preprocessing stage (b). The segmentation considers the previously determined subsampling.

the efficiency of run length encoding (as e.g. in [1]) with the robustness of region growing. Therefore, the image is traversed linearly in a single pass, and regions can only grow to the right. Also, regions are limited to single rows, so that the approach can be seen as a dynamic programming implementation of region growing. Henceforth, a region within a row is referred to as a *segment*. The similarity criterion used is the comparison with the average color of the hitherto segment. A major advantage of the modified region growing is that the memory is only traversed once in a linear fashion, which significantly reduces random memory accesses and improves the caching efficiency and ultimately the processing time.

Color comparison is performed channel-wise: the absolute color difference of each channel is computed and compared to a predefined threshold, and the pixel is added to the segment if no threshold is exceeded, otherwise a new segment is started. The major advantage of separate thresholds per channel in conjunction with the YUV color space provided by NAO's cameras is that a higher threshold can be chosen for the brightness channel than for the color channels to improve the illumination invariance. Also, the parameter space of the segmentation process is reduced to three single values, compared to the high dimensional configuration space of the common color-table-based approach.

The result of the segmentation process is shown in Figure 2. The amount of data is significantly reduced to an average of 1,250 segments compared to 76,800 pixels ($\approx$ 1.6%). Also, the amount of noise is significantly reduced, because the region growing and the averaging acts as a kind of noise filter. Most importantly, all relevant objects (goal posts, lines, ball) have been clustered into segments within each row. However, the approach has some disadvantages. In some areas, especially around the ball, the object is split into multiple segments due to the shadings of the object. Another disadvantage of this segmentation approach is the inter-line inconsistency that arises from line-wise processing [17]. We can observe this by looking at the goal posts (see Figure 2b). Their expected trapezoid-like shape is far from being perfect and include some awkward bends. However, both problems are treated explicitly in the corresponding object detectors.

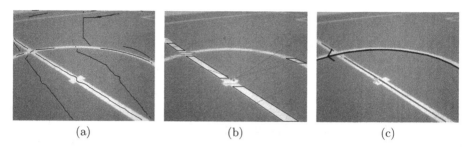

(a)   (b)   (c)

Fig. 3. Stages of the line detection. (a) Adjacent segments of similar color and length are connected. (b) Trapezoid shape fitting and circle center computation. The colors of the trapezoids indicate different shapes. (c) The result of line detection: straight lines, the center circle and line crossings.

3   Object Recognition

After this preprocessing step, distinct object recognition detectors need to be developed. We have created three detectors for lines, goal posts, and the ball.

3.1   Line Recognition

The line recognition is mainly based on region growing. To improve performance this is done in two steps. First, all segments are traversed once, connecting adjacent segments in adjacent lines if their color difference and length ratio is below a threshold. This can be done very fast as the segments are traversed sequentially in a single pass. It narrows down the potential segments processed in the actual region growing, as two segments must fulfill those properties to belong to the same region. The result of this step is shown in Figure 3a. Instead of considering the absolute color of the segment, this considers color difference only, so that the connected components do not only include lines but all types of objects.

The adjacent segments are then connected to actual regions, the average color of which is compared to the color prototype[1]. The result is a set of arbitrarily shaped whitish regions. To identify line segments among those region, the shape of the regions is taken into account. A line in world space projects to a line in the image [7], and therefore a field line, which is bordered by two lines, projects to a trapezoid in general. A trapezoid shape fitting is then applied to the regions, which basically consists of two simultaneous linear regressions for the segments' slope and width along the image's y-axis. In particular, the regions are split as soon as they deviate too much from the trapezoid shape, which is particularly useful to split the arc-like shaped regions of the center circle into short straight line segments. The result of this step is visualized in Figure 3b. The resulting regions are then assessed based on the shape, exploiting that the expected shape

---

[1] The color prototypes for the different object classes are currently calibrated manually, as an approximate calibration is sufficient.

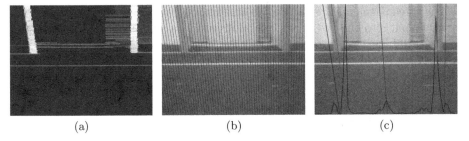

(a)   (b)   (c)

Fig. 4. Goal detection. (a) Visualization of segment rating. (b) Layout of the Hough space. (c) Resulting Hough responses and 5 highest local maxima (the colors of the lines indicate the order of the maxima).

of a line can be computed based on the known camera pose. The different categories for the shapes are *line segments, center circle segments* and *non-line segments*. This step reliably sorts out white segments in other robots, since they have a distinctive shape.

After potential line segments have been determined, a simple sanity check tests whether the segments next to the line segment are sufficiently green by comparison with the corresponding color prototype. Then, the line segments are projected onto the field plane for further processing under consideration of the rolling shutter effect (cf. [15]). Collinear line segments are merged by clustering in the Hesse normal form parameter space. Afterwards, the center circle is detected. Therefore, line segments that potentially lie on the center circle are determined based on their length and shape. Assuming that those segments are tangential to the center circle, hypothetical center points are built by computing two points orthogonal to the segment in a distance of the known center circle radius (see Figure 3b). Afterwards, those points are clustered, and if a sufficiently large cluster with at least three supporting segments is found, its centroid is treated as the center circle's center point. Those segments are disregarded in further processing. Finally, field line intersections are computed based on the line segments found, since they are valuable features for the self-localization. The result of the line detection including the center circle and intersections is visualized in Figure 3c.

3.2 Goal Recognition

Due to the complex shape of the goal as a whole, goal detection is simplified to detecting the goal posts; the goal bar only serves as an indicator for the laterality and as a sanity check. The general idea for detecting goal posts is a combination of Hough transformation [4] and region growing. This is related to *histogram-based recognition*, as presented in [19], extended to work with the segments generated in the preprocessing step. The first step is to assign a rating to the segments, i.e. the color difference (sum of channel-wise color differences) to the color prototype, as visualized in Figure 4a. This is implemented efficiently exploiting the MMX instruction set extension available on NAO's CPU.

The rated segments are accumulated in a one-dimensional Hough space. The image is split into bins (as a compromise of detection quality and processing time we use 120 bins), the borders of which are lines passing through the vanishing point of the world's vertical axis (cf. Figure 4b). This is motivated by the prior knowledge that the goal posts are cylindrical objects oriented parallel to the world's z-axis. A linearized version of the motion compensation presented in [15] is applied to the Hough space to compensate for the rolling shutter effect. The Hough space is filled by iterating over the segments, determining the bin that corresponds to the center of the segment, and adding its rating to the bin. If a goal post is present in the image, its segments have a high rating and accumulate in few bins, resulting in distinct peaks in the Hough space, as seen in Figure 4c. Finally, the Hough space is smoothed with a 1D Gaussian filter to decrease noise, and the $N = 5$ highest local maxima are determined. No hard thresholds (besides the number of local maxima) have been involved so far to increase robustness.

After promising parts of the image have been identified, region growing is applied. Therefore, the local maxima are processed separately. First, the segments belonging to a local maximum bin are determined. Those are traversed from top to bottom to build regions of similar color (the same similarity criterion as in the run length encoding step is used). To identify goal posts among the regions, several sanity checks are applied. First, it is tested whether the region exceeds a certain minimum height (30 pixels) to ensure not to process arbitrary artifacts. Then, the average color of the region is compared to the color prototype of the goal. Afterwards, the average color of the region below is compared to the color prototype of the field because the goal post resides on the field plane. Then, the relative position of the (potential) goal post is computed based on the border between those two regions. Based on the relative position, the potential goal post is projected into the image, and the projected width is compared to the actual width of the region. Finally, the top of the goal post region is checked for the goal bar by examining the uppermost segments resp. those next to them, which can be used to determine the laterality of the goal post.

After a set of goal posts has been determined, it is finally checked whether the constellation of goal posts is valid, i.e., no goal posts of same laterality, no differently colored goal posts, at most two goal posts.

3.3   Ball Recognition

Ball detection is based on region growing using the formerly detected segments, and a subsequent post-processing step. A simple approach for region growing would be to use all segments for initiating a region. This however, has two major drawbacks regarding the processing time. First, the processing time would be too high if no preselection is performed, and secondly, the variance of the processing time would be very high, since it would highly depend on the actual image contents. Taking this into consideration, the first step of the ball detection is a selection of the segments that are used as a seed for the actual region growing.

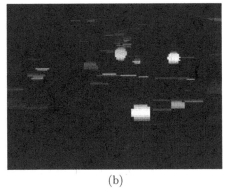

(a)   (b)

Fig. 5. Preprocessing for ball detection. (a) Test scene including real SPL balls (marked with arrows) along with other objects. (b) The resulting weighted segments. The two actual balls get the highest rating.

This step rates segments based on two measures: the first measure is the sum of the channel-wise color differences to the color prototype for the ball color. The second measure is the deviation of a segment's length from the projected diameter of the ball at that position. This way the center segment of the actual ball (if it is in the image) should get the highest rating. This step does not invoke any hard thresholds, but sorts the segments by an objective measure. The result of the step is visualized in Figure 5 using a complex test scene. It can be seen in Figure 5b that the highest rating is assigned to the actual RoboCup SPL balls. The other objects are downgraded because of the size and/or the color. The number of segments processed is bound to 1,500 beginning at the lower boundary of the image to limit the maximum processing time and its variance.

The $N = 5$ best-rated segments are then used as a seed for region growing. The difficulties that arise from the shades and reflections on the ball motivates the use of an alternative color space for the segmentation that focuses on the color *hue*, similar to the HSV or HSL color space [20]. We decided against HSV because the conversion of YUV to HSV is quite expensive due to the use of trigonometric functions resp. case distinction. Also, a lookup table is not well suited due to its space requirement of 16M entries, which makes cache misses very likely. Therefore, a simpler approximation is used. The color model used consists of four (partly redundant) channels y', s, cb', cr', which are defined as follows:

$$y' = y \quad\quad (2) \quad\quad\quad s = \sqrt{cb^2 + cr^2} \quad\quad (3)$$

$$cb' = \frac{cb}{s} \quad\quad (4) \quad\quad\quad cr' = \frac{cr}{s} \quad\quad (5)$$

The y' channel is the same as the original y channel. s is comparable to the saturation channel of HSV. cb' and cr' can be seen as normalized color channels, which together have similar properties as the hue channel of HSV. The advantage of this color space is twofold. Firstly, it allows for the effective use of a lookup table, since the conversion function is now $\mathbb{R}^2 \to \mathbb{R}^3$, yielding a lookup table of 64K entries, which significantly benefits from caching. Secondly, the cyclic property of the hue [20] is avoided, simplifying the actual comparison. Also, the introduction of an additional color channel does not have a negative influence on the processing time because of the implementation using MMX instructions, where it makes no difference whether three or four channels are processed simultaneously.

Based on this color model, the region growing takes place. The segments above and below are processed. To check whether a segment is added to the region, the color difference of the segment's color to the average color of the region is computed and compared to thresholds for each channel. For the ball detection it is reasonable to choose a high threshold for the saturation and luminance considering the shading on and reflections of the ball, and a low threshold for the (normalized) color channels to exploit the uniform color of the ball. Also, due to the known size of the ball, the region growing process can be terminated early if the bounding box of the region exceeds the projected size of the ball. This process is repeated for the $N = 5$ best segments found in the preprocessing stage.

After some potential ball regions have been found, they are post-processed to find the actual ball. First, the average color is compared to the color prototype. Then, the convex hull is determined. Based on the convex hull, a circle fitting is performed to reduce the region found to a circle. We use the *algebraic fit* since it has a closed form solution and the results are neglectably worse compared to the *geometric fit* in this scenario [3]. Due to the convex hull, the circle fitting also handles partially visible balls properly. The RMS error of the data points in the circle fitting is used as a sanity check, since it indicates the "roundness" of the region. Based on the circle center found, the robot-relative position is computed by intersecting the viewing ray with the plane that is one ball radius above the field. Afterwards, the (hypothetical) ball is projected back into the image, and the radius of the projection is compared to the radius found as another sanity check. The first potential ball region that passes all sanity checks is finally selected as the ball.

4 Experiments and Results

The focus of the experiments conducted is twofold. On one hand, the perception rate and quality are of interest. On the other hand, since the processing power of the NAO V3.2 is limited and the resulting system is supposed to be real-time capable, the run time of the modules is examined in greater detail.

For the evaluation of the perception rate and quality, different log files were created and manually annotated, which allows for an automated evaluation. The log files contain 190 frames on average, and were created while the default robot soccer behavior was executed (approaching the ball and kicking). To simulate different lighting conditions, three different scenarios were created, each of which used different light sources, as seen in Figure 6. For comparison purposes the same set of log files was evaluated with a well-established color-table-based vision system that was used by the SPL world champion 2009–2011 in all competitions [16]. For each scene a specialized color table was created which was evaluated

Fig. 6. Different lighting conditions used for the evaluation

Table 1. Detection rate and quality of the newly developed vision system compared to a color-table-based system. The right part of the table represents the color-table-based system. The table shows the true positive rate (TP), the false positive rate (FP) and the average error in image coordinates (err.).

percept type	new system			matching color table			non-matching color table		
	TP	FP	err.	TP	FP	err.	TP	FP	err.
ball	86.4 %	0.3 %	1.538 px	77.1 %	0 %	1.292 px	68.1%	0.2 %	1.328 px
goal post	84.3 %	0.5 %	2.18 px	81.1 %	0 %	1.377 px	7.7 %	6.4 %	2.223 px
lines	19 %	0.4 %	0.94 px	44.7 %	2.9 %	1.524 px	6 %	3.1 %	1.193 px
intersections	6.7 %	2.7 %	2.517 px	18.6 %	1 %	4.456 px	1.4 %	0 %	3.77 px
center circle	64.5 %	1.8 %	5.323 px	75 %	0.5 %	12.635 px	6 %	3.1 %	1.193 px

with each log file to simulate changing lighting conditions. The new system used a single configuration for all scenes.

The results are summarized in table 1. The evaluation is split into three sections: the newly developed system, the color-table-based system using matching color tables (representing static illumination), and the color-table-based system using non-matching color tables[2] (representing changing lighting conditions). The line detection rate refers to the coverage of lines rather than the line count. The performance of the new system is comparable to the color table system under static lighting conditions; the ball and goal rate is slightly better whereas the line detection has a lower detection rate but is more reliable[3]. The average error of both systems is adequate and similar. Under changing illumination, however, the new system significantly outperforms the color-table-based system with regard to every percept type. The color-table-based system is basically unusable in this scenario except for the ball detection.

Besides the perception *quality*, the run time was evaluated as well. It was measured when the robot was standing while the head control was active, which ensures representative diversity in the camera images that are processed. Again, the new system as well as the color-table-based system are evaluated. The results are shown in table 2. As expected, the run time of the new system is higher compared to the color-table-based system. However, the total run time of approx.

---

[2] Non-matching color tables refer to color tables that were created for a different lighting situation than they are used in.

[3] Note that the overall line detection rate is low because all lines, even those barely perceivable for a human, were marked for reference

Table 2. Runtime of the new system in comparison with the color-table-based system

	new system		CT based system	
	mean	std. dev.	mean	std. dev.
preprocessing	4.605 ms	0.077 ms	2.65 ms	0.565 ms
ball detection	0.342 ms	0.05 ms	0.05 ms	0.035 ms
goal detection	0.448 ms	0.036 ms	0.085 ms	0.048 ms
line detection	0.6 ms	0.151 ms	0.345 ms	0.092 ms
total	5.995 ms	0.314 ms	3.13 ms	0.74 ms

6 ms is still comparably low and especially allows for real-time operation. More importantly, the variance of the new system is significantly lower, despite the higher total run time. The latter is a crucial property for real-time systems, as it must be able to perform the task in a given time regardless of the actual input.

5   Conclusion and Outlook

We presented an object recognition system based on region growing. Hard decisions based on the color are delayed to the end of processing to increase robustness. Using different modifications and optimizations it is able to process camera images in real-time on the SPL NAO. In our evaluation, we showed that its performance is comparable under static lighting conditions, and far superior under changing lighting conditions compared to a color-table-based system.

Further work on this system may include a robot recognition module to enable reliable obstacle avoidance. To cope with the deficiency in the line detection, the preprocessing step could be modified to scan the image vertically instead of horizontally, which would ease and improve the detection of (almost) horizontal line segments. Due to the comparably low number of parameters compared to a color table, an automatic parameter optimization is very promising to further improve the detection rate. Finally, the system could easily be transferred to other RoboCup humanoid leagues, as they provide similar preconditions such as an upright upper body and a color-coded environment.

Acknowledgements. The authors would like to thank the RoboCup team *B-Human* for providing the software basis of the developed system, as well as the color-table-based vision system used for comparison. Special thanks go to Tim Laue and Udo Frese for supervising respectively reviewing the thesis this paper is based upon. This work has been partially funded by DFG through SFB/TR 8 "Spatial Cognition".

References

1. Bruce, J., Balch, T., Veloso, M.: Fast and Inexpensive Color Image Segmentation for Interactive Robots. In: 2000 IEEE/RSJ International Conference on Intelligent Robots and Systems, vol. 3, pp. 2061–2066 (2000)
2. Canny, J.: A Computational Approach to Edge Detection. IEEE Transactions on Pattern Analysis and Machine Intelligence (6), 679–698 (1986)

3. Chernov, N., Lesort, C.: Least Squares Fitting of Circles. Journal of Mathematical Imaging and Vision 23(3), 239–252 (2005)
4. Duda, R.O., Hart, P.E.: Use of the Hough Transformation To Detect Lines and Curves in Pictures. Communications of the ACM 15(1), 11–15 (1972)
5. Gevers, T., Smeulders, A.W.M.: Color-based object recognition. Pattern Recognition 32(3), 453–464 (1999)
6. Härtl, A.: Robuste, echtzeitfähige Bildverarbeitung für einen humanoiden Fußballroboter. Master's thesis, Universität Bremen (2012)
7. Hartley, R., Zisserman, A.: Multiple View Geometry in Computer Vision, 2nd edn. Cambridge University Press (2004)
8. Hojjatoleslami, S.A., Kittler, J.: Region Growing: A New Approach. IEEE Transactions on Image Processing 7(7), 1079–1084 (1998)
9. Jähne, B.: Digital Image Processing, 6th edn. Springer (2005)
10. Jain, R.C., Kasturi, R., Schunck, B.G.: Machine vision. McGraw-Hill (1995)
11. Jamzad, M., Sadjad, B.S., Mirrokni, V.S., Kazemi, M., Chitsaz, H., Heydarnoori, A., Hajiaghai, M.T., Chiniforooshan, E.: A Fast Vision System for Middle Size Robots in RoboCup. In: Birk, A., Coradeschi, S., Tadokoro, S. (eds.) RoboCup 2001. LNCS (LNAI), vol. 2377, pp. 71–80. Springer, Heidelberg (2002)
12. Otsu, N.: A Threshold Selection Method from Gray-Level Histograms. IEEE Transactions on Systems, Man, and Cybernetics 9(1), 62–66 (1979)
13. Press, W., Teukolsky, S., Flannery, B., Vetterling, W.: Numerical Recipes in C: The Art of Scientific Computing. Cambridge University Press (1992)
14. Reinhardt, T.: Kalibrierungsfreie Bildverarbeitungsalgorithmen zur echtzeitfähigen Objekterkennung im Roboterfußball. Master's thesis, Hochschule für Technik, Wirtschaft und Kultur Leipzig (2011)
15. Röfer, T.: Region-Based Segmentation with Ambiguous Color Classes and 2-D Motion Compensation. In: Visser, U., Ribeiro, F., Ohashi, T., Dellaert, F. (eds.) RoboCup 2007. LNCS (LNAI), vol. 5001, pp. 369–376. Springer, Heidelberg (2008)
16. Röfer, T., Laue, T., Müller, J., Fabisch, A., Feldpausch, F., Gillmann, K., Graf, C., de Haas, T.J., Härtl, A., Humann, A., Honsel, D., Kastner, P., Kastner, T., Könemann, C., Markowsky, B., Riemann, O.J.L., Wenk, F.: B-Human Team Report and Code Release 2011 (2011),
http://www.b-human.de/downloads/bhuman11_coderelease.pdf
17. Scharstein, D., Szeliski, R.: A Taxonomy and Evaluation of Dense Two-Frame Stereo Correspondence Algorithms. International Journal of Computer Vision 47(1-3), 7–42 (2002)
18. Swain, M.J., Ballard, D.H.: Color Indexing. International Journal of Computer Vision 7(1), 11–32 (1991)
19. Volioti, S., Lagoudakis, M.G.: Histogram-Based Visual Object Recognition for the 2007 Four-Legged RoboCup League. In: Darzentas, J., Vouros, G.A., Vosinakis, S., Arnellos, A. (eds.) SETN 2008. LNCS (LNAI), vol. 5138, pp. 313–326. Springer, Heidelberg (2008)
20. Zhang, C., Wang, P.: A New Method of Color Image Segmentation Based on Intensity and Hue Clustering. In: 15th International Conference on Pattern Recognition, vol. 3, pp. 613–616 (2000)
21. Zucker, S.W.: Region Growing: Childhood and Adolescence. Computer Graphics and Image Processing 5(3), 382–399 (1976)

Automatic Generation of Humanoid's Geometric Model Parameters

Vincent Hugel and Nicolas Jouandeau

LISV, University of Versailles and LIASD, University of Paris 8

Abstract. This paper describes a procedure that automatically generates parameters for the geometric modeling of kinematic chains. The convention of modeling used is the Denavit Hartenberg convention modified by Khalil Kleinfinger, noted DHKK. The procedure proposed here has two advantages. First the user does not need to calculate the geometric parameters by himself. He simply has to give the directions of the successive joint axes, and for each joint axis, a point that belongs to the axis. The second advantage deals with the use of model-generic matrices for the beginning and the end of the kinematic chains, and not only for the joint axes. This prevents the user from doing specific calculation to connect the joint matrices derived from the model with the initial and the final coordinate frames of the chain. Due to its unified formalism, the procedure allows to save time when the kinematics of the robot has to be changed. This paper includes the application of this procedure to the geometric modeling of legs and arms of two versions of the NAO humanoid robot, the one used in the RoboCup 3D Simulation League, and the other one used in the RoboCup Standard Platform League.

1 Introduction

Usually the calculation of the geometric model of a kinematic chain starts with the selection of a convention of parameterization to define the successive coordinate frames to go from one extremity to the other extremity of the chain. The user has to master the convention rules and has to manually calculate distances and angles necessary for the parameterization [1]. This is often a tedious task that requires a certain amount of time for the calculation itself, the checking and the validation with simulations. The task is even harder when there are multiple kinematic chains. If we take the example of a humanoid robot, there are five kinematic chains, the head, the arms and the legs. The number of parameters to determine is at least equal to the total number of degrees of freedom of the robot.

This paper proposes an automatization procedure for determining the parameters related to the modeling convention. This procedure simplifies the job of the robotics engineer since he does not have to do any calculation any more. It is straightforward and can be applied to any robot model or kinematic chain. It can be useful at the design stage and at the simulation stage to modify the dimensions of the bodies of the kinematic chain, the layout and the directions of

S. Behnke et al. (Eds.): RoboCup 2013, LNAI 8371, pp. 408–419, 2014.

the joint axes [2] [3], or the order of the joints without having to calculate all parameters manually.

To our knowledge such an automatization of the calculation of geometric model parameters has never been reported.

2   Modeling Convention

Khalil and Kleinfinger [4] proposed a modified convention for geometric modeling from the Denavit-Hartenberg convention [5]. The first advantage of this modified convention is its suitability for the kinematic description of both open-loop and closed-loop structures. The second advantage lies in the convenient definition of the z_i axis as the axis of rotary joint i. The angle of rotation about axis z_i is denoted by θ_i, with same index i. This rotation is the fourth transformation[1] among the four successive transformations needed to go from coordinate frame $i-1$ linked to joint $i-1$ to coordinate frame i linked to joint i. The first three transformations can be used to position the rotation axis z_i.

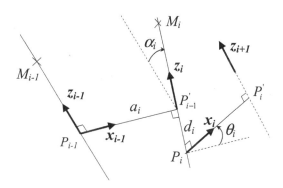

Fig. 1. DHKK convention parameters

There are four DHKK parameters required to go from R_{i-1} to R_i, one for each transformation. Parameters are denoted by a_i, α_i, d_i and θ_i. a_i and d_i are distances, α_i and θ_i are angles. Figure 1 shows the four parameters at stage i. They involve the three axes z_{i-1}, z_i, and z_{i+1}.

- a_i is the signed distance $P_{i-1}P'_{i-1}$ along x_{i-1}.
- α_i is the rotation angle about x_{i-1} between z_{i-1} and z_i.
- d_i is the signed distance $P'_{i-1}P_i$ along z_i axis.
- θ_i is the rotation angle about z_i between x_{i-1} and x_i.

---

[1] Each notation involves four successive transformations, twice one translation followed by one rotation.

where x_{i-1} is along the segment that is orthogonal to z_{i-1} and z_i axes, from z_{i-1} to z_i. x_i is along the segment that is orthogonal to z_i and z_{i+1} axes, from z_i to z_{i+1}. P_i is the intersection point of x_i with z_i. P_i' is the intersection point of x_i with z_{i+1}. M_i is a point that belongs to z_i axis.

The transformation from R_i to R_{i-1} is written as follows, where R stands for rotation and T for translation $(R_i = (P_i, x_i, y_i, z_i))$:

$$_{i-1}^{i}M = Tx_{i-1}(a_i)Rx_{i-1}(\alpha_i)Tz_i(d_i)Rz_i(\theta_i) \tag{1}$$

$$_{i-1}^{i}M = \begin{pmatrix} c\theta_i & -s\theta_i & 0 & a_i \\ c\alpha_i s\theta_i & c\alpha_i c\theta_i & -s\alpha_i & -d_i s\alpha_i \\ s\alpha_i s\theta_i & s\alpha_i c\theta_i & c\alpha_i & d_i c\alpha_i \\ 0 & 0 & 0 & 1 \end{pmatrix} \tag{2}$$

3   Generic Procedure for the Calculation of DHKK Parameters

To determine the DHKK parameters, it is necessary to choose an initial configuration for the kinematic chain. In this configuration the user has to give the coordinates of the joint axes, i.e. the coordinates of (z_i) vectors relative to an external reference frame linked to a body of reference like the trunk in a humanoid robot. The user must also give the coordinates of points that belong to the joint axes, one point for each axis. These points are the (M_i) points. Usually (M_i) are points located at the joint axis intersections, for example hip center, knee center, or ankle center for a humanoid leg.

The first step of the procedure consists of calculating the (x_i) vectors. x_i must be orthogonal to z_i and z_{i+1}. In the general case where (M_i, z_i) is not parallel with (M_{i+1}, z_{i+1}), we can define x_i from the cross-product of z_i with z_{i+1}:

$$x_i = \frac{z_i \times z_{i+1}}{\|z_i \times z_{i+1}\|}$$

There are two particular cases. The first one involves $(M_i, z_i) = (M_{i+1}, z_{i+1})$. Here we choose to keep the same vector as previously defined, i.e. $x_i = x_{i-1}$. The second case implies that (M_i, z_i) is parallel but not equal to (M_{i+1}, z_{i+1}). Here we can use $M_i M_{i+1}$ to obtain the direction orthogonal to both parallel axes inside the plane containing these axes:

$$\frac{z_i \times M_i M_{i+1}}{\|z_i \times M_i M_{i+1}\|} \times z_i$$

In summary:

$$(M_i, z_i) = (M_{i+1}, z_{i+1}) \Rightarrow x_i = x_{i-1}$$
$$(M_i, z_i) \parallel (M_{i+1}, z_{i+1}) \Rightarrow x_i = \frac{z_i \times M_i M_{i+1}}{\|z_i \times M_i M_{i+1}\|} \times z_i$$
$$(M_i, z_i) \not\parallel (M_{i+1}, z_{i+1}) \Rightarrow x_i = \frac{z_i \times z_{i+1}}{\|z_i \times z_{i+1}\|}$$

Then the three DHKK parameters are defined[2] as:

$$a_i = M_{i-1}M_i.x_{i-1} \tag{3}$$
$$\alpha_i = atan2\left([z_{i-1}, z_i, x_{i-1}], z_{i-1}.z_i\right) \tag{4}$$
$$\theta_i = atan2\left([x_{i-1}, x_i, z_i], x_{i-1}.x_i\right) + \Delta\theta_i \tag{5}$$

where $\Delta\theta_i$ represents the command angle of joint i from the initial configuration. The last DHKK parameter is expressed as:

$$d_i = P'_{i-1}P_i.z_i \tag{6}$$

with:

$$P_i = M_i + r_i.z_i$$
$$P'_i = M_{i+1} + s_i.z_{i+1}$$

To calculate r_i and s_i, we use the following property of the convention:

$$P_iP'_i \times (z_i \times z_{i+1}) = 0$$

This equation is useful only in the case where $z_i \times z_{i+1} \neq 0$. Taking the scalar product of this equation with z_i for the one part, and with z_{i+1} for the other part, gives:

$$\left[P_iP'_i, z_i \times z_{i+1}, z_i\right] = 0$$
$$\left[P_iP'_i, z_i \times z_{i+1}, z_{i+1}\right] = 0$$

After replacement:

$$[M_iM_{i+1} + s_i.z_{i+1}, z_i \times z_{i+1}, z_i] = 0 \Leftrightarrow$$
$$s_i.\|z_i \times z_{i+1}\|^2 = [M_iM_{i+1}, z_i, z_i \times z_{i+1}]$$
$$[M_iM_{i+1} - r_i.z_i, z_i \times z_{i+1}, z_{i+1}] = 0 \Leftrightarrow$$
$$r_i.\|z_i \times z_{i+1}\|^2 = [M_iM_{i+1}, z_{i+1}, z_i \times z_{i+1}]$$

In the case where $z_i \times z_{i+1} = 0$, we choose to place P_i at M_i, which means that r_i is set to 0. Then it comes $s_i = M_{i+1}M_i.z_{i+1}$.

In summary,

$$z_i \parallel z_{i+1} \Rightarrow r_i = 0$$
$$s_i = M_{i+1}M_i.z_{i+1}$$
$$z_i \nparallel z_{i+1} \Rightarrow r_i = [M_iM_{i+1}, z_{i+1}, z_i \times z_{i+1}] / \|z_i \times z_{i+1}\|^2$$
$$s_i = [M_iM_{i+1}, z_i, z_i \times z_{i+1}] / \|z_i \times z_{i+1}\|^2$$

---

[2] $[a, b, c]$ denotes the triple scalar product $(a \times b).c$.

4   Application to NAO Humanoids

The NAO robot (Fig. 2) manufactured by the French company Aldebaran-Robotics has a coupled yaw-pitch joint with a 45[deg] inclination with respect to the vertical [6]. Table 1 gives the offsets and lengths related to two versions of NAO, i.e. the version used in the 3D Soccer Simulation League (3D-SSL) and the version used in the Standard Platform League (SPL). The trunk center named T is taken as a reference. The coordinate frame linked to the trunk is $R_T = (T, i_T, j_T, k_T)$, with i_T being the longitudinal axis, j_T the lateral axis, and k_T the vertical axis. Figure 3 illustrates the notations used.

Fig. 2. Nao schematics in reference configuration where all joint angles are equal to zero

Table 1. Leg and arm lengths and offsets of NAO models [m]

Offset/Length	3D-SSL leg	SPL leg	Offset/Length	3D-SSL arm	SPL arm
Hip Offset Z: H_z^o	0.115	0.085	Shoulder Offset Z: S_z^o	0.075	0.1
Hip Offset Y: H_y^o	0.055	0.05	Shoulder Offset Y: S_y^o	0.098	0.098
Hip Offset X: H_x^o	0.01	0.0	Elbow Offset Y: E_y^o	0.0	0.015
Knee Offset X: K_x^o	0.005	0.0	Elbow Offset Z: E_z^o	0.009	0.0
Femur length: L_f	0.12	0.1	Arm length: L_a	0.09	0.105
Tibia length: L_t	0.1	0.1029	Forearm length: L_{fa}	0.105	0.13
Foot Height: F_z^h	0.05	0.04519			

4.1   Leg Geometric Model

Table 2 contains the inputs needed for the DHKK-parameter procedure. It includes the joint axes and points that belong to them. Figure 3 shows the points H (hip), K (knee), A (ankle) and Ah (ankle projection) of each leg, and the joint axes numbered 1 to 6. All these points and axes are given for legs in the

Table 2. Input parameters for the NAO leg DHKK kinematics modeling

Point M_i belonging to joint axis z_i	Joint axis z_i
$M_{s1} = T = [0,0,0]^T$	$z_{s1} = k_T = [0,0,1]^T$
$M_{s2} = T = [0,0,0]^T$	$z_{s2} = k_T = [0,0,1]^T$
$M_1 = H = [-H_x^o, -\xi.H_y^o, -H_z^o]^T$	$z_1 = [0, \cos(45), \xi\sin(45)]^T$
$M_2 = H = [-H_x^o, -\xi.H_y^o, -H_z^o]^T$	$z_2 = i_T = [1,0,0]^T$
$M_3 = H = [-H_x^o, -\xi.H_y^o, -H_z^o]^T$	$z_3 = j_T = [0,1,0]^T$
$M_4 = K = [-K_x^o, -\xi.H_y^o, -H_z^o - L_f]^T$	$z_4 = j_T = [0,1,0]^T$
$M_5 = A = [-K_x^o, -\xi.H_y^o, -H_z^o - L_f - L_t]^T$	$z_5 = j_T = [0,1,0]^T$
$M_6 = A = [-K_x^o, -\xi.H_y^o, -H_z^o - L_f - L_t]^T$	$z_6 = i_T = [1,0,0]^T$
$M_{e1} = Ah = [-K_x^o, -\xi.H_y^o, -H_z^o - L_f - L_t - F_z^h]^T$	$z_{e1} = k_T = [0,0,1]^T$
$M_{e2} = Ah = [-K_x^o, -\xi.H_y^o, -H_z^o - L_f - L_t - F_z^h]^T$	$z_{e2} = -j_T = [0,-1,0]^T$

reference position, i.e. completely stretched vertically. $\xi = 1$ for the right leg and -1 for the left leg.

The first two axes z_{s1}[3] and z_{s2} are used to go from $R_T = (T, i_T, j_T, k_T)$ to the first coordinate frame centered at the hip, and therefore take into account the translation from T to H. Axes z_i refer to the joint rotation axes of angle θ_i. Joints are ordered from hip to ankle as hip-yaw, hip roll, hip pitch, knee pitch, ankle pitch and ankle roll. The last two axes z_{e1}[4] and z_{e2} are used to orient the last coordinate frame so that it matches the axes of R_T. It is centered at Ah which is the projection of the ankle center on the foot sole.

The introduction of two axes at the beginning and at the end of the kinematic chain prevents the user from making connections with the initial and the final coordinate frames. These additional axes are not *joint* axes but are also considered as z-axes in order to apply the same DHKK formalism. Actually each pair of axes – at the beginning for the one part and at the end for the other part – allows to apply an additional DHKK homogeneous matrix at the beginning, and at the end of the kinematic chain.

Once points and axes are given as inputs, the procedure can be applied to get the DHKK parameters of the kinematic chain. Table 3 gives the different values for the kinematic parameters that are related to the NAO leg. Angles $\Delta\theta$ are the angles related to joint moves. They can represent the joint commands or the joint sensor readings. They are zero in the reference position (legs vertically stretched and arms horizontally stretched).

4.2   Arm Geometric Model

Table 4 contains the inputs needed for the DHKK-parameter procedure. It includes the joint axes and the points that belong to them. Figure 3 shows the points S (shoulder), E (elbow), and H_a (hand/wrist) of each arm, and the joint

---

[3] s stands for start.
[4] e stands for end.

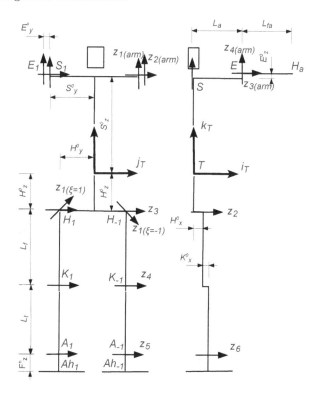

Fig. 3. Humanoid skeleton with notations. Frontal view on the left-hand side and sagittal view on the right-hand side.

axes numbered 1 to 4. All these points and axes are given for arms in the reference position, i.e. completely stretched horizontally. $\xi = 1$ for the right arm and -1 for the left arm.

The first two axes z_{s1}[5] and z_{s2} are used to go from $R_T = (T, i_T, j_T, k_T)$ to the first coordinate frame centered at the shoulder, and therefore take into account the translation from T to S. Joints are ordered from shoulder to hand as shoulder pitch, shoulder yaw, elbow roll and elbow yaw. The last two axes z_{e1}[6] and z_{e2} are used to orient the last coordinate frame so that it matches the axes of R_T. The last two points are used to set the origin at the extremity of the arm H_a.

Table 5 gives the different values for the kinematic parameters that are related to the NAO arm.

5   Inverse Geometric Modeling

The inverse geometric modeling consists of finding analytical solutions for joint angles θ_i given the trajectory of the end effector in the Cartesian space. The

---

[5] s stands for start.
[6] e stands for end.

Table 3. DHKK parameters for the NAO Humanoid leg kinematics model

i	$a_i[m]$	$\alpha_i[deg]$	$d_i[m]$	$\theta_i[deg]$
s	0	0	$-(H_z^o - H_y^o)$	180
1	H_x^o	$45(1+\xi)/2 + 135(1-\xi)/2$	$-\xi.\sqrt{2}.H_y^o$	$-90.0 + \Delta\theta_{yaw-pitch}^{hip}$
2	0	90	0	$135\xi + \Delta\theta_{roll}^{hip}$
3	0	90	0	$\pi - tan^{-1}\frac{H_x^o - K_x^o}{L_f} + \Delta\theta_{pitch}^{hip}$
4	$\sqrt{L_f^2 + (H_x^o - K_x^o)^2}$	0	0	$tan^{-1}\frac{H_x^o - K_x^o}{L_f} + \Delta\theta_{pitch}^{knee}$
5	L_t	0	0	$\Delta\theta_{pitch}^{ankle}$
6	0	90	0	$-90 + \Delta\theta_{roll}^{ankle}$
e	0	90	$-F_z^h$	90

Table 4. Input parameters for the NAO arm DHKK kinematics modeling

Point M_i belonging to joint axis z_i	joint axis z_i
$M_{s1} = T = [0,0,0]^T$	$z_{s1} = k_T$
$M_{s2} = T = [0,0,0]^T$	$z_{s2} = k_T$
$M_1 = S = [0, -\xi.S_y^o, S_z^o]^T$	$z_1 = j_T$
$M_2 = S = [0, -\xi.S_y^o, S_z^o]^T$	$z_2 = k_T$
$M_3 = E = [L_a, -\xi.(S_y^o + E_y^o), S_z^o + E_z^o]^T$	$z_3 = i_T$
$M_4 = E = [L_a, -\xi.(S_y^o + E_y^o), S_z^o + E_z^o]^T$	$z_4 = k_T$
$M_{e1} = H_a = [L_a + L_{fa}, -\xi.(S_y^o + E_y^o), S_z^o + E_z^o]^T$	$z_{e1} = k_T$
$M_{e2} = H_a = [L_a + L_{fa}, -\xi.(S_y^o + E_y^o), S_z^o + E_z^o]^T$	$z_{e2} = -j_T$

product of DHKK homogeneous matrices to go from the end effector coordinate frame to the torso coordinate frame can be used to express the relationship between the position/orientation of the end effector with the joint displacements. Considering the NAO's humanoid leg, we can write the geometric model as:

$$M^0 = M^s . M^{\theta_1} . M^{\theta_2} . M^{\theta_3} . M^{\theta_4} . M^{\theta_5} . M^{\theta_6} . M^e \tag{7}$$

where

- M^s involves the first three z vectors,
- M^e involves the last three z vectors,
- M^{θ_i} involves joint rotation of angle θ_i, and vectors z_{i-1}, z_i, z_{i+1},
- M^0 is the homogeneous matrix given by the user, that represents the wanted trajectory (rotation + translation) of the end effector (position of the projection of ankle on the sole and orientation of the sole).

Joint angles θ_i are the unknowns and, to invert the direct geometric model, we can use the following notation:

$$M^x = \begin{pmatrix} R_x & T_x \\ 0 & 1 \end{pmatrix}$$

Table 5. DHKK parameters for the NAO Humanoid arm kinematics model

i	$a_i[m]$	$\alpha_i[deg]$	$d_i[m]$	$\theta_i[deg]$
s	0	0	S_z^o	180
1	0	90	$\xi.S_y^o$	$180 + \Delta\theta_{pitch}^{shoulder}$
2	0	90	E_z^o	$90 + \Delta\theta_{yaw}^{shoulder}$
3	$-\xi E_y^o$	90	L_a	$180 + \Delta\theta_{roll}^{elbow}$
4	0	90	0	$90 + \Delta\theta_{yaw}^{elbow}$
e	L_{fa}	0	0	0

T_0 is the position of the end effector in the torso coordinate frame. R_0 is the matrix whose columns contain the coordinates of the end effector frame expressed in torso coordinate frame. R_0 can be interpreted as the rotation matrix of the end effector frame inside torso frame. It can also be interpreted as the transformation matrix to pass from the end effector coordinate frame into the torso coordinate frame.

Equation 7 can be developed as:

$$\begin{pmatrix} R_s & T_s \\ 0 & 1 \end{pmatrix}\begin{pmatrix} R_{123} & T_1 \\ 0 & 1 \end{pmatrix}\begin{pmatrix} R_4 & T_4 \\ 0 & 1 \end{pmatrix}\begin{pmatrix} R_{56} & T_5 \\ 0 & 1 \end{pmatrix}\begin{pmatrix} R_e & T_e \\ 0 & 1 \end{pmatrix} = \begin{pmatrix} R_0 & T_0 \\ 0 & 1 \end{pmatrix}$$

$$\Leftrightarrow \begin{pmatrix} R_{123} & 0 \\ 0 & 1 \end{pmatrix}\begin{pmatrix} R_{456} & T_4 + R_4T_5 \\ 0 & 1 \end{pmatrix} =$$

$$\begin{pmatrix} {}^tR_s & -{}^tR_sT_s - T_1 \\ 0 & 1 \end{pmatrix}\begin{pmatrix} R_0({}^tR_e) & T_0 - R_0({}^tR_e)T_e \\ 0 & 1 \end{pmatrix}$$

$$\Leftrightarrow \begin{pmatrix} R_{123} & 0 \\ 0 & 1 \end{pmatrix}\begin{pmatrix} R_{456} & T_4 + R_4T_5 \\ 0 & 1 \end{pmatrix} = \begin{pmatrix} R' & M' \\ 0 & 1 \end{pmatrix}$$

with

$$R' = ({}^tR_s)R_0({}^tR_e)$$
$$T' = -T_1 + ({}^tR_s)(T_0 - Ts - R_0({}^tR_e)T_e)$$

which gives the following system to solve:

$$R_{123}(T_4 + R_4T_5) = T' \qquad (8)$$
$$R_{123}R_{456} = R' \qquad (9)$$

Squaring the first equation gives $(T_4 + R_4T_5)^2 = T'^2$ that allows to solve for the knee angle θ_4 and then $\Delta\theta_4$. Then it is possible to solve either for the hip angles then for the ankle angles, or for the ankle angles then the hip angles. The solving for the hip or the ankle angles is traditional.

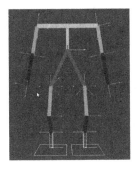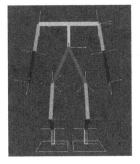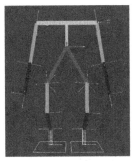

Fig. 4. Stick models of the 3D-SSL version of NAO displayed in the qt-opengl environment for the validation stage. The model on the left-hand side is related to the default model with the femur length equal to $0.12[m]$. The model in the middle is related to a change of the femur length, that is $0.14[m]$ instead of $012[m]$. The model on the right-hand side features a greater distance between hips – $0.065[m]$ instead of $0.055[m]$, and longer arms – $0.13[m]$ instead of $0.09[m]$. DHKK parameters are calculated automatically at the init stage, and are used in the direct geometric model for the display.

Fig. 5. Snapshots of the 3D-SSL NAO robot in standing position and making a step forward, achieved with the 3D-SSL official simulation server [7] [8] combined with the RoboViz monitor [10]

6   Results

The procedure was implemented with success for the participation in the 3D-SSL of the 2013 RoboCup German Open. The DHKK parameters are calculated at the initialization stage when the C++ instance of the robot'smodel class is built. The procedure does not modify the computing time required for the generation of joint commands.

DHKK parameters are used when the direct or the inverse geometric model is needed. The direct geometric model is used at the validation stage to display the robot's skeleton (Fig. 4). The validation stage allows checking the movements of the different limbs of the robot.

The inverse geometric model is used in real time in the 3D-SSL software (SimSpark application for the RoboCup 3D-SSL [7] [8], namely *rcssserver3d*) to calculate the joint command angles to be sent to the server. Walking patterns were designed using the model of the 3D linear inverted pendulum and the Zero Moment Point technique [9]. Figure 5 shows two snapshots of the NAO robot in the 3D-SSL environment. The left-hand side picture shows the robot in standing position with flexed knees after initialization of the model's parameters. The right-hand side picture shows the robot making a forward step.

7 Conclusion

This paper presented a procedure to automatically generate parameters for the geometric modeling of kinematic chains. The convention of modeling used is the Denavit Hartenberg convention modified by Khalil Kleinfinger. This procedure can be used for all kinds of kinematic chains and especially for NAO humanoid models. The user must choose a reference position and give couples (M_i, z_i) that define the joint rotation axes. Then in the initialization stage, the procedure calculates the DHKK model parameters according to the reference position. These parameters are fixed and only depend on the reference position of the chains. They can be used for direct geometric modeling but also for inverse modeling to calculate the joint command angles in real time. The procedure can deal with the entire kinematic chain to switch from a starting coordinate frame to an end coordinate frame and vice-versa.

References

1. Graf, C., Hartl, A., Röfer, T., Laue, T.: A Robust Closed-Loop Gait for the Standard Platform League Humanoid. In: Proc. 4th Workshop on Humanoid Soccer Robots, pp. 30–37 (2009)
2. Zorjan, M., Hugel, V., Blazevic, P.: Influence of hip joint axes change of orientation on power distribution in humanoid motion. In: Proc. IEEE ICARA: The 5th Int. Conf. on Automation, Robots and Applications, pp. 271–276 (2011)
3. Zorjan, M., Hugel, V.: Generalized Humanoid Leg Inverse Kinematics to Deal With Singularities. In: Proc. IEEE Int. Conf. on Robotics and Automation (2013)
4. Khalil, W., Kleinfinger, J.-F.: A new geometric notation for open and closed-loop robots. In: Proc. IEEE Int. Conf. on Robotics and Automation, pp. 1174–1180 (1986)
5. Denavit, J., Hartenberg, R.S.: A kinematic notation for lower-pair mechanisms based on matrices. Trans. ASME J. Appl. Mech. 23, 215–221 (1955)
6. Gouaillier, D., Hugel, V., Blazevic, P., Kilner, C., Monceaux, J., Lafourcade, P., Marnier, B., Serre, J., Maisonnier, B.: Mechatronic design of NAO humanoid. In: Proc. IEEE Int. Conf. on Robotics and Automation, pp. 769–774 (2009)
7. Obst, O., Rollmann, M.: Spark - A Generic Simulator for Physical Multi-Agent Simulations. In: Lindemann, G., Denzinger, J., Timm, I.J., Unland, R. (eds.) MATES 2004. LNCS (LNAI), vol. 3187, pp. 243–257. Springer, Heidelberg (2004)

8. SimSpark, a generic physical multiagent simulator system for agents in three-dimensional environments, http://simspark.sourceforge.net/

9. Hugel, V., Jouandeau, N.: Walking patterns for real time path planning simulation of humanoids. In: Proc. IEEE Int. Symp. on Robot and Human Interactive Comm (RO-MAN), pp. 424–430 (2012)

10. Stoecker, J., Visser, U.: RoboViz: Programmable Visualization for Simulated Soccer. In: Röfer, T., Mayer, N.M., Savage, J., Saranlı, U. (eds.) RoboCup 2011. LNCS, vol. 7416, pp. 282–293. Springer, Heidelberg (2012)

Modification of Foot Placement for Balancing Using a Preview Controller Based Humanoid Walking Algorithm

Oliver Urbann and Matthias Hofmann

Robotics Research Institute,
Section Information Technology,
TU Dortmund University,
44221 Dortmund, Germany

Abstract. Lunges are an important utility to regain balance under strong perturbed biped walking motions. This paper presents a method to calculate the modifications of predefined foot placements with the objective to minimize deviations of the Zero Moment Point from a reference. The modification can be distributed over different points in time to execute smaller lunges, and an arbitrary point in time can be chosen. The calculation is in closed-form, and is embedded into a well-evaluated preview controller with observer based on the 3D Linear Inverted Pendulum Mode (3D-LIPM).

Keywords: 3D-LIPM, ZMP, observer, humanoid robot, reactive stepping, walking algorithm.

1 Introduction

Research in the area of biped robots has become increasingly important over the last years. Especially in environments that contain obstacles and barriers such as stairs, or small objects lying on the ground, biped robots are advantageous in comparison to wheeled robots. Bipedal locomotion in such surroundings is naturally susceptible to disturbances that may occur when the robot collides with those objects. As a precedent step, biped robots must be capable to regain balance even on flat grounds to prevent from damage.

1.1 Related Work

A widely used criterion to determine the stability of a humanoid robot is the Zero Moment Point (ZMP). It was invented by Miomir Vukobratović [1] based on the term support polygon which is the convex hull of all contact points of the feet with the ground. If the ZMP is inside the support polygon, the ground does not exert torques around the roll and pitch axes of the feet, and thus the robot can be considered as stable.

The ZMP can additionally be exploited to define the locomotion of the robot. For this purpose, the reference ZMP is chosen to stay always inside the support

S. Behnke et al. (Eds.): RoboCup 2013, LNAI 8371, pp. 420–431, 2014.

polygons which result from the planned foot positions. As this approach implies a computational solution of non-linear equations which are runtime-intensive, Kajita et al. introduced a linear model of the robot that associates the movement of the center of mass (CoM) of the robot to the ZMP, the 3D linear inverted pendulum mode (3D-LIPM) [2].

There are many approaches to derive a CoM trajectory from the desired ZMP employing the 3D-LIPM, e.g. using a convolution sum [3], or an analytical solution [4]. Kajita et al. proposed a method for an arbitrary ZMP trajectory based on linear-quadratic regulator [5]. To incorporate measurements from different sensors, an observer can be integrated [6,7,8]. While a stabilizing effect of this approach can be shown[1] [8], it is limited due to the fact that major disturbances cannot be compensated without a lunge[2].

In the literature, there are various approaches to determine a step that fits best to the desired motion and dynamical state of the robot. A reasonable way for this is to take the orbital energy into account. Doing so, the switching time of the stance foot can be solved analytically [9]. While this is in fact not a lunge where the desired foot position is modified, Urata et al. proposed a technique based on a preview controller that optimizes the foot placement numerically [10]. Other approaches are based on a concept called Capture Point (CP) [11] which is defined as the foot position where the robot comes to a full stop with the CoM directly above. The CP can be calculated using an extension of the 3D-LIPM, and this can be used to generate a walking motion [12]. While the steps are calculated in a closed form, a major drawback is the validity of the CP which is not constant over time [13]. Moreover, the CP approach does not consider the point in time when the stance foot is changed. This both drawbacks are the motivation to invent the foot placement estimator (FPE) and the foot placement indicator (FPI) [14]. However, all points (CP, FPE and FPI) have the drawback in common that generating a walking motion using these points is counter-intuitive as they were originally introduced to immediately stop the robot after disturbances like a push.

Another approach to simultaneously plan the motion and the desired foot position according to the state of the robot including disturbances is to formulate it as a quadratic program (QP) solved using Model Predictive Control (MPC) [15]. Also Morisawa et al. use a quadratic program to find appropriate foot placements [16]. However, both solutions can reveal large CoM and ZMP deviations under certain circumstances. In these cases, the disturbances should be balanced by other means than stepping [17]. Furthermore, the lunge can only be calculated by numerical optimization in both approaches.

---

[1] A video of the closed-loop performance can be watched here:
https://www.youtube.com/watch?v=ZuebspajdZU

[2] The term for a step that is needed to balance disturbances that cannot be balanced in another way. Normally a lunge is a modification of a planned foot position. In the literature also referred to as reactive stepping or just stepping.

1.2   Objectives

This paper proposes a method to modify the predefined foot placement in order to regain balance after disturbances were observed. To this end, the following required properties of the system are defined:

1. The system should deal with a reference ZMP of an arbitrary shape to avoid oscillations of the body. E.g. a constant ZMP speed in walking direction ensures a constant body speed.
2. The system should ensure that the actual position of the feet is equal to the desired if needed. In case of disturbances, instabilities must then be balanced conventionally. Otherwise it would be difficult to stop next to an object without a collision.
3. The system must always provide and guarantee a solution for the modification of the foot placement.
4. As the controller is supposed to run on platforms with limited computing power, a calculation in closed-form is needed.
5. Because kinematical constraints must be considered to avoid self-collision, it is important to be able to select a random point in time for the lunge execution.
6. To avoid large steps, it should be possible to distribute a single lunge over multiple steps.

Since a reference ZMP of an arbitrary shape is desired, the preview controller with observer approach as presented in [6,7,8] is the basis for this work.

1.3   Structure

The remainder of the paper is structured as follows: Section 2 briefly summarizes the main concepts of the used preview controller and observer, and reflects the output of the balancing system in case of disturbances. Based on this, section 3 presents a method to suppress these disturbances and outlines how the modification of the foot placement is processed. The approach is evaluated accompanied by some examples in section 4 by using a simulation environment. The paper concludes in section 5 including future work. Formulas and algorithms presented in this paper can be downloaded as described in section 6.

2   Preview Controller and Observer

This section describes the basic concepts that are used in the preview controller. Afterwards an observer is presented that is utilized to incorporate sensor measurements. Please note that all derivations are for the sagittal plane but also hold for the coronal plane. The final walking algorithm therefore consists of two controllers, one for each plane, working in parallel. For details about the overall algorithm see [8].

A fast way to determine the target position of the CoM is a preview controller as proposed by Kajita et al. [5]. It is a linear-quadratic regulator (LQR) that

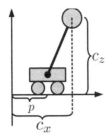

Fig. 1. Cart model of a linear inverted pendulum

requires a preview of the reference ZMP, denoted here as a vector $\boldsymbol{p}^{ref}$ of all reference points for the walk. A detailed description is given in [7], and proofs can be found in [18].

The LQR requires a linear model of the robot. Hence, the 3D-LIPM which abstracts the robot to a single point mass is used to construct the controller. This mass is supported by a single pole, and can only move in a plane with a constant height over ground (see Figure 1). To keep the mass at the height c_z, the ZMP p must be at the x position of the pivot joint. The acceleration of the CoM in x direction $\ddot{c}_x$ must then satisfy:

$$p = c_x - \frac{c_z}{g}\ddot{c}_x, \tag{1}$$

where g is the gravity.

A discrete-time system (with a sampling frequency of $\frac{1}{\Delta t}$) describes the physical behavior, and can be formulated as:

$$\mathbf{x}(k+1) = \mathbf{A}_0\mathbf{x}(k) + \mathbf{b}u(k) \tag{2}$$

$$p(k) = \mathbf{C}\mathbf{x}(k), \mathbf{b} = (0,0,\Delta t)^T \tag{3}$$

where $u(k)$ denotes the controller output which is the time derivative of the ZMP $\dot{p}$, and $\mathbf{x}(k) = (c_x(k), \dot{c}_x(k), p(k))^T$ is the state of the system. Matrix $\mathbf{C} = (0,0,1)$ selects the ZMP as the system output and therefore as the reference. The matrix

$$\mathbf{A}_0 = \begin{bmatrix} 1 & \Delta t & 0 \\ \frac{g}{c_z}\Delta t & 1 & -\frac{g}{c_z}\Delta t \\ 0 & 0 & 1 \end{bmatrix} \tag{4}$$

incorporates the 3D-LIPM equation 1. To derive the controller $u(k)$ in equation 2, a performance index is defined that meets the following requirements: First, the expected ZMP should be equal to the reference. Second, the resulting trajectory of the CoM has to be smooth. Third, the output $\dot{p}$ of the controller should be low as long as the reference is tracked well. Thus:

$$J = \sum_{j=0}^{\infty} \left\{ Q_e \left[p(j) - p_j^{ref} \right]^2 + \Delta\mathbf{x}^T(j)Q_x\Delta\mathbf{x}(j) + R\Delta u^2(j) \right\}, \tag{5}$$

where $\Delta\mathbf{x}$ and Δu are the incremental state vector and controller output respectively. The gains R, Q_x, Q_e are platform-dependent parameters and $\boldsymbol{p}^{ref}$ is the preview of the desired ZMP with a length of $N+1$.

The controller minimizing J is given by:

$$u(k) = -G_I \sum_{i=0}^{k} \left[\mathbf{C}\mathbf{x}(i) - \boldsymbol{p}_i^{ref} \right] - \mathbf{G}_x \mathbf{x}(k) - \sum_{j=0}^{N} G_d(j) \boldsymbol{p}_{k+j}^{ref}, \tag{6}$$

where the gains can be calculated as follows:

$$\mathbf{B} = \begin{bmatrix} \mathbf{Cb} \\ \mathbf{b} \end{bmatrix}, \mathbf{I} = \begin{bmatrix} 1 \\ 0 \\ 0 \\ 0 \end{bmatrix}, \mathbf{F} = \begin{bmatrix} \mathbf{CA_0} \\ \mathbf{A_0} \end{bmatrix}, \mathbf{Q} = \begin{bmatrix} Q_e & 0 \\ 0 & \mathbf{C}^T Q_x \mathbf{C} \end{bmatrix}, \mathbf{A} = [\mathbf{I}, \mathbf{F}] \tag{7}$$

$$G_I = \left[R + \mathbf{B}^T \mathbf{PB} \right]^{-1} \mathbf{B}^T \mathbf{PI}, \quad \mathbf{G}_x = \left[R + \mathbf{B}^T \mathbf{PB} \right]^{-1} \mathbf{B}^T \mathbf{PF} \tag{8}$$

$$G_d(j) = - \left[R + \mathbf{B}^T \mathbf{PB} \right]^{-1} \mathbf{B}^T \left[\mathbf{A}_c^T \right]^j \mathbf{PI}, j = 0, 1, ..., N. \tag{9}$$

The closed-loop matrix $\mathbf{A}_c$ is defined as

$$\mathbf{A}_c = \mathbf{A} - \mathbf{B} \left[R + \mathbf{B}^T \mathbf{PB} \right]^{-1} \mathbf{B}^T \mathbf{PA}, \tag{10}$$

where $\mathbf{P}$ is the solution of the discrete-time matrix Riccati equation that can be solved offline before walking:

$$\mathbf{P} = \mathbf{A}^T \mathbf{PA} - \mathbf{A}^T \mathbf{PB} \left[R + \mathbf{B}^T \mathbf{PB} \right]^{-1} \mathbf{B}^T \mathbf{PA} + \mathbf{Q}. \tag{11}$$

In this paper it is assumed that the actual ZMP and the current CoM position can be measured [8], e.g. by making use of the Force Sensitive Resistors (FSR) in the feet alongside with measured joint angles, or the attitude of the body. Hence, Matrix $\mathbf{C}_m$ is defined as

$$\mathbf{C}_m = \begin{bmatrix} 1 & 0 & 0 \\ 0 & 0 & 1 \end{bmatrix} \tag{12}$$

which is used to select the measurable part $\mathbf{y}$ of the state vector $\mathbf{x}$:

$$\mathbf{y}(k) = \mathbf{C}_m \cdot \mathbf{x}(k) \hat{=} \begin{bmatrix} c_x^{snr}(k) \\ p^{snr}(k) \end{bmatrix}. \tag{13}$$

The discrete time system with sensor feedback is defined slightly different from the open-loop case:

$$\hat{\mathbf{x}}(k+1) = \mathbf{A}_0 \hat{\mathbf{x}}(k) + \mathbf{L} \left[\mathbf{y}(k) - \mathbf{C}_m \hat{\mathbf{x}}(k) \right] + \mathbf{b}\hat{u}(k). \tag{14}$$

Equation 14 incorporates the measurements by adding $e(k) := \mathbf{L} \left[\mathbf{y}(k) - \mathbf{C}_m \hat{\mathbf{x}}(k) \right]$ to equation 2: The summand expresses the difference between the estimated state $\hat{\mathbf{x}}$ and the measurements $\mathbf{y}$ multiplied by a matrix $\mathbf{L}$ that is constructed by executing the LQR procedure again. The latter task can be performed with built-in functions of Matlab[3] or Octave[4]. Figure 2 illustrates an example walk where an error is balanced by the controller. In the left part of the figure an error of $\delta(1.77) := \mathbf{y} - \mathbf{C}_m \hat{\mathbf{x}} = (0.005\,\text{m}, 0\,\text{m})^T$ (a measured error in the CoM position) is added at time $t \approx 1.8\,\text{s}$. One frame after the error was measured, the observed

---

[3] http://www.mathworks.com
[4] http://www.gnu.org/software/octave/

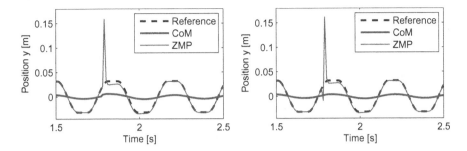

Fig. 2. Lateral plot of disturbed walk balanced by the controller. "Reference" denotes the reference ZMP $\hat{p}^{ref}$, "CoM" the observed CoM position $\hat{c}_x$ and "ZMP" the observed ZMP position $\hat{p}$. On the left, a CoM position error is induced, on the right a ZMP error.

ZMP reveals a large deviation. This is the reaction of the controller that accelerates the CoM towards the CoM trajectory that is optimal under the given reference ZMP. Starting with the next frame, the CoM is then decelerated until the CoM position and velocity reaches optimal values. This deceleration yields to a ZMP error into the opposite direction.

On the right a ZMP error of $\delta(1.77) = (0\,\text{m}, -0.05\,\text{m})^T$ is added. As can be seen, a similar reaction of the controller is the result.

3   Integration of Foot Placement Modification

The behavior of the controller presented in the previous section is not optimal and leads to further deviations. Under circumstances this can result in a ZMP outside the support polygon and a fall down. The central idea presented in this paper is the modification of the future foot positions, and consequently the reference ZMP by carrying out a lunge instead of forcing back the erroneous CoM to the original trajectory. A positive side effect is that the additional ZMP deviation due to the subsequent deceleration after the large acceleration also disappears. In order to calculate the modification, we postulate that the expected ZMP of the closed-loop system is equal to an open-loop system. The latter is not disturbed (and therefore reveals no ZMP deviation) at the time of the first reaction of the controller. The closed-loop system then also does not reveal ZMP deviations.

The case where a ZMP error is measured must be handled slightly different. In this case, a significant larger error is added to the CoM velocity since the CoM acceleration is directly related to the ZMP by the 3D-LIPM equation. If this is ignored, it would result in a similar deceleration deviation as can be seen in figure 2. If the cancellation of any ZMP deviation is desired as soon as possible, an opposite ZMP deviation must be applied in the frame after the error occured to instantaneously decelerate the CoM.

The described deliberations lead to the following requirement:

$$C\hat{\mathbf{x}}(T+2) - C\mathbf{x}(T+2) \stackrel{!}{=} -Ce(T), \tag{15}$$

where T is the time when the error is measured. Please note that $T + 1$ is the time when the error appears in the state vector. The controller $\widehat{u}(T + 1)$ then induces the ZMP deviation into $\widehat{\mathbf{x}}(T + 2)$.

In order to comply with the system requirements defined in section 1.2, the objective is to find a map S calculating modifications in closed-form fulfilling equation 15:

$$S : \{t_0, \ldots, N\} \times \mathbb{R}^2 \to \mathbb{R}, (t, e) \mapsto m \tag{16}$$

where $t \geqslant t_0$ is the point in time where the adjustment m of the reference ZMP is commenced (relative to the point in time where e is measured). The application of the scalar value m can be derived by observing human beings performing a lunge. It is a modification applied to the position of all foot steps starting at t. For this purpose, a ZMP modificator $\boldsymbol{p}^m$ that can be added to the reference ZMP is defined as:

$$\forall k < T + t : \boldsymbol{p}_k^m = 0 \tag{17}$$

$$\forall k \geq T + t : \boldsymbol{p}_k^m = m, \tag{18}$$

Consequently, the modified reference ZMP is:

$$\widehat{\boldsymbol{p}}^{ref} = \boldsymbol{p}^{ref} + \boldsymbol{p}^m. \tag{19}$$

Obviously, the first possible adaption time t_0 of the reference ZMP is when the disturbance occurs in the state vector at $T + 1$. However, equation 15 implicitly includes the reference ZMP $\widehat{\boldsymbol{p}}_{T+2}^{ref}$ but assumes $\boldsymbol{p}_{T+2}^m = 0$. Therefore the adjustment can start at $t_0 = 3$.

To derive S, we first examine the influence of the measured error on the state vector at $T + 2$ by recursively applying equation 14 and 6 :

$$C\widehat{\mathbf{x}}(T + 2) = C\left(\mathbf{A}_0\left(\mathbf{A}_0\widehat{\mathbf{x}}(T) + e(T) + b\widehat{u}(T)\right) + e(T + 1) + b\widehat{u}(T + 1)\right) \tag{20}$$

$$\widehat{u}(T + 1) = \widehat{I}(T + 1) + \widehat{X}(T + 1) + \widehat{D}\left(\left(\widehat{\boldsymbol{p}}_{T+1}^{ref}, \ldots, \widehat{\boldsymbol{p}}_{T+1+N}^{ref}\right)\right) \tag{21}$$

$$\widehat{I}(T + 1) = -G_I \sum_{i=0}^{T+1} \left[C\left(\mathbf{A}_0\widehat{\mathbf{x}}(i - 1) + e(i - 1) + b\widehat{u}(i - 1)\right) - \widehat{\boldsymbol{p}}_i^{ref}\right] \tag{22}$$

$$\widehat{X}(T + 1) = -G_x\left(\mathbf{A}_0\widehat{\boldsymbol{x}}(T) + e(T) + b\widehat{u}(T)\right) \tag{23}$$

$$\widehat{D}(\widehat{\boldsymbol{p}}^{ref}) = -\sum_{i=0}^{N}\left[G_d(i)\widehat{\boldsymbol{p}}_i^{ref}\right] = -\sum_{i=0}^{N}\left[G_d(i)\boldsymbol{p}_i^{ref}\right] - \sum_{i=0}^{N}\left[G_d(i)\boldsymbol{p}_i^m\right] \tag{24}$$

$$= -\sum_{i=0}^{N}\left[G_d(i)\boldsymbol{p}_i^{ref}\right] - m\sum_{i=t}^{N}G_d(i), \tag{25}$$

From 24 to 25, equations 17 and 18 are applied. Since the modification is constant starting at t, the sum of the mapping $\boldsymbol{p}^m$ can be replaced by the corresponding scalar value m.

Equations 20-25 are simplyfied by exploiting the following statements:

$$\forall k \leq T : \widehat{\mathbf{x}}(k) = \mathbf{x}(k), \widehat{u}(k) = u(k) \tag{26}$$

$$\forall k \neq T : e(k) = 0 \tag{27}$$

$$\forall k \leq T + t, 2 < t : \widehat{\boldsymbol{p}}_k^{ref} = \boldsymbol{p}_k^{ref}, \tag{28}$$

Equations 27 and 26 hold since this derivation is based on an impulse like error at time T. It is important to recognize that no disturbance separates the open-loop case from the closed-loop case before T. Hence, equations 20-25 are similar in the open-loop case except for the error terms e and the ZMP modification $\boldsymbol{p}^m$. On the left side of equation 15 this cancels out every term that does not contain e or m:

$$-\boldsymbol{C}e(T) = \boldsymbol{C}\widehat{\mathbf{x}}(T + 2) - \boldsymbol{C}\mathbf{x}(T + 2) \tag{29}$$

$$= \boldsymbol{C}\left(\boldsymbol{A}_0 e(T) + \boldsymbol{b}\left(-\boldsymbol{G}_I \boldsymbol{C}e(T) - \boldsymbol{G}_x e(T) - m\sum_{i=t}^{N}\boldsymbol{G}_d(i)\right)\right) \tag{30}$$

$$\Leftrightarrow m = \underbrace{\left(\boldsymbol{C}\boldsymbol{b}\sum_{i=t}^{N}\boldsymbol{G}_d(i)\right)^{-1}\boldsymbol{C}\left(\boldsymbol{A}_0 + \boldsymbol{I}_{3\times 3} - \boldsymbol{b}\boldsymbol{G}_I\boldsymbol{C} - \boldsymbol{b}\boldsymbol{G}_x\right)e(T)}_{=:\boldsymbol{G}_e(t)}. \tag{31}$$

Function S can now be defined as:

$$S(t, e) = \boldsymbol{G}_e(t)e, \quad t \in \{3, \ldots, N\}. \tag{32}$$

Therefore the modification for the reference ZMP and the foot steps can be calculated by a simple multiplication. The gains $\boldsymbol{G}_e$ can be calculated in closed-form together with the gains described in section 2 offline before walking.

In some cases it is not possible to apply only one lunge to balance error e, e.g. if the sidestep would be too large for the given leg length. Therefore it is possible to split the modification of a single step to n modifiers that can be applied at any desired point. To do so, the error has to be distributed over different smaller modifications according to a desired ratio: $e = \alpha_1 e + \ldots + \alpha_n e$ with $\alpha_1 + \ldots + \alpha_n = 1$. The modifiers $m_1, \ldots, m_n$ can then be computed by $m_k = S(t_k, \alpha_k e)$.

To show that this kind of system has the same stabilizing effect, its influence to the controller must be examined. As can be seen in the controller equation 6, the only term incorporating the future reference ZMP is $\widehat{D}$ as formulated in equation 24. The error can be split between different lunges since $\widehat{D}$ is a linear map (as shown in equations 24 and 25). The resulting modifiers can be applied at random future points in time since S calculates the modifier for the arbitrary time t by definition.

4   Examples and Evaluation

This section starts with an example showing the stabilizing effect on a robot using the 3D-LIPM. It depicts the input ($\widehat{\boldsymbol{p}}^{ref}$) and output ($c_x$ and p) of a walk that

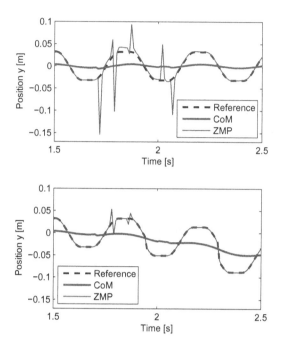

Fig. 3. Example walk with measured errors. "Reference" denotes the reference ZMP $\hat{p}^{ref}$, "CoM" the observed CoM position $\hat{c}_x$ and "ZMP" the observed ZMP position $\hat{p}$. In the top graph the errors are balanced by the controller only, in the bottom graph lunges are included.

is disturbed by different impulse-like errors. As shown in Figure 3, the following errors are measured: $\delta(1.7) = (-0.005, 0)^T, \delta(1.77) = (0, 0.05)^T, \delta(1.85) = (0, -0.02)^T, \delta(2) = (0.003, 0)^T$ and $\delta(2.05) = (-0.003, 0)^T$. In the upper graph, the controller balances the disturbance by forcing the CoM back to the initial movement without a modification of the foot positions.

The lower graph displays an example where lunges are planned for the double support phases at 1.95 s and 2.3 s. In this example, we define that 80% of a measured error has to be corrected by the first lunge, and 20% by the second under the assumption that the error occurs before the first lunge. The later errors are compensated by the second lunge exclusively. As can be seen, all measured CoM errors (at 1.7 s, 2 s and 2.05 s) can be balanced without any ZMP disturbance. The measured ZMP errors at 1.77 s and 1.85 s exhibit the intended ZMP deviations as desired in equation 15 within the subsequent frame. Nevertheless, even with the desired deviations the errors are balanced with far less ZMP deviations compared to the case without modifications.

However, the requirement in equation 15 is based on the 3D-LIPM. Since this is a strong abstraction, the proposed approach has to be evaluated on more complex models. It is also unclear if the inverted ZMP deviation at $T+1$ ($-Ce(T)$ in

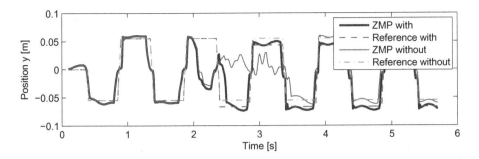

Fig. 4. Multibody simulation of a disturbed walk comparing closed-loop balancing with modification of the foot position ("ZMP/Reference with") with balancing without modification ("ZMP/Reference without")

equation 15) has the desired effect on a real robot. Therefore a walking algorithm incorporating the results of section 2 and 3 is evaluated using SimRobot [19] with a model of the Nao by Aldebaran Robotics[5]. The implementation used for this experiments executes lunges only for larger disturbances and only for the next step since later the lunge would be significantly larger.

In this experiment, the robot walks at a speed of $5\frac{cm}{s}$ with a step duration of 1 s. As can be seen in Figure 4, at time ~ 2 s a force of $1000\,\mathrm{N}$ is applied for 0.01 s. This leads to a disturbance that is compensated by the system after 3 additional steps without carrying out a lunge. When a lunge is executed, stabilization can be achieved in the next step.

It has to be emphasized that results of experiments using a multibody simulation or a real robot strongly depend on the implementation of the overall walking algorithm. For example, the algorithm must not move the swinging foot any further when it accidentally hits the ground caused by tipping. The swinging foot must also be in parallel to the ground at that moment. However, a more detailed evaluation of those algorithmic details is out of the scope of this paper.

5   Conclusion and Future Works

This paper depicts the drawbacks of balancing without modification of footsteps. It can lead to further ZMP deviations besides the measured errors. To overcome this, a requirement for the controller output is formulated and rearranged to a set of gains that can be utilized to calculate the modification in closed-form. This closed-form calculation fulfils objectives 2 (by simply not applying any modification), 3 and 4 (since function S can be calculated by a simple matrix multiplication). Objective 1 is fulfilled by utilizing a preview controller and objectives 5 and 6 are fulfilled since function S can be applied on fractions of the measured error at a random point in time (see end of section 3).

---

[5] http://www.aldebaran-robotics.com

As shown in the example, balancing by performing lunges results into a walk with far less ZMP errors. It can also help stabilising a walk that relies on a more complex model of the robot.

However, this strongly depends on the realization of the entire walking algorithm. The implementation of an adequate walking algorithm that exploits the advantages of the proposed method to stabilize the walk of a real robot is of particular importance for future work. Moreover, the current requirement of the immediate correction of the CoM speed to ensure a correct ZMP leads to the term $-Ce(T)$ in equation 15. While this still results in far less ZMP errors, it would be beneficial to recover the desired speed by a different modification of the reference ZMP. First studies show that it is possible to avoid any ZMP error till an arbitrary point in time, even in presence of false CoM speeds. Currently, this is only possible by applying the modifications in every succeeding frame.

6   Supplementary Material

The algorithms and formulas presented in this paper can be downloaded as Matlab/Octave scripts: `http://irf.tu-dortmund.de/nao-devils/download/ZMPIPController2012.zip`. The examples shown in Figure 2 and 3 are also included.

References

1. Vukobratović, M., Borovac, B.: Zero-moment point – Thirty five years of its life. International Journal of Humanoid Robotics 1(1), 157–173 (2004)
2. Kajita, S., Kanehiro, F., Kaneko, K., Yokoi, K., Hirukawa, H.: The 3d linear inverted pendulum mode: a simple modeling for a biped walking pattern generation. In: Proceedings of the 2001 IEEE/RSJ International Conference on Intelligent Robots and Systems, vol. 1, pp. 239–246 (2001)
3. Kim, J.H.: Walking pattern generation of a biped walking robot using convolution sum. In: 2007 7th IEEE-RAS International Conference on Humanoid Robots, November 29-December 1, pp. 539–544 (2007)
4. Harada, K., Kajita, S., Kaneko, K., Hirukawa, H.: An analytical method for real-time gait planning for humanoid robots. International Journal of Humanoid Robotics 3(01), 1–19 (2006)
5. Kajita, S., Kanehiro, F., Kaneko, K., Fujiwara, K., Harada, K., Yokoi, K., Hirukawa, H.: Biped walking pattern generation by using preview control of zero-moment point. In: ICRA, pp. 1620–1626. IEEE (2003)
6. Czarnetzki, S., Kerner, S., Urbann, O.: Applying dynamic walking control for biped robots. In: Baltes, J., Lagoudakis, M.G., Naruse, T., Ghidary, S.S. (eds.) RoboCup 2009. LNCS (LNAI), vol. 5949, pp. 69–80. Springer, Heidelberg (2010)
7. Czarnetzki, S., Kerner, S., Urbann, O.: Observer-based dynamic walking control for biped robots. Robotics and Autonomous Systems 57(8), 839–845 (2009)
8. Urbann, O., Tasse, S.: Observer based biped walking control, a sensor fusion approach. Autonomous Robots, 1–13 (2013)

9. Alcaraz-Jiménez, J.J., Missura, M., Martínez-Barberá, H., Behnke, S.: Lateral disturbance rejection for the nao robot. In: Chen, X., Stone, P., Sucar, L.E., van der Zant, T. (eds.) RoboCup 2012. LNCS (LNAI), vol. 7500, pp. 1–12. Springer, Heidelberg (2013)
10. Urata, J., Nshiwaki, K., Nakanishi, Y., Okada, K., Kagami, S., Inaba, M.: Online decision of foot placement using singular lq preview regulation. In: 2011 11th IEEE-RAS International Conference on Humanoid Robots (Humanoids), pp. 13–18 (October 2011)
11. Pratt, J.E., Carff, J., Drakunov, S.V., Goswami, A.: Capture point: A step toward humanoid push recovery. In: Humanoids, pp. 200–207. IEEE (2006)
12. Englsberger, J., Ott, C., Roa, M., Albu-Schaffer, A., Hirzinger, G.: Bipedal walking control based on capture point dynamics. In: 2011 IEEE/RSJ International Conference on Intelligent Robots and Systems (IROS), pp. 4420–4427 (September 2011)
13. Yun, S.K., Goswami, A.: Momentum-based reactive stepping controller on level and non-level ground for humanoid robot push recovery. In: 2011 IEEE/RSJ International Conference on Intelligent Robots and Systems (IROS), pp. 3943–3950 (September 2011)
14. van Zutven, P., Kostic, D., Nijmeijer, H.: Foot placement for planar bipeds with point feet. In: 2012 IEEE International Conference on Robotics and Automation (ICRA), pp. 983–988 (May 2012)
15. Herdt, A., Diedam, H., Wieber, P.B., Dimitrov, D., Mombaur, K., Diehl, M.: Online walking motion generation with automatic footstep placement. Advanced Robotics 24(5-6), 719–737 (2010)
16. Morisawa, M., Harada, K., Kajita, S., Kaneko, K., Sola, J., Yoshida, E., Mansard, N., Yokoi, K., Laumond, J.P.: Reactive stepping to prevent falling for humanoids. In: 9th IEEE-RAS International Conference on Humanoid Robots, Humanoids 2009, pp. 528–534 (December 2009)
17. Morisawa, M., Kanehiro, F., Kaneko, K., Mansard, N., Sola, J., Yoshida, E., Yokoi, K., Laumond, J.: Combining suppression of the disturbance and reactive stepping for recovering balance. In: 2010 IEEE/RSJ International Conference on Intelligent Robots and Systems (IROS), pp. 3150–3156 (October 2010)
18. Katayama, T., Ohki, T., Inoue, T., Kato, T.: Design of an optimal controller for a discrete-time system subject to previewable demand. International Journal of Control 41(3), 677–699 (1985)
19. Laue, T., Spiess, K., Röfer, T.: SimRobot - A General Physical Robot Simulator and Its Application in RoboCup. In: Bredenfeld, A., Jacoff, A., Noda, I., Takahashi, Y. (eds.) RoboCup 2005. LNCS (LNAI), vol. 4020, pp. 173–183. Springer, Heidelberg (2006), http://www.springer.de/

Iterative Snapping of Odometry Trajectories for Path Identification

Richard Wang, Manuela Veloso, and Srinivasan Seshan

Computer Science Department,
Carnegie Mellon University,
5000 Forbes Ave, Pittsburgh, PA, USA
{rpw,mmv,srini}@cs.cmu.edu

Abstract. An increasing number of mobile devices are capable of automatically sensing and recording rich information about the surrounding environment. Spatial locations of such data can help to better learn about the environment. In this work, we address the problem of identifying the locations visited by a mobile device as it moves within an indoor environment. We focus on devices equipped with odometry sensors that capture changes in motion. Odometry suffers from cumulative errors of dead reckoning but it captures the relative shape of the traversed path well. Our approach will correct such errors by matching the shape of the trajectory from odometry to traversable paths of a known map. Our algorithm is inspired by prior vehicular GPS map matching techniques that snap global GPS measurements to known roads. We similarly wish to snap the trajectory from odometry to known hallways. Several modifications are required to ensure these techniques are robust when given relative measurements from odometry. If we assume an office-like environment with only straight hallways, then a significant rotation indicates a transition to another hallway. As a result, we partition the trajectory into line segments based on significant turns. Each trajectory segment is snapped to a corresponding hallway that best maintains the shape of the original trajectory. These snapping decisions are made based on the similarity of the two curves as well as the rotation to transition between hallways. We will show robustness under different types of noise in complex environments and the ability to propose coarse sensor noise errors.

1 Introduction

Sensor-equipped mobile devices can sense and record information about their surrounding environments. Mobile devices are presented with incredible opportunities to collect data where static devices cannot. For example, a stationary thermometer is rather uninteresting because it only measures temperature at a single location. In contrast, a mobile device equipped with a thermostat could create a temperature map to reveal patterns and perhaps even identify drafty windows. Interesting spatial maps could be created from many other sensors including UV, air quality, radiation, and Wifi. Devices that move can help reveal

S. Behnke et al. (Eds.): RoboCup 2013, LNAI 8371, pp. 432–443, 2014.
© Springer-Verlag Berlin Heidelberg 2014

rich information about our surrounding environments. One can create these spatial maps if the spatial locations for each corresponding sensor measurement are known. If we knew the path traversed by the device, then we can estimate such spatial locations. In this work, we will address the offline problem of estimating the path traversed by a device in an indoor environment from odometry and a known map.

There exist many techniques to identify the path traversed from advanced sensors like Kinect, LIDAR, Sonar and GPS. However, these sensors are typically limited to a subset of mobile devices like cars and robots. In addition, extrinsic sensors are challenged by scenarios in which there are varying light conditions and dynamic objects and GPS is challenged by indoor environments. Odometry is advantageous because it captures high-resolution changes in motion while also being low cost, occupying a small form factor, and consuming little power. This makes it accessible to a large number of very mobile devices including pedometers, health monitors, cell phones, and blind robots. Odometry faces unique challenges because its measurements are relative. While each measurement alone is a relatively insignificant change in motion, aggregating a lengthy sequence of odometry measurements can reveal a unique shape traversed by the device. In contrast, a single GPS measurement can reveal the global position of the device.

The process of computing the path traversed by odometry is called dead reckoning. Errors in odometry accumulate and typically result in wildly incorrect position estimates. While not ideal for computing the exact path traversed, odometry is very good at capturing the relative shape of the path traversed. Figure 1 is an example of odometry collected from wheel encoders of an omnidirectional robot. Most of the cumulative errors are a result of drift errors that occurred when the robot turned at intersections. With the map, it becomes obvious that the path from odometry does not travel on any traversable paths. Our approach will use the given map in order to correct these significant cumulative errors by matching the shape of the trajectory to traversable paths.

We address offline path identification because the position at a particular point is unclear until one considers the subsequent trajectory that follows. For example, one cannot disambiguate two right turns in the same hallway until the subsequent trajectory reveals a shape that is unique to only one of the two choices. To utilize the shape of the trajectory, we borrow techniques from vehicular GPS map matching. GPS does not consider the infinite space of exact positions of the vehicle on a road. Its primary concern is to match the vehicle to the correct road segment. GPS measurements are snapped to roads by using one of several geometric map matching techniques. Most relevant to our approach is curve-to-curve matching because it computes the distance between an entire sequence of GPS measurements to that of an entire road segment. One can add topological map matching to exploit the connectivity constraints when transitioning between different road segments. Instead of considering all nearby roads for snapping, topological map matching only snaps to road segments connected to the road that the vehicle is currently matched to. The combination of these techniques leads to very robust navigation algorithms for GPS [7] in real world scenarios.

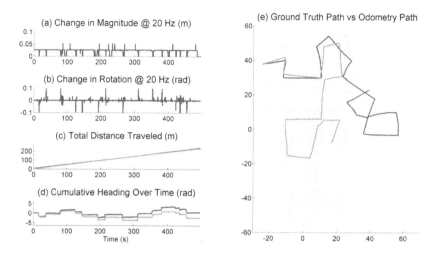

Fig. 1. Wheel encoder odometry (red) collected from a particular traversal of an indoor environment compared to ground truth collected by localization with a Kinect (green). The (a) changes in forward motion are (c) accumulated over time to reveal that the device has traveled over 200 m. The (b) changes in rotation are accumulated to reflect its (d) cumulative heading. At the end of the path, the distance traveled deviated from ground truth by only 6 m while there is much more significant drift in the device's heading. This is (e) visualized by comparing the actual truth path to the path constructed from dead reckoning.

Our approach has many parallels to these GPS map matching algorithms. Instead of snapping to roads, we will be snapping to hallways. In this work, we will assume indoor, office-like environments with straight hallways. To evaluate how well a trajectory segment matches a hallway segment, we perform curve-to-curve matching by computing their similarity instead of distance. Distance is ideal for GPS because better matching roads will be closer to the global position measurements. A similarity metric is better suited for comparing the relative shapes of two curves, which is the case with odometry. We use topological map matching to both enforce connectivity between hallways and also to compare the change in heading required for the device to transition to another hallway. A good candidate hallway will require a change in heading that matches very well to the change in rotation measured by odometry. The combination of these two relative metrics makes our approach more robust in identifying paths that better match the shape of the trajectory from odometry.

We will discuss related work in Section 2, introduce relevant definitions in Section 3, formalize our snapping algorithm in Section 4, and evaluate our algorithm in Section 5.

2   Related Work

Prior works have attempted to improve odometry with careful calibration [1] [4]. While one can localize over short distances [2], odometry will eventually succumb to dead reckoning errors and require corrections to fix these cumulative errors.

Correcting dead reckoning errors has been addressed in prior work with cell phones by using walls of a known map [5] [9] [12] [8]. Particle filters were used to sample the space of reachable paths. Odometry measurements were used to update the motion of the particles. Uncertainty is added to these motion updates because odometry is noisy. The walls of a known map help to constrain uncertainty of the particles because the device should not be able to pass through a wall. The problem addressed is similar to our work but our approach does not attempt to recover the exact positions traversed. As a result, we can focus on high level decisions at hallway intersections to ensure more robust global paths identified and not require being given an accurate sensor model.

Other work attempts to automatically collect unique signatures in an environment from various sensors. These signatures include Wifi [3] [6] as well as a combination of different sensors including magnetic signatures [10]. Odometry is unique because it can capture high-resolution changes in motion with a level of detail that unique signatures cannot capture. There are opportunities in the future to complement position estimates from these unique signatures with the motion trajectory from odometry.

Our algorithm borrows many techniques from GPS map matching [7] [11]. We focus on taking advantage of basic geometric and topological map matching techniques because odometry excels at capturing the relative shape of the traversed path. While distance metrics are ideal for global GPS measurements, some modifications are required so that these map matching techniques are robust for relative sensors. Our approach focuses on the shape of the trajectory by measuring similarity to make better snapping decisions for odometry.

3   Definitions

We consider two dimensional path identification in a Cartesian map. Odometry sensor data is assumed to be captured on the same plane. We consider devices traveling directly towards their destination. We require a few definitions:

Odometry update u is composed of a change in forward motion and rotation since the last update. A trajectory T is a sequence of odometry updates. A pose p is a position and direction on the given map. A path P is a sequence of poses.

$$u = \{dm, dr\}$$
$$T = \{u_1, u_2, u_3, ...\}$$
$$p = \{x, y, a\}$$
$$P = \{p_1, p_2, p_3, ...\}$$

Computing the path of a trajectory, also called dead reckoning, can be computed recursively given an initial pose p_0. A pose p_i is computed from the pose of the previous step p_{i-1} and odometry update u_i.

$$DeadReckoning(p_{i-1}, u_i) \rightarrow p_i :$$
$$p_i^a = u_i^{dr} + p_{i-1}^a$$
$$p_i^y = p_{i-1}^y + sin(p_i^a) * u_i^{dm}$$
$$p_i^x = p_{i-1}^x + cos(p_i^a) * u_i^{dm}$$

Odometry updates arriving at approximately 20 Hz tend to be very small changes in motion. As a result, we will aggregate sequences of odometry updates to form trajectory segments. We consider environments with only straight hallways. Trajectory segments can then be partitioned by identifying significant rotations. A trajectory segment approximates a sequence of odometry updates as a single, large change in motion dM and rotation dR. This will result in a straight trajectory segment that ignores the minor rotations of the aggregated odometry updates. A trajectory T is now a sequence of trajectory segments s. The dead reckoning computation remains the same except the updates are now over these more significant trajectory segments s.

$$s = \{u_i, u_{i+1}, u_{i+2}, ...\} \approx \{dM, dR\}$$
$$T = \{s_1, s_2, s_3, ...\}$$

$$DeadReckoning(p_{i-1}, s_i) \rightarrow p_i :$$
$$p_i^a = s_i^{dR} + p_{i-1}^a$$
$$p_i^y = p_{i-1}^y + sin(p_i^a) * s_i^{dM}$$
$$p_i^x = p_{i-1}^x + cos(p_i^a) * s_i^{dM}$$

4   Snapping

Our algorithm snaps trajectory segments to their corresponding hallways. We will segment the trajectory so that each trajectory segment corresponds to an entire hallway segment. If we assume an environment with only straight hallways, then the trajectory segment should be partitioned into straight segments. We will first explain how the hallways and trajectories are segmented, then how these are processed by the snapping algorithm, and then the relative metrics to best use trajectories from odometry.

4.1   Marking Hallway Segments

Our assumption of straight hallway means that we do not need to consider curved hallways. Examples of the types of indoor environments that we consider are shown in Figure 2. If hallway intersection points are known, then hallways can be identified by connecting all hallway intersection points that do not violate a map constraint. Automatic identification of intersection points is difficult because they need to be carefully placed to have high visibility with surrounding adjacent hallways. This is a challenge especially for open areas. As a result, we manually mark the hallway intersection points as a one-time process for a given map.

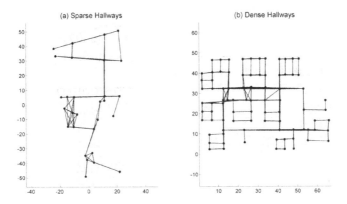

Fig. 2. Indoor environments following our assumption of straight hallways marked by the darker lines

4.2   Extracting Trajectory Segments

Odometry sensors capturing measurements at 20 Hz capture very small changes in motion that individually are quite insignificant. We want to aggregate sequences of odometry updates into trajectory segments $s = \{u_i, u_{i+1}, u_{i+2}, ...\}$ to match with hallway segments. Given that the environment is composed of only straight hallway segments, we can represent each trajectory segment as a large odometry update $s \approx \{dM, dR\}$. Odometry updates are partitioned based on identification of large turns that should indicate when the device is transitioning to another hallway.

Significant rotation peaks are identified by applying a sliding window of 30 frames over the changes in rotation from odometry. This will identify the peak rotations as seen in Figure 1(b) that correspond to significant heading changes in Figure 1(d). The trajectory will be segmented based on these peak rotations. dR is assigned the value of the windowed peak rotation. dM is the total change in motion forward over the sequence of odometry updates occuring since the last significant peak rotation. We have found that this simple approach is sufficient for odometry captured by wheel encoders of an omni-directional robot.

4.3   Snapping Trajectories

Our snapping algorithm iteratively snaps each trajectory to the best matching hallway. As shown in Algorithm 1, the device's poses are recorded at each hallway intersection and appended to form the hallways intersections traversed. These poses can then be connected by their corresponding trajectory segments to form the path traversed. This reflects our algorithm's focus on making robust, high-level decisions at hallway intersection points as opposed to attempting to estimate the exact poses of the device. It also means that our algorithm is efficient because it does not need to consider modifications to the intermediate odometry measurements that form the trajectory segment.

We use both geometric and topological map matching from vehicular GPS algorithms. The function *GetAdjacentHalls* uses topological map matching by limiting the candidate hallways for the device to transition to. The function *GetBestMatch* uses geometric map matching to compute the similarity of the trajectory update to the candidate hallways. The combination of these two techniques means that our algorithm will prefer identifying paths that have a similar shape with the noisy trajectory. This is a benefit because odometry is good at capturing the relative shape of the path traversed.

Algorithm 1. Trajectory Snapping

 1: **function** TRAJECTORYSNAP(Hallways, TrajSegments, PoseInit)
 2:     $PoseCurr \leftarrow PoseInit$;
 3:     $IntersectionsTraversed \leftarrow [PoseCurrent]$;
 4:
 5:     **for** $TrajUpdate$ in $TrajSegments$ **do**
 6:         $Candidates \leftarrow$ GetAdjacentHalls($Hallways, PoseCurr$)
 7:         $SnappedHallway \leftarrow$ GetBestMatch($Candidates, PoseCurr, TrajUpdate$)
 8:         $PoseNext \leftarrow$ GetNextIntersection($SnappedHallway, PoseCurr$)
 9:
10:         $IntersectionsTraversed \leftarrow [IntersectionsTraversed; PoseNext]$;
11:         $PoseCurr \leftarrow PoseNext$)

4.4   Deciding between Candidate Hallways

Our algorithm should prefer hallway candidates that best follow the relative shape of the noisy trajectory. Each trajectory segment is a rotation and forward motion. Applying this trajectory update to the device's current pose should move it from the current hallway intersection to the next hallway intersection. However, there can be many hallways to choose from. We need a metric that will prefer candidate hallways that best matches the measured trajectory update.

Distance is ideal for GPS because global measurements are closer to roads that are better candidates. With relative measurements, drift can result in significant distances between the best hallway candidate and the trajectory. Our relative metric considers the current pose of the device and computes the relative change in motion to transition to each of the candidate hallways. The hallway that best matches the trajectory update will be chosen for snapping. Currently, we use a simple similarity function that computes the square root of the sum of squared differences between the percentage difference in forward motion and the difference in angular rotation. While future work can better engineer these error components, it is most important that the relative metric measures these components separately and then combines them.

As we can see in Figure 3(a), the distance metric would incorrectly snap to the shorter hallway candidate (0,5). In contrast, the relative metric Figure 3(b) would recognize that the change in rotation is equivalent for both candidates but the magnitude of the trajectory matches the longer hallway candidate (0,10) much better. There are many other situations where distance will making puzzling decisions while our relative metric snaps to more intuitive hallways. Because relative snapping cannot recover once a poor decision is made, improving robustness is very important for the success of our algorithm.

Fig. 3. Deciding which candidate hallway (green) to snap the trajectory update (red). A distance metric would select the shorter hallway because the sum of sampled distances (dashed) is smaller. The relative metric separates rotation and motion errors and selects the longer hallway because it only differs from the trajectory update by rotational error while the shorter hallway differs by both rotational and motion errors.

Algorithm 2. Best Matching Hallway

1: **function** GETBESTMATCH(Candidates, PoseCurr, TrajUpdate)
2:      $BestSnapChoice \leftarrow \emptyset$
3:      **for** $Hallway$ in $Candidates$ **do**
4:          $hM \leftarrow$ ComputeRelativeMotion($Hallway$, $PoseCurr$)
5:          $hR \leftarrow$ ComputeRelativeRotation($Hallway$, $PoseCurr$)
6:          $DiffMotion \leftarrow TrajUpdate.dM - hM$
7:          $DiffRotation \leftarrow TrajUpdate.dR - hR$
8:          $PercentageDiffMotion \leftarrow DiffMotion/hM$
9:
10:          $Similarity \leftarrow \sqrt{PercentageDiffMotion^2 + DiffRotation^2}$
11:          **if** isMoreSimilar($Similarity$, $BestSnapChoice$) **then**
12:              $BestSnapChoice \leftarrow Hallway$

4.5   Estimating Coarse Odometry Noise

An important property of our algorithm is that it does not require a given sensor noise model. This is possible because we do not attempt to consider the infinite space of exact positions traversed by the device. Instead, we only consider a finite number of hallway segments and force the trajectory to fit to these hallways. As a result, the relative metrics we use only need to select the best choice among the finite hallway candidates. This allows our algorithm to compute the modifications required to snap each trajectory update to its corresponding hallway. These modifications can be interpreted as the coarse odometry noise estimates of our algorithm. We can then use these noise estimates to recognize when a potentially unsuccessful snap has occurred by looking for modifications that are outliers of normal behavior.

5   Results

We wish to show that our algorithm is robust and operates in real-world scenarios. We will use an extensive simulation in a challenging office environment to evaluate robustness. We will then show that our algorithm can correct errors in wheel encoder odometry from an omni-direction robot and successfully proposes coarse sensor noise errors.

5.1   Evaluating Robustness

We took the map of a real indoor office environment as shown in Figure 2(b) and performed simulations to evaluate the robustness of our snapping algorithm. This environment is especially challenging because there are many similar turns that our algorithm can incorrectly snap to. In our simulation, we take a single ground truth path, extract its trajectory segments, and then add random noise to evaluate our algorithm. Uniform noise is added by taking each trajectory segment $\{dM, dR\}$, stretching the forward motion of the device by a percentage error $\eta = uniform(-\alpha, \alpha)$, and adding rotational noise $\epsilon = uniform(-\beta, \beta)$. This results in a noisy trajectory segment $\{dM(1 + \eta), dR + \epsilon\}$.

We want to show how the success rate of snapping evolves as more noise is injected into the simulation. In our simulation, we perform 100 iterations of each combination of noise. We can see in Figure 4(a-d) the types of paths that our algorithm can successfully identify the correct path traversed. We can see that our algorithm can robustly handle fairly significant noise from the gradient of success rates in Figure 4(e). Our snapping algorithm fails when it makes an incorrect snapping decision. This results from our algorithm making greedy decisions from local information and potentially can discard the globally optimal solution. Instead of maintaining a single guess, we could maintain a set of unique guesses. Figure 4(f) shows the increased success rate when our algorithm is augmented with a set of 5 guesses. When incorrect decisions are made, it is more likely that another guess will have made the correct decision. As we increase the number of unique guesses, the success rate of our algorithm will increase along with the size of the explored search space.

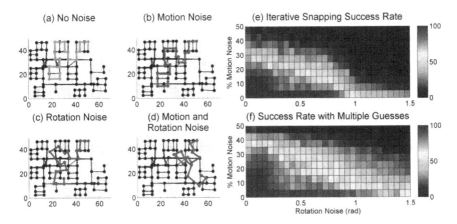

Fig. 4. Testing robustness of our snapping algorithm by injecting noise into a (a) ground truth path. (b) Motion only noise $\eta = \pm 25\%$. (c) Rotation only noise $\epsilon = \pm .8rad$. (d) Both motion and rotation noise $\eta = \pm 25\%, \epsilon = \pm .8rad$. (e) Success rate of snapping injected with various combinations of noise. (f) Improved success when the iterative algorithm is augmented by mantaning a set of 5 unique guesses.

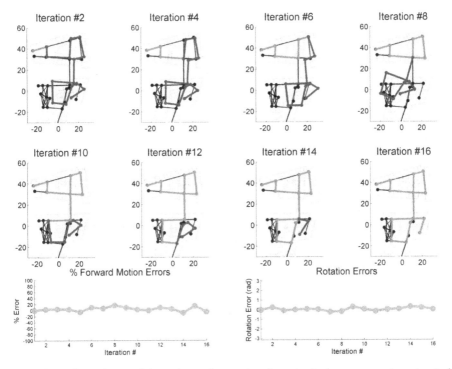

Fig. 5. Snapshot of several iterations of snapping from trajectory segments extracted from real wheel encoder odometry. The snapped segments (green) and unsnapped segments (red). Snapping results in modifications to the trajectory segments and these can be perceived as coarse noise estimates that are shown at the bottom.

5.2 Success with Real-World Data

We evaluate odometry from wheel encoders of an omni-directional robot collected from the path traversed in Figure 1 that travels almost 250 m. This environment is not nearly as challenging as the simulated environment because it requires fewer difficult decisions. In Figure 4, we show how our snapping algorithm iteratively corrects the error in odometry. Each decision is a simple comparison between the trajectory segment and its corresponding hallways because the prior trajectory segments have already been snapped. The noise model discovered reveals that corrections to the wheel encoder odometry resulted in motion noise within 20% and rotation noise within .5 radians. These coarse noise estimates are not as accurate when compared to the ground truth in Figure 1 because they are influenced by the marked positions of the hallway intersections. Nevertheless, these coarse estimates can help to reveal patterns in the normal behavior of snapping and outliers could help to suggest when incorrect snapping decisions are made.

6 Conclusion

Our snapping algorithm can recover paths with similar shapes to the trajectories captured by odometry. Forcing the trajectory to fit the marked hallways allows us to focus on important decisions at intersections to ensure global robustness. In addition, it allows us to recover coarse noise estimates from the modifications required to fit the trajectory to the hallways. Many opportunities remain to consider curved hallways, combining odometry with other sensors, and recovery from incorrect snapping decisions.

References

1. Borenstein, J., Feng, L.: Measurement and correction of systematic odometry errors in mobile robots. IEEE Transactions on Robotics and Automation 12(6), 869–880 (1996)
2. Doh, N., Choset, H., Chung, W.K.: Accurate relative localization using odometry. In: Proceedings of the IEEE International Conference on Robotics and Automation, ICRA 2003, vol. 2, pp. 1606–1612. IEEE (2003)
3. Huang, J., Millman, D., Quigley, M., Stavens, D., Thrun, S., Aggarwal, A.: Efficient, generalized indoor wifi graphslam. In: 2011 IEEE International Conference on Robotics and Automation (ICRA), pp. 1038–1043. IEEE (2011)
4. Kelly, A.: Fast and easy systematic and stochastic odometry calibration. In: Proceedings of the IEEE/RSJ International Conference on Intelligent Robots and Systems, IROS 2004, vol. 4, pp. 3188–3194. IEEE (2004)
5. Kourogi, M., Ishikawa, T., Kameda, Y., Ishikawa, J., Aoki, K., Kurata, T.: Pedestrian dead reckoning and its applications. In: Proceedings of Lets Go Out Workshop in Conjunction with ISMAR, vol. 9 (2009)
6. Liu, H., Gan, Y., Yang, J., Sidhom, S., Wang, Y., Chen, Y., Ye, F.: Push the limit of wifi based localization for smartphones. In: Proceedings of the 18th Annual International Conference on Mobile Computing and Networking, pp. 305–316. ACM (2012)

7. Quddus, M.A., Ochieng, W.Y., Noland, R.B.: Current map-matching algorithms for transport applications: State-of-the art and future research directions. Transportation Research Part C: Emerging Technologies 15(5), 312–328 (2007)
8. Rai, A., Chintalapudi, K.K., Padmanabhan, V.N., Sen, R.: Zee: Zero-effort crowdsourcing for indoor localization. In: Proceedings of the 18th Annual International Conference on Mobile Computing and Networking, pp. 293–304. ACM (2012)
9. Serra, A., Carboni, D., Marotto, V.: Indoor pedestrian navigation system using a modern smartphone. In: Proceedings of the 12th International Conference on Human Computer Interaction with Mobile Devices and Services, MobileHCI, vol. 10, pp. 397–398. Citeseer (2010)
10. Wang, H., Sen, S., Elgohary, A., Farid, M., Youssef, M., Choudhury, R.R.: No need to war-drive: Unsupervised indoor localization. In: Proceedings of the 10th International Conference on Mobile Systems, Applications, and Services, pp. 197–210. ACM (2012)
11. White, C.E., Bernstein, D., Kornhauser, A.L.: Some map matching algorithms for personal navigation assistants. Transportation Research Part C: Emerging Technologies 8(1), 91–108 (2000)
12. Woodman, O., Harle, R.: Pedestrian localisation for indoor environments. In: Proceedings of the 10th International Conference on Ubiquitous Computing, pp. 114–123. ACM (2008)

Evaluation of Recent Approaches
to Visual Odometry from RGB-D Images

Sergey Alexandrov[1] and Rainer Herpers[1,2,3]

[1] Department of Computer Science, Hochschule Bonn-Rhein-Sieg, Germany
[2] Department of Computer Science and Engineering, York University, Canada
[3] Faculty of Computer Science, University of New Brunswick, Canada
sergey.alexandrov@smail.inf.h-brs.de, rainer.herpers@h-brs.de

Abstract. Estimation of camera motion from RGB-D images has been an active research topic in recent years. Several RGB-D visual odometry systems were reported in literature and released under open-source licenses. The objective of this contribution is to evaluate the recently published approaches to motion estimation. A publicly available dataset of RGB-D sequences with precise ground truth data is applied and results are compared and discussed. Experiments on a mobile robot used in the RoboCup@Work league are discussed as well. The system showing the best performance is capable of estimating the motion with drift as small as 1 cm/s under special conditions, though it has been proven to be robust against shakey motion and moderately non-static scenes.

1 Introduction

The objective of a visual odometry system is to compute a continuous camera trajectory through examination of the changes that the motion induces on the images. Research in this area has a long history in robotics [1]. Initially motivated by the NASA Mars exploration program in the 80s, over the years it has yielded systems that were applied on in- and out-door wheeled robots, cars, aerial and underwater vehicles, and even quadruped dog-like robots [2].

The wide choice of vision systems that were used to implement visual odometry includes monocular, stereo, omni-directional and multi-camera setups. Recently low-cost RGB-D cameras were introduced to the market. These visual sensors combine a conventional RGB video camera with a depth sensor, and deliver color images with pre-registered per-pixel depth information at the standard video frame rate. RGB-D cameras have relatively low power consumption and weight which made them suitable and effective sources of information about the environment (and thus the robot's motion). This motivated many researches to explore the possibility of implementing an RGB-D visual odometry system.

Over the past few years a number of such systems were reported in the literature [3,4,5,6,7]. A closely related topic: mapping with RGB-D cameras, also enjoyed considerable attention [8,9,10,11]. The latter emphasizes globally consistent alignment of all captured data, but often involves an odometry subsystem that performs preliminary local alignment. A number of distinct approaches with various benefits and deficits, constraints, and computational complexities exist.

S. Behnke et al. (Eds.): RoboCup 2013, LNAI 8371, pp. 444–455, 2014.

A comparative evaluation of the recent state of the art RGB-D visual odometry systems has been conducted in this contribution. The scope of the evaluation is limited to the software released open-source and capable of real-time performance on the CPU core of a common laptop. Our interest in this area is motivated by participation in the Robocup@Work league. Our target platform is an omni-directional mobile base equipped with an RGB-D camera on which a visual odometry system should be applied. A publicly available RGB-D dataset with various environments and 6 degrees of freedom motions is used for evaluation as well as a dataset captured by ourselves on the target platform.

The remainder of the contribution is organized as follows. Section 2 summarizes the state of the art in RGB-D visual odometry and closely related fields. The evaluated open-source systems are introduced in more detail in Section 3. Section 4 describes the datasets, types of experiments, and metrics that were used to evaluate the systems. The evaluation results and discussion are provided in Section 5. Section 6 summarizes the results and draws the conclusions.

2   Related Work

One core problem of visual odometry is the frame-to-frame transform computation. That is, given two data frames captured by a moving camera at two different time points, compute the rigid transform that relates the positions of the camera. The existing approaches to solve this problem for an RGB-D camera are broadly classified into three groups based on the way they treat the data.

The methods of the first group, often called sparse or feature-based methods, are straightforward adaptations of the classical Structure from Motion stereo odometry pipeline [12]. They compute the motion between frames by solving the absolute orientation problem for a sparse set of 3D points, each of which is a salient and repeatable feature of the environment. This is usually embedded inside a robust estimation framework to tolerate outlier points, which result from incorrect correspondences and independently moving objects. Huang et al. [3] demonstrated such a system for a quadrotor micro aerial vehicle. It is capable of running in real time and performs sufficiently well to allow for autonomous flight in static indoor environments with rich visual features. Du et al. [9] designed an odometry system for interactive environment modeling. It employs a novel hypothesis grading approach which computes visibility conflict criterion for the dense depth data. Recently Domínguez et al. [5] presented a system with an explicit feature filtering stage where a set of rules is applied to keep only the most reliable and consistent in terms of relative 3D positions features.

The second group often referred to as dense or direct methods, solve the motion estimation problem by iteratively aligning the frames to minimize a certain error function. For example: usage of geometrical error between 3D surfaces as defined by the depth images yields a well-understood Iterative Closest Point (ICP) algorithm. Pomerleau et al. [4] presented a modular and efficient ICP library and odometry system. It does not consider color information and is capable of running in real time. Alternatively, Steinbrücker et al. [6] minimize photometric error between intensity images. Their algorithm excels at small displacements but assumes small camera displacements and a static environment. Recently Kerl et al. [7] presented a generalized and extended version which is more accurate and can tolerate moderate dynamics in the environment.

The third group contains hybrid methods. In the pioneering work on RGB-D visual odometry Henry et al. [8] presented a two-stage algorithm. Feature-based method produces an initial motion estimate which is then used as a starting point for the ICP algorithm. The authors reported experiments that showed that the proposed algorithm outperforms its components, however the computational complexity prevents it from running in real time. A similar combined approach was exercised later by by Endres et al. [10] and Hu et al. [11].

3   Evaluated RGB-D Visual Odometry Systems

3.1   Dense Visual Odometry

Dense Visual Odometry (DVO) was developed by Kerl et al. [7] and released open-source[1] as a ROS package. The approach is based on the photo-consistency assumption. It states that the same scene point observed by a camera at two consecutive time instants should have the same intensity in both images. The difference in the intensity between the first and the warped second image is defined as the residual and is regarded as a function of the camera motion. The algorithm proceeds by finding the transform that maximizes the posterior probability of the camera motion given the residual image.

The authors focused exclusively on frame-to-frame motion computation and the system maintains the pose by merely multiplying in estimated transforms. Errors in each estimation are accumulated as well. A technique that is often applied to reduce the drift is keyframe insertion. For an incoming frame instead of estimating transform between it and the previous one, the system computes the transform between it and the keyframe. The latter is periodically updated according to a pre-defined rule. This slows down drift accumulation as only the estimation errors at the moments of keyframe insertion are summed. To allow a fair comparison with other systems, DVO was augmented with a simple keyframing technique. A keyframe is inserted when the transform estimation failed or when the output transform exceeds 4 cm or 0.05 radians. Sometimes DVO produces erroneous estimations without reporting a failure. Such cases are recognized when the transform is greater than 8 cm or 0.08 radians. These numbers are based on the empirical evaluation detailed in Section 5.

3.2   Fast Odometry from Vision

The Fast Odometry from Vision (FOVIS) system was developed by Huang et al. [3] and released open-source[2]. It is based on the standard stereo odometry pipeline and adopts a number of optimization techniques reported in the

---

[1] https://github.com/tum-vision/dvo
[2] https://code.google.com/p/fovis

literature. The FAST detector with adaptive threshold is used for feature detection. Uniform distribution of features in the frame is achieved using bucketing technique. An initial estimation of rotation is produced through direct minimization of the sum of squared pixel errors between frames. This accelerates feature matching and reduces the fraction of invalid matches by restricting the search window. Nevertheless, a certain amount of incorrectly associated features remain, and need to be pruned by the inlier detection algorithm. It proceeds by computing the graph of consistent (in terms of pair-wise spatial relations) feature matches and finds the maximum clique in it. This is considered to be the set of inliers. Finally the motion is estimated by solving the absolute orientation problem and then iteratively refining the solution by minimizing the re-projection error by discarding the outliers which have survived all the previous processing.

3.3   PointMatcher

PointMatcher (PM) is a modular ICP library developed by Pomerleau and Magnenat and released open-source[3]. The transform computation between two frames is regarded as registration of two point clouds. This process is implemented as a reconfigurable chain of modules: data filters, matchers, error minimizers, etc. The authors supply default configuration obtained as a result of extensive tests [4].

Usage of full-resolution depth images is computationally prohibitive and does not allow real-time performance. There are two options how to reduce the amount of data and hence the running time of the algorithm. Either the depth images should be down-scaled, or the point clouds obtained from the full-resolution depth images should be aggressively downsampled. Our initial experiments showed that the former yields better results, so for the evaluation the depth images were downscaled to 160×120 pixels in a pre-processing step.

4   Evaluation Strategy

4.1   RGB-D Datasets

A quantitative evaluation of visual odometry systems requires a dataset of image sequences for which the intrinsic camera parameters and the ground truth trajectories are known. One such dataset was recently published by Sturm et al. [13]. It contains a large number of RGB-D image sequences captured with a Microsoft Kinect camera in a typical office and in a large industrial hall. The data are recorded at a resolution of 640×480 pixels and at a frame-rate of 30 Hz. Sequences are accompanied by time-synchronized trajectories of the camera estimated with a high-precision motion capture system. The authors state that the relative error on the frame-to-frame basis is below 1 mm and $0.5°$. Thus, these trajectories could be used as ground truth data for visual odometry evaluation.

The KUKA youBot [14] is currently the most widely used robot platform in the RoboCup@Work league. It has an omni-directional base with four swedish

---

[3] https://github.com/ethz-asl/libpointmatcher

wheels. Our team has customized it by mounting a sensor tower on the back platform. It hosts a Microsoft Kinect camera which is pointed forwards and is slightly tilted (see Fig. 1a). The robot localizes itself using a laser scanner and a pre-built map of the environment. We drove the robot around a lab and a campus corridor to record several data sequences with trajectories up to 27 m long. The corridor is challenging for visual odometry as the texture is scarce and the structures are ambiguous there, as demonstrated in Fig. 1b and Fig. 1c.

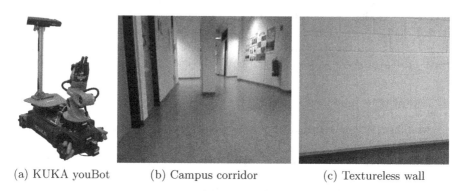

(a) KUKA youBot   (b) Campus corridor   (c) Textureless wall

Fig. 1. The robot and the environment used for evaluation

4.2   Frame-to-Frame Transform Computation

The principal component of a visual odometry system is the module that estimates the relative transform between two given frames. The quality of this estimation is the limiting factor for the accuracy of the system as a whole [1]. In order to analyze the estimation errors the transform computation module of each system is isolated and a large number of RGB-D frame pairs with known ground truth transforms between them are applied.

We isolate the transform computation module of each system, feed it a large number of RGB-D frame pairs with known ground truth transforms between them and analyze the estimation errors.

Formally, given a pair of consecutive RGB-D frames, ground truth transform between them $\mathbf{T}_g$, and the transform estimated by an odometry module $\mathbf{T}_e$, the estimation error is defined as $\mathbf{E} = \mathbf{T}_g^{-1}\mathbf{T}_e$. $Trans\,(\mathbf{T})$ and $rot\,(\mathbf{T})$ are used to denote the translational and rotational components of a transform $\mathbf{T}$.

Analysis of the collected data yields helpful insights. First, we studied how the magnitude of errors $trans\,(\mathbf{E})$ and $rot\,(\mathbf{E})$ depend on the actual transforms $\mathbf{T}_g$. This information allows to understand the limitations the odometry systems impose on the maximum motion velocities. Secondly, we studied how the relative estimation error which is defined as $\eta_{trans} = \frac{trans(\mathbf{E})}{trans(\mathbf{T}_g)}$ and $\eta_{rot} = \frac{rot(\mathbf{E})}{rot(\mathbf{T}_g)}$ (for translational and rotational components) is related with the estimated transforms $\mathbf{T}_e$. In other words, for each possible output transform the average error that this transform bears is empirically calculated. Knowing this allows to make informed choices of the strategy for keyframe insertion.

In order to have representative statistics about the errors, a large number of sample frame pairs that cover the spectrum of possible transforms is required. We boost the number of sample frame pairs by considering frames in the image sequences that are not strictly adjacent. In specific, for each frame a subset of its successors are selected so that the ground truth transform between the frames is limited by 30 cm and 0.15 radians. This way the number of examples is increased up to almost a million and cover the whole spectrum of translation and rotation combinations (of course limited by the mentioned numbers).

4.3   Odometry for Unconstrained Motion

In this group of experiments the visual odometry systems are evaluated with the sequences from the RGB-D dataset and the performance is assessed with the Relative Pose Error (RPE) metric proposed by Sturm et al. [13].

Formally, given an estimated trajectory consisting of a set of poses $\mathbf{P}_1, \ldots, \mathbf{P}_n$ and a ground truth trajectory consisting of a set of poses $\mathbf{Q}_1, \ldots, \mathbf{Q}_n$ the RPE error at a time step i is defined as follows:

$$\mathbf{E}_i = \left(\mathbf{Q}_i^{-1}\mathbf{Q}_{i+\Delta_t}\right)^{-1}\left(\mathbf{P}_i^{-1}\mathbf{P}_{i+\Delta_t}\right), \tag{1}$$

where Δ_t is a fixed time interval which is set to 1 s here, so that the value of RPE could be interpreted as drift per second, a natural and comprehensible metric. Following Sturm et al. we summarize RPE distributions by computing root-mean-square error (RMSE). This is in contrast with commonly used median or mean values which give less influence to gross errors yielding optimistic results.

4.4   Odometry for Planar Motion

In this group of experiments, the in-house dataset is applied to evaluate the performance of the visual odometry systems on a ground mobile robot. Both intrinsic and extrinsic camera parameters are known. This allows us to project the pose estimates computed by the systems on the ground plane and zero out pitch and roll rotations. The resulting trajectories were compared with the output of the laser-based localization system. The latter is not precise enough to make quantitative assessments. However, it is globally accurate so the amount of drift could be visually assessed.

5   Evaluation Results

5.1   Frame-to-Frame Transform Computation

First the evaluation of frame-to-frame transform computation was conducted. The sequences from the "Testing and Debugging" and "Handheld SLAM" groups were used with exclusion of "large_with_loop", "large_no_loop" sequences because they do not have entire ground truth trajectory. This provided us with nearly

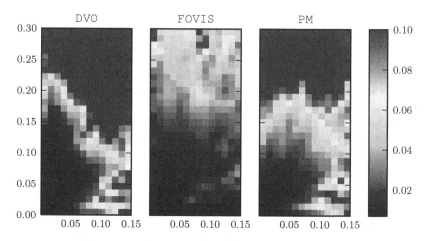

Fig. 2. Magnitude of the translational component of the estimation error (in meters, color-coded) versus ground truth transform rotation (in radians along x-axis) and translation (in meters along y-axis) for the three transform computation modules DVO (left), FOVIS (middle), and PM (right).

one million frame pairs. For each sample the magnitudes of the translational and rotational components of the estimation error were computed.

Figure 2 displays the relation between translational components of estimation errors $trans\,(\mathbf{E})$ and ground truth transforms. All the samples are distributed in a 2D histogram and the mean of each cell is encoded with color. To ensure the same color mapping in all histograms errors above 10 cm were truncated. Thus dark brown corresponds to an error of 10 cm or more. Note that the samples for which the estimation module declared a failure are not included. We observed that FOVIS can handle the whole spectrum of examined transforms with a reasonable accuracy. However, its performance significantly degrades when the translation is more than 15 cm. Both DVO and PM are more restricted. The range of transforms that result in good estimates is limited to about 5° and 15 cm, and in this region DVO is consistently move accurate than PM. The distribution of rotational components of estimation errors $rot\,(\mathbf{E})$ is very similar and is not reproduced here due to space constraints.

Next the relative error in estimated transform versus its magnitude was considered. Figure 3 presents the translational η_{trans} and rotational η_{rot} components. The data are again distributed in a 2D histogram and the mean of each cell is encoded with color. Relative errors above 30% were truncated, thus dark brown corresponds to a relative error of 30% or more. This value could be interpreted as the measure of reliability of an estimated transform. For example: when the translational component of a transform output by DVO is greater than 4 cm and is less than 8 cm and the rotational component is less than 0.05 radians the translational error will be about 15% on average. Therefore, it makes sense to insert a keyframe when the estimated transform falls in the region with the smallest average relative error which will lead to the slowest drift accumulation.

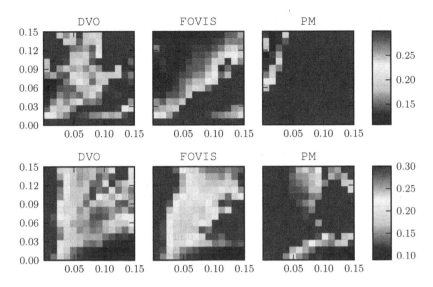

Fig. 3. Relative error η_{trans} in the translational component (top row) and η_{rot} in the rotational component (bottom row) versus estimated transform rotation (in radians along x-axis) and translation (in meters along y-axis) for the three transform computation modules DVO (left), FOVIS (middle), and PM (right).

5.2 Odometry for Unconstrained Motion

The visual odometry systems were evaluated on all sequences from the RGB-D dataset and the summarized results are presented in Table 1.

The sequences in "Handheld SLAM" group contain free flying camera motions with various average velocities and environments. It has been shown that DVO is typically marginally better than FOVIS in terms of translational drift, however in several cases ("fr1/360" and "fr1/desk2") it has a larger drift. The rotational drift, on the other hand, is slightly less for FOVIS. PM consistently gives considerably larger drift. One notable case ("fr2/360_hemisphere") is where the performance of all systems significantly degrades. We attribute this to the fact that in a high percentage of frames in the sequence only the ceiling of the industrial hall is visible. It is distant and hence the depth data is inaccurate. The distribution of RPE of all sequences of the group excluding the aforementioned sequence is presented in Fig. 4. DVO and FOVIS performance is almost identical. However, the latter has a smaller spread but marginally larger median value.

In the "Robot SLAM" sequences the camera is mounted on a Pioneer robot which is driven around in an industrial hall. Wires scattered over the concrete ground produce a severe jittering in the video. Furthermore, the structures are often found at a large distances from the camera. It has been shown that DVO is consistently better than FOVIS. PM completely fails in these sequences as the

Table 1. Evaluation results for the three visual odometry systems on the sequences from RGB-D dataset. The second group column shows the average translational (m/s) and rotational (deg/s) velocities in a sequence according to the ground truth data. The third and fourth group columns show RMSE of translational drift (m/s) and rotational drift (deg/s) respectively for each odometry system.

Sequence name	Avg. velocity		Translational drift			Rotational drift		
	trans	rot	DVO	FOVIS	PM	DVO	FOVIS	PM
Testing and Debugging								
fr1/xyz	0.244	8.92	**0.025**	0.027	0.082	1.53	**1.44**	3.84
fr1/rpy	0.062	50.15	**0.036**	0.053	0.080	10.47	16.42	**6.07**
fr2/xyz	0.058	1.72	0.005	**0.004**	0.075	**0.30**	0.31	2.74
fr2/rpy	0.014	5.77	0.011	**0.005**	0.058	0.54	**0.36**	2.45
Handheld SLAM								
fr1/360	0.210	41.60	0.129	**0.083**	0.113	4.18	**2.53**	4.93
fr1/floor	0.258	15.07	**0.054**	**0.054**	0.163	2.40	**2.12**	6.90
fr1/desk	0.413	23.33	**0.049**	0.055	0.210	7.17	**6.86**	12.07
fr1/desk2	0.426	29.31	0.066	**0.057**	0.109	4.77	**4.65**	6.38
fr1/room	0.334	29.88	**0.054**	0.064	0.115	2.53	**2.23**	6.29
fr2/360_hemisphere	0.163	20.57	**0.111**	0.132	0.685	**2.62**	6.81	11.39
fr2/desk	0.193	6.34	**0.011**	0.013	0.065	0.56	**0.54**	2.49
fr3/long_office_household	0.249	10.19	**0.013**	0.014	0.052	**0.54**	0.64	2.22
Robot SLAM								
fr2/pioneer_360	0.225	12.05	**0.063**	0.163	0.649	**4.01**	5.80	8.11
fr2/pioneer_slam	0.261	13.38	**0.071**	0.142	0.412	**2.86**	3.17	6.74
fr2/pioneer_slam2	0.190	12.21	**0.061**	0.132	0.462	**3.31**	3.61	5.49
fr2/pioneer_slam3	0.164	12.34	**0.091**	0.102	0.324	2.75	**1.92**	6.41
Structure vs. Texture								
fr3/nostruct_notext_far	0.196	2.71	0.200	**0.183**	0.220	6.30	3.67	**3.13**
fr3/nostruct_notext_near	0.319	11.24	0.300	**0.295**	0.321	**9.89**	10.70	12.95
fr3/nostruct_text_far	0.299	2.89	0.186	**0.092**	0.301	2.37	**1.70**	2.55
fr3/nostruct_text_near	0.242	7.43	0.048	**0.017**	0.254	1.46	**0.86**	8.41
fr3/struct_notext_far	0.166	4.00	0.110	0.109	**0.066**	2.69	3.09	**0.73**
fr3/struct_notext_near	0.109	6.25	0.142	0.105	**0.021**	7.75	5.21	**0.97**
fr3/struct_text_far	0.193	4.32	**0.012**	0.014	0.028	**0.45**	0.50	0.99
fr3/struct_text_near	0.141	7.68	0.040	**0.014**	0.051	1.60	**0.71**	1.91
Dynamic objects								
fr2/desk_with_person	0.121	5.34	**0.013**	0.033	0.082	**0.45**	0.87	2.59
fr3/sitting_static	0.011	1.70	**0.008**	0.014	0.040	**0.25**	0.33	1.21
fr3/sitting_xyz	0.132	3.56	**0.014**	0.030	0.055	**0.52**	0.97	1.88
fr3/sitting_halfsphere	0.180	19.09	**0.036**	0.056	0.108	**1.31**	6.98	4.92
fr3/sitting_rpy	0.042	23.84	**0.045**	0.075	0.145	**0.88**	5.13	4.84
fr3/walking_static	0.012	1.39	0.194	**0.153**	0.268	3.26	**2.18**	5.29
fr3/walking_xyz	0.208	5.49	0.405	**0.256**	0.419	7.38	**4.76**	6.71
fr3/walking_halfsphere	0.221	18.27	0.325	**0.240**	0.380	**5.30**	5.31	8.02
fr3/walking_rpy	0.091	20.90	0.406	**0.306**	0.429	**7.01**	8.66	8.51

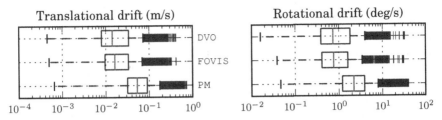

Fig. 4. Aggregated translational drift (left) and rotational drift (right) in the "Handheld SLAM" sequences. DVO and FOVIS demonstrated similar performance, PM has significantly larger drift.

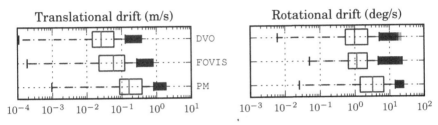

Fig. 5. Aggregated translational drift (left) and rotational drift (right) in the "Robot SLAM" sequences. DVO outperforms the other systems in terms of translational drift.

translational drift that it exposes is twice as large as the average velocity. The aggregate RPE of all the sequences of the group is presented in Fig. 5.

"Structure vs. Texture" sequences expose environments with different amount of structure and visual texture. In the most challenging environment with no texture and no structure all systems completely fail producing drift that is of the same or larger magnitude than the average velocity. In the setup with no structure and visual texture FOVIS is superior to DVO and PM fails as is has virtually no data to compute on. In the sequences with structure and no visual features PM as expected outperforms other systems. It is interesting to note that in the sequences with structure PM performs significantly better than in the other groups. The difference is that here the structures are within approximately one meter and always contain several intersecting planes. Therefore, we conclude that PM is particularly suitable when there is always a structure at the close distance.

The sequences from the "Dynamic Objects" group have people moving around in the scene. In the first five sequences the amount of spurious motion is small and DVO performs significantly better. In the last four sequences people move faster and occupy most of the image area. FOVIS gives much less drift, nevertheless we hold that all the systems failed to produce useful results.

5.3 Odometry for Planar Motion

Furthermore, a number of experiments with the youBot platform as described in Section 4.1 were performed. Figure 6 presents the estimated planar trajectories of the three visual odometry systems for one of the sequences (in red). The ground truth trajectory as output by the laser scanner based localization system is plotted in green. The trajectory estimated by DVO is similar to the actual path. One noticeable deviation happens in the bottom-right corner where the robot rotated in front of a flat textureless wall (see Fig. 1c). FOVIS also showed a good estimation with a similar flaw in the same spot which, however, manifested itself in a wrong turn. The estimate of PM also follows the real path on the whole, however is more erroneous. In general, we observed that the output trajectories mostly preserve the distances and are precise in rotations as well.

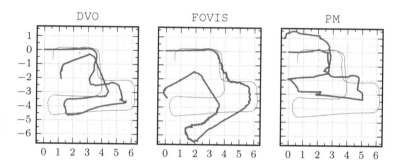

Fig. 6. Planar trajectories of the robot estimated by the odometry systems. Ground truth and estimated trajectories are plotted in thin green and thick red respectively.

6   Conclusions

An evaluation of the three state of the art visual odometry systems (DVO, FOVIS, and PM) was presented. Their frame-to-frame transform estimation components were considered in isolation. The feature-based approach implemented in FOVIS is able to handle the widest spectrum of motions. We presented the relative error in transform estimation as a function of estimated transform which provides an insight for devising a strategy of keyframe insertion for odometry systems.

The algorithms were evaluated on the publicly available RGB-D dataset with precise ground truth. When the motion is smooth and the environment has rich texture and structure both FOVIS and DVO performed equally well giving translational and rotational drift at the level of 1 cm/s and 0.5 deg/s. When the motions are faster and abrupt the performance degrades to nearly 5 cm/s and 5 deg/s. In a particularly challenging group of sequences with excessive jitter DVO clearly outperformed the other methods. It also demonstrated robustness against moderately dynamic scenes where its drift is significantly lower than the others.

The representative of direct methods that minimize geometric error (PM) has performed significantly worse. In the favourable conditions when rich structure is present at the close distance from the camera, it gives drift at the level of 2 cm/s to 6 cm/s and 1 deg/s to 2.5 deg/s. In more challenging conditions it mostly failed producing drift of the same magnitude as the average camera motion. As expected, it outperformed the other systems in a textureless environment.

The systems were evaluated on in-house data collected on an omni-directional robot. The estimated trajectories follow the path of the robot accurately enough.

We conclude that both DVO and FOVIS could be equally well used as odometry systems. The former performs slightly better in normal conditions and is more robust against shakey camera motion and moderately non-static scenes. They both have the same failure mode when the environment has poor visual texture. If this is the target environment for a robot then PM should be considered.

Acknowledgments. Financial support of the BMBF in the FHprofUnt program line, project 6-MIG, grant No 1759X08 and the DAAD PPP project, grant No 50750255, "FPGA based Computer and Machine Vision" is gratefully acknowledged.

References

1. Scaramuzza, D., Fraundorfer, F.: Visual Odometry (Tutorial). IEEE Robotics & Automation Magazine 18(4), 80–92 (2011)
2. Howard, A.: Real-time Stereo Visual Odometry for Autonomous Ground Vehicles. In: Proc. of IROS (2008)
3. Huang, A.S., Bachrach, A., Henry, P., Krainin, M., Maturana, D., Fox, D., Roy, N.: Visual Odometry and Mapping for Autonomous Flight Using an RGB-D Camera. In: Int. Symposium on Robotics Research, ISRR (2011)
4. Pomerleau, F., Magnenat, S., Colas, F., Liu, M., Siegwart, R.: Tracking a Depth Camera: Parameter Exploration for Fast ICP. In: Proc. of IROS (2011)
5. Domínguez, S., Zalama, E., García-Bermejo, J.G., Worst, R., Behnke, S.: Fast 6D Odometry Based on Visual Features and Depth. In: Lee, S., Yoon, K.-J., Lee, J. (eds.) Frontiers of Intelligent Auton. Syst. SCI, vol. 466, pp. 5–16. Springer, Heidelberg (2013)
6. Steinbrücker, F., Sturm, J., Cremers, D.: Real-Time Visual Odometry from Dense RGB-D Images. In: Workshops at ICCV (2011)
7. Kerl, C., Sturm, J., Cremers, D.: Robust Odometry Estimation for RGB-D Cameras. In: Proc. of ICRA (2013)
8. Henry, P., Krainin, M., Herbst, E., Ren, X., Fox, D.: RGB-D Mapping: Using Depth Cameras for Dense 3D Modeling of Indoor Environments. In: Proc. of the Int. Symposium on Experimental Robotics (2010)
9. Du, H., Henry, P., Ren, X., Cheng, M., Goldman, D., Seitz, S., Fox, D.: Interactive 3D Modeling of Indoor Environments with a Consumer Depth Camera. In: Proc. of the Int. Conf. on Ubiquitous Computing (2011)
10. Endres, F., Hess, J., Engelhard, N., Sturm, J., Cremers, D., Burgard, W.: An Evaluation of the RGB-D SLAM System. In: Proc. of ICRA (2012)
11. Hu, G., Huang, S., Zhao, L., Alempijevic, A., Dissanayake, G.: A Robust RGB-D SLAM Algorithm. In: Proc. of IROS (2012)
12. Sünderhauf, N., Protzel, P.: Stereo Odometry - A Review of Approaches. Electrical Engineering (2007)
13. Sturm, J., Engelhard, N., Endres, F., Burgard, W., Cremers, D.: A Benchmark for the Evaluation of RGB-D SLAM Systems. In: Proc. of IROS (2012)
14. Bischoff, R., Huggenberger, U., Prassler, E.: KUKA youBot - A Mobile Manipulator for Research and Education. In: Proc. of ICRA (2011)

Spontaneous Reorientation for Self-localization

Markus Bader and Markus Vincze

Automation and Control Institute (ACIN), Vienna University of Technology,
Gusshausstrasse 27-29 / E376, A-1040 Vienna, Austria
{markus.bader,markus.vincze}@tuwien.ac.at
http://www.acin.tuwien

Abstract. Humanoid robots without internal sensors (e.g. compasses) tend to lose their orientation after a fall or collision. Furthermore, artificial environments are typically rotationally symmetric, causing ambiguities in self-localization. The approach proposed here does not alter the measurement step in the robot's self-localization. Instead it delivers confidence values for rotationally symmetric poses to the robot's behaviour controller, which then commands the robot's self-localization. The behaviour controller uses these confidence values and triggers commands to rearrange the self-localization's pose beliefs within one measurement cycle. This helps the self-localization algorithm to converge to the correct pose and prevents the algorithm from getting stuck in local minima. Experiments in a symmetric environment with a simulated and a real humanoid NAO robot show that this significantly improves the system.

1 Introduction

Mobile robots have to localise themselves in order to navigate reliably and efficiently. Because of this, robot systems are designed to handle multi-modal distribution in self-localization scenarios [1], [2]. However, this leads to ambiguities in rotationally symmetric environments. Estimating the correct viewing direction solves these pose ambiguities in various ways. The estimated viewing direction can be integrated into the self-localization filter [3] or it can augment the Behaviour Controller (BC) with additional knowledge in order to trigger a reorientation behaviour.

Recent experiments conducted by psychologists on humans and animals in symmetric environments [4–6] have proven the existence of a spontaneous reorientation mechanism and the importance of geometric knowledge. In the mid 90s, Hermer and Spelke [5] showed that geometric knowledge is more important than non-geometric information in orientation. After the having been disoriented, young children (at a mean age of 20.9 months) had to find a toy which had been hidden under their observation in a corner. The room was rectangular with two wall lengths, and even when one wall was painted differently, children looked in the geometrically appropriate corners equally often and ignored the non-geometric information. Recently, Lee et al. [6] showed that the geometric impression of a room can be altered by using printed 2D shapes (dots of two sizes)

S. Behnke et al. (Eds.): RoboCup 2013, LNAI 8371, pp. 456–467, 2014.

on walls, to support or suppress the subjective geometric impression. 3-year-old children spontaneously reoriented in his experiments to the correct corner when the bigger dots were on the longer wall, emphasising the 3D impression.

Inspired by these psychological findings, a room-awareness module was developed that captures the subjective geometric impression of a room, without detecting any objects. This was realised by simplifying the environment with a virtual surrounding wall composed of tiles holding trained colour histograms and a filter to estimate the robot's view direction. The colour histograms are linked to observed colours in that specific area and the filter incorporates past perceptions and recent head movements. This makes the filter used sensitive to changes in the perceived background, similar to the natural orientation of the living test subjects in Lee et al. [6].

The room-awareness was integrated as a separate module for computing independent confidence for various poses by evaluating the perceived visual background (see Fig. 1). We also extended the self-localization module to allow the

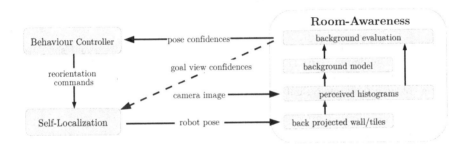

Fig. 1. Room-awareness and its internal sub-modules

BC to rearrange internal pose beliefs spontaneously within one measurement cycle, based on the situation and the results of the room-awareness module.

As a testing environment, the RoboCup *Standard Platform League* (SPL)[1] field was used, which, since 2012 has been a symmetrical playing field with identical goals, as shown in Fig. 2(d). The humanoid used, NAO v3, has limited computational power and acts autonomously, but the visual background used is still able to be processed at the full camera frame rate of 30Hz. Experiments were carried out both in a simulated environment using the B-Human Code Release 2011 [7] and on a real robot.

The scientific contributions of this paper are, on the one hand, the psychologically inspired integrated room-awareness module as an independent module and, on the other hand, the computation of a subjective geometric impression mimicking a human-like belief of one's orientation.

The next section presents related work, followed by a detailed description of the approach in Section 3. Results of experiments are given in Section 4 and a conclusion summarises the findings.

---

[1] Standard Platform League: www.tzi.de/spl/

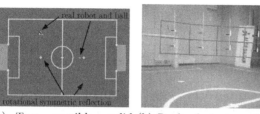

(a) Two possible valid robot poses.

(b) Real robot view of opponent goal.

(c) Real robot view of own goal.

(d) Simulated environment.

(e) Simulated robot view of opponent goal.

(f) Simulated robot view of own goal.

Fig. 2. These figures represent the robot's scene perception in a real and simulated environment. The figures (b,c,e,f) show back-projected tiles, and the tiles in (b,c) and (e,f) appear the same, respectively, due to the rotational symmetry. However, the differing colour histograms disambiguate the pose.

2 Related Work

Due to the SPL's introduction in 2012 of a symmetric playing field, different strategies have been presented over the last two years for estimating the correct pose of a robot in a symmetric environment. However, it should be kept in mind that some of those strategies are specially-designed for RoboCup competition scenarios and that the approach presented in this paper attempts to be more general. Of the strategies presented over the last two years, two types of approaches were distinguishable:

– Strategies using **non-static features** in the environment
– Strategies using **static features** in the background

2.1 Non-static Features

This strategy is related to multi-robot localization [8]. Non-static features, like other robots or, in SPL, the game ball, are observed by multiple robots and used to break the rotational symmetry. Fig. 2(a) shows two possible poses on an SPL playing field, and the robot can determine its correct pose by knowing the ball position. This idea works with single robots tracking objects of interest, but it works even better using team communication to share perceived ball and robot locations. The system starts to fail if the SPL team has only one player left and this robot falls, because the stability of the non-static features cannot be guaranteed after a certain period of time. Some teams (e.g. the Nao-Devils Dortmund) have already integrated non-static features into their Kalman-Filter-based localization [9]. The results of the RoboCup-WC 2012 proved that this strategy is the most competitive one for the SPL.

2.2   Static Features

Similar to humans, this strategy uses features beyond the symmetric environment. The idea is to identify outstanding features in the background and map them using *Self Localization and Mapping* (SLAM) [2] approaches. Typical visual features are based on interest points, i.e. salient image regions, and a descriptor. For example the most popular interest points are Lowes' SIFT [10] and Bays' SURF [11]. Anati et al. [12] trained a vision system to recognize objects such as clocks and trash cans and used this information to create hypotheses about poses and to refine them through particle filtering. But because of the limited computational power of the NAO robot, it is impossible to use such object detection algorithm during a soccer game. Alternatively, Anderson [13] presented a simplified 1D SURF descriptor to map the background. Bader et al. [3] used colour histograms as a descriptor. But they did not propose a reliable strategy for matching and training those descriptors over time, and more importantly, no strategy was presented for integrating the new features into an existing localization system.

Another approach which uses static features in the background uses colour information as well for localization. Sturm et al. [14,15] presented a visual compass and a localization approach which is purely based on the detected colour classes above the horizon. They used a segmented image discretized into colour classes and mapped vertical changes among these classes. In contrast to the work presented here, Sturm used a static map of the background which was trained once in advance, and incorporated the result of the matching into the robot's localization algorithm. Therefore, he was able to build a localization algorithm based on the background and the robot's odometry only by using multiple static maps of the background from different positions. The disadvantage is that the workspace is limited to the area between the positions of the mapped locations. He demonstrated his work on an AIBO robot and published his source code, which works in real-time with the AIBO's camera frame rate.

The paper here proposes instead a colour-histogram-based descriptor linked to virtual tiles surrounding the environment for mapping, and a strategy for reliably matching those descriptors by involving the robot's pose and motion. The information gained is then used to enhance the BC knowledge base in order to control the robot's self-localization and does not alter the measurement step of the self-localization.

3   Room-Awareness

The proposed technique improves existing self-localization algorithm by allowing the robot's BC to trigger reorientation controls. These controls were integrated into a particle-filter, but it would have also been possible to enhance a Kalman-filter-based self-algorithm in order to obtain a similar result. Fig. 1 shows the integration of the room-awareness module and the control channels into an existing self-localization algorithm.

The combination of the external placement of the room-awareness module with the BC as decision making unit would allow further integration of other techniques, e.g. those using non-static features like the game ball, without changing the underlying localization technique by triggering reorientation controls. The room-awareness module tries to capture the subjective geometric room impression by mapping the surrounding colours with colour histograms on a virtual wall. The matching algorithm uses a particle-filter involving the robot's pose and motion, thus relying on geometric and structural properties of the environment for orientation. The approach itself requires initial knowledge like the robot's pose or a previously learned visual background model. The following sections will now describe all of the sub-modules needed to realize the room-awareness module.

3.1   Reorientation Commands

The existing particle-filter-based self-localization algorithm [7] was enhanced in order to rearrange hypotheses according to three possible commands triggered by the robot's BC.

- **flip pose**
  This command changes the robot's belief in being in a symmetric reflection, like a human who realises they are wrong.
- **purge reflection**
  This command triggers the self-localization algorithm to remove beliefs in symmetric reflections, like a human who is sure of their position.
- **reset orientation**
  This command resets the self-localization algorithm's belief in the robot's orientation, and the robot's position belief remains untouched, e.g. after internal sensors detect a fall.

Therefore, the BC is able to optimise the particle distribution of the self-localization algorithm within one measurement cycle. An incorrect particle cluster, for example, is removed using the *purge reflection* command if the confidence for the current pose wins over the reflected pose's confidence. A *flip pose* is triggered if the reflected pose wins. Both cases can be observed in Fig. 6.

3.2   Perceived Colour Histograms and Background Tiles

The *perceived colour histograms* are linked to tiles on a virtual surrounding wall, marked in Fig. 4(a), similar to in [3]. In contrast to [3], which uses a single row of tiles, the virtual wall is modelled as a cylinder with multiple rows and columns of quadrangular tiles. Due to symmetry, the back-projected tiles appear equal in the camera image for all rotationally symmetric poses, see Fig. 2(b,c,e,f), but colour histograms linked to each *background tile* disambiguate the viewing direction. Fig. 3 shows an example of the colour histogram used in this approach, and the same type of chart is used in figures throughout this paper. The histogram has two bins for black and white and 12 colour bins. This allows for a fast computation because only three thresholds c_1, c_2, c_3 and the sign of the colour channel are required to divide the YCbCr colour space into 14 regions, as shown in Fig. 3. The histogram can be blurred, assuming that every bin has four neighbours, in order to suppress noise. Perceived histograms are shown in the viewing direction outside of the background model in Fig. 4(a), but since it is a perception, they are shown without variance.

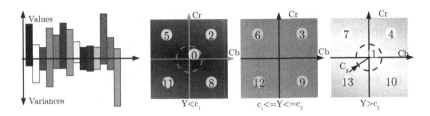

Fig. 3. A colour histogram and bin values are drawn upwards. Histograms used to model the background are augmented with a variance value drawn downwards. The bins are defined by the regions, drawn on the three YCbCr colour space slices.

3.3   The Background Model

The *background model* is trained online with perceived histograms by using a moving average update strategy. In order to stabilize the model, tiles which are too close to the robot, blocked by other robots/obstacles[2], or observed from too steep of a viewing angle, are ignored. The moving average update strategy for each colour bin allows the variance to be computed, thus detecting unstable areas. The equations for computing the moving average μ and variance σ are shown in Eq. (1) and (2), but if a tile is seen for the first time, the perception is copied. Increasing N leads to a more stable model but to a lower rate of adaptation to environmental changes. μ_{new} and σ_{new} represent the new computed average and distribution values, and μ_{last} and σ_{last} are the values before the update with the new measurement x_{mess}.

$$\mu_{new} = \frac{N\mu_{last} + x_{mess}}{N + 1} \tag{1}$$

---

[2] The presence of robots/obstacles is computed by the B-Human Code Release 2011 [7].

$$\sigma^2_{new} = \frac{1}{N+1}\left(N\sigma^2_{last} + \frac{N}{N+1} * (\mu_{last} - x_{mess})^2\right) \tag{2}$$

Training can be interrupted by the robot's BC to avoid learning from incorrect perceptions, for example in the case of a robot falling or during a penalty in the soccer game. A trained background model with colour histograms around the playing field is shown in Fig. 4(a). The two circles of histograms indicate the two rows of tiles cylindrically arranged around the playing field. A subdivided half icosahedron-shaped wall with triangle tiles to cover the room's ceiling was considered, but since the robot looks primarily horizontally, it was decided to use a cylinder for simplicity's sake.

3.4   The Background Evaluation

This sub-module uses robot pose information and perceived colour histograms to estimate current viewing direction based on the background. This estimation is done by using a particle-filter where each particle describes a viewpoint hypothesis on the cylindrically modelled wall. The particle weights are computed by comparing the colour histograms of the perceived tiles with model histograms of viewpoints estimated by a specific particle. For the histogram comparison, the Sum of Squared Differences (SSD) and Jeffrey/Jensen-Shannon Divergence (JSD) distance functions, as suggested in [16], were tested, but since multiple histograms for one view are compared, the best results were achieved by applying the SSD function to the three most dominant histogram bins. All particles are updated by linearising the current rotational motion at the current view centre, plus some additional white noise to compensate for model discrepancy. New particles are injected at the current best matching position to prevent the filter from getting stuck in local minima. Fig. 4(a) indicates particles with magenta dots around the playing field, and the detected cluster centre is drawn as a gray ellipse. The same particles are projected onto the current viewpoint, visualized in Fig. 4(b) (magenta).

Multiple confidence values are computed by counting the number of background particles within specific angle ranges. The confidence for the current pose is computed by counting the number of particles within a virtual field of view around the current view centre, as shown in Fig. 4(a). The opposite pose confidence is computed similarly, with the opposite view centre. A moving average algorithm, the same as Eq. (1), was used in both cases to suppress incorrect estimates and to assure a smoother confidence value curve over time. But in contrast to the background model, the values were initialized with zero. The average confidence values are always set to zero if the BC triggers a signal, in order to build new confidences from scratch. Fig. 4(c) shows a sequence of confidence values and the smoothed mean values over time, together with the BC command signals.

In addition, confidence values indicating which goal was perceived were computed by counting particles within two static areas behind the goals, also shown

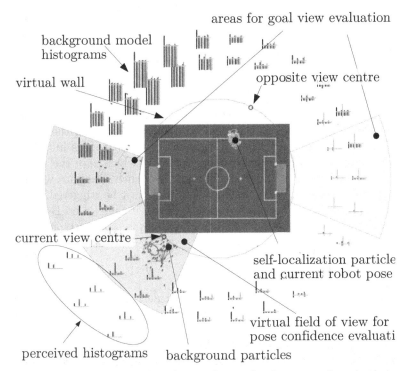

areas for goal view evaluation

background model histograms

virtual wall

opposite view centre

current view centre

self-localization particle and current robot pose

virtual field of view for pose confidence evaluati

perceived histograms   background particles

(a) Background model and background particles drawn on the robot's internal world view around the playing field. Particles of the self-localization are drawn in gray on the playing field.

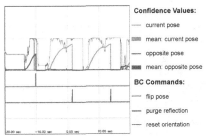

Confidence Values:
— current pose
▬ mean: current pose
— opposite pose
▬ mean: opposite pose

BC Commands:
— flip pose
— purge reflection
— reset orientation

(b) Camera image with particles and tiles with colour histograms. Detected landmarks like field lines and goal posts are drawn as overlay.

(c) History of past confidence values and BC command signals. The BC triggered a *flip* because the robot had been mistakenly placed in the rotational reflective pose. This was followed by a *purge reflection* because the awareness module then indicated a high current pose confidence relative to the reflected pose confidence.

Fig. 4. The robot's internal world view, the corresponding camera image and computed confidence values

in Fig. 4(a) as areas for goal view evaluation. These *goal view confidences* are unlike the pose confidences directly used in the self-localization to generate new pose hypotheses for particle injection. This supports the self-localization in managing kidnapped robot situations.

3.5 Orientation Behaviours

The purpose of a robot is not only to localise itself, but also to fulfill certain tasks. It is the role of the robot's BC to juggle these tasks by taking the appropriate actions. For example, the BC has to recognize if the system is initialised with a pose and not with a pre-learned background, in which case the background evaluation process needs to be stopped, and a sequence of actions which brings the robot into a position to train the background must be triggered. The room-awareness approach presented here does not work if there is no BC or similar module because the interaction with other modules is too intertwined.

4 Experiments

Two test scenarios were used on a simulated and real robot to measure the improvement of this system over a system without a room-awareness module[3]. For both tests the robot was placed on the default soccer start position next to the playing field, but the particle-filter-based self-localization was initialized with the incorrect rotationally symmetric pose, and switched sides after half of the trials.

A system which uses only a symmetric playing field for localization converges only coincidentally to the correct pose, and normally fails. In the *first test* the robot was only allowed to move its head. In the *second test* the robot had to walk from one penalty position to the other and vice versa, based on its own localization. During the first round of tests, the self-localization injected new particles if a goal was detected. In the second round of tests, new injections were allowed only if the room-awareness module indicated clear confidence values to identify which goal had been detected. During all of the trials, the time it took for the BC to trigger a control command to optimize or correct the self-localization's particle distribution was measured. For both the simulation and the real robot, the same room-awareness parameters were used to generate comparable results. One can see in Table 1 that the system tends to fail in up to 15% of the trials if the robot is not moving. This happens because the background is trained online and the wrong background is assumed as correct after a certain period of time. In Fig. 6 we can see two ways in which the BC corrects the self-localization's particle distribution. If the particles are on the wrong pose, the system triggers a flip. A purge is called for if the filter accidentally forms a correct growing particle cluster. Since clusters are primarily the result of new

---

[3] A video with experiments was submitted in addition. It shows a training sequence, an incorrectly initialized robot, a kidnapped robot scenario and a long-term test of 6 minutes.

Table 1. Evaluation of the test scenarios

Robot	Test	Goal	trails	flip	purge	failed
Simulation	head only	unknown	20 trials 18.1 sec	70% 12.9 sec	15% 57.5 sec	15% > 200 sec
		known	20 trials 22.1 sec	80% 16.0 sec	10% 42.0 sec	10% > 200 sec
	moving	unknown	10 trials 20.2 sec	80% 17.1 sec	20% 34.5 sec	0% > 200 sec
		known	10 trials 27.0 sec	80% 19.6 sec	20% 56.5 sec	0% > 200 sec
Robot	**Test**	**Goal**	trails	flip	purge	failed
Real Robot	head only	unknown	20 trials 23.1 sec	90% 23.1 sec	0% -	10% > 200 sec
		known	20 trials 34.3 sec	65% 23.1 sec	25% 63.4 sec	10% > 200 sec
	moving	unknown	10 trials 33.5 sec	100% 33.5 sec	0% -	0% > 200 sec
		known	10 trials 45.4 sec	50% 37.2 sec	40% 55.7 sec	10% > 200 sec

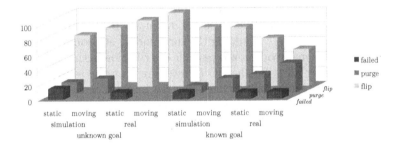

Fig. 5. This chart shows the selected command in different test scenarios, related to Table 1. One can see the increase in purges when the goal was identified using the goal confidence values.

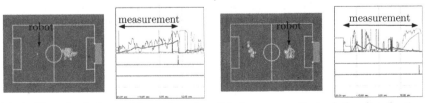

(a) A *flip* solved the problem of an incorrect pose, recorded on a real robot.   (b) A *purge* optimized the distribution, recorded on a simulated robot.

Fig. 6. Two instances of particle distribution optimization observed after an incorrect initialization

injected particles, and the system only creates new particles if a goal has been seen, the tests with goal confidences differ from those without. Without any indication of which goal has been perceived, the system injects poses which might be rotationally-symmetrically incorrect. The room-awareness module identifies goal views and the system is able to inject a most likely hypothesis. We can see this effect in Fig. 5 because the BC triggered a purge more often, indicating that the self-localization was able to recover the correct pose on its own and the purge cleared possible incorrect clusters. The results also show that if a purge was the selected signal, the time it took for this signal to be triggered was significantly longer. This was caused by the self-localization because the estimated pose starts to jump if there are two or more nearly equally-sized clusters. In general, smoother confidence values were experienced with the real robot than with the simulated one, as visible in Fig. 6(a), due to the image being noisier and smoother. Overall, the proposed room-awareness module improves an existing self-localization algorithm.

5   Conclusion

A psychologically-inspired room-awareness module has been presented here which mimics a human-like belief in current pose and triggers a so-called spontaneous reorientation. Experiments with a humanoid robot in a rotationally symmetric environment prove the effectiveness of room-awareness in recognising an incorrectly-estimated pose by mapping the surrounding environment with colour histograms. In addition, an optimized particle distribution in the self-localization used was obtained by allowing the BC to interfere with the self-localization by selectively moving or removing particle clusters caused by a symmetric environment. The room-awareness module was also able to support the self-localization using the visual background to estimate which goal was seen, in order to break the symmetry. However, multiple issues remain, e.g. the optimal number of colour histograms to model environments, the virtual wall shape used, and the implementation of an optimal search and training pattern to force the robot to look at areas with the most distinctive background.

Acknowledgment. The research leading to these results has received funding from the Austrian Science Foundation under grant agreement No. 835735 (TransitBuddy).

References

1. Thrun, S., Burgard, W., Fox, D.: Probabilistic Robotics (Intelligent Robotics and Autonomous Agents). The MIT Press (2005)
2. Siegwart, R., Nourbakhsh, I.R., Scaramuzza, D.: Introduction to Autonomous Mobile Robots. MIT Press (2011)

3. Bader, M., Brunner, H., Hamböck, T., Hofmann, A., Vincze, M.: Colour Histograms as Background Description: An approach to overcoming the Uniform-Goal Problem within the SPL for the RoboCup WC 2012. In: Proceedings of the Austrian Robotics Workshop, ARW 2012 (2012)
4. Landau, B., Gleitman, H., Spelke, E., et al.: Spatial Knowledge and Geometric Representation in a Child Blindfrom Birth. Science 213(4513), 1275–1278 (1981)
5. Hermer, L., Spelke, E.S.: A Geometric Process for Spatial Reorientation in Young Children. Nature 370, 57–59 (1994)
6. Lee, S.A., Winkler-Rhoades, N., Spelke, E.S.: Spontaneous Reorientation Is Guided by Perceived Surface Distance, Not by Image Matching Or Comparison. PLoS ONE 7, e51373 (2012)
7. Röfer, T., Laue, T., Müller, J., Fabisch, A., Feldpausch, F., Gillmann, K., Graf, C., de Haas, T.J., Härtl, A., Humann, A., Honsel, D., Kastner, P., Kastner, T., Könemann, C., Markowsky, B., Riemann, O.J.L., Wenk, F.: B-Human Team Report and Code Release (2011), Only available online: http://www.b-human.de/downloads/bhuman11_coderelease.pdf
8. Fox, D., Burgard, W., Kruppa, H., Thrun, S.: A Probabilistic Approach to Collaborative Multi-Robot Localization. Autonomous Robots 8(3), 325–344 (2000)
9. Czarnetzki, S., Rohde, C.: Handling Heterogeneous Information Sources for Multi-Robot Sensor Fusion, pp. 133–138 (Sepetmber 2010)
10. Loy, G., Eklundh, J.-O.: Detecting Symmetry and Symmetric Constellations of Features. In: Leonardis, A., Bischof, H., Pinz, A. (eds.) ECCV 2006. Part II. LNCS, vol. 3952, pp. 508–521. Springer, Heidelberg (2006)
11. Bay, H., Tuytelaars, T., Van Gool, L.: SURF: Speeded Up Robust Features. In: Leonardis, A., Bischof, H., Pinz, A. (eds.) ECCV 2006, Part I. LNCS, vol. 3951, pp. 404–417. Springer, Heidelberg (2006)
12. Anati, R., Scaramuzza, D., Derpanis, K., Daniilidis, K.: Robot Localization Using Soft Object Detection. In: 2012 IEEE International Conference on Robotics and Automation (ICRA), pp. 4992–4999 (May 2012)
13. Anderson, P., Yusmanthia, Y., Hengst, B., Sowmya, A.: Robot Localisation Using Natural Landmarks. In: Chen, X., Stone, P., Sucar, L.E., van der Zant, T. (eds.) RoboCup 2012. LNCS (LNAI), vol. 7500, pp. 118–129. Springer, Heidelberg (2013)
14. Sturm, J., van Rossum, P., Visser, A.: Panoramic Localization in the 4-Legged League. In: Lakemeyer, G., Sklar, E., Sorrenti, D.G., Takahashi, T. (eds.) RoboCup 2006. LNCS (LNAI), vol. 4434, pp. 387–394. Springer, Heidelberg (2007)
15. Sturm, J., Visser, A.: An appearance-based visual compass for mobile robots. Robotics and Autonomous Systems 57(5), 536–545 (2009)
16. Deselaers, T., Keysers, D., Ney, H.: Features for Image Retrieval: An Experimental Comparison. Information Retrieval 11, 77–107 (2008), doi:10.1007/s10791-007-9039-3

Multi-sensor Mobile Robot Localization for Diverse Environments

Joydeep Biswas and Manuela Veloso

School of Computer Science, Carnegie Mellon University,
Pittsburgh, Pennsylvania 15213, USA
{joydeepb,veloso}@cs.cmu.edu

Abstract. Mobile robot localization with different sensors and algorithms is a widely studied problem, and there have been many approaches proposed, with considerable degrees of success. However, every sensor and algorithm has limitations, due to which we believe no single localization algorithm can be "perfect," or universally applicable to all situations.

Laser rangefinders are commonly used for localization, and state-of-the-art algorithms are capable of achieving sub-centimeter accuracy in environments with features observable by laser rangefinders. Unfortunately, in large scale environments, there are bound to be areas devoid of features visible by a laser rangefinder, like open atria or corridors with glass walls. In such situations, the error in localization estimates using laser rangefinders could grow in an unbounded manner. Localization algorithms that use depth cameras, like the Microsoft Kinect sensor, have similar characteristics. WiFi signal strength based algorithms, on the other hand, are applicable anywhere there is dense WiFi coverage, and have bounded errors. Although the minimum error of WiFi based localization may be greater than that of laser rangefinder or depth camera based localization, the maximum error of WiFi based localization is bounded and less than that of the other algorithms.

Hence, in our work, we analyze the strengths of localization using all three sensors - using a laser rangefinder, a depth camera, and using WiFi. We identify sensors that are most accurate at localization for different locations on the map. The mobile robot could then, for example, rely on WiFi localization more in open areas or areas with glass walls, and laser rangefinder and depth camera based localization in corridor and office environments.

Keywords: Localization, Mobile Robots, Sensor Fusion.

1 Introduction and Related Work

Localization is an ability crucial to the successful deployment of any autonomous mobile robot, and has been a subject of continuous research. As technologies have evolved, we have seen a parallel evolution of algorithms to use new sensors, from SONAR, to infrared rangefinders, to vision, laser rangefinders, wireless

S. Behnke et al. (Eds.): RoboCup 2013, LNAI 8371, pp. 468–479, 2014.

beacons, vision, and most recently depth cameras. The wide-scale deployment of autonomous robots has for a long time been hampered by the limitations of the sensors and the algorithms used for robot autonomy. This has led to work on robots like Minerva [1] that used a "coastal planner" to avoid navigation paths with poor information content perceivable to the robot. Exploration and navigation approaches that account for perceptual limitations of robots [2] have been studied as well. These approaches acknowledge the limitations in perception, and seek to avoid areas where perception is poor.

A number of robotics projects have been launched over the years in pursuit of the goal of long-term autonomy for mobile service robots. Shakey the robot [3] was the first robot to actually perform tasks in human environments by decomposing tasks into sequences of actions. Rhino [4], a robot contender at the 1994 AAAI Robot Competition and Exhibition, used SONAR readings to build an occupancy grid map [5], and localized by matching its observations to expected wall orientations. Minerva [1] served as a tour guide in a Smithsonian museum. It used laser scans and camera images along with odometry to construct two maps, the first being an occupancy grid map, the second a textured map of the ceiling. For localization, it explicitly split up its observations into those corresponding to the fixed map and those estimated to have been caused by dynamic obstacles. Xavier [6] was a robot deployed in an office building to perform tasks requested by users over the web. Using observations made by SONAR, a laser striper and odometry, it relied on a Partially Observable Markov Decision Process (POMDP) to reason about the possible locations of the robot, and to reason about the actions to choose accordingly. A number of robots, Chips, Sweetlips and Joe Historybot [7] were deployed as museum tour guides in the Carnegie Museum of Natural History, Pittsburgh. Artificial fiducial markers were placed in the environment to provide accurate location feedback for the robots. The PR2 robot at Willow Garage [8] has been demonstrated over a number of milestones, where the robot had to navigate over 42 km and perform a number of manipulation tasks. The PR2 used laser scan data along with Inertial Measurement Unit (IMU) readings and odometry to build an occupancy grid map using GMapping [9], and then localized itself on the map using KLD-sampling [10]. The Collaborative Robots (CoBots) project [11] seeks to explore the research and engineering challenges involved in deploying teams of autonomous robots in human environments. The CoBots autonomously perform tasks on multiple floors of our office building, including escorting visitors, giving tours and transporting objects. The annual RoboCup@Home competition [12] aims to promote research in mobile service robots deployed in typical human home environments.

Despite the localized successes of these robotic projects, we still are short of the goal of having indefinitely deployed robots without human intervention. To have a robot be universally accessible, it needs to be able to overcome individual sensor limitations. This has led to the development of a number of algorithms for mobile robot localization by "fusing" multiple sensor feeds. There are a number of approaches to sensor fusion [13] for robot localization, including merging multiple sensor feeds at the lowest level before being processed homogeneously, and hierarchical approaches to fuse state estimates derived independently from multiple sensors.

Our approach, instead of merging sensor feeds into a single algorithm, is to select one out of a set of localization algorithms, each of which independently processes different sensory feeds. The selection is driven by the results of evaluating the performance of the individual algorithms over a variety of indoor scenarios that the mobile robot will encounter over the course of its deployment. We use three localization algorithms, each using a different sensor: a laser rangefinder based localization algorithm that uses a Corrective Gradient Refinement [14] particle filter, a depth camera based localization algorithm [15], and a WiFi localization algorithm that is a cross between graph-based WiFi localization [16] and WiFi localization using Gaussian Processes [17]. Algorithm selection [18] is a general problem that has been applied to problems like SAT solving [19], and spawned the field of "meta-learning" [20]. Here we apply the problem of algorithm selection to the problem of mobile robot localization.

By running each localization algorithm independently, we can decide to pick one over the others for different scenarios, thus avoiding the additional complexity of fusing dissimilar sensors for a single algorithm. To determine which algorithm is most robust in which scenario, we collected sensor data over the entire map, and evaluated each of the algorithms (with respect to ground truth) over the map and over multiple trials. This data is used to determine which algorithm is most robust at every location of the map. Additionally, by comparing the performance of the algorithms while the robot navigates in corridors and while it navigates in open areas, we show that different algorithms are robust and accurate for different locations on the map.

2   Multi-sensor Localization

We use three sensors for localization - a laser rangefinder, a depth camera, and the WiFi received signal strength indicator (RSSI). Each sensor, along with robot odometry, is processed independently of the others. For the laser rangefinder, we use a Corrective Gradient Refinement (CGR) [14] based particle filter. The depth camera observations are processed by the Fast Sampling Plane Filtering [15] algorithm to detect planes, which are then matched to walls in the map to localize the robot using a CGR particle filter. The WiFi RSSI values observed by the robot are used for localization using a Gaussian Processes-Learnt WiFi Graph. This WiFi localization algorithm combines the strengths of graph-based WiFi localization [16] with Gaussian Processes for RSSI based location estimation [17]. We first review each of these sensor-specific localization algorithms.

2.1   Corrective Gradient Refinement

For localization using a laser rangefinder sensor, the belief of the robot's location is represented as a set of weighted samples or "particles", as in Monte Carlo

Localization (MCL)[21]: $Bel(x_t) = \{x_t^i, w_t^i\}_{i=1:m}$. The Corrective Gradient Refinement (CGR) algorithm iteratively updates the past belief $Bel(x_{t-1})$ using observation y_t and control input u_{t-1} as follows:

1. Samples of the belief $Bel(x_{t-1})$ are evolved through the motion model, $p(x_t|x_{t-1}, u_{t-1})$ to generate a first stage proposal distribution q^0.
2. Samples of q^0 are "refined" in r iterations (which produce intermediate distributions $q^i, i \in [1, r-i]$) using the gradients $\frac{\delta}{\delta x}p(y_t|x)$ of the observation model $p(y_t|x)$.
3. Samples of the last generation proposal distribution q^r and the first stage proposal distribution q^0 are sampled using an acceptance test to generate the final proposal distribution q.
4. Samples x_t^i of the final proposal distribution q are weighted by corresponding importance weights w_t^i, and resampled with replacement to generate $Bel(x_t)$.

Thus, given the motion model $p(x_t|x_{t-1}, u_{t-1})$, the observation model $p(y_t|x)$, the gradients of the observation model $\frac{\delta}{\delta x}p(y_t|x)$ and the past belief $Bel(x_{t-1})$, the CGR algorithm computes the latest belief $Bel(x_t)$.

2.2 Depth Camera Based Localization

Depth cameras provide, for every pixel, color and depth values. This depth information, along with the camera intrinsics (horizontal field of view f_h, vertical field of view f_v, image width w and height h in pixels) can be used to reconstruct a 3D point cloud. Let the depth image of size $w \times h$ pixels provided by the camera be I, where $I(i,j)$ is the depth of a pixel at location $d = (i,j)$. The corresponding 3D point $p = (p_x, p_y, p_z)$ is reconstructed using the depth value $I(d)$ as $p_x = I(d)\left(\frac{j}{w-1} - 0.5\right)\tan\left(\frac{f_h}{2}\right)$, $p_y = I(d)\left(\frac{i}{h-1} - 0.5\right)\tan\left(\frac{f_v}{2}\right)$, $p_z = I(d)$.

With limited computational resources, most algorithms (e.g. localization, mapping etc.) cannot process the full 3D point cloud at full camera frame rates in real time with limited computational resources on a mobile robot. The naïve solution would therefore be to sub-sample the 3D point cloud for example, by dropping (say) one out of N points, or sampling randomly. Although this reduces the number of 3D points being processed by the algorithms, it ends up discarding information about the scene. An alternative solution is to convert the 3D point cloud into a more compact, feature - based representation, like planes in 3D. However, computing optimal planes to fit the point cloud for every observed 3D point would be extremely CPU-intensive and sensitive to occlusions by obstacles that exist in real scenes. The Fast Sampling Plane Filtering (FSPF) algorithm combines both ideas: it samples random neighborhoods in the depth image, and in each neighborhood, it performs a RANSAC based plane fitting on the 3D points. Thus, it reduces the volume of the 3D point cloud, it extracts geometric features in the form of planes in 3D, and it is robust to outliers since it uses RANSAC within the neighborhood.

Fast Sampling Plane Filtering (FSPF) [15] takes the depth image I as its input, and creates a list P of n 3D points, a list R of corresponding plane

---

Algorithm 1 Fast Sampling Plane Filtering

1: **procedure** PLANEFILTERING(I)
2:     $P \leftarrow \{\}$ ▷ Plane filtered points
3:     $R \leftarrow \{\}$ ▷ Normals to planes
4:     $O \leftarrow \{\}$ ▷ Outlier points
5:     $n \leftarrow 0$ ▷ Number of plane filtered points
6:     $k \leftarrow 0$ ▷ Number of neighborhoods sampled
7:     **while** $n < n_{max} \wedge k < k_{max}$ **do**
8:         $k \leftarrow k + 1$
9:         $d_0 \leftarrow (\mathrm{rand}(0, h - 1), \mathrm{rand}(0, w - 1))$
10:        $d_1 \leftarrow d_0 + (\mathrm{rand}(-\eta, \eta), \mathrm{rand}(-\eta, \eta))$
11:        $d_2 \leftarrow d_0 + (\mathrm{rand}(-\eta, \eta), \mathrm{rand}(-\eta, \eta))$
12:        Reconstruct p_0, p_1, p_2 from d_0, d_1, d_2
13:        $r = \frac{(p_1 - p_0) \times (p_2 - p_0)}{||(p_1 - p_0) \times (p_2 - p_0)||}$ ▷ Compute plane normal
14:        $\bar{z} = \frac{p_{0z} + p_{1z} + p_{2z}}{3}$
15:        $w' = w \frac{S}{\bar{z}} \tan(f_h)$
16:        $h' = h \frac{S}{\bar{z}} \tan(f_v)$
17:        $[\mathrm{numInliers}, \hat{P}, \hat{R}] \leftarrow \mathrm{RANSAC}(d_0, w', h', l, \epsilon)$
18:        **if** numInliers $> \alpha_{in}l$ **then**
19:            Add $\hat{P}$ to P
20:            Add $\hat{R}$ to R
21:            numPoints $\leftarrow$ numPoints + numInliers
22:        **else**
23:            Add $\hat{P}$ to O
24:        **end if**
25:    **end while**
26:    **return** P, R, O
27: **end procedure**

---

normals, and a list O of outlier points that do not correspond to any planes. Algorithm1 outlines the plane filtering procedure. It uses the helper subroutine $[\mathrm{numInliers}, \hat{P}, \hat{R}] \leftarrow \mathrm{RANSAC}(d_0, w', h', l, \epsilon)$, which performs the classical RANSAC algorithm over the window of size $w' \times h'$ around location d_0 in the depth image, and returns inlier points and normals $\hat{P}$ and $\hat{R}$ respectively, as well as the number of inlier points found. The configuration parameters required by FSPF are listed in Table1.

FSPF proceeds by first sampling three locations d_0, d_1, d_2 from the depth image (lines 9-11). The first location d_0 is selected randomly from anywhere in the image, and then d_1 and d_2 are selected from a neighborhood of size η around d_0. The 3D coordinates for the corresponding points p_0, p_1, p_2 are then computed (line 12). A search window of width w' and height h' is computed based on the mean depth (z-coordinate) of the points p_0, p_1, p_2 (lines 14-16) and the minimum expected size S of the planes in the world. Local RANSAC is then performed in the search window. If more than $\alpha_{in}l$ inlier points are produced as a result of running RANSAC in the search window, then all the inlier points are added to the list P, and the associated normals (computed using a least-squares

Table 1. Configuration parameters for FSPF

Parameter	Value	Description
n_{max}	2000	Maximum total number of filtered points
k_{max}	20000	Maximum number of neighborhoods to sample
l	80	Number of local samples
η	60	Neighborhood for global samples (in pixels)
S	0.5m	Plane size in world space for local samples
ϵ	0.02m	Maximum plane offset error for inliers
α_{in}	0.8	Minimum inlier fraction to accept local sample

fit on the RANSAC inlier points) to the list R. This algorithm is run a maximum of m_{max} times to generate a list of maximum n_{max} 3D points and their corresponding plane normals.

With the list P and R, non-vertical planes are ignored, and the remaining planes are matched to a 2D vector map. These remaining matched points are then used for localization using CGR.

2.3 WiFi Based Localization

WiFi localization using a graph based WiFi map [16] provides fast, accurate robot localization when constrained to a graph. In this approach, the mean and standard deviations of WiFi RSSI observations are approximated by linear interpolation on a graph. This leads to a computationally efficient observation likelihood function, with the computation time of each observation likelihood being independent of the number of training instances. The disadvantage is that to construct the WiFi graph map, the robot needs to be accurately placed at every vertex location of the graph to collect WiFi RSSI training examples.

RSSI based localization using Gaussian Processes [17], on the other hand, does not require training observations from specific locations. Given an arbitrary location x_* and the covariance vector k_* between the training locations and x_*, the expected mean μ_{x_*} and variance is given by,

$$\mu_{x_*} = k_*^T (K + \sigma_n^2 I)^{-1} y \tag{1}$$

$$\sigma_{x_*}^2 = k(x_*, x_*) - k_*^T (K + \sigma_n^2 I)^{-1} k_*. \tag{2}$$

Here, σ_n^2 is the Gaussian observation noise, y the vector of training RSSI observations, and $k(\cdot, \cdot)$ the kernel function used for the Gaussian Process (most commonly a squared exponential). The drawback with this approach is that the computational complexity for Eq.1-2 grows quadratically with the number of training samples, which is more than $100,000$ samples per access point for our single floor map. We introduce an approach that combines the advantages of both algorithms while overcoming each of their limitations. In our approach, training examples are first collected across the map while driving the robot around (not from any specific locations). Next, using Gaussian Processes, the WiFi mean and

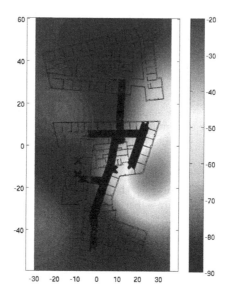

Fig. 1. Gaussian Process-learnt WiFi graph for a single access point. The mean RSSI values are color-coded, varying from $-90dBm$ to $-20dBm$. The locations where the robot observed signals from the access point are marked with crosses.

variance for the nodes of a WiFi Graph (with regularly spaced sample locations on the map) are learnt offline. Once the WiFi graph is learnt, WiFi mean and variance at locations during run time are estimated by bicubic interpolation across nodes of the graph. Figure 1 shows a Gaussian Process-learnt WiFi graph for a single access point on the floor.

3   Comparing Localization Algorithms

To evaluate the robustness of the different localization algorithms over a variety of environments, we collected laser scanner, depth camera, and WiFi sensor data while our mobile robot autonomously navigated around the map. The navigation trajectory covered each corridor multiple times, and also covered the open areas of the map. The sensor data thus collected was then processed offline for 100 times by each of the algorithms (since the algorithms are stochastic in nature), and the mean localization error across trials evaluated for each sensor. Figure 2 shows the errors in localization when using the three different sensors. It is seen that in corridors, the laser rangefinder based localization algorithm is the most accurate, with mean errors of a few centimeters. The depth camera (Kinect) based localization has slightly higher errors of about ten centimeters while the robot travels along corridors. The WiFi localization algorithm consistently had errors of about a meter across the map.

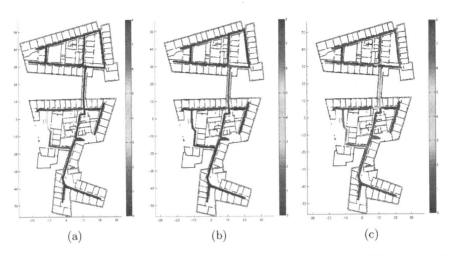

(a)   (b)   (c)

Fig. 2. Errors in localization using (a) a laser rangefinder, (b) the Kinect, and (c) WiFi, while autonomously driving across the map. The error in localization for every true location of the robot is coded by color on the map. All units are in meters.

We then ran additional trials in a large open area of the map, where the dimensions of the area exceed the maximum range of the laser rangefinder and depth image sensors. Figure 3 shows a photograph of this open area. In these trials, the robot started out from the same fixed location, and was manually driven around the open area. The robot was then driven back to the starting location, and the error in localization noted. Figure 4a shows part of such a trajectory as the robot was driven around the open area. This was done 20 times for each localization algorithm, and the combined length of all the trajectories was over 2Km, with approximately 0.7Km per algorithm. Figure 4b shows a histogram of the recorded errors in localization at the end of the driven trajectories for each

Fig. 3. The open area in which localization using different sensors were compared

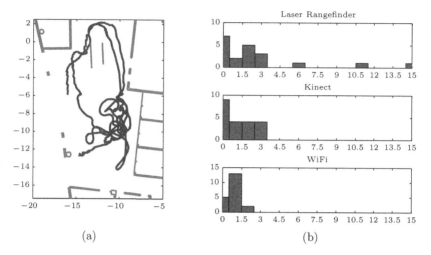

Fig. 4. Trials in the open area: (a) part of a test trajectory of the robot, and (b) histogram of errors in localization across trials for each algorithm

localization algorithm. The following table tabulates the minimum, maximum, and median errors for the different algorithms:

	Laser Rangefinder	Kinect	WiFi
Min Error	0.01	0.17	0.20
Max Error	14.77	3.47	1.88
Median Error	1.83	1.08	0.76

The laser rangefinder has the smallest minimum error during trajectories that the algorithm tracked successfully, but also has the largest maximum error. The WiFi localization algorithm, however, was consistently able to track the robot's trajectory for every trial with a median error of less than a meter.

4   Choosing between Sensors for Localization

A robot equipped with multiple sensors and running different localization algorithms for each sensor needs to decide when to rely on each of the algorithms. There are three broad approaches to choosing between sensors and algorithms for localization: computed uncertainty based selection, sensor mismatch based selection, and data-driven selection.

4.1   Computed Uncertainty Based Selection

Every localization algorithm maintains a measure of uncertainty of the robot's pose estimates. The uncertainty of a particle filter based localization algorithm is captured by the distribution of its particles, while in an Extended Kalman Filter (EKF) based localization algorithm the covariance matrix captures the uncertainty of the robot's pose. Given a measure of the uncertainty of localization using the different sensors, the robot could choose the algorithm (and hence the sensor) with the least uncertainty at any given point of time. However, it is not possible to directly compare the uncertainty measures from different algorithms because of the assumptions and simplifications specific to each algorithm. For example, an EKF based algorithm assumes a unimodal Gaussian distribution of the robot's pose, which is not directly comparable to the sample based multi-modal distribution of poses of a particle filter.

4.2 Sensing Mismatch Based Selection

A more algorithm-agnostic approach to computing a measure of uncertainty for localization algorithms is to compute the mismatch between the observation likelihood function $P(l|s)$ and the belief $Bel(l)$. Sensor Resetting Localization (SRL) [22] evaluates this mismatch by sampling directly from the observation likelihood function $P(l|s)$, and adds the newly generated samples to the particle filter proportional to the mismatch. KLD Resampling [10] similarly evaluates the mismatch by sampling from the odometry model $P(l_i|l_{i-1}, u_i)$ and computing the overlap with the belief $Bel(l)$. With the estimates of sensor mismatch between each localization algorithm and their respective belief distributions, the robot could select the algorithm with the best overlap between the observation likelihood function and the belief. Such an approach would work well for choosing between different algorithms that use similar sensors, like depth cameras and laser rangefinders. However, algorithms using different types of sensors like WiFi signal strength measurements as compared to rangefinder based measurements, cannot be compared directly based on their sensor meismatch due to the very different forms of the observation likelihood functions for the different sensors.

4.3 Data - Driven Selection

If neither uncertainty based nor sensor mismatch based selection is feasible, a data-driven selection of the localization algorithms might be the only recourse. In order to select the best localization algorithm for a given location, the data-driven approach requires experimental evaluation of every localization algorithm (as described in Section 3) over all areas in the environment. During deployments, the robot would then select (at each location) that algorithm which exhibited least variance during the experimental evaluation at the same location. Such an approach would work for any localization algorithm, and does impose any algorithmic limitations on the individual localization algorithms, unlike uncertainty or sensing mismatch based selection. However, this approach does require experimental evaluation of all localization algorithms in all the areas of the environment, which might not be feasible for large areas of deployment.

5 Conclusion and Future Work

In this work, we tested the robustness of localization using three different sensors over an entire map. We showed that in areas where there are features are visible to the laser rangefinder and Kinect, they are both more accurate than WiFi. However, in open areas, localization using a laser rangefinder or Kinect is susceptible to large errors. In such areas, the maximum error of WiFi is bounded, and its median error is lower than that using a laser rangefinder or a Kinect. This indicates that although no single sensor is universally superior for localization, by selecting different sensors for localization for different areas of the map, the localization error of the robot is bounded. Based on these results, we are now interested in learning how to predict a priori which sensors would be most robust for localization for different areas of the map.

References

1. Thrun, S., Bennewitz, M., Burgard, W., Cremers, A., Dellaert, F., Fox, D., Hahnel, D., Rosenberg, C., Roy, N., Schulte, J., et al.: Minerva: A second-generation museum tour-guide robot. In: 1999 IEEE International Conference on Robotics and Automation, vol. 3 (1999)
2. Romero, L., Morales, E., Sucar, E.: An exploration and navigation approach for indoor mobile robots considering sensor's perceptual limitations. In: 2001 IEEE International Conference on Robotics and Automation, vol. 3, pp. 3092–3097. IEEE (2001)
3. Nilsson, N.: Shakey the robot. Technical report, DTIC Document (1984)
4. Buhmann, J., Burgard, W., Cremers, A., Fox, D., Hofmann, T., Schneider, F., Strikos, J., Thrun, S.: The mobile robot rhino. AI Magazine 16(2), 31 (1995)
5. Elfes, A.: Using occupancy grids for mobile robot perception and navigation. Computer 22(6), 46–57 (1989)
6. Koenig, S., Simmons, R.: Xavier: A robot navigation architecture based on partially observable markov decision process models. In: Artificial Intelligence Based Mobile Robotics: Case Studies of Successful Robot Systems, pp. 91–122 (1998)
7. Nourbakhsh, I., Kunz, C., Willeke, T.: The mobot museum robot installations: A five year experiment. In: 2003 IEEE/RSJ International Conference on Intelligent Robots and Systems, vol. 4, pp. 3636–3641 (2003)
8. Oyama, A., Konolige, K., Cousins, S., Chitta, S., Conley, K., Bradski, G.: Come on in, our community is wide open for robotics research? In: The 27th Annual Conference of the Robotics Society of Japan, vol. 9 (2009)
9. Grisetti, G., Stachniss, C., Burgard, W.: Improved techniques for grid mapping with rao-blackwellized particle filters. IEEE Transactions on Robotics 23(1), 34–46 (2007)
10. Fox, D.: KLD-sampling: Adaptive particle filters and mobile robot localization. In: Advances in Neural Information Processing Systems, NIPS (2001)
11. Rosenthal, S., Biswas, J., Veloso, M.: An effective personal mobile robot agent through symbiotic human-robot interaction. In: 9th International Conference on Autonomous Agents and Multiagent Systems, pp. 915–922 (2010)
12. Van der Zant, T., Wisspeintner, T.: Robocup@ home: Creating and benchmarking tomorrows service robot applications. Robotic Soccer, 521–528 (2007)

13. Kam, M., Zhu, X., Kalata, P.: Sensor fusion for mobile robot navigation. Proceedings of the IEEE 85(1), 108–119 (1997)
14. Biswas, J., Coltin, B., Veloso, M.: Corrective gradient refinement for mobile robot localization. In: 2011 IEEE/RSJ International Conference on Intelligent Robots and Systems, pp. 73–78. IEEE (2011)
15. Biswas, J., Veloso, M.: Depth camera based indoor mobile robot localization and navigation. In: 2012 IEEE International Conference on Robotics and Automation, pp. 1697–1702 (2012)
16. Biswas, J., Veloso, M.: Wifi localization and navigation for autonomous indoor mobile robots. In: 2010 IEEE International Conference on Robotics and Automation, pp. 4379–4384 (2010)
17. Ferris, B., Hähnel, D., Fox, D.: Gaussian processes for signal strength-based location estimation. In: Proc. of Robotics Science and Systems. Citeseer (2006)
18. Rice, J.: The algorithm selection problem. Advances in Computers 15, 65–118 (1976)
19. Xu, L., Hutter, F., Hoos, H., Leyton-Brown, K.: Satzilla: portfolio-based algorithm selection for sat. Journal of Artificial Intelligence Research 32(1), 565–606 (2008)
20. Smith-Miles, K.: Cross-disciplinary perspectives on meta-learning for algorithm selection. ACM Computing Surveys (CSUR) 41(1), 1–25 (2008)
21. Dellaert, F., Fox, D., Burgard, W., Thrun, S.: Monte carlo localization for mobile robots. In: 1999 IEEE International Conference on Robotics and Automation, vol. 2, pp. 1322–1328. IEEE (1999)
22. Lenser, S., Veloso, M.: Sensor resetting localization for poorly modelled mobile robots. In: IEEE International Conference on Robotics and Automation 2000, pp. 1225–1232 (2000)

High-Level Commands in Human-Robot Interaction for Search and Rescue

Alain Caltieri and Francesco Amigoni

Artificial Intelligence and Robotics Laboratory, Politecnico di Milano, Milano, Italy
alaincaltieri@gmail.com, francesco.amigoni@polimi.it

Abstract. Successful search and rescue operations require an appropriate inter-action between human users and mobile robots operating on the field. In the literature, use of waypoints for driving the robots has been identified as the main approach to trade-off between fully autonomous robotic systems, which can exclude human users from the control cycle, and completely tele-operated robotic systems, which can excessively burden human users. In this paper, we propose an intermediate level between full autonomy and waypoint guidance. Specifically, human users can issue *high-level commands* to the robots, like "explore along a direction" and "explore in this area", which do not explicitly specify the target locations, but introduce a bias over the autonomous target selection performed by the robots. Experimental results show that high-level commands are effective, provided that notification messages coming from the robots are filtered.

Keywords: human-robot interaction, search and rescue, multirobot systems.

1 Introduction

In many search and rescue settings, the interaction between human users and mobile robots operating in harsh environments is fundamental, given the time constraints and the importance of rescuing victims. Although some fully-autonomous multirobot systems for search and rescue have been developed in the last years, they are not expected to be widely employed in real scenarios, mainly because human users tend to be excluded from the control cycle [1, 2]. As a consequence, autonomous and human components should be balanced. In most of the current multirobot systems for search and rescue, this trade-off is obtained by allowing human users to directly issue waypoint commands (by which the user specifies a target location, often close to robot current position, and the robot autonomously plans a path to reach it) or lower-level tele-operation commands.

In this paper, we propose to introduce an intermediate level between full autonomy and waypoint guidance. Operating at this level, human users can issue *high-level commands* to the robots, like "explore along a direction" and "explore in this area", which do not explicitly specify the target locations, but introduce a bias over the autonomous target selection performed by the robots. High-level commands give the robots more autonomy than waypoint guidance and require an exploration technique that supports them. Experimental results show that using high-level commands increases both the area explored and the number of victims found, but increases also user's *workload*,

S. Behnke et al. (Eds.): RoboCup 2013, LNAI 8371, pp. 480–491, 2014.

namely the cognitive effort for controlling the multirobot system, decreasing user's *situation awareness*, namely the ability to figure out the global status of the system. We show that, introducing an appropriate filtering of notification messages coming from the robots, workload is reduced, keeping only the positive effects of high-level commands on performance of multirobot systems for search and rescue.

We consider a search and rescue setting in which a number of robots coordinate in order to explore an initially unknown environment and find victims. The robots communicate with a base station, which features a Graphical User Interface (GUI) through which a human user can interact with the robots (issuing commands and receiving notifications). Our human-robot interface is embedded in the *PoAReT (Politecnico di Milano Autonomous Robotic Rescue Team)* system that has been developed for the Virtual Robot Competition of the RoboCup Rescue Simulation League. Full information about the system, including a link to the source code is available at `http://home.deib.polimi.it/amigoni/research/PoAReT.html`. The PoAReT system operates in environments simulated in USARSim [3].

2 Human-Robot Interaction in Search and Rescue

The role of the human component in a robotic system can vary depending on the extension of human influence and on the level of autonomy of the system. In the system presented in this paper, the human user interacts both with the whole robotic system and with the single robots (see [4] for a complete taxonomy of forms of human-robot interaction). Specifically, the human user can act as *supervisor* and as *operator* [5]. In the first role, the human user is focused on high-abstraction-level information and tasks. In the role of operator, the human user must also comply with lower-level tasks, ranging from direct tele-operation of robots to their manual coordination. Supervision requires some autonomy of the system in executing complex actions [6]. The most common solution is to implement a dynamic level of autonomy, modifiable by the human operator (*adjustable autonomy*), by the robotic system (*adaptive autonomy*), or by both (*mixed initiative*). Results in [7] show that a mixed initiative approach yields a better global performance in some urban search and rescue tasks.

In the following of this section, we survey some of the most significant systems for interaction between human users and autonomous multirobot systems specifically developed for search and rescue and employed in realistic settings (e.g., competitions).

In the Steel system [8] every function is carried out by a separate module and modules are linked together by the Machinetta framework. The GUI is map-centric (namely, the map built by the robots and representing the environment is the central element of the GUI) and allows the operator to interact with many robots simultaneously. The Steel system allows three different levels of interaction (and autonomy). With full autonomy, the human operator can select some interest areas to be explored with higher priority by the whole multirobot system. Autonomous coordination and exploration modules then accomplish the task. With manual specification of the path to be followed by the robots, the operator can set a number of waypoints. Finally, with tele-operation, the operator directly controls the motion of the robots.

The Hector project [9] aims at developing a fully-autonomous robotic system in which the human user acts as a supervisor. The human user interacts with the system by mean of some high-level policies. For example, a policy could specify what actions should be performed when an exception is raised during a navigation task. The system considers such policies, together with the information about the surrounding environment and the goals, and autonomously decides the action to execute. The human supervisor can cancel or override the decision of the system.

The system in [10] allows a single operator to control a team of robots, interacting with them at three abstraction levels: full autonomy with coordination between robots, waypoint selection, and tele-operation.

Finally, Team Michigan [1] won the MAGIC 2010 competition with a system composed of 14 robots and 2 human operators, in charge of sensor data cleaning and integration and of the allocation of tasks to robots, respectively. Team Michigan features a rich feedback interface that presents information filtered and ordered by priority on a 3D map of the explored environment. The task operator can interact with the system by adjusting the level of autonomy, from full autonomy (the robots autonomously decide what to do and coordinate without human intervention), to semi-autonomy (the human operator assigns to a robot a target point and the robot automatically plans and follows a path to it), to tele-operation.

As it emerges from the discussion above, in most of the current systems, there is no intermediate level between full autonomy and waypoint guidance. In this paper, we propose an approach that contributes to fill this gap.

3   The PoAReT Multirobot System

In this section, we overview the PoAReT system architecture (as reported in [11], to which the reader is referred for full description), highlighting some of its features in order to set the context for illustrating the Human-Robot Interaction (HRI) system proposed in this paper.

The PoAReT system is a controller for simulated robots in USARSim. In particular, it controls mobile platforms (usually Pioneer All Terrain P3AT), each one equipped with laser range scanners, sonars, and a camera. Laser range scanners are used to build a two-dimensional geometrical map of the environment that is represented with two sets of line segments. The first set contains the line segments that represent (the edges of) perceived obstacles. The second set contains the line segments that represent the *frontiers*, namely the boundaries between the known and the unknown portions of the environment. The free space already explored is a polygon (generally with holes) whose edges are either obstacles or frontiers.

When operating autonomously, the main cycle of activities of the PoAReT system is: (a) building a map of the environment composed of line segments, (b) selecting the most convenient frontiers to reach, and (c) coordinating the allocation of robots to the frontiers. At the same time, the system detects victims on the basis of the images returned by the onboard cameras and interacts with the human operator via the GUI.

The architecture of the system is organized in two different types of processes, one related to the base station and one related to the mobile robots. The *base station* embeds the HRI system, which is described in the next section. The base station process can spawn new robots in the USARSim environment: for each robot, a new independent process is created and started. The processes of the base station and of the robots communicate only through Wireless Simulation Server (WSS) [12]. A distance vector routing protocol [13] is implemented to deliver messages.

The *robot* process is structured in different modules, each one related to a high-level functionality: motion control, path planning, SLAM, exploration, coordination, and victim detection. The main functionalities of the above modules are described in the following. The motion control module is straightforward, given the locomotion model of P3AT. The path planning module is invoked to reach a location with a path that lies entirely in the known space (e.g., the location can be on a frontier between known and unknown space). The algorithm of the path planning module is a variant of the Rapidly-exploring Random Tree (RRT) algorithm [14].

The Simultaneous Localization And Mapping (SLAM) problem is tackled using a feature-based method. The SLAM module associates the line segments of a laser scan (points of a scan are approximated with line segments by using the split-and-merge algorithm [15]) to the the linear features in the map, with respect to a distance measure, such as that in [16]. Then, the module executes an Iterative Closest Line (ICL) algorithm (like [16]) to align the scan and the map. All the line segments of a scan are added to the map; periodically a test is carried out to determine whether there is enough evidence to support the hypothesis of two previously associated line segments being in fact the same; if so, they are merged.

The exploration module evaluates new frontiers to explore, in order to discover the largest possible amount of the environment, and calls the coordination module to find an allocation of robots to the frontiers. In our system, we use Multi-Criteria Decision-Making (MCDM) [17], a framework to compose different criteria and obtain the utility $u(f,r)$ of a frontier f for a robot r. Criteria we consider include distance between f and the current position of r, expected information gain at f, and probability of communicating with the base station once in f. One of the main features of MCDM is that adding further criteria is particularly easy. Please refer to [17] for further details and full mathematical treatment of MCDM.

The coordination module is responsible of allocating tasks to the robots. The mechanism we use is market-based and sets up auctions in which tasks (i.e., frontiers to reach) are auctioned to robots [18] that bid according to the evaluation returned by their exploration modules. Market-based mechanisms provide a well-known mean to bypass problems like unreliable wireless connections or robot malfunctions.

Finally, the victim detection module is responsible for searching victims inside the environment. It analyzes images coming from a robot camera (using a skin detector in HSV (Hue, Saturation, Value) color space) and classifies them according to the presence or absence of victims (using a version of the Viola-Jones algorithm [19]). In the positive case, the victim detection module signals the human operator.

4   The PoAReT Human-Robot Interaction System

The PoAReT Human-Robot Interaction (HRI) system allows a single human operator to effectively control a relatively large group of robots. It displays data to the operator and accepts commands from the operator to control the spawned robots. It reduces the workload of the operator and increases her/his situation awareness through a mixed-initiative approach. On the one hand, the PoAReT HRI system allows a single operator to control the system by issuing high-level commands to robots. On the other hand, the PoAReT HRI system filters out notifications arriving from the robots, based on the operator's preferences, past behavior, and situation parameters.

4.1   The PoAReT GUI

The GUI of the PoAReT system (Fig. 1) is map-centric, modular, and with fully-reconfigurable windows to allow the operator to better use the space on the screen and possibly discard unwanted components. The main components of the interface are:

Fig. 1. GUI of the PoAReT system

- The map (Fig. 1, left center). The local line segment maps sent by the robots are merged and the resulting map is displayed to the operator. The map is zoomable, scrollable, and interactive: by clicking on the map the operator can issue commands to the robots and select them. Commands are: operate fully autonomously, high-level commands (see Section 4.2), waypoint guidance, and tele-operation.
- Message manager (Fig. 1, left bottom). Beyond maps, every robot sends to the base station information regarding its internal status (e.g., battery level and wireless connectivity) and events like faults or victims detected. All these data are classified by the HRI system (see Section 4.3) that presents the relevant information to the operator as messages. A pop-up window signals particularly important notification, like a victim detection (as in Fig. 1).
- Camera manager (Fig. 1, left top). Small boxes containing the compressed (reduced framerate and resolution) video streams captured by the cameras of the robots are always shown to the operator.
- Selected robot's camera view (Fig. 1, right top). It shows the video stream of the selected robot, with full framerate and resolution.

- Selected robot's info (Fig. 1, right center). Information includes connectivity, battery status, and internal modules activation (like victim detection module or autonomous exploration module).
- Tele-operation (Fig. 1, center top). This widget allows the operator to directly control the movement of the selected robot.

Despite its modularity, the GUI has a centralized logic and all information is always coherently presented to the operator. Internally, the information flows are managed by a single module, called Base Station Core, which delivers the information to the proper component (internal modules or robots).

4.2 High-Level Commands

As discussed, workload and situation awareness are two main factors that strongly affect the performance of a human operator when controlling a multirobot system. A trade-off between workload and situation awareness by balancing autonomy and direct control of robots has been found in literature mainly using waypoint guidance (Section 2).

We introduce a new kind of High-Level Commands (HLCs) that combine the positive effects of robot's autonomy with the need to preserve the human operator in the control loop. HLCs allow an operator to indicating a preference about the exploration policies of a robot, specifically a preferred direction to explore (a vector $\alpha = (\alpha_x, \alpha_y)$ with origin in the robot's current position) and a specific area of interest (a point $p = (p_x, p_y)$ around which the robot should explore). In the PoAReT GUI, the human operator can issue a HLC through the map component: by drawing an arrow α and by double-clicking a point p. These commands are then integrated in the exploration strategy of the robot that explores autonomously (without human intervention) but following human directives.

In detail, the preferred direction α impacts on the MCDM-based exploration strategy by introducing a new criterion $\mathcal{D}_\alpha(f, r)$ that evaluates a frontier f with respect to a robot r and that returns the smaller (better) values the more the vector $r - f_c$, from the current position of robot r to the centroid f_c of frontier f, lies along the direction of α. Namely, $\mathcal{D}_\alpha(f, r) = \theta(\alpha, r - f_c)$, where $\theta(\cdot, \cdot)$ returns the angle between the directions of two vectors. The areas of interest impact on MCDM in a similar way, by introducing a new criterion $\mathcal{A}_p(f)$ that returns the smaller values the more f lies close to p. Namely, $\mathcal{A}_p(f) = d(p, f_c)$, where $d(\cdot, \cdot)$ returns the Euclidean distance between two points. Criteria $\mathcal{D}_\alpha(f, r)$ and $\mathcal{A}_p(f)$ are merged by MCDM with the other criteria discussed in Section 3 in order to evaluate the utility $u(f, r)$ of frontier f for robot r.

With HLCs the human operator can control robots at a higher level of abstraction than waypoint guidance, without losing control over the single robots (as it could happen in case of full autonomy). The higher level of abstraction is relative to the fact that the human operator does not explicitly tell the robots where to go, but gives them some bias about what locations they should (autonomously) choose. Since our HLCs are issued to single robots, they differ from the approach in [8], in which the operator selects interesting areas for the *whole* multirobot system that autonomously coordinates the motion of the robots. Moreover, the "explore along a direction α" command is not considered in [8].

4.3   Message Filtering

The human operator, to effectively control a multirobot system, needs not only to issue commands to the robots, but also to receive appropriate feedback information. The
robots, during their missions, can encounter difficulties or detect interesting situations,
in which cases they send a message to the human operator. When the number of robots
is large, missions are critical, and time is scarce, the number of notifications can easily
overwhelm the attention capabilities of human operators.

We then introduce a message filtering system in order to display to the operator only
important information and limit the workload due to message handling. Messages that
robots can send to the base station and that can be displayed to the operator, include:
victim detection messages coming from the victim detection modules, feedback messages coming from the motion control modules that signal that an assigned motion task
(e.g., reach a waypoint) has been completed, and fault messages that can come from
motion modules (e.g., a collision) or from path planning modules (e.g., a frontier cannot be reached).

For each message M coming at time t from module m of robot r, the priority of M
is calculated according to three factors:

- The inverse of the current operator workload, $W^{-1}(t) = 100 - \frac{100}{1+100 \cdot \exp^{-n(t)}}$,
  where $n(t)$ is the number of messages currently displayed in the message manager
  to the operator (see Fig. 1); the larger $n(t)$, the smaller $W^{-1}(t)$.
- The current reliability of module m of robot r, $R_{m,r}(t) = \frac{100}{1+\exp^{-z_{m,r}(t)/4}}$, where
  $z_{m,r}(t)$ is the difference between the number of messages coming from m that have
  been archived positively and the number of those that have been archived negatively
  by the operator, from the beginning of the mission (initially, $z_{m,r}(t = 0) = 0$ for
  all modules m and robots r).
- The relevance I_M of the message M, which is related to the specific mission. For
  example, we considered relevance $I_M = 100$ for messages M about victim detection, $I_M = 80$ for messages M about faults, $I_M = 50$ for feedback messages, and
  $I_M = 10$ for other messages. These values can be dynamically set by the operator
  during the mission.

The priority of M is then calculated as $p(M) = \frac{W^{-1}(t)+R_{m,r}(t)+I_M}{3}$. Obviously, $0 \leq$
$p(M) \leq 100$. Messages are displayed in the message manager ordered by their priority,
after elimination of messages with priority less than 10 (this threshold can be set by the
operator), which are assumed to be not relevant.

The rationale of our approach is that, when the workload is high, only very important
messages are displayed to the operator. Moreover, the reliability of the sending module
m depends on the previous interaction of the operator with the system and accounts
for the fact that some modules might be sending unreliable information (e.g., in the
case of the victim detection module, due to malfunctioning or to adverse environmental
conditions, like scarce light). Finally, the nature of the message (e.g., detection of a
victim or successful completion of motion) influences the priority in an obvious way,
namely the higher the importance, the higher the priority.

Messages can be removed from the message manager windows in two ways. Firstly, messages are automatically archived after a given time interval (25 seconds in our experiments, but the value can be set by the operator). In this case, the reliability balance $z_{m,r}$ of the module that sent the message is left untouched. Secondly, messages can be removed by the operator, explicitly marking them as positive or negative. In this case, the corresponding reliability balance $z_{m,r}$ is updated accordingly ($+1$ and -1 for positive and negative markings, respectively).

5 Experimental Activity

5.1 Experimental Setup

We implemented our PoAReT HRI system in C++ and we experimentally tested three different system configurations with an increasing number of functionalities. In the *standard* configuration the robots are tele-operated or controlled by waypoint commands. This configuration represents a basic level of functionalities available in many of the works discussed in Section 2 and is used as a baseline for comparison. *High-level commands* (HLC) configuration adds autonomous exploration, driven by high-level commands issued by the operator (Section 4.2). Still no message filtering capabilities is used. Finally, *full functionalities* (FF) configuration includes both improvements proposed in this paper: high-level commands and message filtering (Section 4.3).

We tested our approach in three indoor environments simulated in USARSim (shown in Fig. 2). The environments are created in order to be fully mapped in 5 minutes with almost perfect coordination among a given number of robots: the small environment has 8 victims for runs with 2 robots, the bigger one has 10 victims for runs with 5 robots, and the huge one has 10 victims for runs with 8 robots. Victims are randomly placed and can be detected automatically by the robots or by the operator looking at camera views of the GUI (Section 4.1).

Experimental tests are carried out by 14 volunteers, 7 are expert users of the system (with at least 1 full-day training) and 7 are non-expert users (a similar number of testers

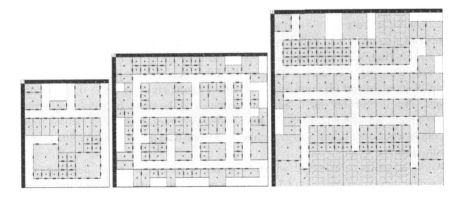

Fig. 2. Test environments (sizes are 27 m × 33 m, 51 m × 42 m, and 57 m × 57 m; corridors are in green, rooms in blue, inaccessible areas in white, and obstacles in black)

has been used also in [1,7,10,20]). Testers are aged between 18 and 27 and have expertise at least in using personal computers. Each tester performs 9 runs (corresponding to
three system configurations in the three environments) of 5 minutes each. To limit bi-
ases caused by learning, which can be relevant for non-expert users, runs are randomly
ordered for each tester.

For performance evaluation we use some well-known metrics. *Fan out* is defined as
number of robots properly controlled by a single operator. We calculate it by measuring
the amount a of area covered (mapped) in a run by all robots. As a benchmark we
considered the area $\bar{a}$ mapped by a single robot manually operated by an average expert
operator. Fan out is then defined as follows: $fan\ out = \frac{a}{\bar{a}}$. A similar measure is used for
fan out relative to distance travelled. Since the main goal of a search and rescue mission
is to find as many victims as possible, the percentage of *victims found* is also considered
(as in [7]). Finally, we measure the workload (e.g., see [21]) considering the number of
cognitive events the operator must face. A cognitive event can be a command issued to
a robot, a message received, or an interaction with the GUI.

5.2 Experimental Results

Let start from fan out results (Fig. 3(a)). Fan out relative to area explored and distance travelled shows a relevant grow when HLCs are introduced. These results are
statistically significant according to an ANOVA analysis with a threshold for significance p-value < 0.05 [22]: for example, the difference between fan out of standard
and HLC configurations with 5 robots has p-value $= 0.019$ for area mapped and p-
value $= 0.0037$ for distance travelled. With 8 robots, the p-values are 0.011 and 0.040,
respectively. These performance improvements are more evident with 5 and 8 robots
because controlling more robots is more challenging. Experts testers perform slightly
better than non-expert ones in terms of both area mapped and distance travelled (data
not shown here due to space constraints). The difference is more relevant when the
number of robots grows: performance of non-experts users in controlling 8 robots is
similar to that in controlling 5 robots.

Percentage of victims found (Fig. 3(b)) slightly grows when introducing HLCs but
the increase is not statistically significant (for example, p-value $= 0.37$ for the difference between percentage of victims found of standard and HLC configurations with 5
robots and p-value $= 0.83$ with 8 robots). A more evident improvement is obtained by
message filtering, leading to statistically significant better performances in FF configuration (for example, p-value $= 0.0010$ for the difference between percentage of victims
found of standard and FF configurations with 5 robots and p-value < 0.0001 with 8
robots). This is probably due to a better awareness of the status of every single robot.
Expert testers perform slightly better than non-expert ones in terms of victims found,
especially when controlling a large number of robots (data not shown here due to space
constraints).

Workload grows, as expected, with the number of robots to be controlled (Fig. 3(c)).
An evident increase comes with HLCs: robots explore the environment by themselves,
sending feedbacks to the operator and asking for help when facing some difficulty. Feedback messages, together with a more intense action of the robots (as highlighted by the
results on fan out), cause an increase of the workload of the operator. Message filtering

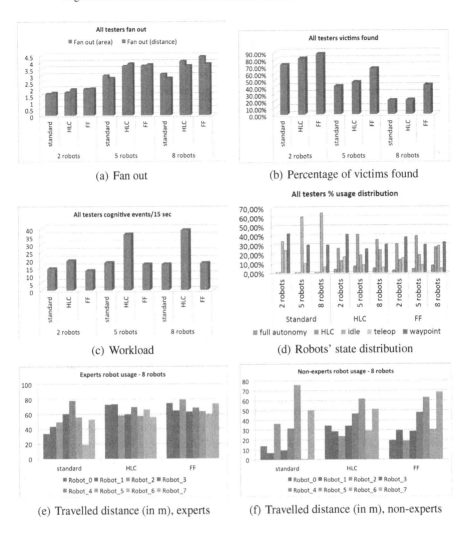

Fig. 3. Experimental results (averages are over runs and testers)

effectively limits this drawback of autonomous exploration, reducing the workload to a level comparable to a standard configuration (p-value < 0.0001 for the difference between workload of HLC and FF configurations with both 5 and 8 robots; p-value $= 0.72$ and p-value $= 0.77$ for the difference between workload of standard and FF configurations with 5 and 8 robots, respectively). This is a relevant result because workload is not effectively reduced for large numbers of robots in some other systems, for example in the Steel system [8]. We noticed that expert testers could work with a consistently higher workload with respect to non-expert testers. This can explain the performance gap between the two groups of testers.

We now show some results that provide insights on the impact of our approach on the control of robots during a mission. Fig. 3(d) shows the percentage distribution of mission time (5 minutes) over the possible states of a robot: operating fully autonomously, or with high-level commands, waypoint guidance, or tele-operation, or being idle. At any time, a robot can be in only one of these states. As expected, a drastic reduction of idle time is observed thanks to HLCs (p-value < 0.001 for the difference between idle time of standard and HLC configurations with both 5 and 8 robots). This is a relevant result because idle time is relevant for some other systems, like the Team Michigan system [1]. Even more important is the fact that expert users strongly prefer to issue HLCs instead of letting robots explore fully autonomously. We guess this is due to the better level of control experienced by users with HLCs.

During the runs we noticed that expert users were able to use almost all robots, while non-experts used just a subset of robots, presumably in order to feel more confident with the system. This behaviour is related to the differences in workload we already noticed, and it is more evident when the number of robots grows. In Figs. 3(e) and 3(f) we show the distance travelled (in meters) by single robots for the two groups of testers with 8 robots. In general, robots are used more homogeneously by expert testers. HLCs and message filtering further improve the homogeneity in controlling the robots.

6 Conclusion

In this paper, we have presented the PoAReT human-robot interaction system that allows operators to issue high-level commands to the robots. Experiments show that our approach increases the performance of the system (in terms of fan out and victims found), without worsening the cognitive effort of the operators (in terms of workload) thanks to a mechanism that filters out messages coming from the robots. The PoAReT HRI system allows to effectively control a large number of robots exploiting and driving the potentialities of autonomous exploration. This is one of the reasons that led the PoAReT system to win the 2012 edition of the Virtual Robot competition within the RoboCup Rescue Simulation League, effectively controlling up to a dozen of robots.

Future work will mainly focus on moving from a realistic simulation (like that provided by USARSim) to a robotic system operating in the real world. This will allow to further assess the significance of our approch.

References

1. Olson, E., Strom, J., Morton, R., Richardson, A., Ranganathan, P., Goeddel, R., Bulic, M., Crossman, J., Marinier, B.: Progress towards multi-robot reconnaissance and the MAGIC 2010 competition. Journal of Field Robotics 29(5), 762–792 (2012)
2. Petersen, K., von Stryk, O.: Towards a general communication concept for human supervision of autonomous robot teams. In: Proc. ACHI, pp. 228–235 (2011)
3. Carpin, S., Lewis, M., Wang, J., Balakirsky, S., Scrapper, C.: USARSim: A robot simulator for research and education. In: Proc. ICRA, pp. 1400–1405 (2007)
4. Yanco, H.: Classifying human-robot interaction: An updated taxonomy. In: Proc. IEEE SMC, pp. 2841–2846 (2004)
5. Scholtz, J.: Theory and evaluation of human robot interactions. In: Proc. HICSS (2003)
6. Kaber, D., Endsley, M.: The effects of level of automation and adaptive automation on human performance, situation awareness and workload in a dynamic control task. Theoretical Issues in Ergonomics Science 5(2), 113–153 (2004)

7. Wang, J., Lewis, M.: Human control for cooperating robot teams. In: Proc. HRI, pp. 9–16 (2007)
8. Velagapudi, P., Kleiner, A., Brooks, N., Scerri, P., Lewis, M., Sycara, K.: RoboCup Rescue Simulation League 2010 Team STEEL (USA). Technical report, Carnegie Mellon University, Pittsburgh (2010)
9. Graber, T., Kohlbrecher, S., Meyer, J., Petersen, K., von Stryk, O., Klingauf, U.: RoboCup Rescue Robot League 2012 Team Hector Darmstadt (Germany). Technical report, Technische Universität Darmstadt (2012)
10. Nevatia, Y., Stoyanov, T., Rathnam, R., Pfingsthorn, M., Markov, S., Ambrus, R., Birk, A.: Augmented autonomy: Improving human-robot team performance in urban search and rescue. In: Proc. IROS, pp. 2103–2108 (2008)
11. Amigoni, F., Caltieri, A., Cipolleschi, R., Conconi, G., Giusto, M., Luperto, M., Mazuran, M.: PoAReT Team Description Paper. In: RoboCup 2012 CD (2012)
12. WSS Wireless Simulation Server, http://sourceforge.net/apps/mediawiki/usarsim/index.php?title=Wireless_Simulation_Server
13. Comer, D.: Internetworking with TCP/IP, vol. 1. Addison-Wesley (2006)
14. LaValle, A., Kuffner, J.: Rapidly-exploring random trees: Progress and prospects. In: Donald, B., Lynch, K., Rus, D. (eds.) Algorithmic and Computational Robotics: New Directions, pp. 293–308. CRC Press (2001)
15. Nguyen, V., Gächter, S., Martinelli, A., Tomatis, N., Siegwart, R.: A comparison of line extraction algorithms using 2d range data for indoor mobile robotics. Autonomous Robots 23(2), 97–111 (2007)
16. Li, Q., Griffiths, J.: Iterative closest geometric objects registration. Computers & Mathematics with Applications 40(10-11), 1171–1188 (2000)
17. Basilico, N., Amigoni, F.: Exploration strategies based on multi-criteria decision making for searching environments in rescue operations. Autonomous Robots 31(4), 401–417 (2011)
18. Zlot, R., Stentz, A., Dias, M.B., Thayer, S.: Multi-robot exploration controlled by a market economy. In: Proc. ICRA, pp. 3016–3023 (2002)
19. Viola, P., Jones, J.: Robust real-time face detection. International Journal of Computer Vision 57(2), 137–154 (2004)
20. Kane, B., Velagapudi, P., Scerri, P.: Asking for help through adaptable autonomy in robotic search and rescue. In: Bai, Q., Fukuta, N. (eds.) Advances in Practical Multi-Agent Systems. SCI, vol. 325, pp. 339–357. Springer, Heidelberg (2010)
21. Prewett, M., Johnson, R., Saboe, K., Elliott, L., Coovert, M.: Managing workload in human-robot interaction: A review of empirical studies. Computers in Human Behavior 26(5), 840–856 (2010)
22. Pestman, W.: Mathematical Statistics: An Introduction. de Gruyter (1998)

Optimizing Energy Usage through Variable Joint Stiffness Control during Humanoid Robot Walking

Ercan Elibol[1], Juan Calderon[1,3], and Alfredo Weitzenfeld[2]

[1] Dept. of Electrical Engineering, University of South Florida, Tampa, FL, USA
{ercan,juancalderon}@mail.usf.edu
[2] Div. of Information Technology, University of South Florida, Tampa, FL, USA
aweitzenfeld@usf.edu
[3] Dept. of Electronic Engineering, Universidad Santo Tomás, Bogotá, Colombia

Abstract. The objective of this paper and our current research is to optimize energy usage in a humanoid robot during diverse tasks such as basic walking by dynamically controlling individual joint stiffness. In the current work we analyze individual and total usage of current, voltage and power in a NAO V4 humanoid robot joints during short walks around a circle at different speeds and under varying control of joint stiffness. We perform experimental studies to understand the main factors affecting power consumption and energy usage and look at ways to improve overall energy usage. We describe experiments and corresponding results. We discuss the state of advancement of our research.

Keywords: Energy usage, power consumption, joint stiffness, motor control, humanoid robots.

1 Introduction

A critical challenge in mobile robots is the optimization of energy usage during specific robot tasks [1-3]. This is particularly relevant to humanoid robots where the increased number of joints and corresponding DOFs make energy usage hard to analyze and optimize during basic tasks such as walking [4]. In general, battery life is one of the main constraints in the use of robots for extended time. In the context of RoboCup soccer, games last only a few minutes where batteries are usually recharged at half time. As battery usage improves we expect future games to last longer or have power restrictions imposed on teams.

As part of our goal to better understand energy usage in robots and develop appropriate energy optimization algorithms, we present in this paper our initial study on power consumption in the NAO V4 humanoid robot in the context of the Standard Platform League (SPL). We analyze current, voltage and power consumption at individual joints and improve their usage by dynamically modifying motor stiffness without sacrificing task performance. We analyze the effect of variations on walking step frequency and joint stiffness on overall energy usage during simple short walks.

In contrast to other work we develop more in depth analysis of power consumption and energy usage in both individual joint and overall system at various speeds. In the

S. Behnke et al. (Eds.): RoboCup 2013, LNAI 8371, pp. 492–503, 2014.
© Springer-Verlag Berlin Heidelberg 2014

study by Kormushev et al. [5] electric energy consumption is optimized on the COMAN humanoid robot during walking by varying-height and robot's center of mass. The authors applied reinforcement learning algorithm to reduce the energy utilization with respect to variations in the robot's center of mass position. In the study by Kulk and Welsh [6-8], the authors compare various optimization algorithms to reduce overall energy usage while increasing walking speed. The authors analyze energy usage in an older NAO V3 by reading current from joints and voltage and analyzing primarily overall power usage in the system.

In the rest of the paper we briefly discuss in Section 2 the humanoid biped walking in the NAO that will be used for our experiments; in Section 3 we describe the basis for the power consumption and energy usage analysis; in Section 4 we describe the experimental results; and we finish by conclusions where we discuss our findings and future work.

2   Humanoid Robot Walking

While there are numerous approaches to humanoid robot biped walking, we will concentrate in this paper on the NAO open-loop walk engine [9]. The NAO V4 robot includes 25 degrees of freedom (DOF) including five DOF in each leg (3 in the hip, 1 in the knee, and 2 in the ankle) each controlled by a brushed DC motor with magnetic rotary encoders for position feedback. The NAO walk patterns are generated from a Zero-Moment Point (ZMP) [10] trajectory that is calculated from user specified step parameters that include walking step frequency, walking step width and walking step length. The ZMP trajectory is transformed into a center of gravity (CoG) trajectory using an inverted pendulum model [11-12].

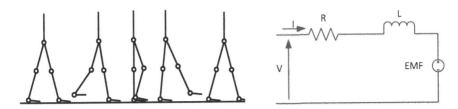

Fig. 1. (Left) Diagram illustrating the humanoid robot joints side connectivity schematics (hip pitch, knee pitch and ankle pitch). (Right) Magnetic brush DC motor used in and equivalent circuitry for the NAO DC motors with EMF representing the Electro-Magnetic Force [13].

3   Energy Usage during Humanoid Robot Walking

The energy usage of the humanoid robot joints can be computed from the corresponding motors controlling each joint. Figure 2 (left) shows the electrical circuit, while (middle) shows the corresponding joint location in the NAO V4, and (right) shows the joint connectivity schematics. To compute full energy usage in the system additional components in the robot need to be considered including CPU, sensors, communications, etc. Note that the NAO uses different types of motors depending on version and joint location [14]. To compute the energy usage of each motor in the robot we need to first compute power defined by motor voltage and current. In the following equations we describe the basic equations defining motor power and energy usage [15].

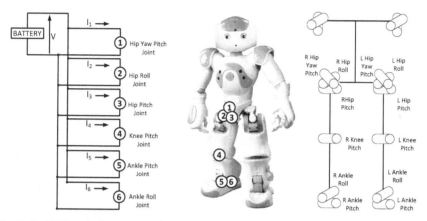

Fig. 2. (Left) Electric diagram of lower body NAO joints: Hip Yaw Pitch, Hip Pitch, Hip Roll, Knee Pitch, Ankle Pitch, Ankle Roll. (Middle) Corresponding location of joints in NAO robot. (Right) Diagram showing joints relative location.

In Eq (1), voltage V is equal to the change in motor inductance L, the current I through the motor windings with resistance R, and the motor torque constant k_b corresponding to the motor's back-EMF multiplied by the motor's rotor angular velocity $\dot{\theta}$,

$$V = L\frac{dI}{dt} + RI + k_b\dot{\theta} \tag{1}$$

In Eq (2), the motor's moment of inertia M multiplied by the robot's acceleration $\ddot{\theta}$ is equivalent to the motor's torque constant k_t multiplied by the electric current minus the motor's viscous friction constant v times motor angular speed minus the torque τ being applied from the external load,

$$M\ddot{\theta} = k_t I - v\dot{\theta} - \tau \tag{2}$$

At steady state the voltage V and torque τ are given by Eqs (3-7)

$$V = RI + k_b\dot{\theta} \tag{3}$$

$$\tau = k_t I - v\dot{\theta} \tag{4}$$

$$V = \tau\frac{R}{k_t} + \frac{Rv\dot{\theta}}{k_t} + k_b\dot{\theta} \tag{5}$$

$$\dot{\theta} = \left(v + \frac{k_b k_t}{R}\right)^{-1}\left(V\frac{k_t}{R} - \tau\right) \tag{6}$$

$$\tau = V\frac{k_t}{R} - \left(v + \frac{k_b k_t}{R}\right)\dot{\theta} \tag{7}$$

Stall torque τ_s at zero angular velocity is given when $\dot{\theta} = 0$ as shown by Eq (8)

$$\tau_s = V\frac{k_t}{R} \tag{8}$$

And the corresponding stall current I_s is given by Eq (9)

$$I_s = \frac{\tau_s}{k_t} = \frac{V}{R} \tag{9}$$

As current increases in the joint, torque also increases until it gets to a maximum level, i.e. the stall torque. The "no load" max speed $\dot{\theta}_n$ is defined when torque $\tau = 0$ given by Eq (10)

$$\dot{\theta}_n = \tau_s \left(v + \frac{k_b k_t}{R}\right)^{-1} \tag{10}$$

At steady state, the motor angular velocity $\dot{\theta}$, the torque τ and current I_s can be defined by Eqs (11-13),

$$\dot{\theta} = \dot{\theta}_n - \left(v + \frac{k_b k_t}{R}\right)^{-1}\tau \tag{11}$$

$$\tau = \tau_s - \left(v + \frac{k_b k_t}{R}\right)\dot{\theta} \tag{12}$$

$$I = I_s - \left(\frac{v}{k_t} + \frac{k_b}{R}\right)\dot{\theta} \tag{13}$$

Mechanical power P delivered by the motor is given by multiplying torque by motor angular velocity as shown by Eq (14)

$$P = \tau\dot{\theta} = \tau_s\dot{\theta} - \left(v + \frac{k_b k_t}{R}\right)\dot{\theta}^2 \tag{14}$$

Total power consumption P_{total} at time t is the summation of power P_i from all joints where i denotes the i-th joint as given by Eq (15),

$$P_{total} = V\sum_{i=i}^{n} I_i = \sum_{i=i}^{n} P_i \tag{15}$$

Finally, energy usage integrates power in time as described by Eq (16)

$$E = \int P_{total}\, dt \tag{16}$$

4   Experimental Results

In order to analyze power consumption and energy usage in the NAO V4 humanoid robot, we developed a simple walking experiment where the robot walks counter-clockwise around a small circle having 1 meter in diameter. We read electric current and voltage data from individual joints while walking at different speeds for about 1.5 minutes corresponding to 4-5 circles at max walking speed. We collected during each experiment approximately 2000 sets of data for each joint under different values of stiffness, walking step length and walking step frequency for both left and right joints: Hip Yaw Pitch, Hip Roll, Hip Pitch, Knee Pitch, Ankle Pitch, Ankle Roll. Note that a single motor controls in the NAO both Left and Right Hip Yaw Pitch joints. We performed each experiment three times and produced individual joint averages from those runs. Due to slow readings from the NAO motor joints, individual joint data was read at each experiment with stiffness and walking parameters changed at different runs. For data reading we used the NAO SDK platform version 12.3 under Python version 2.7. Current and voltage data was collected and saved into a text file for offline analysis. The following section of code illustrates the portion used to collect the data:

```
ALMEMORY_KEY_NAMES = [ "De-
vice/SubDeviceList/LAnklePitch/ElectricCurrent/Sensor/Value",
"Device/SubDeviceList/Battery/Charge/Sensor/CellVoltageMin",]

for key in ALMEMORY_KEY_NAMES:
    value = memory.getData(key)
```

Table 1 shows robot defaults and range of values used during the various experiments.

4.1   Hip Pitch Joint

Figures 3, 4, and 5 show Left Hip Pitch joint current, voltage and power for three stiffness values (1, 0.8 and 0.6) walking at maximum speed.

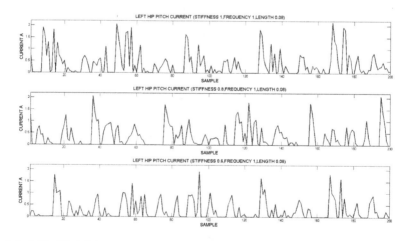

Fig. 3. Left Hip Pitch joint current usage at walking speed frequency 1 and step length 0.8 with the following stiffness (top) stiffness 1, (middle) stiffness 0.8, (bottom) stiffness 0.6

Table 1. The values to be changed during the experiment. Joint stiffness, walking step frequency and length were used as input to the humanoid system in order to obtain the output. Default values, their maximum and minimum values are presented in this table.

Name	Minimum	Maximum	Default
Joint Stiffness	0	1	1
Walking Step Frequency	0 (1.667 Hz)	1 (2.382 Hz)	1
Walking Step Length	0.001 (m)	0.08 (m)	0.04 (m)

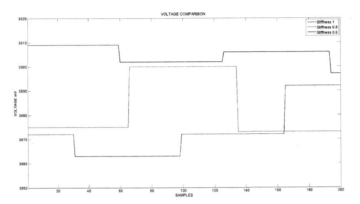

Fig. 4. Battery voltage usage for various Left Hip Pitch joint stiffness at walking speed frequency 1 and step length 0.8: (black) stiffness 1, (red) stiffness 0.8, (blue) stiffness 0.6.

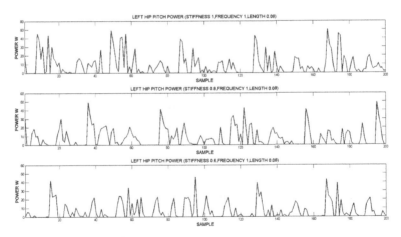

Fig. 5. Left Hip Pitch joint power usage at walking speed frequency 1 and step length 0.8 with the following stiffness (top) stiffness 1, (middle) stiffness 0.8, (bottom) stiffness 0.6.

4.2   Knee Pitch Joint

Figures 6, 7, and 8 show Left Knee Pitch joint current, voltage and power for three stiffness values (1, 0.8 and 0.6) walking at maximum speed.

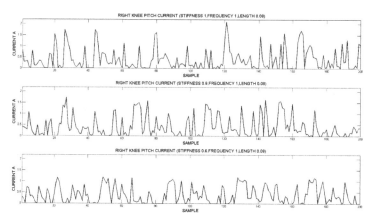

Fig. 6. Left Knee Pitch joint current usage at walking speed frequency 1 and step length 0.8 with the following stiffness (top) stiffness 1, (middle) stiffness 0.8, (bottom) stiffness 0.6

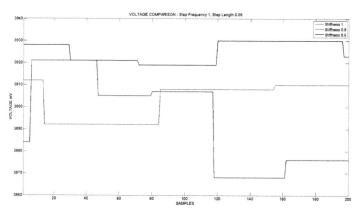

Fig. 7. Battery voltage usage for various Left Knee Pitch joint stiffness at walking speed frequency 1 and step length 0.8: (black) stiffness 0.6, (red) stiffness 0.8, (blue) stiffness 1

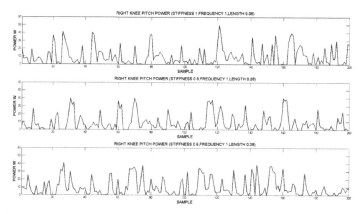

Fig. 8. Left Knee Pitch joint power usage at walking speed frequency 1 and step length 0.8 with the following stiffness (top) stiffness 1, (middle) stiffness 0.8, (bottom) stiffness 0.6

4.3   Ankle Pitch Joint

Figures 9, 10, and 11 show Ankle Pitch joint current, voltage and power for three stiffness values (1, 0.8 and 0.6) walking at maximum speed.

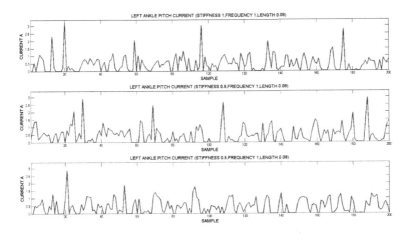

Fig. 9. Left Ankle Pitch joint current usage at walking speed frequency 1 and step length 0.8 with the following stiffness (top) stiffness 1, (middle) stiffness 0.8, (bottom) stiffness 0.6

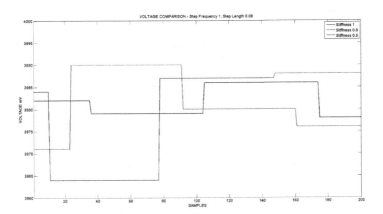

Fig. 10. Battery voltage usage for various Left Ankle Pitch joint stiffness at walking speed frequency 1 and step length 0.8: (black) stiffness 0.6, (red) stiffness 0.8, (blue) stiffness 1

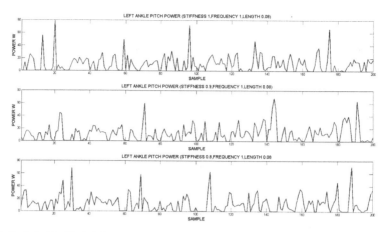

Fig. 11. Left Ankle Pitch joint power usage at walking speed frequency 1 and step length 0.8 with the following stiffness (top) stiffness 1, (middle) stiffness 0.8, (bottom) stiffness 0.6

4.4   Power and Energy Usage

Tables 2, 3, and 4 show max values of individual joint current and power for various stiffness values for Left Hip Pitch, Left Knee Pitch and Left Ankle Pitch. It is interesting to note that lowest current and power varies for various joints depending on stiffness. Note that when applying stiffness value under 0.6 walking becomes unstable with robot falling.

Table 2. Max current and power for Left Hip Pitch (Step Frequency 1, Step Length 0.08)

Stiffness	1	0.9	0.8	0.7	0.6
Current (A)	2.72	2.75	2.68	3.34	3.31
Power (W)	63.77	64.24	62.52	78.99	78.67

Table 3. Max current and power for Left Knee Pitch (Step Frequency 1, Step Length 0.08)

Stiffness	1	0.9	0.8	0.7	0.6
Current (A)	2.19	1.87	1.87	1.71	2.44
Power (W)	51.28	43.38	43.1	39.81	56.75

Table 4. Max current and power for Left Ankle Pitch (Step Frequency 1, Step Length 0.08)

Stiffness	1	0.9	0.8	0.7	0.6
Current (A)	3.31	3.77	3.00	2.78	2.84
Power (W)	78.71	88.51	70.47	65.44	67.94

Table 5 shows the accumulated power, i.e. total energy usage in Joules (J), for individual and total (grey) left and right joints according to the different stiffness values. Note the difference in energy usage for left and right side joints, whereas the Left and Right Hip Yaw Pitch have a unique motor controlling them both, hence their equal values.

Table 5. Total energy usage for individual joints and totals in Joules (J)

Stiffness	1	0.9	0.8	0.7	0.6
Left Hip Yaw Pitch (J)	5770	5668	5421	5540	5365
Left Hip Pitch (J)	15447	14751	14622	14188	14482
Left Hip Roll (J)	13358	13472	14197	14625	14486
Left Knee Pitch (J)	18236	18546	18675	18507	19105
Left Ankle Pitch (J)	22590	22017	21872	21166	21163
Left Ankle Roll (J)	4669	4906	4810	4956	5367
Total Left Joints (J)	80070	79360	79597	78982	79968
Right (Left) Hip Yaw Pitch (J)	5770	5668	5421	5540	5365
Right Hip Pitch (J)	12004	11380	11325	10956	11351
Right Hip Roll (J)	11557	11543	11306	11446	11394
Right Knee Pitch (J)	14954	14379	13315	12766	11619
Right Ankle Pitch (J)	14703	14964	14906	14882	15398
Right Ankle Roll (J)	5097	5043	5178	5112	4996
Total Right Joints (J)	64085	62977	61451	60702	60123
Total Joints (J)	144155	142337	141048	139684	140091

Tables 6 shows the resulting total energy usage per joint in Joules (J) when apply-ing best stiffness values from Table 5 although keeping similar values for correspond-ing left and right joint. The total energy usage is reduced from 139684 Joules (Table 5 – Stiffness 0.7) to 138073 Joules (Table 6). This is a small 1% total energy usage reduction for a 1.5 min walk but exemplifies the use of variable stiffness control to reduce individual and total joint energy usage according to specific task.

Table 6. Total Energy Usage per Joint in Joules (Step Frequency 1, Step Length 0.08) for Left and Right Joints under variable stiffness values

Joint	Stiffness	Left Joints	Right Joints	Total Energy (J)
Left and Right Hip Yaw Pitch (J)	0.6	5448	5476	10924
Left and Right Hip Pitch (J)	0.7	14388	11487	25875
Left and Right Hip Roll (J)	0.7	13211	11501	24712
Left and Right Knee Pitch (J)	1	17460	14835	32295
Left and Right Ankle Pitch (J)	0.6	21261	14309	35570
Left and Right Ankle Roll (J)	1	4048	4677	8725
Total Joints (J)		75816	62285	138101

Table 7. Total Energy Usage per Joint in Joules (Step Frequency 1, Step Length 0.08) for Left and Right Joints for different stiffness values

Stiffness	1	0.9	0.8	0.7	0.6
Left Hip Yaw Pitch (J)	2893	8555	5925	3189	3300
Left Hip Roll (J)	21458	21464	21485	22571	20379
Left Hip Pitch (J)	10541	10345	9376	10065	12703
Left Knee Pitch (J)	22454	23508	18973	27085	29802
Left Ankle Pitch (J)	32743	21586	16827	13391	12877
Left Ankle Roll (J)	6961	4882	7560	5352	6131
Total Left Joints (J)	97050	90340	80146	81653	81925

Table 7 shows the total energy usage per joint in Joules (J) while the robot is standing for 1 minute. It is interesting to note that energy usage is relatively large when standing with stiffness values affecting the results.

Table 8 show the total energy usage per joint in Joules (J) for 1 minute walking at different step frequencies (0.1, 0.5 and 1) that proportionally control resulting walking speed. Stiffness is set to 1 for all experiments. Note that walking at 10% of max speed (0.1) results in about 80% of max speed energy usage.

Table 8. Total Energy Usage per Joint in Joules for various Walking Step Frequencies (0.1, 0.5, 1) while keeping Walking Step Length at 0.08 and Stiffness at 1

Step Frequency	0.1	0.5	1
Left Hip Yaw Pitch (J)	4186	4813	6812
Left Hip Pitch (J)	9613	10294	14365
Left Hip Roll (J)	14735	14462	14011
Left Knee Pitch (J)	12445	12643	15793
Left Ankle Pitch (J)	20004	21782	25147
Left Ankle Roll (J)	3012	2712	3936
Total Left Joints (J)	63995	66706	80065

5   Conclusions and Discussion

We have described in this paper results from our current research in analyzing during open-loop humanoid robot walking the effect of variations in motor stiffness over individual joints in terms of current, battery voltage, and power. During experiments we set the stiffness and walking parameters manually and performed readings on single joints to achieve the best possible sampling rate considering the logic restrictions when reading out data from the NAO. Data readings show relatively good correspondence to walking cycles. The results are very interesting in that power consumption and energy usage can be decreased by dynamically setting specific stiffness values at individual robot joints. The current analysis has been applied only to the NAO V4 robot while we are currently performing similar experiments to NAO V3 and looking to test on other humanoids and walking algorithms. In the current paper we have presented initial results of energy usage optimization. The long-term goal of our research is to better understand the relationship between stiffness control and energy usage depending on the particular humanoid configuration and task. In future work we will be extending the analysis to include other factors that affect energy usage such as battery drainage, joint load, friction and temperature.

Acknowledgements. This work has been funded in part by NSF IIS Robust Intelligence research grant #1117303 at USF entitled "Investigations of the Role of Dorsal versus Ventral Place and Grid Cells during Multi-Scale Spatial Navigation in Rats and Robots." We thank Ralph Fehr at USF for the discussions and suggestions to improve the paper.

References

1. Mei, Y., Lu, Y.-H., Hu, Y.C., Lee, C.S.: A Case Study of Mobile Robot's Energy Consumption and Conservation Techniques. In: Proceedings of 12th International Conference on Advanced Robotics, ICAR 2005, Seattle, WA, July 18-20, pp. 492–497 (2005)
2. Zhang, W., Hu, J.: Low Power Management for Autonomous Mobile Robots Using Optimal Control. In: Proceedings of 46th IEEE Conference on Decision and Control, New Orleans, LA, December 12-14, pp. 5364–5369 (2007)
3. Ogawa, K., Kim, H., Mizukawa, M., Ando, Y.: Development of the Robot Power Management System Adapting to Tasks and Environments-The design guideline of the Power Control System Applied to the Distributed Control Robot. In: Proceedings of the SICE-ICASE International Joint Conference, Busan, Korea, October 18-21, pp. 2042–2042 (2006)
4. Yamasaki, F., Hosoda, K., Asada, M.: An Energy Consumption Based Control for Humanoid Walking. In: Proceedings of IEEE/RSJ International Conference Intelligent Robots and Systems, IROS 2002, Lausanne, Switzerland, September 30-October 4, vol. 3, pp. 2473–2477 (2002)
5. Kormushev, P., Ugurlu, B., Calinon, S., Tsagarakis, N.G., Caldwell, D.G.: Bipedal walking energy minimization by reinforcement learning with evolving policy parameterization. In: Proceedings of IEEE/RSJ International Conference on Intelligent Robots and Systems, IROS 2011, San Francisco, CA, September 25-30, pp. 318–324 (2011)
6. Kulk, J., Welsh, J.S.: A low power walk for the NAO robot. In: Proceedings of the Australian Conference on Robotics and Automation, Canberra, Australia, December 3-5 (2008)
7. Kulk, J., Welsh, J.S.: Autonomous optimisation of joint stiffnesses over the entire gait cycle for the NAO robot. In: Proceedings of the International Symposium on Robotics and Intelligent Sensors, IRIS 2010, Nagoya, Japan, March 6-11 (2010)
8. Kulk, J., Welsh, J.S.: Evaluation of walk optimisation techniques for the NAO robot. In: Proceedings of the 11th IEEE-RAS International Conference onHumanoid Robots, Bled, Slovenia, October 26-28, pp. 306–311 (2011)
9. Gouaillier, D., Hugel, V., Blazevic, P., Kilner, C., Monceaux, J., Lafourcade, P., Mariner, B., Serre, J., Maisonnier, B.: The NAO humanoid: A combination of performance and affordability. In: Proceedings of IEEE International Conference on Robotics and Automation, ICRA 2009, Kobe, Japan, May 12-17, pp. 769–774 (2009)
10. Vukobratovic, M., Borovac, B.: Zero-Moment Point- Thirty five years of its life. International Journal of Humanoid Robotics 1(1), 157–173 (2004)
11. Miura, H., Shimoyama, I.: Dynamic walk of a biped. The International Journal of Robotics Research 3(2), 60–74 (1984)
12. Kajita, S., Kanehiro, F., Kaneko, K., Yokoi, K., Hirukawa, H.: The 3D linear inverted pendulum mode: A simple modeling for a biped walking pattern generation. In: Proceedings of IEEE/RSJ Intelligent Robots and Systems, IROS 2001, Maui, Hawaii, USA, October 29-November 3, vol. 1(4), pp. 239–246 (2001)
13. Liu, X., Yang, R.: The simulated technology research of DC motor load. In: Proceedings of International Conference on Electronics and Optoelectronics, ICEOE 2011, vol. 3, July 29-31, pp. 375–377 (2011)
14. Aldebaran-Robotics, H25 Motor Documentation (2013), http://www.aldebaran-robotics.com/documentation/family/nao_h25/motors_h25.html
15. Movellan, J.: DC Motors (2010), http://mplab.ucsd.edu/tutorials/DC.pdf

A System for Building Semantic Maps
of Indoor Environments
Exploiting the Concept of Building Typology

Matteo Luperto, Alberto Quattrini Li, and Francesco Amigoni

Politecnico di Milano, Milano, Italy
{matteo.luperto,alberto.quattrini,
francesco.amigoni}@polimi.it

Abstract. Semantic mapping of indoor environments refers to the task of building representations of these environments that associate spatial concepts with spatial entities. In particular, semantic labels, like 'rooms' and 'corridors' are associated to portions of an underlying metric map, to allow robots or humans to exploit this additional knowledge. Usually, the classifiers that build semantic maps process data coming from laser range scanners and cameras and do not consider the specific type of the mapped building. However, in architecture it is well known that each building has a specific typology. The concept of *building typology* denotes the set of buildings that have the same function (e.g., being a school building) and that share the same structural features. In this paper, we exploit the concept of building typology to build semantic maps of indoor environments. The proposed system uses only data from laser range scanners and creates a specific classifier for each building typology, showing good classification accuracy.

Keywords: semantic mapping, building typology, line segment maps.

1 Introduction

A *semantic map* of an environment provides human-level knowledge about its structure, for example about the type of the rooms. Automatic building of semantic maps by means of mobile robots has received significant attention in the last years (e.g., [1, 2]). Usually semantic maps are built from metric maps, which represent the physical structure of environments (e.g., the locations of obstacles), and by labeling specific areas (e.g., as 'kitchen' or as 'bathroom'). The knowledge embedded in a semantic map could be exploited in several robotic tasks; for example in a domestic scenario, a semantic map could be used by a robot to find an object or to reason about what actions can be performed in a specific room (e.g., kitchen, bedroom). The labeling is often performed by a classifier that has been previously trained and that is fed with data coming from a metric map and from robot sensors. To the best of our knowledge, no approach attempts to exploit information on the typology of the buildings in classifying places of indoor environments.

In this paper, we contribute in this direction by proposing a semantic mapping system for indoor environments that considers the concept of building typology. The *building*

S. Behnke et al. (Eds.): RoboCup 2013, LNAI 8371, pp. 504–515, 2014.

typology, as studied in architecture [3], denotes a set of buildings that have the same function (e.g., being a school or an office building) and that share the same structural features (e.g., the fact that classrooms are not directly connected to gymnasiums). The system proposed in this paper exploits these common features to improve the classification of places of buildings with specific typologies.

More precisely, in our approach we create and train classifiers specific for each building typology. These classifiers are used for labeling rooms of indoor environments with five semantic labels: 'small room', 'medium room', 'big room', 'corridor', 'hall'. The input data for the classifiers are a number of features that characterize a room and that are extracted from a metric map represented as a set of line segments. Such a metric map could be easily built by mobile robots using laser range scanners.

The main original contribution of this paper is to provide an initial study and evaluation of the intuition of considering building typology for semantic labeling of rooms. In particular, we propose to exploit *a priori* knowledge on a building for performing a sort of "informed" semantic mapping that uses classifiers specific for each building typology. Our experimental results show that these classifiers present a better classification accuracy than a generic classifier that does not refer to any building typology. Moreover, they show performance comparable with that of state-of-the-art classifiers that require much richer input data (for example, coming from cameras) and that use a smaller number of semantic labels.

2 Related Work

Semantic mapping refers to the "process of building a representation of the environment which associates spatial concepts with spatial entities" [4]. A semantic map usually associates semantic labels to places and objects present in an underlying metric map. In this paper, we are interested in labeling portions of a metric map with the corresponding tag, like 'corridor' or 'room'. This problem is known as *place classification* and can be further specialized as place recognition, when the classifier is used in the same environment in which it was trained, and place categorization, when the classifier is used in previously unseen environments [4]. The system presented in this paper performs place categorization.

Approaches to classification of places can be organized according to the sensor modalities they employ, basically laser range scanners and cameras. Since the system we propose in this paper relies on data that are obtained from laser range scanners, in the following we analyze in detail the systems belonging to this class.

The system in [1] extracts a topological map of the environment, which is a graph where every node represents a room and every edge represents a passage connecting two rooms. This method is based on partitioning the environment into a set of open spaces connected by narrow passages. The partitioning is done using a fuzzy grid map of the emptiness of the environment, from which the rooms (assumed to be rectangular) are extracted using a fuzzy mathematical morphological technique.

In [5], a topological map of the environment, where edges have semantic values ('corridor', 'room', 'door'), is used by a behavior-based robot. However, the construction of the map is not illustrated in detail.

The system of [6] classifies a single laser range scan as belonging to a 'room', to a 'corridor', or to a 'hallway'. The classifier is based on AdaBoost and considers both features that are extracted from raw sensor data (e.g., average beam length) and features that are calculated after approximating raw data with a polygon (e.g., area of the approximating polygon). Experimental results show that the system is able to correctly classify scans taken in environments different of those used for learning the classifier. A similar approach, which uses also features extracted from cameras, is presented in [7].

Authors in [8] describe a method for integrating the robot map with a human representation of the environment. This goal is reached using a semantic map based on data obtained from a laser range scanner and using the concept of region and locations, statistically derived from sensor data. Every room is characterized by features, such as its area, its major and minor axes, and its excentricity. These features are used to affix a semantic label ('small', 'medium size', or 'large') to rooms.

In [9], a topological map of the environment is extracted using Voronoi Random Fields (VRF). To build this map, a Voronoi graph is extracted from an occupancy grid map. Each point of the Voronoi graph is a node of a conditional random field, and the resulting VRF estimates the label ('room', 'doorway', 'hallway', 'junction', 'other') of each node using features from both the grid map and the Voronoi graph.

The system proposed in [10] extracts semantic information from 2D and 3D maps of outdoor environments, using Hidden Markov Models (HMM) and Support Vector Machines (SVM) classifiers, and labels terrains as 'traversable' or 'non-traversable' and parts of urban environments as 'street' or as 'sidewalk'.

Basically, all these approaches, which use only data coming from laser range scanners, have been shown to have good performance on the same data set used for training and with a small number of labels (often two or three).

Semantic labeling of places has been also addressed using visual data obtained from cameras. Some works from computer vision make exclusive use of images (e.g., [11]). In robotics, a combination of camera and laser range scanner is usually employed to improve performance of place classification. For example, the system in [2] uses SVMs to combine multiple visual cues and laser range data. Another example is [12], which integrates a line segment-based metric map (built using data from laser range scanners) with higher-level maps that contain information about places and objects. These maps are built on the basis of the metric map, of the perceived images, and of an ontology representing knowledge about indoor environments (e.g., the fact that fridges are usually located in kitchens).

The system we propose in this paper exploits a metric map composed of a set of line segments, which can be easily built from points acquired by laser range scanners. We do not use any visual information. However, we exploit the *a priori* knowledge about the building typology of the indoor environment being mapped. To the best of our knowledge, no other approach attempts to distinguish between different building typologies. Assuming that this background information is available, we show that it can improve place classification.

3   The Proposed System

3.1   Building Typology

A building is an artifact created for a specific function, its purpose, because it is built by people for people who inhabit it. This simple observation is often neglected when designing mobile robots that operate in indoor environments. Indeed, these environments are usually treated as natural, fixed, immutable entities, rather than as a cultural product, the result and summary of centuries of social evolution, as the modern buildings are. As a consequence, robotics designers consider the environments in which the robots move as structured, without fully exploiting the implications of their structure.

The function of a building imposes its structure, its floor plan, and the structure of its rooms. Each building, having a precise function, shares some structural features with all other buildings with the same purpose. A *building typology* is a set of buildings that have the same function [3]. A building typology can be associated to a model that represents the structural features shared by buildings belonging to the typology.

Main features of this model are well known (as cultural facts) to humans, who use them for localization and orientation, as largely studied in architecture [13]. For example, office buildings are structured in the following way: at the ground floor, there are the rooms for public relations and social spaces, and possibly a canteen and conference rooms; at the upper floors, there are office rooms for the back-end activities.

In this paper, we propose a system that attempts to formalize and build the model of a building typology, in order to provide it to a mobile robot. We consider three building typologies: *House* (residential buildings and houses), *School* (school buildings), and *Office* (office buildings and open spaces). These three building typologies have been chosen because they are particularly significant in the real world and because they are usually considered for experimental activities of autonomous mobile robots. Note that, when comparing two school buildings, they are, at an initial sight, different from each other. However, they share a common model that, for example, represents the fact that rooms are typically connected in certain ways. Recognizing this common model helps humans in acting properly in schools they never entered before.

Finally, it is worth noting that the concept of building typology is debated in architecture and is considered part of an analytical approach to architecture that has produced a huge amount of data on structures of buildings [14].

3.2   Construction of the Classifiers

Our system semantically labels portions of a metric map using different classifiers that are trained on data belonging to different building typologies. In this way, by knowing the typology of the building where a robot operates, the correct classifier can be selected. Note that having different classifiers for different building typologies allows to have explicit models of the specific building typologies, which could be used for different applications. In this section, we illustrate how we construct and train the classifiers for place classification.

We represent indoor environments by a line segment-based metric map, built using data from laser range scanners, as shown in the next section. From this metric map, our system derives a topological map, represented as a graph. A node represents a room. Doorways are associated to edges that connect two nodes (rooms). A description of the method for the extraction of the topological map from the metric map, including doorway identification, can be found in the next section. Each room is associated to a *semantic label*. We consider the following set of semantic labels: *'small room'*, *'medium room'*, *'big room'*, *'corridor'*, *'hall'*. They have been chosen to be simple enough to be immediately understandable by humans, but sufficiently descriptive of types of environments operated by robots. The number of labels we use is higher than those typically used in state-of-the-art semantic mapping approaches using only laser range scanners (e.g., [6] uses 'corridor', 'hall', and 'room' and [8] uses 'small', 'medium size', and 'large' room).

To associate a semantic label, each room is described by a vector of features, used by classifiers. The features are chosen to capture (some of) the characteristics of the model underlying a specific building typology. The features can be divided in two groups. The first group of features captures the shape of a room and consists of the area a of the room and the axes ratio $rt = M/m$ of the major axis M and minor axis m. The details on how the latter ratio is calculated are given in the next section. Note that these features are similar to those used in [8] and constitute a subset of those used in [6]. The second group of features represents the structure of the building and the connections of the room with the rest of the environment (in particular, with adjacent rooms), as this is one of the distinguishing characteristics of a building typology. The features of this second group include the number d of doorways present in a room r and the labels of the rooms directly connected to r. More precisely, for each semantic label s (s could be S, M, B, C, H for 'small room', 'medium room', 'big room', 'corridor' and 'hall', respectively), we consider the number of doorways l_s that connect r to a room with semantic label s. If one of the adjacent rooms has not yet a semantic label, a temporary label is associated to that room with a naïve classifier that uses only the features $\langle a, rt, d \rangle$ and that has been manually trained. A room r is hence described by the feature vector $F_r = \langle a, rt, d, l_S, l_M, l_B, l_C, l_H \rangle$.

According to these features, rooms are semantically labeled by using classifiers specifically trained for each building typology. We consider four well-known classification algorithms: rule induction (RI), multi-layer perceptron (MLP), decision tree (DT), and k-nearest neighbor (K-NN) [15], in order to compare their performance. These algorithms have been chosen because they are general and not tailored for the task of classifying places. In this way, we do not introduce any bias in the evaluation of our approach.

All the classifiers are trained using a supervised learning approach starting from four data sets, that have been manually built. Each entry in these data sets consists of a vector of features F_r (describing the structure of a room r) and of a corresponding semantic label. Three data sets refer to specific building typologies: D_H refers to House, D_S refers to School, D_O refers to Office. The idea is that the classifiers trained using one of these data sets, called *typology classifiers*, embed an implicit model of the corresponding building typology. The fourth data set D_{H+S+O} is composed of all the elements in D_H, D_S, and D_O and does not refer to any specific building typology, as commonly assumed in literature. The classifier trained with D_{H+S+O} is called *standard classifier*.

The data sets have been built as follows. We selected the floor plans of dozens of buildings belonging to the typologies House, School, and Office from eleven monographic books used for the design and analysis of buildings in architecture (for example, [16] is one of the books used for the School building typology). In these books, rooms are labeled according to their function and these labels are assigned by the architects who designed the buildings. The labeled floor plans have been digitally represented using a custom CAD-like software program that allows, for each floor plan, to identify rooms and to semantically label them, as shown, for example, in Fig. 1.

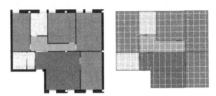

Fig. 1. An example of a floor plan of a house made with AutoCAD (left) and with our CAD-like software (right). Colors indicate semantic labels (green is corridor, yellow is small room, red is big room, brown is medium room).

All the above steps have been manually performed. Then, each labeled floor plan is fed into another software program that automatically extracts the features F_r for each room r. These features F_r and the corresponding label for r are used to create entries to populate the data sets D_H, D_S, and D_O. In the following table, some entries of data set D_H with features and semantic labels are shown.

Label	a	d	rt	l_S	l_M	l_B	l_C	l_H
S	3.0	1	1.0	0	0	0	1	0
B	44.0	2	1.4	1	1	0	0	0
C	13.0	8	3.0	0	1	5	2	0
M	15.0	1	1.2	1	0	0	0	0
C	9.0	6	2.0	0	0	4	1	1

These three building typology-specific data sets are composed of approximately 2,000 rooms (entries) each. Hence, D_{H+S+O} contains around 6,000 rooms.

In summary, we have 16 different classifiers (12 typology classifiers and 4 standard classifiers) obtained from all the possible combinations of algorithms (RI, MLP, DT, and K-NN) and of data sets (D_H, D_S, D_O, and D_{H+S+O}) used for training. These classifiers, given a vector of features as input, return a semantic label.

3.3 Use of the Classifiers

The classifiers are used to associate a semantic label to rooms detected by a robot moving in the environment. In this section, we illustrate how the vector of features for a room to be classified is constructed from sensor data.

We assume to have a mobile robot equipped with two laser range scanners that are mounted back-to-back so that they cover an area of $360°$ around the robot. The point-based scans acquired by the sensors are approximated with line segments using the split-and-merge algorithm [17]. The set of line segments representing a scan is then merged in the metric map, starting from an initial guess provided by odometry, by using a simple scan matching approach similar to that of [18]. The free space already explored can be represented by a polygon (possibly with holes) whose edges are either line segments representing obstacles or line segments representing frontiers, namely the boundaries between the known and the unknown portions of the environment. We use line segments instead of more common occupancy grids to have a more compact representation of the environment.

The metric map built as illustrated above is then used to extract the vectors of features. Initially, we eliminate short line segments (in our experiments, shorter than 30 cm) to remove from the map line segments representing parts of furniture and small obstacles and to keep only the line segments representing the structure of the rooms. Fig. 2 shows the filtered metric map of an environment, where objects and furniture were present, as built by our mapping system. The map preserves the main walls and, thus, the structure of the environment.

Fig. 2. Line segment map built by our system (white area: known free space, gray area: unknown space, black line segments: obstacles, red line: path of the robot)

Then, the filtered metric map is partitioned in rooms in the following way. Walls are identified by finding out collinear line segments. Two line segments are considered collinear if their angular coefficients are equal, up to a threshold value (10 degrees). Doorways are identified as passages in a wall, which connect rooms. In particular, a doorway is found by identifying the gaps between the collinear line segments of a wall or between the endpoints of a wall and another wall. If the gaps have the size of a doorway (e.g., more than 30 cm and less than 150 cm), they are recognized as such. Note that, although more complex situations could be handled, we consider static environments, in which doors can be either open or closed (as many semantic mapping systems do, with few notable exceptions [19]). Portions of area delimited by walls and doorways are identified as rooms using a Monte Carlo method. A random sample of points is thrown within the space. For each point p, the set of line segments $\{L_p\}$ visible (in straight line) from p is calculated. Line segments $\{L_p\}$ are tentatively considered as delimiting the room in which p is placed. The points are checked for straight line visibility between them. Mutually visible points p, p' belong to the same room, and $\{L_p\}$ and $\{L_{p'}\}$ are merged accordingly (see Fig. 3).

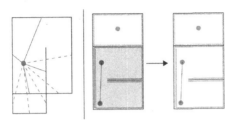

Fig. 3. On the left, line segments directly visible from the point (red lines) and not directly visible from the point (dotted lines). On the right, two mutually visible points in the same room.

A graph representing the topological map is then created by considering each room as a node and connecting nodes through edges that represent doorways. Finally, for each identified room r, the corresponding vector of features F_r is calculated. In particular, the area a is calculated as the average of the area of the inner and outer quadrilaterals that best approximate walls of the room. The major axis M is calculated as the longest wall of the room. The minor axis m is calculated as the longest wall of the room perpendicular to M. The rest of the features, representing the connection graph, are directly derived from the graph. Semantically unlabeled rooms in the graph are temporarily labeled using a naïve rule classifier, as described in the previous section.

Each feature vector F_r becomes an input for the classifiers, which associate a semantic label to the corresponding room r; the result of this last step is a topological map with semantically labeled nodes, namely a semantic map.

4 Experimental Activity

4.1 Experimental Setting

The implementation of our semantic mapping system is divided in two parts. The first one is responsible for the extraction of the features from the metric map. This part has been developed in C++ and tested in environments simulated in USARSim [20] using a Pioneer P3AT robot equipped with two laser range scanners, as described before. Results are satisfactory, but are not shown here because this paper focuses on the classifiers. The second part refers to the training and evaluation of the classifiers. The classifiers have been trained using algorithms implemented in OpenCV [21] and Rapid-Miner [22]. In particular, the training phase and the testing phase of the classifiers are carried out on the same data set, using 10-fold cross-validation. The results of the experimental evaluation of the classifiers are presented in the following. We measured the accuracy of classification comparing typology classifiers with standard classifiers. The accuracy of classification is defined, according to [2, 6, 8], as the percentage of rooms that have been properly classified with the right semantic label.

Our data sets D_H, D_S, D_O, and D_{H+S+O} are composed of simulated data. This allowed us to use data sets with thousands of rooms, an amount of data larger than those typically used in state-of-the-art approaches, usually limited to few buildings. Available real data sets (like those in Radish [23]) do not cover all the building typologies we consider (e.g., houses are largely underrepresented). With our data sets, we preliminarily assess the usefulness of the concept of building typology for semantic mapping by evaluating it under ideal conditions, without measurement and mapping errors.

4.2 Experimental Results

Fig. 4 shows the accuracy of classification of the 12 typology classifiers compared with the classification accuracy of the 4 standard classifiers and Table 1 shows the confusion matrices of each building type. The average (over algorithms) accuracy of the typology classifiers is 87.8% (4.9 is the standard deviation) for building typology House, 88.3% (4.8) for building typology School, and 81.6% (4.6) for building typology Office. The lower performance on the Office building typology with respect to classifiers trained and tested on the other two building typologies may be due to a further division that can be made internally in this category, namely between small offices, located in mixed office/residential buildings, and large open spaces. By contrast, standard classifiers have a lower average accuracy of 73.6% (4.7). These results seem to suggest the effectiveness of using building typology in semantic mapping. The performance of the standard classifiers is rather good because they exploit the several rooms labeled as S and C, whose features (large number of doors and small area) are similar in all the building typologies. Analyzing the algorithms, it can be seen that, for all building typologies, their results are similar, with the RI algorithm performing slightly better than the others over all the typologies.

A statistical analysis of some of the features of the rooms in the data sets provides an explanation and a motivation for the above results. In this regard, mean and standard

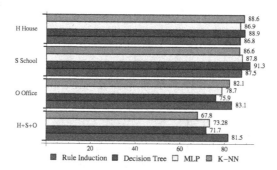

Fig. 4. Accuracy of classification (average and standard deviation) of the typology and standard classifiers

Table 1. Confusion matrices of the typology and the standard classifiers (averaged over the algorithms), where all numbers are in percentage, rows are true labels, columns are predicted labels (C 'corridor'; H 'hall'; S 'small room'; M 'medium room'; B 'big room'), TP are true positives, FP false positives, TN true negatives, and FN false negatives

Label	C	H	S	M	B	TP	FP
H House							
C	92.3	1.2	1.7	3.5	1.3	92.3	7.7
H	3.0	88.0	0.0	0.7	8.2	88.0	12.0
S	3.8	0.3	91.4	4.5	0.0	91.4	8.5
M	3.7	0.4	8.1	83.4	4.3	83.4	16.6
B	1.0	7.1	0.0	5.4	86.5	86.5	13.5
TN	75.7	98.5	96.3	96.0	97.5		
FN	24.2	1.5	3.7	3.9	2.5		

Label	C	H	S	M	B	TP	FP
S School							
C	92.0	3.3	4.2	0.3	0.1	92.0	8.0
H	7.4	80.0	1.3	5.2	6.1	80.0	20.0
S	3.1	0.3	94.0	2.6	0.0	94.0	6.0
M	0.6	3.5	6.5	85.5	3.9	85.5	14.5
B	0.0	21.7	2.9	18.8	56.5	56.5	43.5
TN	97.4	98.2	42.4	97.4	98.8		
FN	2.6	1.8	57.6	2.6	1.2		

Label	C	H	S	M	B	TP	FP
O Office							
C	91.9	3.6	0.2	2.6	1.7	91.9	8.1
H	7.8	80.9	0.0	6.5	4.8	80.9	19.1
S	2.0	0.0	92.3	5.4	0.3	92.3	7.7
M	2.6	1.2	5.6	70.9	19.6	70.9	29.1
B	3.6	5.9	0.1	25.3	65.1	65.1	34.8
TN	79.3	98.1	96.8	92.4	90.1		
FN	20.7	1.9	3.2	7.6	9.9		

Label	C	H	S	M	B	TP	FP
H+S+O							
C	85.7	6.9	2.1	2.8	2.4	85.7	14.2
H	7.9	75.8	1.0	7.1	8.0	75.8	24.1
S	4.0	0.2	75.4	17.6	2.7	75.4	24.6
M	2.3	3.0	11.9	63.3	19.6	63.3	36.7
B	4.6	9.3	3.6	15.8	66.7	66.7	33.3
TN	96.4	96.9	64.7	87.9	91.1		
FN	3.6	3.1	35.3	12.1	8.9		

Table 2. Characteristics of the rooms in the data sets, where % is the percentage of semantic labels (C 'corridor'; H 'hall'; S 'small room'; M 'medium room'; B 'big room') present in the data set and μ and σ are the mean and the standard deviation of the corresponding feature, respectively

		Area a		Doorways d		rt=M/m	
Label	%	μ	σ	μ	σ	μ	σ
H House							
C	23.9	14.0	11.4	4.0	1.8	3.4	3.6
H	8.0	61.5	25.1	3.3	1.5	1.8	0.8
S	28.0	6.0	2.6	1.0	0.2	1.6	0.6
M	26.3	17.4	4.8	1.3	0.4	2.0	1.0
B	13.7	33.7	7.8	1.8	1.0	1.5	0.4

		Area a		Doorways d		rt=M/m	
Label	%	μ	σ	μ	σ	μ	σ
S School							
C	15.4	36.1	42.2	5.7	3.8	7.4	7.6
H	5.6	205.1	172.9	6.9	4.5	2.0	1.3
S	55.6	10.0	9.7	1.1	0.4	1.6	0.9
M	21.3	54.8	14.8	1.6	0.8	1.3	0.4
B	2.0	102.9	22.3	2.1	0.7	1.7	0.7

		Area a		Doorways d		rt=M/m	
Label	%	μ	σ	μ	σ	μ	σ
O Office							
C	20.6	30.5	25.7	6.7	4.8	8.1	7.3
H	6.6	106.9	84.5	6.5	3.6	2.4	1.6
S	23.2	4.5	2.3	1.0	0.1	1.3	0.7
M	34.5	18.5	14.4	1.2	0.3	1.9	0.8
B	14.8	31.3	12.2	1.4	0.9	1.9	1.0

		Area a		Doorways d		rt=M/m	
Label	%	μ	σ	μ	σ	μ	σ
H+S+O							
C	20.3	26.4	27.9	5.7	4.1	6.5	6.8
H	6.8	110.9	109.2	3.7	3.7	2.2	1.3
S	32.3	7.1	6.9	1.0	0.4	1.5	0.8
M	29.3	24.5	18.9	1.3	0.6	1.8	0.8
B	11.5	35.0	18.0	1.6	0.9	1.8	0.9

deviation of some of these features have been calculated. Table 2 shows this analysis. Some of the characteristics of the rooms of the same type are rather constant within a building typology. For example, for houses, all the rooms (except corridors and halls) have a standard deviation of the number of doorways not larger than one. The same happens for other building typologies. Rooms with the same label belonging to buildings of different typologies are consistent. For example, a medium room in a house has an area a, on average, of 17.4 square meters, while a medium room in a school building has an average area a of 54.8 square meters. Moreover, a corridor in a house has an average axes ratio ($rt = M/m$) of 3.4, while a corridor in a school or in an office has a ratio of 7.4 and 8.1, respectively. The distinctive characteristics of each building typology are effectively exploited by the typology classifiers.

To further assess the use of building typology, we classified rooms belonging to school buildings (D_S) using House typology classifiers (M_H) and vice versa (D_H/M_S).

As shown in the following table, the classification accuracy is low. This implies that the models embedded in the typology classifiers are actually specific to the corresponding building typology. Classic semantic mapping approaches, not using building typologies, are implicitly developed for a unique typology (e.g., university campus). Their use for a different typology (e.g., houses) may result in a deterioration of their performance by 50%, similarly to our results.

Algorithm	D_H/M_S	D_S/M_H
RI	40.0(3.2)	50.7(2.5)
MLP	35.7(1.1)	56.1(0.5)
DT	43.0(1.8)	53.0(2.2)
K-NN	41.2(0.4)	51.0(0.3)
average	40.0(3.25)	52.7(2.75)

Finally, we qualitatively compare the results obtained by the proposed system with those of the related literature. The use of the concept of building typology in semantic mapping seems to provide better performance compared to systems that use the same type of sensor (i.e., a laser range scanner) as the only source of information, like [2, 8]. The classification accuracy of these systems is usually in a range from 40% and 80%, and using a smaller number of semantic labels for place classification. The classification accuracy we achieve is similar to that of systems that use a camera coupled with a laser range scanner (e.g., [2, 7]). This fact can be explained by saying that knowing *a priori* the building typology (namely, by using the right typology classifier) turns out to be as significant as the use of additional sensors like the camera, but without introducing the computational burden arising from their use. In a sense, the use of information on the structure of the building enriches the data obtained by the laser range scanner.

5 Conclusions

In this paper we have presented a semantic mapping system that classifies rooms of indoor environments considering typology of buildings where a robot is operating. More precisely, assuming that a robot is moving in a building with a known typology, the proposed system employs classifiers specific for that typology to semantically label rooms identified from data acquired by laser range scanners. Experimental results show that considering building typology is effective for place classification. In particular, availability of *a priori* information about the building typology of an environment seems to have an impact similar to that of using a camera on improving classification accuracy.

Future work will address the further assessment of the proposed system, by considering other building typologies and larger data sets (also acquired by real robots). This will allow to evaluate effects of perception and mapping noise on the classifiers. Moreover, we will investigate the system performance when *a priori* knowledge is not provided, but the system itself has to infer the building typology according to the best classifier. Finally, possible uses of the system will be investigated. For example, semantic maps built by our system could be used to predict labels of rooms not yet discovered during an exploration task or to validate realism of synthetic environments used in robotic simulations, like those used in the RoboCup Rescue Simulation League.

References

1. Buschka, P., Saffiotti, A.: A virtual sensor for room detection. In: Proc. IROS, pp. 637–642 (2002)
2. Pronobis, A., Mozos, O., Caputo, B., Jensfelt, P.: Multi-modal semantic place classification. Int. J. Robot. Res. 29(2-3), 298–320 (2010)
3. Rossi, A.: The architecture of the city. MIT Press (1984)
4. Pronobis, A.: Semantic mapping with mobile robots. PhD thesis. KTH (2011)
5. Althaus, P., Christensen, H.: Behavior coordination in structured environments. Adv. Robotics 17(7), 657–674 (2003)
6. Mozos, O., Stachniss, C., Burgard, W.: Supervised learning of places from range data using AdaBoost. In: Proc. ICRA, pp. 1730–1735 (2005)
7. Mozos, O., Triebel, R., Jensfelt, P., Rottmann, A., Burgard, W.: Supervised semantic labeling of places using information extracted from sensor data. Robot. Auton. Syst. 55(5), 391–402 (2007)
8. Topp, E., Christensen, H.: Topological modelling for human augmented mapping. In: Proc. IROS, pp. 2257–2263 (2006)
9. Friedman, S., Pasula, H., Fox, D.: Voronoi random fields: Extracting the topological structure of indoor environments via place labeling. In: Proc. IJCAI, pp. 2109–2114 (2007)
10. Wolf, D., Sukhatme, G.: Semantic mapping using mobile robots. IEEE T. Robot. 24(2), 245–258 (2008)
11. Quattoni, A., Torralba, A.: Recognizing indoor scenes. In: Proc. CVPR, pp. 413–420 (2009)
12. Zender, H., Mozos, O., Jensfelt, P., Kruijff, G., Burgard, W.: Conceptual spatial representations for indoor mobile robots. Robot. Auton. Syst. 56(6), 493–502 (2008)
13. Lynch, K.: The image of the city. MIT Press (1960)
14. Neufert, E., Neufert, P.: Architects' data. Wiley-Blackwell (2012)
15. Bishop, C.: Pattern recognition and machine learning. Springer (2006)
16. Perkins, B.: Building type basics for elementary and secondary schools. Wiley (2001)
17. Nguyen, V., Gächter, S., Martinelli, A., Tomatis, N., Siegwart, R.: A comparison of line extraction algorithms using 2d range data for indoor mobile robotics. Auton. Robot. 23(2), 97–111 (2007)
18. Lu, F., Milios, E.: Robot pose estimation in unknown environments by matching 2d range scans. J. Intell. Robot. Syst. 18(3), 249–275 (1997)
19. Anguelov, D., Koller, D., Parker, E., Thrun, S.: Detecting and modeling doors with mobile robots. In: Proc. ICRA (2004)
20. Carpin, S., Lewis, M., Wang, J., Balakirsky, S., Scrapper, C.: USARSim: A robot simulator for research and education. In: Proc. ICRA, pp. 1400–1405 (2007)
21. OpenCV: OpenCV (2012), http://opencv.org/
22. RapidMiner: RapidMiner (2012), http://www.rapidminer.com/
23. Howard, A., Roy, N.: The robotics data set repository (Radish) (2003), http://radish.sourceforge.net/

Semantic Object Search Using Semantic Categories and Spatial Relations between Objects

Patricio Loncomilla, Marcelo Saavedra, and Javier Ruiz-del-Solar

Advanced Mining Technology Center & Dept. of Elect. Eng., Universidad de Chile
{ploncomi,jruizd}@ing.uchile.cl

Abstract. In this work, a novel methodology for robots executing informed object search is proposed. It uses basic spatial relations, which are represented by simple-shaped probability distributions describing the spatial relations between objects in space. Complex spatial relations can be defined as weighted sums of basic spatial relations using co-occurrence matrices as weights. Spatial relation masks are an alternative representation defined by sampling spatial relation distributions over a grid. A Bayesian framework for informed object search using convolutions between observation likelihoods and spatial relation masks is also provided. A set of spatial relation masks for the objects "monitor", "keyboard", "system unit" and "router" were estimated by using images from Label-Me and Flickr. A total of 4,320 experiments comparing six object search algorithms were realized by using the simulator *Player/Stage*. Results show that the use of the proposed methodology has a detection rate of 73.9% that is more than the double of the detection rate of previous informed object search methods.

Keywords: Semantic search, Informed search, Co-occurrence matrix.

1   Introduction

Object search is an important ability for a mobile robot. Some previous work on object search are based on the use of spatial object-place relations, assuming that in a scene the searched object will be readily available to the robot's field of vision. However, there are sets of objects inside a setting that tend to appear near of each other as they have a particular spatial relation, making it possible to infer the existence of an object A, given an object B.

For example, when looking inside offices, it can be noted that the object "keyboard" is often very near the object "monitor". For a human, the ability to deduce that object A tends to be "near", "very near" or "far" from object B is a trivial task, since human beings can learn spatial relations between objects by observing a large number of similar settings. In this paper, we focus on giving robots the ability to find objects using existing semantic relations with other objects such as "near" or "far," in a given environment.

We conducted an exhaustive analysis of spatial relations between real-world objects that share a common use. For this objective, we created a database of spatial co-occurrences where distance values between objects are represented by the linguistic

S. Behnke et al. (Eds.): RoboCup 2013, LNAI 8371, pp. 516–527, 2014.
© Springer-Verlag Berlin Heidelberg 2014

variables "very near," "near," "far," and "very far". We say that two objects "co-occur" when they appear together in several images showing a particular spatial relation.

The main contribution of this work is the use of convolutions for computing the probability of the presence of an object from positive and negative detections of all the objects on the map in a unified way. There are two secondary contributions. The first one is the representation of spatial relations as spatial relation masks, and the use of basic semantic categories such as "very near," "near," "far," and "very far" for generating basic spatial relation masks that can be combined to generate complex spatial relation masks by using a set of weights named co-occurrence values. The second contribution is the creation of a methodology for computing a co-occurrence matrix associated with a set of basic relation masks from a database of labeled images containing real world objects.

This paper is organized as follows. In Section 2 some related work is presented. In Section 3 the proposed methodology for object search is described. In Section 4 an experimental validation of the methodology is presented. Finally, in Section 5 some conclusions of this work are given.

2   Related Work

The search for objects in a real environment is a very complex task for robots. Garvey in [1] recognized this problem and proposed the idea of indirect search, i.e. searching for another intermediate object that maintains a particular spatial relation with the object being searched for. Wixson et al. materialize the idea of indirect search, demonstrating greater efficiency both theoretically and empirically [2]. But the problem with indirect search is that the spatial relationship between the object sought and the intermediate object does not always exist. Furthermore, the detection of the intermediate object may be more difficult than the detection of the desired/primary object itself. In fact, Ye and Tsotsos demonstrated that the search for an arbitrary object in 3D space is NP-complete [3]. Shubina and Tsotsos propose an algorithm that considers the cost and effect of different actions with different types of prior knowledge and different spatial relations between objects [4]. Kollar et al. perform the search for an object in a known map of the environment, using object-object and object-scene context [5]. They obtain the co-occurrence of the existence of objects in two-dimensional images for learning correlations between object categories, and between objects and place labels (semantic labels such as "kitchen"). These images were taken from the Flickr website, and they do not take the distances between objects into consideration. Viswanathan et al. propose an approach using existing resources: common-sense knowledge of machine learning of object relations [6]. They use marked images from the LabelMe database, designed by Russell et al. [7]. They train an automatic classifier of places based on the presence of the detected objects to infer the probability that the other objects exist, and the kind of place (e.g., kitchen or office) is seen in the setting. Kasper et al. perform a study on spatial relations in three-dimensional images using a Kinect sensor [8]. They created a database using nine different office space settings with a total of 168 objects in 35 object classes. Then, they found the distances between different objects. They also made predictions about the location of unfound objects by detecting their surrounding objects.

Galindo et al. performed a study based on 2D data, combining metric, topological, and semantic aspects on a map [9]. In addition, they proposed a method for learning these semantic representations from sensory data. Vasudevan et al. attempted to create a spatial representation in terms of objects, by encoding typical household objects and doors within a hierarchical probabilistic framework [10]. They used a SIFT [11] based object recognition system and a door detection system based on lines extracted from range scans. They also proposed a conceptualization of different places, based on the objects that were observed inside them.

Aydemir et al. developed a method for object search using explicit spatial relationships between objects in order to perform an efficient visual search [12]. They presented a computational model using several random views to guide the robot's camera to the points where the objects have a high probability of being found by using the spatial-relation term "on" between objects, in an indoor environment since the objects are mostly on horizontal surfaces. The work presented in this paper is an extension and improvement of the models proposed in [8] and [12].

3   Methodology

3.1   Map Update Using Spatial Relation Masks

The proposed methodology is designed for finding an object using informed search, i.e., by using information about other objects, which have a spatial relation with the object to be found. In [12] spatial relations are defined as functions from a space of a pair of poses π_A, π_B of two objects to the interval [0,1], where 1 indicates that the relation is completely fulfilled by these pose combinations, and 0 that the relation does not apply at all:

$$Rel_{A,B} : \{\pi_A, \pi_B\} \rightarrow [0,1] \tag{1}$$

In this work spatial relations will be defined in a probabilistic sense. A spatial relation between two objects is defined as the probability distribution of the pose of the first object given the known pose of the second object:

$$Rel_{A,B}(\pi_A, \pi_B) = p(\pi_A \mid \pi_B) \tag{2}$$

As the relation is treated as a probability distribution, the sum over all possible poses of A for a fixed pose of B is equal to one:

$$\int_{\pi_A} p(\pi_A \mid \pi_B) = 1 \tag{3}$$

If the spatial relation is invariant to translations and rotations, i.e. it only depends on the relative pose $\pi_{A/B}$ of object A with respect to object B, then the expression for the probability can be rewritten as:

$$Rel_{A/B}(\pi_{A/B}) = p(\pi_{A/B}) \tag{4}$$

In our case, the robot is in a two-dimensional space parameterized by using coordinates (x,y). The space is quantized into squared cells with size k, and parameterized by using indexes (i,j). By using the index notation, a spatial relation can be written as:

$$R_{A/B}(i,j) = K_{norm} * Rel_{A/B}(ki,kj) \tag{5}$$

$$\sum_i \sum_j R_{A/B}(i,j) = 1 \tag{6}$$

where K_{norm} is a normalizing constant.

The term $p(a_{i,j})$ represents the probability that the center of the main object A is in the cell (i,j). Both positive and negative observations z_A provide valuable information for the object search process, and can be used to compute an updated probability $p(a_{i,j}|z_A)$. The probability $p(a_0)$ is treated as a special case, and it represents the probability of the object being outside the search region.

Positive detections $z_A=true$ provide information about the places where the object has high probability of being, by means of a likelihood $p(z_A=true|a_{i,j})$, which is defined over the cell (i,j). The likelihood has a high value over the cell where the object was detected, and a low value in the other cells. Negative detections $z_A=false$ provide information $p(z_A=false|a_{i,j})$ about the cells where the objects have low probability of being, which are those cells visible from the current viewpoint that have a low probability of containing the object.

The problem addressed in this work is to find a main object, A, by moving the robot appropriately. The robot search process is applied until the main object, A, is found. In consequence, the search process includes only negative detections of the main object, A, before the object is found. Two cases are considered:

$$p(a_{i,j} \mid z_A = false) = \frac{p(z_A = false \mid a_{i,j})p(a_{i,j})}{p(a_o) + \sum_{i,j} p(z_A = false \mid a_{i,j})p(a_{i,j})} \tag{7}$$

$$p(a_0 \mid z_A = false) = \frac{p(a_0)}{p(a_o) + \sum_{i,j} p(z_A = false \mid a_{i,j})p(a_{i,j})} \tag{8}$$

The secondary object, B, can produce positive and negative detections z_B, which can be used to compute an updated probability:

$$p(a_{i,j} \mid z_B) = \frac{p(z_B \mid a_{i,j})p(a_{i,j})}{p(z_B \mid a_o)p(a_0) + \sum_{i,j} p(z_B \mid a_{i,j})P(a_{i,j})} \tag{9}$$

$$p(a_0 \mid z_B) = \frac{p(z_B \mid a_o)p(a_0)}{p(z_B \mid a_o)p(a_0) + \sum_{i,j} p(z_B \mid a_{i,j})P(a_{i,j})} \tag{10}$$

The terms $p(z_B|a_{i,j})$ and $p(z_B|a_0)$ will be called cross-likelihoods, as they relate the detection of a secondary object, B, with the presence of the main object, A, on the map.

These probabilities can be derived by considering probabilities $p(b_{u,v})$ for the presence of a secondary object, B, at locations (u,v) in the grid:

$$p(z_B \mid a_{i,j}) = \sum_u \sum_v p(z_B \mid b_{u,v}) p(b_{u,v} \mid a_{i,j}) \tag{11}$$

$$p(z_B \mid a_0) = \sum_u \sum_v p(z_B \mid b_{u,v}) p(b_{u,v} \mid a_0) \tag{12}$$

The term $p(b_{uv}\mid a_0)$ is considered a constant over (u,v) whose sum has a value of 1 because B is supposed to be on the map. The term $p(b_{u,v}\mid a_{i,j})$ corresponds to the spatial relation between the main object, A, at location (i, j) and a secondary object, B, at location (u, v). By replacing this term with the spatial relation $R_{B/A}$, there is no need for storing a map for the secondary object; only the map for the main object and the likelihoods of the detections of the secondary object are needed:

$$p(z_B \mid a_{i,j}) = \sum_u \sum_v p(z_B \mid b_{u,v}) R_{B/A}(u-i, j-v) \tag{13}$$

$$p(z_B \mid a_0) = \frac{1}{n_U n_V} \sum_u \sum_v p(z_B \mid b_{u,v}) \tag{14}$$

where $n_U n_V$ is the size of the map.

Equation (13) can be implemented as a convolution in the (i,j) space between a likelihood image and a mask $R_{B/A}(i,j)$ describing the spatial relation between the main and secondary objects, which will be named a *spatial relation mask*:

$$p(z_B \mid a_{i,j}) = p(z_B \mid b_{i,j}) * R_{B/A}(i, j) \tag{15}$$

The proposed system is highly versatile because any spatial relation can be represented by an appropriate mask. It must be noted that extra secondary objects can be added to the system by creating additional spatial relation masks. In case these relations are chained, as an example object A is near B, and object B is near C, then the mask of the chained relation can be obtained by convolution of the original masks:

$$P(z_C \mid a_{i,j}) = P(z_C \mid b_{i,j}) * R_{B/A}(i, j) \tag{16}$$

$$P(z_C \mid a_{i,j}) = P(z_C \mid c_{i,j}) * R_{C/B}(i, j) * R_{B/A}(i, j) \tag{17}$$

$$\Rightarrow R_{C/A} = R_{C/B} * R_{B/A} \tag{18}$$

A path for searching for the object can be created by generating optimal viewpoints at each iteration. The optimal viewpoint is generated from a set of k random poses reachable in a fixed time, and selecting the one that maximizes the probability of finding the object in the visible area, as shown in [8]:

$$\arg\max_{k=1..N} \sum_{i=1}^{n} \sum_{j=1}^{n} p(a_{i,j}) V(a_{i,j}, k) \tag{19}$$

where N is the number of candidate poses, and $V(a_{i,j},k)$ is defined as:

$$V(a_{i,j},k) = \begin{cases} 1, & \text{if } a_{i,j} \text{ is inside the } k^{th} \text{ view cone} \\ 0 & \text{otherwise} \end{cases} \qquad (20)$$

The navigation algorithm has an impact on the reliability of the object search algorithm. For generating random poses, random sets of parameters for the navigation algorithm must be selected. Simple navigation strategies like going forward and rotating have an easily computable navigation time for a given final pose, and more complex navigation algorithms are not easily parameterized, then there is a trade-off between simplicity of the navigation algorithm (which allows simple and accurate generation of trajectories), and complex navigation approaches (which enable the robot to reach better poses in a given time). In the experiments reported in Section 4 we choose to use a simple navigation strategy, which every time selects the execution of a composed sequence of movements consisting on one initial rotation, followed by a translation and a final rotation. The parameter space associated to this sequence of movements is explored randomly, and the best sequence is chosen as the one that maximize the probability of finding the searched object.

For computing the convolutions, a way of generating appropriate masks from examples of real-world images is needed. This will be explained in the following section.

3.2   Creating Spatial Relation Masks from Co-occurrences

In this section, a procedure for approximating a complex spatial relation as a weighted sum of basic spatial relations is presented. Each basic spatial relation corresponds to a semantic category meaningful to humans. In this work, we focus on four simple spatial relations: "very near" (VN), "near" (N), "far" (F), and "very far" (VF). The use of these spatial relations is useful as it enables the system to estimate a set of basic probability distributions from samples of relative positions of the objects in the real world. The masks for each of the spatial relations are defined in two versions, *hard masks* and *soft masks*. Hard masks are defined by two thresholds and have a rectangular profile, while soft masks are defined by four numbers and have a trapezoid-shaped profile. Each basic mask is normalized to sum one over all of the cells on the map, thus a normalization constant is added to the formulas

Equations for hard masks are defined in equations (21) to (25).

$$R_{B/Ahard}(i,j;a_1,a_2) = K * \begin{cases} 1 & a_1 \leq \sqrt{(ki)^2 + (kj)^2} < a_2 \\ 0 & other \end{cases} \qquad (21)$$

$$R_{B/Ahard}^{VN}(i,j) = R_{B/Ahard}(i,j;0,u_1) \qquad (22)$$

$$R_{B/Ahard}^{N}(i,j) = R_{B/Ahard}(i,j;u_1,u_2) \qquad (23)$$

$$R_{B/Ahard}^{F}(i,j) = R_{B/Ahard}(i,j;u_2,u_3) \qquad (24)$$

$$R_{B/Ahard}^{VF}(i,j) = R_{B/Ahard}(i,j;u_3,\infty) \qquad (25)$$

Equations for soft masks are similar, but have a smooth transition between values 0 and 1, that is regulated by a gap parameter δ. The equations for soft masks are shown in equations (26) to (30).

$$R_{B/A soft}(i,j;a_1,a_2,a_3,a_4) = K * \begin{cases} \dfrac{\sqrt{(ki)^2+(kj)^2}-a_1}{a_2-a_1} & a_1 \le \sqrt{(ki)^2+(kj)^2} < a_2 \\ 1 & a_2 \le \sqrt{(ki)^2+(kj)^2} < a_3 \\ \dfrac{a_4-\sqrt{(ki)^2+(kj)^2}}{a_4-a_3} & a_3 \le \sqrt{(ki)^2+(kj)^2} < a_4 \\ 0 & other \end{cases} \tag{26}$$

$$R_{B/A soft}^{VN}(i,j) = R_{B/A soft}(i,j;0,0,u_1-\delta,u_1+\delta) \tag{27}$$

$$R_{B/A soft}^{N}(i,j) = R_{B/A soft}(i,j;u_1-\delta,u_1+\delta,u_2-\delta,u_2+\delta) \tag{28}$$

$$R_{B/A soft}^{F}(i,j) = R_{B/A soft}(i,j;u_2-\delta,u_2+\delta,u_3-\delta,u_3+\delta) \tag{29}$$

$$R_{B/A hard}^{VF}(i,j) = R_{B/A hard}(i,j;u_3-\delta,u_3+\delta,\infty,\infty) \tag{30}$$

In this work, basic masks defined by a circle of radius u_1 in the case of "very near", a circular ring of radii u_1 and u_2 in the case of "near," a circular ring of radii u_2 and u_3 in the case of "far," and a circular ring of internal radius u_3 and an external radius that cover the whole map in the case of "very far". The radius values are selected by considering statistics of the distances between objects A and B, and by modeling their selection process as a classification problem. Thus, the optimal radius value between two categories, e.g., "near" and "far", is the one that generates the same mean classification error in both classes.

A complex mask can be created as a weighted sum of basic hard or soft masks:

$$R_{B/A}(x,y) = C_{B/A}^{VN}R_{B/A}^{VN}(x,y) + C_{B/A}^{N}R_{B/A}^{N}(x,y) + \\ + C_{B/A}^{F}R_{B/A}^{F}(x,y) + C_{B/A}^{VF}R_{B/A}^{VF}(x,y) \tag{31}$$

An example of a complex mask sampled over a grid with pixel size 0.1[m] is shown in Figure 1.

The four coefficients $C_{B/A}^{VN}$, $C_{B/A}^{N}$, $C_{B/A}^{F}$ and $C_{B/A}^{VF}$ are called co-occurrences because they indicate the relative frequency of occurrence of a pair of objects for each spatial relation. They can be constructed from samples of positions of both objects by computing the number of occurrences of each basic spatial relation. If a set of samples is divided into basic semantic categories and the count is n_{VN} for "very near," n_N for "near," n_F for "far," and n_{VF} for "very far," the co-occurrences can be computed by using equations (32) to (35):

$$C_{B/A}^{VN} = \frac{n_{VN}}{n_{VN}+n_N+n_F+n_{VF}} \tag{32}$$

$$C_{B/A}^{N} = \frac{n_N}{n_{VN}+n_N+n_F+n_{VF}} \tag{33}$$

$$C_{B/A}^{F} = \frac{n_F}{n_{VN} + n_N + n_F + n_{VF}} \qquad (34)$$

$$C_{B/A}^{VF} = \frac{n_{VF}}{n_{VN} + n_N + n_F + n_{VF}} \qquad (35)$$

A co-occurrence matrix is the set of co-occurrence values of two or more objects with a particular spatial relation.

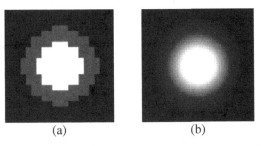

(a)   (b)

Fig. 1. (a) Hard complex mask composed by summing hard masks "very near," "near," and "far," sampled over a grid. (b) The equivalent soft complex mask (see main text for details).

4 Results

4.1 Experimental Setup

In order to characterize the proposed object search methodology and to compare it with existing methodologies, a simulation environment that uses data of real-world objects was developed. We believe that a simulation environment is an appropriate tool to evaluate object search algorithms, because it allows obtaining repeatable experiments and to better quantify the performance of the different methodologies.

Experiments were performed on maps containing four objects, by using the *Player/Stage*, a robot simulation tool [13]. Each map contains a main object, named A (monitor), to be searched for, and three secondary objects named B (keyboard), C (system unit), and D (router). The size of each map is 6[mt] x 6[mt]. A total of 20 maps were created by picking a random position for the main object, A, and then picking a random position for the objects B, C, and D following a distribution that represents the co-occurrences shown in Table 1. The 20 maps are used to perform the experiments, each map being used the same number of times as the others. A laser sensor from *Player/Stage* is used for avoiding collisions. Observations of the objects are obtained by using a fiducial sensor included in Player/Stage that is able to measure the position of a detected object inside a view cone whose size depends on the object: the main object, A, can be detected up to 1[mt], and the secondary objects, B, C, and D can be detected up to 2[mt].

In each experiment the goal is to find the main object A before 1,500 views have been processed. In the search process, only negative detections of the main object need to be processed because a positive detection causes the search process to finish.

Six algorithms of object search are compared on the same set of maps. The first 4 correspond to different variants of the proposed methodology, the fifth algorithm is the one proposed by Aydemir in [12], and the sixth one corresponds to the baseline algorithm where no information from secondary objects is used. The algorithms are:

1. *Informed search with convolutions using positive and negative information with hard masks*: A probability map $P(a_{i,j})$ for the main object A is estimated by using positive and negative detections of objects B, C, and D, negative detections of object A, and spatial relation masks. Then object A is searched by finding viewpoints that maximize the probability of containing it. The relation masks $R_{A/B}$, $R_{A/C}$ and $R_{A/D}$ needed for updating the probability map $P(a_{i,j})$ from detections of secondary objects are constructed by using co-occurrence matrices.

2. *Informed search with convolutions using only positive information with hard masks*: Similar to algorithm 1, but in this case only positive detections of objects B, C, and D are used.

3. *Informed search with convolutions using positive and negative information with soft masks*: Similar to algorithm 1, but in this case soft spatial relation masks are used.

4. *Informed search with convolutions using only positive information with soft masks*: Similar to algorithm 2, but in this case soft spatial relation masks are used.

5. *Informed search using particles*: The algorithm of Aydemir [12] is used for constructing a probability map $P(a_{i,j})$ for object A by using negative detections of that object. Then object A is searched for by finding viewpoints that maximize the probability of containing A. When a secondary object is detected, a set of particles is generated around the detection inside the current view cone, and the ones which fulfill the spatial relation computed from Table 1 are used to select the next optimal viewpoint. A spatial relation is considered fulfilled when its current value is equal or greater than half of its maximum possible value.

6. *Uninformed search*: A probability map $P(a_{i,j})$ for object A is estimated by using negative detections of that object, then object A is searched for by finding viewpoints that maximize the probability of containing A. This is the baseline algorithm used in [12], and no information from secondary objects is used.

4.2   Creation of Co-occurrence Matrices

A set of 243 images from LabelMe [7] and captions of photos on the Flickr website were used for generating co-occurrence matrices. In each image, instances of the objects "monitor," "system unit," "keyboard," and "router" were labeled. As the sizes of the objects and the parameters of the camera are known, it is possible to compute the pose of each object in space. Several instances of the objects on the set of images and their poses were used to construct co-occurrence matrices for the categories "very near," "near," "far," and "very far" for each of the objects with respect to the others.

Given the poses of a pair of objects, a distance was computed and used for selecting whether the sample belongs to the categories "very near," "near," or "far". If an object is detected alone in an image, the sample belongs to the category "very far". Only the depth and horizontal axis were used to compute the distances, as differences in the vertical direction do not affect the position of the object when it is transformed onto the 2D grid. The statistics of the distances between the object "monitor" and the objects "keyboard," "system unit," and "router," as well as the delimitation between the basic spatial relations are shown in Figure 2. The final co-occurrences for the object, *Monitor,* as the main object are shown in Table 1.

The basic hard and soft masks are defined by equations (21)-(25) and (26)-(30), respectively. The thresholds that separate the categories very near, near, far and very far are $u_1=60$[cm], $u_2=100$[cm] and $u_3=150$[cm]. The gap parameter for the soft masks is $d=20$[cm].

Table 1. Final co-occurrences of objects around the object "monitor"

Main object Monitor	Secondary objects		
Semantic categories	*keyboard*	*system unit*	*router*
Very Near	0.773	0.178	0.061
Near	0.143	0.491	0.151
Far	0.046	0.258	0.485
Very Far	0.038	0.074	0.303

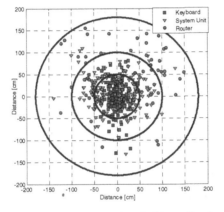

Fig. 2. Relative poses between object "monitor" and objects "keyboard," "system unit," and "router". Thresholds for basic spatial relations are shown as circles.

4.3   Experiments

A total of 4,320 experiment trials were executed for comparing the six search algorithms explained in Section 4.1. For each algorithm, a total of 720 experiments considering different maps and different initial robot poses were executed. Each experiment is considered successful if the object is found before 1,500 sensor frames. This considers that all algorithms are able to run at 15 fps, which is the frame rate used in the simulations. The results from the experiments are shown in Table 2.

From the results of the experiments, it is evident that the algorithms that use convolutions for integrating information about secondary objects into the probability distribution of the main object have the best performance. This happens because the integrated information can be used several frames after the moment when the object was seen. The integration of positive detections of secondary objects using hard masks improves the detection rate from 28.9% up to 49.6%. The use of soft masks is also a factor that improves the detection rate up to a 53.2%. Finally, the use of positive and negative information, in addition to soft masks, generates an important improvement in the detection rate that rises up to a 73.9%. The integration of information using masks, the comparison between different kind of masks and the ability of using negative information about secondary objects are all contributions of this work. Aydemir's particle based informed search algorithm [12] performs better than the baseline; however, the best algorithm described in this paper has a detection rate that is more than the double of the detection rate of Aydemir's algorithm.

The methods can be optimized by observing that the cross-likelihood images are constant except on the detection area in the case of positive detections, and on the view cone in the case of negative detections. Then, focalized convolutions can be applied on these areas. Informed search using positive and negative information scales linearly with the number of total secondary objects. Informed search using only positive information scales linearly with the number of observed secondary objects. As the mean amount of detected objects is low, the methods that use only positive information run as fast as uninformed search the most of the time. Both kind of methods scale quadratically with the size of the mask when using focalized convolutions, and they scale linearly with the size of the mask when using convolutions with the full map. Time of focalized convolutions do not depend on the size of the map. In the experiments, a small map (100x100) and mask (32x32) were used for representing a 6m x 6m environment and the processing time of informed search is lower than the processing time of object detection algorithm.

Table 2. Results from the experiments comparing six variants of object search algorithms. In each variant, a total of 720 experiments were performed. Each experiment is successful if the main object is found before 1,500 sensor frames. DR: Detection Rate.

Algorithm	Number of searches	Successful searches	DR
Informed search using positive and negative information and hard masks	720	496	68.9%
Informed search using only positive information and hard masks	720	357	49.6%
Informed search using positive and negative information and soft masks	720	532	73.9%
Informed search using only positive information and soft masks	720	383	53.2%
Informed search using particles(Aydemir et al. [12])	720	248	34.4%
Uninformed search	720	208	28.9%

5   Conclusions

In this work, a novel methodology for performing informed search of objects was proposed and tested. The methodology is based on integrating information provided by secondary objects into the probability distribution of the main object to be found. Spatial relations between objects are estimated by using a set of basic spatial relations which are mixed by using co-occurrence values as weights. Six algorithms of object search were compared by using spatial relations estimated from real-world data by performing a total of 4,320 simulations in *player/stage*. The results show that the detection rate of the search process increases from 28.9% to 73.9% when integrating positive and negative detections from the secondary objects into the probability distribution of the main object using soft masks. The integration of positive and negative detections of secondary objects, as well as the use of soft masks increases the detection rate. The obtained detection rate is more than the double of the one obtained by previous informed search algorithms.

Future work includes the creation of a full 3D model of the informed search system, the management of false detections, the comparison with extra object search methods, and the realization of experiments with a real robot for validating the results in the real world.

Acknowledgments. This work was partially funded by FONDECYT under Project Number 1130153.

References

[1] Garvey, T.D.: Perceptual strategies for purposive vision, Technical report, SRI International, vol. 117 (1976)

[2] Wixson, L., Ballard, D.: Using intermediate object to improve efficiency of visual search. Int. J. Comput. Vis. 18(3), 209–230 (1994)

[3] Ye, Y., Tsotsos, J.K.: Sensor Planning for 3D Object Search. Computer Vision and Image Understanding 73-2, 145–168 (1999)

[4] Shubina, K., Tsotsos, J.: Visual search for an object in a 3d environment using a mobile robot. Computer Vision and Image Understanding 114(5), 535–547 (2010)

[5] Kollar, T., Roy, N.: Utilizing object-object and object-scene context when planning to find things. In: Proc. of the 2009 IEEE Int. Conf. on Robotics and Automation, ICRA 2009 (2009)

[6] Viswanathan, P., Meger, D., Southey, T., Little, J.J., Mackworth, A.: Automated Spatial-Semantic Modeling with Applications to Place Labeling and Informed Search. In: Proc. Canadian Conf. on Computer and Robot Vision (2009)

[7] Russell, B., Torralba, A., Murphy, K., Freeman, W.: Labelme: A database and web-based tool for image annotation. International Journal of Computer Vision 77, 157–173 (2008)

[8] Kasper, A., Jäkel, R., Dillmann, R.: Using spatial relations of objects in real world scenes for scene structuring and scene understanding. In: Proc.15th Int. Conf. on Advanced Robotics, ICAR 2011, Tallinn, Estonia (2011)

[9] Galindo, C., Saffiotti, A., Coradeschi, S., Buschka, P., Fernandez-Madrigal, J.A., Gonzalez, J.: Multi-hierarchical semantic maps for mobile robotics. In: Proc. 2005 IEEE/RSJ Int. Conf. on Intelligent Robots and Systems, IROS 2005, pp. 2278–2283 (2005)

[10] Vasudevan, S., Gachter, S., Nguyen, V., Siegwart, R.: Cognitive maps for mobile robots-an object based approach. Robot. Auton. Syst. 55, 359–371 (2007)

[11] Lowe, D.G.: Distinctive Image Features from Scale-Invariant Keypoints. Int. Journal of Computer Vision 60, 91–110 (2004)

[12] Aydemir, A., Sjöö, K., Jensfelt, P.: Object search on a mobile robot using relational spatial information. In: Proc. 11th Int Conf. on Intelligent Autonomous Systems, IAS 2011 (2010)

[13] The Player Project, http://playerstage.sourceforge.net/

HELIOS Base: An Open Source Package for the RoboCup Soccer 2D Simulation

Hidehisa Akiyama[1] and Tomoharu Nakashima[2]

[1] Faculty of Engineering, Fukuoka University, Japan
akym@fukuoka-u.ac.jp
[2] Department of Computer Science and Intelligent Systems,
Osaka Prefecture University, Japan
tomoharu.nakashima@kis.osakafu-u.ac.jp

Abstract. To promote the research of multiagent systems, several base codes have been released for the RoboCup soccer 2D simulation community. As described herein, we present HELIOS base, currently the most popular base codes for 2D soccer simulation. HELIOS base involves a common library, a sample team, a visual debugger, and a formation editor, which help us to develop a simulated soccer team.

1 Introduction

In this paper, we present a base code, named HELIOS base, for the RoboCup soccer 2D simulation. The RoboCup Soccer Simulation 2D League is a long-running competition among the RoboCup leagues. It is based on the RoboCup Soccer 2D Simulator [8,11], which enables two teams of 11 autonomous player agents and an autonomous coach agent to play a game of soccer with highly realistic rules and real-time game play. Because of its stability, the 2D soccer simulator is extremely useful for research and education related to multiagent systems, artificial intelligence, and machine learning.

The soccer simulation league has devoted more attention to team work techniques than to robot control techniques. The 2D soccer simulator adopts a discrete timer model and an abstract and simple kinematic model, although its virtual soccer field has a continuous space. Therefore, we can avoid the burdens of developing and maintaining mechanical devices and also developing complex robot control tasks such as bipedal walking. These characteristics enable us to concentrate on research efforts related to multiagent systems. However, developing an agent program from scratch is as difficult a challenge as ever because other complex modules, such as a stable network communication, synchronization, world modeling, and so on, are necessary to produce an agent program that fully functions in the soccer simulator. We must resolve these technical problems before progressing with research of multiagent systems. The base code presented in this paper provides a framework that enables us to concentrate on teamwork techniques.

The remainder of this paper is organized as follows. Section 2 introduces the base code released by other teams. Section 3 introduces an outline of our

S. Behnke et al. (Eds.): RoboCup 2013, LNAI 8371, pp. 528–535, 2014.
© Springer-Verlag Berlin Heidelberg 2014

base code. Section 4 describes our base code components. Section 5 describes the impact of our base code on the 2D soccer simulation community. Section 6 concludes this report.

2   Related Works

In the 2D soccer simulation community, several teams have released (parts of) their respective source codes to promote the research of multiagent systems using the soccer simulator. Especially, the champion teams often release their source codes after competitions.

CMUnited [10,6] is an important code release in the early 2000s. This team was the RoboCup champion of 1998 and 1999. Their released code was widely used as a base by teams around the world. Its concept, such as Locker Room Agreement, Layered Learning, and so on, still have an impact on the development of simulated soccer agents. The base code released by TsinghuAeolus [14], the champion team of RoboCup 2001 and 2001, provides excellent skills such as ball kicking and dribbling. However, because the code was designed for a Windows environment at first, it did not capture the global popularity. UvA Trilearn [7,12], the champion team of RoboCup 2003, has released an extremely successful base code. Their code provides a sophisticated design and extremely rich documentation. Consequently, a large number of new teams have managed to participate in the competition. One team still uses this base code in 2013. Recent champion teams such as Brainstormers [9,5] and WrightEagle [4,13], also released their base code. Although their codes have sophisticated implementation and provide satisfactorily high performance, none is widely used yet because it is difficult for new users to use them.

3   Outline of HELIOS Base

HELIOS base is a base code and related development tools released by HELIOS, which is the champion team of RoboCup 2010 and 2012 [1].

3.1   History

HELIOS is a simulated soccer team for the RoboCup soccer 2D simulation league. The team, a joint team of Fukuoka University and Osaka Prefecture University since 2010, has been participating in RoboCup competitions since 2000. The team has won 2 championships and 2 runner-up places to date, and has remained among the top 3 in the world championships since 2007.

The first release of HELIOS base was in 2006. The code is still being maintained continually according to the change of competition rules. HELIOS base is said to be the most popular base code in 2012. All source codes are available at our project site http://sourceforge.jp/projects/rctools/.

3.2   Features

HELIOS base provides a sample source code for developing a team and a data set that can run as a simple but competitive team, for new teams to participate to the competition of the 2D soccer simulation easily. HELIOS base consists of several software components designed to reduce the maintenance cost. an overview of each component is described in Section 4.

All software components included in HELIOS base are written using Standard C++ and implemented from scratch without the source codes of other simulated soccer teams. The code depends on the POSIX API, the boost C++ libraries[1], and Qt[2], but never includes environment specific dependencies. Therefore, HELIOS base has high portability to various operating systems. Now, HELIOS base supports Linux, Mac OSX, and Windows (Cygwin).

The users of HELIOS base can use it freely if they follow its license. The common library of HELIOS base is licensed under GNU Lesser General Public License[3]. The code of the sample team and development tools are licensed under GNU General Public License[4].

4   Components

HELIOS base provides the following components:

- librcsc
- agent2d
- soccerwindow2
- fedit2

This section presents a description of the overview of each component.

4.1   librcsc: The Common Library

librcsc (LIBrary for the RoboCup Soccer simulation Client) is a basic and common library for developing a 2D soccer simulation software. librcsc contains several library files such as geometry, network interface, communication and synchronization with simulator, world model, basic actions, log parser, debug message management, formation model, and so on.

librcsc encapsulates almost all things related to the communication between the simulator and agent programs. The users of HELIOS base need not consider the synchronization problem related to the network programming and the timing of decision making, which are usually out of focus from the viewpoint of multiagent research.

---

[1] http://boost.org/
[2] http://qt-project.org/
[3] http://www.gnu.org/copyleft/lesser.html
[4] http://www.gnu.org/licenses/gpl.html

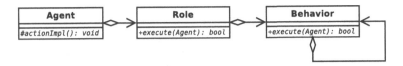

Fig. 1. The UML diagram showing the decision making architecture of agent2d

4.2 agent2d: A Sample Team

agent2d is a sample team that utilizes librcsc and also contains a data set of the simulated soccer team that can run as a simple but competitive team. The implemented behavior is more complicated than that of UvA base code [7]. Each player can intercept, dribble, pass and shoot by judging from the field situations. Although the team strategies remain simple, the team performance is better than any other sample teams. We assume that this sample team is used as a template when starting team development.

The decision making process of agent2d comprises three layers: agent class, role class and behavior class. Any decision making originates from the agent class. The agent class decides the current strategy and the player's role in the team. The role class is responsible for determining strategic behavior. The role class first gets the current situation and then executes tactical behavior according to the current strategy. Finally, the behavior class performs the actual action. Figure 1 shows the relation among these classes.

Developers of a team never need to implement their own agent class, but might need to implement their own role classes or behavior classes. The default implementation provides several role instances, but in agent2d there is no difference among them. If developers would like to step into more detailed development such as improving or adding new role classes or behavior classes, then they require some knowledge and experience of C++ because agent2d is written entirely in C++.

In agent2d, our formation framework [3] and online multiagent planning framework [2] have already been implemented. These frameworks enable us to change the characteristics of team behavior by modifying the team formation and the evaluation function. Consequently, we can concentrate on improving the team strategy without considering a complicated decision tree. For example, we can focus on the optimization of team formation only by changing parameters in configuration files. The multiagent planning framework was introduced into agent2d in 2010. This framework brings up several research issues related to the online search approach in a continuous state and action space.

4.3 soccerwindow2: A Visual Debugger

soccerwindow2, a viewer program for the 2D soccer simulation, has many useful features. For example, the following functions help us to develop a team:

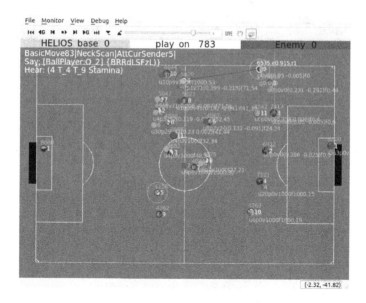

Fig. 2. soccerwindow2 is useful not only as a compatible monitor client but also as a visual debugger. This image shows the actual simulation state and one agent's internal state on the virtual soccer field.

- It can work as a monitor client that is compatible with the official monitor client. The timeshift function is also available. We can always replay any recorded scene during the game online.
- It can function as a stand-alone log player. The game log files recorded by the 2D simulator can be replayed.
- It can function as a visual debugger: not only as an online debug server but also as an offline debug message viewer. The agents' internal state can be examined during the game online if agents send their internal state to the integrated visual debug server. Moreover, if agents record their internal state with the specified format as their own log files, then soccerwindow2 can load and replay them.

Figure 2 portrays a snapshot of soccerwindow2 in which the online debug server mode is active. As the figure shows, soccerwindow2 can visualize not only the actual simulation state but also the agents' internal state on the virtual soccer field. This feature helps us to observe a gap separating the actual simulation state from the agents' internal state. This debugging information can be sent from agent programs to the visual debug server integrated to soccerwindow2 via UDP/IP communication during the game. The debug server function facilitates our development process to a considerable degree.

Figure 3 presents a snapshot of the debug message window, which can show agents' more detailed internal states by unlimited length text message with arbitrary format. We assume that the text messages are loaded from files recorded

Fig. 3. Snapshot of the debug message window that can show the agents' internal state by text with greater detail. We can check all agents' respective states by changing the tab window. The left pain shows log level buttons that can toggle the information type shown in the main message panel.

by agents. Therefore, this feature cannot be used during the game. However, this feature helps us to recognize more details related to agents' decision making process.

soccerwindow2 is developed using Qt, which is a cross-platform application development library. Therefore, we can use soccerwindow2 in several systems.

4.4   fedit2: A Formation Editor

fedit2 is a GUI application to edit the formation data for agent2d. agent2d supports the formation framework [3] provided by librcsc. This framework enables the definition of the team formation through external configuration files. Here, agent2d can change the team formation easily by loading different configuration files. However, this framework defines the team formation using Delaunay triangulation. Therefore, it is difficult for us to modify the team formation by editing the text data. In addition, fedit2 helps us to modify the team formation.

Figure 4 portrays a snapshot of fedit2, which can not only display the ball and 11 player agents on the soccer field, but can also enable them to be edited intuitively by the mouse and keyboard. The recorded ball positions are used as vertices of Delaunay triangulation. The agents' positions are calculated using a linear interpolation algorithm, resembling the Gouraud shading algorithm if the input ball position is unknown and if it is contained by one triangle. For more details of this framework, another report of the literature is helpful [3].

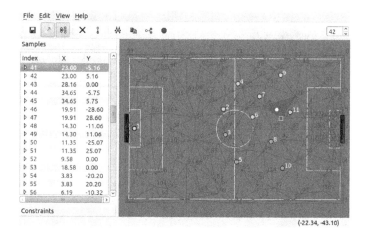

Fig. 4. Snapshot of fedit2. The left panel shows a list of recorded data sets. Each row contains the ball position and ideal agent positions according to the ball position. The right pane visualizes Delaunay triangulation constructed from the recorded data. Each vertex represents the ball position of input data. We can modify them easily using a mouse and keyboard on the soccer field.

Table 1. Transition of the number of teams that use HELIOS base

	# of participating teams	# of HELIOS based teams	Percentage(%)
RoboCup2007	15	3	20.0
RoboCup2008	15	4	26.7
RoboCup2009	19	7	36.5
RoboCup2010	19	8	42.1
RoboCup2011	17	8	47.1
RoboCup2012	18	15	83.3
RoboCup2013	24	20	83.3

5   Impact on the Community

Until the release of HELIOS base, the most popular base code in the 2D soccer simulation community was UvA Trilearn. After releasing HELIOS base, the number of teams that use HELIOS base has increased year by year. Table 1 shows the transition of the number of teams that use HELIOS base. Results show that HELIOS base became the most popular base code in 2012.

6   Conclusions and Future Works

This paper described HELIOS base, currently the most popular base code for the RoboCup soccer 2D simulation league. HELIOS base involves a common library, a sample team, and development tools for the 2D soccer simulation to promote the research of multiagent systems.

An important subject for future work is to prepare comprehensive documentation. It is still necessary to improve the design of the decision making architecture.

References

1. Akiyama, H., Nakashima, T.: HELIOS2012: RoboCup 2012 Soccer Simulation 2D League Champion. In: Chen, X., Stone, P., Sucar, L.E., van der Zant, T. (eds.) RoboCup 2012. LNCS (LNAI), vol. 7500, pp. 13–19. Springer, Heidelberg (2013)
2. Akiyama, H., Nakashima, T., Aramaki, S.: Online cooperative behavior planning using a tree search method in the robocup soccer simulation. In: Proceedings of Fourth IEEE International Conference on Intelligent Networking and Collaborative Systems, INCoS-2012 (2012)
3. Akiyama, H., Noda, I.: Multi-agent positioning mechanism in the dynamic environment. In: Visser, U., Ribeiro, F., Ohashi, T., Dellaert, F. (eds.) RoboCup 2007. LNCS (LNAI), vol. 5001, pp. 377–384. Springer, Heidelberg (2008)
4. Bai, A., Chen, X., MacAlpine, P., Urieli, D., Barrett, S., Stone, P.: Wright Eagle and UT Austin Villa: RoboCup 2011 simulation league champions. In: Röfer, T., Mayer, N.M., Savage, J., Saranlı, U. (eds.) RoboCup 2011. LNCS, vol. 7416, pp. 1–12. Springer, Heidelberg (2012)
5. Brainstormers Public Source Code Release, http://sourceforge.net/projects/bsrelease/
6. CMU RoboSoccer RoboCup Simulator Team Homepage, http://www.cs.utexas.edu/~pstone/RoboCup/CMUnited-sim.html
7. Kok, J.R., Vlassis, N., Groen, F.: UvA Trilearn 2003 team description. In: Polani, D., Browning, B., Bonarini, A., Yoshida, K. (eds.) Proceedings CD RoboCup 2003. Springer, Padua (2003)
8. Noda, I., Matsubara, H.: Soccer server and researches on multi-agent systems. In: Kitano, H. (ed.) Proceedings of IROS-96 Workshop on RoboCup, pp. 1–7 (November 1996)
9. Riedmiller, M., Gabel, T., Knabe, J., Strasdat, H.: Brainstormers 2d - team description 2005. In: Bredenfeld, A., Jacoff, A., Noda, I., Takahashi, Y. (eds.) Proceedings CD RoboCup 2005. Springer (2005)
10. Stone, P., Riley, P., Veloso, M.: The CMUnited-99 champion simulator team. In: Veloso, M., Pagello, E., Kitano, H. (eds.) RoboCup 1999. LNCS (LNAI), vol. 1856, pp. 35–48. Springer, Heidelberg (2000)
11. The RoboCup Soccer Simulator, http://sserver.sourceforge.net/
12. UvA Trilearn (2003), - Soccer Simulation Team, http://staff.science.uva.nl/~jellekok/robocup/2003/
13. WrightEagle 2D Soccer Simulation Team, http://wrighteagle.org/2D/
14. Yao, J., Chen, J., Cai, Y., Li, S.: Architecture of tsinghuaeolus. In: Birk, A., Coradeschi, S., Tadokoro, S. (eds.) RoboCup 2001. LNCS (LNAI), vol. 2377, pp. 491–494. Springer, Heidelberg (2002)

The Open-Source TEXPLORE Code Release for Reinforcement Learning on Robots

Todd Hester and Peter Stone

Department of Computer Science,
University of Texas at Austin,
Austin, TX, 78712
{todd,pstone}@cs.utexas.edu

Abstract. The use of robots in society could be expanded by using reinforcement learning (RL) to allow robots to learn and adapt to new situations on-line. RL is a paradigm for learning sequential decision making tasks, usually formulated as a Markov Decision Process (MDP). For an RL algorithm to be practical for robotic control tasks, it must learn in very few samples, while continually taking actions in real-time. In addition, the algorithm must learn efficiently in the face of noise, sensor/actuator delays, and continuous state features. In this paper, we present the TEXPLORE ROS code release, which contains TEXPLORE, the first algorithm to address all of these challenges together. We demonstrate TEXPLORE learning to control the velocity of an autonomous vehicle in real-time. TEXPLORE has been released as an open-source ROS repository, enabling learning on a variety of robot tasks.

Keywords: Reinforcement Learning, Markov Decision Processes, Robots.

1 Introduction

Robots have the potential to solve many problems in society by working in dangerous places or performing unwanted jobs. One barrier to their widespread deployment is that they are mainly limited to tasks where it is possible to hand-program behaviors for every situation they may encounter. Reinforcement learning (RL) [19] is a paradigm for learning sequential decision making processes that could enable robots to learn and adapt to their environment on-line. An RL agent seeks to maximize long-term rewards through experience in its environment.

Learning on robots poses at least four distinct challenges for RL:

1. The algorithm must learn from very few samples (which may be expensive or time-consuming).
2. It must learn tasks with continuous state representations.
3. It must learn good policies even with unknown sensor or actuator delays (i.e. selecting an action may not affect the environment instantaneously).
4. It must be computationally efficient enough to take actions continually in real-time.

S. Behnke et al. (Eds.): RoboCup 2013, LNAI 8371, pp. 536–543, 2014.
© Springer-Verlag Berlin Heidelberg 2014

Addressing these challenges not only makes RL applicable to more robotic control tasks, but also many other real-world tasks.

While algorithms exist that address various subsets of these challenges, we are not aware of any that are easily adapted to address all four issues. RL has been applied to a few carefully chosen robotic tasks that are achievable with limited training and infrequent action selections (e.g. [11]), or allow for an off-line learning phase (e.g. [13]). However, to the best of our knowledge, none of these methods allow for continual learning on the robot running in its environment.

In contrast to these approaches, we present the TEXPLORE algorithm, the first algorithm to address all four challenges at once. The key insights of TEXPLORE are 1) to learn multiple domain models that generalize the effects of actions across states and target exploration on uncertain and promising states; and 2) to combine Monte Carlo Tree Search and a parallel architecture to take actions continually in real-time. TEXPLORE has been released publicly as an open-source ROS repository at: http://www.ros.org/wiki/rl-texplore-ros-pkg.

2   Background

We adopt the standard Markov Decision Process (MDP) formalism for this work [19]. An MDP consists of a set of states S, a set of actions A, a reward function $R(s, a)$, and a transition function $P(s'|s, a)$. In many domains, the state s has a factored representation, where it is represented by a vector of n state variables $s = \langle x_1, x_2, ..., x_n \rangle$. In each state $s \in S$, the agent takes an action $a \in A$. Upon taking this action, the agent receives a reward $R(s, a)$ and reaches a new state s', determined from the probability distribution $P(s'|s, a)$.

The value $Q^*(s, a)$ of a given state-action pair (s, a) is an estimate of the future reward that can be obtained from (s, a) and is determined by solving the Bellman equation: $Q^*(s, a) = R(s, a) + \gamma \sum_{s'} P(s'|s, a) \max_{a'} Q^*(s', a')$, where $0 < \gamma < 1$ is the discount factor. The goal of the agent is to find the policy π mapping states to actions that maximizes the expected discounted total reward over the agent's lifetime. The optimal policy π is then $\pi(s) = \text{argmax}_a Q^*(s, a)$.

Model-based RL methods learn a model of the domain by approximating $R(s, a)$ and $P(s'|s, a)$ for each state and action. The agent can then plan on this model through a method such as value iteration [19] or UCT [10], effectively updating the Bellman equations for each state using their model. RL algorithms can also work without a model, updating the values of actions only when taking them in the real task. Generally model-based methods are more sample efficient than model-free methods, as their sample efficiency is only constrained by how many samples it takes to learn a good model.

3   TEXPLORE

In this section, we describe TEXPLORE [9], a sample-efficient model-based real-time RL algorithm. We describe how TEXPLORE returns actions in real-time in Section 3.1, and its approach to model learning and exploration in Section 3.2.

3.1   Real-Time Architecture

In this section, we describe TEXPLORE's real-time architecture, which can be used for a broad class of model-based RL algorithms that learn generative models. Most current model-based RL methods use a sequential architecture, where the agent receives a new state and reward; updates its model with the new transition $\langle s, a, s', r \rangle$; plans exactly on the updated model (i.e. by computing the optimal policy with a method such as value iteration); and returns an action from its policy. Since both the model learning and planning can take significant time, this algorithm is not real-time. Alternatively, the agent may update its model and plan on batches of experiences at a time, but this requires long pauses for the batch updates to be performed. Making the algorithm real-time requires two modifications to the standard sequential architecture: 1) utilizing sample-based approximate planning and 2) developing a novel parallel architecture called the Real-Time Model-Based Architecture (RTMBA) [7].

First, instead of planning exactly with value iteration, RTMBA uses UCT [10], a sample-based anytime approximate planning algorithm from the Monte Carlo Tree Search (MCTS) family. MCTS planners simulate trajectories (rollouts) from the agent's current state, updating the values of the sampled actions with the reward received. The agent performs as many rollouts as it can in the given time, with its value estimate improving with more rollouts. These methods can be more efficient than dynamic programming approaches in large domains because they focus their updates on states the agent is likely to visit soon rather than iterating over the entire state space.

In addition, we developed a Real-Time Model Based Architecture (RTMBA) that parallelizes the model learning, planning, and acting such that the computation-intensive processes (model learning and planning) are spread out over time. Actions are selected as quickly as dictated by the robot control loop, while still being based on the most recent models and plans available.

Since both model learning and planning can take significant computation (and thus wall-clock time), RTMBA places both of those processes in their own parallel threads in the background, shown in

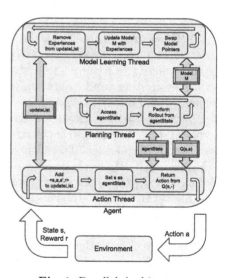

Fig. 1. Parallel Architecture

Figure 1. A third thread interacts with the environment, receiving the agent's new state and reward and returning the action given by the agent's current policy. The threads communicate through shared data structures protected by mutex locks. By de-coupling the action thread from the time-consuming model-learning and planning processes, RTMBA releases the algorithm from the need to complete the model update and planning between actions. The full details of this architecture are described in [7,9].

3.2   Model Learning

While the parallel architecture we just presented enables TEXPLORE to operate in real-time, the algorithm must learn an accurate model of the domain quickly to learn the task with high sample efficiency. While tabular models are a common approach, they require the agent to take every action from each state once (or multiple times in stochastic domains), since they learn a prediction for each state-action separately. If we assume that the transition dynamics are similar across state-action pairs, we can improve upon tabular models by incorporating *generalization* into the model learning. TEXPLORE achieves high sample efficiency by combining this generalization with targeted exploration to improve the model as quickly as possible.

TEXPLORE approaches model learning as a supervised learning problem with (s, a) as the input and s' and r as the outputs the supervised learner is predicting. The supervised learner can make predictions about the model for unseen or infrequently visited states based on the transitions it has been trained on. TEXPLORE uses C4.5 decision trees [15] as the supervised learner to learn models of the transition and reward functions. The algorithm learns a model of the domain by learning a separate prediction for each of the n state features and reward. For continuous domains, the algorithm uses the M5 regression tree algorithm [16], which learns a linear regression model in each leaf of the tree, enabling it to better model continuous dynamics by building a piecewise linear model.

Each tree makes predictions for the particular feature or reward it is given based on a vector containing the n features of the state s along with the action a: $\langle x_1, x_2, ..., x_n, a \rangle$. To handle robots, which commonly have sensor and actuator delays, we provide the model with the past k actions, so that the model can learn which of these past actions is relevant for the current prediction.

Using decision trees to learn the model of the MDP provides us with a model that can be learned quickly with few samples. However, it is important that the algorithm focuses its exploration on the state-actions most likely to be relevant to the the task. To drive exploration, TEXPLORE builds multiple possible models of the domain in the form of a random forest [3]. The random forest model is a collection of m decision trees, where each tree is trained on only a subset of the agent's *experiences* ($\langle s, a, s', r \rangle$ tuples). Each tree in the random forest represents a hypothesis of what the true domain dynamics are. TEXPLORE then plans on the average of these predicted distributions, so that TEXPLORE balances the models predicting overly positive outcomes with the ones predicting overly negative outcomes. More details, including an illustrative example of this exploration, are provided in [9].

4   ROS Code Release

The TEXPLORE algorithm and architecture presented in this paper has been fully implemented, empirically tested, and released publicly as a Robot Operating System (ROS) repository at: http://www.ros.org/wiki/rl-texplore-ros-pkg.

ROS[1] is an open-source middleware operating system for robotics. It includes tools and libraries for commonly used functionality for robots such as hardware abstraction and messaging. In addition, it has a large ecosystem of users who release software packages for many common robot platforms and tasks as open source ROS repositories. With TEXPLORE released as a ROS repository, it can be easily downloaded and applied to a learning task on any robot running ROS with minimal effort. The goal of this algorithm and code release is to encourage more researchers to perform learning on robots using state-of-the-art algorithms.

The code release contains five ROS packages:

1. **rl_common**: This package includes files that are common to both reinforcement learning agents and environments.
2. **rl_msgs**: The **rl_msgs** package contains a set of ROS messages (http://www.ros.org/wiki/rl_msgs) that we have defined to communicate with reinforcement learning algorithms. The package defines an *action* message that the learning algorithm publishes, and a *state/reward* message that it subscribes to. To interface the learning algorithm with a robot already running ROS, one only needs to write a single node that translates *action* messages to actuator commands for the robot and translates robot sensor messages into *state/reward* messages for the learning algorithm. The advantages of this design are that we can apply RL on a wide variety of tasks by simply creating different RL interface nodes, while the actual learning algorithms and the robot's sensors/actuators interfaces are cleanly abstracted away.
3. **rl_agent**: This package contains the TEXPLORE algorithm. In addition, it includes several other commonly used reinforcement learning algorithms, such as Q-LEARNING [21], SARSA [17], R-MAX [2], and DYNA [18]. The agent package also includes a variety of model learning and planning techniques for implementing different model-based methods.
4. **rl_env**: This package includes a variety of benchmark RL domains such as Fuel World [8], Taxi [4], Mountain Car [19], Cart-Pole [19], Light World [12], and the simulated car velocity control domain presented in Section 5.
5. **rl_experiment**: This package contains code to run RL experiments without ROS message passing, by compiling both the experiment and agent together into one executable. This package is useful for running simulated experiments that do not require message passing to a robot.

5   Example Application

In this section, we present more details on how to interface the TEXPLORE algorithm with a robot already running ROS. In particular, we will demonstrate the ability of TEXPLORE to learn velocity control on our autonomous vehicle [1]. This task has a continuous state space, delayed action effects, and requires learning that is both sample efficient (to learn quickly) and computationally efficient (to learn on-line while controlling the car).

---

[1] www.ros.org

The task is to learn to drive the vehicle at a desired velocity by controlling the pedals. For learning this task, the RL agent's 4-dimensional state is the desired velocity of the vehicle, the current velocity, and the current position of the brake and accelerator pedals. The agent's reward at each step is -10.0 times the error in velocity in m/s. Each episode is run at 10 Hz for 10 seconds. The agent has 5 actions: one does nothing (no-op), two increase or decrease the desired brake position by 0.1 while setting the desired accelerator position to 0, and two increase or decrease the desired accelerator position by 0.1 while setting the desired brake position to 0. While these actions change the desired positions of the pedals immediately, there is some delay before the brake and accelerator reach their target positions.

The vehicle was already using ROS [14] as its underlying middleware. The **rl_msgs** package defines an *action* and *state/reward* message for the agent to communicate with an environment. To connect the agent with the robot, we wrote a ROS node that translates actions into actuator commands and sensor information into state/reward messages. The agent's action message is an integer between 0-4 as the action.

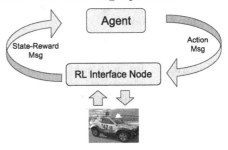

Fig. 2. ROS RL Interface

The interface node translates this action into messages that provide commanded positions for the brake and throttle of the car. The interface node then reads sensor messages that provide the car's velocity and the true positions of the pedals, and uses these to create a state vector and reward for the agent and publishes a state/reward message. Figure 2 visualizes how the learning algorithm interfaces with the robot.

We ran five trials of Continuous TEX-PLORE (using M5 regression trees) with $k = 2$ delayed actions on the physical vehicle learning to drive at 5 m/s from a start of 2 m/s. Figure 3 shows the average rewards over 20 episodes. In all five trials, the agent learned the task within 11 episodes, which is less than 2 minutes of driving time. This experiment shows that our TEXPLORE code release can be used to learn a robotic task that has continuous state and actuator delays in very few samples while selecting

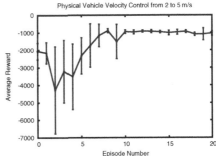

Fig. 3. Average rewards of TEXPLORE learning to control the physical vehicle from 2 to 5 m/s

actions continually in real-time. In addition to learning to control the velocity of an autonomous vehicle, a variant of TEXPLORE has also been used to learn how to score penalty kicks in the Standard Platform League of RoboCup [6]. More results with comparisons against other state-of-the-art algorithms are available in [9,5].

6 Related Work

Since TEXPLORE is addressing many challenges, there is ample related work on each individual challenge, although no other methods address all four challenges. Work related to the algorithmic components of TEXPLORE is detailed in [9,5]. In this section, we look at related reinforcement learning code releases.

Similar to our **rl_msgs** package which defines ROS messages for an agent to communicate with an environment, RL-GLUE [20] defines similar messages for general use. The RL-LIBRARY[2] builds off of this as a central location for sharing RL-GLUE compatible RL projects. However, currently the library only contains the SARSA(λ) [17] algorithm, and does not have any algorithms which focus on learning on robots.

RLPARK[3] is a Java-based RL library that includes both learning methods and methods for the real-time display of learning data. RLPARK contains a variety of algorithms for both control and prediction, both on-line or off-line. It does contain some algorithms that target learning on robots, however it does not provide a ROS interface for connecting with other robots.

The York Reinforcement Learning Library (YORLL)[4] focuses on multi-agent learning, however it works for single-agent learning as well. It contains a few basic algorithms like Q-LEARNING [21] and SARSA [17], and has various options for handling multiple agents. However, it does not have any algorithms which focus on learning on robots.

Teaching Box[5] is a learning library that is focused on robots. It contains some algorithms for reinforcement learning as well as learning from demonstration. It uses RL-GLUE to interface the agent and environment, rather than using ROS.

7 Conclusion

We identify four properties required for RL to be practical for continual, on-line learning on a broad range of robotic tasks: it must (1) be sample-efficient, (2) work in continuous state spaces, (3) handle sensor and actuator delays, and (4) learn while taking actions continually in real-time. This article presents the code release of TEXPLORE, the first algorithm to address all of these challenges. Note that there are other challenges relevant to robotics that TEXPLORE does not address, such as *partial observability* or *continuous actions*, which we leave for future work.

The code release provides the TEXPLORE algorithm along with a variety of other commonly used RL algorithms. It also contains a number of common benchmark tasks for RL. The release includes a set of ROS messages for RL which define how an RL agent can communicate with a robot. Using these defined messages, it is easy to interface TEXPLORE or the other algorithms provided in the code release with robots already running ROS.

---

[2] `library.rl-community.org/wiki/Main_Page`

[3] `rlpark.github.com/`

[4] `www.cs.york.ac.uk/rl/software.php`

[5] `amser.hs-weingarten.de/en/teachingbox.php`

Acknowledgements. This work has taken place in the Learning Agents Research Group (LARG) at UT Austin. LARG research is supported in part by NSF (IIS-0917122), ONR (N00014-09-1-0658), and the FHWA (DTFH61-07-H-00030).

References

1. Beeson, P., O'Quin, J., Gillan, B., Nimmagadda, T., Ristroph, M., Li, D., Stone, P.: Multiagent interactions in urban driving. Journal of Physical Agents 2(1), 15–30 (2008)
2. Brafman, R., Tennenholtz, M.: R-Max - a general polynomial time algorithm for near-optimal reinforcement learning. In: IJCAI (2001)
3. Breiman, L.: Random forests. Machine Learning 45(1), 5–32 (2001)
4. Dietterich, T.: The MAXQ method for hierarchical reinforcement learning. In: ICML, pp. 118–126 (1998)
5. Hester, T.: TEXPLORE: Temporal Difference Reinforcement Learning for Robots and Time-Constrained Domains. PhD thesis, Department of Computer Science, University of Texas at Austin, Austin, TX (December 2012)
6. Hester, T., Quinlan, M., Stone, P.: Generalized model learning for reinforcement learning on a humanoid robot. In: ICRA (May 2010)
7. Hester, T., Quinlan, M., Stone, P.: RTMBA: A real-time model-based reinforcement learning architecture for robot control. In: ICRA (2012)
8. Hester, T., Stone, P.: Real time targeted exploration in large domains. In: ICDL (August 2010)
9. Hester, T., Stone, P.: TEXPLORE: Real-time sample-efficient reinforcement learning for robots. Machine Learning 87, 10–20 (2012)
10. Kocsis, L., Szepesvári, C.: Bandit based Monte-Carlo planning. In: Fürnkranz, J., Scheffer, T., Spiliopoulou, M. (eds.) ECML 2006. LNCS (LNAI), vol. 4212, pp. 282–293. Springer, Heidelberg (2006)
11. Kohl, N., Stone, P.: Machine learning for fast quadrupedal locomotion. In: AAAI Conference on Artificial Intelligence (2004)
12. Konidaris, G., Barto, A.G.: Building portable options: Skill transfer in reinforcement learning. In: IJCAI (2007)
13. Ng, A., Kim, H.J., Jordan, M., Sastry, S.: Autonomous helicopter flight via reinforcement learning. In: NIPS (2003)
14. Quigley, M., Conley, K., Gerkey, B., Faust, J., Foote, T., Leibs, J., Wheeler, R., Ng, A.: ROS: An open-source robot operating system. In: ICRA Workshop on Open Source Software (2009)
15. Quinlan, R.: Induction of decision trees. Machine Learning 1, 81–106 (1986)
16. Quinlan, R.: Learning with continuous classes. In: 5th Australian Joint Conference on Artificial Intelligence, pp. 343–348. World Scientific, Singapore (1992)
17. Rummery, G., Niranjan, M.: On-line Q-learning using connectionist systems. Technical Report CUED/F-INFENG/TR 166. Cambridge University Engineering Department (1994)
18. Sutton, R.: Integrated architectures for learning, planning, and reacting based on approximating dynamic programming. In: ICML, pp. 216–224 (1990)
19. Sutton, R., Barto, A.: Reinforcement Learning: An Introduction. MIT Press, Cambridge (1998)
20. Tanner, B., White, A.: RL-Glue: Language-independent software for reinforcement-learning experiments. JMLR 10, 2133–2136 (2009)
21. Watkins, C.: Learning From Delayed Rewards. PhD thesis. University of Cambridge (1989)

NUbugger: A Visual Real-Time Robot Debugging System

Brendan Annable, David Budden, and Alexandre Mendes

School of Electrical Engineering and Computer Science,
Faculty of Engineering and Built Environment,
The University of Newcastle, Callaghan, NSW, 2308, Australia
{brendan.annable,david.budden}@uon.edu.au,
alexandre.mendes@newcastle.edu.au

Abstract. As modern autonomous robots have improved in their ability to demonstrate human-like motor skills and reasoning, the size and complexity of software systems have increased proportionally, with developers actively working to leverage the full processing performance of next-generation computational hardware. This software complexity corresponds with increased difficulty in debugging low-level coding issues, with the traditional methodology of inferring such issues from emergent high-level behaviour rapidly approaching intractability. This paper details the development and functionality of NUbugger: a visual, real-time and open source robot debugging utility that provides the user with comprehensive information regarding low-level functionality. This represents a paradigm shift from corrective to preventative debugging, and concrete examples of the application of NUbugger to the identification of fundamental implementation errors are described. The system implementation facilitates simple and rapid extension or modification, making it a useful utility for debugging any similar complex robotic framework.

Keywords: debugging, robotics, open source, visualisation.

1 Introduction

The problem of developing a team of humanoid robots capable of defeating the FIFA World Cup champion team, coined "The Millennium Challenge" [7], has been a milestone that has driven research in the fields of artificial intelligence, robotics and computer vision for over a decade. Corresponding with the continual improvement in a robot's ability to demonstrate human-like motor skills and reasoning is an exponential blowout in software size and complexity, facilitated by the evolution of robot platforms and subsequent advances in processor performance (from the 384 MHz RISC-based processors of the Sony AIBO ERS-210 (2002) to the 1.6 GHz Intel Atom processors of the Robotis DARwIn-OP [5] platform (2012)); a trend often inferred from Moore's Law [11].

As with any system of software or hardware, exponential increases in system size and complexity necessitate the introduction of hierarchial layers of abstraction, allowing low-level functionality to be handled transparently by higher-level

S. Behnke et al. (Eds.): RoboCup 2013, LNAI 8371, pp. 544–551, 2014.

functions, classes and packages. As the majority of open source libraries are supported by large developer communities, one critical factor is often ignored: the cascading effect of low-level errors, and the difficulty of locating the source of such errors by qualitative observations of emergent system misbehaviour (such as a robot refusing to kick a ball [2]). The probability of such low-level issues is increased by a number of factors encountered in typical RoboCup research environments: few team members handling a large number of issues, strict deadlines and the disjoint nature of development (both in terms of frequent developer turnover and one-developer-per-subsystem strategies [9]).

An analogy may be drawn between complex robotic systems and modern genomics, where abnormalities in an organism's genotype are almost exclusively identified by complex emergent behaviour (e.g. a disease or hereditary trait) at the phenotypic level. Although determining the root cause of some observed abnormalities is a straightforward process, more complex system misbehaviour may be the result of a significantly larger number of low-level contributing factors. Due to the massive complexity of biological systems and the inability to make perfect observations at the genome-level, the exact systematic causes of many common diseases remain largely unknown, despite years of research by very large teams.

Although robotic and biological systems are (to date) fundamentally different in implementation, they exhibit two primary common traits: massive complexity and high-level issues caused by nontrivial combinations of low-level implementation errors. In this sense, the above analogy provides a useful insight to common debugging methodologies; locating low-level issues by observing high-level behaviour is an increasingly intractable problem. However, unlike in biological systems, where network models are constructed and analysed to infer genotype from phenotype, the "genome" of a robotic system (i.e. the low-level software behaviour and environmental response) is directly observable in real-time. Specifically, the following debugging methodologies are possible:

- **Data Logs:** Most robotic systems provide functionality for generating debug logs, commonly enabled by setting the value of some *debug verbosity* parameter at compile-time (such as per the NUbots system [9]). Once an error has been appropriately reproduced, data logs are generated capturing some of (and not limited to) the following information as a function of time:
  - Sensor values (accelerometer, gyroscopes, pressure sensors, etc.)
  - Servo positional values (from which kinematics and pose may be inferred)
  - Current captured image, colour-classified image [1] and recognised features
  - Self-localisation belief and perceived location of dynamic features (such as the ball in robot soccer [2])

  This information may be later analysed (either manually or by applying classification techniques for well-known errors) to discover the sources of error.
- **Visual Monitoring:** Although a significant improvement over high-level inferential debugging, data logs have three major shortcomings: limited

memory resources restricting the amount or resolution of data that can be collected; the requirement for an error to be reproducible, allowing it to be captured in a log once debugging has been enabled; and the adoption of a purely *corrective* (rather than *preventative*) debugging methodology. All of these issues are addressed by a *visual monitoring* methodology, where all critical system information is streamed in real-time to a graphical web client, to be monitored by the user. Concretely, this addresses the aforementioned issues in the following ways:

- By comparison to memory, network communication is relatively cheap. Large amounts of information (including real-time video) may be streamed from a robot to a corresponding web client for an effectively unlimited period of time.
- Uncommon issues may be immediately identified by the user, removing the presumption of error reproducibility.
- Abnormalities in low level functionality (such as an incorrectly classified image or reduced video frame-rate) may be identified and corrected before they are able to visually affect high-level behavioural performance.

This paper describes the implementation of *NUbugger* (Newcastle University's Debugger); a visual, real-time and open source robot debugging utility that addresses the aforementioned issues. Firstly, an overview of the system implementation is provided, both in terms of high (system inputs, outputs and architecture) and low-level (languages and library dependencies) application structure. The main functionality of NUbugger is explained, with concrete visual examples of the real-time information provided to the user. Finally, the significance and outcomes of the system are justified by providing examples of actual low-level errors in the NUbots RoboCup source code, that have been identified and corrected as a direct result of NUbugger's implementation.

2   Implementation

NUbugger implements a many-to-many service between robots and web clients, via a single web server. Concretely, an arbitrary number of robots (3-4 in the case of RoboCup humanoid league soccer) are able to stream real-time, low-level system information wirelessly to a single web server (which may be a robot in itself). This information is then distributed to an arbitrary number of web clients, allowing users to monitor performance-critical visualisations of sensory data and emergent high-level behaviour. This process is illustrated in Fig. 1.

Due to the complexity of streaming large quantities of data from a robot (implemented in C++) to web server (implemented in JavaScript), and finally to a lightweight web client for real time display, a number of libraries were utilised to provide abstraction over low-level networking and visualisation mechanisms. The following libraries facilitated rapid application development, in addition to minimising the effort required for future extension or modification:

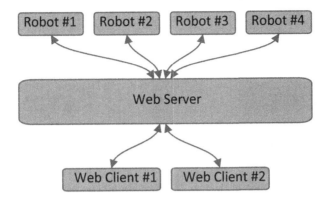

Fig. 1. Visual representation of the many-to-many NUbugger implementation, allowing multiple users to visualise the real-time performance of an arbitrary number of robots. The selected libraries provide sufficient levels of abstraction over underlying network communication that the robots and web clients need not be coded in the same language.

- **∅MQ:** Referred to as "the intelligent transport layer" by developers iMatrix, ∅MQ is a high-performance, asynchronous message library facilitating speed and control over low-level message passing [6]. It was chosen for implementation of the raw transport layer from robot to web server due to the ease of embedding into a pre-existing application; concretely, it provides direct support for C++, in addition to providing simple abstractions over multithreading and automatic reconnection.
- **Protocol Buffers:** Developed by Google, Protocol Buffers are a language-independent, cross-platform and extensible mechanism for serialising structured data [4]. As ∅MQ only provides a transportation mechanism for raw binary data, such structures are critical for the addition of application level information. Protocol Buffers allows for objects to be created and transported across the network in a packed binary form, rather than an inefficient ASCII representation (utilised by JSON or similar transport protocol alternatives). Language independence is realised in a manner that allows C++ structures to be transparently interpreted by the JavaScript browser client.
- **Socket.IO:** Developed by Guillermo Rauch, Socket.IO is a JavaScript library for real-time web applications [10], and was implemented as the raw transport layer from web server to web client. Socket.IO was chosen due to its native support for WebSockets (the only streaming technology natively supported by most modern web browsers), in addition to its transparent support of automatic reconnection and fall-back transport methods. As not all transports supported by Socket.IO support binary, Base64 encoding was applied.
- **Express:** Developed by VisionMedia, Express is a minimal and flexible node.js web application framework, providing a robust set of features for building single, multi-page and hybrid web applications [13]. Express was chosen for web server implementation due to easy integration with the

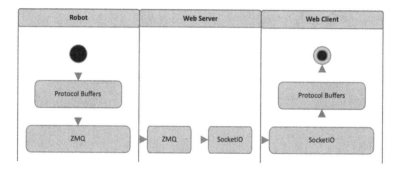

Fig. 2. Visual representation of the flow of information between the various networking packages implemented during the development of NUbugger. The selected libraries provide sufficient levels of abstraction over underlying network communication that the robots and web clients need not be coded in the same language.

JavaScript web client, in addition to its low computational expense, which allows for the option of running the web server directly on the robot (if other options are not available).
– **Three.js:** Developed by Ricardo Cabello, Three.js is a lightweight, cross-browser JavaScript library that allows animated 3-dimensional computer graphics to be displayed directly within a web browser, by providing an abstracted interface over WebGL [3]. It was implemented for the 3D rendering of the main application display.
– **Ext JS:** Developed by Sencha, Ext JS is a pure JavaScript application framework for building interactive web applications, using techniques such as Ajax, DHTML and DOM scripting [12]. It provides several widget and component templates that are commonly used within a web development context, and was used for the implementation of movable, resizable and configuration windows to contain each developed application user interface. Ext JS provides flexibility for future extensions to the user interface by allowing for the simple addition of further displays.

The flow of information between the various networking packages is illustrated in Fig. 2.

3   Functionality

As demonstrated in Sec. 2, the NUbugger implementation allows for critical system information to be streamed in real time from robot (in this case the Robotis DARwIn-OP [5]) to web client (via web server). Although readily modifiable and extensible, the NUbots debugging environment currently consists of the following elements:

– The main display, as demonstrated on the right half of Fig. 3. This interface provides a visualisation the robots self-localisation belief (i.e. the position

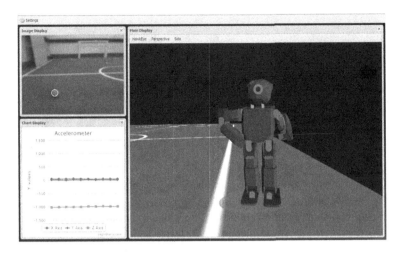

Fig. 3. The NUbots configuration of the NUbugger utility, demonstrating: the main display (right), image display (top-right) and scrolling chart display (bottom-left). These displays are able to be customised by the user to provide any manner of real-time information from the robot.

and orientation of the rendered 3D model [8]) and error (indicated by a transparent purple ellipse), in addition to the current pose of the robot, which utilises real-time accelerometer and servo positional data.

- An image display, as demonstrated on the top-left of Fig. 3 and Fig. 4b. This interface provides an indication of what the robot sees in real time, and includes a number of overlays to allow the user to visualise image classification [1] or detected salient features [2].
- A scrolling chart display, as demonstrated on the bottom-left of Fig. 3 and Fig. 4a. This interface is able to provide real-time data directly from any number of the robot's sensors, including: accelerometers (as demonstrated), gyroscopes, pressure sensors and temperature monitors.

4   Conclusion

In the short time since its development, NUbugger has already been applied with great success to the University of Newcastle's NUbots RoboCup team [9]. Concretely, it has assisted with the following low-level issues:

- **Video latency:** A previous version of the NUbots vision system maintained a buffer of 20 image frames [9], facilitating random access and explicitly enforcing thread-safe execution. Recent redevelopment of the vision system removed the need for this functionality [2], and introduced conflicts preventing buffered frames from being accessed synchronously with sensor data. This resulted in up to a 0.6 second latency in image processing; a significant

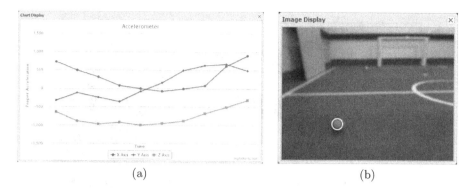

(a)   (b)

Fig. 4. Two examples of NUbugger debugging interfaces: a) the scrolling chart display, demonstrating real-time information from the robot's 3-axis accelerometer; and b) the image display, with an overlay demonstrating the correct identification of the ball [2].

issue which went undiscovered until streamed in real-time to the NUbugger utility, despite causing an observable reduction in self-localisation accuracy and overall system performance.

- **Ball detection accuracy:** An implementation error in the vision system prevented the robot's head pitch and yaw from being considered when projecting ball-localisation coordinates from image to field-plane, causing significant inaccuracy whenever the robot was not facing directly ahead. This was identified in the main NUbugger display, which provides an overlay of the robot's ball positional belief; rotating the robot's head from side-to-side caused the overlay to transcribe an arc about the robot, despite the fact that the ball remained physically stationary.

- **Camera firmware:** The Robotis DARwIn-OP robot platform is equipped with a Logitech C905 camera, which utilises the Linux UVC driver [5]. Although this driver provides control over a large subset of fundamental camera parameters (such as brightness and contrast), a small number of parameters (including auto white balance and a number of proprietary Logitech colour correction values) remain inaccessible. Although this is a known and unresolved issue, NUbugger allows for instant detection of one of the resultant errors: automatic white balance adjustment caused by lighting variations across the soccer field, resulting in a dramatic reduction in classification accuracy and therefore object recognition performance. Once the white balance of the streaming images is observed to have changed, the robot may be instantly substituted and restarted.

Although still in the process of active development, it is hoped that NUbugger will be successfully adopted by other RoboCup teams. This may be possible at a number of levels:

- The adoption of the complete NUbugger system (both server and client applications), with modifications made to the server-robot interface where necessitated by different hardware platforms.

– The adoption of the NUbugger server application, with the functionality and layout of the client modified to suit an individual team's specific needs.

In recent months, it has proven critical to the identification of low-level issues that were plaguing system performance, but near-impossible to identify using traditional debugging methodologies. The latest NUbugger source code is available for download at `https://github.com/nubots/NUbugger`, and the authors are happy to collaborate with other RoboCup teams in the extension of the framework to better incorporate their needs.

References

1. Budden, D., Fenn, S., Mendes, A., Chalup, S.: Evaluation of colour models for computer vision using cluster validation techniques. In: Chen, X., Stone, P., Sucar, L.E., van der Zant, T. (eds.) RoboCup 2012. LNCS (LNAI), vol. 7500, pp. 261–272. Springer, Heidelberg (2013)
2. Budden, D., Fenn, S., Walker, J., Mendes, A.: A novel approach to ball detection for humanoid robot soccer. In: Thielscher, M., Zhang, D. (eds.) AI 2012. LNCS, vol. 7691, pp. 827–838. Springer, Heidelberg (2012)
3. Cabello, R.: Three.js (2013), `https://github.com/mrdoob/three.js/`
4. Google: Protocol Buffers (2012),
   `https://developers.google.com/protocol-buffers/`
5. Ha, I., Tamura, Y., Asama, H., Han, J., Hong, D.: Development of open humanoid platform DARwIn-OP. In: Proceedings of SICE Annual Conference, SICE 2011, pp. 2178–2181. IEEE (2011)
6. iMatrix: ØMQ: The Intelligent Transport Layer (2013), `http://www.zeromq.org/`
7. Kitano, H., Asada, M.: The robocup humanoid challenge as the millennium challenge for advanced robotics. Advanced Robotics 13(8), 723–736 (1998)
8. Michel, O.: Webots: Symbiosis between virtual and real mobile robots. In: Heudin, J.-C. (ed.) Virtual Worlds 1998. LNCS (LNAI), vol. 1434, pp. 254–263. Springer, Heidelberg (1998)
9. Nicklin, S.P., Bhatia, S., Budden, D., King, R.A., Kulk, J., Walker, J., Wong, A.S., Chalup, S.K.: The nubots team description for (2011)
10. Rauch, G.: Socket.IO, `http://socket.io/` (2012)
11. Schaller, R.R.: Moore's law: past, present and future. IEEE Spectrum 34(6), 52–59 (1997)
12. Sencha: Sencha Ext JS: JavaScript Framework for Rich Desktop Apps. (2013), `http://www.sencha.com/products/extjs`
13. VisionMedia: Express: Web application framework for node (2013), `http://expressjs.com/`

The 2012 UT Austin Villa Code Release

Samuel Barrett, Katie Genter, Yuchen He, Todd Hester, Piyush Khandelwal, Jacob Menashe, and Peter Stone

Department of Computer Science,
The University of Texas at Austin
{sbarrett,katie,hychyc07,todd,piyushk,jmenashe,pstone}@cs.utexas.edu
http://www.cs.utexas.edu/~AustinVilla

Abstract. In 2012, UT Austin Villa claimed the Standard Platform League championships at both the US Open and the 2012 RoboCup competition held in Mexico City. This paper describes the code release associated with the team and discusses the key contributions of the release. This release will enable teams entering the Standard Platform League and researchers using the Naos to have a solid foundation from which to start their work as well as providing useful modules to existing researchers and RoboCup teams. We expect it to be of particular interest because it includes the architecture, logic modules, and debugging tools that led to the team's success in 2012. This architecture is designed to be flexible and robust while enabling easy testing and debugging of code. The vision code was designed for easy use in creating color tables and debugging problems. A custom localization simulator that is included permits fast testing of full team scenarios. Also included is the kick engine which runs through a number of static joint poses and adapts them to the current location of the ball. This code release will provide a solid foundation for new RoboCup teams and for researchers that use the Naos.

Keywords: RoboCup, Nao, SPL, UT Austin Villa.

1 Introduction

Developing intelligent behavior for robots is difficult due to the time and effort needed to create and debug code for robots. This code release attempts to reduce that burden for researchers that use humanoid robots. Specifically, it contains the code from the UT Austin Villa team that competes in RoboCup, the Robot Soccer World Cup. The international research initiative behind RoboCup pushes towards advancing robotics and artificial intelligence by using the game of soccer as a test domain. Soccer requires robots to deal with real world issues such as teamwork, localization, motion, and vision all in real time while interacting with an environment that is not under their control. The long-term goal of RoboCup is to build a team of 11 humanoid robot soccer players that can beat the best human soccer team on a real soccer field by the year 2050 [5].

RoboCup is organized into several leagues, including both simulation leagues and leagues that compete with physical robots. This paper describes the code used by the 2012 championship team in the Standard Platform League (SPL)[1].

---

[1] http://www.tzi.de/spl/

S. Behnke et al. (Eds.): RoboCup 2013, LNAI 8371, pp. 552–559, 2014.
© Springer-Verlag Berlin Heidelberg 2014

All teams in the SPL compete with identical Aldebaran Nao humanoid robots robots[2], making it essentially a software competition. This format allows for teams to share ideas and code, such as this code release, to speed the development of capable, intelligent robots.

UT Austin Villa has competed in the Standard Platform League with the Nao robots every year since the Nao was introduced in 2008. Through these years, we have built a substantial code infrastructure for robot soccer that served as the base for our championship in 2012 [2,1]. This paper describes the partial release of that code base, including its software architecture, stream-lined vision processing, localization, localization simulator, kick engine, and debug tool. The high level strategy and behavior code is omitted, but is described in [3,1].

2   Code Release

This section describes the steps necessary to use the released code[3] and adapt it to new situations. More documentation is included in the code release in the form of a README and comments in the code. The vision system is based on colors, which it interprets through the use of color tables. Generating these color tables is done with the provided UTNaoTool, using a combination of painting colors onto the images and annotating objects in the images as described in Section 4. In addition, as the included behavior only directly walks towards and kicks the ball, it will be necessary to create more complex behaviors. Further behaviors can be created using a task hierarchy that is written in Lua for rapid development and testing.

The main components of the code are contained in the `core` directory, with modules separated into their respective directories, but note that the behavior code is contained in the `lua` directory. The main access point to modules is the `processFrame` function, and the main control of the vision and cognition process is primarily handled by the `processFrame` function in the file `lua/init.lua`. The main control of the motion process is handled by `processFrame` in `MotionCore.cpp`. The NaoQi interface is primarily found in `interfaces/nao/src/naointerface.cpp`.

In 2012, UT Austin Villa won the Standard Platform League at the 16th International RoboCup Competition in Mexico City, Mexico. Twenty-five teams entered the competition, and games were played with four robots on each team. Our code release allows new teams to ramp up quickly with modules that have been demonstrated to support a RoboCup championship.

In principle, this code release can be useful for projects other than RoboCup and on robots other than the Nao. The included tools for color-based vision are easy and quick to use. The localization algorithms used are general and can be extended to use other types of landmarks, while coordinating with other robots. Furthermore, the most useful tool included is the general infrastructure of the memory and modules. This infrastructure permits creating and replaying logs

---

[2] http://www.aldebaran.com/
[3] http://www.cs.utexas.edu/~AustinVilla/?p=downloads/
source_code_and_binaries

in a simple and customizable manner, and these logs are vital for understanding and debugging behaviors on robots. In addition, it allows for decoupling the code modules, while still maintaining good computational performance. This infrastructure is described in more depth in the following section.

3   Software Architecture

When dealing with robots, it is important that the code be both easily debugable and testable. The design's key element is to enforce that the environment *interface* and its *logic* are kept distinct (Figure 1), and that all communication is done via shared *memory*. The interface simply extracts sensor data into memory and then sends commands to the joints, whether the system is a robot, simulator, or debug tool. This design allows the same logic to be applied regardless of the underlying system being used. The advantages of this architecture are discussed further in [1].

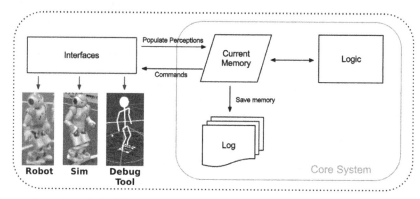

Fig. 1. Overview of the UT Austin Villa software architecture, which separates interface, memory, and logic

Furthermore, for better modularity, the logic is organized into a number of modules that each take care of specific functionality. These modules are discussed in depth in Sections 4–6. In addition, the memory is split into memory blocks that group related pieces of information and allow us to analyze which modules rely on each piece of information. Modules communicate via the memory blocks; for example, the higher level strategy behavior talks to the walk and kick engines in the form of motion request blocks and get information about odometry and the current status of the kick in the form of information blocks.

To take advantage of the modularity of our code, the execution is split into three separate processes:

1. Interface: this code runs at 100 hz, is called by naoqi (or a simulator), and copies the current sensor information to memory and writes out the current motion commands
2. Motion: this code runs at 100 hz and converts high level commands into direct commands for the motors
3. Vision: this code runs at 30 hz and processes the incoming imaging and selects the high level commands

Splitting execution into three processes allows us to restart crashed processes, as well as debug each process separately. In addition, it aids in development as changing code in the motion and vision processes does not require the slow procedure of restarting the NaoQi interface on the robots. However, this design requires us to store our memory structure in shared memory. The majority of the information is kept separate for each of the processes, so mutex locks are only required when copying local information into these shared blocks.

The walk engine included in the release was developed by the B-Human team from the University of Bremen [8], using the code for interfacing with this walk engine released by Bowdoin College's Northern Bites team [6]. All the other major modules, comprising the large majority of the codebase, were developed by the Austin Villa team. These modules are described in the following sections.

4   Vision

The vision system is dedicated to the task of detecting the ball, goals, field lines, and robots in the camera images and reporting their relative distance and bearings to the robot. The team's vision system divides the object detection task into 4 stages, each of which is carried out on both cameras. These stages are listed below:

1. **Segmentation** - The raw image is read and segmented using a color table.
2. **Blob Formation** - The segmented image is scanned using horizontal and vertical scan lines, and "blobs" are formed for further processing.
3. **Object Detection** - The blobs are merged into different objects.
4. **Transformation** - The information given by the pose of the robot is used to generate ground plane transformations of the line segments detected.

In addition, our tool's vision analysis system provides a set of key features for debugging and tuning color calibrations. In particular, the log annotation and color table generation features are useful for creating and analyzing color tables with measurable accuracy. Since the vision system is based on segmented images, proper construction of color tables is key to its effectiveness.

The overall framework for these processing steps is outlined in detail along with high level descriptions of the color table analysis system in our previous papers [2,1]. We therefore focus on new implementation techniques that are found in this code release, the first of which is the arrangement of object detection methods into segmentation and detector modules.

The vision system relies on a handful of primary classes:

- `TransformationProvider` - This class converts body position estimates into transforms for mapping pixel values into world locations and vice versa.
- `Classifier` - Classifies pixels by color and constructs vertical and horizontal runs of contiguous color. This is the segmentation phase.
- `BlobDetector` - Merges adjacent runs of common color into a number of blobs. This is the blob formation phase.
- `[Line,Ball,Goal,Cross,Robot]Detector` - These detector classes run in sequence to transform blobs into complete object detections. Each detector

is responsible for finding relevant blobs, performing sanity checks, and indicating results to the global world object array. Various statistics such as bearing and image position are stored as well.

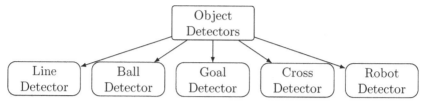

It should be noted that the cross detector, which is responsible for discerning the cross directly in front of the goal box, was used only for the goalie as a localization aid. In general it can be difficult to discern the cross from other lines on the field due primarily to occlusion, however in the setting of the goalie we are able to enforce more restrictive assumptions on what qualifies as the cross. Other field lines are too ambiguous to provide the goalie with enough certainty in its position. Goal posts provide a good alternative to field lines as localization anchors, however they often lie outside the goalie's field of view and would thus require a complete head turn to maintain positional certainty. The cross overcomes both of these drawbacks by providing a unique localization anchor that is consistently between the goalie and the main area of play.

The vision system also implements selective high-resolution scanning based on subsampled pixel values, which is used most effectively for locating the ball across large distances. In normal operation, each of the listed object detectors relies on a 320x240 subsampled segmented image to ensure that vision processing runs at the hardware-enforced framerate of 30 Hz. At any point a detector may request a high-resolution scan of the top camera image for a particular color from the Classifier. The Classifier then reclassifies all regions surrounding subsampled pixels of the detected color using the full 640x480 resolution allowed by our implementation. Generally only small regions of the image are reclassified, thus providing the benefits of full resolution images with virtually no hit to performance. By using this technique we are able to reliably detect the ball at distances up to 5 meters, up by a factor of 2 over detections with subsampled images. High resolution scans of the goal are also enabled, which similarly improves goal detection distances.

5   Localization

Keeping track of the location of the robot is important in all robot tasks, especially in the RoboCup domain. The goal of the localization system is to take in the detected objects from vision, and output the location of the robot, the ball, the robot's teammates, and the robot's opponents. Having a good world model like this enables the robot to make smart strategic decisions during the game. Like many other teams at the competition [7,4], since 2011 we have used a multi-modal 7-state Unscented Kalman Filter (UKF) for localization [2]. The

UKF provides a number of benefits compared to other approaches such as Monte Carlo localization as it is computationally efficient and enables easy sharing and integration of ball information between teammates.

One of the challenges of the RoboCup Standard Platform League in 2012 is that there are no unique markers anywhere on the field. Therefore, the robots must do a good job of integrating both odometry and vision estimates of landmarks while recovering from noise added by bumps, falls, and kidnapping. Since the field is symmetrical, the robot has no information to recover its orientation on its own. Instead, the robot takes advantage of shared information from teammates to resolve which direction it is facing. Our solution to resolving this issue was to have robots check whether their teammates thought the ball was in the same location or in the symmetrically opposite location. It is important to listen to more than one teammate, as we do not want one teammate that is going the wrong way to convince the entire team to start going the wrong direction. This functionality is described in the **processSharedBall** and **checkSharedBallForFlips** functions in the file **localization/UKFModule.cpp**.

However, testing and debugging localization with full teams of robots is quite challenging. Therefore, we wrote a localization simulator to implement, test, and debug our solutions, shown in Figure 2. The simulator can be run from the world window of the tool included in the code release. The simulator takes advantage of the modularity of our code. Instead of interfacing with the code at the perception and motor command layers, it populates the code with simulated output

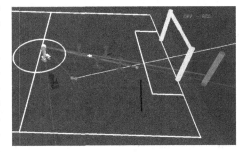

Fig. 2. The localization simulator

from the vision module, and then uses the code's output to the kicking and walking modules to move the robots in the simulation. The simulator proved useful not just for localization, but also for testing various strategies and behaviors.

The main interface into the localization module is in the **processFrame** function in the file **localization/UKFModule.cpp**, and the updates from observations is handled by the **processObservations** function. The UKF has many parameters that affect how it handles observations and odometry updates. These parameters were initially optimized using a number of long logs of the robot's observations and movements. Additional tuning was performed by hand to handle issues that were discovered during test games.

6   Kick Engine

The code release includes our static standing kick engine, which is found in the file **motion/KickModule.cpp**. This kick engine utilizes a series of keyframe states that the robot moves through at specified times. For each state, we define a state time, a joint movement time, the desired center of mass in x,y,z space, and the desired kicking foot position in x,y,z space. These values are defined for each state in the file `lua/cfgkick.lua`. The robot's joints are controlled using inverse kinematics to reach each desired state in the allocated joint movement time. In particular, our kick engine utilizes the states shown in Table 1.

Table 1. States utilized by our static kick engine

State	Description
Stand	stand at the same height as when walking
Shift	shift the center of mass away from the kicking foot
Lift	lift the kicking foot
Align	align the kicking foot
Spline	swing the kicking foot through the ball
ResetFoot	reset the kicking foot such that it is centered under the body
FootDown	place the kicking foot back on the ground
ShiftBack	shift the center of mass back towards the center of the body
FinishStand	stand at the same height as when walking

In the Spline state, splines are used to compute the path for the kicking foot to follow as it moves forward through the ball. We use splines in this state to obtain a smooth path for the kicking foot to follow, as informal experimentation showed that a smooth kicking foot trajectory resulted in a more consistent kick. You might note that two additional states — Kick1 and Kick2 — are defined in the file `lua/cfgkick.lua`. The state time for these states is set to 0 though, so these states are not currently utilized by the kick engine (although the target for the end of the Spline state uses the desired swing foot position defined for the Kick2 state). However, one could remove the Spline state by setting its state time to 0, and give positive state times to the Kick1 and Kick2 states, to obtain a kick that does not use splines to compute the path for the foot to follow as it moves through the ball.

The kick engine obtains the desired kick distance by controlling the amount of time needed to move the foot during the Spline state. For RoboCup 2012, the only static standing kicks that the robots used were straight kicks ranging between 1.5 meters and 3.5 meters. Hence, 190 ms were spent in the Spline state for the 3.5 meter kicks, while 300ms were spent in the Spline state for the 1.5 meter kicks.

The kick engine alters the desired swing foot position for the Align and Spline states of the kick engine based on the desired kick distance and the current ball position with respect to the kicking foot. The complete details of this kick engine are available in [2], and the improvements to it for 2012 are available in [1].

7 Conclusion

This paper describes the partial release of 2012 RoboCup SPL champions UT Austin Villa's 2012 code base. The code release includes UT Austin Villa's software architecture, stream-lined vision processing, localization, localization simulator, motion engine, and debug tool. This release omits the strategy and high level behaviors for playing soccer. Importantly, all of the code was developed on a flexible architecture developed previously, that enables easy testing and debugging of code. This code release is intended to serve as a useful foundation for any team entering the RoboCup SPL competition as well as any researchers using the Nao robots or similar platforms.

Acknowledgments. This code release builds upon the work of all of the past UT Austin Villa teams. We thank all the previous team members for their important contributions, especially Michael Quinlan and Mohan Sridharan. This work has taken place in the Learning Agents Research Group (LARG) at UT Austin. LARG research is supported in part by NSF (IIS-0917122), ONR (N00014-09-1-0658), and the FHWA (DTFH61-07-H-00030).

References

1. Barrett, S., Genter, K., He, Y., Hester, T., Khandelwal, P., Menashe, J., Stone, P.: UT Austin Villa 2012: Standard Platform League world champions. In: Chen, X., Stone, P., Sucar, L.E., van der Zant, T. (eds.) RoboCup 2012. LNCS (LNAI), vol. 7500, pp. 36–47. Springer, Heidelberg (2013)
2. Barrett, S., Genter, K., Hester, T., Khandelwal, P., Quinlan, M., Stone, P., Sridharan, M.: Austin Villa 2011: Sharing is caring: Better awareness through information sharing. Technical Report UT-AI-TR-12-01, The University of Texas at Austin, Department of Computer Sciences, AI Laboratory (January 2012)
3. Barrett, S., Genter, K., Hester, T., Quinlan, M., Stone, P.: Controlled kicking under uncertainty. In: The Fifth Workshop on Humanoid Soccer Robots at Humanoids (December 2010)
4. Jochmann, G., Kerner, S., Tasse, S., Urbann, O.: Efficient multi-hypotheses unscented Kalman filtering for robust localization. In: Röfer, T., Mayer, N.M., Savage, J., Saranlı, U. (eds.) RoboCup 2011. LNCS, vol. 7416, pp. 222–233. Springer, Heidelberg (2012)
5. Kitano, H., Asada, M., Kuniyoshi, Y., Noda, I., Osawa, E.: RoboCup: The robot world cup initiative. In: Proceedings of The First International Conference on Autonomous Agents. ACM Press (1997)
6. Neamtu, O., Dawson, W., Googins, E., Jacobel, B., Mamantov, L., McAvoy, D., Mende, B., Merritt, N., Ratner, E., Terman, N., Zalinger, J., Morrison, J., Chown, E.: Northern Bites code release (2012), https://github.com/northern-bites
7. Quinlan, M.J., Middleton, R.H.: Multiple model Kalman filters: A localization technique for RoboCup soccer. In: Baltes, J., Lagoudakis, M.G., Naruse, T., Ghidary, S.S. (eds.) RoboCup 2009. LNCS (LNAI), vol. 5949, pp. 276–287. Springer, Heidelberg (2010)
8. Röfer, T., Laue, T., Müller, J., Fabisch, A., Feldpausch, F., Gillmann, K., Graf, C., de Haas, T.J., Härtl, A., Humann, A., Honsel, D., Kastner, P., Kastner, T., Könemann, C., Markowsky, B., Riemann, O.J.L., Wenk, F.: B-Human team report and code release (2011), http://www.b-human.de/downloads/bhuman11_coderelease.pdf

Ground Truth Acquisition of Humanoid Soccer Robot Behaviour

Andrea Pennisi, Domenico D. Bloisi, Luca Iocchi, and Daniele Nardi

Dept. of Computer, Control, and Management Engineering,
Sapienza University of Rome, via Ariosto 25, 00185, Rome, Italy
{pennisi,bloisi,iocchi,nardi}@dis.uniroma1.it

Abstract. In this paper an open source software for monitoring humanoid soccer robot behaviours is presented. The software is part of an easy to set up system, conceived for registering ground truth data that can be used for evaluating and testing methods such as robot coordination and localization. The hardware architecture of the system is designed for using multiple low-cost visual sensors (four Kinects). The software includes a foreground computation module and a detection unit for both players and ball. A graphical user interface has been developed in order to facilitate the creation of a shared multi-camera plan view, in which the observations of players and ball are re-projected to obtain global positions. The effectiveness of the implemented system has been proven using a laser sensor to measure the exact position of the objects of interest in the field.

1 Introduction

A ground truth system is a necessary tool to evaluate and improve algorithms dealing with a series of challenging tasks in the RoboCup competitions. In particular, visual data collected from a global view of the environment can be used for validating innovative methods concerning the following aspects:

- Coordination;
- Localization;
- Game strategies;
- Feedback for adaptive methods;
- Quantitative measurements;
- Debugging.

Indeed, since humanoids robots have only a local and limited point of view of the environment in which they operate, tasks like coordination and localization are still open problems [6] and there is the need of creating and adopting publicly available benchmarks in order to validate and quantitatively evaluate different solutions.

In this paper, an open source software for monitoring humanoid soccer robot behaviours in the Standard Platform League (SPL) is described. The software

S. Behnke et al. (Eds.): RoboCup 2013, LNAI 8371, pp. 560–567, 2014.
© Springer-Verlag Berlin Heidelberg 2014

is part of an easy and fast to set up system, designed to give a global point of view of the observed scene that is useful in order to recognize and evaluate the behaviour of the robots.

The system aims at providing:

1. A simple and fast calibration set up;
2. The foreground mask for each captured frame;
3. 3D information about each player in the field as well as the position of the ball in each frame;
4. A multi-camera data fusion scheme;
5. The set of the tracks representing the objects of interest.

The proposed approach is robust to the presence of people on the field (e.g., the referees), illumination changes, shadows, and modifications in the background geometry (e.g., the audience around the field).

The remainder of the paper is organized as follows. Related work is analysed in Section 2, while Section 3 describes both the hardware architecture and the software modules of the system. The software tools as well as the evaluation of the results coming from a first data set are described in Section 4. Conclusions are drawn in Section 5.

2   Related Work

Zickler *et al.* in [12] propose a shared vision system for RoboCup competition and use it to estimate a ground truth for NAO robots. Two RGB cameras are used to span the field and a set of makers on the top of the robots are used to identify them. Since markers are not allowed in the competition games, such an approach is limited to collecting ground truth information for testing purposes.

De Morais *et al.* in [4] describe a method for estimating the position of indoor soccer players by using multiple cameras in real indoor soccer games. The proposed set up consists of four cameras positioned around the soccer court. A simple object detector runs on each camera and projects the detections onto a world plan representing the soccer court. The main drawback of such a method is the need of a long calibration time for each camera.

Niemüller *et al.* in [9] present a data collection system for NAO robots. Reflective markers attached to NAO's body are used in combination to a 6D professional vision-based body motion tracking system. The position and the orientation of the observed robots are tracked by using 15 pulsed infra-red cameras placed on the field. Although the acquired data have a high accuracy, the system is difficult to set up and calibrate. Moreover, markers are not allowed in the competition games and infra-red cameras are quite expensive sensors.

A low-cost and portable system using a single Kinect has been proposed by Khandelwal and Stone in [5]. The system is designed to find the location of robots and ball in the SPL environment. No special identifiers on the robots are required, although 22 known landmarks must be placed on the field to carry out the calibration. The correspondences between the point cloud acquired from the

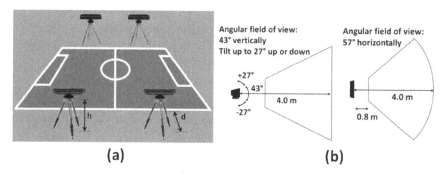

Fig. 1. Hardware architecture of the proposed system. a) Each Kinect is mounted on a tripod at an height h from the ground (in our setup $h = 1.05\ m$) and at a distance d from the long side line (in our setup $d = 0.8\ m$). b) Vertical and Horizontal field of view for the Kinect.

Kinect and the position of the landmarks are obtained through a graphical user interface (GUI). The user clicks a pixel in the image and the system performs the association. Only the points above 0.30 m of height are considered and a Euclidean method for cloud clusterization is used. However, the system cannot provide the orientation of the robots and it allows to monitor only a part of the field.

Compared to related work, the main improvements of our approach are: 1) A network of low-cost sensors to cover the entire field is used; 2) No markers are needed on the field or on the robots; 3) An accurate foreground mask is extracted for each frame; 4) The orientation of each robot is estimated from its 3D point cloud. The details of our method are described in the remainder of the paper.

3   System Description

In this section the hardware architecture of the proposed system is described, followed by the details of each software module involved in the ground truth registration.

3.1   Hardware Architecture

The hardware architecture of the system is shown in Fig. 1a. Four Kinects [7] are placed along the long side lines of the field, two for each side. The choice of this setting derives from the physical limitations of the sensors [8], that are reported in Fig. 1b in terms of the vertical and the horizontal angular field of view. Given the above limitations and the dimensions of the field to be monitored ($9 \times 6\ m^2$ for the SPL [10]), we find that a reasonable set up is the following. Each Kinect is mounted at about 1 m height, with the tilt angle set to the maximum achievable down value (-27°), at a distance of 0.8 m from the side line. This is an easy

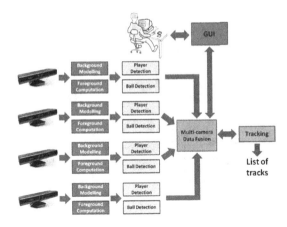

Fig. 2. Software modules

and reproducible set up, specially with respect to other approaches that require to place the sensor(s) on the top of the field and/or at a considerable height (e.g., [9]). Furthermore, the given distances are only suggested and can vary from sensor to sensor, because the system can receive those measures as input (see Section 4). In order to acquire the images, we connected the two Kinects on the same side of the field to a commercial notebook (CPU Intel Core i7, 2.4 GHz, 8 GB RAM).

3.2 Software Modules

The software architecture of the system is shown in Fig. 2. For each captured frame, the foreground mask is extracted and the positions of the ball and players are estimated (Fig. 3). A data fusion module re-projects all the detected positions on a shared plan view of the field (Fig. 4), while a tracking module generates the list of tracks in output.

To compute an accurate foreground mask both RGB and depth information are used [3]. An RGB image (Fig. 3a) and a 16 bit depth map are stored for each captured frame; they are labelled with a timestamp and a sequence number (for synchronization purposes). A statistical approach, called IMBS [1], is used to create the background model that is updated every 30 seconds for dealing with illumination changes.

The positions of the players are obtained by using the foreground mask (Fig. 3b) in combination with the depth map (Fig. 3c). A set of blobs (*observations*) is extracted by means of a height filter that considers as valid observations only the blobs under 0.70 m (in this way, humans on the field are filtered out). Moreover, the pixels that belong to shadows are also suppressed since they are recognized as ground points on the basis of their height values.

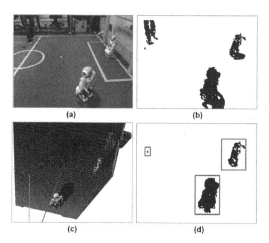

Fig. 3. Player and ball position estimation. a) Original frame. b) Foreground extraction. c) 3D scene reconstruction. d) Player and ball detection.

An example of player detection is shown in Fig. 3d. In order to estimate the robot position to be re-projected on the plan view, the lower point of the blob is considered and it is transformed in world coordinates.

A surface normal approach [11] is used to detect the orientation of the robot. Given the set of 3D points of each blob, the problem of determining the normal to a point on the surface is approximated by estimating the normal of a plane tangent to the surface, thus resulting in a least-square plane fitting estimation problem. Therefore, surface normal estimation is reduced to an analysis of the eigenvectors and eigenvalues of a covariance matrix created from the nearest neighbours of the query point, where the sign of the normal is assigned by knowing the view point of the scene.

A color based approach is used for detecting the ball. The RGB image is converted into the HSV color space and the ball is detected searching for a red coloured blob. Then, the same transformation applied to the players' positions is used.

Multi-camera data fusion is carried out thanks to four homography matrices that are used to generate the plan view of the field (Fig. 4). The homography matrices are calculated through a GUI (that is described in the next section) and are used to generate a warp image for each Kinect. Then, all the four warp images are stitched together to reconstruct the whole field using the Stitcher class provided by OpenCV.

For tracking both the players and the ball a multi-hypothesis approach based on a set of Kalman Filters has been chosen [2]. Data association is used to determine relationships between observations and tracks, but multiple hypotheses are maintained when observations may be associated to more than one track.

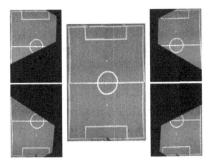

Fig. 4. Plan view. Multiple camera data fusion is carried out merging the four warp images generated using the GUI.

4   System Outcomes

We created the web site `http://labrococo.dis.uniroma1.it/gnao` containing the C++ source code for all the software modules and for the GUI, whose features are described in the following. In addition, we made available for downloading also a first data set used to obtain the measurements discussed in this section.

4.1   Graphical User Interface (GUI)

The GUI supports the calibration of the four Kinects through a two step procedure (Fig. 5). In the first step, the GUI allows to load up to four different Kinect views (Fig. 5a).

Pressing the button "Execute" under each Kinect view it is possible to start the homography based calibration (Fig. 5b). A homography is a relation between points that belong to two different spaces S and S', in which each point of S coincides to only one point of S'. Since the problem of 2D homography consists

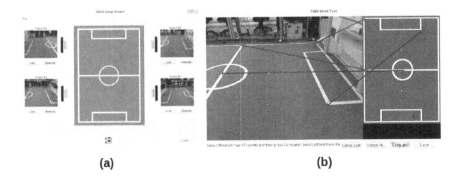

(a)   (b)

Fig. 5. Graphical User Interface. a) Camera assignment. b) Point selection.

in finding a transformation that converts the points from an image plane to the points of another plane, the GUI allows the user to select a point in the image coming from a Kinect and its corresponding point in a virtual representation of the field (right side of Fig. 5b). The user is required to select at least four correspondences in order to compute the homography, although selecting more points can produce a more accurate transformation. When the user ends the selection of the points, a text file that contains the homography related to the selected Kinect is saved by pressing the button "Compute". The point selection procedure must be repeated for each Kinect and it is also possible to save the tilt angle and the height of each sensor.

4.2   Results on a First Data Set

We registered a first data set of ground truth data capturing three NAOs by using four Kinects on a $4 \times 3\ m^2$ training field. We made three runs (each one about 20 minutes long) obtaining a computational speed of about 15 frames per second acquiring 640×480 images. In Fig. 6 two frames from the collected data set show the output of the robot orientation extraction routine.

Fig. 6. Extraction of the robot orientation

Table 1. Quantitative evaluation for the player estimation method

Experiment	Ground Truth	Estimated Position	Error
	3.398 m	3.489 m	0.09
1	4.078 m	4.1268 m	0.049
	2.932 m	2.681 m	0.25
	3.349 m	3.357 m	0.008
2	4.156 m	4.125 m	0.03
	2.761 m	2.578 m	0.182
	3.566 m	3.344 m	0.221
3	4.002 m	4.08 m	0.078
	2.794 m	2.578 m	0.215
	2.784 m	2.658 m	0.125
4	2.76 m	2.562 m	0.197
	2.914 m	2.635 m	0.278
	3.627 m	3.781 m	0.154
5	2.85 m	3.008 m	0.158
	3.133 m	3.609 m	0.476
	2.56 m	2.54 m	0.018
6	3.135 m	3.015 m	0.119
	4.101 m	3.952 m	0.148

To quantitatively evaluate the accuracy of the proposed set up, we measured the exact position of the players in the field using a laser range finder. Then, we compared the results obtained by using the laser with the measurements provided by the system. The results, obtained monitoring three NAOs placed in six different relative positions, are reported in Tab. 1 (units in meters).

5   Conclusions

In this paper, an open source software for creating a ground truth acquisition system for evaluating humanoid soccer robot behaviours is described. The software is divided in modules and it is designed to work in a low-cost multi-camera architecture. We used four Kinect sensors that can be calibrated using a specifically designed GUI (also made available). Both the hardware architecture and all the software modules needed to achieve the goal have been discussed in detail. The main features of the proposed approach are as follows: 1) Fast and easy multi-camera set up; 2) No markers needed; 3) Real-time foreground mask computation; 4) Orientation of each robot estimated from its 3D point cloud. The proposed approach can be easily adapted to be suitable for other RoboCup leagues.

References

1. Bloisi, D., Iocchi, L.: Independent Multimodal Background Subtraction. In: CompIMAGE, pp. 39–44 (2012)
2. Bloisi, D.D., Iocchi, L.: ARGOS - A Video Surveillance System for Boat Traffic Monitoring in Venice. IJPRAI 23(7), 1477–1502 (2009)
3. Bloisi, D.D., Iocchi, L., Monekosso, D.N., Remagnino, P.: A Novel Segmentation Method for Crowded Scenes. In: VISAPP, pp. 484–489 (2009)
4. De Morais, F.E., Goldenstein, S., Roch, A.: Automatic Localization of Indoor Soccer Players from Multiple Cameras. In: VISAPP, 205–212 (2012)
5. Khandelwal, P., Stone, P.: A Low Cost Ground Truth Detection System Using the Kinect. In: Röfer, T., Mayer, N.M., Savage, J., Saranlı, U. (eds.) RoboCup 2011. LNCS, vol. 7416, pp. 515–527. Springer, Heidelberg (2012)
6. Li, X., Lu, H., Xiong, D., Zhang, H., Zheng, Z.: A Survey on Visual Perception for RoboCup MSL Soccer Robots. Int. J. Adv. Robotic Sy. 10, 1–10 (2013)
7. Microsoft Corp. Redmond WA: Kinect for Xbox 360
8. MSDN Library: Skeletal Tracking, http://msdn.microsoft.com/en-us/library/hh973074.aspx#ID4ENB
9. Niemüller, T., Ferrein, A., Eckel, G., Pirro, D., Podbregar, P., Kellner, T., Rath, C., Steinbauer, G.: Providing Ground-truth Data for the Nao Robot Platform. In: Ruiz-del-Solar, J., Chown, E., Plöger, P.G. (eds.) RoboCup 2010. LNCS (LNAI), vol. 6556, pp. 133–144. Springer, Heidelberg (2010)
10. RoboCup Technical Committee: RoboCup Standard Platform League (Nao) Rule Book, http://www.tzi.de/spl/pub/Website/Downloads/Rules2013.pdf
11. Rusu, R.B.: Semantic 3D Object Maps for Everyday Manipulation in Human Living Environments. Articial Intelligence (KI - Kuenstliche Intelligenz) (2010)
12. Zickler, S., Laue, T., Birbach, O., Wongphati, M., Veloso, M.: SSL-Vision: The Shared Vision System for the RoboCup Small Size League. In: Baltes, J., Lagoudakis, M.G., Naruse, T., Ghidary, S.S. (eds.) RoboCup 2009. LNCS (LNAI), vol. 5949, pp. 425–436. Springer, Heidelberg (2010)

Humanoid TeenSize Open Platform NimbRo-OP

Max Schwarz, Julio Pastrana, Philipp Allgeuer, Michael Schreiber,
Sebastian Schueller, Marcell Missura, and Sven Behnke

Autonomous Intelligent Systems, Computer Science, Univ. of Bonn, Germany
{schwarzm,schuell1,behnke}@cs.uni-bonn.de,
{pastrana,allgeuer,schreiber,missura}@ais.uni-bonn.de
http://ais.uni-bonn.de/nimbro/OP

Abstract. In recent years, the introduction of affordable platforms in
the KidSize class of the Humanoid League has had a positive impact on
the performance of soccer robots. The lack of readily available larger
robots, however, severely affects the number of participants in Teen-
and AdultSize and consequently the progress of research that focuses
on the challenges arising with robots of larger weight and size. This paper presents the first hardware release of a low cost Humanoid TeenSize
open platform for research, the first software release, and the current
state of ROS-based software development. The NimbRo-OP robot was
designed to be easily manufactured, assembled, repaired, and modified.
It is equipped with a wide-angle camera, ample computing power, and
enough torque to enable full-body motions, such as dynamic bipedal locomotion, kicking, and getting up.

1   Introduction

Low-cost and easy to maintain standardized hardware platforms, such as the
DARwIn-OP [1], have had a positive impact on the performance of teams in
the KidSize class of the RoboCup Humanoid League. They lower the barrier
for new teams to enter the league and make maintaining a soccer team with
the required number of players easier. Out-of-the-box capabilities like walking
and kicking allow the research groups to focus on higher-level perceptual or
behavioral skills, which increases the quality of the games and, hence, the attractiveness of RoboCup for visitors and media. In the competition classes with
larger robots, teams so far are forced to participate with self-constructed robots.
Naturally, this severely affects the number of participants willing to compete,
and, in consequence, the progress of the research that attempts to solve the
challenges arising with robots of larger weight and size.

Inspired by the success of DARwIn-OP, we developed a first prototype of a
TeenSize humanoid robot and released it as an open platform. Our NimbRo-OP
bipedal prototype is easy to manufacture, assemble, maintain, and modify. The
prototype can be reproduced at low cost from commonly available materials and
standard electronic components. Moreover, the robot is equipped with configurable actuators, sufficient sensors, and enough computational power to ensure
a considerable range of operation: from image processing, over action planning,
to the generation and control of dynamic full-body motions. These features and

S. Behnke et al. (Eds.): RoboCup 2013, LNAI 8371, pp. 568–575, 2014.

the fact that the robot is large enough for acting in real human environments make NimbRo-OP suitable for research in relevant areas of humanoid robotics.

2   Related Work

In the KidSize class, a number of robust and affordable off-the shelf products suitable for operation on the soccer field are available. Entry-level construction kits including Bioloid [2] are a very cost effective way to enter the competitions. However, the limited capabilities of these construction kits are an obstacle for achieving the high performance required by soccer games.

With a body height of approximately 58 cm, the Nao robot [3], produced by Aldebaran Robotics, replaced Sony's quadruped Aibo as the robot used in the RoboCup Standard Platform League. Nao is an attractive platform because it offers a rich set of features and reliable walking capabilities. However, the fact that it is a proprietary product is a disadvantage because one has restricted possibilities for customizing and repairing the hardware.

The recently introduced DARwIn-OP [1] is very popular in the KidSize class. In 2011 and 2012, team DARwIn won the KidSize competitions of RoboCup and successfully demonstrated its potential. DARwIn-OP has been designed to be assembled and maintained by the owner, but a fully operational version can be ordered from Robotis. Most importantly, DARwIn-OP has been released as an open platform. Software and construction plans are available for public access.

The limitations of the KidSize robots lay in their size. Falling—an undesired but inevitable consequence of walking on two legs—is a negligible problem for small robots. Larger robots, however, can suffer severe damage as result of a fall. A number of commercial platforms are available with sizes larger than 120 cm. The most prominent examples include Honda Asimo [4], the HRP [5] series, the Toyota Partner Robots [6], and Hubo [7]. The extremely high acquisition and maintenance costs of these robots and their lack of robustness to falls make them unsuitable for use in soccer games. NimbRo-OP closes the gap between large, expensive robots and affordable small robots.

3   Hardware Design

The design of the NimbRo-OP hardware takes into consideration the following criteria: affordable price, low weight, readily available parts, robot appearance, and reproducibility in a basic workshop. Fig. 1 gives an overview of the main components.

NimbRo-OP is 95 cm tall and weighs 6.6 kg, including the battery. Selecting this specific size allowed the use of a single actuator per joint. All 20 joints are driven by configurable Robotis Dynamixel MX series actuators [8]. MX-106 are used in the 6 DoF legs and MX-64 in the 3 DoF arms and the 2 DoF neck. All Dynamixel actuators are connected with a single TTL one-wire bus. The servo motors, as well as all other electronic components are powered by a rechargeable 14.8 V 3.6 Ah lithium-polymer battery.

Keeping the weight as low as possible is achieved by using lightweight materials like carbon composite and aluminum. Arms and legs are constructed using milled carbon-composite sheets which are connected with U-shaped aluminum parts cut from sheets and bent on two sides. The feet are made of flexible carbon composite sheets. The torso is composed of a milled rectangular aluminum profile. The head and the connecting pieces in the hands are 3D printed using ABS+ polymer.

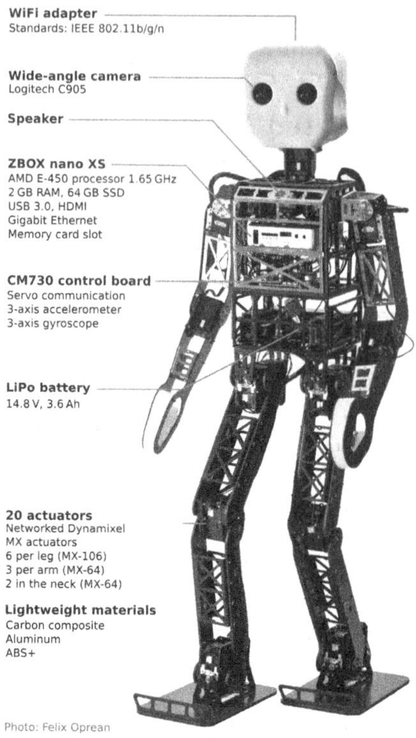

NimbRo-OP is equipped with a Zotac Zbox nano XS PC featuring a dual-core AMD E-450 1.65 GHz processor, 2 GB RAM, 64 GB solid state disk, and memory card slot. Communication interfaces are USB 3.0, HDMI, and Gigabit Ethernet. The PC is embedded into the torso without modifications to facilitate upgrades. The head contains a USB WiFi supporting IEEE 802.11b/g/n.

A Robotis CM730 board mediates

Fig. 1. NimbRo-OP hardware

communication between the PC and the actuators. It also contains three-axes accelerometers and three-axes gyroscopes for attitude estimation.

The robot head contains a Logitech C905 USB camera with fish eye wide-angle lens. Its extremely wide field of view (ca. 180°) allows for simultaneously having multiple objects in sight, e.g. the ball and the goal.

The NimbRo-OP CAD files are available [9] under Creative Commons Attribution-NonCommercial-ShareAlike 3.0 license. This allows research groups to reproduce the robot and to modify it to their needs. The University of Bonn [10] also offers fully assembled and tested robots.

4   First Software Release

The hardware release of the first NimbRo-OP prototype is supported by a software package that provides a set of fundamental functionalities, such as bipedal walking and ball tracking. Out of the box, the robot is able to walk up to a uniformly colored ball and to kick it away. As a simple fall protection mechanism, the robot relaxes all its joints when it detects an inevitable fall. Moreover, the robot is able to stand up from a prone and a supine position. We used the freely available DARwIn-OP software framework [11] as a starting point for development and made only the modifications necessary for providing the aforementioned functionalities. More specifically, we added

- correction of the wide-angle lens distortion,
- attitude estimation based on CM730 acceleration and turning rate sensors,
- a feedback stabilized bipedal gait configuration,
- an instability detection and simple fall protection mechanism, and
- get-up and kicking motions.

Please refer to Schwarz et al. [12] for more details. This software release has been made available [9] as a list of small patches against the original DARwIn-OP framework. For easier traceability of changes, each added or modified feature resides in an own patch file with a header describing the change.

5   ROS-Based Software Development

The first version of the NimbRo-OP software is based on the freely available DARwIn-OP framework, making NimbRo-OP easily accessible to researches familiar with the DARwIn-OP robot, but it includes only basic skills.

We aim for more advanced functionality, including cognitive perceptual and action capabilities. To this end, we started a new software development based on the Robot Operative System framework (ROS) [13,14]. This popular middleware makes it easy to implement multiple processes (nodes) that communicate with each another using data streams (topics)—facilitating modular software development. Additionally, effective development tools are provided that perform tasks such as data logging and serialization, specifying launch configurations, unit testing, and visualization. Most importantly, the ROS community includes thousands of robotic researchers and enthusiasts who are constantly contributing their experiences and results.

Our ROS-based software will incorporate accumulated experience, scientific achievements, and technologies developed throughout the successful history of team NimbRo. Fig. 2 depicts the initial structure of the soccer robot software that is being implemented. At the bottom of the figure are nodes that interface the robot and abstract from its hardware. The top of the figure, Perception, is composed of nodes that interpret all available sensory data to estimate the robot and environment state. The Behavior Control modules, shown in the middle of the figure, decide the robot actions based on the perceived state. The Motion Control modules generate walking, kicking, and other full-body motion and send joint targets to the robot.

5.1   Configuration Server and Visualization

A significant part of our development are the visualization and parameter tuning tools. For development and debugging, it is important to be able to capture and inspect the information flow through the system. ROS offers two fundamental software packages for this purpose: RVIZ and RQT. The former provides a 3D visualization of the robot and its environment. The latter permits the visualization

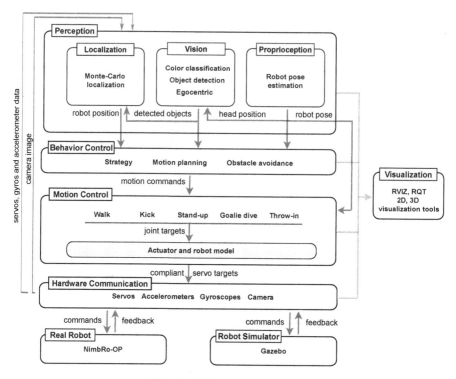

Fig. 2. Robot control diagram

of 2D data such as plots and images. Fig. 3 shows two examples of visualization tools that we implemented.

The software modules that we develop are highly configurable and, hence, contain many parameters. In order to allow for and keep track of parameter changes, we developed a configuration server. It manages parameters in a hierarchical structure and notifies subscribed nodes about changes. This server also makes robot configurations persistent for future use.

5.2   High-Precision Timer

Some of the previously mentioned procedures require execution with high-precision periodic timing. In order to ensure this, the Linux native *timerfd* interface is being used. Furthermore, the communication thread uses the realtime FIFO scheduler instead of the non-realtime default scheduler. The entire motion layer and hardware communications including sending servo commands, position feedback and IMU readings are executed every 8 ms.

5.3   Model-Based Feed-Forward Position Control

Basis for our current actuator interface is a servo model that facilitates compliant robot motion [15]. This model is used for feed-forward position control to compensate gravity and dynamic forces known in advance. The model parameters are identified using Iterative Learning Control. Our approach tracks desired trajectories with low control gains, yielding smooth, energy efficient motions that comply to external disturbances.

5.4   Soccer Vision

Visual perception of the game situation is a necessary prerequisite for playing soccer. The robot camera supports resolutions up to 1600×1200. We process 800×600 YUYV images to achieve a frame rate above 24 fps. We classify all pixels by color and create lower-resolution images for the individual color classes (see Fig.4(a)). Using smaller, color classified images facilitates the fast and simple detection of ball, goals, obstacles, field lines, and field line crossings. Fig. 4(b) shows example detections. Please refer to Schulz and Behnke [16] for more detail.

5.5   Behavior Control

For initial testing, we implemented a central-pattern generated gait, based on the gait engine of earlier NimbRo robots [17]. We also implemented a simple ball approach behavior that makes the robot walk to the ball while simultaneously aiming at the goal [18]. The robot executes kicking and getting-up [19] motions based on trajectories designed in a motion editor that we developed.

6   Impact

The initial release triggered quite some interest from the RoboCup community and the general public. Several RoboCup teams expressed interest to acquire the robot from the University of Bonn or to manufacture it in their own lab using the released CAD files. There was even a vivid interest from resellers. Many news outlets reported on the release of the NimbRo-OP robot, which yielded

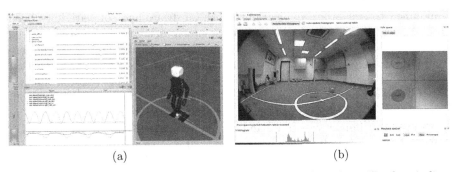

(a)   (b)

Fig. 3. NimbRo-OP visualization tools. a) 3D model of the robot, 2D plots indicating joint positions, gait parameter tuning tool and battery status notifier. b) Color calibration tool for generating YUV color look-up tables.

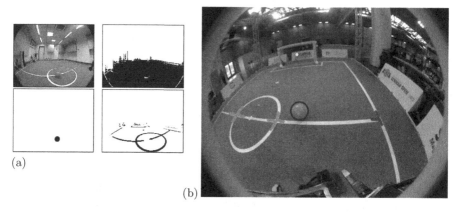

Fig. 4. Soccer vision. (a) Color segmentation (orange, green, white); (b) Identification of field, ball, goal, and line crossings (T and X).

more than 32,000 views on its release video. The first software release of our robot was demonstrated at the IROS 2012 and Humanoids 2012 conferences. The newer ROS-based software was demonstrated at ICRA 2013 and RoboCup German Open 2013, where NimbRo-OP received the HARTING Open-Source Award.

Our robot contributed to an increased interest in the Humanoid TeenSize class. For RoboCup 2013, the number of qualified teams in the TeenSize class reached an all-time high of six.

7   Conclusions

NimbRo-OP is a bipedal TeenSize robot prototype for research. It is expected that it will be of particular interest to those research groups that want to work with a larger robotic platform able to act in everyday environments. The robot has comparably low hardware costs, and it is easy to assemble and operate. Moreover, if necessary, the robot can be repaired by non-specialists and can be modified to suit other applications. Both hardware and software have been released open-source to support low-cost manufacturing and community-based improvements. The robot is equipped with high torque actuators, a dual-core PC for processing power, a wide-angle camera, and a software package that implements numerous fundamental robot skills.

Further development is planned in order to improve the hardware and software. With respect to the hardware, special attention will be paid to appearance, durability, impact resistance and easy maintenance. The NimbRo-OP ROS framework, which is still a work in progress, will receive additional functionality. For example, a self-localization node that uses information from the vision detection and the robot state to feed a particle filter model [16] and a simulator, where the user will be able to test behaviors and new algorithms before deployment on the real robot, will be among the next steps.

Acknowledgements. This work has been partially funded by grant BE 2556/6 of German Research Foundation (DFG). We also acknowledge support of Robotis Inc. and RoboCup Federation for the first robot prototype.

References

1. Ha, I., Tamura, Y., Asama, H., Han, J., Hong, D.W.: Development of open humanoid platform DARwIn-OP. In: SICE Annual Conference (2011)
2. Wolf, J., Hall, P., Robinson, P., Culverhouse, P.: Bioloid based humanoid soccer robot design. In: Proc. of Second Workshop on Humanoid Soccer Robots (2007)
3. Gouaillier, D., Hugel, V., Blazevic, P., Kilner, C., Monceaux, J., Lafourcade, P., Marnier, B., Serre, J., Maisonnier, B.: Mechatronic design of NAO humanoid. In: IEEE International Conference on Robotics and Automation, ICRA (2009)
4. Hirai, K., Hirose, M., Haikawa, Y., Takenaka, T.: The Development of Honda Humanoid Robot. In: Int. Conf. on Robotics and Automation, ICRA (1998)
5. Kaneko, K., Kanehiro, F., Morisawa, M., Miura, K., Nakaoka, S., Kajita, S.: Cybernetic human HRP-4C. In: IEEE-RAS Int. Conf. on Humanoid Robots (2009)
6. Takagi, S.: Toyota partner robots. J. of Robotics Society of Japan 24(2), 62 (2006)
7. Park, I.-W., Kim, J.-Y., Lee, J., Oh, J.-H.: Mechanical design of humanoid robot platform KHR-3 (KAIST Humanoid Robot 3: HUBO). In: IEEE-RAS Int. Conf. on Humanoid Robots (Humanoids), pp. 321–326 (2005)
8. Robotis Dynamixel MX actuators, http://www.robotis.com/xe/dynamixel_en
9. NimbRo-OP Hardware/Software, https://github.com/NimbRo/nimbro-op
10. NimbRo-OP Website, http://www.nimbro.net/OP/
11. DARwIn-OP Robotis Software, http://sourceforge.net/projects/darwinop/
12. Schwarz, M., Schreiber, M., Schueller, S., Missura, M., Behnke, S.: NimbRo-OP Humanoid TeenSize Open Platform. In: IEEE-RAS Int. Conference on Humanoid Robots Proceedings of 7th Workshop on Humanoid Soccer Robots, Osaka (2012)
13. ROS Website, http://www.ros.org
14. Quigley, M., Conley, K., Gerkey, B.P., Faust, J., Foote, T., Leibs, J., Wheeler, R., Ng, A.Y.: ROS: An open-source robot operating system. In: ICRA Workshop on Open Source Software (2009)
15. Schwarz, M., Behnke, S.: Compliant robot behavior using servo actuator models identified by iterative learning control. In: Behnke, S., Veloso, M., Visser, A., Xiong, R. (eds.) RoboCup 2013. LNCS (LNAI), vol. 8371, pp. 207–218. Springer, Heidelberg (2014)
16. Schulz, H., Behnke, S.: Utilizing the structure of field lines for efficient soccer robot localization. Advanced Robotics 26, 1603–1621 (2012)
17. Behnke, S.: Online trajectory generation for omnidirectional biped walking. In: IEEE Int. Conference on Robotics and Automation (ICRA), pp. 1597–1603 (2006)
18. Stückler, J., Behnke, S.: Hierarchical reactive control for humanoid soccer robots. International Journal of Humanoid Robotics 5(3), 375–396 (2008)
19. Stückler, J., Schwenk, J., Behnke, S.: Getting back on two feet: Reliable standing-up routines for a humanoid robot. In: 9th International Conference on Intelligent Autonomous Systems (IAS-9), pp. 676–685 (2006)

Modular Development of Mobile Robots with Open Source Hardware and Software Components

Martino Migliavacca, Andrea Bonarini, and Matteo Matteucci

Politecnico di Milano, Dipartimento di Elettronica, Informazione e Bioingegneria,
Piazza Leonardo Da Vinci 32, 20133, Milano, Italy
{migliavacca,bonarini,matteucci}@polimi.it

Abstract. Prototyping and engineering robot hardware and low-level control often require time and efforts thus subtracted to core research activities, such as SLAM or planning algorithms development, which need a working, reliable, platform to be evaluated in a real world scenario. In this paper, we present Rapid Robot Prototyping (*R2P*), an open source, hardware and software architecture for the rapid prototyping of robotic applications, where off-the-shelf embedded modules (e.g., sensors, actuators, and controllers) are combined together in a plug-and-play fashion, enabling the implementation of a complex system in a simple and modular way. R2P makes people involved in robotics, from researchers and designers to students and hobbyists, dramatically reduce the time and efforts required to build a robot prototype.

1 Introduction

In recent years, several development frameworks [6, 4, 11, 8] have been proposed to assist researchers in the design of robotic applications. While these projects really boosted the development of high-level software, hardware design and low-level firmware development are still critical tasks. To develop a new mobile robot, designers always face the problem of selecting hardware devices, controlling them, and interfacing them with the high-level software. This slows down the progress of robotic research, as prototyping and engineering often requires more time and resources than tasks strictly related to the target application.

To simplify the development of new robotic applications, we developed Rapid Robot Prototyping (*R2P*) [2, 1], an open source hardware and software framework focused on speeding up the prototyping of robotic systems. R2P provides hardware modules that implement basic functionalities needed by common robotic applications, and a lightweight, real-time, middleware to easily write low-level control software. R2P targets span from mobile autonomous robots used for research purposes to entertainment and service applications, such as games, telepresence, and rescue. The limits of R2P, at the actual stage of development, are only imposed by the modules already available; moreover, as R2P is an open

S. Behnke et al. (Eds.): RoboCup 2013, LNAI 8371, pp. 576–583, 2014.

source, modular, framework, it can be extended by users with additional modules to cover other application fields.

2 Modular Hardware and Software Development

When a new robotic application is investigated, the first steps involve selecting the hardware devices, e.g., sensor and actuators, and building the platform needed to validate the overall idea. Looking at today's possibilities, we can pick devices either from the automation market or from the hobby market. Components from automation market are often expensive and offer overkilling performance with respect to the requirements of a robotic application prototype. Moreover, automation devices often require power supplies not suitable for battery powered systems like mobile robots. On the other hand, devices from hobby market are usually cheap, but they show poor performance, low reliability, and no real-time capabilities making impossible any distributed control loop. Having selected hardware devices, here it comes the problem of interfacing them with each other, and with the high-level control software. Different manufacturers generally use different data links and protocols, increasing wiring complexity and requiring specialized device drivers. As a consequence, resulting platforms are commonly based on custom setups, which are hardly reusable in different projects. Although mobile robots have been built for decades by integrating heterogeneous devices, or implementing custom solutions, we firmly believe that a modular approach based on off-the-shelf components would strongly help robot designers in developing new applications. To the best of our knowledge, the only available modular robotic platforms, such as the E-puck educational robot [10], the Kephera robot [7], and a few others, are aimed at developing small mobile robots for applications like swarm robotics and their usage is restricted to control the platform they are designed for.

With R2P, we aim at fulfilling the lack of hardware components focused on robot prototyping, pushing design strategies commonly exploited in software development – such as modular, component-based, software engineering – down to the hardware level. R2P relies on the principle that the requirements of a generic robot application can be implemented by modules not only at software level, as it is common in most frameworks, but also at hardware level. Basic functionalities such as motor control, distance measurement, inertial navigation are implemented by specific, standardized hardware modules, with corresponding firmware, that can be plugged on a common bus and can interact in real-time. Firmware development tools, and a middleware to foster distributed, reusable, software development, are provided, supporting users in writing code on resource-constrained devices. Using R2P, robot designers can build generic platforms by choosing the modules they need, configuring them, and easily developing the control software, implementing complex systems in a plug-and-play fashion. Integration with high-level robotics frameworks, such as ROS [11], is provided by a gateway module.

3   R2P: The Rapid Robot Prototyping Framework

In this section, we introduce R2P design choices and architecture. Then, a review of some of the already available hardware modules are presented.

3.1   Power and Data Link

R2P uses a single connector to transport both power and data. Power consumption is limited to $5V$, $200mA$, for each module, which suites the requirements of most electronic devices, while modules needing higher power, such as motor drivers, must rely on auxiliary connections. Modules exchange data using the CAN-Bus, which has been designed to work in harsh environments and is available on many microcontrollers. Its maximum data rate of $1Mbps$ is generally enough for a distributed system of smart devices, where only high level information needs to be sent over the network (i.e., no raw sensor data is exchanged), thus needing a relatively small bandwidth [3]. As part of R2P, we developed RTCAN [9], a CAN-Bus protocol targeted at robotic applications that supports both sporadic, event-triggered, and periodic, time-triggered communication, with soft and hard real-time constraints. To reduce wires, a daisy chain wiring schema is adopted: each module has two ports to connect to the previous and the next component, as shown in Figure 2(a). This also supports easy connection of new modules to an existing system.

3.2   Embedded Firmware Development

Writing code for resource-constrained devices, such as microcontrollers used to interface with sensors and actuators, requires specific knowledge and competence. Most robot designers are used to write software on desktop-level computer systems, and they have to spend time and efforts to start developing code targeted to embedded devices. To reduce this effort, the use of an operating system can significantly support software development even for small embedded systems as it features threads, memory management, message passing primitives, and other services programmers are commonly used to deal with. Moreover, an operating system with real-time capabilities is important to manage critical, high-priority tasks, which are often involved in robotic systems, e.g., for closed-loop control. For the mentioned reasons, R2P relies on ChibiOS/RT [12], a real-time operating system designed for deeply embedded real time applications. ChibiOS/RT has been preferred to other alternatives for its portability, ease of use, rich features set, and extremely high efficiency; anyway, a review of available embedded operating systems is out of the scope of this paper. ChibiOS/RT also includes a Hardware Abstraction Layer (HAL), which abstracts the hardware implementation of different low level peripherals, relieving the developer from acquiring specific competence on each specific platform and making easier the port of existing code to different targets.

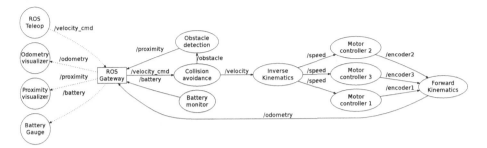

Fig. 1. The distributed architecture of the embedded software controlling Triskar2

3.3   Publish/Subscribe Middleware

To support the development of modular software components on embedded target, R2P features a lightweight communication middleware. R2P middleware main goals are software reuse, real-time communication, efficient implementation, and ease of use. It follows the *publish/subscribe* paradigm [5]: data producers *publish* messages on a *topic*, i.e., a communication channel, while data consumers *subscribe* to the corresponding topic to receive messages. Identifying data by its content, i.e., the topic it is published on, instead of by its producer, also promotes loosely-coupled software design and, thus, code reuse. The middleware provides concepts common to most robotics frameworks used on computer systems, such as software nodes, topics, publishers, subscribers, and message queues.

R2P middleware is written in a subset of C++, to take advantage of some object-oriented programming features without compromising performance on embedded targets. Its implementation is focused on code efficiency and messaging performance. Software nodes can subscribe to both local and remote publishers, with no difference from the user point of view. The middleware supports both periodic and sporadic publishers, which can specify real-time communication constraints: update period for time-triggered messages, and delivery deadline for event-triggered ones. Finally, a simple API, which reminds the ROS syntax, enables developers to write embedded, distributed code as they are used to do on computer systems, fostering code reuse through different projects.

3.4   Integration with ROS

While R2P supports rapid development of robotic systems using off-the-shelf hardware and software components, applications involving computation-intensive tasks such as computer vision, localization, and complex planning, must also rely on a computer system and, eventually, a software framework. Among the many available development frameworks for robotics software, ROS [11] is currently the most widely adopted in academia and research laboratories, and, recently, it has been considered also by industrial developers. To natively integrate resource-constrained devices within ROS, we developed *μROSnode*, a lightweight, open

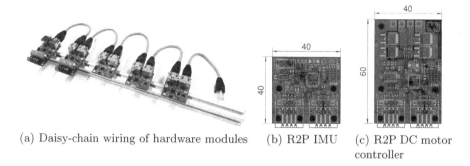

(a) Daisy-chain wiring of hardware modules   (b) R2P IMU   (c) R2P DC motor controller

Fig. 2. R2P hardware modules

source, ANSI C ROS client library. R2P provides a gateway module (see Section 3.5), which acts as a proxy between the R2P middleware and ROS systems. Topics published on the R2P network can be accessed from ROS nodes, and, at the same way, R2P modules can subscribe data published by ROS software.

3.5   Off-the-Shelf Hardware Components

We have designed and built, as part of the R2P framework, a set of plug-and-play hardware modules that implement basic functionalities required by common robotics applications. Modules are based on STM32 Cortex-M3 microcontrollers with $20Kb$ of RAM and $128Kb$ of Flash memory, running the ChibiOS/RT and the R2P middleware. Each module has two RJ45 ports for daisy-chain connection to the bus, a serial port to download new firmware and for debugging purposes, and a JTAG header for advanced users who want to directly access the microcontroller. An overview of the currently available modules follows.

PSU Module. This is the power supply unit, which powers all the modules connected to the bus. Input voltage range is from $5.5V$ to $36V$ DC. A DC-DC converter produces a $5V$ regulated output with maximum current supply of $4A$ and short circuit protection. Both battery voltage and current drain can be published over the network to monitor power consumption and to estimate the residual battery life.

DC Motor Module. This high-power motor controller board can drive DC motors up to $36V$, delivering a continuous $20A$ current. It features closed loop control, with position feedback from a quadrature encoder and current measurement from the on-board Hall-effect sensor. The DC motor module accepts position, speed, and torque set points, and can publish position and speed messages, exploiting data from the encoder, and the measured current drawn.

IMU Module. A 10-DoF Inertial Measurement Unit featuring MEMS accelerometer, gyroscope, magnetometer and pressure sensor. An additional serial port to acquire GPS coordinates from an external GPS receiver is also provided on this module. The on-board sensor fusion algorithm produces heading, attitude, and position messages.

Proximity Module. A module to interface with proximity sensors such as the Sharp IR rangers or *MaxBotix* ultrasonic sensors. Each module connects to up to 4 sensors. Calibration and data filtering algorithms run on the microcontroller, which produces distance measurements.

Gateway Module. This is the gateway module mentioned in Section 3.4. It features an Ethernet port and a more powerful, Ethernet-enabled, microcontroller to handle the TCP/IP stack. R2P messages can be forwarded from the CAN-Bus to the IP network, and the other way around. The gateway module runs μROSnode, which enables a direct integration of R2P modules with ROS systems.

3.6 Open Source Development

R2P is fully open source, both hardware and software, to encourage its adoption and to take advantage of community-driven improvements to became a mature and widespread project. The design of the boards, the code they run and the middleware are available on the R2P repository: `http://github.com/openrobots-dev`. At the moment of writing, R2P has reached its maturity (see, e.g., the use case in the next section), but its development is still actively progressing, thus, the repository is frequently updated.

4 Use Case: An Omnidirectional Robot

We used R2P to develop the omnidirectional wheeled robot *Triskar2*, shown in Figure 3(a). The robot sports 3 R2P DC modules, a PSU module, a Proximity module, and the Gateway module to interface with a computer running ROS. The low-level control software embedded on the modules, which exploits the R2P publish/subscriber middleware, is reported in Figure 3. Software components are enclosed in R2P nodes, which implement basic functionalities, performing a specific task. Then, nodes are composed as a distributed architecture, implementing a complex system from basic, reusable, components. This design strategy is not innovative, being commonly used in software development; the main contribution of R2P middleware is to bring the same approach, and, thus, the same advantages, to embedded firmware development, with the same programming interfaces known to most robot developers.

Software nodes have been deployed on the modules as shown in Figure 3(b). Some nodes have to run on specific boards (e.g., those that are directly connected to the hardware like motor controller nodes), while others can run on any connected module. For example, in our tests, the inverse kinematics model to compute wheel speeds was run on the *Motor 1* module, while the odometry node was deployed on *Motor 2*. In this way, we can balance processor load and reduce latency, easily moving nodes from an hardware module to another.

Thanks to the R2P gateway, Triskar2 can be controlled by any ROS application publishing native ROS topics. We firstly teleoperated the robot by using standard ROS *teleop* messages, then we developed a robotic game, involving the Triskar2 robot and a quadricopter, both controlled by ROS software.

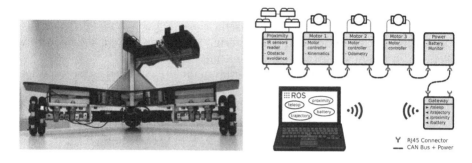

Fig. 3. The Triskar2 omnidirectional platform (a) and the R2P hardware modules controlling the robot (b)

5   Conclusions

In this paper, we presented R2P, an open source hardware and software framework for the rapid prototyping of robots. Bringing design strategies such as modular development, and components reuse, down to hardware level, R2P enables robot designers to build and control a robotic platform using off-the-shelf modules. Exploiting the R2P framework, generic mobile robots can be built bottom-up in a distributed plug-and-play fashion by simply selecting the hardware modules to satisfy the needed functional requirements and easily programming their interaction. Integration with high-level software frameworks, e.g., ROS, allows to develop complex application, while low-level control is implemented by means of a modular distributed architecture, with real-time performance, without the need for advanced domain-specific knowledge. We are exploiting R2P to design new robots in our laboratory, as shown by the use case presented in Section 4, and to upgrade our previous platforms, the first being a balancing wheeled robot, a differential drive heavy-duty robot, and an autonomous wheelchair. The open source license encourages robot designers to adopt existing R2P modules to control their platforms, and to develop new hardware modules and software components that implement new functionalities.

Acknowledgements. This work has been partially supported by the research grant "Robotics for the Masses" from ST Microelectronics and Regione Lombardia, and by the Italian Ministry of University and Research (MIUR) through the PRIN 2009 grant "ROAMFREE: Robust Odometry Applying Multi-sensor Fusion to Reduce Estimation Errors".

References

[1] Bonarini, A., Matteucci, M., Migliavacca, M., Rizzi, D.: R2P: An open source hardware and software modular approach to robot prototyping. Robotics and Autonomous Systems

[2] Bonarini, A., Matteucci, M., Migliavacca, M., Rizzi, D.: R2P: an open source modular architecture for rapid prototyping of robotics applications. In: Proceedings of 1st IFAC Conference on Embedded Systems, Computational Intelligence and Telematics in Control, CESCIT 2012 (2012)

[3] Bonarini, A., Matteucci, M., Migliavacca, M., Sannino, R., Caltabiano, D.: Modular low-cost robotics: What communication infrastructure? In. In: Proceedings of 18th World Congress of the International Federation of Automatic Control (IFAC), pp. 917–922 (2011)

[4] Bruyninckx, H.: Open robot control software: the OROCOS project. In: Proceedings 2001 ICRA, IEEE International Conference on Robotics and Automation, pp. 2523–2528 (2001)

[5] Eugster, P.T., Felber, P.A., Guerraoui, R., Kermarrec, A.-M.: The many faces of publish/subscribe. ACM Computing Surveys 35(2), 114–131 (2003)

[6] Gerkey, B.P., Vaughan, R.T., Howard, A.: The player/stage project: Tools for multi-robot and distributed sensor systems. In: Proceedings of the 11th International Conference on Advanced Robotics, pp. 317–323 (2003)

[7] Harlan, R.M., Levine, D.B., McClarigan, S.: The khepera robot and the krobot class: a platform for introducing robotics in the undergraduate curriculum. ACM SIGCSE Bulletin 33, 105–109 (2001)

[8] Huang, A., Olson, E., Moore, D.: LCM: Lightweight communications and marshalling. In: IEEE/RSJ International Conference on Intelligent Robots and Systems (IROS), pp. 4057–4062 (2010)

[9] Migliavacca, M., Bonarini, A., Matteucci, M.: RTCAN: a real-time CAN-Bus protocol for robotic applications. In: 2013 International Conference on Informatics in Control, Automation and Robotics, ICINCO (2013)

[10] Mondada, F., Bonani, M., Raemy, X., Pugh, J., Cianci, C., Klaptocz, A., Magnenat, S., Zufferey, J.-C., Floreano, D., Martinoli, A.: The e-puck, a robot designed for education in engineering. In: Proceedings of the 9th Conference on Autonomous Robot Systems and Competitions, vol. 1, pp. 59–65 (2009)

[11] Quigley, M., Conley, K., Gerkey, B.P., Faust, J., Foote, T., Leibs, J., Wheeler, R., Ng, A.Y.: ROS: an open-source robot operating system. In: ICRA Workshop on Open Source Software (2009)

[12] Sirio, G.D.: ChibiOS/RT real time operating system, http://www.chibios.org

Sharing Open Hardware through ROP, the Robotic Open Platform

Janno Lunenburg, Robin Soetens, Ferry Schoenmakers, Paul Metsemakers, René van de Molengraft, and Maarten Steinbuch*

Eindhoven University of Technology,
Den Dolech 2, P.O. Box 513, 5600 MB Eindhoven, The Netherlands
rop@tue.nl
www.roboticopenplatform.org

Abstract. The robot open source software community, in particular ROS, drastically boosted robotics research. However, a centralized place to exchange open hardware designs does not exist. Therefore we launched the Robotic Open Platform (ROP). A place to share and discuss open hardware designs. Among others it currently contains detailed descriptions of Willow Garage's TurtleBot, the NimbRo-OP created by the University of Bonn and the AMIGO robot of Tech United Eindhoven. Eventually, ROP will contain a collection of affordable hardware components, allowing researchers to focus on cutting-edge research on a particular component instead of having to design the entire robot from scratch. As an example of how the Robotic Open Platform is able to facilitate this knowledge transfer, we introduce TURTLE-5k: A redesign of an existing soccer robot by a consortium of our university and companies in the wider Eindhoven area. Cooperating with industrial partners resulted in a significant cost reduction.

Keywords: Open Hardware, RoboCup, Low-Cost Robot Design, Robotic Open Platform, TURTLE-5k, Robotics Hardware.

1 Introduction

In recent years, increasing research efforts have been devoted to domestic service robots. Creating robotic agents able to autonomously assist humans in a highly unstructured environment is a huge step in research and engineering, too big for any university, research institute or company alone. Collaboration and knowledge exchange is therefore of utmost importance. Within robotics, the open source community has given software developments a tremendous boost, with the Robot Operating System (ROS) as its best-known exponent. Open hardware, on the other hand, is less common.

Nevertheless, some examples of open robotic hardware designs already exist. Willow Garage's PR2 [15] has been introduced as an Open Platform: Users are encouraged to modify the system and open interfaces enable the use of different

---

* This researched received funding from the Dutch ministry of Economic affairs.

S. Behnke et al. (Eds.): RoboCup 2013, LNAI 8371, pp. 584–591, 2014.

grippers, forearms, whole arms or sensors. A second open hardware design by Willow Garage is TurtleBot. Open hardware designs for humanoid robots are, e.g, NimbRo-OP [11], iCup [8] and DARwIn-OP [5]. Finally, Thymio II [6] is a low-cost open hardware educational robot.

Besides having individual robots released under an open hardware licensce, there are also community efforts to promote the exchange of hardware designs. Robotsource[1] is an online initiative on which some of the designs mentioned above are released, but it is aimed at marketing and productizing robotic solutions rather than sharing open hardware. RoboCup on the other hand, being the largest annual open source robotics event, provides a huge pool of knowledge on open hardware design.

However, unlike ROS provides for software, there is no centralized online platform for hardware designs yet. Furthermore, there is no standard format to publish open robot hardware, causing the information that is published to be insufficient to actually build the robots.

Therefore this work presents the Robotic Open Platform (ROP): An online place to share open hardware designs for robots. Inspired by ROS, embedded wiki pages provide specifications and explanations but also link to a repository containing CAD files, electric drawings and information on open source hardware licenses. The ROP 'Questions and Answers' page provides a place people can go to ask others in the ROP community for help. The Robotic Open Platform is a place where people can share, discuss and follow up on hardware designs, which eventually should be as easy as sharing ROS packages is now.

In the following section, we will discuss some examples of hardware knowledge exchange within and between RoboCup leagues. In Section 3 we discuss ROP in more detail, and elaborate on how ROP currently facilitates hardware knowledge exchange. In the final section, we introduce TURTLE-5k, an example of ROP facilitating knowledge transfer between industry and the RoboCup community.

2   Sharing Hardware Designs Within RoboCup

RoboCup, being the biggest annual open source robotics event, provides a huge pool of knowledge on hardware design. In this section we will highlight some examples of knowledge exchange within the RoboCup community.

2.1   RoboCup Middle-Size League

The effects and benefits of sharing hardware designs is clearly visible within the RoboCup Middle Size League (MSL). According to the rulebook, a team is almost completely free in designing their soccer playing robots. Only restrictions on size, weight and color exist. However, the developments over the past years have led to solutions commonly used throughout the league. In this section vision

---

[1] http://www.robotsource.org/

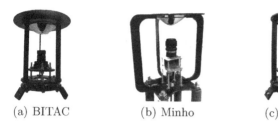 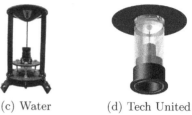

(a) BITAC   (b) Minho   (c) Water   (d) Tech United

Fig. 1. Omnivision systems of several MSL teams

units, the motion base, shooting mechanisms and ball handling mechanisms will be discussed respectively[2].

A robot needs to be able to perceive its environment and as much of the soccer field as possible. While in the past sometimes tilting camera systems were used, nowadays most teams use an omnidirectional vision module [9]: A single camera pointed at a hyperbolic mirror on top of the robot (Figure 1).

MSL soccer robots need to be both fast, agile and preferably holonomic. In the past, steering wheels and differential drives were common. However, since these drive concepts are non-holonomic or semi-holonomic, most MSL teams currently use omni wheels [1] (Figure 2).

A third example is found in the adoption of the shooting mechanism. Many systems can be used to shoot a ball, such as spring driven devices [3], pneumatic actuators [12] or electromechanical systems using a solenoid [7]. Of these solutions, the solenoid has proven to be superior since the shooting power can be fully controlled [7] and is therefore currently used by most MSL teams.

A final example concerns ballhandling mechanisms. Solutions using only rubber bands to control the ball have been used, but more recently most teams use a system consisting of two movable levers in front of the robot with rotating wheels attached [2], enabling the robots to move in any direction while keeping possession of the ball.

 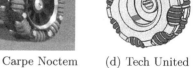

(a) Water   (b) NuBot   (c) Carpe Noctem   (d) Tech United

Fig. 2. Omni wheels of several MSL teams

---

[2] The figures depicted in this section are taken from the team description papers or websites of the corresponding teams.

(a) Carpe Noctem   (b) MRL   (c) Tech United

Fig. 3. Ballhandling systems of several MSL teams

2.2 RoboCup@Home League

In the previous section we discussed knowledge exchange within a RoboCup league. However, knowledge can also be shared between leagues. As an example we will discuss the design of the AMIGO robot, participating in RoboCup@Home.

The @Home league is aimed at the development of autonomous service robots for domestic applications. Compared to MSL, this league is relatively new. Therefore less solutions used throughout the league exist.

Various types of base-platforms are used, ranging from differential drives [13,16] and steering wheels [4,14] to fully holonomic platforms with omni wheels or Mecanum wheels [10]. The concepts for the upper bodies and manipulators are also very different. Some robots have a human-like concept with a movable torso and two anthropomorphic manipulators [10,14], while others only have a single (industrial) manipulator without a torso joint [4,13,16].

While designing the AMIGO robot for the RoboCup@Home league, team Tech United Eindhoven used hardware knowledge obtained in MSL as a blueprint. Although the purpose of a soccer robot and a domestic service robot are different, they do share some requirements:

- Both robots need powerful actuators and amplifiers. Either to accelerate quickly and reach a high velocity (soccer robots), or to drive over small thresholds such as doorsteps (domestic service robot).
- Various sensors and actuators need to be connected to on-board computers.
- A (semi-) holonomic base platform is convenient, either for object manipulation (service robot) or for agility (soccer robot).
- Both platforms need an internal power source.

To design a base, an omni wheel platform which was originally developed to be an MSL soccer robot is used (Figure 4). This design has been scaled up for stability reasons and to increase space available for peripheral equipment.

The diameter of the omni wheels has been increased to 15 cm, which is sufficient to drive over small thresholds such as doorsteps. The wheels are actuated by the same motors and amplifiers as used for the MSL robot. Only the gearboxes are different, since these require a larger reduction due to the larger mass, larger wheels and smaller velocities of a domestic service robot.

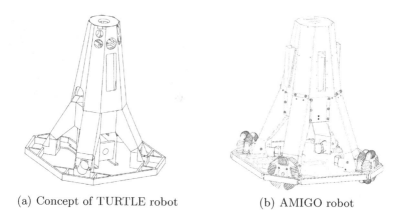

(a) Concept of TURTLE robot   (b) AMIGO robot

Fig. 4. Using mechanical base design of a soccer robot for a domestic service robot

The PCs on AMIGO are connected to sensors and actuators using Beckhoff EtherCAT stacks. This is also directly copied from the TURTLE soccer robots. A final similarity between the two robots concerns the power supply. In both cases, power tool batteries are placed on the base platform.

Since other than handling a ball, no object manipulation is required for an MSL robot, this is where the comparison ends. The torso of AMIGO is a custom design with a ball screw actuator for vertical movement and two commercially available anthropomorphic robotic arms attached. For vision purposes a commonly used Kinect camera is mounted on a pan-tilt unit made out of Dynamixel servomotors.

3   Robotic Open Platform

In the sections above we have shown that sharing knowledge on hardware design decreases the development efforts required to build a robot. However, such knowledge transfer is not always as efficient as it should be, mainly due to the lack of a centralized platform to do so.

Inspired by the impact ROS had on creating and sharing open source robot software, we launched the Robotic Open Platform (ROP). This website with embedded wiki pages is designed to provide a central place for the exchange of hardware designs. Comparable to open source software, companies and research institutions can release their robot hardware design on ROP under an open hardware license. Users can post a description of their robot or component on the ROP wiki, including figures and videos. By providing a link to a repository containing detailed lists of specifications, CAD drawings, electric schemes, part-lists and any additional documentation required to build the robot, ROP grants access to the hardware-equivalent of source code. These documents are accompanied by the license under which the designs are released.

Furthermore, a Questions and Answers section has been included to allow contributors to ask for assistance.

With the information provided, all users should be able to build their own robot. Similar to open source software, possible modifications and improvements to designs also have to be released. Having ROP as a central platform to share hardware designs can enhance the collaboration within RoboCup, but also within the entire international robotics community. The first robots that have been published on ROP were discussed in previous sections: The TURTLE soccer robot and service robot AMIGO, followed by the TurtleBot, Thymio II and NimbRo-OP.

Although having a central place to share hardware designs is a nice step in what we envision to be the right direction, there is still a limitation: The hardware designs discussed in Section 2 are fully integrated designs, making it difficult to combine different components in a new robot. Therefore modular design principles must be applied and standard electro-mechanical interfaces must be defined. When this is achieved, a collection of highly configurable and affordable hardware components will become available. Eventually, using hardware modules will be as easy as using ROS packages is now.

4   TURTLE-5k, a Low-Cost MSL Robot

In the previous section, ROP has been described as a platform to share open hardware designs. TURTLE-5k is a first example of ROP leading to better hardware: Together with industry an existing MSL soccer robot has been redesigned to significantly reduce costs.

4.1   Small Series versus Mass Production

Within RoboCup, we are building a maximum of six robots of a specific design, which is much less than the large series of mass-produced service robots we all envision. This quantitative difference is of huge influence on choices made while designing a robot, since the optimal ratio between development costs and production costs depends on the size of the series. For example, using folded sheet metal instead of milling parts out of an expensive aluminum monolith can reduce costs but requires more development effort. When only one or two robots of a specific design are built, it does not pay off to put additional effort in making the design itself as low-cost as possible.

4.2   Cooperation with Industry

While a lot of expertise on smart robot design exists at RoboCup teams, knowledge on how to turn these ideas into something that can be produced at low-cost is lacking. This knowledge, however, is abundantly present in industry. With this in mind, a consortium of three companies[3] and Eindhoven University of Technology was founded. By building on specifications and CAD drawings of the original TURTLE soccer robot, previously published on ROP, the consortium came up with a prototype design of a much more affordable MSL robot. Since this robot design will be on ROP as well, it can be improved by people all over the world. Assuming a minimum of ten teams chooses to buy these low-cost soccer robots, we think it is possible to reach a cost-per-robot of five thousand euros, which is one-fifth of the original cost. Hence, the name of this robot will be TURTLE-5k.

At the time of writing this paper, a prototype is being built, including a ball handling and shooting mechanism. It will be tested, evaluated and redesigned if necessary, such that it can be presented at RoboCup 2013.

After completion, this newly designed robot should provide a base-platform, recapitulating hardware knowledge gained over the past fifteen years of MSL's existence as much as possible. It should be easily adjustable and extensible such that the league maintains innovative on the hardware level. Teams focussing on a specific part of the robot, e.g., vision, should be able to add their own equipment or replace existing units with their own. Therefore, TURTLE-5k will not be a standard-platform for the MSL. Instead, it gives teams the opportunity to buy an affordable base, already able to play soccer but requiring extensions in order to stay competitive. It allows teams to spend a larger part of their budget on cutting-edge research by reducing start-up costs and time.

We consider the TURTLE-5k project to be a first example of ROP facilitating knowledge transfer between different areas of robotics research. In industry much of the knowledge on cost-focussed design we are looking for, is readily available. Through ROP, we are able to reach it.

5   Conclusions

In robotics research, well-designed hardware is just as important as smart software. Open hardware releases will pave the way towards such well-designed hardware. Sharing CAD files, parts lists and electrical schemes reduces start-up costs for anybody willing to start research in the robotics field. It boosts progress towards better hardware design. Although initiatives such as RoboCup promote knowledge exchange, a central platform to share knowledge on hardware designs was still lacking.

Through the Robotic Open Platform, such knowledge on mechanical and electrical design can be centralized, which stimulates modular design principles, standardization of interfaces and facilitates knowledge exchange within the worldwide robotics community. Having a collection of highly configurable and affordable hardware components allows researchers to spend their effort and budget on cutting-edge research instead of having to 'reinvent the wheel' once again.

---

[3] ACE, project management and mechanical design. VEDS, electrical design. Frencken Group, manufacturing and assembly.

The TURTLE-5k project is a perfect example of ROP providing knowledge exchange between different communities. Based on specifications of an existing soccer robot, published on the ROP wiki, industry is able to improve the design. As a result, knowledge from industry is blended with knowledge accumulated in the RoboCup community.

References

1. D'Andrea, R., Kalmár-Nagy, T., Ganguly, P., Babish, M.: The Cornell RoboCup team. In: Stone, P., Balch, T., Kraetzschmar, G. (eds.) RoboCup 2000. LNCS (LNAI), vol. 2019, pp. 41–51. Springer, Heidelberg (2001)
2. De Best, J., Van de Molengraft, R., Steinbuch, M.: A novel ball handling mechanism for the RoboCup middle-size league. Mechatronics 21(2), 469–478 (2011), special Issue on Advances in intelligent robot design for RoboCup MSL
3. Dirkx, F.B.: Philips CFT RoboCup team description. In: RoboCup 2004: Robot Soccer World Cup VIII. Springer (2004)
4. Dwiputra, R., Füller, M., Hegger, F., Hochgeschwender, N., Paulus, J., Schneider, S., Bierbrauer, A., Shpieva, E., Ivanovska, I., Deshpande, N., Gaier, A., Hagg, A., Sanches Loza, J., Ozhigov, A., Ploeger, P., Kraetzschmar, G.: The b-it-bots RoboCup@Home team description paper (2013)
5. Ha, I., Tamura, Y., Asama, H.: Development of open platform humanoid robot DARwIn-OP. Advanced Robotics (2013) (article in press)
6. Magnenat, S., Riedo, F., Bonani, M., Mondada, F.: A programming workshop using the robot 'Thymio II'. pp. 24–29 (2012)
7. Meessen, K.J., Paulides, J.J.H., Lomonova, E.A.: A football kicking high speed actuator for a mobile robotic application. In: IECON 2010 (2010)
8. Metta, G., Sandini, G., Vernon, D., Natale, L., Nori, F.: The iCub humanoid robot: An open platform for research in embodied cognition. In: PerMIS 2008, pp. 50–56 (2008)
9. Nadarajah, S., Sundaraj, K.: Vision in robot soccer: A review. Artificial Intelligence Review, pp. 1–23 (2013) (article in press)
10. Okada, H., Omori, T., Watanabe, N., Shimotomai, T., Iwahashi, N., Sugiura, K., Nagai, T., Nakamura, T.: Team eR@sers 2012 in the @Home league team description paper (2012)
11. Schwarz, M., Schreiber, M., Schueller, S., Missura, M., Behnke, S.: NimbRo-OP humanoid teensize open platform. In: IEEE-RAS International Conference on Humanoid Robots (2012)
12. Searock, J., Browning, B., Veloso, M.: Segway CMBalance robot soccer player. Technical report, DTIC Document (2004)
13. Seib, V., Kathe, F., McStay, D., Manthe, S., Peters, A., Jöbchen, B., Memmesheimer, R., Jakowlewa, T., Vieweg, C., Stümper, S., Günther, S., Müller, S., Veith, A., Kusenback, M., Knauf, M., Paulus, D.: RoboCup - homer@UniKoblenz, Germany (2013)
14. Stückler, J., Dröschel, D., Gräve, K., Holz, D., Schreiber, M., Behnke, S.: NimbRo@Home 2011 team description (2011)
15. Wyrobek, K.A., Berger, E.H., Van Der Loos, H.F.M., Salisbury, J.K.: Towards a personal robotics development platform: Rationale and design of an intrinsically safe personal robot. In: Proceedings of IEEE ICRA 2008 (2008)
16. Ziegler, L., Wittrowski, J., Schöpfer, M., Siepmann, F., Wachsmuth, S.: ToBI - Team of Bielefeld: The human-robot interaction system for RoboCup@Home (2013)

Enabling Codesharing in Rescue Simulation with USARSim/ROS

Zeid Kootbally[1], Stephen Balakirsky[2], and Arnoud Visser[3]

[1] Department of Mechanical Engineering, University of Maryland,
College Park, Maryland, USA
[2] Georgia Tech Research Institute, Atlanta, Georgia, USA
[3] Intelligent Systems Laboratory Amsterdam, Universiteit van Amsterdam,
Amsterdam, NL

Abstract. The Robot Operating System (ROS) has been steadily gaining popularity among robotics researchers as an open source framework for robot control. The Unified System for Automation and Robot Simulation (USARSim) has been used for many years by robotics researchers and developers as a validated framework for simulation. This paper presents a new ROS node that is designed to seamlessly interface between ROS and USARSim. It provides for automatic configuration of ROS transforms and topics to allow for full utilization of the simulated hardware. The design of the new node as well as examples of its use for mobile robot inside the RoboCup Rescue Simulation League are presented.

1 Introduction

With the development of advanced but also more complex algorithms one cannot expect that a robotic control system will be developed from scratch. With the aid of open source projects such as the Robot Operating System (ROS) [17] allow anyone with a Linux computer to download and run some of the most advanced robotic algorithms that exist. This is essential for the RoboCup mission; to accelerate the developments of intelligent and dexterous robots. With working modules that cover all basic capabilities needed for a functional robot, developers can concentrate on improving the aspects needed for their application.

Most programmers have access to a single robot or small sensor suite, but are missing access to some of the robotic hardware needed for the job. Simulators exist to fill this void and allow both experts and novices to experiment with robotic algorithms in a safe, low-cost environment. However, to truly provide valid simulation, the simulator must provide noise models for sensors and must be validated [2, 13–15, 21]. One modern robotic simulator, known as the Unified System for Automation and Robot Simulation (USARSim) [2] provides such a simulation platform. This simulator has been used by the expert robotics community for several years and has played an important role in developing robotics applications [3].

This paper examines how a new interface to the ROS control framework (introduced by [5]) can be used inside the RoboCup Rescue Simulation League [1]. The ROS framework allows for easy sharing of modules [9], allowing fast progress.

S. Behnke et al. (Eds.): RoboCup 2013, LNAI 8371, pp. 592–599, 2014.

2   Background

2.1   The USARSim Framework

USARSim [7] is a high-fidelity physics-based simulation system based on the Unreal Developers Kit (UDK)[1] from Epic Games. Through its usage of UDK, USARSim utilizes the PhysX physics engine [6] and high-quality 3D rendering facilities to create a realistic simulation environment that provides the embodiment of, and environment for a robotic system. The current release of USARSim consists of various environmental models, models of commercial and experimental robots, and sensor models. High fidelity at low cost is made possible by building the simulation on top of a game engine. By loading the most difficult aspects of simulation to a high volume commercial platform (available for free to academic users) which provides superior visual rendering and physical modeling, full user effort can be devoted to the robotics-specific tasks of modeling platforms, control systems, sensors, interface tools and environments. These tasks are in turn accelerated by the advanced editing and development tools integrated with the game engine. This leads to a virtuous spiral in which a wide range of platforms can be modeled with greater fidelity in a short period of time.

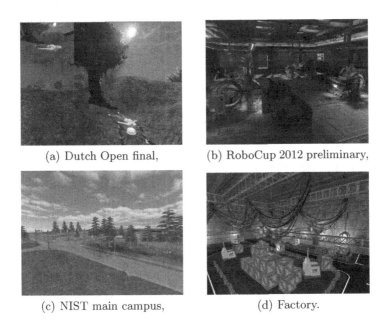

(a) Dutch Open final,   (b) RoboCup 2012 preliminary,

(c) NIST main campus,   (d) Factory.

Fig. 1. Sample of 3D environments in USARSim

USARSim was originally based upon simulated environments in the USAR domain [20]. Realistic disaster scenarios as well as robot test methods were created (Figure 1(a) and 1(b)). Since then, USARSim has been used worldwide and

---

[1] http://www.unrealengine.com/udk

more environments have been developed for different purposes, including human robot interaction and educational games [3]. Other environments such as the NIST campus (Figure 1(c)) and factories (Figure 1(d)) have been used to test the performance of robotic algorithms in different circumstances [2, 21]. The simulation is also widely used for the RoboCup Virtual Robot Rescue Competition [1], the IEEE Virtual Manufacturing and Automation Challenge [4].

2.2   The ROS Framework

ROS [17] is an open source framework designed to provide an abstraction layer to complex robotic hardware and software configurations. It provides libraries and tools to help software developers create robot applications and has found wide use in both industry and academia [8]. Examples of ROS applications include Willow Garage's Personal Robots Program [22], the Stanford University STAIR project [16] and the European Nifti project [11]. Developers of ROS code are encouraged to contribute their code back to the community and to provide documentation and maintenance of their algorithms.

ROS consists of a large range of tools and services that both users and developers alike can benefit from. The philosophical goals of ROS include an advanced set of criteria and can be summarized as: peer-to-peer, tools-based, multi-lingual, thin, and free and open-source [17]. Furthermore, debugging at all levels of the software is made possible with the full source code of ROS being publicly available. Thus, the main developers of a project can benefit from the community and vice-versa.

3   The ROS/USARSIM Interface

USARSim is designed to communicate over a TCP/IP socket with a computer hosting the controller of the robot. The robot is "spawned" into the simulated world running on the game server. A robot's configuration is controlled by an initialization file that resides on the simulation system's computer, comparable with the launch file from Gazebo. This file controls aspects such as sensor configuration, battery life, and simulated noise models. One socket connection is established per simulated robot, with both commands and sensor data being transmitted over the socket. An additional separate socket is established for high-volume sensors such as camera systems.

ROS stacks are designed to "bottom out" at a hardware abstraction layer that provides the interface to sensors and the motors of the robot; publishing and subscribing to the basic topics of the robot. For example, the mobility stack expects to be able to control a platform by writing commands to low-level topics that control items such as the velocity. In addition, the mobility stack expects feedback from sensors, such as the movement detected by the inertia sensor. In order to close this low-level loop between ROS and USARSim, a USARSim package was created inside ROS[2]. This package contains a node called *RosSim*

---

[2] http://sourceforge.net/projects/usarsimros/

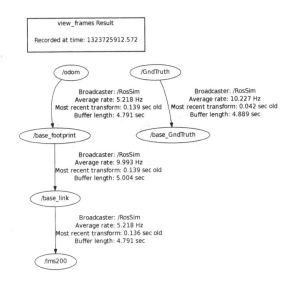

Fig. 2. Auto-generated `tf` transform tree for P3AT robot

that publishes a ROS transform tree `tf` and sensor messages, and also accepts platform and actuator motion commands. When run, it provides a mechanism for spawning a robot in USARSim, and then auto-discovering the robot's sensors, actuators, and drive configuration in order to provide the necessary ROS topics.

The *RosSim* node relies on several parameters for its configuration. These are necessary for the creation of a robot in USARSim and a transform tree `tf` in ROS. A transform tree for the P3AT robot is shown in Figure 2. This transform tree is built automatically from data obtained from the USARSim GEO and CONF messages. Since USARSim supports multiple sensors on a robot, it should be specified which sensor should be preferred for an initial localization estimation (later updated in a SLAM algorithm). That sensor's name is automatically changed to *odom*. The *base_footprint*, representing the robot platform and the *base_link* representing robot sensor mounting points are also automatically generated.

Vehicle movement commands into USARSim vary depending on the robot type. For example, skid-steered vehicles require left and right wheel velocities while Ackerman steered vehicles require steering angle and linear velocity. ROS provides a *cmd_vel* topic that includes both linear and angular velocities. The *RosSim* node automatically converts these velocities into the appropriate commands and values for the USARSim simulator based on the robots steering type, wheelbase, and wheel separation. Vehicle speeds are also clamped to not exceed maximum velocities that are set in the simulation.

3.1   Sensor Interface

ROS provides a rich vocabulary of sensor interface messages. The *RosSim* node strives to automatically match simulated sensors to the appropriate ROS topic.

Fig. 3. Kenaf in RoboCup 2012 final environment

Currently, a wide variety of sensors are supported; ground truth, inertial navigation, GPS, range-scanners, sonar, CO2 and acoustic sensors are supported. It is our ambition to support all sensors which are used in the RoboCup Rescue competition [1]. This presence of sensors is queried via USARSim's CONF and GEO messages and automatically included the robot transform tree used by ROS. The corresponding sensor messages are published at the rate that the *RosSim* node receives the sensor output.

3.2   Mobile Robot Control with the ROS Navigation Stack

Control of mobile robots through the ROS/USARSim interface is performed with the ROS navigation stack[3]. The navigation stack provides for 2D navigation and takes in information from odometry, sensor streams, and a goal pose to plan the velocity commands that does not lead to collisions. The planning for collision free paths is performed on two levels: one on a local costmap and a global costmap.

Although different models of mobile robot are developed in USARSim, the Kenaf has proven its worth in the Rescue League [14]. The Kenaf robot (see Figure 3) is a challenge to be used the navigation stack with its elongated rectangular form (so not square or circular) and its flippers. As configured in our experiments, it includes a laser scanner mounted on his base.

Low-level Navigation. The ROS/USARSim interface allows for the start-up and control of the default robot base controllers by directly sending velocity commands to the base. This task was performed using the following commands:

1. Bring up an environment in USARSim.
2. $roscore
3. $roslaunch usarsim usarsim.launch
4. $rosrun teleop_twist_keyboard teleop_twist_keyboard.py
5. $rosrun gmapping slam_gmapping scan:=lms200 _odom_frame:=odom

---

[3] http://www.ros.org/wiki/navigation

In step 1. an environment is started on the server side (USARSim). If an environment is not up and running, passing messages between ROS and US-ARSim will fail. Step 2. starts *roscore*, a collection of nodes and programs that are pre-requisites of a ROS-based system for ROS nodes to communicate. Step 3. launches the *usarsim.launch* file. This launch file contains the parameters to connect the game server and starts the *RosSim* node that provides a connection between ROS and USARSim. Step 4. starts the *teleop_twist_keyboard* node which sends velocity commands to the *RosSim* node through the computer keyboard. At this point, the Kenaf can be controlled by keyboard teleop in the USARSim environment. Step 5. starts the node *slam_gmapping* which transforms each incoming scan from the laser into the odometry tf frame to build a map. Here, the topic *scan* is used to create the map with the parameter *_odom_frame*, the frame attached to the odometry system.

Fig. 4. Mobile robot control using *teleop*

Figure 4 is a graph generated by `rxgraph` with the option "quiet". The graph illustrates the communication between the nodes *RosSim*, *teleop_twist_keyboard*, and *slam_gmapping*. The keyboard inputs are converted in velocity commands and then communicated to the *RosSim* node on the topic *cmd_vel*. *slam_gmapping* uses the topics (*lms200*) and (*tf*) as inputs to build the map. To save the generated map, the following command is used:

$rosrun map_server map_saver

Figure 5(a) is a bird's eye view of the environment used to run the `teleop` command to steer a robot through the environment and Figure 5(b) is the map generated by the *map_saver* utility-command.

(a) USARSim environment.

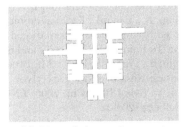

(b) Map of the environment.

Fig. 5. Environment in USARSim and the corresponding map

High-level Navigation. Several teams [18, 19] are currently building high-level navigation software by combining existing ROS algorithms with algorithms from their institutes. Those teams perform simultaneous localization and mapping to generate online maps, and use those maps to plan path (for instance with the RRT algorithm [19]) to replace `teleop` with autonomous navigation.

4   Conclusion and Future Work

This paper has presented a new ROS package that allows for the seamless interface of USARSim with ROS. The package provides for auto-discovery of robots and sensors, and produces the standard ROS topics that one would expect from a physical platform. Several researchers from outside the RoboCup community [10, 12] have already tried this interface. Still, further work is needed to provide the control interface for legged robots (as the Nao robot) and flying robots (such as the AirRobot and the AR.Drone). In addition, the whole sensor suite must have their USARSim interfaces wrapped to be supported in the ROS environment. On the positive site, as the Darpa Robotics Challenge is based on the ROS framework, progress in this competition could be an advance for the RoboCup, and vice versa.

References

1. Akin, H., Ito, N., Kleiner, A., Pellenz, J., Visser, A.: Robocup rescue robot and simulation leagues. AI Magazine 34(1), 78–86 (2013)
2. Balaguer, B., Balakirsky, S., Carpin, S., Lewis, M., Scrapper, C.: USARSim: A Validated Simulator for Research in Robotics and Automation. In: IEEE/RSJ IROS 2008 Workshop on Robot Simulators: Available Software, Scientific Applications and Future Trends (2008)
3. Balakirky, S., Carpin, S., Lewis, M.: Workshop on robots, games, and research: Success stories in usarsim. In: Proceedings of the International Conference on Intelligent Robots and Systems, IROS 2009 (October 2009)
4. Balakirsky, S., Chitta, S., Dimitoglou, G., Gorman, J., Kim, K., Yim, M.: Robot challenge. IEEE Robotics Automation Magazine 19(4), 9–11 (2012)
5. Balakirsky, S., Kootbally, Z.: USARSim/ROS: A combined framework for robot control and simulation. In: Proceedings of the ASME 2012 International Symposium on Flexible Automation, ISFA 2012 (June 2012)
6. Boeing, A., Bräunl, T.: Evaluation of real-time physics simulation systems. In: Proceedings of the 5th International Conference on Computer Graphics and Interactive Techniques in Australia and Southeast Asia. pp. 281–288. ACM (2007)
7. Carpin, S., Wang, J., Lewis, M., Birk, A., Jacoff, A.: High Fidelity Tools for Rescue Robotics: Results and Perspectives. In: Bredenfeld, A., Jacoff, A., Noda, I., Takahashi, Y. (eds.) RoboCup 2005. LNCS (LNAI), vol. 4020, pp. 301–311. Springer, Heidelberg (2006)
8. Cousins, S.: Is ros good for robotics? IEEE Robotics Automation Magazine 19(2), 13–14 (2012)
9. Cousins, S., Gerkey, B., Conley, K., Garage, W.: Sharing software with ros. IEEE Robotics Automation Magazine 17(2), 12–14 (2010)

10. Haber, A., McGill, M., Sammut, C.: Jmesim: An open source, multi platform robotics simulator. In: Proceedings of Australasian Conference on Robotics and Automation (December 2012)
11. Larochelle, B., Kruijff, G.J., Smets, N., Mioch, T., Groenewegen, P.: Establishing human situation awareness using a multi-modal operator control unit in an urban search & rescue human-robot team. In: 2011 IEEE RO-MAN, pp. 229–234 (2011)
12. Meyer, J., Sendobry, A., Kohlbrecher, S., Klingauf, U., von Stryk, O.: Comprehensive simulation of quadrotor uavs using ros and gazebo. In: Noda, I., Ando, N., Brugali, D., Kuffner, J.J. (eds.) SIMPAR 2012. LNCS (LNAI), vol. 7628, pp. 400–411. Springer, Heidelberg (2012)
13. van Noort, S., Visser, A.: Validation of the dynamics of an humanoid robot in usarsim. In: Proceedings of Performance Metrics for Intelligent Systems Workshop, PerMIS 2012 (March 2012)
14. Okamoto, S., Kurose, K., Saga, S., Ohno, K., Tadokoro, S.: Validation of simulated robots with realistically modeled dimensions and mass in usarsim. In: IEEE International Workshop on Safety, Security and Rescue Robotics, SSRR 2008, pp. 77–82. IEEE (2008)
15. Pepper, C., Balakirsky, S., Scrapper, C.: Robot simulation physics validation. In: Proceedings of the 2007 Workshop on Performance Metrics for Intelligent Systems, PerMIS 2007, New York, NY, USA, pp. 97–104 (2007)
16. Quigley, M., Berger, E., Ng, A.Y.: et al.: Stair: Hardware and software architecture. In: AAAI 2007 Robotics Workshop, Vancouver, BC, pp. 31–37 (2007)
17. Quigley, M., Conley, K., Gerkey, B.P., Faust, J., Foote, T., Leibs, J., Wheeler, R., Ng, A.Y.: Ros: An open-source robot operating system. In: ICRA Workshop on Open Source Software (2009)
18. Takahashi, T., Nishimura, H., Shimizu, M.: Hinomiyagura team description paper for robocup 2013 virtual robot league (February 2013)
19. Taleghani, S., Shayesteh, M.H., Samizade, S., Sistani, F., Hashemi, S., Hashemi, A., Najafi, J.: Mrl team description paper for virtual robots competition 2013 (February 2013)
20. Wang, J., Lewis, M., Gennari, J.: A Game Engine Based Simulation of the NIST Urban Search and Rescue Arenas. In: Proceedings of the 2003 Winter Simulation Conference. vol. 1, pp. 1039–1045 (2003)
21. Wang, J., Lewis, M., Hughes, S., Koes, M., Carpin, S.: Validating USARSim for use in HRI Research. In: Proceedings of the Human Factors and Ergonomics Society 49th Annual Meeting. pp. 457–461 (2005)
22. Wyobek, K., Berger, E., der Loos, H.V., Salisbury, K.: Towards a Personal Robotics Development Platform: Rationale and Design of an Intrinsically Safe Personal Robot. In: Proceedings of the IEEE International Conference on Robotics and Automation (ICRA), Pasadena, CA, pp. 2165–2170 (2008)

FUmanoids Code Release 2012

Daniel Seifert and Raúl Rojas

Institut für Informatik, Arbeitsgruppe Intelligente Systeme und Robotik,
Freie Universität Berlin, Arnimallee 7, 14195 Berlin, Germany
{dseifert,rojas}@inf.fu-berlin.de
http://www.fumanoids.de

Abstract. Code re-use and sharing between teams are important factors in the advancement of a RoboCup league. This paper presents the code release of the Humanoid League KidSize team *Berlin United - FUmanoids*. We describe the underlying frameworks and their design principles. Abstraction of hardware and independence of the platform allow the use for different robot types and usage scenarios.

1 Introduction

In this paper we describe the software release of the Humanoid League (HL) KidSize team *Berlin United - FUmanoids*. The first section covers the motivation for code releases in RoboCup, as well as a short history of our team. Section 2 discusses the current code release and its main features, as well as plans for the next code release. Relevant design decisions, the overall architecture and the architecture of the underlying frameworks are presented in section 3. Section 4 covers the debug and analysis tool *FUremote*. Finally, section 5 provides an overview of the use of the release and offers some conclusions.

1.1 Motivation for Code Release

Interoperability between teams in a RoboCup league is an important factor for the advancement of the league. Being able to start with a working system allows new participants to quickly focus on their areas of research. And having the possibility to integrate another team's code allows teams to accelerate their progress without having to reinvent the wheel.

The benefits of interoperability are evident in the Standard Platform League (SPL), where each team has the same robots and several teams regularly provide full releases of their code. A notable example is B-Human, who won first place in 2009-2011. They provide their source code on a yearly basis including an extensive team report [10,11]. As a result of being able to share code easily, the SPL was able to strongly push their rules towards the 2050 goal by increasing team size, reducing color-dependency and enlarging their playing field [9].

On the other hand, teams in the HL usually build their own robots, or at least modify existing platforms. Significant time is spent on hardware development and a team's software is usually incompatible with other teams' robots due to

S. Behnke et al. (Eds.): RoboCup 2013, LNAI 8371, pp. 600–607, 2014.
© Springer-Verlag Berlin Heidelberg 2014

the close ties to the specific hardware platform. This severely slows down progress in the HL. Only in the last two years have the release of the DARwIn-OP [2] and the recent release of the NimbRo-OP [13] provided means to build upon the experience of established teams more easily.

1.2   Berlin United - FUmanoids

Berlin United - FUmanoids has been participating in the Humanoid KidSize League since 2007. Since 2010, the team has cooperated with the SPL team *Nao Team Humboldt*, resulting in the multi-league joint-research group *Berlin United*.

In the first year of the team's existence, a commercially available robot platform was used. In 2008 this platform was slightly modified and finally completely replaced by a fully custom-built robot in 2009 [8]. Since then, a variety of robot models have been designed and successfully[1] used in RoboCup competitions.

Over time, it became obvious how important code sharing is and how little has happened in the Humanoid League in this regards. As a result, we decided to release our code after RoboCup 2011. The first release was published on the team's website in December 2011, thus being one of the first and few code releases in the league. In December 2012, the updated code from RoboCup 2012 was released [1].

2   Current State and Outlook

Our release includes the sources for *FUmanoid* and *FUremote*, a Java application handling data display, debugging and many additional tasks (section 4).

FUmanoid is the robot control software which runs on the robots. At its heart is a modular framework developed by *Berlin United - NaoTH* [5]. It is incorporated as part of the *Berlin United Framework*, consisting of hardware and operating system abstractions as well as a number of service classes. In the current code release [1] the *Berlin United Framework* has not yet been separated. However a standalone version is available online [1]. Utilizing both frameworks, modules separated into cognition and motion blocks are responsible for handling robot specific tasks.

Section 3 highlights the technical aspects of the *FUmanoid* source code. It is the result of several years of development which gradually changed the code from being highly specific for the team's robots to a more generic approach. This has allowed the *Berlin United - Racing Team*[2] to use the software as the basis for their autonomous RC car. The *Berlin United Framework* will also be used in two other robots at our working group, namely in a quadcopter and in a robot participating in the DLR's SpaceBot Cup, supporting inter-project exchange of code and knowledge.

---

[1] The team won 3rd place in 2007, was runner-up in 2009 and 2010, and placed 4th in 2011.

[2] The *Racing Team* is a student project participating in the CaroloCup competition, where an autonomous RC car has to drive on a modelled street and park itself.

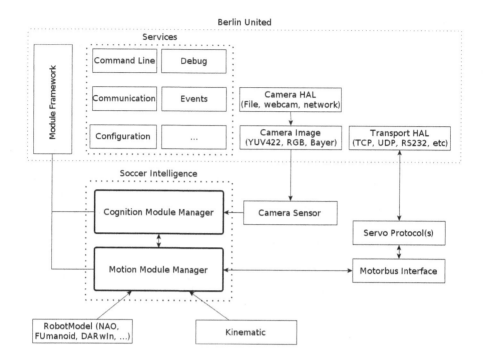

Fig. 1. Overall structure of the control software

For the next release, we are actively working on additional hardware abstractions and increased functionality of the framework. Development of a generic actuator interface is almost finished. It will add full support for the SPL's NAO robot as well as the DARwIn-OP robot to our software. Work has also started on proper documentation outside the source code to aid new students to get a better overall grasp of the system as well as to allow new projects to easily adapt the source code to their needs.

3   Robot Control Architecture

The robot control software *FUmanoid* presented in this section consists of multiple parts. Figure 1 gives an overview of the important components.

3.1   Module Framework

The *NaoTH Module Framework* [5,6] is a clean standalone implementation of a blackboard architecture. The historical roots go back to the GermanTeam's module framework [12], which results in some similarities of the concepts.

The blackboard holds and manages instances of objects, which are *representations* of data that *modules* can either *require* (read) or *provide* (write). Read and write access to the blackboard objects is granted based solely on the declared requirements of the module. The resulting dependency chain permits to automatically calculate an execution order of the modules. As modules are independent from each other they can be exchanged easily, which supports to have concurrent solutions for the same problem.

The modules are executed serially and access data through their module manager's dedicated blackboard only. Thus, concurrency issues are avoided without the overhead of using critical sections throughout the code.

The framework simplifies the development process, especially for new students joining the team. With this approach, no extensive knowledge of the whole software stack is required. Instead work can focus on a single, contained module.

3.2 Multi-platform and Multi-sensor Support

A core design principle of the code in the last few years was the requirement to support different hardware platforms and sensors. This principle arose naturally out of the yearly changing hardware of the team. Additionally it became necessary to support testing the software outside the robots to accelerate development and to improve software quality.

Rather than providing a platform interface which acts as a hardware abstraction layer (HAL) to a pre-defined platform (e.g. a certain robot model), our software is heavily based on the concept of exchangeable input and output devices, i.e. sensors and actuators. Instead of a single HAL, there are specific hardware abstraction interfaces for specific types of sensors and actuators. Being able to easily combine these different sensors allows to quickly adjust to changes in the platform or to assemble new platforms based on pre-existing components. Furthermore, sensor information can be retrieved from multiple sources[3]. Representations can be pre-filled with ground truth data aquired via e.g. simulators[4].

Currently the software only supports robots running the Linux operating system. As CPU-specific functionality is avoided, the code works on both ARM and x86 based CPUs. Due to the multi-sensor approach the code also runs seamlessly on Linux desktop computers. However, supporting other common operating systems would be easy. All Linux-specific components are in wrapper classes which can be easily extended if required.

3.3 Berlin United Framework

The *Berlin United Framework* incorporates various service classes as well as the module framework discussed in 3.1. All functionality included in this framework is independent of the type of robot used, and as such can be used as the basis for

---

[3] One example would be to mix sensors from an actual robot and a simulated one.

[4] The code release includes support for our multi-level sensor emulator *Sim** [3,7], which is not part of the code release due to limited resources for its continued development. However, support for the Simspark simulator has been added and will be available in the next release of the code.

one's own robot software. Standard hardware functionality is included but not mandatory to use. The service classes offer all common functionality a mobile platform requires in an easy to use way to simplify development. For example:

- A dedicated thread handles incoming and outgoing **communication**, supporting multiple remote clients. All communication is based on Protobuf[5], which is an efficient [14] data serialization format that can be extended easily and accessed with a variety of computer languages. Modules are able to register for specific incoming Protobuf messages.
- **Configuration** parameters can be defined in the code and stored and retrieved from a Protobuf-serialized file or via network (ref. section 4), as well as set via the command line.
- An extensive **debug** interface allows the registration of debug messages of several types, including text, table, 2D plot and various image representations. These can be selectively enabled. While disabled, run-time overhead is negligible. Debug output, including images, can be streamed to several remote instances, allowing multiple users to debug or inspect different aspects of the robot simultaneously.
- **Log files** can be created, consisting of all or select representations from each execution cycle. These log files can be used in subsequent runs to populate representations, allowing to debug modules or compare algorithms against each other for the same input data set.

3.4   Motion and Cognition

So far, we discussed the generic capabilities included in our code release. However the code release also includes the HL specific soccer code of our team. We separated our modules into a cognition and a motion block (figure 2).

The *Cognition* module manager is executed when a new camera image is received and thus runs at most 30 times per second. At first the image processing modules run and retrieve information like the position of the ball, the goal and the obstacles from the image, as well as line segments. This data is then classified, e.g. field lines and field line features (crossings, etc) are extracted. Various modelling modules try to model the world and finally behavior modules based on XABSL [4] decide on the actions the robot should take. This sequence follows the Sense-Think-Act paradigm.

The *Motion* module manager runs at 100 Hz. It is responsible for hardware control, primarily interfacing with sensors (except camera) and actuators. An execution cycle starts with reading out sensors and actuators and managing the various information to handle sensor fusion and related tasks, e.g. calculation of camera height based on the robot's kinematics. Depending on the type of motion requested by the cognition, modules are selected which perform keyframe motions like standing up, or which calculate and set the walking trajectories.

Communication between the two module blocks is achieved by event notifications. At the end of each manager's cycle, specific representations are sent and

---

[5] https://code.google.com/p/protobuf

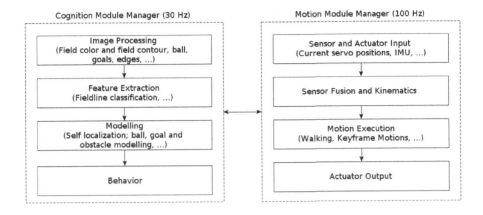

Fig. 2. Core Tasks in the Module Managers

the other manager injects them into its own representations at the next execution cycle. This is a simple yet effective method of sharing data and distributing commands with negligible overhead.

4   Remote Control and Debugging

The code release also includes our debug and control tool *FUremote*. It is based on Eclipse RCP and provides most of its features by means of plugins which allows to switch out features or extend functionality easily, for example when the software is used by other projects.

For the most part, *FUremote* is developed in a way that is agnostic to the specific hardware and environment being used. Plugins retrieve the relevant information from the robots, e.g. the list of debug options, the configuration keys or the available camera settings. This makes it a versatile tool supporting changes to the robot's hardware and software.

Figure 3 shows a possible setup of the tool. On the top-left, the robots currently online are shown including relevant status information like battery level. Below, the Debug plugin lists the available debug options, four of which are active and displaying information in various ways.

Other notable plugins available allow to configure the camera settings, modify configuration parameters, change the robot's game state, visualize the behaviour status of the robot, interact with the servos and record and preview keyframe motions.

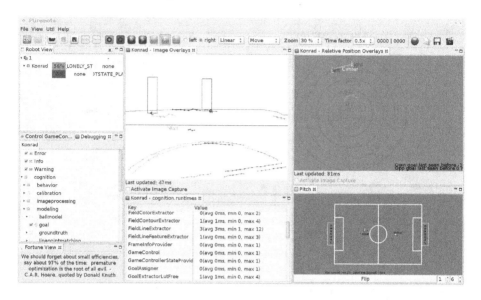

Fig. 3. Screenshot of the *FUremote* tool

5   Conclusion

We presented our code release consisting of the *FUremote* tool, as well as *FU-manoid* and its frameworks. The *Berlin United Framework* is actively used in our humanoid RoboCup team as well as in an autonomous RC car. It will also be used for a quadcopter and a robot participating in the SpaceBot Cup. Its platform-independence and modularity makes it suitable for a variety of robots within and outside the RoboCup community. We expect increased impact when the NAO und DARwIn-OP support is fully integrated in the next code release.

The code release can be downloaded at http://www.fumanoids.de/code.

Acknowledgements. The code released is based on many years of development by the members and students of the *Intelligent Systems and Robotics* group at Freie Universität Berlin. More information about the team and its members, as well as publications and details to the robots is available online at http://www.fumanoids.de.

References

1. FUmanoids: FUmanoids Code Release (2012),
   http://www.fumanoids.de/publications/coderelease
2. Ha, I., Tamura, Y., Asama, H., Han, J., Hong, D.: Development of open humanoid platform DARwIn-OP. In: Proceedings of SICE Annual Conference, SICE 2011, pp. 2178–2181 (2011)

3. Heinrich, S.: Development of a Multi-Level Sensor Emulator for Humanoid Robots. Diploma thesis, Freie Universität Berlin, Institut Mathematik und Informatik, Deutschland (2012)

4. Loetzsch, M., Risler, M., Jungel, M.: Xabsl - a pragmatic approach to behavior engineering. In: 2006 IEEE/RSJ International Conference on Intelligent Robots and Systems, pp. 5124–5129 (2006)

5. Mellmann, H., Xu, Y., Krause, T.: Common Platform Architecture - A Simple and Clean Architecture for Participation in SPL and Simulation 3D (2011), `http://www.naoth.de/projects/multi-platform-robot-controller`

6. Mellmann, H., Xu, Y., Krause, T., Holzhauer, F.: NaoTH Software Architecture for an Autonomous Agent. In: Proceedings of the International Workshop on Standards and Common Platforms for Robotics (SCPR 2010), pp. 316–327 (2010)

7. Mielke, S.: Kamerabildrekonstruktion der Simulationsumgebung für humanoide Fußballroboter. Diploma thesis, Freie Universität Berlin, Institut für Mathematik und Informatik, Deutschland (2012)

8. Mobalegh, H.: Development of an Autonomous Humanoid Robot Team. Ph.D. thesis, Freie Universität Berlin (2011)

9. RoboCup SPL Technical Committee: RoboCup Standard Platform League (Nao) Rule Book (2013), `http://www.tzi.de/spl/pub/Website/Downloads/Rules2013.pdf`

10. Röfer, T., Laue, T., Müller, J., Bartsch, M., Batram, M.J., Böckmann, A., Lehmann, N., Maß, F., Münder, T., Steinbeck, M., Stolpmann, A., Taddiken, S., Wieschendorf, R., Zitzmann, D.: B-Human Team Report and Code Release 2012 (2012), `http://www.b-human.de/wp-content/uploads/2012/11/CodeRelease2012.pdf`

11. Röfer, T., Laue, T., Müller, J., Fabisch, A., Feldpausch, F., Gillmann, K., Graf, C., de Haas, T.J., Härtl, A., Humann, A., Honsel, D., Kastner, P., Kastner, T., Könemann, C., Markowsky, B., Riemann, O.J.L., Wenk, F.: B-Human Team Report and Code Release 2011 (2011), `http://www.b-human.de/downloads/bhuman11_coderelease.pdf`

12. Röfer, T., Brose, J., Göhring, D., Jüngel, M., Laue, T., Risler, M.: GermanTeam 2007 - The German national RoboCup team. In: RoboCup 2007: Robot Soccer World Cup XI Preproceedings, RoboCup Federation (2007)

13. Schwarz, M., Schreiber, M., Schueller, S., Missura, M., Behnke, S.: NimbRo-OP Humanoid TeenSize Open Platform. In: IEEE-RAS International Conference on Humanoid Robots Proceedings of 7th Workshop on Humanoid Soccer Robots (2012)

14. Sumaray, A., Makki, S.K.: A comparison of data serialization formats for optimal efficiency on a mobile platform. In: Proceedings of the 6th International Conference on Ubiquitous Information Management and Communication, ICUIMC 2012, pp. 48:1–48:6. ACM, New York (2012), `http://doi.acm.org/10.1145/2184751.2184810`

Extensions of a RoboCup Soccer Software Framework

Stephen G. McGill, Seung-Joon Yi, Yida Zhang, and Daniel D. Lee

University of Pennsylvania
{smcgill3,yiseung,yida,ddlee}@seas.upenn.edu

Abstract. The RoboCup soccer leagues have greatly benefitted Team DARwIn and the UPennalizers. The stiff competition has hardened our code into a robust framework, and the community has allowed it to flourish as an open source project used by many teams. Working with the open source DARwIn-OP hardware allows even more clairvoyance into the inner workings on the low level code the builds to state machines. We show how our codebase performs in the Webots simulation and on the Open Source DARwIn-OP platform. From these beginnings, we wish to apply our codebase to new scenarios for humanoids including human robot interaction and manipulation tasks. Many of these scenarios are explored by other RoboCup leagues including @Home and Rescue, where we see a new avenue for our codebase. New human robot interaction features are described in our framework, and example performances are demonstrated. Finally, we describe added standards compliance and open source tool usage that will give our codebase more accessibility.

1 Introduction

As a brief background, Team DARwIn and the UPennalizers are teams from the Humanoid KidSize league and the Standard Platform League, respectively, for RoboCup soccer. These teams work on the same codebase for playing software, and have released open source versions of this code since 2011. The current releases can be obtained online[1]. Because these teams operate on two totally different humanoid robots in different leagues, the focus with each software release has always been on portability and compatibility with a variety of humanoid platforms.

As such, this code has been tested, and utilized in competition, on the ubiquitous Nao and DARwIn-OP platforms, and, additionally, on custom DARwIn-XOS and CHARLI platforms. This code has performed well, pushing the DARwIn and CHARLI teams to victory in each of the past two years, and running the UPennalizers standard platform team in the same period. Furthermore, many teams in the humanoid kid-size league have used our code in competition; we have received bug reports from these teams to improve its quality.

This software has been used in-house on humanoid robot experiments outside the realm of soccer, including teleoperative control and recently, the DARPA

---

[1] http://seas.upenn.edu/~robocup

S. Behnke et al. (Eds.): RoboCup 2013, LNAI 8371, pp. 608–615, 2014.

robotics challenge[4]. With these new directions, it became clear that we needed to extend the platform in a few crucial directions, many of which will directly impact the usability in RoboCup.

In summary, many teams have begun to examine or use our codebase as a starting point for their own RoboCup entries, and we have found it useful in working with external colleagues and visiting scholars on humanoid research. While RoboCup soccer does not focus on these extensions, the RoboCup@Home, RoboCupRescue, and the new RoboCup@Work leagues may find it applicable. With applications suitable for a wide audience, we document our recent developments for an open source humanoid robot framework.

2   Related Work

Open source robotic frameworks are not new; a good survey of original formulations of them can be found here [8]. There are many examples, from ROS to OROCOS, from the Microsoft Robotics suite to the B-Human RoboCup soccer release. Each framework, however, has its own focus based on its history. The ROS operating system [12] tries to be all things to all people, and has broad support in the robotics community. However, its history with humanoid robots in particular has not been in as much focus. With our experience stemming from RoboCup and its humanoids, we wish to prioritize humanoid control, sensing, and planning.

OROCOS, on the other side of the coin, focuses on its niche application of realtime control [2], and does this well. However, it cannot support a variety of device drivers that are required for building complex robots. Thus, OROCOS is not sufficient for running a soccer playing robot. Contrasting the niche focus, the Microsoft Robotics studio [6] provides a high level design suite for writing robotics programs. However, this system is restricted to use on the Windows operating system. We wish our framework to be supported on multiple platforms, with a focus on UNIX based systems.

Specific to RoboCup, the B-Human code release is tailored to the Nao robot, which many teams in the Standard Platform league choose to use. While the code has been proven very effective at winning championships, it will not port to other humanoids. Most recently, accompanying the release of the NimBro-OP robot was open source software. While this release is promising, it has just begun and is not tested on other platforms. Other frameworks originating from RoboCup competitions include the Fawkes system[10] which uses the Lua language like ours. However, there is no included walk engine - crucial for teams in RoboCup soccer and for humanoid research.

Given this representative sample of robot frameworks, what we try to accomplish with our framework is a focus on humanoids and a low dependency, small footprint, code base. Most importantly for RoboCup, we wish to keep compliant with the latest rules and regulations for all four humanoid robot devisions - Standard Platform league and each of the three humanoid leagues.

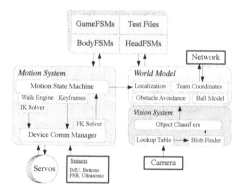

Fig. 1. The system architecture is divided into low level drivers (boxed in red), task specific motion and perception modules, and high level finite state machines (lavender boxes), communicating over shared memory and a message passing system.

3 Soccer Playing Framework

Our soccer framework has focused on providing a general purpose walk controller, a simple vision system, and a set of extensible state machines. The code is written in a combination of the Lua scripting language and the widely used C and C++ languages.

Shown in Figure 1 is a high level overview of how data flows in our system. There are three processes that execute the vision system the device communications manager (DCM) and the motion system state machines. Data is shared using a shared memory segment on the system, where the world model, vision system, motion system, and DCM have their own memory segments. These segments can be read by any other process. For instance, the localization system requires odometry information from the motion system, which is shared using this memory segment.

The benefit of this structure is that the DCM and Camera modules are able to abstract away the platform dependent drivers. We have a different DCM for the Nao, OP, simulated hardware, and any other robot we operate. For instance, the DARwIn-OP and Nao have Linux V4L2 cameras, while Webots has a memory segment. Each driver, however, interfaces the shared memory in the same way.

3.1 Locomotion

We provide a well-documented implementation of the popular Zero Moment Point (ZMP) walking engine. This implementation uses the analytic solution of the ZMP equation, assuming that the ZMP is piecewise linear. The main advantage of this approach is that it is very simple, and it does not require any preview interval. After the foot and center of mass (COM) trajectories are calculated, the inverse kinematics solver generates joint angle values for the leg actuators. It is important for open source software to have a simple and understandable

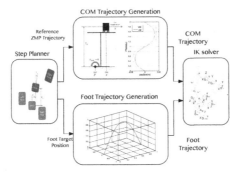

Fig. 2. The motion system provides a abstractions from desired footsteps through ZMP trajectories to inverse kinematics

base system. Shown in Figure 2 is a detailed look at how commanded walking velocities are transformed into joint commands.

3.2 Vision

In our provided vision system, we use a look up table to first categorize raw camera pixels into one of 8 color labels. The goal of the colortable generating utility is to provide a supervised learning approach to classify raw camera pixels into a set of 8 color class labels - possibly green for the field, orange for the ball, etc. Typically, we receive a YCbCr color formatted pixel array from the camera, and convert this to RGB for displaying to the user. Figure 3 shows the colortable generating tool for a real robot camera frame, and how the generated lookup table performs in simulation.

The user clicks on pixels that belong to each of these labels to generate both positive and negative examples of the color. We then apply a gaussian mixture model to generate color segments. For each mixture, we perform a threshold to find pixels with a high probability of belonging to that mixture?s color class. With a high enough score, we add that pixel to a lookup table mapping to the color class. This lookup table from the mixture model to color class saves on per pixel computation, since the probability does not need to be computed for each pixel on every incoming frame; downsampling further increases speed. This labeled image is fed into high level blob detection routines and object classifiers.

Given a labeled image it is downsampled, where labeled pixel are grouped into 4 pixel by 4 pixel bitwise OR-ed blocks for faster execution in later image analysis. This downsampled image is fed into high level blob detection routines and object classifiers. Results are show in Figure 3.

3.3 World Model

Our World model collects information about odometry from the motion system, teammate positions from network, and observed objects from the vision system.

Fig. 3. Colortables can be made with a QT user interface for classifying ball colors (left). The generated lookup table can be monitored in MATLAB (right).

With this information the world model can assign roles for the robot, determine the robot's pose, and identify obstacles. Localization is performed using acoustic triangulation or particle filters based on landmark observations. Our obstacle avoidance methods is described in [13].

3.4   Interprocess Communication

While we use the Boost shared memory system to synchronize data across processes, we needed to add event based information sharing for more responsive processing loops. To ensure compatibility, we choose to implement a widely used standard that supports a broad array of features. Our framework has adopted the ZeroMQ system [5], which focuses on being lightweight with low latency. An additional benefit of the ZeroMQ system is that it is an open source, standard, definition with bindings in many languages.

Since we use Lua objects to represent information like poses and ball position, we must serialize these objects before sending over a ZeroMQ channel. We have two serialization libraries - our own custom library and a MessagePack library. With MessagePack, Lua objects are reconstituted easily in other languages, since tables, strings, and numbers, for instance, are well defined.

3.5   Debugging Tools

Figure 3 shows our monitoring program that receives frames from the Webots simulator or the real robot, and displays the raw camera frame and the labeled image side by side. Recently, we have added open source tensor manipulation libraries using the Torch7 project [3], which runs under Lua. We have extended the available Torch system as a packaged module. For graphical interfaces, we provide QT windows for displaying information. These moves allow team members, and other institutions, who do not have access to MATLAB to be able to use our system and contribute to it readily. An additional benefit of having two methods for visualization is that it pushes the codebase to be more modular. If a better interface becomes available in the future, we can port more easily our code.

Fig. 4. Left: Teamplay is simulated using the Robotstadium competition, where we can rapidly prototype behaviors which perform similarly on real hardware. Right: The open source ATLAS model used in the Webots simulator to execute our teleoperated manipulation routine.

4   Evaluation

We utilize the freely available Webots [9] Robotstadium platform to evaluate our teamplay code. Using a simulator allows for rapid prototyping, with reasonable speeds, even for full physics simulation. Shown in Figure 4 is an example of two teams of four robots running our code. Each team (or even each player) can be slightly modified to run a different parameter set. We test all of our software on Mac and Linux operating systems, and provide initial support for the Gazebo simulation platform [7].

Most importantly, we evaluate our platform on the DARwIn-OP, the Aldebaran Nao, and custom humanoid hardware. Our software platform is extensively tested in competition each year during RoboCup.

We evaluate our platform on the DARwIn-OP, the Aldebaran Nao, and custom humanoid hardware. Our software platform is extensively·tested in competition each year during RoboCup.

5   Beyond RoboCup

Outside of robot soccer, humanoids perform manipulation tasks and interact with humans. While manipulation tasks are gaining traction in soccer competitions, they are not a focus; human interaction is limited to off field training. Thus, these aspects have not become a focus for many teams.

However, outside RoboCup soccer, these tasks are immeasurably important. In RoboCup@Home, for instance, robots must manipulate objects on tables, as well as interpret commands from a human. In RoboCupRescue puts an emphasis on exploration using sensors not allowed in the soccer competitions. However, the recent DARPA Robotics competition has actually merged RoboCup soccer and the Rescue league by encouraging humanoid robots to be used in disaster relief scenarios. Many people involved in our RoboCup effort are working on this DARPA project, and so our software must scale to achieve this broader impact. For instance, shown in Figure 4 is our Webots version of the open source ATLAS model picking up an object.

Our first foray into manipulation and human interface requires a set of device drivers to be present in our a code base. To control our robots by teleoperation, we have focused on providing drivers for some human interface devices. While UVC cameras have well defined drivers in out framework, we needed to add more devices We now interface a Spacemouse, originally developed for robots [1], for 6D control of the manipulator, a Kinect for skeleton tracking using the NiTE [11] libraries, and recently a LEAP sensor for gripper control.

5.1 Challenges and Opportunities

While providing support for a multitude of humanoids and devices enhances the usability of the codebase, it presents challenges for maintaining concise abstractions. For instance, allowing more operation modes that the DARPA challenge and other RoboCup leagues require that more motor interfaces must be specified, and that the higher level state machines need to amend certain API calls. By using using Lua in place of C/C++ for dynamically adding function calls, extra operating modes become less intrusive.

Additionally, sensors must be able to interpret a wider scope of environments than that of the well defined soccer field. While our framework does not solve these problem, it can provide a way to approach it. This particular aspect is where many new state machines, sensor drivers, and other aspects of the framework are evolving. We hope to release these new features as they become more mature.

Overall, we are are attempting to target humanoid researchers who have a background in RoboCup, or an understanding of the RoboCup competition. Because RoboCup provides a concrete evaluation of software, the particular state machines and abstractions become easier to understand precisely their is some intuition of the nature of soccer. In the near term, we hope that our general humanoid implementation can be applied to leagues outside of soccer.

References

1. Space mouse, the natural man-machine interface for 3d-cad comes from space. In: ECUA 1996 Conference for CATIA and CADAM users in Europe (1996)
2. Bruyninckx, H., Soetens, P., Koninckx, B.: The real-time motion control core of the Orocos project. In: IEEE International Conference on Robotics and Automation, pp. 2766–2771 (2003)
3. Collobert, R., Kavukcuoglu, K., Farabet, C.: Torch7: A matlab-like environment for machine learning. In: BigLearn, NIPS Workshop (2011)
4. Defense Advanced Research Projects Agency DARPA. DARPA robotics challenge (April 2012)

5. Hintjens, P.: ZeroMQ: The Guide (2010)
6. Jackson, J.: Microsoft robotics studio: A technical introduction. IEEE Robotics and Automation Magazine 14, 82–87 (2007)
7. Koenig, N., Howard, A.: Design and use paradigms for gazebo, an open-source multi-robot simulator. In: Proceedings of the 2004 IEEE/RSJ International Conference on Intelligent Robots and Systems, IROS 2004, vol. 3, pp. 2149–2154 (2004)
8. Kramer, J., Scheutz, M.: Development environments for autonomous mobile robots: A survey. Auton. Robots 22(2), 101–132 (2007)
9. Michel, O.: Webots: Professional mobile robot simulation. Journal of Advanced Robotics Systems 1(1), 39–42 (2004)
10. Niemüller, T., Ferrein, A., Lakemeyer, G.: A Lua-based Behavior Engine for Controlling the Humanoid Robot Nao. In: Baltes, J., Lagoudakis, M.G., Naruse, T., Ghidary, S.S. (eds.) RoboCup 2009. LNCS (LNAI), vol. 5949, pp. 240–251. Springer, Heidelberg (2010)
11. PrimeSense Inc. Prime Sensor NITE 2 Algorithms notes, 2012 (last viewed March 21, 2013)
12. Quigley, M., Conley, K., Gerkey, B.P., Faust, J., Foote, T., Leibs, J., Wheeler, R., Ng, A.Y.: Ros: An open-source robot operating system. In: ICRA Workshop on Open Source Software (2009)
13. Vadakedathu, L., Sreekumar, A., Yi, S.-J., McGill, S., Zhang, Y., Lee, D.: Comparison of obstacle avoidance behaviors for a humanoid robot in real and simulated environments. In: The 7th Workshop on Humanoid Soccer Robots (2012)

Inexpensive Image Processing Solution for the RoboCupJunior Soccer Scenario

Marek Šuppa and Ján Ďurkáč*

XLC Team, Topoľčany, Slovakia
{marek,durikon}@xlc-team.info
http://www.xlc-team.info

Abstract. It has been a while since image processing has been introduced to RoboCupJunior. Since then, computational power has become cheaper and tools have become more mature. It is time to evolve again.

In this paper we present a simple and inexpensive system which is capable of processing images in real time and can provide information about the situation on the playing field and about the robot itself.

Inspired by many teams in the RoboCup Senior Soccer leagues which use similar image processing systems, we believe that it can serve as a framework for other RoboCupJunior teams who are trying to use more logic than force in their strategy.

Keywords: robot, image processing, RoboCupJunior Soccer.

1 Introduction

The RoboCupJunior Soccer is being played on a field which resembles the football pitch that is being used in human soccer. Robots are required to fit into a cylinder 22 cm in diameter and 22 cm in height. Depending on the league, there is also a weight limit. As defined in the rules, total dimensions of this field are 182 cm by 243 cm [1].

Since 2009 the rules specify that "the interior walls and the cross-bar of each goal are painted, one goal yellow, the other goal blue. The exterior is painted black". Additionally, any interference that might happen with other robots is prevented ("Robots are not allowed to be coloured yellow or blue in order to avoid interference with the goal colour") as well as interference with the outer environment ("Any person close to the playing field is not allowed to wear any yellow or blue clothes that can be seen by the robots (to avoid interference with the goal colour)").

This set of rules allows for easy detection of the goal. Even though there still might be some interference from the environment (e.g. walls might be painted in yellow) we can expect it to be minor.

---

* We would like to thank our mentor Rastislav Gaži whose energy and passion is crucial for anything we do at XLC Team, our patient and supportive parents and everyone who broaden our horizons.

S. Behnke et al. (Eds.): RoboCup 2013, LNAI 8371, pp. 616–623, 2014.

Image processing systems are being commonly used in all RoboCup Soccer leagues. In many of them they are used as the primary source of information about the field since objects like the yellow ball would be difficult to detect effectively using different methods [3]. Image processing has already been introduced to RoboCupJunior [2] and it is gaining popularity in robot designs.

We present a simple and inexpensive image processing system that can detect the goal and provide information that might be used to enhance robot's strategy.

2 Setup

In one of the first applications of image processing in RoboCupJunior Soccer, the use of USB camera was refuted due to its low frame rate [2]. However, since then computational power has become cheaper and thus more usable in RoboCupJunior scenario.

We use a PlayStation Eye webcam which is capable of taking images with 640×480 pixels at the rate of 60Hz [4]. Its dimensions (80 mm × 55 mm × 65 mm) make it also well suited for the usual design of RoboCupJunior Soccer robots.

For the actual image processing we use the Raspberry Pi, Model B . This single-board computer has a 700 MHz processor and 512 MB of RAM[1]. It does not include a built-in storage but uses an SD card and runs the Linux operating system. Raspberry Pi is described as "credit-card-sized" (85.60 mm × 53.98 mm) which means that it can fit within the size requirements for robots competing in RoboCupJunior Soccer. [5]

The PlayStation Eye webcam is connected to the Raspberry Pi via an USB cable. Even thought the Raspberry Pi could serve as the only processing unit for the whole robot, we rather have a separate Arduino Mega board which controls motors and gathers other sensory data. The Arduino board is connected via USB cable to the Raspberry Pi. After the processing happens, the results are sent to the Arduino board.

Raspberry Pi provides many possible ways of interfacing so it is possible to use this system with any micro controller that is able to read digital values (Raspberry Pi can provide the results via GPIO pins for instance).

The aim of robots in the RoboCupJunior Soccer competition is to win the match by shooting goals. Usually, this is done using a kicker device that pushes the ball in the forward direction. Since this system tries to enhance the effectiveness of scoring a goal, the camera is located just above the kicker and dribbler (a device, which holds the ball in robot's possession) as shown in Figure 1.

3 Image Processing

The software part of this image processing system is based on the freely available OpenCV library[6]. It is implemented in the C programming language,

---

[1] Model B originally had 256 MB of RAM but was later upgraded to 512 MB. See [5] for more details.

Fig. 1. The robot with the webcam mounted above the dibbler and the kicker device

in from of a library called `visy`. This library is open-source, available at `http://xlc-team.info/visy` and licensed under the terms of the BSD License.

3.1 Camera Calibration

In order for this system to work reliably we need to make sure that the camera does not distort images. For this, we use the standard OpenCV camera calibration procedure [7].

As a result, we obtain configuration constants, such as distortion matrix and focal length of the camera. The distortion matrix lets us remove distortion from the image. The focal length lets us compute the real size of an object given its size measured in pixels.

3.2 Goal Detection

The ability to detect the opponent's goal is crucial for a mobile robot competing in the RoboCupJunior Soccer. Many teams use a compass sensor to steer the robot in the direction of the opponent's goal side. Even though this setup is easy to use it is ineffective, because it does not take into account the situation in which the robot is not located exactly in front of the goal or when there are obstacles which could prevent the ball from entering the goal (e.g. goalkeeper).

The aim of our system is to detect the goal and any possible obstacles to increase effectiveness of scoring a goal from any point on the field. For this we use these four steps:

Fig. 2. Example of input image

Step 1: Goal Extraction. Since 2009 the rules define colors of goals as blue and yellow. However, the exact color of these goals is not defined.

In order to locate the goal in the field we extract its color from the image. To make this system resistant to lighting changes and also as error free as possible we use the HSV color space instead of the RGB model. The process of this step is described as follows: The image is loaded from the webcamera in the RGB format. It is then converted to the HSV color space. The system then marks pixels which are within the defined range. The results are then used in the next step.

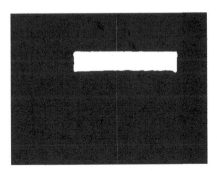

Fig. 3. Example of extracted goal

Step 2: Histogram Generation. Using the results from previous step, in this step the system creates a histogram which simplifies the analysis of results. For any given column in the hiostogram the value equals the number of pixels marked in Step 1 in this column.

Moreover, it computes statistical variables of this graph such as the mean and the mode of all non-zero heights. This histogram and the statistical variables are then used in following steps.

Fig. 4. Example of the generated histogram

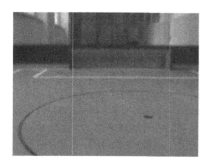

Fig. 5. Example of the image with marked boundaries

Fig. 6. Example of the image with an obstacle in the goal with marked boundaries

Step 3: Histogram Analysis. In this step, the system looks for the places in histogram with significant difference. By significant difference we mean, that the value of one pixel is higher than the mode value and the value of the other pixel is smaller than the mode value. These places are marked as boundaries. If the distance between two boundaries is within the defined configuration value it is removed. This results in a small set of boundaries. These boundaries then allow us to compute the position of the goal and the position of any obstacle.

Aditionally, the system splits the histogram into two equal parts. For both of these parts it computes the sum of all values and call them $lsum$ and $rsum$, respectively.

Step 4: Results. Given the data gathered in previous steps, the system tries to detect whether there is enough free space in the goal for the ball. If not, it tries to suggest what action should the robot take to successfully score.

In section 3.4 the rules specify that "The goal inner space is 60 cm width, 10 cm high and 74 mm deep, box shaped. It has a cross-bar on top". The height of the cross-bar is not specified in the rules but it is usually around 2 cm. The

RoboCupJunior Soccer competition uses a ball that emits IR signal and has 74 mm in diameter. [1]

The mode value computed in previous step represents the height of the goal. We have chosen the mode value to assure correct results in cases when there is an obstacle in the goal.

In this step the system computes the pixel:centimeter ratio using the mode and the known height of the goal (approximately 12 cm) and looks if there is enough room in the centre of the image for the ball to fit in. If there is not, it compares the *lsum* and *rsum* values and advises the robot to move to the left or the right side. If *lsum* and *rsum* both equal to zero (e.g. when no goal has been detected) the system advises the robot to rather rely on other sensory data. The result is stored in the *status* variable which can have four states: K (kick), L (move to the left), R (move to the right) or N (nothing). [2]

3.3 Pose Estimation

Using the values computed while detecting the goal we can also find the approximate distance from the goal. For this we need to know the focal length of the camera. It can either be found as the result of the camera configuration process or in a simpler process described as follows:

Consider an object with known height H in known distance d from the camera. We measure its apparent height in pixels and call it h. From the pinhole camera model we see that the focal length of the camera f is then given by the following:

$$f = \frac{hd}{H} \tag{1}$$

If we want to compute the approximate distance from this object (the opponents goal in our case), given its apparent height in pixels h', then by applying triangle similarity we get:

$$\frac{f}{h'} = \frac{D}{H}$$
$$D = \frac{Hf}{h'} \tag{2}$$

Where D is the approximate distance from this object (the Y coordinate).

We have also explored the possibility applying a similar approach in order to find the horizontal distance from the goal centre (the X coordinate of the robot). However, it proved to be impractical since there are many situations in which it this value cannot be computed properly[3]. Furthermore, its usefulness in the strategy was minor so the system does not compute it.

---

[2] In Fig. 5 the system would tell the robot to kick. In Fig. 6, however, it would advise the robot to move to the right.

[3] For instance, when there is a robot on each end of the goal.

3.4   System Calibration

In order to achieve precision and good results during the gameplay, it is important to calibrate the system before each halftime, mostly due to the fact that "After the first half, teams will switch sides." [1] but also because the lighting conditions might change.

The calibration procedure consists of changing the range in which the goal color should be. This means changing values for the minimal and maximal hue, saturation and value in the HSV model. These values are represented by variables $minH$, $minS$, $minV$ and $maxH$, $maxS$, $maxV$, respectively.

To make it as easy and straightforward as possible, we use the OpenCV High-GUI library, which provides a way of changing these variables via trackbars and immediately seeing the results.

After calibration, the values are stored in a config file which uses the YAML format. This file is then loaded on the next run of the system.

4   Usage in Strategy

After each iteration of the Image Processing System two variables are sent to the Arduino controller: the goal status and the approximate distance from the goal. The goal status variable is being used more, as it can be used to enhance shooting efficiency of the robot.

The following code shows a possible usage implementation of this value:

Example usage of the results provided by the Image Processing System

```
switch (status) {
    case 'K':
        kick();
        break;
    case 'R':
        move_right();
        break;
    case 'L':
        move_left();
        break;
    default:
        break;
}
```

Even thought the code is very simple, the improvement in the robots strategy is significant because rather than blindly shooting the ball and thus giving the opponent a chance go gain its possession, it moves until there is enough room in the goal to score. Moreover, since it is possible to find the robot's distance from the goal it is possible to create more complex and effective strategies.

5   Conclusion

We have presented a simple and inexpensive Image processing system which can be used on robots in the RoboCupJunior Soccer scenario. It can locate the opponent's goal and provide useful data which can be then used by the robot to decide whether it is better to kick or to move in some direction.

The system has been able to process 10 images per second which we have found sufficient to be used in the RoboCupJunior Soccer. The cost of this system did not exceed $60 which makes it even more usable since it removes the financial barrier teams in the RoboCupJunior competition often face. The software this system uses is based on the OpenCV library which supports a variety of devices so there is still room for reducing its cost.

Even though the system is very basic in the current state, we believe that it can be helpful to the whole RoboCupJunior community and can serve as the framework for other teams that are interested in using image processing in their designs. While being inexpensive, it can be easily extended to for instance detect white borders around the field or even distinguish robots on the field which can lead into effective team play.

References

1. RoboCupJunior Soccer Rules, http://rcj.robocup.org/rcj2013/soccer_2013.pdf (accessed February 23, 2013)
2. Siedentop, C., Schwarz, M., Pfülb, S.: Introducing Image Processing to RoboCupJunior. In: Iocchi, L., Matsubara, H., Weitzenfeld, A., Zhou, C. (eds.) RoboCup 2008. LNCS (LNAI), vol. 5399, pp. 308–317. Springer, Heidelberg (2009)
3. Gönner, C., Rous, M., Kraiss, K.-F.: Real-time adaptive colour segmentation for the robocup middle size league. In: Nardi, D., Riedmiller, M., Sammut, C., Santos-Victor, J. (eds.) RoboCup 2004. LNCS (LNAI), vol. 3276, pp. 402–409. Springer, Heidelberg (2005)
4. Wikipedia contributors, "PlayStation Eye," Wikipedia, The Free Encyclopedia, http://en.wikipedia.org/w/index.php?title=PlayStation_Eye&oldid=526514281 (accessed February 23, 2013)
5. Wikipedia contributors, "Raspberry Pi," Wikipedia, The Free Encyclopedia, http://en.wikipedia.org/w/index.php?title=Raspberry_Pi&oldid=539924790 (accessed February 23, 2013)
6. Bradski, G.: The opencv library. Doctor Dobbs Journal 25(11), 120–126 (2000)
7. OpenCV dev team: Camera calibration With OpenCV, http://docs.opencv.org/doc/tutorials/calib3d/camera_calibration/camera_calibration.html#camera-calibration-with-opencv

Hector Open Source Modules for Autonomous Mapping and Navigation with Rescue Robots

Stefan Kohlbrecher[1], Johannes Meyer[2], Thorsten Graber[2], Karen Petersen[1],
Uwe Klingauf[2], and Oskar von Stryk[1]

[1] Department of Computer Science, TU Darmstadt, Germany
[2] Department of Mechanical Engineering, TU Darmstadt, Germany
http://www.gkmm.tu-darmstadt.de/rescue

Abstract. Key abilities for robots deployed in urban search and rescue tasks include autonomous exploration of disaster sites and recognition of victims and other objects of interest. In this paper, we present related open source software modules for the development of such complex capabilities which include *hector_slam* for self-localization and mapping in a degraded urban environment. All modules have been successfully applied and tested originally in the RoboCup Rescue competition. Up to now they have already been re-used and adopted by numerous international research groups for a wide variety of tasks. Recently, they have also become part of the basis of a broader initiative for key open source software modules for urban search and rescue robots.

1 Introduction

While robots used for Urban Search and Rescue (USAR) tasks will remain mainly tele-operated for the immediate future when used in real disaster sites, increasing the autonomy level is an important area of research that has the potential to vastly improve the capabilites of robots used for disaster response in the future.

The RoboCup Rescue project aims at advancing research towards more capable rescue robots [1]. Rescue robotics incorporates a vast range of capabilities needed to address the challenges involved, e.g. resulting from a degraded environment. The availability of re-useable and adaptable open source software can significantly reduce development time and increase robot capabilities while simultaneously freeing resources and, thus, accelerating progress in the field.

In this paper, we present open source modules that provide the building blocks for a system capable of autonomous exploration in USAR environments. Different modules have been applied with great success in RoboCup Rescue and other applications, both by Team Hector (Heterogeneous Cooperating Team of Robots) of TU Darmstadt and numerous other international research groups.

Robot Operating System (ROS) [2] is used as the robot middleware for the software modules. It has been widely adopted in robotics research and can be considered a de-facto standard. The provided modules have also become part of a recently established, broader initiative of the RoboCup Rescue community for providing standard software modules useful for USAR tasks [3].

S. Behnke et al. (Eds.): RoboCup 2013, LNAI 8371, pp. 624–631, 2014.
© Springer-Verlag Berlin Heidelberg 2014

At the RoboCup competition, we mainly use the Ackermann-steered Hector UGV vehicle (Figure 1)[4]. While this method is in many ways more challenging than differential steering, we do not focus on these challenges in this paper, instead providing a simulated skid-steered vehicle based on the Hector Lightweight UGV (Figure 1) that bears more similarity to differential drive vehicles commonly used for USAR tasks.

Fig. 1. Robots used by Team Hector. Left: Hector UGV based on Kyosho Twin Force chassis. Right: Hector Lightweight UGV based on "Wild Thumper" robot kit.

1.1 Related Work

Research in Simultaneous Localization and Mapping (SLAM) and exploration of unknown environments received a lot of attention in recent years, with impressive results being demonstrated. Many of these results often cannot be reproduced due several reasons, like a lack of standardized interfaces, closed source software and limited robustness to different (e.g. environmental) conditions.

Evaluation of state-of-the-art visual SLAM approaches [5], [6] in the standardized RoboCup Rescue setting showed promising results, but consistent localization/mapping as with the system described in this paper could not be achieved so far, as ramps and other obstacles lead to jerky vehicle motion and pose significant challenges to any SLAM system.

The RoboCup Rescue Robot League competition provides especially challenging scenarios, as the competition setting enforces strict constraints on the time and environment for robot operation.

2 System Overview

This paper covers many of the higher-level nodes originally developed and tested for the Hector UGV system, which can be used and adapted for other platforms without or with only slight modifications (Fig. 2). Hardware dependent modules like camera and motor drivers or low-level controllers are not within the scope of this work. It is assumed that robots intended to use the described modules provide the necessary sensor data according to existing ROS standards and are steerable by publishing velocity commands. All nodes holding some sort of state information are subscribing the command topic which is primarily used to reset the system whenever necessary.

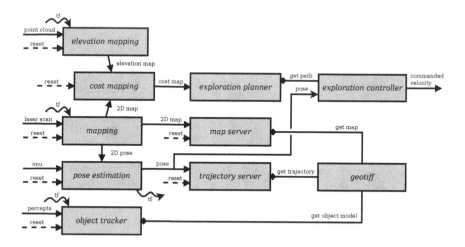

Fig. 2. System overview schematic. ROS nodes are represented by rectangles, topics by arrow-headed and services by diamond-headed lines. Services are originated at the service caller.

The following sections describe the ROS nodes provided[1]. Section 3 presents the open source software for 2D and 3D mapping, perception of objects of interests and the generation of GeoTIFF maps to visualize the relevant information according to the RoboCup Rescue rules. The subsequent Section 4 introduces the modules required for planning and autonomous exploration. While not directly related to autonomous robots being able to test individual modules and the robots overall behavior in simulation in a close-to-reality scenario is crucial in order to detect bugs or possible failure cases earlier and allows shorter development cycles. We present our simulation environment in Section 5.

3   Localization and Mapping

Creating maps of the environment is important for two reasons: Allowing first responders to both perform situation assessment and localize themselves inside buildings and for path planning and high level autonomous behaviors of robot systems.

---

[1] For details see http://www.ros.org/wiki/tu-darmstadt-ros-pkg

While purely geometric maps such as occupancy grid maps are useful for navigation and obstacle avoidance, additional semantic information like the location of objects of interest is very important for first responders and required for intelligent high level autonomous behavior control.

3.1 Simultaneous Localization and Mapping (SLAM)

As disasters can significantly alter the environment compared to a pre-disaster state, USAR robots have to be considered as operating in unknown environments as to be most robust against changes. This means the SLAM problem has to be solved to generate sufficiently accurate metric maps useful for navigation of first responders or a robot system.

For this task we provide *hector_slam*, consisting of *hector_mapping*, *hector_map_server*, *hector_geotiff* and *hector_trajectory_server* modules. As odometry is notoriously unreliable in USAR scenarios, the system is designed to not require odometry data, instead purely relying on fast LIDAR data scanmatching at full LIDAR update rate. Combined with an attitude estimation system and an optional pitch/roll unit to stabilize the laser scanner, the system can provide environment maps even if the ground is non-flat as encountered in the RoboCup Rescue arena. A comprehensive discussion of *hector_slam* is available in [7].

3.2 Pose Estimation

The estimation of the full 6 degrees of freedom robot pose and twist is realized in the *hector_pose_estimation* node that implements an Extended Kalman Filter (EKF) and fuses measurements from an inertial measurement unit (IMU), the 2D pose error from the laser scan matcher and optionally from additional localization sensors like satellite navigation receivers, magnetometers and barometric pressure sensors if available. The filter is based on a generic motion model for ground vehicles and is primarily driven by the IMU, without using the control inputs or wheel odometry as they typically are unreliable due to wheelspin or side drift on uneven or slippery ground.

3.3 Elevation and Cost Mapping

In addition to a two-dimensional world representation obtained by the *hector_slam* package, USAR robots have to take the traversability of the environment into account. To this end we developed *hector_elevation_mapping*. This package fuses point cloud measurements obtained by a RGBD-camera such as the Microsoft Kinect into an elevation map. The elevation map is represented by a 2D grid map storing a height value with a corresponding variance for each cell. The cell measurement update is based on a local Kalman Filter and adapted from the approach described in [8].

Finally, *hector_costmap* fuses the 2.5D elevation map with the 2D occupancy grid map provided by *hector_mapping* and computes a two-dimensional cost map for the exploration task.

Fig. 3. Examples for autonomous exploration. Left: Simulated Thailand Rescue Robot Championship 2012 arena. Right: Simulated random maze.

3.4 Objects of Interest

Plain occupancy grid maps provide information about the environment geometry, but do not contain semantic information. We track information about objects of interest in a separate module, using a Gaussian representation for their position. The *hector_object_tracker* package is based on an approach described comprehensively in [9]. It subscribes to percept messages from victim, QR code or other object detectors, projects them to the map frame based on the robot's pose, camera view angle and calibration information and solves the association and tracking problem for subsequent detections.

3.5 GeoTIFF Maps

To achieve comparability between environment maps generated by different approaches, the GeoTIFF format is used as standard map format in the RoboCup Rescue League competition. Using geo-reference and scale information, maps can be overlaid over each other using existing tools and accuracy can be compared. The *hector_geotiff* package allows generating RoboCup Rescue rules compliant GeoTIFF maps which can be annotated through a plugin interface. Plugins for adding the path travelled by the robot, victim and QR code locations are provided. The node can run onboard a robot system and save maps to permanent storage based on a timer, reducing the likelihood of map loss in case of connectivity problems. All map shown in figures in this paper have been generated using the *hector_geotiff* node.

4 Planning and Exploration

While a plethora of research results are available for exploration using autonomous robots, there are very few methods readily available for re-use as open source software. We provide the *hector_exploration_planner* that is based on the exploration transform approach presented in [10]. In our exploration planner, frontiers towards the front of the robot are weighted favorably, to prevent frequent costly turning of the robot. Inspired by wall following techniques used by firefighters [11], a "follow wall" trajectory can also be generated using the exploration planner. The planned trajectory is generated based on map data and thus does not exhibit weaknesses associated with reactive approaches that only consider raw sensor data [12]. High level behaviors can thus switch between using the exploration transform and wall follow approach at any time. In case the environment has been completely explored, the planner has been extended to start an "inner exploration" mode. Here, the traversed path of the robot containing a discrete set of past robot poses is retrieved from the *hector_trajectory_server* node. These positions are sampled based on distance from each other and added to a list. This list is passed to the exploration transform algorithm as a list of goal points. An exhaustive search for the exploration transform cell with the highest value then yields a point that is farthest away from the previous path and safe to reach for the robot.

5   Simulation

Experiments using real robots are time-consuming and costly as availability of appropriate scenarios and wear and tear of robot systems have to be considered. This holds especially for USAR environments (like the RoboCup Rescue arena) as those also put high strain on robot hardware and lab space is often limited.

5.1   Environments

To conveniently be able to create simulated environments for experiments, the *hector_nist_arenas_gazebo* stack provides the necessary tools that allow the creation of scenarios by composition of provided NIST standard test arena elements. Users can also easily add further elements. The *hector_nist_arena_worlds* package provides example arenas, including both models for the RoboCup German Open 2011 and the Thailand Robot Championship 2012 Rescue arenas. Gazebo does not support multispectral sensor simulation originally. To enable simulation of thermal images often used for the detection of victims that emit body heat, the *hector_gazebo_thermal_camera* package provides a gazebo camera plugin that can be used for this task.

5.2   Ground Vehicles

The *hlugv_gazebo* package provides a model of the Hector Lightweight UGV system (Fig. 1 left). The robot uses differential drive for its six wheels and thus behaves similar to tracked robot systems commonly applied in USAR scenarios.

6   Application and Impact

6.1   RoboCup

Within less than two years *hector_slam* has become the de-facto standard SLAM system used by many teams with great success in RoboCup competitions. With

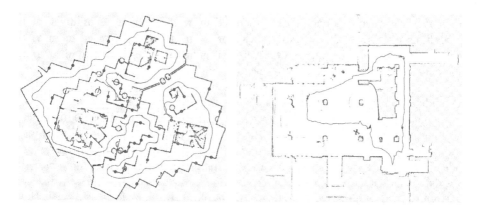

Fig. 4. Maps learned using the provided Hector modules. Left: Map learned using *hector_slam* and *hector_exploration_planner* at the RoboCup 2012 final mission with the Hector UGV robot. The robot started at the right middle position and autonomously explored the majority of the arena, finding 3 victims (red markers). The fourth victim was found using tele-operation for the last 4 meters of the travelled path. Blue markers indicate the positions of 35 QR codes that were detected autonomously by the robot. Right: Application of *hector_slam* to the ccny quadrotor lobby dataset [13].

Team Hector winning the Best in Class Autonomy award both at the RoboCup German Open 2012 and RoboCup Mexico 2012 and Team BARTlab winning the award at the Thai Rescue Robot Championship 2012, the applicability and adaptability of the system to challenging environments and different robot platforms has clearly been demonstrated. Fig. 4 left shows a real-world map learned using the presented modules with the Hector UGV system.

6.2 Other Applications

Hector open source modules have been re-used for both research and commercial purposes[2]. *hector_mapping* was succesfully deployed in different applications such as mapping of littoral areas using a unmanned surface vehicle, mapping different environments using a handheld mapping system and building radio maps for wireless sensor networks [14]. Fig. 4 right shows results when applied to the quadrotor datasets provided in [13]. The resulting map is consistent and comparable to the results in the original paper, showing the flexibility of the system.

7 Conclusion

A collection of open source modules has been presented for providing urban search and rescue robots with abilities like mapping and exploration of disaster sites and tracking of objects of interest. Many of the presented modules have already been adopted by other research groups for RoboCup Rescue and beyond.

---

[2] http://www.youtube.com/playlist?list=PL0E462904E5D35E29

Acknowledgments. This work was supported by the DFG Research Training Group 1362. We thank contributing past and present team members, notably Florian Berz, Florian Kunz, Mark Sollweck, Johannes Simon, Georg Stoll and Laura Strickland.

References

1. Jacoff, A., Sheh, R., Virts, A.M., Kimura, T., Pellenz, J., Schwertfeger, S., Suthakorn, J.: Using competitions to advance the development of standard test methods for response robots. In: Proc. Workshop on Performance Metrics for Intelligent Systems, PerMIS 2012, pp. 182–189. ACM, New York (2012)
2. Quigley, M., Gerkey, B., Conley, K., Faust, J., Foote, T., Leibs, J., Berger, E., Wheeler, R., Ng, A.: ROS: An open-source Robot Operating System. In: ICRA Workshop on Open Source Software, vol. 3 (2009)
3. Kohlbrecher, S., Petersen, K., Steinbauer, G., Maurer, J., Lepej, P., Uran, S., Ventura, R., Dornhege, C., Hertle, A., Sheh, R., Pellenz, J.: Community-Driven Development of Standard Software Modules for Search and Rescue Robots. In: IEEE Intern. Symposium on Safety, Security and Rescue Robotics, SSRR (2012)
4. Graber, T., Kohlbrecher, S., Meyer, J., Petersen, K., von Stryk, O., Klingauf, U.: RoboCupRescue 2013 - Robot League Team Hector Darmstadt (Germany). Technical report. Technische Universität Darmstadt (2013)
5. Geiger, A., Ziegler, J., Stiller, C.: StereoScan: Dense 3d Reconstruction in Real-time. In: IEEE Intelligent Vehicles Symposium, Baden-Baden, Germany (2011)
6. Huang, A.S., Bachrach, A., Henry, P., Krainin, M., Maturana, D., Fox, D., Roy, N.: Visual odometry and mapping for autonomous flight using an RGB-D camera. In: International Symposium on Robotics Research, ISRR (2011)
7. Kohlbrecher, S., Meyer, J., von Stryk, O., Klingauf, U.: A Flexible and Scalable SLAM System with Full 3D Motion Estimation. In: IEEE International Symposium on Safety, Security and Rescue Robotics (2011)
8. Kleiner, A., Dornhege, C.: Real-time Localization and Elevation Mapping within Urban Search and Rescue Scenarios. Journal of Field Robotics (8-9), 723–745 (2007)
9. Meyer, J., et al.: A Semantic World Model for Urban Search and Rescue Based on Heterogeneous Sensors. In: Ruiz-del-Solar, J., Chown, E., Plöger, P.G. (eds.) RoboCup 2010. LNCS (LNAI), vol. 6556, pp. 180–193. Springer, Heidelberg (2010)
10. Wirth, S., Pellenz, J.: Exploration transform: A stable exploring algorithm for robots in rescue environments. In: IEEE International Workshop on Safety, Security and Rescue Robotics (SSRR), pp. 1–5 (2007)
11. International Association of Fire Chiefs and National Fire Protection Association: Fundamentals of fire fighter skills. Jones & Bartlett Learning (2008)
12. Van Turennout, P., Honderd, G., Van Schelven, L.: Wall-following control of a mobile robot. In: IEEE International Conference on Robotics and Automation (ICRA), pp. 280–285 (1992)
13. Dryanovski, I., Morris, W., Xiao, J.: An open-source pose estimation system for micro-air vehicles. In: IEEE International Conference on Robotics and Automation (ICRA), pp. 4449–4454 (2011)
14. Scholl, P.M., Kohlbrecher, S., Sachidananda, V., van Laerhoven, K.: Fast Indoor Radio-Map Building for RSSI-based Localization Systems. In: Demo Paper, International Conference on Networked Sensing Systems (2012)

SimSpark: An Open Source Robot Simulator Developed by the RoboCup Community

Yuan Xu[1] and Hedayat Vatankhah[2]

[1] DAI-Labor, Technische Universität Berlin,
Ernst-Reuter-Platz 7, Berlin, D-10587 Germany
yuan.xu@dai-labor.de
[2] Department of Computer Engineering and IT,
Amirkabir University of Technology, Tehran, Iran
hedayatv@gmail.com

Abstract. SimSpark is an open source robot simulator developed by RoboCup Community. This paper briefly describes the development of SimSpark since 2008. Furthermore, some new features are proposed and implemented for the next RoboCup, including realistic motor, heterogeneous robots, and agent proxies. As a powerful tool to state different multi-robot researches, SimSpark has been successfully used in RoboCup simulation league, standard platform league and humanoid league.

1 Introduction

The development on robots may be severely limited by the constrained resources. This is especially true in the research of multi-robot systems in areas such as RoboCup. Using simulation for algorithm development and testing makes thing easier.

SimSpark, a multi-robot simulator based on the generic components of the Spark[7] physical multi-agent simulation system, has been used in the RoboCup Soccer Simulation League since 2004. The project was registered as open source project in SourceForge in 2004, it has an established code base with development increasing year-over-year. As the result, RoboCup soccer simulations have changed significantly over the years, going from rather abstract agent representations to more and more realistic humanoid robot games[2,3]. Thanks to the flexibility of the Spark system, these transitions were achieved with little changes to the simulator's core architecture.

In this paper we describe the recent development of SimSpark project, which make the SimSpark possible to simulate 11 vs. 11 humanoid robot soccer games in real time. In section 2, we will give an overview of the SimSpark project since 2008. After that, we will describe the development of the Spark simulation platform in section 3 and the implementation of RoboCup 3D Soccer Simulator in section 4. We will introduce some new features for RoboCup 2013 in section 5. Furthermore, we will give the application examples of SimSpark in section 6. Finally, we will outline future development plans in section 7.

S. Behnke et al. (Eds.): RoboCup 2013, LNAI 8371, pp. 632–639, 2014.

2 Project Overview

Until 2008, soccer simulation and SimSpark simulator were developed and released as a single project called rcssserver3d. In late 2008, the SimSpark project migrated from CVS repository at http://sserver.sourceforge.net to the new project Subversion repository at http://simspark.sourceforge.net. As part of this process, the project was broken to two main projects (Spark simulation platform and RCSSServer3D soccer simulation server) and two auxiliary projects (RSGEdit and simspark-utilities). The separation clarifies SimSpark is a generic simulation environment rather than only a robot soccer simulator.

SimSpark was never a single-platform project, but it practically didn't support Windows. In recent years, Windows support was added after migration from Autotools to CMake, and windows installers are now released.

3 Spark

As a generic simulation environment, Spark provides a rich set of features to create, debug and modify multi-robot simulations. It has three main components, including the simulation engine, the object and memory management system, and the physics engine. Details about architecture and concepts of Spark have been described in [2,7]. In this section, we describe the necessary changes for simulating 11 vs. 11 humanoid robot soccer game. However changes to the simulator core were never specialized for the soccer simulation.

Sensor Plugins. Sensors of a robot allows awareness of the robot's state and the environment. Humanoid robots like the NAO usually have many different sensors installed. We have implemented new plugins to simulate sensors of humanoid robots. Some of the sensors delivers information from physics engine, such as joint position, gyro, accelerometer, and force resistance. Furthermore, lines can be sensed by virtual vision. Additionally, a more realistic camera which delivers images rendered via OpenGL hardware accelerated offscreen buffers is implemented. Details of these sensors can be found in [3].

Multi-threads Supporting. One great feature of SimSpark is switching between single thread mode and multi-threads mode. The multi-threads mode can improve the performance in computer with multi-cores processor, but the single thread mode is also useful for developing the simulator.

The implementation of multi-threads loop is based on two conditions. First, different tasks are assigned to different *SimControlNode*s in SimSpark. For example, *AgentControl* is a node that manages the communication with agents. For each simulation cycle, *SimControlNode*s are executed one by one in the single thread mode, but they can run in parallel. Second, all data of simulation state is stored in a tree called *active scene*, the physics engine and *SimControlNodes* interact through the *active scene*. As we know, the physics computation is the most time-consuming, and the physics engine does not need to access the active scene during physics computation. So the physics computation and *SimControlNodes* can run in parallel. Furthermore, the physics engine, e.g. Open Dynamic Engine[9], is modified to run in parallel by using Intel Threading Building Blocks[1].

In RoboCup 2012, the 11 vs. 11 humanoid robot (22 degree of freedom) soccer games were simulated in real time by computer with Intel Core i7-975 @3.33GHz CPU and 4G DDR3 RAM.

4   RCSSServer3D

As the competition environment for the Soccer Simulation 3D League at RoboCup, RCSSServer3D simulates the soccer field where two team of robots play soccer game. In its initial version players were modeled as spheres in a physical three dimensional world. Now it supports humanoid players with articulated bodies.

Soccer Simulation. Most rules of the soccer game are judged by an automatic rule set that enforces the basic soccer rule set. However more involved situations like detection unfair behavior still require a human referee. Noticeably, the size of soccer field is increasing every year, since the number of robots in each team is increasing. In RoboCup 2012, each team has 11 robots. This makes coordination between multi-robots more important.

Robot Model. The NAO robot is currently used in the competitions of the Soccer Simulation 3D League at RoboCup. Its height is about 57cm and its weight is around 4.5kg. Its biped architecture with 22 degrees of freedom allows NAO to have great mobility. The NAO robot model is also equipped with a powerful selection of the sensors described in section 3, to provide a widespread information base for agent development. There are differences between simulated NAO and real NAO. For example, the hip joints are physically connected by one motor in the real NAO. Furthermore, the simulated robot gets positions of objects via virtual vision sensor, while the real robot has to process images to understand the world. Nevertheless, this NAO model enables RCSSServer3D to be a good humanoid robot research platform.

5   Experimental Features

For getting more realistic simulation, some new features are proposed and implemented, including realistic motor, heterogeneous robots, and agent proxies. These experimental features probably will be used in RoboCup 2013 for the first time.

5.1   Realistic Motor

NAO robot has twenty-one motor joints as its actuators. The simulation engine, i.e. ODE, provides a simple model of real life servos: the motor brings the body up to speed in one step; and provides force that is not more than is allowed. The simple motor model is one reason for the unrealistic simulation results. In this section, we proposed and implemented a realistic motor model. More details and experimental results are described in [11].

Stiffness. The stiffness determines how strong the motor is. In the real robot, this percentage is the maximum electric current applied to the motor. For a DC motor, the electric current, I, determines the output torque, $\tau = K_\tau I$; where K_τ is the torque constant of the motor, and can be found in the specification. The simulation engine can specify the maximum torque of the servo, therefore the stiffness control can be easily implemented by setting the maximum torque of the simulated servo:

$$\tau_{max}(t) = k_s(t)T_{max} \tag{1}$$

where $\tau_{max}(t)$ is the maximum torque set in the simulated servo at time t; $k_s(t)$ is the stiffness at time t; and T_{max} denotes the maximum torque of the servo when stiffness is 1.

Power Consumption. Another important aspect of real motor is power consumption. Because the robot is powered by a battery with limited energy, and has to walk during the game. Furthermore, the motor can overheat if it consumes too much energy and becomes too hot. In real robots, the temperature of each motor is measured, and the motor shuts down when the temperature is too high.

DC motors can be modeled by the following equation: $U = U_e + IR = K_e\dot{\theta}$; where U is the voltage of input, U_e is the back electromotive force (EMF), I is the electric current, R is resistance, $\dot{\theta}$ is the speed, and K_e is the speed constant of the motor. The value of R and K_e can be found in the specifications of motor again; and the simulation engine provides the value of τ and $\dot{\theta}$; therefore we can calculate the power consumption:

$$P = UI = U_eI + I^2R = \frac{K_e}{K_\tau}\dot{\theta}\tau + \frac{R}{K_\tau^2}\tau^2 \tag{2}$$

And the total energy used by the motor is:

$$E = \sum_t P_t \Delta t \tag{3}$$

where Δt is the time step of the simulation, and P_t is the power consumed at time t. For the overall power consumption, the energy consumed by devices other than motors, e.g. main board, CPU, camera, etc. has to be counted.

Temperature Regulation. We model the temperature and heat of the motor with the following equations:

$$\Delta Q = \Delta Q^+ + \Delta Q^- = I^2R\Delta t - \lambda(T - T_e)\Delta t \tag{4}$$

$$\Delta Q = C\Delta T \tag{5}$$

where T is the temperature of the motor, T_e is the environment temperature inside the motor, so it is higher than outside and differs from motor to motor, ΔQ is the heat changing, ΔQ^+ is the heat produced by the motor, ΔQ^- is the heat transferred from the motor to the environment, λ is thermal conductivity which indicates the ability of a motor to conduct heat, and C is the heat capacity of the motor, which can be seen as constant. Finally, the temperature of the motor at time $t + \Delta t$ can be solved as:

$$T_{t+\Delta t} = T_t + \Delta T = T_t + \frac{[I^2 R - \lambda(T_t - T_e)]\Delta t}{C} \tag{6}$$

In this model, we need to determine T_e, λ, and C by experiments. A sequence values of Δt, T_t, and $I^2 R$ can be measured by experiment, therefore the optimum parameters of eq. (6) is determined.

The whole process of joint simulation is summarized in Fig. 1. When the battery is empty, the maximum torque τ_{max} is set to 0 to turn off the motor.

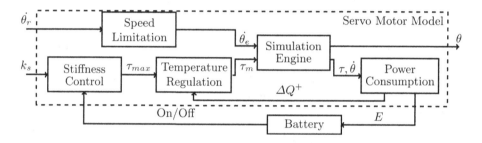

Fig. 1. Pipeline of the servo motor simulation

5.2   Heterogeneous Robots

Heterogeneous systems are emerging as effective solutions to many multi-robot tasks, it is also true for the soccer team. The physical difference between robots are important when they serve diverse roles in a cooperating team. In simulation, the physical properties of the robot can be easily changed, so it is a good platform for researching on heterogeneous systems.

Heterogeneity also creates new challenges for robot AI developers. The robots with different mechanical properties will be generated at running time, different robots are provided to teams, so teams have to adopt their developed behaviors to unknown robots.

SimSpark already provided the ability to define different robot models and use them in its simulation environment. However, each model was fixed and if you wanted to have several variations of a robot, you had to defined several separate robot models. To better support games with heterogeneous robots, we have implemented the parametric robot model. This model has a number of parameters and can have different values for different robot variations. For example, the length of the legs can be different for different robots. In order to launch this model, only these parameters have to be specified additionally.

5.3 Agent Proxies

Initially, SimSpark used SPADES[8] for managing external agents. It managed the simulation time and the time spend in agents for thinking. Later it was dropped because of its complexities and agents communicated with the server directly. The cycle duration was managed by the server without considering agents and the network latency. Therefore, if agents were unable to deliver their commands to the server in a cycle because of network latency, their commands were executed by the simulator in the next simulation cycle. As the number of soccer players in soccer simulation increased, this problem became more apparent.

To solve this problem, the concept of agent proxies was proposed. Agents communicate with the proxies instead of communicating directly with the simulator. In this model, the cycle time is managed by the proxies on the client side, and they communicate with the simulator in Sync Mode. Therefore, the simulator waits for all proxies to signal the end of a cycle and then proceeds with the next cycle. This model ensures that agents can use the allowed cycle time to think in a cycle when new information is delivered to them.

This model is not perfect because the agents can have some spare time if network latency occurs, but it is more predictable and fairer than the old model. The network latency can be reduced by using binary network protocol between proxies and simulator. Nevertheless, no new information is arrived to agents in that spare time and they cannot send new commands for that cycle.

Currently, an initial version of agent proxies is under development outside SimSpark project using Java.

6 Applications

In addition to be used in RoboCup Soccer Simulation 3D League, SimSpark is gaining popularity in other leagues, including Standard Platform League and Humanoid League. It is also widely used to teach artificial intelligence and robotic lectures.

RoboCup Soccer Simulation 3D League. Since 2004, SimSpark has been used as a official competitions environment. Research teams have developed useful research tools based on SimSpark. For example, RoboViz[10] is designed to assess and develop agent behaviors in SimSpark. And these tools are also released in open source. Combined with these tools, SimSpark becomes a featured platform to develop and test new algorithms for multi-robot systems.

RoboCup Soccer Standard Platform League. The special situation between Standard Platform League and 3D Simulation League is that both leagues use the same robot model — NAO from Aldebaran. So it appears to be natural to reuse the work which has already been done. Nao Team Humboldt developed a software architecture[6] which enables their control software can run both in real NAO and simulated NAO with SimSpark. This helps them to participate in both Simulation League and Standard Platform League, and achieve some good

Fig. 2. Prototype of the extended SimSpark for Standard Platform League. The bottom of screen are images of robot cameras.

results. Furthermore, they also promotes the usage of SimSpark in the Standard Platform League by implementing its rules, see Fig. 2.

RoboCup Soccer Humanoid League. Of course the simulator can also be extended for other leagues by adding new robot models. For example, in RoboCup Humanoid Kid Size League, FUmanoid[4] uses SimSpark to perform multi-level testing methods for archiving higher quality in each module of their robot control software and unlink the module test from the robotic hardware.

7   Conclusion and Future Work

SimSpark is a powerful tool to state different multi-robot researches. The introduction of a humanoid robot model increase the realism of SimSpark, and gains popularity in research teams who have real robots. The increased number of players per team makes research slowly get back to the high level behaviors based on solid low level behavior for realistic humanoid robot teams.

SimSpark has undergone continuous development driven by the requirement of continuous research in multi-robot system. The following extentions are planned:

Integration with Existing Robotic Frameworks. To enhance the usability of SimSpark for wider usage, and also to make it easier to develop and test their robotic software on SimSpark and to use the same code on real hardware, we aim to integrate SimSpark to an existing robotic framework, such as ROS or Player.

Physics Abstraction Layer. Relying on a single physics engine, i.e. ODE, hampers Simspark's flexibility. Thus, the physics abstraction layer has be developed[5], ODE and Bullet can be used as plugins of SimSpark. However, current implementation has many drawbacks and difficult to maintenance.

Robot Model Importers. The usage of SimSpark will be extended when it has more robot models, however creating robot models is a time consuming task. The idea is to create model importers for different model format, so SimSpark will be able to use the robot model which is available in other simulators.

Acknowledgments. We wish to thank other members in the Maintenance Committee of the RoboCup Soccer Simulation League, especially the original authors of SimSpark: M. Kögler, M. Rollmann, and O. Obst. Furthermore, thanks go to J. Boedecker for supporting us to work in the SimSpark project.

References

1. Intel ® threading building blocks, http://threadingbuildingblocks.org/
2. Boedecker, J., Asada, M.: Simspark – concepts and application in the robocup 3d soccer simulation league. In: Proceedings of the SIMPAR-2008 Workshop on The Universe of RoboCup Simulators, Venice, Italy (November 2008)
3. Boedecker, J., Dorer, K., Rollmann, M., Xu, Y., Xue, F., Buchta, M., Vatankhah, H., Glaser, S.: SimSpark User's Manual, 1.3 edn. (August 2010)
4. Donat, H.: Evaluation of Simulators for Humanoid Soccer Playing Robots and Integration in an Existing System. Bachelor thesis. Department of Computer Science, Freien Universität Berlin (2012)
5. Held, A.: Creating an abstract physics layer for simspark. Tech. rep. University of Koblenz-Landau (2010)
6. Mellmann, H., Xu, Y., Krause, T., Holzhauer, F.: Naoth software architecture for an autonomous agent. In: International Workshop on Standards and Common Platforms for Robotics (SCPR 2010), Darmstadt (November 2010)
7. Obst, O., Rollmann, M.: SPARK – A Generic Simulator for Physical Multiagent Simulations. Computer Systems Science and Engineering 20(5), 347–356 (2005)
8. Riley, P.: Spades: A system for parallel-agent, discrete-event simulation. AI Magazine 24(2), 41 (2003)
9. Smith, R.: Open Dynamic Engine User Guide (2006), http://www.ode.org (last visited at July 02, 2013)
10. Stoecker, J., Visser, U.: Roboviz: Programmable visualization for simulated soccer. In: Röfer, T., Mayer, N.M., Savage, J., Saranlı, U. (eds.) RoboCup 2011. LNCS, vol. 7416, pp. 282–293. Springer, Heidelberg (2012)
11. Xu, Y.: From Simulation to Reality: Migration of Humanoid Robot Control. Ph.D. thesis. Humboldt-Universität zu, Berlin (2012)

Visualizing and Debugging Complex Multi-Agent Soccer Scenes in Real Time

Justin Stoecker and Ubbo Visser

Department of Computer Science,
University of Miami, Coral Gables FL
{justin,visser}@cs.miami.edu

Abstract. The RoboCup Soccer environment is one of the most difficult scenarios for autonomous agents. With the potential for so many
things to go wrong, debugging and analyzing agents' behaviors becomes
a significant task. We propose RoboViz, an open-source program for integrating agent-driven visualizations into a real-time, 3D rendered environment; the scene becomes a shared, interactive whiteboard for all agents,
and the user can moderate by filtering drawings they are interested in.
Visualization is an effective tool for tracking down errant behaviors and
explaining algorithms. RoboViz is embraced by the RoboCup Soccer
Simulation 3D sub-league as the de facto monitor application, and the
latest revision makes it useful for other leagues as well. We are currently
testing RoboViz in the Standard Platform League (SPL).

1   Introduction

A significant challenge in developing a robotic agent is debugging and evaluating its behavior. The RoboCup Soccer scenario is one of the most difficult
environments for intelligent agents, and presents several hurdles: an uncertain
and dynamic world, multiple competitive and cooperative agents, physics, and
the need for high-level strategy. With such a challenging environment, it is inevitable that teams developing soccer agents for RoboCup will stumble over
bugs and struggle with solutions. As might be expected, teams competing in
the various RoboCup leagues develop their own specialized tools, in isolation,
to optimize and debug agent code. However, there is generally a lack of tools
that are truly beneficial to all teams participating in a league. In this paper, we
focus on the development and debugging issues shared by teams in the RoboCup
Soccer Simulation 3D sub-league; however, the same issues are relevant to other
RoboCup Soccer sub-leagues.

There are several challenges to overcome when interpreting robot behaviors,
such as localization or task planning, and simple approaches are often inadequate
in diagnosing problems. The real-time nature of the environment is the most notable complication: agents generally process input and act within milliseconds, so
outputting values on a console provides an incomprehensible amount of information. Logging this information and parsing it later has two serious drawbacks: the
data can fill volumes very quickly, and it can be difficult to synchronize one agent's

S. Behnke et al. (Eds.): RoboCup 2013, LNAI 8371, pp. 640–647, 2014.

view of the world with an external ground truth. Because of the dynamic environment, inspecting a single agent using an interactive debugger, such as the GNU Debugger, can be disruptive to a simulation or game and change the outcome.

Our answer to these problems is to visually integrate agent state and behavior information into a 3D representation of the environment. The design of these visualizations is delegated to individual agents; the soccer scene becomes a shared whiteboard for all agents, and the user can moderate by filtering drawings they are interested in. The primary contribution we present is a software program, called RoboViz, which enables real-time visualization that is accessible through a simple drawing protocol. RoboViz is also an implementation of the SimSpark [2] soccer simulation monitor, and it provides additional functionality not previously available. We expand on our recent work in [10] by redesigning the software to be usable in environments outside 3D simulation.

In the next section, we cover some of the previous work that has been done with regard to debugging or visualization in the RoboCup Soccer scenario. In section 3, we describe our visualization approach. We comment on how RoboViz has affected the 3D simulation sub-league in section 4, and we end with some ideas for future progression and applicability to other leagues and areas.

2   Related Work

Several teams competing in the simulation leagues of RoboCup develop their own tools specific to their team's agent architecture. For the most part, these tools are described in team description papers. The 2D simulation team *Mainz Rolling Brains* developed a debug and visualization tool called *FUNSSEL* [1]. *FUNSSEL* acts as a layer between the server and agents to intercept and process communication. The primary features of this software include filtering data, agent training, and graphic overlays for the 2D field in a secondary monitor. Other examples of tools for the 2D league can be found, for example, in the Portuguese team *FC Portugal* [9] and many other team description papers not mentioned here for brevity. For the 3D simulation league, the team *Virtual Werder 3D* utilized an evaluation tool [5] to analyze agent performance. The program also supported basic drawings in a simplified 2D monitor; however, all analysis and visualization was done on server and agent generated logs.

There have been few attempts at providing useful analysis and debugging tools for the community. The *logfile player and analyzer* [8] provided improvements to the log-playing capabilities of the 3D monitor; for example, it allows agents to record behaviors as simple drawings displayed in the 3D scene when the log is replayed. Additional features of this program included some basic filtering of logfile data and an improved graphical user interface.

Simulators for multi-agent systems often include some manner of visualization. Usually, these visualizations are for modeling the robots and environment. However, a few simulators do provide additional functionality. In Webots [7], for example, user code can initiate the drawing of primitives to further model the robot components. Breve [4] is another simulation program that supports the modeling and simulation of large multi-agent environments. Breve also exposes simple drawing routines to add shapes to the scene. Yet another simulator that has been used in the SPL would be SimRobot [6].

Tools mentioned in team description papers are, unfortunately, not well documented, obsolete, or unsuitable for general use as they may be tightly bound to a particular agent framework. Simulators such as Webots and Breve, while providing more advanced visualization capabilities, are unwieldy or impossible to integrate into the SimSpark simulation; additionally, the drawing routines exposed by these simulators are secondary and not integrated with the interface in a meaningful way.

3   Approach

We frame our technical solution to visualizing agent state and behavior by using the RoboCup Soccer Simulation 3D sub-league as an example. Other environments, such as the Standard Platform League, are less complicated to work with: there is no server component to interface with. Games in the RoboCup Soccer Simulation 3D sub-league consists of three components: (1) the simulation server, implemented by *rcssserver3d*, orchestrates the simulation by computing physics, enforcing rules, and managing the update cycle; (2) agents connect to the simulation server to receive *perceptors* (sensor input and game state changes) and send *effectors* (desired joint changes); (3) a monitor, implemented by *rcssmonitor3d*, communicates with the simulation server to receive a *scene graph* for rendering, and allows a user to referee the game. The simulation server, monitor, and each agent runs as an individual processes that communicate using TCP and UDP sockets.

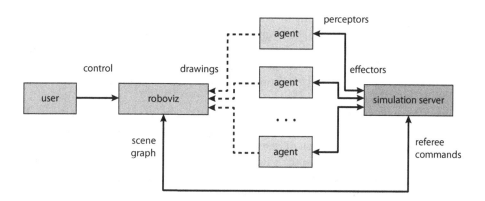

Fig. 1. Interaction between the components of a simulation-3D game, with RoboViz replacing the default monitor. Each component runs as a separate process, with communication handled by either TCP or UDP sockets. The dotted arrows between agents and RoboViz indicate the optional visualization commands.

RoboViz is our implementation of an advanced monitor, and it effectively replaces the default implementation provided by *rcssmonitor3d*. The simulation server communicates the environment in *scene graph* and *game state* objects; these objects are sent to any monitor connected to the server when a change occurs. The scene graph stores only the information necessary to render the environment: geometric primitives, model names, transform matrices, lights, and so forth. The game state contains information about the time, scores, team names, rules, play mode, and field dimensions. It is up to the monitor to decide how to parse this information and display it to a user. The monitor can also send messages to the simulation server, which is useful for referee purposes.

In the physical sub-leagues, there is (clearly) no server to provide a scene graph. The 3D scene must be rendered by a module in RoboViz that is specific to the environment. For example, a Standard Platform League module is implemented that draws a field, field lines, and goals according to the dimensions outlined in the rules. These modules require much less effort to implement because there is no network communication or parsing required; these scenes contain mostly static geometry, and the agents can visualize themselves using the drawing protocol which is discussed next.

Efficiently rendering a scene is a non-trivial task, and RoboViz provides modernized graphics in comparison to *rcssmonitor3d*; however, the rest of this paper will focus on the agent-driven visualization capabilities of RoboViz. A protocol is provided that allows external processes to submit 3D drawings, in real time, that are integrated directly into soccer scene. Figure 2 shows an example where agents on the blue team are drawing circles, lines, and points that represent believed orientations, path-planning, localization particles, and so forth.

Fig. 2. This rendered scene highlights some of the graphics and visualization capabilities of RoboViz. The robots, field, ball, and goals are constructed entirely from the simulation server's scene graph message. Colored drawings, like the circles and lines, are shapes submitted directly from individual agents in the RoboCup Simulation 3D sub-league.

3.1 Drawing Protocol

A simple network protocol enables clients, such as agents, to draw shapes in RoboViz. This visualization interface is intentionally low level, and it provides great flexibility and is easy to integrate into existing agent architectures. Clients interact with RoboViz by issuing commands. An example of a command is *draw line*. Commands are serialized into bytes[1], and each command has a unique format so it can be tightly packed in a buffer. When a client is ready to submit drawings, it can fire off a UDP packet to RoboViz that contains one or more commands. We chose UDP primarily because we expect for RoboViz and agents to be run on the same host or LAN, where packet loss is highly uncommon. UDP simplifies the connections between RoboViz and agents, which may crash unexpectedly, and makes it easier for teams to add drawing functionality to their agents.

To make the visualization interface flexible, all drawing is accomplished by working with a small set of shapes: circles, points, lines, spheres, and convex polygons. More complex shapes can be constructed from these primitives. In addition to geometric shapes, strings can be rendered by RoboViz. We refer to both geometric shapes and text drawings as *draw commands*. Draw commands includes properties such as position, color, and scaling, and they also contain a name.

The name assigned to a drawing is very important, because all drawings with the same name will be grouped into a *shape set* within RoboViz. Grouping shapes into sets is useful so drawings can be filtered inside RoboViz. For example, each robot may have its own set of shapes, so the user can choose to render only that particular agent's shapes and hide everything else. A robot may also have sets based on behaviors or algorithms. How the shape names are chosen is entirely up to the agent programmer; however, we recommend a hierarchical naming pattern, as the shapes are stored as a tree in RoboViz. Each shape set is a node in the tree. For instance, a shape set name such as *RoboCanes.1.PathPlanning* could be used to indicate all shapes belonging to the first agent on the RoboCanes team that pertain to path planning; this shape set is the child of *RoboCanes.1*, which is the child of *RoboCanes*. The advantage of this convention is that the user can filter rendered drawings very easily.

3.2 Rendering Control

Let's say an agent wants to draw a circle to represent its current location. This position changes constantly, so a new circle is drawn every cycle. Very quickly, the scene is flooded with circles because none of the old ones have been removed. The drawing protocol has another command, *swap buffers*, that will clear out all of the existing shapes whose names begin with a given string. For example, the shape sets *RoboCanes.1.PathPlanning* and *RoboCanes.1.Localization* would

---

[1] For detailed command formats refer to the documentation:
    https://sites.google.com/site/umroboviz/drawing-api/draw-commands

both be cleared by issuing a *swap buffer* with *RoboCanes.1* as the argument. This allows the programmer to avoid lots of extra commands if they organize their shape sets in a hierarchical fashion.

The command is called *swap buffers* instead of *clear* or *reset* because it addresses another problem. During rendering, it's not possible for RoboViz to iterate over a set of shapes while new shapes are being added: the rendering and network communication of RoboViz are handled by separate threads. Each shape set has two buffers: a front and back buffer. When a client sends a draw command, RoboViz parses the shape and adds it to the back buffer. These shapes will not be visible until the client sends a swap command, at which point the back and front buffers exchange roles. This behavior enables asynchronous reading and writing of shapes. In other words, the front buffer always contains a set's shapes which may be rendered in RoboViz. The *swap buffers* command is synchronized inside of RoboViz, and will block while shapes are being rendered.

It is important to note that as soon as RoboViz executes a buffer swap, the shape set's new back buffer is cleared of all shapes. If multiple agents try to swap the buffers of the same shape set, flickering will occur in RoboViz. For this reason it is expected that agents will submit drawings relevant to their own state or behavior within their own shape sets, but this is not strictly required. A team may, for instance, designate a captain that sends drawings relevant to the team as a whole. We did not wish to impose a set of rules on how drawings should be assigned to RoboViz, so there is nothing to prevent agents from sending drawings identified by names other than their own.

3.3   Implementation

RoboViz was programmed in Java to mitigate cross-platform issues and reduce the number of libraries needed: the Java runtime environment provides solutions for networking, threading, windowing, and GUI development. This makes RoboViz especially easy to deploy on any platform, and the source code can be compiled on any Linux, Windows, or OS X system that has Java and recent graphics drivers. Rendering is done entirely with OpenGL through the Java Bindings for OpenGL (JOGL) library [3]. We have made RoboViz open source under the Apache 2.0 License.

4   Results

RoboViz is particularly useful for visualizing belief states and high-level strategies. We have seen drawings for localization routines, path planning, decision-making, and behavior modeling. Figure 3 shows visualizations from robots in the Simulation 3D and SPL environments.

RoboViz has been adopted as a monitor for the RoboCup Soccer Simulation 3D sub-league since it was announced: the source code was released in February 2011, and its first public use in competition was at the 2011 RoboCup German Open held in Magdeburg, Germany. This was followed by official use at

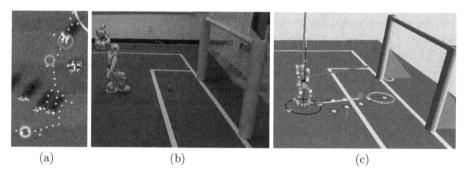

(a)   (b)   (c)

Fig. 3. In (a), a Sim3D agent's search tree is visualized to show path planning from the blue circle (current state) to the yellow circle (desired state). The chosen path is highlighted in blue, with other options in white, and the pink circle is the first destination to reach without collision. In (b), a NAO robot kicks a ball towards the goal; this scene is visualized in RoboViz in (c): the red and blue dots indicate the robot part centers and orientations; the red vectors are the robot's orientation; and the blue circles indicate the path.

RoboCup 2011 in Istanbul, Turkey and RoboCup 2012 in Mexico City, Mexico. The visualization feature of RoboViz has been popular with many teams in the 3D simulation sub-league. In particular, we would like to note that many of the teams in the RoboCup 2012 semi-finals used RoboViz during development. We have also seen agent frameworks released that explicitly provide support for RoboViz [11].

Outside of RoboCup, we have seen RoboViz incorporated into courses in multi-agent systems at the University of Miami and the University of Texas at Austin[2]. RoboViz has also been used at the University of Miami as a demonstration tool to promote awareness of the RoboCup events and attract students to computer science.

5   Conclusion and Future Work

Our initial goal with RoboViz was to present a tool that was accessible and useful to the entire simulation 3D sub-league. We believe this has been clearly demonstrated over the past two years: several teams make use of the visualization capabilities, and RoboViz is used at the major competitions as the monitor. As a bonus, RoboViz has been used for education purposes outside of RoboCup.

Moving forward, we would like to see RoboViz used in other RoboCup soccer leagues, such as the Standard Platform League. It would be especially interesting to see teams that write cross-league agents that can reuse the same tools. While the primary release of RoboViz[3] is still focused on the Simulation 3D sub-league, a branch[4] is under development that makes the tool usable for many other

---

[2] http://www.cs.utexas.edu/~todd/cs344m/

[3] http://sourceforge.net/projects/rcroboviz/

[4] https://github.com/jstoecker/roboviz

environments; our team, RoboCanes, is using this for use with our NAO robots. We expect this branch to become the primary version as it is polished for use in RoboCup 2013. Adding support for heterogeneous robot types would also be useful for 3D simulation.

Acknowledgments. The authors would like to thank Klaus Dorer of mag-maOffenburg[5] and Drew Noakes for contributing to the source code of RoboViz. Thanks to Andreas Seekircher for the images used in figure 3.

References

1. Arnold, A., Flentge, F., Schneider, C., Schwandtner, G., Uthmann, T., Wache, M.: Team Description Mainz Rolling Brains 2001. In: Birk, A., Coradeschi, S., Tadokoro, S. (eds.) RoboCup 2001. LNCS (LNAI), vol. 2377, pp. 531–534. Springer, Heidelberg (2002)
2. Bödecker, J., Dorer, K., Rollmann, M., Xu, Y., Xue, F.: SimSpark User's Manual (June 2008)
3. Java Bindings for OpenGL (JOGL), http://www.jogamp.org
4. Klein, J., Spector, L.: 3D Multi-Agent Simulations in the breve Simulation Environment. In: Komosinski, M., Adamatzky, A. (eds.) Artificial Life Models in Software, pp. 79–106. Springer, London (2009)
5. Lattner, A.D., Rachuy, C., Stahlbock, A., Warden, T., Visser, U.: Virtual Werder 3D Team Documentation 2006. Technical Report 36, TZI, Universität Bremen (August 2006)
6. Laue, T., Röfer, T.: Simrobot-development and applications. In: The Universe of RoboCup Simulators-Implementations, Challenges and Strategies for Collaboration. Workshop Proceedings of the International Conference on Simulation, Modeling and Programming for Autonomous Robots (SIMPAR 2008). LNCS (LNAI). Springer, Heidelberg, Citeseer (2008)
7. Michel, O.: Webots: Professional Mobile Robot Simulation. Journal of Advanced Robotics Systems 1(1), 39–42 (2004)
8. Planthaber, S., Visser, U.: Logfile Player and Analyzer for RoboCup 3D Simulation. In: Lakemeyer, G., Sklar, E., Sorrenti, D.G., Takahashi, T. (eds.) RoboCup 2006. LNCS (LNAI), vol. 4434, pp. 426–433. Springer, Heidelberg (2007)
9. Reis, L.P., Lau, N.: FC Portugal Team Description: RoboCup 2000 Simulation League Champion. In: Stone, P., Balch, T., Kraetzschmar, G. (eds.) RoboCup 2000. LNCS (LNAI), vol. 2019, pp. 29–40. Springer, Heidelberg (2001)
10. Stoecker, J., Visser, U.: RoboViz: Programmable Visualization for Simulated Soccer. In: Röfer, T., Mayer, N.M., Savage, J., Saranlı, U. (eds.) RoboCup 2011. LNCS, vol. 7416, pp. 282–293. Springer, Heidelberg (2012)
11. TinMan. C-Sharp framework for 3D simulation league, http://code.google.com/p/tin-man/

---

[5] http://robocup.fh-offenburg.de/html/index.htm

On B-Human's Code Releases
in the Standard Platform League –
Software Architecture and Impact

Thomas Röfer and Tim Laue

Deutsches Forschungszentrum für Künstliche Intelligenz,
Cyber-Physical Systems, Enrique-Schmidt-Str. 5, 28359 Bremen, Germany
{Thomas.Roefer,Tim.Laue}@dfki.de

Abstract. In RoboCup, source code releases after a competition are
one important part of the competition's overall progress. In particular
teams in the Standard Platform League can strongly benefit from other's
software, as everybody shares the same robot platform. Therefore several
releases by different teams exist. In this paper, we describe the past code
releases of our team B-Human, particularly focusing on its underlying
software architecture, which is set in relation to concepts of the currently
popular Robot Operating System (ROS), as well as on its impact on the
league's progress.

1 Motivation

B-Human started as a team in the Humanoid League. We adapted the German-
Team framework for our Bioloid-based robots [10] before we switched to the
Standard Platform League (SPL) in 2009 and thereby to the NAO. Our history
in the Humanoid League gave us a certain advantage in the Standard Platform
League, because our main weakness in the Humanoid League was the robot
hardware, which – in contrast – is identical for all teams in the SPL. B-Human
won all official games it played so far except for the final in RoboCup 2012. As a
result, the team won the RoboCup German Open five times (2009–2013) and the
RoboCup World Championship three times (2009, 2010, 2011). Although code
releases are not mandatory in the SPL, B-Human has been releasing its software
since the league started to use the NAO as standard platform in 2008 [11] at
www.b-human.de/publications. We believe that providing software as Open
Source is the best way to foster research and push a RoboCup league forward.

The remainder of this paper is as follows: Section 2 presents the technical
aspects of the B-Human framework which are compared to the popular Robot
Operating System in the following Section 3. The impact of the B-Human code
releases is described in Section 4. Finally, the paper is concluded in Section 5.

2 B-Human's Software Framework

B-Human's software framework is based on the framework of the GermanTeam
in the Four-Legged League [10]. The GermanTeam was a joint team of four (later

S. Behnke et al. (Eds.): RoboCup 2013, LNAI 8371, pp. 648–655, 2014.

three) universities that competed separately at local events, but participated as a single team at the RoboCup. Some elements of the software framework are motivated by these circumstances, namely that several modules which provide very similar functionality can exist in parallel, because they were originally developed by members of different university teams for a local competition.

Since the GermanTeam framework always aimed at being as hardware-independent as possible, porting it from AIBO to Bioloid to NAO was not very difficult. Such transitions would have been a lot more complicated if we had directly used Sony's Open-R on the AIBO that does not exist anywhere else. Therefore, we also abstract from Aldebaran Robotics' NaoQi, because the league might switch to a different robot platform in the future.

Since 2011, the architecture that is now called *B-Human framework* supports the three most popular operating systems as development platforms. Microsoft Windows running Visual Studio, OS X running Xcode, and Linux are all treated as first class citizens in the development process. To reduce the logistic overhead that comes with this variety, we recently switched to llvm/clang as C++ compiler. It can cross-compile the code for the robots on all three platforms, it is Xcode's native compiler, and it works with any development environment on Linux. Only when targeting Windows, Microsoft's compiler is still used. In addition, we increasingly use the expressiveness of C++ and its preprocessor to add new features rather than employing external tools such as pre-compilers. Such tools complicate the build process and they always need to be available for all three platforms. In 2013, even the modeling of the robots' behavior using hierarchical state machines is done in C++, still following the same general approach that B-Human and the GermanTeam used since 2002 [7], but avoiding the overhead of coupling two different programming languages together.

2.1 Processes

A robot control program reacts to external events, i. e. when new sensor data was acquired or when the robot requires new target values for its actuators. In our framework, *processes* are employed to promptly react to these events. These framework processes can be either implemented by using operating system processes or by using threads. The former was the case on the GermanTeam's AIBOs and on our robots in the Humanoid League, the latter is the case in B-Human's current implementation on the NAO.[1] The NAO offers two kinds of sensor information: camera images from *Video for Linux* and everything else from NaoQi's *Device Control Manager*, i. e. actual joint angles, IMU and sonar measurements, etc. The latter is offered at the same speed as target joint angles are requested. Therefore, the framework uses two processes, namely *Cognition* that runs at the speed at which camera images are taken (30 Hz for NAO V3.x, 60 Hz for NAO V4) and *Motion* that runs at the speed the other sensor measurements are provided, i. e. 100 Hz. A third process *Debug* offers a network connection to an external PC and is used for debugging purposes only (cf. Fig. 1).

---

[1] To keep consistent between different instantiations of the framework, we still use the term *process*, although the current implementation is based on threads.

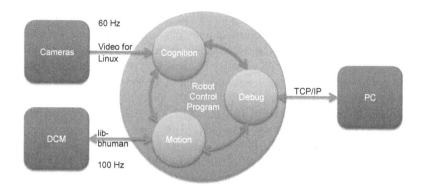

Fig. 1. The processes used on the NAO

2.2   Data Representation

B-Human's software architecture consists of *modules* that implement the functionality (cf. next section). These modules exchange data in the form of *representations*. Representations are instances of C++ classes that mainly store data and provide nearly no functionality. For instance, there are representations that store objects detected in an image, e. g. the ball, models of objects in the environment, e. g. the location of the ball on the field (Fig. 2 contains a simplistic example), or requests from one module to another one, e. g. to walk with a certain speed.

All representations are stored in a blackboard [5]. Each process has its own blackboard that contains all representations that are used in this process. The representations are either provided by modules that run in the same process or received from another process. Since each process has a copy of each representation that is processed by its modules, there is no concurrent access to the representations, which guarantees consistency, i. e. the contents of a representation

```
class BallModel : public Streamable {
  void serialize(In* in, Out* out) {
    STREAM_REGISTER_BEGIN
    STREAM(position)
    STREAM(wasLastSeen)
    STREAM_REGISTER_FINISH
  }
public:
  Vector2<float> position;
  unsigned wasLastSeen;
};
```

Property	Value
▼ position	
x	906
y	-0.31742
wasLastSeen	67411

robot3.data.representation:BallModel

Fig. 2. Simple example of the definition of a representation on the left (only #include statements omitted) and its live visualization on the right

provided in a different process will always be the same through a whole processing cycle. At the end of each processing cycle, representations that are required by the other process are simply sent to it, which it receives at the beginning of its next processing cycle.

All representations are streamable, i. e. they can be written to and read from a data stream. The C++ macro (cf. Fig. 2 left) used to stream attributes of representations does not only handle reading and writing, it also provides the name of the attribute and its data type at runtime. On the one hand, this is used to read and write structured, human-readable configuration files without any further code necessary. On the other hand, any representation can be visualized and edited at runtime through a debugging tool on a PC (cf. Fig. 2 right).

2.3 Components Performing Computations

In the B-Human framework, computations are performed by *modules*. Each module requires a number of representations as input to perform its job and it provides one or more representations as output to other modules. The dependencies between requirements and outputs define an execution sequence for all modules that is automatically determined, i. e. the execution of the modules is automatically ordered in a way that it is guaranteed that before a module is executed, all of its requirements have already been updated by other modules. Thereby, inconsistent configurations are detected and reported to the programmer. It is also automatically computed, which of the representations have to be exchanged between the different processes. Figure 3 shows an example of the dependencies between modules and representations.

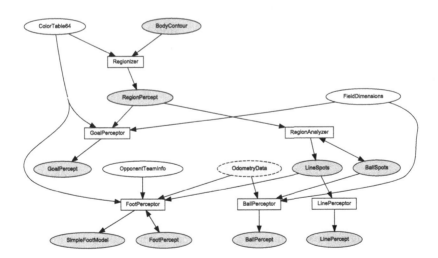

Fig. 3. A small subset of the modules (shown as rectangles) and representations (shown as ellipses – with a dashed border if received from another process) used. Dependencies on representations that are not depicted have been removed for a better readability.

The architecture allows that several modules exist that provide the same representation, but only one of them can be active at the same time. The set of active modules is specified in a configuration file. When a new module is developed, it can either be activated by changing the configuration file or interactively while the B-Human software is running, and thereby recomputing the execution sequence of the modules and rejecting inconsistent configurations. This eases software development, because different implementations can exist in parallel.

Since it is the intention of the architecture to simplify replacing existing implementations with new and hopefully better ones, it is an important rule that both representations and modules should be small and coherent. Small representations such as the position of the ball relative to the robot allow implementing a module that computes exactly this information, which can easily be re-implemented in a new module that tries to follow a different approach. In contrast, a representation that would consist of everything soccer-relevant that can be found in an image (ball, lines, goals, robots, etc.) would require for a single module to provide all this information, making it highly complex and hard to replace by a different approach. Therefore, B-Human's software consists of a large number of small representations and rather small modules. For instance, currently there are nearly 100 modules in the system that exchange approximately 120 representations.

3   Comparison to the Robot Operating System (ROS)

A robotics framework that gained much attention in recent years and has also been used on the NAO robot by some SPL teams is the Robot Operating System (ROS) [9]. Similarly to the B-Human framework, it provides a clear encapsulation of computing components and the data representation exchanged among these components. However, the underlying concepts and intentions – i. e. *nodes* and *messages* versus *modules* and *representations* – differ from each other.

In ROS, computations are carried out by the so-called *nodes* with each node being a single (operating system) process. The corresponding *modules* in B-Human's framework do not run in parallel to each other but are grouped in a fixed number (currently two) of processes. This is – of course – less flexible and less suitable given a scenario that involves a number of available CPU cores and associated enormous computations. However, the NAO has only a single CPU core (with Hyper-Threading in case of the V4) that is not even used to full capacity by our software. We do not expect this situation to change significantly in the near future. Our current module layout focuses on a very fine-grained modularization which allows to put even trivial functionalities (such as computing the overall walked distance based on odometry offsets provided by another module) into separate modules. This can be done without losing reactivity and efficiency due to scheduling and communication overheads which would occur in case of a direct mapping of our current modules to ROS nodes.

The ROS counterpart to B-Human's *representations* are *messages* (also referred to as *topics* after instantiation). Both concepts are quite similar and provide equal features such as streaming or the possibility of replacing senders and receivers during run-time. The main difference is the way of defining the data structures. ROS messages are specified in external files, which are used to generate actual code. This allows ROS to support different programming languages. As aforementioned, we aim to keep the number of external tools, languages, and compilers as low as possible. That is why B-Human's representations are specified directly in C++. In theory, this allows to add computational functionality to the representations, but in general, we refuse to use this option, because in contrast to modules, the implementations of representations cannot be switched at runtime.

4   Impact

Since 2008, we released a documented software package after each year's RoboCup world championship. They found broad interest in the RoboCup community and many teams use the B-Human software or certain parts of it.

4.1   Release Procedure and Impact on Our Own Development

The B-Human software is released every fall, shortly before the winter term starts. Each release consists of two parts: a software archive and a team report that includes detailed installation and operating instructions. We do not maintain any public repository as some other teams do. Hence, we do not publish any bug fixes for our released code but so far also never received any bug fixes from teams that use our code.

A code release always contains all components that have been used in the previous competition (without any modifications beyond bug fixes) except for the behavior, some special kicking motions, or components that are part of a thesis in progress. We always release our code under a BSD-style license as this grants the users the most liberties. However, we added some special RoboCup clauses to the license: Each team that uses B-Human code has to a) announce this on the league's mailing list prior to the next competition it takes part in and b) replace more than one major part of the module stack. As the B-Human team has been very successful in the past years, we consider this combination of a slightly trimmed code release and a demanding license to be the best trade-off between fostering the league's progress on the one hand and preventing a domination of B-Human clones on the other hand.

In addition to contributing to the league's progress as described in the following section, these code releases also benefit our own progress. On the one hand, the team reports serve as final reports of the current student project. On the other hand, we use our team reports internally to brief the new team members who join us every year. Instead of referencing a collage of research papers, README files, and Wiki pages as starting points, the latest team report serves as *one* single source including all necessary information for installing, using and understanding the B-Human system. Even experienced team members refer to the team report as reference as the system has grown to a complexity that makes it hard for anyone to know "everything". Therefore, we can recommend all teams to provide code releases and corresponding team reports.

4.2 Impact on Other Teams

In recent years, several other SPL teams used B-Human's software. This has been done in two different ways: Using the whole B-Human system and replacing single parts or using an own framework and integrating specific B-Human components. From a software engineering point of view, the former is quite easy due to the highly modular framework which allows a straight forward addition and replacement of new modules and representations. The latter is comparatively difficult as most modules have – despite their algorithmic and semantic enclosure – many dependencies to B-Human-specific implementations for streaming, mathematical computations, and debugging.

At least seven teams have based their development on the B-Human framework, including *NTU Robot PAL* [12] (reaching the third place at RoboCup 2011), *BURST* [6], *MRL* [4], and *NimbRo* (the runner-up at the RoboCup German Open 2010 and 2011). The latter also based research on the framework and the output of certain B-Human modules [8].

At least another seven teams use parts of our code. The most popular one used without the framework appears to be the implementation of the walking gait that has been presented by Graf and Röfer [3]. Amongst others, the current SPL world champion *UT Austin Villa* [2] uses the B-Human 2011 gait, relying on a version provided by the *Northern Bites* team at `github.com/northern-bites`.

Applications and connections outside the SPL include the usage of the walking gait by the NUbots in the Humanoid League [1] as well as the integration of a client for the Small Size League's standard vision system SSL-Vision [13] in the B-Human software. This allows the usage of SSL-Vision as a source for ground truth data for experiments with NAO robots.

5   Conclusions and Future Works

B-Human's software framework is well suited for controlling soccer robots that are equipped with embedded computers of rather low performance. It supports the rather dynamic development necessary to successfully participate in a robot soccer competition with continuously evolving rules. Due to its yearly releases, the software has been used by quite a number of other RoboCup teams as well.

We recently considered the option to switch to ROS but its strengths and preferences do not fit the way we currently want to lay out our software. In addition, the ROS support for some of our development platforms is still experimental, whereas our framework supports Linux, Windows, and OS X in equal measure without preferring any of these. However, in the near future, we plan to integrate other Open Source software such as *Eigen*.

References

1. Annable, B., Budden, D., Calland, S., Chalup, S.K., Fenn, S.K., Flannery, M., Fountain, J., King, R.A., Mendes, A., Metcalfe, M., Nicklin, S.P., Turner, P., Walker, J.: The NUbots team description paper. In: RoboCup 2013: Robot Soccer World Cup XVII Preproceedings, RoboCup Federation, Eindhoven, Netherlands (2013)

2. Barrett, S., Genter, K., He, Y., Hester, T., Khandelwal, P., Menashe, J., Stone, P.: UT Austin Villa 2012: Standard platform league world champions. In: Chen, X., Stone, P., Sucar, L.E., van der Zant, T. (eds.) RoboCup 2012. LNCS (LNAI), vol. 7500, pp. 36–47. Springer, Heidelberg (2013)

3. Graf, C., Röfer, T.: A center of mass observing 3D-LIPM gait for the RoboCup Standard Platform League humanoid. In: Röfer, T., Mayer, N.M., Savage, J., Saranlı, U. (eds.) RoboCup 2011. LNCS (LNAI), vol. 7416, pp. 102–113. Springer, Heidelberg (2012)

4. Hashemi, E., Ghiasvand, O.A., Jadidi, M.G., Karimi, A., Hashemifard, R., Lashgarian, M., Shafiei, M., Farahani, S.M., Zarei, K., Faraji, F., Harandi, M.A.Z., Mousavi, E.: Mrl team description 2010 standard platform league. In: Chown, E., Matsumoto, A., Ploeger, P., del Solar, J.R. (eds.) RoboCup 2010: Robot Soccer World Cup XIV Preproceedings. RoboCup Federation, Singapore (2010)

5. Jagannathan, V., Dodhiawala, R., Baum, L.S. (eds.): Blackboard Architectures and Applications. Academic Press, Boston (1989)

6. Keidar, M., Aharon, I., Barda, D., Kamara, O., Levy, A., Polosetski, E., Ramati, D., Shlomov, L., Shoshan, J., Yakir, A., Zilka, A., Kaminka, G.A., Kolberg, E.: Robocup 2010 standard platform league team burst description. In: Chown, E., Matsumoto, A., Ploeger, P., del Solar, J.R. (eds.) RoboCup 2010: Robot Soccer World Cup XIV Preproceedings, RoboCup Federation, Singapore (2010)

7. Lötzsch, M., Risler, M., Jüngel, M.: XABSL - A pragmatic approach to behavior engineering. In: Proceedings of IEEE/RSJ International Conference of Intelligent Robots and Systems (IROS), Beijing, China, pp. 5124–5129 (2006)

8. Metzler, S., Nieuwenhuisen, M., Behnke, S.: Learning visual obstacle detection using color histogram features. In: Röfer, T., Mayer, N.M., Savage, J., Saranlı, U. (eds.) RoboCup 2011. LNCS, vol. 7416, pp. 149–161. Springer, Heidelberg (2012)

9. Quigley, M., Conley, K., Gerkey, B.P., Faust, J., Foote, T., Leibs, J., Wheeler, R., Ng, A.Y.: ROS: An open-source robot operating system. In: Proceedings of the Open-Source Software Workshop of the International Conference on Robotics and Automation (ICRA), Kobe, Japan (2009)

10. Röfer, T., Brose, J., Göhring, D., Jüngel, M., Laue, T., Risler, M.: GermanTeam 2007. In: Visser, U., Ribeiro, F., Ohashi, T., Dellaert, F. (eds.) RoboCup 2007: Robot Soccer World Cup XI Preproceedings, RoboCup Federation (2007)

11. Röfer, T., Laue, T., Burchardt, A., Damrose, E., Gillmann, K., Graf, C., de Haas, T.J., Härtl, A., Rieskamp, A., Schreck, A., Worch, J.H.: B-Human team report and code release 2008 (2008), only available online: http://www.b-human.de/file_download/16/bhuman08_coderelease.pdf

12. Wang, C.C., Wang, S.C., Yen, H.C., Chang, C.H.: NTU robot PAL 2009 team report (2009), only available online: http://www.csie.ntu.edu.tw/~bobwang/RoboCupSPL/NTU_Robot_PAL_09Report.pdf

13. Zickler, S., Laue, T., Birbach, O., Wongphati, M., Veloso, M.: SSL-Vision: The shared vision system for the RoboCup Small Size League. In: Baltes, J., Lagoudakis, M.G., Naruse, T., Ghidary, S.S. (eds.) RoboCup 2009. LNCS (LNAI), vol. 5949, pp. 425–436. Springer, Heidelberg (2010)

Five Years of SSL-Vision –
Impact and Development

Stefan Zickler[1], Tim Laue[2], José Angelo Gurzoni Jr.[3],
Oliver Birbach[2], Joydeep Biswas[4], and Manuela Veloso[4]

[1] iRobot Corp.,
8 Crosby Dr., Bedford, MA, 01730, USA
szickler@irobot.com
[2] Deutsches Forschungszentrum für Künstliche Intelligenz GmbH,
Cyber-Physical Systems, Enrique-Schmidt-Str. 5, 28359 Bremen, Germany
{Tim.Laue,Oliver.Birbach}@dfki.de
[3] Centro Universitário da FEI, AI and Robotics Lab.,
Av. Humberto A.C. Branco, 3972, S. Bernardo do Campo, SP, 09850, Brazil
jgurzoni@ieee.org
[4] Carnegie Mellon University, Robotics Institute, Computer Science Department,
5000 Forbes Ave., Pittsburgh, PA, 15213, USA
joydeepb@ri.cmu.edu, mmv@cs.cmu.edu

Abstract. Since its start in 1997, the setup of the RoboCup Small Size
Robot League (SSL) enabled teams to use their own cameras and vision
algorithms. In the fast and highly dynamic SSL environment, researchers
achieved significant algorithmic advances in real-time complex colored-
pattern based perception. Some teams reached, published, and shared
effective solutions, but for new teams, vision processing has still been
a heavy investment. In addition, it became an organizational burden to
handle the multiple cameras from all the teams. Therefore, in 2008, the
league started the development of a centralized, shared vision system,
called SSL-Vision, which would be provided for all teams. In this paper,
we discuss this system's successful implementation in SSL itself, but
also beyond it in other domains. SSL-Vision is an open source system
available to any researcher interested in processing colored patterns from
static cameras.

1 Introduction

The RoboCup Small Size League (SSL) has evolved to a fast-paced and dynamic
environment for cooperative multi-robot research. In the SSL, robots are able
to traverse the game field in merely two seconds, which demands algorithms
capable of decision making, team coordination and motion control in fractions
of a second. In part, this dynamism is due to the adoption of global vision
systems in the league. Not having to be concerned with robot localization and
mapping problems, teams can focus more on intelligent software algorithms and
more precise hardware and control engineering.

S. Behnke et al. (Eds.): RoboCup 2013, LNAI 8371, pp. 656–663, 2014.
© Springer-Verlag Berlin Heidelberg 2014

However, what should have been an advantage turned out to be a considerable bottleneck. Most of the time available before a competition was devoted to the assembly of multiple cameras brought by each team, and to the iterative adjustment of their vision algorithms. Also, any new team interested in joining the league had to develop its own computer vision system before they could place robots on the field.

An open-source software[1], SSL-Vision [17], changed this scenario five years ago by providing a shared vision system that could be used by all teams. SSL-Vision not only resolved the organizational obstacles and provided a high performance common ground teams could rely on, it exceeded and served purposes outside the SSL domain. In this paper, we review the ramifications of the introduction of SSL-Vision and the impact it had both within and outside of the SSL domain. We also look at the ongoing and future developments that can further improve SSL-Vision.

This paper is organized as follows: Section 2 provides a brief overview of the overall system. Its impact on the SSL and other domains is described in Sect. 3, followed by a summary of ongoing and future work on SSL-Vision in Sect. 4.

2   Structure of SSL-Vision

SSL-Vision [17] was designed with the goal of being flexible and robust enough to meet the demands of all SSL teams. The system's architecture employs an extendable design that reflects the collaborative spirit of the League and that aims to foster use of SSL-Vision as a research tool beyond the SSL and beyond RoboCup. SSL-Vision supports concurrent image processing of multiple cameras in a single integrated application, bringing together all of its functionality, including robot and ball detection, configuration, calibration, and visualization.

SSL-Vision's architecture is described in detail in a previous paper [17]. Briefly, SSL-Vision's processing flow is encoded in a *multi-camera stack* that defines how many cameras are used for capturing, and what particular processing pipeline should be executed. A multi-camera stack consists of several threads, each representing a *single-camera stack*, consisting of multiple *plugins* which are executed in order. The system allows developers to create different stacks for different application scenarios. All configuration parameters of the system are represented in a unified way through a variable management system called *VarTypes* [15], which allows real-time introspection, editing, and XML-based data storage of all stack and individual plugin parameters. Fig. 1 shows a snapshot of SSL-Vision, including the data-tree's visualization on the left-hand pane.

2.1   RoboCup SSL Image Processing Stack

SSL-Vision's default multi-camera stack implements a processing flow for solving the RoboCup Small Size League vision task using a dual-camera setup. In the

---

[1] http://code.google.com/p/ssl-vision/

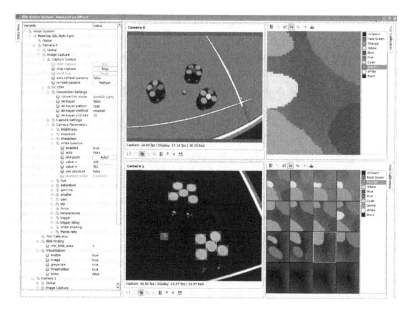

Fig. 1. Screenshot of SSL-Vision, showing the parameter configuration tree (left), live-visualizations of the two cameras (center), and views of their respective color thresholding YUV LUTs (right).

Small Size League, each robot is uniquely identifiable and locatable based on colored markers that are arranged on top of the robot in a standardized pattern. The processing flow of the Small Size League vision stack is as follows:

Image Capture

SSL-Vision supports capture from DC1394-based devices. In the Small Size League, the system typically captures two 780x580 YUV422 streams, each at 60Hz on a separate 1394B bus. We discuss ongoing community efforts to support other capture methods in Section 4.

Color Segmentation

The color segmentation process has been implemented by porting core algorithms of the CMVision library [1] to the SSL-Vision plugin architecture. CMVision's color segmentation is based on a lookup table (LUT) that maps the multi-dimensional camera color space (e.g. raw RGB values) into a discrete 1-dimensional color label (e.g. 'Pink', 'Cyan', 'Yellow', ...). To ease the calibration of this LUT, the SSL-Vision system features a fully integrated GUI which is able to visualize the 3D LUT (Fig. 1) and which allows to directly pick calibration measurements and histograms from the incoming video stream. The plugin then computes the bounding boxes and centroids of all merged regions and finally sorts them by color and size.

Conversion from Pixel Coordinates to Real-World Coordinates

In order to deduce information about the objects on the field from the measurements of the cameras, a calibration defining the relationship between the field geometry and the image plane is needed. SSL-Vision stands apart from many previous Small Size League vision systems, because it includes tools that significantly simplify the calibration process and that do not require any calibration accessories (such as checkerboard patterns). SSL-Vision's calibration routine consists of a few simple steps. First, it is assumed that the physical field dimensions (which are standardized by the League's rules) and the camera height (which can be measured easily) are known. Next, using the video image, the user has to manually annotate a rough estimate of the corners of the field and a rough search boundary for the field's line locations. Given these manually defined constraints, SSL-Vision automatically detects line locations as calibration reference points. It then uses the Levenberg-Marquardt algorithm [8] to find the optimal intrinsic and extrinsic model parameters.

Pattern Detection and Filtering

Once all colored markers have been segmented and all their real-world coordinates have been computed, the processing flow continues with the execution of the pattern recognition plugin that extracts the identities, locations, and orientations of all the robots, as well as the location of the ball. The pattern detection algorithm was adopted from the CMDragons vision system [2].

Delivery of Detection Results

The results of the pattern detection are delivered to participating teams via UDP Multicast. Data packets are encoded using Google Protocol Buffers [6], and contain positions, orientations, and confidences of all detected objects, as well as additional meta-data, such as a timestamp and frame-number.

3 Impact

3.1 Impact on the Small Size League

From the beginning, SSL-Vision was not only intended to be an alternative vision system that teams *could* use but a system that everyone *has* to use during official competitions since RoboCup 2010. Therefore, a significant impact on the league has always been foreseeable.

The process of its introduction, including a one-year trial period starting in 2009, can be considered as essentially flawless and appeared to be much appreciated by the community. The fact that there have been few improvements on SSL-Vision since its release as well as almost no complaints about its usage indicate the maturity of the already existing computer vision approaches for the RoboCup Small Size League domain. However, for anybody who wants to do vision algorithm research in the SSL context, SSL-Vision provides a modular framework to write, test and objectively evaluate new approaches against a well-tested baseline implementation.

One major motivation for the introduction of a shared vision system has always been the impact on the organizational realization of Small Size League competitions. Having each team set up its own vision equipment – as was the case before the official use of SSL-Vision– much of the valuable time during a

RoboCup was lost as teams blocked whole fields for setting up their equipment. This was the case during the setup days before the round robin as well as during the finals. In addition, the consequent field assignment made friendly matches against teams from other round robin groups almost impossible. Having the shared vision system, changing fields became as easy as it is in every other RoboCup robot soccer league.

Although successful in resolving the organizational issue during competitions, the improvements due to having an efficient shared vision system go well beyond that. The time a new team entering the League takes to have a functional system has significantly reduced, as the vision system can be executed on most modern computers and requires no knowledge in the computer vision field.

SSL-Vision has also become a foundation for the development of other long-term goals of the League. One of these is the SSL Autonomous RefBox[2], an initiative started in 2011 with the goal to create an autonomous assistant referee for the league. Another related open-source project used by some teams in the League is the grSim simulator [9], a 3D physics simulator developed on top of the SSL-Vision architecture and communication protocol.

The current vision equipment[3] has been purchased by the RoboCup Federation and is handed over to the following year's organizing committee after each competition. This procedure allows a setup of·the whole equipment even before the first teams arrive at the competition site, as well as the configuration and calibration. To operate SSL-Vision, the League created the role of the *vision experts*, who are members of the teams who help in operating SSL-Vision in the competitions. Their task is to mark the field boundaries, perform the color calibration, as well as adjusting the settings if the field luminance conditions change. The stages described in Sect. 2.1 happen mostly in the background after the operator performed these few interactions.

3.2   Impact beyond the Small Size League

As the system does not make too specific assumptions regarding the setup in detail, it can easily be applied to other scenarios, such as tracking objects that are carrying patterns on their tops. A currently popular application is the usage of SSL-Vision as a source for ground-truth data in RoboCup scenarios that demand local vision systems, such as the Standard Platform League (Fig. 2a) and the Humanoid Leagues. Some examples for SPL self-localization research using SSL-Vision for ground truth are [4], [13], and [3]. The two latter projects have been based on the B-Human software which is released as open source since several years and already includes an interface for SSL-Vision data [12]. SSL-Vision has also been selected and demonstrated as the vision system for a proposed new robot soccer league *SSL-Humanoid* that combines the Small Size League's global vision paradigm with a small humanoid robot platform [10]. In addition, SSL-Vision has been used to track robots and obstacles in real-time motion planning research [16] and in university courses such as *CMRoboBits* [14].

---

[2] `http://code.google.com/p/ssl-autonomous-refbox/`

[3] Four pairs of AVT Stingray F46C cameras with Tamron 12VM412ASIR lenses.

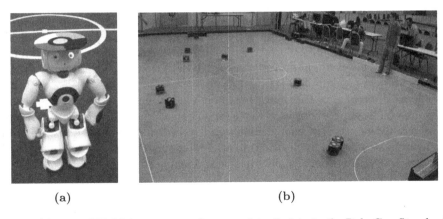

(a)   (b)

Fig. 2. (a) Using SSL-Vision as source for ground truth data in the RoboCup Standard Platform League. Image has been taken from [13].
(b) Demonstration game on a larger field during RoboCup 2011. The field was 9.4 x 6.15 meters and 4 cameras were used.

4   Ongoing and Future Work

The modular plugin-based architecture of SSL-Vision allowed the Small Size League to take important evolutionary steps, such as the addition of a sixth robot to teams without need for vision algorithm adjustments, and the convenient setup of the so called mixed-team games, where two teams play side-by-side, cooperating, each providing a half of the robots in play.

An important step for the League, made possible by SSL-Vision, is the ongoing work towards the creation of a larger field, of almost twice the current official size, for games with more robots playing, perhaps even an 11 vs. 11 soccer match. The first trial of the larger field setup occurred during RoboCup 2011 in Istanbul, and can be seen in Fig. 2b. Two fields were placed side-by-side, and the 4 cameras mounted were used to transmit vision information to the teams participating on the demonstration match. Although field geometry calibration and sensor fusion can become more complex as more cameras are added, and were manually adjusted for the demonstration in 2011, we believe that extending the current capabilities of the software to handle 4 or more cameras is feasible.

There are several extensions to SSL-Vision that are worth to be explored, among them the color calibration system. The current procedure is the most time-consuming configuration task in the software, and requires the operator to calibrate the colors for all different luminance conditions found on the field. The calibration is also static, so changes in the lighting of the field result in need for recalibration. Currently, a few plugins for automated color calibration are under development, one of them employing Artificial Neural Networks and another using Self-Organizing Map (SOM) networks.

An interesting extension would be the change in the robot detection plugin for probabilistic detection of partially detected patterns. This is the case, for instance, when the robot stops under the goal bar, occluding some of its pattern circles, or when a color segmentation error occurs in a small patch of the field. Several algorithms could be used, from naive Bayes classifiers to particle filters (i.e.: [7],[5]). This would improve teams' robot detection and could lead to a fully featured object tracking like [11].

The addition of a capture stack based on Video4Linux, to support USB cameras, as well as support for cameras using Gigabit Ethernet (GigE), would also considerably increase the compatibility and spread the use of the software both for new SSL teams and applications in other domains. A community-contributed stack to add Video4Linux support has been announced[4] as of the writing of this paper.

Another interest of the Small Size League is the distribution of game logs and their corresponding videos in a standard format, therefore a plugin to stream and record camera images via network could improve the qualify of the logs teams use for training and machine learning tasks.

5   Conclusions

In this paper we have shown the vast impact that SSL-Vision has had since its introduction five years ago. By replacing all teams' individual vision systems with a unified open architecture, SSL-Vision solved the biggest organizational challenge of the league and furthermore made it easier for new teams to get involved and immediately focus on the key research aspects of the league. Beyond the Small Size League, SSL-Vision has had significant impact as a research tool, providing a flexible and inexpensive ground-truthing solution for various applications. SSL-Vision continues to thrive on the strong momentum of its active developer community. Hence, looking forward, we expect SSL-Vision's capabilities to further grow and mature, and its applicability to increase to an even larger set of domains, both within and beyond the scope of robot soccer.

Acknowledgments. The authors would like to thank all persons who contributed to SSL-Vision.

References

1. Bruce, J., Balch, T., Veloso, M.: Fast and inexpensive color image segmentation for interactive robots. In: Proceedings of the 2000 IEEE/RSJ International Conference on Intelligent Robots and Systems (IROS 2000), vol. 3, pp. 2061–2066 (2000)
2. Bruce, J., Veloso, M.: Fast and accurate vision-based pattern detection and identification. In: Proceedings of the 2003 IEEE International Conference on Robotics and Automation, ICRA 2003 (2003)
3. Burchardt, A., Laue, T., Röfer, T.: Optimizing particle filter parameters for self-localization. In: Ruiz-del-Solar, J., Chown, E., Plöger, P.G. (eds.) RoboCup 2010. LNCS (LNAI), vol. 6556, pp. 145–156. Springer, Heidelberg (2010)

---

[4] https://github.com/cktben/ssl-vision

4. Coltin, B., Veloso, M.M.: Multi-observation sensor resetting localization with ambiguous landmarks. In: Burgard, W., Roth, D. (eds.) Proceedings of the Twenty-Fifth AAAI Conference on Artificial Intelligence, AAAI 2011. AAAI Press, San Francisco (2011)
5. Fergus, R., Perona, P., Zisserman, A.: Object class recognition by unsupervised scale-invariant learning. In: Proceedings of the 2003 IEEE Computer Society Conference on Computer Vision and Pattern Recognition, 2003, vol. 2, pp. 264–271 (2003)
6. Google Inc.: Protocol Buffers, http://code.google.com/p/protobuf/
7. Isard, M., MacCormick, J.: Bramble: A bayesian multiple-blob tracker. In: Proceedings of the Eighth IEEE International Conference on Computer Vision, ICCV 2001, vol. 2, pp. 34–41 (2001)
8. Marquardt, D.: An algorithm for least-squares estimation of nonlinear parameters. SIAM Journal on Applied Mathematics 11(2), 431–441 (1963)
9. Monajjemi, V., Koochakzadeh, A., Ghidary, S.S.: grSim - RoboCup Small Size robot soccer simulator. In: Röfer, T., Mayer, N.M., Savage, J., Saranlı, U. (eds.) RoboCup 2011. LNCS, vol. 7416, pp. 450–460. Springer, Heidelberg (2012)
10. Naruse, T., Masutani, Y., Mitsunaga, N., Nagasaka, Y., Fujii, T., Watanabe, M., Nakagawa, Y., Naito, O.: Ssl-humanoid. In: Ruiz-del-Solar, J., Chown, E., Plöger, P.G. (eds.) RoboCup 2010. LNCS (LNAI), vol. 6556, pp. 60–71. Springer, Heidelberg (2010)
11. Nguyen, H., Smeulders, A.: Fast occluded object tracking by a robust appearance filter. IEEE Transactions on Pattern Analysis and Machine Intelligence 26(8), 1099–1104 (2004)
12. Röfer, T., Laue, T., Müller, J., Bösche, O., Burchardt, A., Damrose, E., Gillmann, K., Graf, C., de Haas, T.J., Härtl, A., Rieskamp, A., Schreck, A., Sieverdingbeck, I., Worch, J.H.: B-human team report and code release 2009 (2009), only available online: http://www.b-human.de/file_download/26/bhuman09_coderelease.pdf
13. Seekircher, A., Laue, T., Röfer, T.: Entropy-based active vision for a humanoid soccer robot. In: Ruiz-del-Solar, J., Chown, E., Ploeger, P.G. (eds.) RoboCup 2010. LNCS (LNAI), vol. 6556, pp. 1–12. Springer, Heidelberg (2010)
14. Veloso, M., Lenser, S., Vail, D., Rybski, P., Aiwazian, N., Chernova, S.: CMRoboBits: Creating an intelligent AIBO robot. In: 2004 AAAI Spring Symposium on Accessible Hands-on Artificial Intelligence and Robotics Education (March 2004)
15. Zickler, S.: The VarTypes System, http://code.google.com/p/vartypes/
16. Zickler, S.: Physics-Based Robot Motion Planning in Dynamic Multi-Body Environments. Ph.D. thesis, Carnegie Mellon University, Thesis Number: CMU-CS-10-115 (May 2010)
17. Zickler, S., Laue, T., Birbach, O., Wongphati, M., Veloso, M.: SSL-vision: The shared vision system for the RoboCup Small Size League. In: Baltes, J., Lagoudakis, M.G., Naruse, T., Ghidary, S.S. (eds.) RoboCup 2009. LNCS (LNAI), vol. 5949, pp. 425–436. Springer, Heidelberg (2010)

Integration of the ROS Framework in Soccer Robotics: The NAO Case

Leonardo Leottau Forero, José Miguel Yáñez, and Javier Ruiz-del-Solar

Advanced Mining Technology Center, Department of Electrical Engineering,
Universidad de Chile,
Av. Tupper 2007, Santiago, Chile
{dleottau,jruizd}@ing.uchile.cl, jm.yanez.arancibia@gmail.com
http://uchilert.amtc.cl

Abstract. The SPL robot soccer league focuses its efforts on the development of robot control software for standard humanoid robots. Nevertheless, few interchange of software modules are observed in the league, being the B-Human effort an exception. In addition, a large difference in performance is observed between experienced teams and new teams. This situation makes difficult the incorporation of new teams in the league. Therefore, it seems attractive to explore the use of ROS within the SPL soccer robotics community in order to revert the described situation. As a first step, this paper presents some work in this direction, such as the installation of ROS in the new NAO V4 robots, the integration of the B-Human motion engine as a ROS node, and the communication of two robots running a ROS-based control software.

Keywords: Humanoid robots, NAO robot, Robot Operating System (ROS), shared code.

1 Introduction

The SPL (Standard Platform League) is a RoboCup robot soccer league in which all teams compete with identical robots; since 2008 the robots being used are the Aldebaran NAO robots [5]. The research efforts focus on the development of sophisticated control software for fully autonomous operation of a team of robots. With teams participating in the league for 5-10 years, or ever more, some control software solutions have reached a very high level of effectiveness and functionality, which new teams cannot reach easily. In addition, even tough being a league where most of the work is on the development of software, there are very few cases of code sharing. One of these cases is the one of the B-Human team [1], which makes every year its software open to the community. However, even in this case it is not easy to use just some specific parts of the code. Normally teams use the whole code and modify the high-level behaviors. In addition, it is not easy for other teams to develop new components that can be integrated in the B-Human software. Basically, the B-Human software is being developed and extended just by the B-Human team.

S. Behnke et al. (Eds.): RoboCup 2013, LNAI 8371, pp. 664–671, 2014.

On the other hand, the ROS (Robot Operating System) Framework, which can be defined as an open-source, meta-operating system for robots, has emerged as tool that allows the sharing of robot software modules. In addition to provide a proper architecture to build robot applications and to communicate appropriately different modules using the ROS Node concept, the main attractiveness of ROS is that is being used almost by everyone in the robotics research community. In this context it seems attractive to explore the use of ROS within the SPL soccer robotics community. The use of ROS within the SPL soccer community could allow:

- to share easily software module between teams,

- to encourage the development of very specialized solutions which can be shared among the teams,

- to facilitate the incorporation of new teams in the league,

- to attract new students and new researchers to the leagues teams,

- to encourage the specialization of some teams in some robot control areas (e.g. motion control or perception), and

- to facilitate the comparison and benchmarking of specific software modules.

In order to explore the use of ROS within the SPL league, the first step requires being able to compile ROS in the NAO robots, and to build a basic architecture for these robots. The main goal of this paper is to present such developments. The article is organized as follows: Section 2 describes the use of the ROS framework in the NAO robots. This includes a description of how to install ROS into the NAO robots and the integration of the B-Human Motion Engine as a ROS Node. Section 3 presents as a proof of concept, the communication between two NAO robots using a ROS-based core. Finally, some conclusions are drawn in section 4.

2   ROS Use in NAO Robots

2.1   ROS Installation on NAO Robots

There are few research groups and labs working with NAO robots and ROS [3] [7] [4]. The Humanoid Robots Lab at the Albert-Ludwigs-Unversitaet that have developed the *nao-robot* stack which contains some useful nodes to integrate the NAO robot into ROS [8]. Within its activities of robotic soccer, the Mobile Robotics Group of the University of Chile is currently using the ROS framework as part of their NAO robot software architecture [3]. Recently the group has uploaded to the ROS community a detailed tutorial to build, install and run ROS natively onto the NAO V4 [6]. To the best on our knowledge, this is the first tutorial that provides a step-by-step guide to build, install and run ROS Fuerte embedded onto the Atom CPU of the latest NAO robot. In Figure 1 are shown the main steps of this installation.

1. Requirements

- Download and uncompress the appropriate Naoqi toolchain version.

- Download ROS BASE from source code.

- Setup paths, environment variables, and directories such as: ROS-NAO root path ($TARGETDIR), ROS fuerte installation path, the CTC directory where Aldebaran's cross-compilation toolchain was uncompressed, the CTC boost root path, among others.

2. Preliminary Steps

- To include NAO's toolchain within ROS build toolchain.

- Since NAO's toolchain does not contains "log4cxx.so" nor "aprutil*.so" libraries, these should be copied.

3. Compiling ROS Base from source

- To compile ROS Base from source, generating the project using cmake, including the Aldebarans Nao Cross toolchain withing the ROS toolchain file. It is important to set the CMake-Install-Path in the same absolute path where ROS will be run in the Nao`s CPU.

4. Copying some files, scripts and missing libraries

- It is necessary to copy some files, scripts and libraries that are missing on the NAO. These files are needed for launching ROS on the NAO's CPU. For example the *log4cxx*, *boost* and *apr-util* libraries, python packages "yaml", among others.

5. Installing and running ROS onto NAO's CPU

- To install ROS in the NAO simply copy the installation root folder to the robot via ssh, remembering that the $TARGETDIR necessarily must match the installation path indicated in the robot (/home/nao/ros-nao for example).

- After that it is necessary to specify to ROS the path where files, scripts and libraries were added in the previous step. This should be done by exporting the environment variables $LD_LIBRARY_PATH and $PYTHONPATH. These paths indicate to ROS an alternative directory to find libraries and python files respectively. It must be set on each terminal where some ROS task will be run.

- To run ROS onto the Nao`s CPU, it is necessary run *roscore* command and executes each needed process-node.

Fig. 1. ROS installation procedure in NAO V4 robots. A detailed tutorial can be found in [6].

2.2 Integration of the B-Human Walking and Motion Engine as a ROS Node

Currently, UChile Robotics Team [3] is using the B-Human walking and motion engine [2]. In order to carry out the isolation and integration of this motion engine, two key aspects are important to mention: the implementation of a shared memory block for inter-communicating the motion process directly with NAOs hardware interface (NaoQi-DCM), and the implementation of a blackboard for reading and writing the motion engine inputs and outputs. A basic B-Human actuation block diagram is showed in Figure 2.

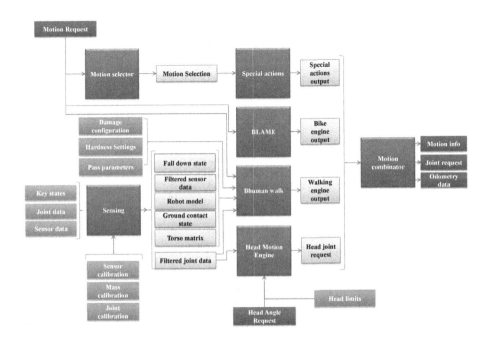

Fig. 2. B-Human actuation block diagram. The blue blocks are the main motion engine modules and the orange blocks some of their configuration parameters. Green blocks are sensors and encoders readings which are the inputs obtained each frame. The outputs (purple blocks) are mainly a joint request and odometric information. Red blocks are the high level requests. (Based on B-Human 2012 team description paper and code release [1]).

In order to integrate the B-Human walking and motion module as a ROS node (*bh-motion*), several types of mesages have been defined, maintaining as much as possible the compatibility between the original B-Human representations and ROS mesagges. Table 1 shows some of these messages; a complete documentation about their descriptions and functions can be found in [1] and [2].

The basic procedure to subscribe, execute and publish the *bh-motion* node is as follows:

Table 1. Some relevant ROS messages defined for the B-Human motion node

Msg. Name	First Level Attributes	Second Level Attributes
HeadRequest (input)	float pan	-
	float tilt	-
	float speed	-
MotionRequest (input)	motionType	-
	specialActionRequest	specialActionType
		bool mirror
	walkRequest	mode
		Pose2D speed
		Pose2D target
	bikeRequest	bMotionType
		bool mirror
FilteredJointDataPrev (output)	float[] angles	-
	int timeStamp	-
	float cycleTime	-
FilteredJointData (output)	float[] angles	-
	int timeStamp	-
	float cycleTime	-
FilteredSensorData (output)	float[] data	-
	int[] currents	-
	uint[] temperatures	-
	uint timeStamp	-
	uint usActuatorMode	-
	uint usTimeStamp	-
GroundContactState(output)	bool contactSafe	-
	bool contact	-
	bool noContactSafe	-
HeadJointRequest (output)	float tilt	-
	float pan	-
	bool reachable	-
	bool moving	-
JointRequest (output)	jointData	filteredJointData
	jointHardness	int[] hardness
OdometryData (output)	pose2D	-
RobotModel (output)	Pose3D[] limbs	-
	Vector3 centerOfMass	-
	float totalMass	-
TorsoMatrix (output)	Pose3D offset	-

1. The node subscribes to the correspondent ROS topics waiting for a motion request.

2. If there is a callback, the input messages are copied to the blackboard.

3. Every time NaoQi-DCM updates the sensors, the motion engine checks the blackboard for pending jobs and computes the motion engine to obtain its outputs (mainly actuator requests and odometric information).

4. Obtained outputs are put in the blackboard and published on its correspondent ROS topics.

5. Naoqi-DCM receives the actuator requested values and updates the values of the sensors again.

The source code of *bh-motion* ROS package can be downloaded in [9], soon it will be shared on the ROS community website.

3 Proof of Concept

3.1 Communication between Two NAO Robots

The objective of this proof of concept is communicating two robots by using the ROS framework, and its distributed computing capability. The first robot (Robot 1 - Goalie) has the UChile Robotics Team code with an interface for sharing the most important representations as ROS messages (perceptors, odometry, robot pose, etc). The second robot (robot 2 - blind), only has the B-Human motion engine ROS node, and it does not have any vision or perception module.

Robot 1 is the master, it is static, knows its pose on the field and can see the ball. That robot publishes the global position of the seen ball in a topic which is subscribed by blind robot. Robot 2 knows its initial pose on the field, it receives a callback from robot 1 with the ball position, transforms it to an adequate navigation message (walk request), and goes to the ball.

The procedure for sharing the ball position from robot 1 to robot 2 can be described as following:

1. Robot 1 receives a READY flag from Robot 2.

2. Robot 1 publishes the ball position.

3. Robot 2 disables the READY flag while navigates up to the ball position on the field.

4. Robot 2 enables the READY flag once it is on the indicated position.

5. Go to step 1.

Results of this test can be watched on video in this link [10]. The experiment has worked as expected taking into a account the purpose of this proof of concept. No communication failures have been observed neither wrong shared ball positions. Since robot 2 goes to the ball only using its odometry to navigate, reached target positions are inaccurate due to cumulative odometry errors.

Table 2 shows average of CPU and RAM consumption of the NAO V4 (ATOM Z530 1.6GHz CPU and 1 GB RAM) taken while BHmotion node is running. It can be noticed that average consumption of all ROS processes is about 1% for CPU and 2.6% for RAM.

Table 2. Average of CPU and RAM consumption of the NAO V4

Process	%CPU Avg.	%CPU Std.dev.	%RAM Avg	%RAM Std.dev.
naoqi-bin	10.69	0.53	7.81	0.10
nao_hald	5.10	0.29	0.30	0.00
BHmotion	3.22	0.89	0.70	0.00
rosout	0.36	0.15	0.50	0.00
roscore	0.33	0.11	1.20	0.00
rosmaster	0.33	0.11	0.90	0.00

4   Conclusions

This article described the use of the ROS middleware in NAO robots. A guide to cross-compile, install and run ROS onto the CPU of the latest NAO robot was presented. Also, a brief description of how B-Human motion engine was integrated as a ROS node has been presented. In addition, a proof of concept which two NAO robots have been successfully communicated by using ROS framework was presented.

Since this is our first approach towards integration of ROS framework with NAOs and soccer robotics, our main objective was not to measure or compare the ROS computational-performance into the Atom CPU. This is part of our future work.

A line of work that needs to be explored by the comunity is the integration of the *bh-motion* node with simulation, visualization, debugging, and other useful tools currently available for ROS.

Acknowledgments. This work was partially funded by FONDECYT under Project Number 1130153. We thank B-Human for their contribution to SPL.

References

1. Röfer, T., Laue, T., Müller, J., Bartsch, M., Batram, M.J., Böckmann, A., Lehmann, N., Maaß, F., Münder, T., Steinbeck, M., Stolpmann, A., Taddiken, S., Wieschendorf, R., Zitzmann, D.: B-Human Team Report and Code Release (2012), http://www.b-human.de/wp-content/uploads/2012/11/CodeRelease2012.pdf
2. Graf, C., Härtl, A., Röfer, T., Laue, T.: A Robust Closed-Loop Gait for the Standard Platform League Humanoid. In: Zhou, C., Pagello, E., Menegatti, E., Behnke, S., Röfer, T. (eds.) Proceedings of the Fourth Workshop on Humanoid Soccer Robots in Conjunction with the 2009 IEEE-RAS International Conference on Humanoid Robots in Conjunction with the 2009 IEEE-RAS International Conference on Humanoid Robots, Paris, France, pp. 30–37 (2009)
3. Yáñez, J.M., Leottau, L., Cano, P., Mattamala, M., Vidal, A., Celedón, W., Mardones, J., Silva, C., Ruiz-del-Solar, J., UChile, R.T.: Team Description Paper, RoboCup 2013 - Standard Platform League. In: RoboCup 2013: Robot Soccer World Cup XVII Symposium Preproceedings (2013)

4. Austrian-Kangaroos 2013 Team Description Paper. In: RoboCup 2013: Robot Soccer World Cup XVII Symposium Preproceedings (2013)
5. Gouaillier, D., Hugel, V., Blazevic, P., Kilner, C., Monceaux, J., Lafourcade, P., Marnier, B., Serre, J., Maisonnier, B.: Mechatronic design of NAO humanoid. In: 2009 IEEE Int. Conf. Robot. Autom., pp. 769–774. IEEE (2009), http://ieeexplore.ieee.org/articleDetails.jsp?arnumber=5152516
6. UChile Robotics Team: Ros fuerte cross-compiling and installation for the NAO v4, http://www.ros.org/wiki/nao/Tutorials/Cross-Compiling_NAO-V4
7. Hornung, A., Dornbush, A., Likhachev, M., Bennewitz, M.: Anytime search-based footstep planning with suboptimality bounds. In: 2012 12th IEEE-RAS International Conference on Humanoid Robots (Humanoids), pp. 674–679 (2012)
8. Nao-Robot ROS Stack, http://www.ros.org/wiki/nao_robot
9. UChile Robotics Team: B-Human Motion ROS Package (Source Code), https://github.com/uchile/bh-motion_ros-pkg
10. UChile Robotics Team: Communication between two robots (Video), http://www.youtube.com/watch?v=UgR6JxY1R8A&feature=youtu.be

Development of RoboCup@Home Simulation towards Long-term Large Scale HRI

Tetsunari Inamura[1,2], Jeffrey Too Chuan Tan[1], Komei Sugiura[3],
Takayuki Nagai[4], and Hiroyuki Okada[5]

[1] National Institute of Informatics, Japan
[2] The Graduate University for Advanced Studies, Japan
[3] National Institute of Information and Communication Technology, Japan
[4] The Univ. of Electro-communications, Japan
[5] Tmagawa Univ., Japan

Abstract. Research on high level human-robot interaction systems that aims skill acquisition, concept learning, modification of dialogue strategy and so on requires large-scaled experience database based on social and embodied interaction experiments. However, if we use real robot systems, costs for development of robots and performing many experiments will be too huge. If we choose virtual robot simulator, limitation arises on embodied interaction between virtual robots and real users. We thus propose an enhanced robot simulator that enables multiuser to connect to central simulation world, and enables users to join the virtual world through immersive user interface. As an example task, we propose an application to RoboCup@Home tasks. In this paper we explain configuration of our simulator platform and feasibility of the system in RoboCup@Home.

1 Introduction

An important task in the field of human-robot interaction (HRI) is elucidating the mechanisms of social and physical interactions and then embodying them into the design of robots. Completing this task requires the evaluation and modification of a hypothetical interaction model based on long-term and large-scale social interaction between people and robots. However, conventional HRI experiments are limited due to their laboratory setting, while large-scale experiments are quite costly in terms of people and time, especially if the aim of HRI is learning from demonstration or instruction such as was investigated by Sugiura et al. [1].

One of the purposes of the RoboCup@Home laps over the above goal. Typical tasks in the competition are designed on the basis of the assumption that robots must possess advanced social and embodied interaction functions. However, due to huge cost of developing robot hardware and executing experiments, benchmark tasks tend to focus on basic recognition functions and physical functions such as grasping, navigation, object recognition, and face recognition. The RoboCup@Home tasks are performed in a kitchen or living room environment,

S. Behnke et al. (Eds.): RoboCup 2013, LNAI 8371, pp. 672–680, 2014.
© Springer-Verlag Berlin Heidelberg 2014

where a higher level of natural interaction is required. Such interaction includes dialogue management, intention understanding via gestures and eye gaze, and clarification of vague user instructions. Additionally, machine learning techniques for adapting to unknown environments and situations should be evaluated on the basis of past experience. These tasks shold be designed for RoboCup@Home not only from the viewpoint of competition but also from the viewpoint of academic interest. We believe that the RoboCup@Home competition should incorporate high-level intuitive and natural interaction tasks; however, the current rules requiring the use of real robots prevent the incorporation of such tasks.

Here, we present a RoboCup@Home simulation system with an immersive interface based on our SIGVerse simulation platform and discuss its feasibility. This platform was used in an official competition at the RoboCup Japan Open in 2013.

2   Current Mechanism of RoboCup@Home Competition

The current RoboCup@Home competition mechanism was designed on the basis of a benchmarking approach in which a set of functional abilities corresponding to the target robot technical abilities [2]. These functional abilities are weighted for use in the scoring of the yearly competitions. Table 1 shows the function weightings for the 2010 competition [2]. The cognition ability, which reflects the robot's high-level intuitive and natural interaction capability, was first introduced in 2010 and was given a weight of 13%, lower than the weighting for the two physical abilities, navigation and object manipulation. Cognition ability is mainly tested in the GPSR (General Purpose Service Robot) challenge. For the top five teams in 2012, the average achievement percentage for the GPSR challenge was only 8.2% (Table 2). This reflects the weakness of the cognition ability development at the latest RoboCup@Home competition.

Table 1. Function weighting for 2010 competition [2]

Ability	Weight(%)
Navigation	22
Mapping	9
Person Recognition	12.5
Person Tracking	3
Object Recognition	7.5
Object Manipulation	14
Speech Recognition	15
Gesture Recognition	3.5
Cognition	13

Table 2. Average score for top 5 teams in 2012 competition

Test	Max. Score	Score Weight	Avg. Score	Percent Achvd (%)
RIPS	1000	7.69	721.8	72.18
FM	1000	7.69	195	19.5
WW	2000	15.38	655	32.75
CU	1000	7.69	366.6	36.66
OC	2000	15.38	1249.8	62.49
GPSR	2500	19.23	205	8.2
DC	1500	11.54	644	42.93
Res	2000	15.38	260	13
Total	13000	100	4297.2	33.06

The above analysis clearly shows that the latest RoboCup@Home competition did not sufficiently stimulate cognition ability development. Due to the requirement of using a real robot in the competition, only very basic and limited interaction was possible. Focusing on the balance between cognition ability and the physical functional abilities is far more important than focusing on high-level robot intelligence.

We have developed a new approach to addressing these limitations: robot simulation. Moving the robot in a simulated world reduces the complexity of the low-level hardware issues, making it easier to focus on the high-level cognition issues. However, maintaining a rich interaction medium requires that people in the real world participate naturally through embodiment and multimodal interaction. We thus use various techniques to immerse a real-world person into the 3D virtual world and thereby enable human interaction with the virtual robot. This human participation (without a physical presence) supports our higher aim of having many human participants interact with an easy-to-maintain virtual robot in much deeper, longer term, and larger scale HRI. We believe this will significantly facilitate high-level robot intelligence development through the RoboCup@Home competitions.

3   Platform for RoboCup@Home Simulation

3.1   SIGVerse: SocioIntelliGenesis Simulator

We developed a platform for RoboCup@Home simulation on the basis of our SIGVerse [3] system, which enables easy development of HRI experiments in which people and virtual robot agents can interact socially and physically. This system enables arbitrary users to join virtual HRI experiments through the Internet with log-on to the central virtual world. It has three basic simulation modules: dynamics, perception, and communication.

Dynamics Simulation. The Open Dynamics Engine (ODE)[1] is used for dynamic simulation of interactions between agents and objects. Basically, the motions of each agent and object are calculated by the dynamics engine, and the user controls the calculations to reduce simulation costs. A switch flag can be set for each object and agent to turn off the dynamics calculations if required.

Perception Simulation. The perception simulation module provides the senses of vision, sound, force, and touch. OpenGL is used for visual simulation; it provides each agent with a pixel map that is a visual image derived from the agent's viewpoint and field of view. A distance sensor is also emulated. A robot agent can get the distance vector as a single dimension if the laser range finder is selected. A distance image as a two-dimensional matrix is available if stereo vision is selected.

For the sense of touch, it is possible to acquire information on the force and torque between objects, which are calculated mainly by ODE.

---

[1] http://www.ode.org

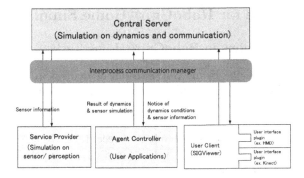

Fig. 1. Software configuration of SIGVerse

Communication Simulation. The sense of hearing is simulated by enabling every agent to communicate with audio data. The effect of sound volume is simulated by attenuating the sound in inverse relationship with the square of the distance under the assumption that the voice of an agent becomes more difficult to hear as the distance increases. It is possible to set the system so that only voices within a certain distance are audible.

With this system, not only can agents within the virtual environment communicate with each other but the virtual environment can interact with users in the real world.

There have been several studies [4][5] in which human agents act in a virtual social world based on the Second Life framework. However, it is difficult to implement dynamics and perception simulation.

3.2 Configuration of Simulator Software

SIGVerse is a client/server system consisting of a Linux server and a Windows client application. Dynamics calculations are mainly performed on a central server system. The behaviors of the robot and human avatars are controlled by an 'agent controller,' which is a dynamic link library on the Linux SIGVerse server. Autonomous actions and sensing functions of the agents are written using C++ APIs. Since the software libraries for ROS (Robot Operating System)[2] and OpenRTM[3] are also available, source code compatibility is supported. The avatars' behaviors are programmed using the APIs; they can also be controlled by user operation through the user interface in real time. Each user connects to the server system through the Windows client system. The configuration of the SIGVerse software is shown in Fig. 1.

---

[2] http://www.ros.org/wiki/

[3] http://www.openrtm.org/

4   User Interface for RoboCup@Home Simulation

4.1   Frequent Functions in RoboCup@Home Tasks

The tasks in the RoboCup@Home competition, as defined in the rule-book [6], are *Follow Me, Clean Up, Enhanced Who is Who,* and so on. In the *Follow Me* task, the robot has to recognize the facial image of the user, track the image, and follow the walking user. In the *Clean Up* task, the robot has to grasp a piece of trash targeted by the user and place it in a receptacle. In the *Enhanced Who is Who* task, the robot has to bring a drink in a cup or bottle to the user and hand it to the user. The implementation of these tasks requires the development of many basic functions such as receiving the user's instructions via speech and gestures, understanding the meaning of the instructions by speech recognition, image processing of pictures captured with a camera, grasping an object using end effectors, navigation by wheels, and dialogue management. These basic functions can be simulated in the SIGVerse world.

4.2   Immersive Interface for RoboCup@Home Simulation

We developed an inexpensive, flexible, and immersive interface for SIGVerse to enable social and embodied interaction with the virtual agents that frequently appear in RoboCup@Home tasks.

Projection of SIGVerse World to HMD. The interface uses a head mounted display (HMD) (eMagin Z800 3DVisor) that has a motion detector for the pan and tilt directions. Detected motion is transferred to the SIGVerse system for use in controlling the user's avatar. The HMD can display the sequence of images captured by the avatar's eye in accordance with the head motion. An overview of a user and the scene image projected on the HMD is shown in Fig. 4.2.

Control of Avatar Body Motion. Microsoft's Kinect controller can be connected to the latest SIGVerse client terminal. The motion pattern measured with the Kinect controller is transferred to the server for use in controlling the motion of the avatar's body. This enables virtual pointing gestures. In addition, the avatar can be made to grasp virtual objects by displaying a grasping motion in front of the Kinect controller. Since finger motion is difficult to measure with the Kinect controller, the system provides a grasp/release command for controlling avatar grasping. An image of avatar control using the Kinect controller is shown in Fig. 4.2.

These immersive interfaces can be connected to the SIGVerse client system as plug-in modules. There are also text-based chat and speech recognition and synthesis functions provided as standard user interfaces. Future developments in the RoboCup@Home simulation competition should lead to additional proposals from the participants for immersive interfaces such as a joystick controller and a motion capture system. Therefore, the SIGVerse client has a plug-in system

Fig. 2. A user joining a virtual world through HMD

Fig. 3. Whole body control of an avatar by Kinect sensor

that accepts arbitrary interface systems such as users' original speech recognition systems, sound source detection systems, face recognition systems, and haptic interfaces.

4.3 Limitations of the RoboCup@Home Simulator Interface

In the *Follow Me* task, the robot has to track the user's facial image and follow the walking user; however, the SIGVerse user interface is fixed on a client computer. Therefore, the user has to interact in a limited area covered by fixed sensors, such as the Kinect sensors. Although SIGVerse can be connected with motion capture systems, such systems are difficult to use as an interface for the RoboCup@Home competition.

Due to the huge computational cost of simulation physical grasping due to the need to consider the friction coefficient of soft and flexible material, the proposed system does not currently simulate grasping. Instead, it uses a binary status: grasping/not grasping. Likewise, the handing over of an object from a robot agent to a user avatar is only reflected as a change of status. The user does not feel the reaction force.

We do not aim to implement all of the RoboCup@Home simulation tasks in the SIGVerse simulator. Instead, we aim to implement advanced interaction scenarios that require high-level agent functionality such as decision making, machine learning from conversation, and inference of future events. Examples of such scenarios are presented in the next section.

5 Implementation of RoboCup@Home Simulation Tasks

5.1 Clean Up Task

A simulated version of the Clean Up task introduced in the 2011 RoboCup@Home competition [6] is practical (Figure 5.1). This task tests the robot's abilities to detect, recognize, and manipulate objects, to navigate, and to systematically search and explore. The robot is directed to explore a room and determine

Fig. 4. Screen-shot of Clean Up task   **Fig. 5.** Screen-shot of Cooperative Cooking task

whether known and unknown objects are items to be discarded. In this simulation, techniques similar to those used by real robots can be simulated on a virtual robot, for instance, the use of computer vision (e.g., OpenCV) to perform image processing for object detection and recognition.

Another function that can be simulated is using natural language and gesture instruction to recognize which object is being referred to by the user. A user might ask the robot to move and/or manipulate an object by saying something like, "Please bring that dish to the dining table" while pointing to the dish. If the pointing and/or speech are vague, the robot should be able to ask appropriate questions to remove the uncertainty. Such dialogue management is a high-level interaction function inherent in high-level HRI.

5.2   Cooperative Cooking Task (Future Candidate)

Tasks not described in the rulebook can also be simulated in the SIGVerse system and fit within the scope of the RoboCup@Home competition. One such task is a Cooperative Cooking task, which is a very high-level HRI task [7]. The task requires recognition of human behavior, real-time planning in accordance with recipe, dialogue management, and so on. Here, we describe an implementation of the Cooperative Cooking task as a future representative task. Since typical cooking tasks are cutting foodstuffs, grasping dishes and cups, operating microwaves, and so on, observation of such behaviors can focus on upper-body motions, and it is easily accomplished with a Kinect controller. Figure 5.1 shows a screen shot of this task.

The evaluation target is an effective human-robot interaction strategy. For example, if the user is not taking a cooking action, the robot should ask the user to do something complementary to what the robot is doing. If the user is taking a cooking action, the robot should search for, find, and take a suitable complementary action. Implementation of the required action selection algorithm would make the system more effective.

We performed preliminary experiments corresponding to two cases: one in which the user performed all of the actions through the interface, and one in which the robot performed complementary tasks as well. In the first case, the task took 194 [sec], and in the second case it took 58 [sec]. The task completion time would be an evaluation criterion in the RoboCup@Home competition.

6   Conclusion

Social and embodied interaction and adaptation and learning methods based on big data will be important issues in the next stage of human-robot interaction. Implementation of a simulation platform on which many people can easily participate in social interaction experiments will be a breakthrough in promoting human-robot interaction research.

We have developed an immersive interface for our SIGVerse simulation platform. Effective tasks for the RoboCup@Home competition can be developed by using a multi-user connection environment on the Internet. The interaction is currently one-to-one; however, it is easy to develop many-to-many human-agent interaction scenarios based on the multi-user connection system. Use of this simulation platform should lead to various contributions such as an autonomous judge system and machine learning competition for intelligent behavior through long-term large-scale interaction between robots and humans.

The proposed multi-user connection system has limitations, so its application is limited to simulating interactions that use kinematic motion, eye direction, and speech. Nevertheless, the tasks still have a moderate level of complexity.

The first official RoboCup@Home simulation competition was held in RoboCup Japan Open in May 2013. Since our team, the eR@sers, took first prize at the RoboCup@Home competitions in 2008 and 2010, our proposed use of simulation for the competition was accepted by the Japanese RoboCup@Home community. In the Japan Open, Cleaan Up task was chosen as a competition task and 5 teams participated. Since autonomous judge system was introduced at the competition, scoring procedure progressed smoothly.

The SIGVerse client application can be downloaded at http://www.sigverse. org. The source code for a sample application is available at[4]. The tutorial has more than 15 sample programs such as 1) controller for mobile robots, humanoid robots and human avatar, 2) sensor emulators such as distance sensor, and 3) usage of Kinect and head mounted display for immersive virtual reality system.

References

1. Sugiura, K., Iwahashi, N., Kawai, H., Nakamura, S.: Situated spoken dialogue with robots using active learning. Adv. Robotics 25(17), 2207–2232 (2011)
2. van der Zant, T., Iocchi, L.: Robocup @home: Adaptive benchmarking of robot bodies and minds. In: Mutlu, B., Bartneck, C., Ham, J., Evers, V., Kanda, T. (eds.) ICSR 2011. LNCS (LNAI), vol. 7072, pp. 214–225. Springer, Heidelberg (2011)

---

[4] http://www.sociointelligenesis.org/SIGVerse/index.php?Tutorial

3. Inamura, T., et al.: Simulator platform that enables social interaction simulation –SIGVerse: SocioIntelliGenesis simulator–. In: IEEE/SICE Int'l Symp. on System Integration, pp. 212–217 (2010)
4. van der Kapri, A., Ullrich, S., Brandherm, B., Prendinger, H.: Global lab: An interaction, simulation, and experimentation platform based on "second life" and "open-simulator". In: Proc. Pacific-Rim Symp. on Image and Video Technology (2009)
5. Johansson, M., Verhagen, H.: Massively multiple online role playing games as normative multiagent systems. In: Normative Multi-Agent Systems. Dagstuhl Seminar Proceedings, vol. 09121 (2009)
6. Robocup @home rules & regulations, Version, Revision 286:288 (2012), http://www.ais.uni-bonn.de/~holz/2012_rulebook.pdf
7. Bollini, M., Tellex, S., Thompson, T., Roy, N., Rus, D.: Interpreting and executing recipes with a cooking robot. In: Int'l Symp. on Experimental Robotics (2012)

Author Index

Abdennadher, Slim   125
Abouraya, Ahmed   125
Agrawal, Richa   49
Akhtar, Naveed   219
Akiyama, Hidehisa   528
Alers, Sjriek   147
Alexandrov, Sergey   444
Allgeuer, Philipp   56, 568
Amasyalı, M. Fatih   316
Amigoni, Francesco   480, 504
Anderson, John   292
Anderson, Peter   244
Annable, Brendan   544
Asada, Minoru   68

Bader, Markus   456
Balakirsky, Stephen   592
Balcılar, Muhammet   316
Baltes, Jacky   292
Barrett, Samuel   552
Behnke, Sven   56, 135, 207, 568
Birbach, Oliver   171, 656
Biswas, Joydeep   468, 656
Bloisi, Domenico D.   560
Böckmann, Arne   80
Bonarini, Andrea   576
Brahmbhatt, Samarth   49
Budden, David   268, 373, 385, 544

Calderon, Juan   492
Caltieri, Alain   480
Chalup, Stephan K.   268
Chen, Song   37
Chen, Xiaoping   114
Claes, Daniel   147
Cliff, Oliver M.   1
Correa, Mauricio   231
Costa, Anna Helena Reali   256
Cozman, Fábio Gagliardi   256
Cunha, Bernardo   183

Dasagi, Vibhavari   49
da Silva, Valdinei Freire   256
Davletov, Feruz   316
de Denus, Michael   292

Droeschel, David   135
Ďurkáč, Ján   616

Elfring, Jos   13
Elibol, Ercan   492
Ewert, Daniel   336

Fang, Li   92
Ferrein, Alexander   336
Forero, Leonardo Leottau   664
Fossel, Joscha   147
Fountain, Jake   268
Frese, Udo   171

Garcia, João   280
Geihs, Kurt   304
Genter, Katie   552
Graber, Thorsten   624
Gräve, Kathrin   135
Gurzoni Jr., José Angelo   656

Hammer, Tobias   171
Hao, Yue   104
Härtl, Alexander   396
He, Yuchen   552
Helal, Dina   125
Helgadóttir, Lovísa Irpa   326
Hengst, Bernhard   244
Hennes, Daniel   147
Herpers, Rainer   444
Hester, Todd   536, 552
Hirose, Dai   68
Hofmann, Matthias   420
Holz, Dirk   135
Hugel, Vincent   408

Inamura, Tetsunari   672
Iocchi, Luca   560

Jansen, Simon   13
Jeschke, Sabina   336
Jouandeau, Nicolas   408

Kawano, Keisuke   68
Khandelwal, Piyush   552
Khater, Noha   125
Kirchner, Dominik   304

Klingauf, Uwe 624
Kobayashi, Kunikazu 159
Koga, Marcelo Li 256
Kohlbrecher, Stefan 624
Kootbally, Zeid 592
Küstenmacher, Anastassia 219

Lakemeyer, Gerhard 219, 336
Laue, Tim 80, 171, 648, 656
Lee, Daniel D. 49, 608
Li, Binbin 37
Li, Chuan 92
Liang, Zhiwei 104
Lima, Pedro U. 280
Liu, Juan 104
Liu, Wei 37
Liu, Xiaoming 37
Lizier, Joseph T. 1
Loncomilla, Patricio 231, 516
Lunenburg, Janno 584
Luperto, Matteo 504
Lv, Ye 37
Lynch, Charles 37

Mahmoodi, Khudaydad 316
Mankowitz, Daniel Jaymin 195
Matteucci, Matteo 576
McGill, Stephen G. 49, 608
Menashe, Jacob 552
Mendes, Alexandre 268, 373, 544
Metsemakers, Paul 584
Meyer, Johannes 624
Migliavacca, Martino 576
Missura, Marcell 56, 568
Mobalegh, Hamid 326
Müller, Judith 80
Münstermann, Cedrick 56
Murakami, Kazuhito 159

Nagai, Takayuki 672
Nakashima, Tomoharu 528
Nardi, Daniele 560
Naruse, Tadashi 159
Neves, António J.R. 183
Niemczyk, Stefan 304
Niemueller, Tim 336
Niwa, Toshinori 348

Obst, Oliver 1
Okada, Hiroyuki 672

Okaya, Masaru 348
Osama, Salma 125
Oshima, Yuji 68

Pairo, Wilma 231
Pastrana, Julio 56, 568
Pennisi, Andrea 560
Petersen, Karen 624
Plöger, Paul G. 219
Prokopenko, Mikhail 1, 385

Qin, Shibo 68
Quattrini Li, Alberto 504

Ramamoorthy, Subramanian 195
Reis, João C.G. 280
Reis, Luís Paulo 360
Reuter, Sebastian 336
Ribeiro, Fernando 360
Röfer, Thomas 25, 80, 396, 648
Rojas, Raúl 326, 600
Ruiz-del-Solar, Javier 231, 516, 664

Saavedra, Marcelo 516
Sakr, Fadwa 125
Schoenmakers, Ferry 584
Schreiber, Michael 56, 135, 568
Schueller, Sebastian 56, 568
Schwarz, Max 56, 135, 207, 568
Seifert, Daniel 600
Seshan, Srinivasan 432
Shen, Ping 104
Shibata, Kazumasa 68
Soetens, Robin 584
Steinbuch, Maarten 13, 584
Stoecker, Justin 640
Stone, Peter 536, 552
Stückler, Jörg 135
Sugiura, Komei 672
Šuppa, Marek 616
Suzuki, Tomoya 68

Takahashi, Tomoichi 348
Takuma, Takashi 68
Tan, Jeffrey Too Chuan 672
Tian, Ye 37
Tong, Hangjun 92
Topalidou-Kyniazopoulou, Angeliki 135
Toyoyama, Syohei 68
Trifan, Alina 183

Trigueiros, Paulo   360
Tsogias, Alexis   80
Tuyls, Karl   147

Urbann, Oliver   420
Uzun, Yücel   316

Vadakedathu, Larry   49
van de Molengraft, René   13, 584
Vatankhah, Hedayat   632
Veloso, Manuela   432, 468, 656
Verschae, Rodrigo   231
Vincze, Markus   456
Visser, Arnoud   592
Visser, Ubbo   396, 640
von Stryk, Oskar   624

Walker, Josiah   268
Wang, Chenyu   37
Wang, Miao   37
Wang, Peter   1
Wang, Richard   432
Wang, Rosalind X.   1
Wang, Xueyan   37
Weiss, Gerhard   147

Weitzenfeld, Alfredo   492
Wenk, Felix   25

Xiong, Rong   92
Xu, Xinxin   37
Xu, Yuan   632

Yáñez, José Miguel   664
Yang, Feichan   37
Yasui, Kotaro   159
Yavuz, Sırma   316
Yi, Seung-Joon   49, 608

Zhang, Haochong   114
Zhang, Wanjie   37
Zhang, Yida   49, 608
Zhang, Zongyi   37
Zhao, Hecheng   104
Zhao, Liang   37
Zhao, Peng   37
Zhao, Yue   92
Zhou, Jieming   37
Zhu, Di   37
Zhu, Zhe   37
Zickler, Stefan   656